I0280801

Studies in Logic

Mathematical Logic and Foundations

Volume 94

A Lambda Calculus Satellite

Volume 88
Belief Attitudes, Fine-Grained Hyperintensionality and Type-Theoretic Logic
Jiří Raclavský

Volume 89
Essays on Set Theory
Akihiro Kanamori

Volume 90
Model Theory for Beginners. 15 Lectures
Roman Kossak

Volume 91
A View of Connexive Logics
Nissim Francez

Volume 92
Aristotle's Syllogistic Underlying Logic: His Model with his Proofs of Soundness and Completeness
George Boger

Volume 93
Truth and Knowledge
Karl Schlechta

Volume 94
A Lambda Calculus Satellite
Henk Barendregt and Giulio Manzonetto

Volume 95
Transparent Intensional Logic. Selected Recent Essays
Marie Duží, Daniela Glavaničovà, Bjørn Jespersen and Miloš Kosterec, eds

Studies in Logic Series Editor
Dov Gabbay
dov.gabbay@kcl.ac.uk

A Lambda Calculus Satellite

Henk Barendregt
Giulio Manzonetto

© Individual authors and College Publications, 2022
All rights reserved.

Photograph on back cover © Archives of the Mathematisches Forschungsinstitut Oberwolfach.

College Publications and the authors have no responsibilities for the persistence or accuracy of URLs for external or third-party internet websites referred to in this publication.

ISBN 978-1-84890-415-6

College Publications
Scientific Director: Dov Gabbay
Managing Director: Jane Spurr

http://www.collegepublications.co.uk

All rights reserved. No part of this publication may be reproduced, stored in a retrieval system or transmitted in any form, or by any means, electronic, mechanical, photocopying, recording or otherwise without prior permission, in writing, from the publisher.

I conjecture that:

"All conjectures in λ-calculus are false"

– Ugo de'Liguoro

In this book we show that his conjecture is false.

As he conjectured.

This book is dedicated to the women in our lives:
Lidia, Buffee, and Sophia,
Laura and Nadia.

Contributors

Henk Barendregt Institute of Computing & Information Science Radboud University Nijmegen	All parts, except Chapters 5, 7, §1.2
Stefano Guerrini Northern Paris Computer Science Laboratory CNRS & LIPN, University Sorbonne Paris Nord	Chapter 5, §1.2
Giulio Manzonetto Northern Paris Computer Science Laboratory CNRS & LIPN, University Sorbonne Paris Nord	All parts, except Chapters 5, 7, §1.2
Vincent Padovani Research Institute on the Foundations of Computer Science CNRS & IRIF, University Paris Cité	Chapter 7

Contents in short

Contents

IV Theories

V Models

VI Open Problems

VII Appendix

Acknowledgements

We would like to start with a special thanks to Martin Hyland, who suggested the time had come to write a satellite book of "The Lambda Calculus — its syntax and semantics", since most of the problems stated there were nowadays solved. As in that book, we also include some portraits of researchers that have either found these solutions, or contributed to shape the field. We thank them for sending us their photos—in particular Corrado Böhm's family for providing a nice photo of him and consent for its publication.

We are grateful to Vincent Padovani for accepting to publish here his original results and for writing a large part of Chapter 7, and to Stefano Guerrini for writing Chapter 5.

Several people were involved in the later phases of writing the book. Each chapter was assigned to one or two colleagues that kindly accepted to act as reviewers. We warmly thank them for their careful reading and insightful commentaries. We take responsibility for any remaining errors or deficiencies.

Chapter	Reviewers
1	Jakob Grue Simonsen
2	Davide Barbarossa
3	Antonio Bucciarelli
4	Jakob Grue Simonsen
5	James McKinna & Luca Roversi
6	Łukasz Czajka
7	Johannes Waldmann
8	Davide Barbarossa
9	Mariangiola Dezani-Ciancaglini & Enno Folkert
10	Lionel Vaux Auclair
11	Simona Ronchi Della Rocca
12	Andrew Polonsky
13	Ugo de'Liguoro
14	Delia Kesner & Michele Pagani
15	Antonino Salibra

Chapter 9 required a real group effort. Besides Enno Folkert's invaluable help in understanding his results and in writing Section 9.2, we had to gather Davide Barbarossa, Flavien Breuvart, Stefano Guerrini, and Axel Kerinec to fully understand the notion of 'core' in §9.1 and to complete the proof of Theorem 9.40.

Guy McCusker helped with §§3.2, 12.1 and A.2, Mariangiola Dezani-Ciancaglini and Freek Wiedijk with §A.1, and Antonino Salibra with Chapter 3 and §A.4. Some colleagues helped proof-reading the final version of the book: André Batenburg, Flavien Breuvart, Rémy Cerda, Simona Kašterović, Axel Kerinec, Samuel Frontull, Federico Olimpieri, and Paolo Pistone. Mariangiola Dezani-Ciancaglini deserves an honorable mention, as she managed to read the whole manuscript during the summer of 2022. Other help came from Alberto Carraro, Thomas Ehrhard, Jörg Endrullis, Els Hornix, Wilberd van der Kallen, Jan Willem Klop, Jan Kuper, Chris Lambe, Jozef Steenbrink, and Pierre Vial. We thank Saito Teppei for pointing out some typos, discovered after the initial publication of this book.

The Northern Paris Computer Science Laboratory (LIPN) of the University Sorbonne Paris Nord, has financed several meetings between the two authors. We also thank the logical team of LIPN for the scientific discussions and for providing a stimulating environment. The Mathematical Forschungsinstitut at Oberwolfach, Germany, provided hospitality (two weeks) through their 'Research in Pairs' program. The Institut Henri Poincaré, France, provided hospitality (one week) and some fundings via their 'Research in Paris' program. Since February 2020, Manzonetto has benefited from two sabbatical periods that proved to be crucial in the finalization of this volume: one semester of *Congé pour recherches ou conversions thématiques* and one year of *délégation CNRS*.

Nijmegen and Paris 30, September 2022

Henk Barendregt[1]
Giulio Manzonetto[2]

[1]Institute of Computing & Information Science, Radboud University, Nijmegen, The Netherlands.
[2]CNRS & LIPN, University Sorbonne Paris Nord, Villetaneuse, France.

About λ-calculus

The birth of λ-calculus

In 1972 Henk Barendregt visited Alonzo Church at UCLA. Barendregt seems to remember that Church had told him the following story about the invention of the λ-calculus. As a PhD student of Oswald Veblen, Church received the following research assignment:

> *Find a computational method to determine from a simplicial 2-dimensional manifold in 3-dimensional space its Betti numbers.*

While working on his doctorate, completed in 1927, Church could not solve this problem. Because of his failure to settle the question of Veblen, Church started to wonder the following: was it conceivable that the Betti numbers actually *could not* be computed from the description of these surfaces? And if that was the case, how would one even proceed to prove it? Around 1928,[3] Church invented the λ-calculus to settle these questions. Later, in his famous paper Church (1936), he proposed to identify intuitive computability with the precisely defined notion of λ-definability. This equivalence is nowadays known as *Church's Thesis*. In the same paper he showed that undecidable problems, and therefore non-computable functions, do exist.[4] Perhaps Veblen's problem was one of these?[5]

As no one could be found who was able to confirm the above story, it should be considered as an anecdote. But it is quite possible that things happened this way. In the introduction of the same paper where Church's Thesis was proposed, a topological problem[6] was mentioned related to the computational problem that Church allegedly got from Veblen. This is not a typical problem to be mentioned in the introduction of an article about computability.

The first versions of λ-calculus did appear as part of a system dealing with logic and functions. Several of these turned out to be inconsistent, for more details we refer to Section 16.4. After stripping away the notions from deductive logic, the resulting system

[3]The year was deduced by Cardone and Hindley (2009), Section 4.1, from Seldin (2009) in which it is stated that when Curry was in Göttingen in 1928-1929, working on his PhD on Combinatory Logic, he saw a paper of Church that contained many symbols 'λ', that at the time he did not understand as being related to his own work, but only later when seeing Church (1932).

[4]Undecidability was proved for the following problem: given a λ-term, does it have a normal form?

[5]Eventually, that turned out not to be the case. In Chen and Rong (2010) a linear time algorithm was given for computing the Betti numbers of a simplicial 2D surface in 3D space.

[6]To find a complete set of effectively calculable invariants of closed 3D simplicial manifolds under homeomorphisms.

was proved consistent in Church and Rosser (1936)[7] by establishing the confluence of the associated reduction rule. When Kleene (1935) showed how functions like the predecessor could be represented in this consistent system, Church had the intuition that the λ-calculus could be seen as a way to capture the full class of intuitively computable functions and then formulated this as a Thesis,[8] in Church (1936). As mentioned in Davis (1982), Gödel was not convinced of the validity of Church's Thesis, until he saw the two articles Turing (1937a,b) in which Turing Machines were introduced and it was proved that these devices represent the same class of functions as the λ-definable ones.

Lambda terms are elements of a formal language, subject to conversion rules, that deserves to be studied for several reasons. First of all it forms an equational theory that is simple to define, has a clear semantics, and yet is Turing-complete and therefore undecidable.[9] Moreover, the theory has two major (related) applications: functional programming and proof-checking. As argued in Barendregt (2020b), these two features make λ-calculus one of the major tools of logic, even of all of mathematics, for scientific and industrial applications.

Lambda calculus and functional programming

A computational job consists of transforming some data (information) into other data, according to some specification. This can be a one-time job, like computing the value $f(n)$ for a given number n. Or it can be a perpetual job, in which after receiving a value n_0, the value $f(n_0)$ needs to be computed, in order to be ready again for the next input n_1, and so on.[10] Computational jobs can be performed using tools that are helping to get the value. Such tools can be mechanical, like a Pascal calculator, often for *ad hoc* computations, or can be digital computers, usually for universal computations, consisting of the evaluation of a program on some input. In 1937 two approaches to computability existed that claimed to be universal (programmable): λ-calculus, Church (1936), and Turing Machines, Turing (1937a) (submitted in 1936). Moreover, it was proved that both

[7] An early version of this paper was cited in Church (1935), showing that the result was already known in 1935. The Church-Rosser paper is unpleasant to read: for the system that was proved to be consistent, our beloved λ-calculus, reference was made to "rules *I*, *II*, *III* in Church (1932), as modified by Kleene (1934)". These rules are α-conversion, β-reduction, and β-anti-reduction. The modification of Kleene consisted in changing the original $\lambda\mathsf{K}$-terms of Church into $\lambda\mathsf{I}$-terms. (So the $\lambda\mathsf{I}$-calculus was proposed by Kleene, while Church had the $\lambda\mathsf{K}$-version! The preference of Kleene may be understood realizing that the $\lambda\mathsf{K}$-calculus doesn't have an obvious interpretation in the partial combinatory algebra $\mathcal{K}$ of Kleene, while the $\lambda\mathsf{I}$-calculus does. See Barendregt (2020b).) The resulting systems, the $\lambda\mathsf{I}$-calculus and the $\lambda\mathsf{K}$-calculus, were finally described clear cut in Church (1941).

[8] At the celebration of Robin Gandy's retirement in 1986, Stephen Kleene spoke about this episode and mentioned how he showed to Church the possibility of λ-defining the predecessor function. Kleene remarked: "At that moment in time I could have stated the thesis that all intuitively computable functions are λ-definable. But I didn't, Church did."

[9] Even essentially undecidable, if one adds negation and the inequality $\mathsf{S} \neq \mathsf{K}$, as remarked by A. Grzegorczyk (private communication).

[10] Such a perpetual job happens on a calculator with, say, a button for $f(x) = x^2$, or an operating system, patiently ready to evaluate the next pair (program, input).

computational models define the same class of functions, see Turing (1937b).[11] These approaches generated two alternative styles of programming: Functional and Imperative Programming (FP and IP).

Seen abstractly a computational job can be rendered as a λ-term FA, where A represents the input and F is the functional program that expresses what should be computed. In the functional computation model the expression FA has to be rewritten (reduced), while keeping its meaning. Because λ-calculus and hence functional languages have a notation for values (without having to evaluate them) and also for functions that can be given as parameter to other functions, a comfortable level of abstraction is provided. Moreover there exists a natural and computable type assignment (called Hindley-Milner algorithm, but actually due to Curry) for functional programs, that provides a partial warranty that the programming is done correctly, so that many programming bugs (but not all) can be caught in a relatively easy way (types for imperative programs are less elegant). These qualitative factors contribute to bringing the (mathematical) specification of the computational job and the program F very close to each other.

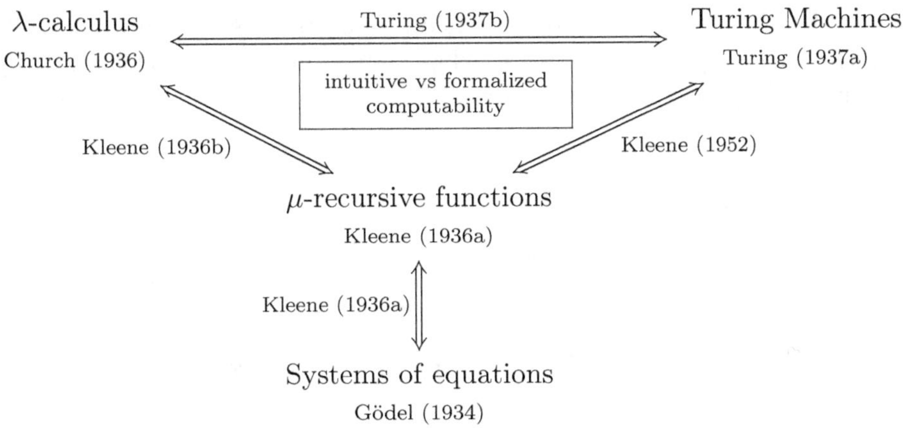

Figure 1: Various models of computation and their equivalences.

Technological considerations

In the imperative paradigm a program indicates in a stepwise fashion how the information in memory must be changed to execute a specific job. In the early 1940s specific hardware (generally called a Von Neumann processor) was developed to realize this process, and an imperative language in fact is tailored towards describing the behavior of such a processor. A functional program can also be executed on a Von Neumann processor, but mature compilers for this only appeared in the 1980s. See Barendregt et al. (2013b) for a short comparison and references.

[11] Also other proposals for universal computing appeared in the 1930s. The most notable ones are Herbrand-Gödel (HG) computability, Gödel (1934), and the μ-recursive computability, Kleene (1936a). All these computational models turned out to be equivalent to Church's and Turing's ones, see Figure 1.

Since a functional program is not concerned with memory, its compiled version may not be efficient, both for space and time reasons. Therefore it is necessary to perform a program transformation that takes into account how the functional language is being compiled in order to optimize performance. This is also the case for an imperative program that needs to be translated into machine language of the CPU, while imperative languages have evolved as well (e.g. with `for` and `while` loops and more complex constructions). In Figure 2 this is indicated by often necessary program transformations, that moreover depend on the underlying hardware.

Programmable parallel hardware makes a difference for specific computations. An FPGA (Field Programmable Gate Array) consists of many 'logic blocks' that can be configured to obtain the functionality of a desired logical gate. This way one creates dedicated parallel hardware for intensive tasks,[12] providing efficiency at relatively low energy costs. Also, here the functional program may need to be optimized by a careful balancing of the space and time resources. The technology is in its infancy but promising, see Érdi (2021). The dotted line with 'HLS' (High Level Synthesis), from IP to FPGA, is only possible using backwards engineering to obtain first the specification and then to proceed to the FPGA; the results are not encouraging because of low time performance. In Figure 2 the various possible paths are depicted to perform a specified computation. Depending on the situation one may need/prefer to follow one of the paths.

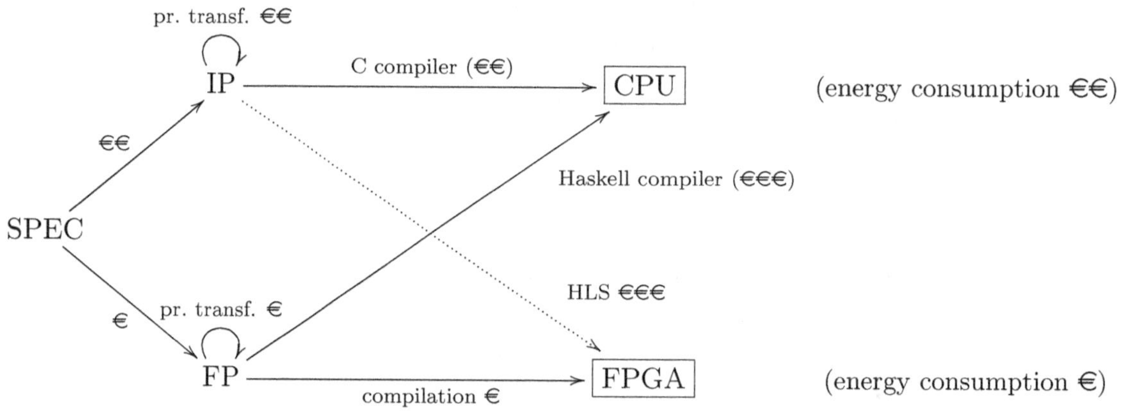

Figure 2: Comparing imperative and functional programming on standard processor and dedicated programmable hardware FPGA. The work to be done for each of the four possibilities is displayed and given an intuitive (possibly subjective) 'price' (number of €s; some of these are in parentheses, as e.g. the Haskell compiler has already been constructed): from the abstract specification, to the program implementing it, and finally to the hardware on which it is running. In addition to compilation one needs to count also the effort to perform program transformations for space and time optimization.

[12]E.g. for audio streaming that needs to be continuously repeated.

Lambda calculus and proof-checking

Formalization and theoretical consequences

The *force, precision, and certainty*[13] of mathematics are known. In the foundations of mathematics, it became the subject of interest to find out what are the mechanisms that enabled these qualities. Mathematicians often have shown little interest in understanding these, just using successfully the mental capacities that made them possible. For example, Poincaré says about Weierstrass something like the following "He was a logician, not able to understand what is outside his vocabulary; but we need logicians", Poincaré (1903). And this was presumably meant to be friendly.

Nevertheless, in the course of several millennia these mechanisms have been described well enough to establish the following.

- There are logical systems in which mathematical proofs can be fully formalized.
- A substantial part of these logical systems use some form of λ-calculus, either as exact notational system for the proofs, or as essential auxiliary system to manipulate these proofs.
- Fully formalized proofs can be checked by a reliable program.
- Interactive Theorem Provers (ITPs) have been built, in which the computer helps the human to create a formal proof. The human provides the intuition and the machine checks each step.
- Part of an ITP consists of Automated Theorem Proving (ATP), making it easier to come to a full formalization.

The quest started with the Greek philosopher Aristotle, who described the axiomatic method. Mathematics consists of notions and established properties of these. In order to avoid an infinite regression one needs to start with primitive notions and by means of definitions more complex notions can be introduced. Similarly, one starts with axioms and theorems are obtained by deductions. Aristotle also started to look for logical rules sufficient for this reasoning. This did repel Aristotle's teacher, Plato, but in the course of time it attracted many philosophers and even mathematicians. More than two millennia later Frege found a formal system for logical deduction, see Frege (1879). In Gödel (1930) it was proved to be sufficient for the derivation of valid mathematical results. This implies that in principle the consequences of some mathematical axioms form a well-defined set of formulas. In this spirit Whitehead and Russell (1910) started in their Principia Mathematica to fully formalize a small part of mathematics. Colleagues made fun of the fact that it took them 321 pages to prove that $1 + 1 = 2$. But formalizing is interesting as it raises the question "Do we have sufficient insight in the mechanism of our (mathematical) thinking?"

[13] Citation from Musil (1930-1943). See Barendregt (2020b) for a precise interpretation of these epithets.

Formalizing is not only interesting for phenomenological reasons, it also has led to the following well-known impressive applications. The set of provable sentences of Peano Arithmetic is incomplete, provided that this theory is ω-consistent, Gödel (1931), and undecidable, Church (1936). In Rosser (1936) both these fundamental results are improved by replacing ω-consistency by simple consistency.

Formalization and technical consequences

In the system of Natural Deduction, Gentzen (1935a,b), the logic of Frege was given a canonical form. These natural deduction proofs turned out to be closely related to typed λ-terms, see Howard (1980). Earlier Curry had noted that the types for the combinators $\mathsf{I}, \mathsf{K}, \mathsf{S}$ have the same form as some axioms of propositional logic, Curry and Feys (1958). Independently the formal proofs in de Bruijn (1970) are also directly described as λ-terms. For more on this so-called Curry-de Bruijn-Howard isomorphism between derivations and λ-terms, and propositions and types, see de Groote (1995) and Wadler (2015), including some of the history.

There are situations in mathematics and computer science, including their applications, in which one would like to be able to issue a predicate of certainty, and in the case of products provide a warranty of their reliability. Certainty comes from proofs. Reliability from proofs that the design of a product meets a partial but precise specification. Formal methods in computer science started to appear in the 70s and 80s of last century: providing full proofs for claims of program correctness. Both the full formalization of mathematics, started with Principia Mathematica and the formalized correctness proofs of hardware and software, did suffer from a methodological flaw: it was not clear how the correctness of these formal proofs could be verified.

Since the 1960s a technology of proof reliable proof-checking was developed, bringing the development, started with Principia in mathematics and Formal Methods in computer science, to a higher level of certainty and reliability. McCarthy (1962) proposed to construct a computer program for checking the correctness of mathematical proofs. In McCarthy's paper it was suggested that proofs in natural deduction form may be useful to this purpose, but this suggestion was not followed up by him. Some years later, in de Bruijn (1970) a precise language, AUTOMATH, was described for writing down mathematical proofs, together with a method for verifying them. This language was some form of typed λ-calculus. The work of McCarthy had as a follow-up the development of systems for Automated Theorem Proving (ATP) for a subset of mathematical theorems, while the work of de Bruijn had the systems for Interactive Theorem Proving (ITP).[14] This resulted in several proof-checking systems, like `Coq` and `Lean`, in which proofs consist of typed λ-terms. See Barendregt (2013) for considerations concerning the choice of formal system in the different systems of ITPs, Blanchette et al. (2016) and Blanchette and Mahboubi (2022) for a survey.

We now mention some state of the art applications of proof-checking.

[14] Also called *Mathematical Assistants.*

Applications in computer science and technology

Proof-checking was developed for mathematics. But as the designers of hardware and software knew from experience that their work is error prone, applications started first with the certification of IT products. This can be done by proof-checking their (partial) specification, see Wupper (2000). The following are impressive examples, collected in Barendregt (2020b).

- The ARM6 processor—predecessor of the ARM7 embedded in the large majority of mobile phones, personal organizers and MP3 players—was certified using the aforementioned method, Fox (2003).
- The seL4 operating system has been fully specified and certified, Klein et al. (2009).
- The same holds for a realistic kernel of an optimizing compiler for the C programming language, Leroy (2009).
- In Dutle et al. (2020) the Compact Position Reporting algorithm, that enables an aircraft to share its position and velocity with other aircraft in its vicinity, has been secured, after correcting some errors in an earlier version.

Applications in mathematics

Substantial proof-checking in mathematics started somewhat later. The methodology became necessary for those theorems requiring computer calculations for their proofs, like the Four Color Theorem, and establishing Kepler's conjecture. What needed to be done was the verification of three things: 1. that a program fits a given specification; 2. that an actual computation was done correctly; and last but not least 3. that the specified program has relevance for the proof of the stated theorem. Apart from this, even Fields medal laureates, realizing that sometimes they could not follow their own previous work, started to be also interested in proof-checking and in designing new formal systems for this. See, e.g., Voevodsky (2014) and the Univalent Foundations Program (2013). Finally, it was discovered that formalizing was an intellectually rewarding effort.

The following are impressive examples of formalized and checked proofs in mathematics, one could say providing certified theorems in mathematics.

- The Four Color Theorem, Gonthier (2007).
- The Prime Number Theorem, both the elementary proof of Selber, see Avigad et al. (2007), and the analytical proof by Hadamard and de la Vallée Pousin, see Harrison (2009).
- The Odd Order Theorem, stating that all finite groups of odd order are solvable, Gonthier et al. (2013).
- The proof of Kepler's conjecture about optimal sphere packing, Hales et al. (2017).

For more examples, we refer to the forthcoming Blanchette and Mahboubi (2022). See also Barendregt (2013) for a view on fundamental choices of various formal systems underlying proof-checkers.

At the moment of this writing (2022) proof-checking requires a steep learning curve. This may change when ATP provides more help to ITP. Machine learning may play a role here, see Urban (2021).

View on λ-calculus in this book

The way λ-calculus is considered in this book is as a type-free theory that can be compared to, say, number theory. In this perspective, λ-terms are studied from various points of view—a principal one being the relation between their applicative properties and their syntactic representation.[15] A prime example of this relation to be found in this book is Folkerts' Proposition 9.78, characterizing λ-terms that act as a bijection in the closed term model of $\boldsymbol{\lambda}\eta$ as finite hereditary permutations. A more sophisticated example is given by observational equivalences on λ-terms that can be characterized in terms of extensional equalities between their Böhm trees. Depending on the concept of 'observation' under consideration, one ends up with different notions of extensionality on Böhm trees—compare Lévy's Proposition 11.9 and Hyland's Proposition 11.12.

Of independent interest is the construction of domains that constitute denotational models of the λ-calculus. These contributed considerably to the investigation of the aforementioned tension between applicative *vs* syntactic properties of terms. As these domains live in Cartesian closed categories, either with enough points or non well-pointed, also some category theory is brought into play. In Chapter 15 a uniform proof is given, using notions from universal algebra, showing that all models considered so far do not cover all possible λ-theories, by taking the set $\mathrm{Th}(\mathcal{M})$ of equations valid in $\mathcal{M}$.

Reduction Systems

Reduction Systems do not fall under the λ-calculus, but are closely related. A Reduction System is a set X endowed with a (one step) reduction relation $\to$. That is, a structure $(X, \to)$, where $\to$ is a binary relation. Prime examples are $(\Lambda, \to_\beta)$ and $(\Lambda, \to_{\beta\eta})$ consisting of λ-terms with β- (and $\beta\eta$-) one-step reduction. But there are more and highly significant classes of reduction systems next to λ-calculus.

Most of these classes of reduction systems consist of pairs $(T, \to)$, where T is a set of linguistic expressions (terms) and $\to$ is a relation generated by some basic rewrite rule $t \to_0 s$ ('contraction'), that is monotonically extended to all terms

$$\begin{array}{rcl} t \to s & \iff & \text{for some context } C[\,] \text{ and terms } t_0, s_0 \text{ one has} \\ & & t = C[t_0] \;\&\; s = C[s_0] \;\&\; t_0 \to_0 s_0. \end{array}$$

[15] In number theory a main point of interest is studying the relation between additive and multiplicative properties of numbers. Goldbach's unresolved conjecture:

"Is every even number $n > 2$ the sum of two primes?"

can be considered as an example of this relation.

In this case one speaks about Rewriting Systems. Those reduction systems $(X, \rightarrow)$, in which the X is not necessarily specified as a set of terms, are also given the more explicit name 'Abstract Reduction Systems' (ARSs).

The general study of Reduction Systems is beyond the scope of this book. We briefly discuss the main classes of these systems in order to give the reader a hint of the wealth of notions and results in this field. These classes are Recursive Programming Schemes (RPS), Term Rewriting Systems (TRS), Combinatory Reduction Systems (CRS), and Abstract Reduction Systems (ARS). Rather than giving formal definitions of these systems, prime examples and some results will be indicated. For a thorough basic treatment of this rich field of still ongoing[16] research the reader is referred to Terese (2003) (abbreviated below as 'Terese').

- Abstract reduction systems (ARSs). In their most general form there is the notion of Abstract Rewrite System (ARS) in which there are elements without any structure, subject to a binary one-step reduction relation.[17]

  A basic result for ARSs is Newman's Lemma, Theorem 1.2.1 in Terese.
  If an ARS is locally confluent and strongly normalizing, then it is confluent.
  The following counterexample from Hindley (1969) shows that the condition of being strongly normalizable is necessary:

  $$a' \longleftarrow a \rightleftarrows b \longrightarrow b'$$

  This ARS is locally confluent, but not confluent; indeed it is not strongly normalizing.

  A much more complex result is the de Bruijn-van Oostrom theorem on Trace-decreasing Diagrams, Theorem 14.2.18 in Terese, and its consequence Strong Confluence Theorem, Theorem 14.2.26 in the same book. See Endrullis and Klop (2013) and Endrullis et al. (2020) for an exposition of the result and its relation with the independent and unpublished work of de Bruijn.

- Recursive programming schemes (RPSs). The most simple form of a Rewrite System is the notion of Recursive Programming Scheme (RPS). Here the objects that can be rewritten[18] are expressions of the form $F(t_1, \ldots, t_n)$, where all the t_i's are required to be variables; these can be rewritten to expressions of the form $G(s_1, \ldots, s_m)$, where now the s_i's may be arbitrary expressions.

  A typical example of an RPS from Terese is

  $$\begin{array}{rcl} F_1(x) & \rightarrow & G_1(x, F_1(x), F_2(x, x)); \\ F_2(x, y) & \rightarrow & G_2(F_2(x, x), F_1(G_3)). \end{array}$$

---

[16]Initially there were two series of recurring conferences: on Lambda Calculus (Typed Lambda Calculi and Applications, TLCA) and on Rewriting (Rewriting Techniques and Applications, RTA). Those two series have been merged in 2016 into Formal Structures for Computation and Deduction (FSCD).

[17]A more general alternative definition is that an ARS is a *digraph*, in which there may be multiple edges from an element a to an element b in the set.

[18]Reducible expressions, or *redexes* for short.

In Terese the following is shown.
Every RPS is confluent. Example 4.1.6(iii) & Theorem 4.3.4 in Terese.
In an RPS it is decidable whether a term has a normal form, Khasidashvili, see Remark 4.6.4 in Terese.

- TERM REWRITING SYSTEMS (TRSs). These are in general stronger than systems in RPS: there may be pattern matching.
  The following rewrite rules are from Dedekind (1965).

$$\begin{array}{rcl} A_+(x,0) & \to & x; \\ A_+(x,S(y)) & \to & S(A_+(x,y)); \\ A_\times(x,0) & \to & 0; \\ A_\times(x,S(y)) & \to & A_+(A_\times(x,y),x). \end{array}$$

Using these rules one can add and multiply numerals $S^k(0) = S(S(\cdots S(0)\cdots))$. Herbrand-Gödel computability extends this system and captures the class of all partial recursive functions, see Mendelson (2009).

Also Combinatory Logic (CL) is a TRS. For example the rule $\mathsf{I}x \to x$ is *de facto* $\mathrm{App}(\mathsf{I},x) \to x$, so that pattern matching is involved.

An interesting example of a system with pattern matching is surjective pairing:

$$\begin{array}{rcll} P_1(Pxy) & \to & x; & \text{(left projection)} \\ P_2(Pxy) & \to & y; & \text{(right projection)} \\ P(P_1x)(P_2x) & \to & x. & \text{(surjectivity)} \end{array}$$

The system with the surjectivity rule is not confluent, as it was proved by Klop (1980a). He discovered that the culprit is the surjectivity rule that is not left-linear (since the variable x occurs twice in the LHS). Without the surjectivity rule the system is confluent, as follows by the general result on TRS:
Every orthogonal[19] *TRS is confluent,* Terese, Theorem 4.3.4.

Another non-trivial result is the Knuth-Bendix completion method. This is a partial algorithm that takes as input a congruence relation on terms, and in case of termination yields a confluent reduction relation, having as conversion relation the initial congruence, see Terese, §7.4. This algorithm can be applied to obtain Buchberger's algorithm constructing Gröbner bases. With the latter, one can test whether a polynomial $p \in R[\vec{x}]$ belongs to the ideal $\{p_1,\ldots,p_n\}$ generated by a finite set of polynomials $\vec{p} \in R[\vec{x}]$.[20]

An entire chapter in Terese, Chapter 6, is devoted to the rich methodology for proving strong normalization of several TRSs.

[19] That is, non-ambiguous and left-linear.
[20] This result has many applications, including determining the reachability of robot arms.

- COMBINATORY REDUCTION SYSTEMS (CRSs). These have been introduced by Klop (1980a) and were inspired by Aczel (1978). In these systems both pattern matching and binding effects are possible.

  The Prawitz-reductions on natural deduction derivations, Prawitz (2006), in which an elimination-rule follows an introduction-rule, obtain an elegant linear notation. For example

$$P\;\frac{Q_i\,\dfrac{\dfrac{z_0}{\phi_i}}{\phi_1 \vee \phi_2} \qquad \dfrac{\begin{array}{c}[\phi_1]\\ z_1\end{array}}{\psi} \qquad \dfrac{\begin{array}{c}[\phi_2]\\ z_2\end{array}}{\psi}}{\psi} \;\to\; \dfrac{\begin{array}{c}\dfrac{z_0}{[\phi_i]}\\ z_i\end{array}}{\psi},$$

  where Q_i and P are 'rule-constants' of $\vee$-introduction and $\vee$-elimination, becomes

$$P(Q_i z_0)([x]z_1(x))([y]z_2(y)) \;\to\; z_i(z_0),$$

  see Klop (1980a), pp. 127–128. This can be done for all the Prawitz-reduction rules.

  Fundamental results about CRSs (an orthogonal CRS is confluent and a left-normal orthogonal CRS satisfies a standardization theorem, see Klop (1980a)) made it possible to prove an important fact observed by Statman (2005) about the syntactic structure of λ-terms, namely that two bound variables do not suffice to express all closed λ-terms; three bound variables are sufficient, due to the close relation between λ-terms and combinators $\mathsf{S}, \mathsf{K}, \mathsf{I}$ from CL.

About this book

This book is a satellite to "The Lambda Calculus, its Syntax and Semantics", Barendregt [1981/1984]. In the second edition of the book, several open problems were stated, many in the form of a conjecture. In the course of forty years most, but not all, open problems have been solved, sometimes establishing and sometimes refuting a conjecture.

One may imagine other satellites of B[1984][21]. For example entering domain theory, or linear logic semantics, with their wealth of λ-calculus models. Also other books on λ-calculus could be embellished by satellites. For example books on typed versions of λ-calculus may obtain satellites emphasizing theory or applications (functional programming and automatized verification of proofs in mathematics and computer science).

Solved open problems

The book B[1984] was meant to be a survey of what was known about untyped λ-calculus at the time. Several open problems concerning semantical and syntactical aspects of λ-calculus were presented, often in the form of conjectures. Most of these have been solved in the subsequent 35 years, but the results are scattered throughout the literature and difficult to piece together. Some of these solutions occupied an entire PhD thesis, e.g. Folkerts (1995), or part of a thesis, e.g. Polonsky (2011b), with the complexity (but not the method) of a proof using priority for results on the degrees of undecidability.

> The main purpose of this book is to give a uniform description of these solutions, and also to present other results relevant to the study of λ-calculus.

We would like to mention that in B[1984] a significant effort was made for handling infinitary objects like Böhm trees in a precise mathematical way. However, in retrospect, the final outcome is not particularly satisfactory. This is due to the fact that Böhm trees possess a coinductive nature, but this was understood only later, hence they are best manipulated using coinductive techniques. A recurring problem in the literature is that proofs relying on coinduction often become awkward and overly complicated, essentially because the authors reassert the principle every time it is used. Since coinduction has been around for decades, we adopt the more informal style of coinductive reasoning introduced by Kozen and Silva (2017). We believe that this approach, explained thoroughly in Section 2.1, greatly improves the readability and benefits the reader.

[21] This is an abbreviation for the reference Barendregt (1984).

Reduction

1. **Leaving the β-plane.** Let $M, N \in \Lambda$. Write $M \circlearrowleft_\beta N$ if $M \twoheadrightarrow_\beta N \twoheadrightarrow_\beta M$. An equivalence class under $\circlearrowleft_\beta$ is called a β-plane. It is possible to *leave a plane $\mathcal{P}$ at point $M \in \mathcal{P}$* if there is an N such that $M \to_\beta N \notin \mathcal{P}$. In 1980, Jan Willem Klop conjectured that if one can leave a plane at one of its points, then such a plane can be left at any of its points. Hans Mulder (1986) and Sekimoto and Hirokawa (1988) have refuted this conjecture (independently).

2. **Deciding normalizability of S-terms.** Consider combinatory terms built up from application and **S** only. Many of these have a weak normal form, like **SSSSSSS**. Others do not, like **S(SS)SSSS**, **SSS(SSS)(SSS)**. The question was raised whether normalizability of **S** is decidable. This was shown by Johannes Waldmann (1998). We also present original results by Vincent Padovani on the **S**-fragment of combinatory logic, including: (i) the termination of head reduction is decidable; (ii) there exist two non-interconvertible terms having the same Berarducci tree.

3. **Optimal reduction.** A consequence of the standardization theorem is that the leftmost-outermost reduction strategy allows to reach β-normal forms, whenever they exist. Due to the duplication of redexes, the reduction sequences obtained in this way may not be *optimal.* In this setting optimality means that the normal form is reached in a minimum number of steps. In his *Thèse d'État*, Jean-Jacques Lévy (1978b) introduced the notion of redex family in order to capture an intuitive idea of optimal sharing between 'copies' of the same redex. By studying the causal history of redexes using a suitable labeled extension of λ-calculus it is possible to define (and implement!) an optimal reduction method for λ-calculus.

4. **Reduction under substitution.** If $M[x:=L] \twoheadrightarrow_\beta P$, what can be said about the reduction of M itself and about the use that is made of L? In van Daalen (1980) a result is proved about this, see B[1984], Exercise 15.4.8. In Endrullis and de Vrijer (2008) several consequences are derived, among which the Genericity Lemma and the non-definability of surjective pairing.

5. **Infinitary reduction.** In λ-calculus there are terms, like Turing's fixed point combinator, generating an infinite reduction sequence. Pushing this reduction to infinity, one generates the infinite term $\lambda f.f(f(f(\cdots)))$, namely, its Böhm tree. Inspired by this phenomenon, Kennaway et al. (1997) introduced the infinitary λ-calculus, whose terms and reductions can possibly be infinite. The resulting infinitary term rewriting system is well defined and enjoys the unicity of its normal forms, but many properties fail like normalization and—most importantly—confluence. Berarducci (1996) showed that collapsing meaningless terms allows to restore confluence and induces a new model of λ-calculus based on Berarducci trees. By modifying the notion of meaningless terms, one also retrieves Böhm trees and Lévy-Longo trees as infinitary normal forms of λ-terms. By adding η-reduction, or a variation of it, one obtains extensional Böhm trees and Nakajima trees. Our presentation of these results mostly adopts the modern approach based on coinduction.

Conversion

1. **Perpendicular lines property.** Given a λ-algebra $\mathcal{M}$, suppose that a λ-term F interpreted as λ-definable map of three arguments is constant on three perpendicular lines:

$$\forall Z \in \mathcal{M} \begin{cases} F\ Z\ M_{12}\ M_{13} =_{\mathcal{M}} N_1 \\ F\ M_{21}\ Z\ M_{23} =_{\mathcal{M}} N_2 \\ F\ M_{31}\ M_{32}\ Z =_{\mathcal{M}} N_3 \end{cases}$$

then F is constant everywhere. This implication is called the perpendicular lines property at dimension 3 (PLP(3)) for the model $\mathcal{M}$. We consider as $\mathcal{M}$ the term model $\mathcal{M}(\mathcal{T})$ of a λ-theory $\mathcal{T}$, or its closed version $\mathcal{M}^o(\mathcal{T})$. For $\mathcal{T} \in \{\boldsymbol{\lambda}, \mathcal{B}\}$, where $\boldsymbol{\lambda}$ is the theory of β-conversion and $\mathcal{B}$ the theory induced by Böhm tree equality, the following has been shown.

$\mathcal{M}(\mathcal{B}) \models \forall n.\mathrm{PLP}(n)$,		by using Berry's theorem; B[1984].
$\mathcal{M}^o(\mathcal{B}) \models \forall n.\mathrm{PLP}(n)$,		by Bethke, adapting Berry's theorem.
$\mathcal{M}(\mathcal{B}) \models \forall n.\mathrm{PLP}(n)$,		by Endrullis and de Vrijer (2008), using reduction under substitution.
$\mathcal{M}^{(o)}(\mathcal{B}) \models \forall n.\mathrm{PLP}(n)$,		in this book, by coinduction.
$\mathcal{M}(\boldsymbol{\lambda}) \models \forall n.\mathrm{PLP}(n)$,		by Endrullis and de Vrijer (2008), using reduction under substitution.
$\mathcal{M}^o(\boldsymbol{\lambda}) \not\models \mathrm{PLP}(2)$,		by Statman and Barendregt (1999), using variants of Plotkin terms.

Another proof of the fact that $\mathcal{M}^{(o)}(\mathcal{B})$ satisfies the perpendicular lines property was found by Barbarossa and Manzonetto (2020), using Ehrhard and Regnier's Taylor expansion. This proof-technique is however outside the scope of this book.

2. **Invertibility and bijectivity.** In set theory it is well known, and easy to verify, that a function $f : X \to X$ is bijective if and only if it is invertible. Now, given a λ-theory $\mathcal{T}$, every closed λ-term $F \in \Lambda^o$ can be considered as function

$$F : \mathcal{M}^o(\mathcal{T}) \to \mathcal{M}^o(\mathcal{T}).$$

Assuming that F is a bijection, can one conclude from this that F is $\mathcal{T}$-invertible? In other words, is there a λ-term $G \in \Lambda^o$ such that $F \circ G =_{\mathcal{T}} G \circ F =_{\mathcal{T}} \mathsf{I}$? For $\mathcal{T} = \boldsymbol{\lambda}$ the answer is positive because the only β-invertible closed λ-term is the identity I. This was shown by Böhm and Dezani-Ciancaglini (1974). The invertibility problem for $\mathcal{T} = \boldsymbol{\lambda}\eta$ was first raised in B[1984], Exercise 21.4.9. More than 10 years later, in his PhD thesis, Enno Folkerts (1995) showed that this correspondence does hold. As a consequence of this, and of combined results by Dezani and Bergstra-Klop, a closed λ-term is $\beta\eta$-invertible if and only if it is a finite hereditary permutator.

Theories

1. **Placing Morris' theory $\mathcal{T}_{\mathrm{NF}}$.** Consider the λ-theories axiomatized as follows.

$$\begin{array}{lcl} \boldsymbol{\lambda} & = & \{M = N \mid M =_\beta N\}. \\ \boldsymbol{\lambda}\eta & = & \{M = N \mid M =_{\beta\eta} N\}. \\ \mathcal{T}_X & = & \{M = N \mid \forall F \in \Lambda^o . [FM \in X \iff FN \in X]\}. \\ \mathcal{H} & = & \{M = N \mid M, N \text{ are unsolvable}\}. \\ \mathcal{H}^+ & = & \mathcal{T}_{\mathrm{NF}}, \text{ where } \mathrm{NF} = \{M \in \Lambda^o \mid M \text{ has a } \beta\text{-nf}\}. \\ \mathcal{H}^* & = & \mathcal{T}_{\mathrm{SOL}}, \text{ where } \mathrm{SOL} = \{M \in \Lambda^o \mid M \text{ is solvable}\}. \\ \mathcal{B} & = & \{M = N \mid \mathrm{BT}(M) = \mathrm{BT}(N)\}. \\ \mathcal{B}\eta & = & \mathcal{B} \text{ closed under the } \eta\text{-rule}. \\ \mathcal{B}\omega & = & \mathcal{B} \text{ closed under the } \omega\text{-rule, which is} \\ & & \dfrac{FZ = GZ \text{ for all } Z \in \Lambda^o}{F = G}\ \omega\text{-rule} \end{array}$$

The λ-theories above are interesting because they capture operational properties of λ-terms. For their origin, we invite the reader to consult Barendregt (2020b).

One has the following inclusions in the lattice of λ-theories.

$$\lambda \subsetneq \mathcal{H} \subsetneq \mathcal{B} \subsetneq \mathcal{B}\eta \subsetneq \mathcal{B}\omega \subsetneq \mathcal{H}^* \ \&\ \mathcal{B}\eta \subsetneq \mathcal{H}^+ \subsetneq \mathcal{H}^*$$

The exact position of $\mathcal{H}^+$ was unknown. Patrick Sallé conjectured that

$$\mathcal{B}\omega \subsetneq \mathcal{H}^+.$$

This conjecture was refuted by Benedetto Intrigila, Giulio Manzonetto, and Andrew Polonsky by showing that $\mathcal{H}^+ = \mathcal{B}\omega$ holds, thus improving our understanding of the 'canonical' λ-theories above. This result first appeared in Intrigila et al. (2017).

2. **Failure of the range property modulo $\mathcal{H}$.** Let $F \in \Lambda^o$. For a λ-theory $\mathcal{T}$ and its closed term model $\mathcal{M} = \mathcal{M}^o(\mathcal{T})$ one can consider $F\colon \mathcal{M} \to \mathcal{M}$ as a map. The range of this map is given by $\mathsf{Range}_{\mathcal{T}}(F) = \{[FM]_{\mathcal{T}} \mid M \in \Lambda^o\}$. For many theories $\mathcal{T}$ the following property holds: the range modulo $\mathcal{T}$ of an arbitrary F is of cardinality either 1 or $\aleph_0$ ('is either a singleton or an infinite set'). For $\mathcal{T} = \mathcal{H}$ this property was resisting. Several steps were made attempting to prove the range property for $\mathcal{H}$. In his PhD thesis, Andrew Polonsky (2011b) cleverly used all these hints to construct a devil's tunnel (in the sense of Barendregt (2008)) and refute the conjecture: there is a λ-term with range modulo $\mathcal{H}$ of cardinality 2.

3. **Number of fixed points modulo $\mathcal{H}$.** Let $F \in \Lambda^o$. The fixed point property states that the number of fixed point of F, again modulo a λ-theory $\mathcal{T}$, is either 1 or $\aleph_0$. Richard Statman conjectured that for sensible $\mathcal{T}$, i.e. such that $\mathcal{H} \subseteq \mathcal{T}$, the fixed point property holds for $\mathcal{T}$. This conjecture was also refuted by Andrew Polonsky, in collaboration with Giulio Manzonetto, Alexis Saurin and Jakob Grue Simonsen. Their result appeared in Manzonetto et al. (2019).

Models

1. **Understanding the notions of λ-algebra and λ-model.** Combinatory algebras are applicative structures that are combinatory complete. A λ-model is obtained if one has a natural interpretation of λ-terms. Dana Scott gave a denotational description of these models. The more relaxed notion of λ-algebra was introduced to accommodate the closed term models. Karst Koymans gave a categorical description of λ-algebras and λ-models differentiating in a categorical way ('Does the categorical model have enough points?') between these two classes. Although technically correct, his result led part of the community to believe that only categorical models having enough points deserve the status of models of λ-calculus. Antonio Bucciarelli, Thomas Ehrhard and Giulio Manzonetto proved that it is sufficient to slightly modify Koymans' construction to obtain a λ-model, even when starting from a categorical model without enough points. See Bucciarelli et al. (2007).

2. **Filter models.** Intersection type assignment systems, introduced by Coppo and Dezani-Ciancaglini (1980), allow to give a logical description of several operational properties of λ-terms, like solvability and various forms of normalization. Moreover, thanks to the celebrated Stone's duality, they correspond to a class of denotational models of λ-calculus, called filter models. This nomenclature derives from the fact that the denotation of a λ-term is given by the filter of its types. We show that classical lattice models, whose construction mimics the one of Scott's $\mathcal{D}_\infty$, can be presented as filter models. Others examples are the original filter model $\mathcal{F}_{\text{BCD}}$ defined by Barendregt et al. (1983) and the model $\mathcal{F}_{\text{CDZ}}$ by Coppo et al. (1987). We mainly focus on the latter since it has been largely overlooked in the literature, with the notable exception of Ronchi Della Rocca and Paolini (2004). Exploiting Tait's reducibility argument, we show that $\mathcal{F}_{\text{CDZ}}$ satisfies an Approximation Theorem. Finally, using Lévy's extensional approximants, we prove that it is (in)equationally fully abstract for $\mathcal{H}^+$, a result first established in Coppo et al. (1987).

3. **Relational models.** In the eighties Jean-Yves Girard realized that the category of sets and relations constitutes a simple quantitative semantics of Linear Logic, where the promotion $!A$ is given by the set of all finite multisets over A. The relational semantics of λ-calculus, obtained by applying the coKleisli construction, has been largely studied in the last decades because of its peculiar properties. First, its quantitative features allow to expose semantically intensional properties of λ-terms, like the amount of head-reduction steps needed to reach their head normal form. This also allows to endow fundamental results, like Approximation Theorems, with easy inductive proofs bypassing the usual techniques based on Tait's computability. Second, relational models can be expressed through tensor type[22] assignment systems whose inhabitation problem is decidable. Finally, the fact that it is a non-well-pointed category contributes to justify, together with categories of games, the interest in categorical models without enough points.

[22]Intuitively, relevant intersection type systems where $\wedge$ is a non-idempotent operator, i.e. $\sigma \wedge \sigma \neq \sigma$.

4. **Indecomposable models.** As remarked in B[1984], page 91, combinatory algebras are algebraically pathological because they are never commutative, associative, finite or recursive. In fact, at first sight, they seem to have little in common with the mathematical structures that are usually considered in universal algebra. Manzonetto and Salibra (2008) viewed these topics from a wider perspective. They introduced the variety of Church algebras, namely algebras possessing two distinguished nullary terms representing the truth values, and a ternary term representing the if-then-else conditional construct, which is ubiquitous in programming languages. Beyond combinatory algebras, this class includes all Boolean algebras, Heyting algebras and rings with unity. They also proved that combinatory algebras satisfy a Representation theorem stating that every combinatory algebra can be decomposed as a direct product of directly indecomposable combinatory algebras. It is therefore natural to study the *indecomposable semantics* of λ-calculus, namely the class of λ-models that are indecomposable in this sense. It turns out that this class is large enough to include all the main semantics of λ-calculus, but also largely incomplete: there is a wealth of λ-theories whose models must be decomposable. This furnishes a uniform algebraic proof of incompleteness for the main semantics.

This book is not about...

When writing a mathematical book, the authors often start from the results they certainly want to present, but are eventually faced with the problem of what topics they need to exclude. We discuss below a non-exhaustive list of topics that would not have been misplaced in this satellite, but that we decided not to treat. We profit from the occasion and suggest some publications that might interest the reader.

1. **Evaluation strategies.** The λ-calculus is considered a call-by-name language. When restricting β-reduction to values, i.e. variables or abstractions, one obtains Plotkin's call-by-value λ-calculus (1975). The monograph "The Parametric Lambda Calculus", by Ronchi Della Rocca and Paolini (2004), presents the state of the art at that time. For decades this calculus has been considered somewhat pathological because some redexes remain stuck along the reduction for silly reasons, giving rise to premature normal forms. A solution arose from the analysis of the proof-nets obtained via Girard's so-called 'boring' translation of the intuitionistic arrow into linear logic. Solvability in call-by-value has been characterized both syntactically, by introducing explicit substitutions to unblock stuck redexes (Accattoli and Paolini (2012)), and semantically (Carraro and Guerrieri (2014)) via Ehrhard's relational model (2012). This clarified the global picture and generated several new results, e.g. De Benedetti and Ronchi Della Rocca (2015), Accattoli and Guerrieri (2016), Guerrieri et al. (2017), Kerinec et al. (2020), and Santo (2020). Other popular evaluation strategies are call-by-need (Accattoli et al. (2014), Kesner (2016), Balabonski et al. (2017) and Kesner et al. (2018)) and call-by-push-value (Ehrhard and Guerrieri (2016), Guerrieri and Manzonetto (2018), Faggian and Guerrieri (2021)).

2. **Variations of λ-calculus.** Over the years the λ-calculus has been extended in a number of different directions. The most common is towards actual programming languages. By adding to the simply typed λ-calculus constants for representing integers and booleans, programming primitives like the conditional test-on-zero, and a fixed point combinator, one obtains Plotkin's PCF (1977) that can be further extended with other features. Other variations maintain unchanged the pure untyped setting, while modifying the nature of the computations that are modeled. For instance, non-deterministic λ-calculi are obtained by adding an inner-choice operator, or a parallel composition (Boudol (1994)), or both (Dezani-Ciancaglini (1996), Bucciarelli et al. (2012b)). Ehrhard and Regnier (2003) enriched the λ-calculus with a syntactic derivative operator having a precise operational meaning, and developed an approximation theory based on the Taylor expansion, rather than Böhm trees (Ehrhard and Regnier (2008)). More recently, the interest has grown within the scientific community towards extensions of λ-calculus with probabilistic choice (Danos and Ehrhard (2011), Di Pierro (2017), Faggian and Ronchi Della Rocca (2019), Dal Lago and Leventis (2019)), quantum data (Dal Lago et al. (2009)) and even quantum control (Sabry et al. (2018), Díaz-Caro et al. (2019)).

3. **Game semantics.** Categories of games proved to be crucial for constructing a fully abstract model of PCF. This celebrated result was achieved by Abramsky et al. (2000), Hyland and Ong (2000) and Nickau (1996), independently. For an introduction to this topic we refer to the lecture notes of Abramsky and McCusker (1997), or to the original papers. Games models of the untyped λ-calculus have been defined by Di Gianantonio et al. (1999) and Ker et al. (2002), they are also fully abstract in the sense that they induce the maximal sensible λ-theory $\mathcal{H}^*$. Subsequently, a variety of categories of games were proposed in the literature as semantics of extensions of PCF, most recently by Castellan et al. (2015), Castellan et al. (2018) and Clairambault and de Visme (2020). Interestingly, games models are also well suited for imperative programming languages like John Reynolds's Idealized Algol and its variations: Abramsky and McCusker (1999), Harmer and McCusker (1999) and Longley (2009).

4. **Generalizations of the relational semantics.** The relational semantics of λ-calculus discussed in Chapter 14, has been generalized along several directions. The simplest variations are obtained by adding more structure, like in non-uniform coherent spaces (Bucciarelli and Ehrhard (2001)) with the free comonad by Boudes (2011) or probabilistic coherence spaces (Danos and Ehrhard (2011)). In all these semantics the underlined set of the denotation of a λ-term coincides with its relational interpretation. Other variations arise by considering finite multisets with possibly infinite multiplicities, as in Carraro et al. (2010), or modifying the construction of the comonad involved in the coKleisli, as in Melliès et al. (2018) or Grellois and Melliès (2015). Starting from the consideration that relations from A to B can be seen as matrices indexed by A and B, and populated by values from the boolean semi-ring, Laird et al. (2013) realized that several weighted relational

models arise by simply varying the semi-ring under consideration. Pagani et al. (2014) modified this construction to build a denotational semantics for quantum PCF, which turned out to be fully abstract (Clairambault and de Visme (2020)). Finally, Fiore et al. (2007) introduced a bicategorical model of linear logic based on profunctors (or distributors), representing the categorical generalization of relations. See Olimpieri (2021) for a profunctorial model of the untyped λ-calculus. We refer to Ong (2017) for a survey on the generalizations of the relational semantics.

5. **Linear logic.** Since its discovery by Jean-Yves Girard (1987), linear logic proved to be an inexhaustible source of inspiration for studying the λ-calculus and its variations. In particular, it allows to expose quantitative properties of λ-terms by creating a bridge between the mathematical notion of linearity (linear map between two vector spaces) and its analogue in computer science (program using its argument exactly once during its execution). The relational semantics of λ-calculus and its generalizations discussed above, arise from a quantitative semantics of differential linear logic—see Ehrhard (2018) for an introduction. We mention this fact, but we decided not to enter into details to avoid superposition with the "Handbook of Linear Logic" written in the same period by a group of logicians collectively called "International Research Network Linear Logic". We warmly welcome their initiative because, for too long, the definition of linear logic proof-nets mainly remained an oral tradition.

6. **Term rewriting.** This area of theoretical computer science already has a book of reference, namely "Term Rewriting Systems" by Terese (2003). In the last decades outstanding advances were made in the study of infinitary term rewriting (see, e.g., Simonsen (2006), Zantema (2008), Ketema and Simonsen (2009), Endrullis et al. (2012), Endrullis et al. (2014), Endrullis et al. (2018)) and graph rewriting (Corradini et al. (2019a), Corradini et al. (2019b), Overbeek and Endrullis (2020), Overbeek et al. (2021)). A thorough presentation of these topics would require separate monographs, therefore we decided to discuss only the infinitary λ-calculus in Chapter 6. Concerning this calculus, we regretfully need to omit its denotational semantics since presenting its infinitary rewriting properties is already a challenge. We make amend here, and mention some classical and recent results:

   (i) a Completeness Theorem by Salibra and Goldblatt (1999) stating that every infinitary λ-theory is the theory of an appropriate lambda abstraction algebra;

   (ii) relational models of the infinitary λ-calculus (equivalently, tensor type assignment systems) were obtained by Grellois and Melliès (2015), Vial (2017) and Vial (2018), by modifying the comonad of finite multisets.

We would like to emphasize that the articles and monographs mentioned above are intended as interesting pointers to the literature, but should by no means be regarded as a comprehensive list of publications in the respective areas. These are, in fact, ongoing research topics and an exhaustive bibliography should include hundreds of papers.

Dependencies between the chapters

The following dependency graph suggests an order in which the chapters can be read, but it should not be taken too literally. The actual logical dependencies between the various chapters are in fact too complex to be completely rendered in the graph.

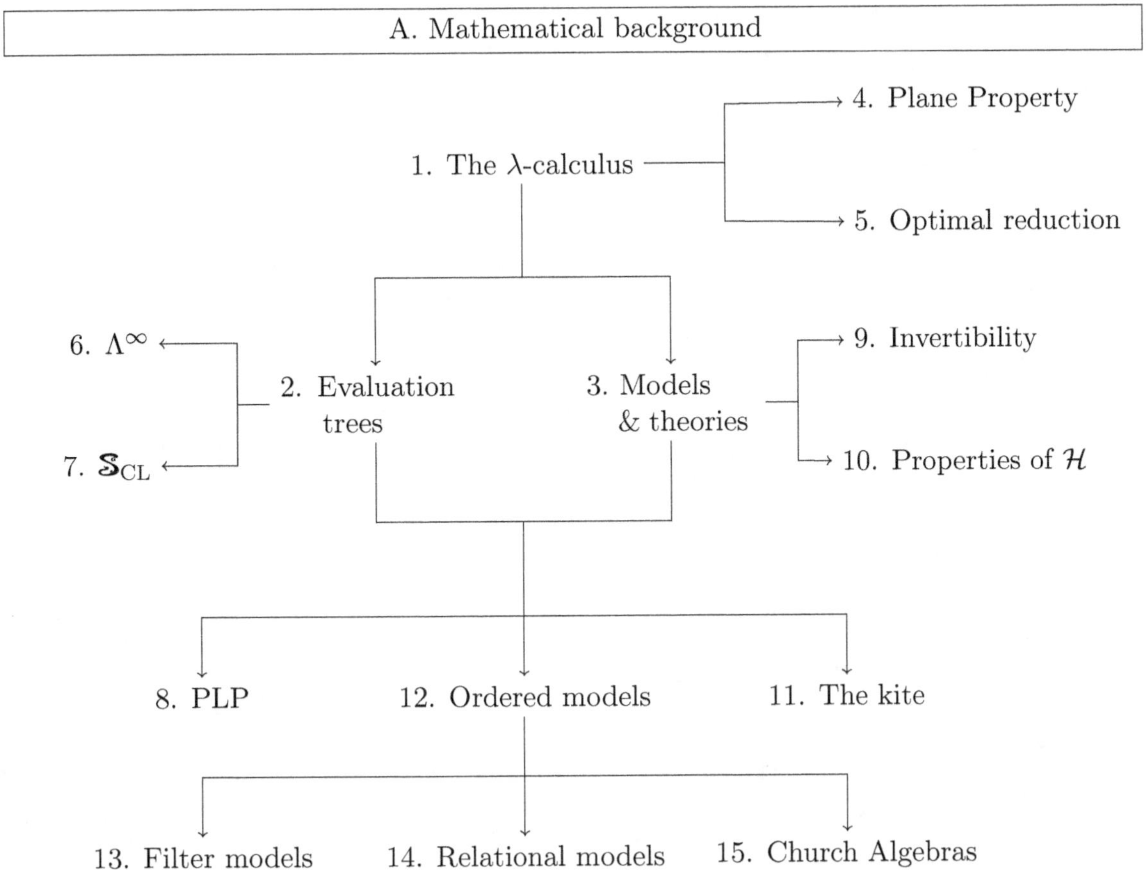

Appendix A presents some mathematical background that is best used 'by need'.

General notations

$\mathbb{N}$	The set of natural numbers.
$\mathbb{R}$	The set of real numbers.
$\|X\|$	The cardinality of a set X.
$\aleph_0$	The cardinality of $\mathbb{N}$.
$\aleph_1$	The cardinality of the set of all countable ordinals.
$\mathscr{P}(X)$	The powerset of X.
$\mathscr{P}_{\mathrm{f}}(X)$	The set of finite subsets of X.
Y^X	The set of functions from X to Y.
$f\colon X \to Y$	The function f belongs to Y^X.
$\mathrm{dom}(f)$	The domain of f.
f^{-1}	The inverse of f.
$\lambda\!\!\lambda x.E_x$	The 'meta' λ-abstraction: $f(x) = E_x$.
$\mu n.\mathrm{P}(n)$	Minimization: least $n \in \mathbb{N}$ satisfying $\mathrm{P}(n)$.
$X \,\&\, Y$	Logical and: X and Y.
$\{\mathrm{X}\|\mathrm{Y}\}$	Meta-operation 'or': either X or Y.

Acronyms

aka	also known as.
wlog	without loss of generality.
resp	respectively.
IH	induction hypothesis.
co-IH	coinduction hypothesis.
LHS	left-hand side.
RHS	right-hand side.

Conventions

In mathematical statements we sometimes denote optional symbols in parentheses. E.g., we may write "$M \in \Lambda^{(o)} \,\&\, M \twoheadrightarrow_\beta N \Rightarrow N \in \Lambda^{(o)}$" to represent the two statements "$M \in \Lambda^o \,\&\, M \twoheadrightarrow_\beta N \Rightarrow N \in \Lambda^o$" and "$M \in \Lambda \,\&\, M \twoheadrightarrow_\beta N \Rightarrow N \in \Lambda$" at once. However, this writing is not intended to represent the statement "$M \in \Lambda \,\&\, M \twoheadrightarrow_\beta N \Rightarrow N \in \Lambda^o$", which is false. That is, optional symbols should be uniformly considered or omitted.

Confluence diagrams are depicted as directed graphs. In these diagrams, solid arrows denote universally quantified reductions, while dashed arrows stand for existentially quantified ones. An analogous graphic convention is adopted for commutative diagrams representing morphisms in categories.

Part I

Preliminaries

Corrado Böhm

Gordon Plotkin

Erwin Engeler

Dana Scott

Jean-Yves Girard

Martin Hyland

DISCLAIMER. In this part we briefly present fundamental definitions and results that are useful in the rest of the book. The reader who is not familiar with untyped λ-calculus should be redirected towards gentler introductions to the subject like Barendregt's first book B[1984], Amadio and Curien (1998), Hindley and Seldin (2008) or Selinger (2008). Some of the definitions and results that are presented here, although not difficult, go beyond the material in these books. For a detailed historical survey on λ-calculus and combinatory logic, we refer to Cardone and Hindley (2009) and Seldin (2009).

Chapter 1

The λ-calculus in a nutshell

For the *λ-calculus* we generally follow the terminology and notations from B[1984] and, when it is not the case, we explicitly mention it.

1.1 The λ-calculus — Its syntax

Let $\mathbb{N}$ be the set of natural numbers. Consider fixed a denumerable set Var of *variables*, whose elements are usually denoted by x, y, w, z, possibly with indices $i, j, k, \ell, m, n \in \mathbb{N}$.

DEFINITION 1.1. The set Λ of *λ-terms* is defined by the simplified[1] grammar:

$$(\Lambda) \qquad M ::= x \mid \lambda x.M \mid MM$$

We use the symbol $\triangleq$ for definitional equality and $=$ for syntactic equality. (In contrast with B[1984], where both are denoted $\equiv$ and the symbol $=$ stands for β-conversion.)

NOTATION. (i) *We use uppercase letters like F, M, N, P, Q as metavariables for λ-terms.*

(ii) *We assume that application associates to the left and has higher priority than abstraction. For example, $\lambda x.\lambda y.\lambda z.xyz$ denotes the λ-term $\lambda x.(\lambda y.(\lambda z.(xy)z))$.*

(iii) *For $n \in \mathbb{N}$, we let $MN^{\sim n} \triangleq MN \cdots N$ and $M^n(N) \triangleq M(M(\cdots(MN)))$ (n times). In particular, we have $MN^{\sim 0} = M$ and $M^0(N) = N$.*

(iv) *Similarly, given $M, N_1, \dots, N_n \in \Lambda$, we let $M\vec{N} \triangleq MN_1 \cdots N_n$.*

(v) *We write $\lambda x_1 \dots x_n.M$, or simply $\lambda\vec{x}.M$, as an abbreviation for $\lambda x_1. \dots \lambda x_n.M$.*

DEFINITION 1.2. Consider a λ-term M.

(i) The number of symbols in M is called its *size* and is denoted by $\mathrm{size}(M)$.

(ii) The set $\mathrm{FV}(M)$ of *free variables of M* is defined inductively:

$$\begin{array}{lcl} \mathrm{FV}(x) & \triangleq & \{x\}, \\ \mathrm{FV}(MN) & \triangleq & \mathrm{FV}(M) \cup \mathrm{FV}(N), \\ \mathrm{FV}(\lambda x.M) & \triangleq & \mathrm{FV}(M) - \{x\}. \end{array}$$

[1] We always consider simplified grammars, this basically means that parentheses are left implicit.

(iii) The variables occurring in M that are not free are called *bound variables*.

(iv) We say that M is *closed*, or *a combinator*, whenever $\mathrm{FV}(M) = \emptyset$. We set

$$\Lambda^o \triangleq \{M \in \Lambda \mid \mathrm{FV}(M) = \emptyset\}.$$

(v) Similarly, given $x_1, \ldots, x_n \in \mathrm{Var}$, we write $\Lambda^o(\vec{x}) \triangleq \{M \in \Lambda \mid \mathrm{FV}(M) \subseteq \{\vec{x}\}\}$.

We now recall the notion of *substitution*, which is central in the theory of λ-calculus. To avoid unintended 'capture' of free variables in the abstraction case, we sometimes need to rename a bound variable choosing from Var a *fresh variable*, namely a variable that has not been used yet. This is always possible because the set Var is infinite.

DEFINITION 1.3. (i) Let $M, N \in \Lambda$ and $x \in \mathrm{Var}$. The *capture free substitution of N for x in M*, denoted $M[x{:=}N]$, is defined by induction on M as follows:

$$y[x{:=}N] \triangleq \begin{cases} N, & \text{if } y = x, \\ y, & \text{if } y \neq x; \end{cases}$$

$$(PQ)[x{:=}N] \triangleq (P[x{:=}N])(Q[x{:=}N]);$$

$$(\lambda y.P)[x{:=}N] \triangleq \begin{cases} \lambda y.P, & \text{if } y = x, \\ \lambda y.P[x{:=}N], & \text{if } x \neq y \ \&\ y \notin \mathrm{FV}(N), \\ \lambda z.P[y{:=}z][x{:=}N], & \text{if } x \neq y \ \&\ y \in \mathrm{FV}(N), \\ & \text{for some fresh variable } z. \end{cases}$$

(ii) More generally, given variables $x_1, \ldots, x_n$ and λ-terms $N_1, \ldots, N_n$, a *substitution* σ is a finite map sending $x_i \mapsto N_i$ for all i $(1 \leq i \leq n)$.

(iii) Given a substitution σ as above and a λ-term M, we may write $M\sigma$ or M^σ for

$$M[x_1{:=}N_1] \cdots [x_n{:=}N_n]$$

When no confusion may arise, this will be abbreviated in $M[x_1 := N_1, \ldots, x_n := N_n]$ or even $M[\vec{x}{:=}\vec{N}]$. In case $N \triangleq N_1 = \cdots = N_n$, we simply write $M[\vec{x}{:=}N]$.

CONVENTION 1.4. *Hereafter, as in* B[1984], *λ-terms are assumed to be* abstract terms, *i.e. they are considered modulo α-*conversion. *We also assume the* Variable Convention *stating that in each subterm of a term the names of free and bound variables are distinct.*

Now that we have introduced the syntax of λ-calculus, we may explain its operational semantics. Indeed, λ-calculus can be thought of as the first higher-order term rewriting system to have appeared in the literature. The reader is invited to consult the book Terese (2003) for a more detailed discussion on this approach.

DEFINITION 1.5. (i) A *notion of reduction* R on Λ is any binary relation $\mathrm{R} \subseteq \Lambda^2$.

(ii) If $(M, N) \in \mathrm{R}$, then M is called an R-*redex* and N its R-*contractum*.

(iii) $\mathrm{R} \subseteq \Lambda^2$ is called *compatible* if it is compatible w.r.t. abstraction and application:

$$\frac{M \mathrel{\mathrm{R}} N}{\lambda x.M \mathrel{\mathrm{R}} \lambda x.N} \qquad \frac{M \mathrel{\mathrm{R}} M'}{MN \mathrel{\mathrm{R}} M'N} \qquad \frac{N \mathrel{\mathrm{R}} N'}{MN \mathrel{\mathrm{R}} MN'}$$

(iv) For $\mathrm{R} \subseteq \Lambda^2$, its *compatible closure* is the least compatible relation including R.

The compatibility of a binary relation R on Λ can equivalently be expressed in terms of a closure under λ-calculus contexts. In λ-calculus a *context* is a λ-term possibly containing occurrences of a *hole* $[\,]$, which should be thought of as an algebraic variable.

DEFINITION 1.6. (i) A *context* $C[\,]$ is generated by the following grammar:

$$C[\,] ::= \ [\,] \mid x \mid \lambda x.C[\,] \mid C[\,]C[\,].$$

(ii) A context $C[\,]$ is called *single hole* if it has a unique occurrence of its hole $[\,]$.
(iii) Given a context $C[\,]$ and a λ-term M, $C[M]$ is the λ-term obtained by substituting M for all occurrences of $[\,]$ in $C[\,]$, possibly with capture of free variables in M.

A relation R is compatible whenever $M \mathrel{\mathrm{R}} N$ entails $C[M] \mathrel{\mathrm{R}} C[N]$, for all single hole contexts $C[\,]$. For this reason, compatible relations are often called *context closed*.

DEFINITION 1.7. Let R be a notion of reduction on Λ.
(i) *One step* R*-reduction*, written $\rightarrow_{\mathrm{R}}$, is the compatible closure of R.
(ii) *Many step* R*-reduction*, in notation $\twoheadrightarrow_{\mathrm{R}}$, is the reflexive, transitive closure of $\rightarrow_{\mathrm{R}}$.
(iii) R*-convertibility*, written $=_{\mathrm{R}}$, is the reflexive, symmetric, transitive closure of $\rightarrow_{\mathrm{R}}$.
(iv) An R*-normal form* (R-nf) is a λ-term M such that for no N one has $M \rightarrow_{\mathrm{R}} N$.
(v) We denote by $M^{\mathrm{R\text{-}nf}}$ the R-normal form of M, if it exists and is unique.

DEFINITION 1.8. (i) On Λ we define the following notions of reductions by writing

$$M \rightarrow N, \ \ldots \tag{R}$$

to denote $\mathrm{R} = \{(M, N) \mid \ldots\}$.
(ii) The notion of *β-reduction* is given by the axiom:

$$(\lambda x.M)N \rightarrow M[x{:=}N]. \tag{β}$$

This uses in addition to (i) pattern matching and means

$$\beta = \{(P, Q) \mid P = (\lambda x.M)N,\ Q = M[x{:=}N]\}.$$

(iii) The notion of *η-reduction* is given by the axiom:

$$\lambda x.Mx \rightarrow M, \qquad \text{if } x \notin \mathrm{FV}(M). \tag{η}$$

These notions of reduction generate reduction relations $\rightarrow_\beta, \twoheadrightarrow_\beta, \rightarrow_\eta, \twoheadrightarrow_\eta, \rightarrow_{\beta\eta}$ and $\twoheadrightarrow_{\beta\eta}$. The corresponding conversion relations are denoted by $=_\beta, =_\eta$ and $=_{\beta\eta}$.

DEFINITION 1.9. A notion of reduction R is called:
(i) *confluent* if, for all $M, N_1, N_2 \in \Lambda$:

$$M \twoheadrightarrow_{\mathrm{R}} N_1 \ \&\ M \twoheadrightarrow_{\mathrm{R}} N_2 \quad \Rightarrow \quad \exists Z \in \Lambda \,.\, N_1 \twoheadrightarrow_{\mathrm{R}} Z \ \&\ N_2 \twoheadrightarrow_{\mathrm{R}} Z.$$

(ii) *Church-Rosser* (*CR*, for short) if, for all $M, N \in \Lambda$:

$$M =_{\mathrm{R}} N \quad \Rightarrow \quad \exists Z \in \Lambda \,.\, M \twoheadrightarrow_{\mathrm{R}} Z \ \&\ N \twoheadrightarrow_{\mathrm{R}} Z.$$

Definition			Informal description
I	$\triangleq$	$\lambda x.x$	identity
$\mathsf{1}$	$\triangleq$	$\lambda xy.xy$	an η-expansion of the identity
Δ	$\triangleq$	$\lambda x.xx$	self-applicator
B	$\triangleq$	$\lambda fgx.f(gx)$	*composition*, i.e. $M \circ N \triangleq \mathsf{B}MN$
K	$\triangleq$	$\lambda xy.x$	K from combinatory logic and first projection, denoted T when representing the value *true*
F	$\triangleq$	$\lambda xy.y$	truth value *false* and second projection
S	$\triangleq$	$\lambda xyz.xz(yz)$	S from combinatory logic
c_n	$\triangleq$	$\lambda fz.f^n(z)$	n-th *Church numeral*
Y	$\triangleq$	$\lambda f.(\lambda x.f(xx))(\lambda x.f(xx))$	*Curry's fixed point combinator*
Θ	$\triangleq$	$(\lambda fx.x(ffx))(\lambda fx.x(ffx))$	*Turing's fixed point combinator*
Ω	$\triangleq$	$(\lambda x.xx)(\lambda x.xx)$	paradigmatic looping combinator
Ω_3	$\triangleq$	$(\lambda x.xxx)(\lambda x.xxx)$	'garbage' producing looping combinator
$\mathsf{K}^\star$	$\triangleq$	$\Theta\mathsf{K}$	'$\lambda x.$' producing looping combinator
J	$\triangleq$	$\mathsf{Y}(\lambda jxz.x(jz))$	Wadsworth's 'infinite' η-expansion of I

Table 1.1: Some well-known combinators.

PROPOSITION 1.10. (i) R *is confluent if and only if it is Church-Rosser.*
(ii) *Suppose* R *is confluent. If M has an* R*-normal form, then it is unique.*

The Church–Rosser Theorem, first demonstrated by Church and Rosser (1936), is a fundamental result having as a consequence the consistency of λ-calculus.

PROPOSITION 1.11 (CHURCH–ROSSER THEOREM).
The notions of reduction (β), (η) and $(\beta\eta)$ are confluent.

We simply write M^{nf} for $M^{\beta\text{-}\mathrm{nf}}$ and let $\mathrm{NF} \triangleq \{M \in \Lambda \mid \exists M^{\mathrm{nf}} \,.\, M \twoheadrightarrow_\beta M^{\mathrm{nf}}\}$.

NOTATION 1.12. (i) In Table 1.1 we fix the notations for some specific combinators.
(ii) Let S^+ be the *successor* satisfying $\mathsf{S}^+\mathsf{c}_n =_\beta \mathsf{c}_{n+1}$, for all $n \geq 0$, and S^- be the *predecessor* satisfying $\mathsf{S}^-\mathsf{c}_0 =_\beta \mathsf{c}_0$ and $\mathsf{S}^-\mathsf{c}_{n+1} =_\beta \mathsf{c}_n$, for all $n > 0$.
(iii) Let ifz be a combinator implementing the programming primitive *if-zero*. Write $\mathsf{ifz}(L, M, N)$ for $\mathsf{ifz}\,LMN$, whence

$$\mathsf{ifz}(\mathsf{c}_n, M, N) =_\beta \begin{cases} M, & \text{if } n = 0, \\ N, & \text{otherwise.} \end{cases}$$

Pairing and finite sequences can be represented in λ-calculus in different ways.

DEFINITION 1.13. Let $M, N, M_0, \dots, M_n \in \Lambda$, for $n \geq 0$, and x be a fresh variable.
(i) The *pairing* $[M, N] \triangleq \lambda x.xMN$ satisfies $[M, N]\mathsf{K} =_\beta M$ and $[M, N]\mathsf{F} =_\beta N$.

(ii) *Finite sequences* $[M_0, \ldots, M_n]$ are defined in terms of $[\cdot, \cdot]$ inductively (for $m \geq 0$):

$$\begin{array}{lcl}[M_0] & \triangleq & M_0, \\ [M_0, \ldots, M_{m+1}] & \triangleq & [M_0, [M_1, \ldots, M_{m+1}]].\end{array}$$

Given a finite sequence $[M_0, \ldots, M_n]$ of length $n+1$, there are projections $\pi_i^n \triangleq \lambda x.x\mathsf{F}^{\sim i}\mathsf{K}$ for all $i\,(0 \leq i < n)$ and $\pi_n^n \triangleq \lambda x.x\mathsf{F}^{\sim n}$.

(iii) A more direct definition of finite sequences is $\langle M_0, \ldots, M_n\rangle \triangleq \lambda x.xM_0 \cdots M_n$.

The Böhm Theorem, due to Böhm (1968), is another important result in λ-calculus stating that two β-normalizable λ-terms that are not η-convertible can be separated.

THEOREM 1.14 (BÖHM THEOREM). *Let $M, N \in \Lambda^o$. Assume M and N have η-distinct β-normal forms. There exists $F \in \Lambda^o$ such that*

$$FM =_\beta \mathsf{K} \qquad\qquad FN =_\beta \mathsf{F}$$

REMARK 1.15. (i) In the above theorem, F can be taken of the form $\lambda x.x\vec{P}$ for $\vec{P} \in \Lambda^o$.

(ii) Two λ-terms M, N satisfying the hypotheses of the theorem above can be sent respectively to two distinct variables y, z using $F' \triangleq \lambda x.Fxyz \in \Lambda^o(x, y)$.

The following families of combinators are central in the Böhm-out technique traditionally used to demonstrate the Böhm Theorem (see B[1984], Chapter 10).

DEFINITION 1.16. The *selector* U_k^n and the *tupler* P_n are defined as follows:

$$\mathsf{U}_k^n \triangleq \lambda x_1 \ldots x_n.x_k, \qquad \mathsf{P}_n \triangleq \lambda x_1 \ldots x_n.\langle x_1, \ldots, x_n\rangle, \qquad (1 \leq k \leq n).$$

The definition of U_k^n is slightly different from the one in B[1984], as in the old book the indices n, k start from 0. The next remark states explicitly the key properties satisfied by the selectors and the tuplers and should clarify the choice of such a terminology.

REMARK 1.17. Let $m, n \geq 0$ and $M_1, \ldots, M_m, N_1, \ldots, N_n \in \Lambda^o$.

(i) $(\mathsf{P}_{m+n}M_1 \cdots M_m)N_1 \cdots N_n =_\beta \langle M_1, \ldots, M_m, N_1, \ldots, N_n\rangle$,

(ii) $\langle N_1, \ldots, N_n\rangle\, \mathsf{U}_i^n =_\beta N_i$ for every $i\,(1 \leq i \leq n)$.

The Böhm Theorem 1.14 can be generalized to tuples of λ-terms.

THEOREM 1.18. *Let $M_0, \ldots, M_{n-1} \in \Lambda^o$ have distinct βη-normal forms. Then*

(i) $\exists F \in \Lambda^o, \forall k < n\,.\,FM_k =_\beta \mathsf{c}_k$.

(ii) $\exists \vec{P} \in \Lambda^o, \forall k < n\,.\,M_k\vec{P} =_\beta \mathsf{c}_k$.

PROOF. (i) See Böhm et al. (1979) or B[1984], Corollary 10.4.14.

(ii) Following the proof of (i). □

There are also generalizations to arbitrary finite sets of λ-terms or even recursively enumerable (*r.e.*, for short) infinite sets of these—see Statman and Barendregt (2005), or Barendregt (2020a) for a survey.

DEFINITION 1.19. (i) Let $M \in \Lambda$. A λ-term X is a *fixed point* of M if $MX =_\beta X$.

(ii) A λ-term Y is a *fixed point combinator* (*fpc*) if $YM =_\beta M(YM)$, for all $M \in \Lambda$.

(iii) An fpc Y is *reducing* whenever $YM \twoheadrightarrow_\beta M(YM)$ holds, for all $M \in \Lambda$.

EXAMPLES 1.20. Both Θ and Y are fpc's: the former is reducing while the latter is not.

Solvability

The λ-terms are classified into solvable and unsolvable, depending on their capability of interaction with the environment.

DEFINITION 1.21. (i) A closed λ-term N is *solvable* if there exist $P_1, \dots, P_n \in \Lambda$ such that $NP_1 \cdots P_n =_\beta \mathsf{I}$. An arbitrary λ-term M is *solvable* if its closure $\lambda\vec{x}.M$ is solvable.
(ii) A λ-term M is called *unsolvable* if it is not solvable.
(iii) We denote by SOL (resp. UNS) the set of all solvable (resp. unsolvable) λ-terms.

It was first noticed by Wadsworth (1976) that solvable λ-terms can be characterized in terms of 'head normalizability'. Indeed, it is well known that every λ-term M has one of these following shapes (for $n, k \geq 0$):

- $\lambda x_1 \dots x_n.yM_1 \cdots M_k$ where y is called the *head variable* of M. In this case we say that M *is in head normal form* (*hnf*, for short).
- $\lambda x_1 \dots x_n.(\lambda y.P)QM_1 \cdots M_k$ where $(\lambda y.P)Q$ is called the *head redex* of M.

DEFINITION 1.22. (i) We let HNF be the set of λ-terms having a head normal form.
(ii) The *head reduction* $\to_h$ is the reduction strategy obtained by contracting the head redex. The corresponding multi-step relation is denoted by $\twoheadrightarrow_h$.
(iii) The *principal hnf of* M, written M^{phnf}, is the hnf (if it exists) obtained from M by head reduction, i.e. $M \twoheadrightarrow_h M^{\mathsf{phnf}} \not\to_h$.
(iv) Two λ-terms M, N in hnf are *similar*, in symbols $M \sim N$, if for some n, n', k, k':

$$M = \lambda x_1 \dots x_n.yM_1 \cdots M_k, \qquad N = \lambda x_1 \dots x_{n'}.yN_1 \cdots N_{k'},$$

where $k - k' = n - n'$ and y is either free in both λ-terms, or bound in both λ-terms.
(v) For arbitrary $M, N \in \Lambda$ we set $M \sim N$ if either they are both unsolvable or $M, N \in \mathrm{SOL}$ and $M^{\mathsf{phnf}} \sim N^{\mathsf{phnf}}$.

Notice that the quantities $n - n'$ and $k - k'$ in Definition 1.22(iv) can be negative.
Head reduction is an effective and deterministic strategy, so the principal hnf of M (if any) is unique. A λ-term M has a head normal form if and only if M^{phnf} exists.

THEOREM 1.23. *A λ-term M is solvable if and only if M has a head normal form.*

COROLLARY 1.24. $\mathrm{NF} \subsetneq \mathrm{HNF} = \mathrm{SOL}$.

EXAMPLES 1.25. (i) Typical examples of unsolvable λ-terms are $\Omega, \Omega_3, \mathsf{YI}, \mathsf{K}^\star$.
(ii) $\mathsf{I}, \mathsf{K}, \mathsf{F}, \mathsf{F}\Omega, \mathsf{Y}, \Theta$ are solvable λ-terms.
(iii) I and $\mathsf{1}$ have similar hnf's, while K and F do not.

The next result appears as Lemma 17.4.4 in B[1984] and shows that any $M \in \Lambda^o$ can be turned into an unsolvable by applying enough Ω's.

LEMMA 1.26. *For all $M \in \Lambda^o$ there exists $k \in \mathbb{N}$ such that $M\Omega^{\sim k}$ is unsolvable.*

PROOF. If $M \in \mathrm{UNS}$ take $k = 0$. Otherwise M has a hnf $\lambda x_1 \dots x_n.x_iM_1 \cdots M_m$ $(n > 0)$. Then $M\Omega^{\sim i} =_\beta \lambda x_{i+1} \dots x_n.\Omega M_1^\sigma \cdots M_m^\sigma$ for $\sigma = [x_1, \dots, x_i := \vec{\Omega}]$, so $M\Omega^{\sim i} \in \mathrm{UNS}$. □

Encoding lambda terms

It is well known that the λ-calculus, despite its very simple syntax, is Turing-complete. A *partial numeric function* is a possibly partial mapping $f : \mathbb{N}^k \to \mathbb{N}$, for some $k \in \mathbb{N}$. *Partial recursive* and *recursive (numeric) functions* are defined as usual (for example, see Rogers (1967)). Given a partial (numeric) function $f : \mathbb{N}^k \to \mathbb{N}$, we write $f(n_1, \ldots, n_k)\downarrow$ whenever $(n_1, \ldots, n_k) \in \mathrm{dom}(f)$ and $f(n_1, \ldots, n_k)\uparrow$ otherwise.

DEFINITION 1.27 (CHURCH). A partial function $f : \mathbb{N}^k \to \mathbb{N}$ is *λ-definable* whenever there exists $F \in \Lambda^o$ satisfying for all $n_1, \ldots, n_k \in \mathbb{N}$:

$$\begin{array}{ll} F\mathsf{c}_{n_1} \cdots \mathsf{c}_{n_k} =_\beta \mathsf{c}_{f(n_1,\ldots,n_k)}, & \text{if } f(n_1, \ldots, n_k)\downarrow, \\ F\mathsf{c}_{n_1} \cdots \mathsf{c}_{n_k} \in \mathrm{UNS}, & \text{otherwise.} \end{array}$$

The following result is due to Church (1933) and Kleene (1936b).

PROPOSITION 1.28. *A function is partial recursive if and only if it is λ-definable.*

Important results in λ-calculus are obtained by applying techniques from recursion theory. The key idea is to exploit the fact that λ-terms can be encoded as natural numbers, and therefore as Church numerals, in an effective way.

DEFINITION 1.29. (i) Let $\# : \Lambda \to \mathbb{N}$ be an effective one-to-one encoding associating with every λ-term M its *code* $\#M$ (the *Gödel number* of M).

(ii) For example we can define # as follows. For $n, m \in \mathbb{N}$, write

$$(n, m) \triangleq \frac{1}{2}(n + m)(n + m + 1) + m.$$

This function is a computable bijection $\lambda\!\!\lambda nm.(n, m) \colon \mathbb{N}^2 \to \mathbb{N}$, with computable inverse $p_1, p_2 \colon \mathbb{N} \to \mathbb{N}$, such that writing $n \cdot i \triangleq p_i(n)$ one has $(n \cdot 1, n \cdot 2) = n$ and $n \cdot i \leq n$.

Fix an enumeration $\mathrm{Var} = \{x_0, x_1, x_2, \ldots\}$. Now, define $\# \colon \Lambda \to \mathbb{N}$ inductively by

$$\begin{array}{rcl} \#(x_k) & \triangleq & (0, k), \\ \#(PQ) & \triangleq & (1, (\#(P), \#(Q))), \\ \#(\lambda x_k.P) & \triangleq & (2, (k, \#(P))). \end{array}$$

(iii) The *quote of* M, written $\ulcorner M \urcorner$, is the corresponding Church's numeral $\mathsf{c}_{\#M}$.

REMARK 1.30. The following operations are effective:
- from $\#M$ compute $\#N$ where $M \to_h N$ (as head reduction is an effective strategy);
- from $\#(\lambda\vec{x}.yM_1 \cdots M_k)$ compute $\#M_i$ for all $i\,(1 \leq i \leq k)$;
- from $\#M$ compute $\#(\lambda x_1 \ldots x_n.M)$ for $x_1, \ldots, x_n \in \mathrm{Var}$.

The next theorem shows that it is possible to reconstruct a closed λ-term M starting from its quote via a suitable combinator E. It does not generalize to arbitrary λ-terms M.

THEOREM 1.31 (B[1984], THEOREM 8.1.6). *There exists* $\mathsf{E} \in \Lambda^o$ *such that*

$$\mathsf{E}\ulcorner M \urcorner \twoheadrightarrow_\beta M, \text{ for all } M \in \Lambda^o.$$

PROOF. See B[1984] for a proof based on the original construction of Kleene, or better Barendregt (2001) for a simpler construction by P. de Bruin. □

It is not possible to generalize Theorem 1.31 to open terms, using # and the corresponding $\ulcorner M \urcorner$. T. Mogensen used another ('higher-order') coding $\#^{\mathrm{M}}$ for which an evaluator E^{M} exists that decodes quoted terms for all $M \in \Lambda$. In Barendregt (2001) the methods are compared. For the code $\ulcorner M \urcorner$ defined above one can discriminate open terms by looking at their free variables, which is not possible for the higher-order coding.

REMARK 1.32. The combinator E is defined in such a way that if M is unsolvable then $\mathsf{E}M$ is unsolvable as well.

The theorem above is related to the theory of self-interpreters in λ-calculi, which is an ongoing subject of study, see e.g. Mogensen (1992); Given-Wilson and Jay (2011); Polonsky (2011a); Brown and Palsberg (2016) and Jay (2018). In Chapters 8, 10 and 11 we present some constructions that exploit such interpreters in a fundamental way. As this book is conceived as a satellite of B[1984], we prefer to keep the same notion of encoding. However, these constructions could be recast using any (effective) encoding, like the one proposed by Mogensen (1992) that can be applied to open λ-terms as well.

EXERCISE 1.33. (i) Show that there exists a combinator $M \in \Lambda^o$ such that

$$M\ulcorner x \urcorner \ulcorner y \urcorner =_\beta \begin{cases} \mathsf{T}, & \text{if } x = y; \\ \mathsf{F}, & \text{else.} \end{cases}$$

This is not possible for the Mogensen coding.

(ii) Show that there is no $\mathsf{E}^* \in \Lambda$ such that $\forall M \in \Lambda \,.\, \mathsf{E}^*\ulcorner M \urcorner =_\beta M$.

(iii) Show that there is a λ-term $\mathsf{E}_{\vec{x}} \in \Lambda^o(\vec{x})$ such that $\forall M \in \Lambda^o(\vec{x}) \,.\, \mathsf{E}_{\vec{x}}\ulcorner M \urcorner =_\beta M$.

(iv) Show that in the previous item one even may require $\forall M \in \Lambda^o(\vec{x}) \,.\, \mathsf{E}_{\vec{x}}\ulcorner M \urcorner \twoheadrightarrow_\beta M$.

(v) Note that in the λI-calculus[2] the set $\Lambda^o(x)$ is closed under $=_\beta$, but decidable. This shows that the theorem of Scott (B[1984], Theorem 6.6.2(ii)) is not valid for the restricted calculus. Where does the proof fail for the λI-calculus?

Effective sequences

The combinator E from Theorem 1.31 can be used to construct λ-terms enumerating along their reduction infinitely many combinators.

DEFINITION 1.34. (i) An enumeration $e \triangleq (M_0, M_1, M_2, \dots)$ of closed λ-terms is called *effective*[3] if there exists an $F \in \Lambda^o$ such that

$$\forall n \in \mathbb{N} \,.\, F\mathsf{c}_n =_\beta M_n.$$

Such an F is called an *enumerator* of e.

[2]The λI-*calculus* is the λ-calculus without erasing: $\lambda x.M$ entails $x \in \mathrm{FV}(M)$. See B[1984], Chapter 9.
[3]Or *uniform* in B[1984], Definition 8.2.1.

(ii) Given an effective enumeration e as above, it is possible[4] to define the *sequence* $[M_n]_{n \in \mathbb{N}}$ as a single λ-term satisfying

$$[M_n]_{n \in \mathbb{N}} =_\beta [M_0, [\ldots, [M_{i-1}, [M_{n+i}]_{n \in \mathbb{N}}] \cdots]].$$

The corresponding *i-th projection* is $\pi_i \triangleq \lambda y.y\mathsf{F}^{\sim i}\mathsf{K}$ since $\pi_i[M_n]_{n \in \mathbb{N}} =_\beta M_i$ (for $i \geq 0$).

(iii) A *stream* is any λ-term of the form $\lambda x_1 \ldots x_k.[M_n x_1 \cdots x_k]_{n \in \mathbb{N}}$ for some sequence $[M_n]_{n \in \mathbb{N}}$ and fresh variables $x_1, \ldots, x_k \in \mathrm{Var}$ (for $k \geq 0$).

Notice that all sequences are streams, while the converse is false.

NOTATION 1.35. For sequences $[M_n]_{n \in \mathbb{N}}$ and streams S, we use the following notations:

$$\begin{array}{rclcl} [M_n]_{n \in \mathbb{N}} & = & [M_0, [M_1, [M_2, \ldots]]] & = & [M_0, M_1, M_2, \ldots]. \\ S x_1 \cdots x_k & = & [M_0\vec{x}, [M_1\vec{x}, [M_2\vec{x}, \ldots]]] & = & [M_0\vec{x}, M_1\vec{x}, M_2\vec{x}, \ldots]. \end{array}$$

EXAMPLES 1.36. (i) The stream $[\mathsf{I}]_{n \in \mathbb{N}} \triangleq [\mathsf{I}, \mathsf{I}, \mathsf{I}, \ldots]$ contains infinitely many copies of I.

(ii) The stream $[\mathsf{Ec}_n]_{n \in \mathbb{N}} \triangleq [\mathsf{Ec}_0, \mathsf{Ec}_1, \mathsf{Ec}_2, \ldots]$ enumerates the whole set Λ^o.

EXERCISE 1.37. Write the actual λ-terms representing the streams $[\mathsf{I}]_{n \in \mathbb{N}}$ and $[\mathsf{Ec}_n]_{n \in \mathbb{N}}$.

1.2 Properties of reduction

We present some more advanced properties of reduction that are needed in Part II of this book, for studying optimal reductions. Most definitions and results are taken from a survey paper by Lévy (2017). We refer the reader to Asperti and Guerrini (1998) for a gentler introduction to the notions of residuals and finite developments. We conclude the section by recalling the Standardization Theorem for β- and $\beta\eta$-reductions.

DEFINITION 1.38. Let $M \in \Lambda$ and R be a notion of reduction.

(i) An R-*reduction sequence* ρ is a finite or infinite sequence of λ-terms such that the $(n+1)$-th term is obtained from the n-th term by contracting an R-redex. I.e.,

$$\rho = M \to_{\mathrm{R}} M_1 \to_{\mathrm{R}} M_2 \to_{\mathrm{R}} \cdots \to_{\mathrm{R}} M_n \to_{\mathrm{R}} \cdots$$

The term M is called the *initial term* of ρ. If ρ is finite, then its last term is called *final*.

(ii) Two R-reduction sequences ρ, ν are called *coinitial* if their initial terms coincide. They are called *cofinal* if they are both finite and their final terms coincide.

(iii) We write $\rho : M \twoheadrightarrow_{\mathrm{R}} N$ to specify any finite R-reduction sequence whose initial and final terms are M and N, respectively.

(iv) The *composition* of two finite R-reduction sequences $\rho : M \twoheadrightarrow_{\mathrm{R}} N$, $\nu : N \twoheadrightarrow_{\mathrm{R}} P$, is denoted by $\rho\,;\nu : M \twoheadrightarrow_{\mathrm{R}} P$ and is defined as their concatenation.

(v) The *empty* R-*reduction sequence* has length 0 and is denoted by o. Note that, for all finite R-reduction sequences ρ, we have $\rho\,; o = \rho = o\,; \rho$.

[4]See B[1984], Corollary 8.2.6.

When there is no possibility of confusion, we say that ρ is an R-reduction, thus omitting the 'sequence'. Moreover, in case $\mathrm{R} = \beta$, we simply say that ρ is a reduction.

DEFINITION 1.39. Let $M \in \Lambda$ and R be a notion of reduction.

(i) The R-*reduction graph of* M is the directed graph defined as follows:

$$\mathcal{G}_{\mathrm{R}}(M) \triangleq (\{N \mid M \twoheadrightarrow_{\mathrm{R}} N\}, \to_{\mathrm{R}})$$

In other words, the nodes of $\mathcal{G}_{\mathrm{R}}(M)$ are all the λ-terms N satisfying $M \twoheadrightarrow_{\mathrm{R}} N$ and there exists an edge from a node P to a node Q (possibly, $Q = P$) if $P \to_{\mathrm{R}} Q$ holds.

(ii) An edge $M \to_{\mathrm{R}} M$ having identical source and target is called *a loop*.

(iii) We say that $\mathcal{G}_{\mathrm{R}}(M)$ is *finite* if it has finitely many nodes, *infinite* otherwise.

(iv) We slightly abuse notations, and write

$$\begin{array}{lcl} N \in \mathcal{G}_{\mathrm{R}}(M) & \iff & M \twoheadrightarrow_{\mathrm{R}} N, \\ \rho \in \mathcal{G}_{\mathrm{R}}(M) & \iff & \text{the sequence } \rho \text{ is a path in the graph } \mathcal{G}_{\mathrm{R}}(M). \end{array}$$

Notice that a reduction $\rho \in \mathcal{G}_{\mathrm{R}}(M)$ can be infinite even when the graph $\mathcal{G}_{\mathrm{R}}(M)$ is finite. For instance, we have

$$\Omega \to_\beta \Omega \to_\beta \Omega \to_\beta \cdots \in \mathcal{G}_\beta(\Omega)$$

where $\mathcal{G}_\beta(\Omega)$ is the graph having one node Ω and a loop $\Omega \to_\beta \Omega$.

Residuals and parallel reduction

Recall that the notion of redex, relative to a notion of reduction R, has been introduced in Definition 1.5(ii). We are now interested in considering the case $\mathrm{R} = \beta$, therefore we write 'redex' to mean β-redex.

DEFINITION 1.40. Let $M \in \Lambda$.

(i) Given a redex (occurrence) R in M, in symbols $R \in M$, we write $M \xrightarrow{R}_\beta N$ if the λ-term N is obtained from M by contracting the redex R.

(ii) A redex $R \in M$ is called a K-*redex* if it is of the form $R = (\lambda x.P)Q$ with $x \notin \mathrm{FV}(P)$.

(iii) Consider a reduction $\rho : M \twoheadrightarrow_\beta N$ and a redex $R \in M$. Now, underline the leftmost symbol 'λ' in R, as in $R = (\underline{\lambda} x.P)Q$. At every step of ρ, the contraction of some redex may erase[5], preserve or copy the underlined λ's. The set R/ρ of *residuals of* R *across the reduction* ρ is the set of redex occurrences in N whose λ remains underlined. In this situation, we also say that R is an *ancestor* of any redex $S \in R/\rho$.

(iv) Let $\rho : M \twoheadrightarrow_\beta N$ and $R \in N$. We say that the redex R *has been created along* ρ, or simply that R *is new*, whenever there is no redex $S \in M$ such that $R \in S/\rho$.

REMARK 1.41. Consider $M \to_\beta N$.

(i) A redex in M may have several residuals in N, or none.

(ii) Every redex in N has at most one ancestor in M.

[5] A $\underline{\lambda}$ can be erased either because it occurs in the argument of a K-redex, as in $\mathsf{F}((\underline{\lambda} x.x)y) \to_\beta \mathsf{I}$, or because the corresponding redex is contracted, as in $(\underline{\lambda} x.x)y \to_\beta y$.

EXAMPLES 1.42. (i) Take $\Delta(\mathsf{F}y)$ and $R = \mathsf{F}y$ as redex occurrence. By underlining the leftmost 'λ' in $\mathsf{F}y = (\lambda x.\mathsf{I})y$ and then contracting the outermost redex, we obtain:

$$\rho : \Delta((\underline{\lambda}x.\mathsf{I})y) \xrightarrow{\Delta(\mathsf{F}y)}_\beta ((\underline{\lambda}x.\mathsf{I})y)((\underline{\lambda}x.\mathsf{I})y)$$

The set R/ρ is a doubleton containing as residuals the two copies of R.

(ii) The reduction ρ above can be continued until no residuals are left:

$$\begin{array}{lllllll} \rho\,;\nu : & \Delta((\underline{\lambda}x.\mathsf{I})y) & \to_\beta & ((\underline{\lambda}x.\mathsf{I})y)((\underline{\lambda}x.\mathsf{I})y) & \to_\beta & \mathsf{I}((\underline{\lambda}x.\mathsf{I})y) & \to_\beta & \mathsf{II} \\ \rho\,;\nu' : & \Delta((\underline{\lambda}x.\mathsf{I})y) & \to_\beta & ((\underline{\lambda}x.\mathsf{I})y)((\underline{\lambda}x.\mathsf{I})y) & \to_\beta & ((\underline{\lambda}x.\mathsf{I})y)\mathsf{I} & \to_\beta & \mathsf{II} \end{array}$$

where $R/(\rho\,;\nu) = R/(\rho\,;\nu') = \emptyset$.

(iii) If $R = \Delta(\mathsf{I}y)$ then we need to underline the λ in Δ. By contracting the rightmost redex, we get:

$$\rho : (\underline{\lambda}x.xx)(\mathsf{I}y) \xrightarrow{\mathsf{I}y}_\beta (\underline{\lambda}x.xx)y$$

In this case, $R/\rho = \{\Delta y\}$ is a singleton. Otherwise, if we contract the outermost redex, then we obtain:

$$\nu : (\underline{\lambda}x.xx)(\mathsf{I}y) \xrightarrow{\Delta(\mathsf{I}y)}_\beta \mathsf{I}y(\mathsf{I}y)$$

whence $R/\nu = \emptyset$.

LEMMA 1.43 (LÉVY). *Let $M, N \in \Lambda$ be such that $M \to_\beta N$. A new redex $R \in N$ can only be created in one of the following three ways.*

1. *A function is passed to the left of an application as in:*

$$(\lambda x.\cdots(xN)\cdots)(\lambda y.M) \to_\beta \cdots((\lambda y.M)(N[x:=\lambda y.M]))\cdots$$

2. *A curried function takes its first argument:*

$$((\lambda xy.M)N)P \to_\beta (\lambda y.M[x:=N])P$$

3. *A function is applied to the identity at the left of an application:*

$$(\lambda x.x)(\lambda y.M)N \to_\beta (\lambda y.M)N$$

PROOF. By case analysis. □

DEFINITION 1.44. Let $M \in \Lambda$, $\mathcal{F}$ be a set of redexes in M and $\rho \in \mathcal{G}_\beta(M)$.

(i) If ρ is finite, i.e. $\rho : M \twoheadrightarrow_\beta N$, then the *residual relation* maps $\mathcal{F}$ to the set $\mathcal{F}/\rho$ of all redexes in N that are residuals of some redex in $\mathcal{F}$ across ρ. The elements of $\mathcal{F}/\rho$ are called *residuals of $\mathcal{F}$ across ρ*.

(ii) The reduction ρ is *relative to $\mathcal{F}$* if every step in the sequence ρ is obtained by contracting a residual of a redex in $\mathcal{F}$.

(iii) We say that ρ is a *(complete) development of $\mathcal{F}$* if it is a maximal reduction relative to $\mathcal{F}$. This means that ρ is finite and there are no residuals of $\mathcal{F}$ across ρ (i.e., $\mathcal{F}/\rho = \emptyset$).

The finiteness of developments was first proved in Schroer's PhD thesis (1965), but his work remained unpublished.

THEOREM 1.45 (FINITE DEVELOPMENTS$^+$). *Let $M \in \Lambda$ and $\mathcal{F}$ be a set of redexes in M. Consider relative reductions that only contract residuals of redexes in $\mathcal{F}$.*
(i) *There is no infinite reduction relative to $\mathcal{F}$.*
(ii) *All developments of $\mathcal{F}$ end at the same term.*
(iii) *The residuals of any redex $R \in M$ (possibly not in $\mathcal{F}$) are the same by all developments of $\mathcal{F}$.*

PROOF. Cf. B[1984], Theorem 11.2.25. See also Lévy (1978b). □

In Chapter 5, we present a proof of the so-called Generalized Finite Developments (Theorem 5.65), using Jean-Jacques Lévy's labeled λ-calculus (see Section 5.3).

The item (iii) of Theorem 1.45 is important as it shows that the notion of residual is consistent with parallel steps. One can simultaneously contract a set $\mathcal{F}$ of redexes and uniquely identify its residuals, since they are independent from the order in which redexes of $\mathcal{F}$ are contracted. This leads to the following definition of parallel reduction.

DEFINITION 1.46. (i) Given a set $\mathcal{F}$ of redexes of M, define the *parallel reduction of M relative to $\mathcal{F}$*, in symbols $\xrightarrow{\mathcal{F}}$, by:

$$\frac{\rho : M \twoheadrightarrow_\beta N \text{ is a development of } \mathcal{F}}{M \xrightarrow{\mathcal{F}} N}$$

(ii) We let $\mathcal{F}_1 ; \cdots ; \mathcal{F}_n$ denote the sequence of parallel reduction steps

$$M_0 \xrightarrow{\mathcal{F}_1} M_1 \xrightarrow{\mathcal{F}_2} \cdots \xrightarrow{\mathcal{F}_{n-1}} M_{n-1} \xrightarrow{\mathcal{F}_n} M_n$$

When the $\mathcal{F}_i$'s are singletons, say $\mathcal{F}_i = \{R_i\}$, the parentheses are omitted. So, we simply write $M_{i-1} \xrightarrow{R_i} M_i$ for $M_{i-1} \xrightarrow{\{R_i\}} M_i$ and $R_1 ; \cdots ; R_n$ for the corresponding reduction.

(iii) Given two coinitial reductions $\mathcal{F}_1$ and $\mathcal{F}_2$, define:

$$\mathcal{F}_1 \sqcup \mathcal{F}_2 \triangleq \mathcal{F}_1 ; (\mathcal{F}_2/\mathcal{F}_1)$$

REMARK 1.47. (i) In Definition 1.46(i) there is no restriction on $\mathcal{F}$, whence $M \xrightarrow{\mathcal{F}} N$ is defined also for $\mathcal{F} = \emptyset$, in which case $M = N$. Notice, however, that the reduction $\emptyset$ is different from o, because the former has length 1 while the latter has length 0.

(ii) The similarity between the notations $M \xrightarrow{R} N$ (Definition 1.46(ii)) and $M \xrightarrow{R}_\beta N$ (Definition 1.40(i)) should not generate any confusion, since their meanings coincide.

(iii) Note that $\mathcal{F}_1 \sqcup \mathcal{F}_2$ is the parallel reduction of length 2 obtained by contracting the redexes of $\mathcal{F}_1$ first, and subsequently the residuals of $\mathcal{F}_2$. In terms of developments, it is not difficult to realize that, appending a development ρ of $\mathcal{F}_2/\mathcal{F}_1$ to a development ν of $\mathcal{F}_1$, we get a development $\nu;\rho$ of $\mathcal{F}_1 \cup \mathcal{F}_2$. Nevertheless, $\mathcal{F}_1 \cup \mathcal{F}_2$ and $\mathcal{F}_1 \sqcup \mathcal{F}_2$ represent different reduction sequences.

(iv) For every set $\mathcal{F}$ of redexes of M and every reduction $\rho_1 ; \rho_2$ starting from M, we have

$$\mathcal{F}/(\rho_1 ; \rho_2) = (\mathcal{F}/\rho_1)/\rho_2$$

There are two well-known consequences of the Finite Developments Theorem that are useful for extending the notion of residual to reductions and for introducing the permutation equivalence on reductions.

LEMMA 1.48. *Let $\mathcal{F}_1, \mathcal{F}_2$ and $\mathcal{F}_3$ be sets of redexes of a λ-term M.*

(i) *Parallel moves: both $\mathcal{F}_1 \sqcup \mathcal{F}_2$ and $\mathcal{F}_2 \sqcup \mathcal{F}_1$ end at the same λ-term;*

(ii) *Cube lemma: the following equivalences hold and the corresponding reductions are cofinal*

$$\begin{aligned} \mathcal{F}_1/(\mathcal{F}_2 \sqcup \mathcal{F}_3) &= \mathcal{F}_1/(\mathcal{F}_3 \sqcup \mathcal{F}_2) \\ \mathcal{F}_2/(\mathcal{F}_1 \sqcup \mathcal{F}_3) &= \mathcal{F}_2/(\mathcal{F}_3 \sqcup \mathcal{F}_1) \\ \mathcal{F}_3/(\mathcal{F}_1 \sqcup \mathcal{F}_2) &= \mathcal{F}_3/(\mathcal{F}_2 \sqcup \mathcal{F}_1) \end{aligned}$$

In diagrammatic form:

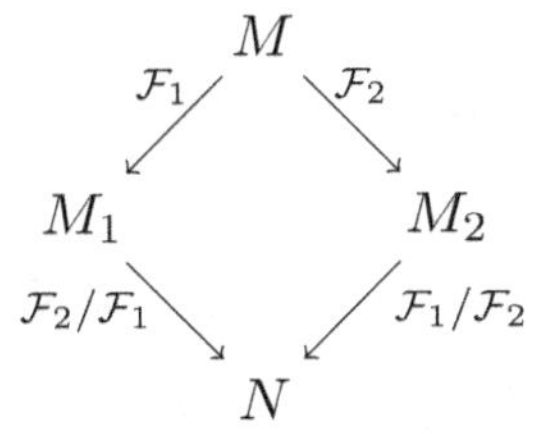

(i) Parallel moves

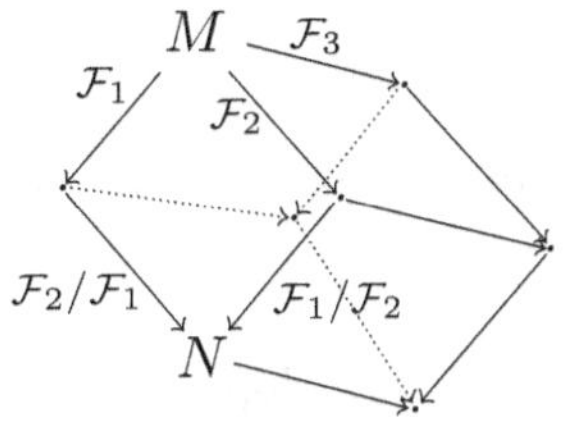

(ii) Cube lemma

PROOF. (i) It is an immediate consequence of Theorem 1.45(ii), since any development of $\mathcal{F}_1 \sqcup \mathcal{F}_2$ and any development of $\mathcal{F}_2 \sqcup \mathcal{F}_1$ are also developments of $\mathcal{F}_1 \cup \mathcal{F}_2$.

(ii) By Theorem 1.45(iii) and the previous observation, the residuals of $\mathcal{F}_3$ across $\mathcal{F}_1 \sqcup \mathcal{F}_2$ and across $\mathcal{F}_2 \sqcup \mathcal{F}_1$ are the same. Therefore, $\mathcal{F}_3/(\mathcal{F}_1 \sqcup \mathcal{F}_2) = \mathcal{F}_3/(\mathcal{F}_2 \sqcup \mathcal{F}_1)$, and similarly for the other two cases. Finally, by observing that the developments of $(\mathcal{F}_1 \sqcup \mathcal{F}_2); (\mathcal{F}_3/(\mathcal{F}_1 \sqcup \mathcal{F}_2))$, of $(\mathcal{F}_1 \sqcup \mathcal{F}_3); (\mathcal{F}_2/(\mathcal{F}_1 \sqcup \mathcal{F}_3))$, and of $(\mathcal{F}_2 \sqcup \mathcal{F}_3); (\mathcal{F}_1/(\mathcal{F}_2 \sqcup \mathcal{F}_3))$ are developments of $\mathcal{F}_1 \cup \mathcal{F}_2 \cup \mathcal{F}_3$ too, so we close the cube by Theorem 1.45(ii). □

Residual of a reduction and permutation equivalence

DEFINITION 1.49. Let $\rho_1 : M \twoheadrightarrow N_1$ and $\rho_2 : M \twoheadrightarrow N_2$ be coinitial parallel reductions. The *residual* ρ_1/ρ_2 *of* ρ_1 *across* ρ_2 is the reduction starting at N_2 inductively defined on ρ_1 by setting:

$$\begin{aligned} o/\rho_2 &\triangleq o, \\ (\nu_1 ; \mathcal{F})/\rho_2 &\triangleq (\nu_1/\rho_2) ; (\mathcal{F}/(\rho_2/\nu_1)). \end{aligned}$$

In order to ensure that the case $\rho_1 = \nu_1 ; \mathcal{F}$ of the previous definition is well-defined, we need to show that $\nu_1/\rho_2 : N_2 \twoheadrightarrow N$ and $\rho_2/\nu_1 : N_1 \twoheadrightarrow N$ reduce to a same λ-term N. Indeed, only in this case the reduction $\mathcal{F}/(\rho_2/\nu_1)$, which starts at the ending term of ρ_2/ν_1, can be composed with ν_1/ρ_2.

LEMMA 1.50. *Let $\rho_1 : M \twoheadrightarrow N_1$ and $\rho_2 : M \twoheadrightarrow N_2$ be two coinitial parallel reductions.*

(i) *ρ_1/ρ_2 and ρ_2/ρ_1 are well-defined and have the same length as ρ_1 and ρ_2, respectively. Moreover:*

$$\begin{array}{lll} a) \ \rho_1 = o & \Rightarrow & \rho_2/\rho_1 = \rho_2/o = \rho_2 \\ b) \ \rho_1 = \nu_1 \,;\mathcal{F}_1 & \Rightarrow & \rho_2/\rho_1 = \rho_2/(\nu_1 \,;\mathcal{F}_1) = (\rho_2/\nu_1)/\mathcal{F}_1 \end{array}$$

(ii) *$\rho_1/\rho_2 : N_2 \twoheadrightarrow N$ and $\rho_2/\rho_1 : N_1 \twoheadrightarrow N$, for some term N.*

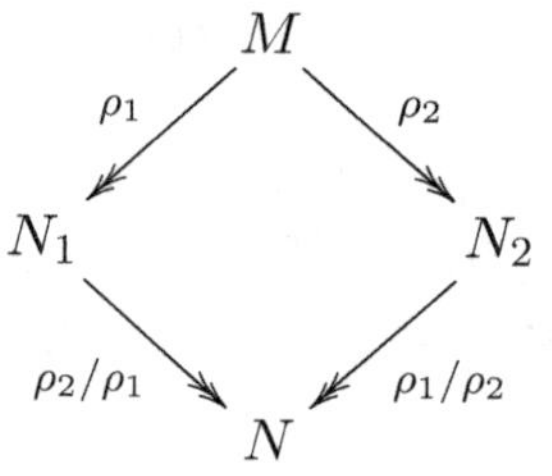

PROOF. First of all, we analyze the case in which one of ρ_1 or ρ_2 is empty. Let us write o_X to denote the empty reduction $o : X \twoheadrightarrow X$.

- We start with the case $\rho_1 = o_M$. In particular, this entails $M = N_1$.

  (i) ρ_1/ρ_2 is well-defined and empty, since $\rho_1/\rho_2 = o_M/\rho_2 = o_{N_2}$ by the first case of Definition 1.49. To prove that ρ_2/ρ_1 is well-defined and that it has the same length as ρ_2, let us directly prove by induction on the length of ρ_2 that $\rho_2/o_M = \rho_2$. The base case $\rho_2 = o_M$ is trivial, as $o_M/o_M = o_M$, by the first case of Definition 1.49. When $\rho_2 = \nu_2 \,;\mathcal{F}_2$, with $\nu_2 : M \twoheadrightarrow M_2$ and $\mathcal{F}_2 : M_2 \to N_2$, we have that $\nu_2/o_M = \nu_2$, by the IH, and that $o_M/\nu_2 = o_{N_2}$, by the first case of Definition 1.49, and that $\mathcal{F}_2/o_{M_2} = \mathcal{F}_2$, by the definition of residual. Therefore, the second item of Definition 1.49 is well-defined in the case $(\nu_2 \,;\mathcal{F}_2)/o_M$, since

  $$(\nu_2 \,;\mathcal{F}_2)/o_M = (\nu_2/o_M)\,;(\mathcal{F}_2/(o_M/\nu_2)) = (\nu_2/o_M)\,;(\mathcal{F}_2/o_{N_2}) = \nu_2 \,;\mathcal{F}_2 = \rho_2.$$

  (ii) Since $\rho_1/\rho_2 = o_M/\rho_2 = o_{N_2}$ and $\rho_2/\rho_1 = \rho_2/o_M = \rho_2$, the statement trivially holds with $N = N_2$.

- Consider now the case $\rho_2 = o_M$, which also implies $M = N_2$. By swapping ρ_1 with ρ_2 in the case $\rho_1 = o_M$ proved above, it is readily seen that $\rho_2/\rho_1 = o_M/\rho_1 = o_{N_1}$ and that $\rho_1/\rho_2 = \rho_1/o_M = \rho_1$ are well-defined and have the same length as ρ_2 and ρ_1, respectively. Moreover, the second item of the lemma trivially holds with $N = N_1$. Finally, the remaining parts follow by a direct application of the first case of Definition 1.49, i.e., $o_M/\rho_1 = o_M$, when $\rho_1 = o_M$, and $o_M/\nu_1 = o_{M_1}$, and $o_{M_1}/\mathcal{F}_1 = o_{N_1}$, when $\rho_1 = \nu_1 \,;\mathcal{F}_1$, with $\nu_1 : M \twoheadrightarrow M_1$ and $\mathcal{F}_1 : M_1 \to N_1$.

- We can now prove the general case of the lemma. We proceed by induction on the sum of the lengths of ρ_1 and ρ_2. Since, we have already proved the cases in which either ρ_1 or ρ_2 is empty, it is left to prove the inductive case in which $\rho_1 = \nu_1 ; \mathcal{F}_1$ and $\rho_2 = \nu_2 ; \mathcal{F}_2$, with $\nu_i : M \twoheadrightarrow M_i$ and $\mathcal{F}_i : M_i \to N_i$ (for $i = 1, 2$).

We start by proving that there are some λ-terms M_0, N_1', and N_2' such that the reduction diagram in Figure 1.1 holds.

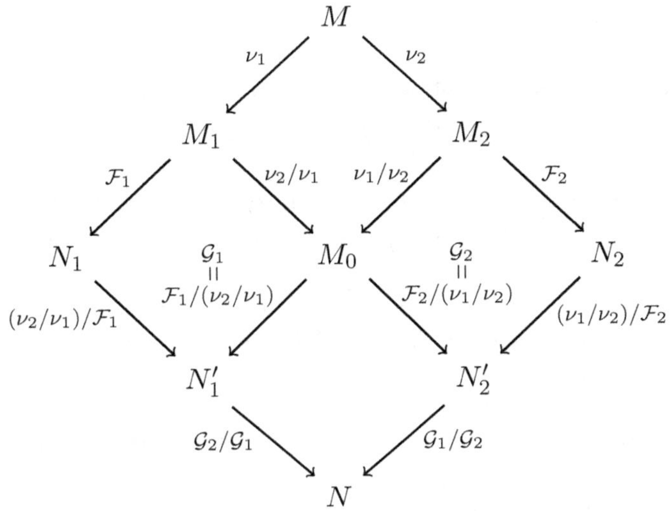

Figure 1.1: Residual of a reduction across another reduction (see Lemma 1.50).

- In the uppermost diamond from M to M_0, the reductions ν_1/ν_2 and ν_2/ν_1 are well-defined, they have the same lengths as ν_1 and ν_2, respectively, and reduce to a same λ-term M_0 by IH.
- In the leftmost diamond from M_1 to N_1', the reduction ν_2/ν_1 has the same length as ν_2 and is then shorter than ρ_2. Moreover, the reduction $\mathcal{F}_1$ is not longer than $\rho_1 = \nu_1 ; \mathcal{F}_1$. We can therefore apply the IH to conclude that $(\nu_2/\nu_1)/\mathcal{F}_1$ and $\mathcal{G}_1 = \mathcal{F}_1/(\nu_2/\nu_1)$ reduce to a same λ-term N_1'. Finally, $(\nu_2/\nu_1)/\mathcal{F}_1$ has the same length as ν_2/ν_1, and therefore as that of ν_2, while $\mathcal{G}_1$ is a set of redexes (the residual of $\mathcal{F}_1$ across ν_2/ν_1).
- By symmetry, all the assertions in the previous item hold for the rightmost diamond from M_2 to N_2', after swapping the indices 1 and 2.
- Since $\mathcal{G}_1$ and $\mathcal{G}_2$ are sets of redexes, the lowermost diamond is an instance of the parallel moves lemma (Lemma 1.48(i))

The residuals of the composed reductions in the diagram of Figure 1.1 can be obtained by composition from the other reduction in the diagram. Indeed, by the IH, we get the following equivalences:

$$\rho_2/\nu_1 = (\nu_2 ; \mathcal{F}_2)/\nu_1 = (\nu_2/\nu_1) ; (\mathcal{F}_2/(\nu_1/\nu_2)) = (\nu_2/\nu_1) ; \mathcal{G}_2$$
$$\nu_2/\rho_1 = \nu_2/(\nu_1 ; \mathcal{F}_1) = (\nu_2/\nu_1)/\mathcal{F}_1$$
$$\mathcal{G}_2/\mathcal{G}_1 = (\mathcal{F}_2/(\nu_1/\nu_2))/\mathcal{G}_1 = \mathcal{F}_2/((\nu_1/\nu_2) ; \mathcal{G}_1) = \mathcal{F}_2/(\rho_1/\nu_2)$$

By swapping the indices 1 and 2, these lead to the dual equivalences:

$$\begin{array}{rcl} \rho_1/\nu_2 & = & (\nu_1/\nu_2)\,;\mathcal{G}_1 \\ \nu_1/\rho_2 & = & (\nu_1/\nu_2)/\mathcal{F}_2 \\ \mathcal{G}_1/\mathcal{G}_2 & = & \mathcal{F}_1/(\rho_2/\nu_1) \end{array}$$

We can finally prove the two items of the lemma.

(i) Since ν_1/ρ_2 and ρ_2/ν_1 are well-defined and reduce to the same λ-term, the second case of Definition 1.49 is well-defined for ρ_1/ρ_2 and ρ_2/ρ_1, with

$$\begin{array}{l} \rho_2/\rho_1 = (\nu_2\,;\mathcal{F}_2)/\rho_1 = (\nu_2/\rho_1)\,;(\mathcal{F}_2/(\rho_1/\nu_2)) \\ \rho_1/\rho_2 = (\nu_1\,;\mathcal{F}_1)/\rho_2 = (\nu_1/\rho_2)\,;(\mathcal{F}_1/(\rho_2/\nu_1)) \end{array}$$

from which, we see that the length of ρ_1/ρ_2 is equal to the length of ρ_1, since the length of ν_1/ρ_2 is equal to the length of ν_1, by the IH. Analogously for ρ_2/ρ_1, whose length is equal to that of ρ_2. Moreover:

$$\begin{array}{rlcl} \rho_2/\rho_1 = & \rho_2/(\nu_1\,;\mathcal{F}_1) & = & (\nu_2\,;\mathcal{F}_2)/\rho_1 \\ = & (\nu_2/\rho_1)\,;(\mathcal{F}_2/(\rho_1/\nu_2)) & = & ((\nu_2/\nu_1)/\mathcal{F}_1)\,;(\mathcal{G}_2/\mathcal{G}_1) \\ = & (\nu_2/\nu_1)/\mathcal{F}_1\,;(\mathcal{G}_2/(\mathcal{F}_1/(\nu_2/\nu_1))) & = & ((\nu_2/\nu_1)\,;\mathcal{G}_2)/\mathcal{F}_1 \\ = & (\rho_2/\nu_1)/\mathcal{F}_1 & & \end{array}$$

(ii) The λ-term N we are looking for is the N in the bottom diamond in Figure 1.1, since we have already seen that

$$\begin{array}{rcl} \rho_2/\rho_1 & = & (\nu_2/\rho_1)\,;(\mathcal{F}_2/(\rho_1/\nu_2)) \\ \rho_1/\rho_2 & = & (\nu_1/\rho_2)\,;(\mathcal{F}_1/(\rho_2/\nu_1)) \end{array}$$ □

The distribution rules already proved for the cases $(\rho_1;\mathcal{F})/\nu$ and $\nu/(\rho_1;\mathcal{F})$ generalize to the cases $(\rho_1;\rho_2)/\nu$ and $\nu/(\rho_1;\rho_2)$ as follows.

LEMMA 1.51. (i) $(\rho_1;\rho_2)/\nu = (\rho_1/\nu);(\rho_2/(\nu/\rho_1))$.
(ii) $\nu/(\rho_1;\rho_2) = (\nu/\rho_1)/\rho_2$.

PROOF. By induction on ρ_2. The case $\rho_2 = o$ is trivial. Now, assume $\rho_2 = \rho;\mathcal{F}$.
(i) We have the following chain of equalities

$$\begin{array}{rcll} (\rho_1;\rho;\mathcal{F})/\nu & = & ((\rho_1;\rho)/\nu);(\mathcal{F}/(\nu/(\rho_1;\rho))) & \\ & = & (\rho_1/\nu);(\rho/(\nu/\rho_1));(\mathcal{F}/((\nu/\rho_1)/\rho)), & \text{by IH,} \\ & = & (\rho_1/\nu);((\rho;\mathcal{F})/(\nu/\rho_1)) & \end{array}$$

(ii) We have $\nu/(\rho_1;\rho;\mathcal{F}) = (\nu/(\rho_1;\rho))/\mathcal{F} \overset{\text{IH}}{=} ((\nu/\rho_1)/\rho)/\mathcal{F} = (\nu/\rho_1)/(\rho;\mathcal{F})$. □

We can now define the *permutation equivalence* which is the least congruence with respect to composition of reductions, satisfying the parallel moves property (Lemma 1.48(i)) and the elimination of empty steps.

DEFINITION 1.52. Let M be a λ-term. The *permutation equivalence* $\sim$ on finite parallel reductions starting from M is defined inductively as follows (where $\nu \neq o$):

$$\frac{}{\mathcal{F}_1 \sqcup \mathcal{F}_2 \sim \mathcal{F}_2 \sqcup \mathcal{F}_1}\ (\text{perm}) \qquad \frac{}{o \sim \emptyset}\ (o_\ell) \qquad \frac{}{\emptyset \sim o}\ (o_r)$$

$$\frac{\rho_1 \sim \rho \sim \rho_2}{\rho_1 \sim \rho_2}\ (\text{trans}) \qquad \frac{\rho_1 \sim \rho_2}{\rho_1 ; \nu \sim \rho_2 ; \nu}\ (\text{cong}_\ell) \qquad \frac{\rho_1 \sim \rho_2}{\nu ; \rho_1 \sim \nu ; \rho_2}\ (\text{cong}_r)$$

REMARK 1.53. (i) Although the relation $\sim$ is defined on parallel reductions, it relates regular reductions as well, since their reduction steps can be considered as contractions of singleton sets of redexes.

(ii) The relation $\sim$ is indeed an equivalence. Reflexivity follows from (cong_ℓ) or (cong_r) by assuming $\rho_1 = \rho_2 = o$; symmetry is a consequence of the symmetry of all the rules in Definition 1.52; transitivity is given by (trans). The rules (cong_ℓ) and (cong_r) ensure that $\sim$ is a congruence w.r.t. composition of reductions—indeed, by combining them, we obtain $\nu_1; \rho_1; \nu_2 = \nu_1; \rho_2; \nu_2$, whenever $\rho_1 \sim \rho_2$. From this property plus rules (o_ℓ) and (o_r), we also have that every empty step can be erased from a reduction.

One of the most important properties of the permutation equivalence is the *diamond property*—a stronger version of confluence showing how to complete two coinitial reductions ρ, ν in order to get the same result. More precisely, if we extend the operator $\sqcup$ to reductions of any length by setting

$$\rho_1 \sqcup \rho_2 \triangleq \rho_1; (\rho_2/\rho_1)$$

then the $\sqcup$ operator is still symmetric, assuming that reductions are equated modulo $\sim$.

THEOREM 1.54 (DIAMOND PROPERTY). *For any pair of coinital reductions ρ_1 and ρ_2, we have that $\rho_1 \sqcup \rho_2 \sim \rho_2 \sqcup \rho_1$. In diagrammatic form:*

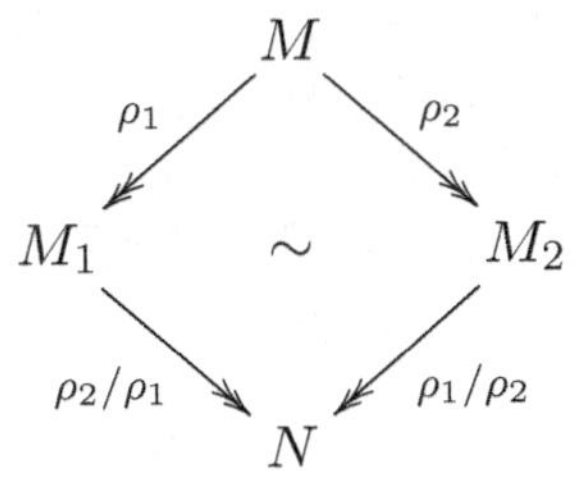

PROOF. We proceed by induction on the sum of the lengths of ρ_1 and ρ_2.

The cases $\rho_1 = o$ or $\rho_2 = o$ are trivial.

Consider $\rho_1 = \nu_1; \mathcal{F}_1$ and $\rho_2 = \nu_2; \mathcal{F}_2$, as in the proof of Lemma 1.50. The four inner diamonds in the diagram in Figure 1.1 commute with respect to $\sim$:

– the upper diamond from M to M_0 and the left and right diagrams from M_i and N_i', with $i = 1, 2$, commute by induction hypothesis;

– the lower diamond from M_0 to N commutes by applying the rule (perm).

By diagram composition, we have then that all the paths from M to N are $\sim$-equivalent. In particular, from the commutation of the leftmost and of the rightmost one, we obtain

$$\nu_1; \mathcal{F}_1; ((\nu_2/\nu_1)/\mathcal{F}_1); (\mathcal{G}_2/\mathcal{G}_1) \sim \nu_2; \mathcal{F}_2; ((\nu_1/\nu_2)/\mathcal{F}_2); (\mathcal{G}_1/\mathcal{G}_2)$$

Since $\rho_1 = \nu_1; \mathcal{F}_1$ and $\rho_2 = \nu_2; \mathcal{F}_2$, and in the proof of Lemma 1.50 we have seen that

$$\rho_2/\rho_1 = ((\nu_2/\nu_1)/\mathcal{F}_1)\,;(\mathcal{G}_2/\mathcal{G}_1) \qquad \text{and} \qquad \rho_1/\rho_2 = ((\nu_1/\nu_2)/\mathcal{F}_2)\,;(\mathcal{G}_1/\mathcal{G}_2)$$

we conclude. □

The residuals of a reduction across another reduction behave well with respect to permutation equivalence. In particular, by replacing a reduction by a $\sim$-equivalent one we get an equivalent residual.

LEMMA 1.55. *Let ρ_1, ρ_2, and ν be coinitial reductions such that $\rho_1 \sim \rho_2$.*
(i) $\nu/\rho_1 = \nu/\rho_2$
(ii) $\rho_1/\nu \sim \rho_2/\nu$

PROOF. We start by considering the particular case where $\rho_1 \sim \rho_2$ can be shown without using the transitivity rule of $\sim$. Wlog, we assume that ρ_1 is shorter than ρ_2, and proceed by induction on the length of ρ_1 and by case analysis on the last rule applied in the proof of the equivalence $\rho_1 \sim \rho_2$.
Case (perm): $\rho_1 = \mathcal{F}_1 \sqcup \mathcal{F}_2$ and $\rho_2 = \mathcal{F}_2 \sqcup \mathcal{F}_1$. First, we prove by induction on ν, that

$$(\mathcal{G}_1 \sqcup \mathcal{G}_2)/\nu = (\mathcal{G}_1/\nu) \sqcup (\mathcal{G}_2/\nu) \tag{*}$$

for every pair of sets of redexes $\mathcal{G}_1$ and $\mathcal{G}_2$ of a λ-term M. The case $\nu = o$ is trivial. To prove the inductive case $\nu = \nu_0; \mathcal{G}$, it is useful to prove first the particular case where $\nu_0 = o$. Indeed, for every set of redexes $\mathcal{G}$ of M, we have

$$\begin{aligned}(\mathcal{G}_1 \sqcup \mathcal{G}_2)/\mathcal{G} &= (\mathcal{G}_1/\mathcal{G}); ((\mathcal{G}_2/\mathcal{G}_1)/(\mathcal{G}/\mathcal{G}_1)) \\ &= (\mathcal{G}_1/\mathcal{G}); (\mathcal{G}_2/(\mathcal{G}_1 \sqcup \mathcal{G})) \\ &= (\mathcal{G}_1/\mathcal{G}); (\mathcal{G}_2/(\mathcal{G} \sqcup \mathcal{G}_1)), \qquad \text{by Lemma 1.48(ii)}, \\ &= (\mathcal{G}_1/\mathcal{G}); ((\mathcal{G}_2/\mathcal{G})/(\mathcal{G}_1/\mathcal{G})) \\ &= (\mathcal{G}_1/\mathcal{G}) \sqcup (\mathcal{G}_2/\mathcal{G}).\end{aligned}$$

Coming back to general case for ν_0, by the induction hypothesis and the particular case just seen (denoted $\stackrel{**}{=}$), we have

$$\begin{aligned}(\mathcal{G}_1 \sqcup \mathcal{G}_2)/(\nu_0; \mathcal{G}) &= ((\mathcal{G}_1 \sqcup \mathcal{G}_2)/(\nu_0))/\mathcal{G} \\ &= ((\mathcal{G}_1/\nu_0) \sqcup (\mathcal{G}_2/\nu_0))/\mathcal{G}, \qquad \text{by IH}, \\ &\stackrel{**}{=} ((\mathcal{G}_1/\nu_0)/\mathcal{G}) \sqcup ((\mathcal{G}_2/\nu_0)/\mathcal{G}) \\ &= (\mathcal{G}_1/(\nu_0; \mathcal{G})) \sqcup (\mathcal{G}_2/(\nu_0; \mathcal{G})).\end{aligned}$$

We can then come back to the proof of the two items of the lemma.

(i) By induction on ν. The case $\nu = o$ is trivial. So let us take $\nu = \nu_0; \mathcal{F}$. We have

$$\begin{aligned}
(\nu_0;\mathcal{F})/(\mathcal{F}_1 \sqcup \mathcal{F}_2) &= (\nu_0/(\mathcal{F}_2 \sqcup \mathcal{F}_1)); (\mathcal{F}/((\mathcal{F}_1 \sqcup \mathcal{F}_2)/\nu_0)), && \text{by Def. 1.49 and IH,}\\
&= (\nu_0/(\mathcal{F}_2 \sqcup \mathcal{F}_1)); (\mathcal{F}/((\mathcal{F}_1/\nu_0) \sqcup (\mathcal{F}_2/\nu_0))), && \text{by (*),}\\
&= (\nu_0/(\mathcal{F}_2 \sqcup \mathcal{F}_1)); (\mathcal{F}/((\mathcal{F}_2/\nu_0) \sqcup (\mathcal{F}_1/\nu_0))), && \text{by Lemma 1.48(ii),}\\
&= (\nu_0/(\mathcal{F}_2 \sqcup \mathcal{F}_1)); (\mathcal{F}/((\mathcal{F}_2 \sqcup \mathcal{F}_1)/\nu_0)), && \text{by (*),}\\
&= (\nu_0;\mathcal{F})/(\mathcal{F}_2 \sqcup \mathcal{F}_1).
\end{aligned}$$

(ii) By (*) plus Lemma 1.48(ii).

Cases (o_ℓ) and (o_r): $\rho_1 = o$ and $\rho_2 = \emptyset$.

(i) $\nu/o = \nu = \nu/\emptyset$.

(ii) $o/\nu = o \sim \emptyset = \emptyset/\nu$.

Case (cong$_r$): $\rho_1 = \rho;\nu_1$ and $\rho_2 = \rho;\nu_2$, with $\nu_1 \sim \nu_2$. By Lemma 1.51 and the IH:

(i) $\nu/(\rho;\nu_1) = (\nu/\rho)/\nu_1 \overset{\text{IH}}{=} (\nu/\rho)/\nu_2 = \nu/(\rho;\nu_2)$

(ii) $(\rho;\nu_1)/\nu = (\rho/\nu);(\nu_1/(\nu/\rho)) \overset{\text{IH}}{\sim} (\rho/\nu);(\nu_2/(\nu/\rho)) = (\rho;\nu_2)/\nu$

Case (cong$_\ell$): $\rho_1 = \nu_1;\rho$ and $\rho_2 = \nu_2;\rho$, with $\nu_1 \sim \nu_2$.

(i) $\nu/(\nu_1;\rho) = (\nu/\nu_1)/\rho \overset{\text{IH}}{=} (\nu/\nu_2)/\rho = \nu/(\nu_2;\rho)$.

(ii) $(\nu_1;\rho)/\nu = (\nu_1/\nu);(\rho/(\nu/\nu_1)) \overset{\text{IH}}{\sim} (\nu_2/\nu);(\rho/(\nu/\nu_2)) = (\nu_2;\rho)/\nu$.

We have then proved the lemma for the particular case in which $\rho_1 \sim \rho_2$ can be shown without using the transitivity rule of the permutation equivalence. But, a proof of $\rho_1 \sim \rho_2$ that contains k applications of the rule (trans) can be split into the proofs of $k+1$ $\sim$-equivalences

$$\rho_1 = \nu_0 \sim \nu_1 \sim \cdots \sim \nu_k \sim \nu_{k+1} = \rho_2$$

that do not contain the transitivity rule and such that:

(i) $\nu/\nu_i = \nu/\nu_{i+1}$;

(ii) $\nu_i/\nu \sim \nu_{i+1}/\nu$;

for $i = 0, 1, \ldots, k$. We conclude by transitivity. □

In the particular case when $\nu = R$ is just a single redex, the first item of the previous lemma implies that, for every $\rho : M \twoheadrightarrow N$ and every $\rho' \sim \rho$, a redex S of N is the residual of a redex R of M across ρ if and only if it is a residual of R across ρ'.

COROLLARY 1.56. *Let R be a redex of a term M and $\rho \sim \rho'$ be two reductions starting at M. Then, $R/\rho = R/\rho'$.*

The diamond property (Theorem 1.54) both generalizes and strengthens the parallel moves Lemma 1.48(i) by replacing coinitial reductions for sets of redexes and by showing that the diamond obtained in this way commutes with the permutation equivalence $\sim$. In a similar way, the following lemma strengthens and generalizes the cube lemma.

LEMMA 1.57. *For every coinitial reductions ρ_1, ρ_2, and ρ_3, we have:*

(i) $\rho_3/(\rho_1 \sqcup \rho_2) = \rho_3/(\rho_2 \sqcup \rho_1)$;

(ii) $(\rho_1 \sqcup \rho_2)/\rho_3 = (\rho_1/\rho_3) \sqcup (\rho_2/\rho_3)$.

PROOF. (i) By Lemma 1.55 and Theorem 1.54.

(ii) We have $(\rho_1 \sqcup \rho_2)/\rho_3 = (\rho_1/\rho_3)\,;((\rho_2/\rho_1)/(\rho_3/\rho_1)) = (\rho_1/\rho_3)\,;(\rho_2/(\rho_1 \sqcup \rho_3))$. By (i), we get $(\rho_1/\rho_3)\,;(\rho_2/(\rho_1 \sqcup \rho_3)) = (\rho_1/\rho_3)\,;(\rho_2/(\rho_3 \sqcup \rho_1)) = (\rho_1/\rho_3) \sqcup (\rho_2/\rho_3)$. □

The two statements in the previous lemma remain obviously true if one permutes the three reductions ρ_1, ρ_2, and ρ_3. We have then the following generalization of the cube in Lemma 1.48.

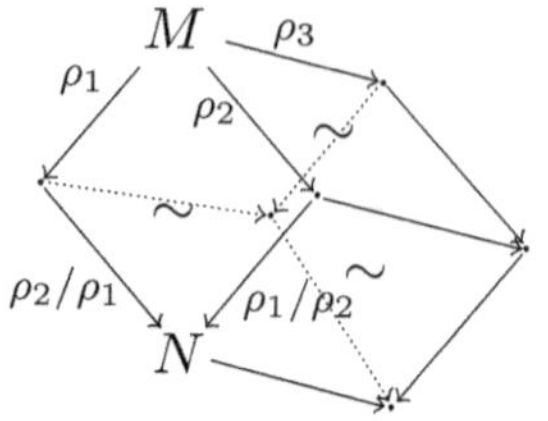

DEFINITION 1.58. A reduction ρ_1 is a *prefix* of a coinitial reduction ρ_2, written $\rho_1 \precsim \rho_2$, if there exists a reduction ν such that $(\rho_1\,;\nu) \sim \rho_2$.

Before proving that the prefix relation is a partial order, we shall see some basic properties of permutation equivalence and prefixes. Among the properties presented in the next lemma, particularly relevant are 'left deletion' (iii) and the fact that two reductions are permutationally equivalent exactly when they are prefixes of each other (vi).

LEMMA 1.59. (i) *If* $\rho_1 \sim \rho_2$ *then* $\rho_1/\rho_2 = \emptyset^n$, *where* n *is the length of* ρ_1.

(ii) $\rho_1 \sim \rho_2 \iff \rho_1/\rho_2 \sim o \sim \rho_2/\rho_1$.

(iii) *(left deletion)* $\nu\,;\rho_1 \sim \nu\,;\rho_2 \iff \rho_1 \sim \rho_2$.

(iv) $\rho_1\,;\nu \sim \rho_2$ *implies* $\nu \sim \rho_2/\rho_1$.

(v) $\rho_1 \precsim \rho_2 \iff \rho_2 \sim \rho_1 \sqcup \rho_2 \iff \rho_1/\rho_2 \sim o$.

(vi) $\rho_1 \sim \rho_2 \iff \rho_1 \precsim \rho_2\ \&\ \rho_2 \precsim \rho_1$.

PROOF. (i) By induction on n, we can easily prove $\rho_1/\rho_1 = \emptyset^n$. We conclude by applying Lemma 1.55.

(ii) ($\Rightarrow$) It follows from (i), since $o \sim \emptyset^n$.

($\Leftarrow$) We have $\rho_1 \sim \rho_1 \sqcup \rho_2 \sim \rho_2 \sqcup \rho_1 \sim \rho_2$.

(iii) ($\Rightarrow$) For every $i \in \{1,2\}$, we have

$$(\nu\,;\rho_i)/\nu \sim (\nu/\nu)\,;(\rho_i/(\nu/\nu)) \sim o\,;(\rho_i/o) \sim \rho_i.$$

Thus, by Lemma 1.55, we obtain

$$\rho_1 \sim (\nu\,;\rho_1)/\nu \sim (\nu\,;\rho_2)/\nu \sim \rho_2.$$

($\Leftarrow$) It follows directly from (cong_r) in Definition 1.52.

(iv) If $\rho_1\,;\nu \sim \rho_2$ then $\rho_2/\rho_1 \sim (\rho_1\,;\nu)/\rho_1 = (\rho_1/\rho_1)\,;(\nu/(\rho_1/\rho_1)) \sim o\,;(\nu/o) \sim \nu$.

(v) By (iv), $\rho_1 \precsim \rho_2$ entails $\rho_2 \sim \rho_1 \sqcup \rho_2$. If moreover $\rho_2 \sim \rho_1 \sqcup \rho_2$ holds, then so does $\rho_1 \precsim \rho_2$ by the definition of the prefix relation. Now $\rho_2 \sim \rho_1 \sqcup \rho_2 \sim \rho_2 \sqcup \rho_1$ which, by left deletion, leads to $\rho_1/\rho_2 \sim o$. Finally, if $\rho_1/\rho_2 \sim o$, then $\rho_1 \sqcup \rho_2 \sim \rho_2 \sqcup \rho_1 \sim \rho_2$.

(vi) We have $\rho_1 \precsim \rho_2 \ \&\ \rho_2 \precsim \rho_1 \iff \rho_1/\rho_2 \sim o \sim \rho_2/\rho_1$, by (v),
$\iff \rho_1 \sim \rho_2$, by (ii).

This concludes the proof. □

LEMMA 1.60. *The relation $\precsim$ is a partial order on reductions equated modulo $\sim$.*

PROOF. The prefix relation is reflexive since $\rho\,; o \sim \rho$. Transitivity holds as $\rho_1\,;\nu_1 \sim \rho_2$ and $\rho_2\,;\nu_2 \sim \rho_3$ imply $\rho_1\,;\nu_1\,;\nu_2 \sim \rho_3$. Antisymmetry holds by Lemma 1.59(vi). □

Standard reductions

DEFINITION 1.61. (i) Given two redexes $R, S \in M$, we say that R *is to the left of* S if the main λ in R is to the left of the main λ in S (equivalently, either R occurs to the left of S in M, or it contains S).

(ii) A redex $R \in M$ is the *leftmost-outermost redex of* M if it is to the left of any other redex $S \in M$.

(iii) A reduction $\rho \in \mathcal{G}_\beta(M)$ of the form

$$\rho : M = M_0 \xrightarrow{R_1}_\beta M_1 \xrightarrow{R_2}_\beta M_2 \xrightarrow{R_3}_\beta \cdots \xrightarrow{R_n}_\beta M_n$$

is called *standard* if, for all i, j such that $0 < i < j \leq n$, the redex R_j is not a residual of a redex S_j to the left of R_i in M_{i-1} across ρ. This definition extends to parallel reductions $\rho = \mathcal{F}_1\,;\cdots;\mathcal{F}_n$ by saying that ρ is *standard* when the previous proviso holds assuming that R_i and R_j are the leftmost-outermost redexes of the respective sets $\mathcal{F}_i$ and $\mathcal{F}_j$.

(iv) We write $\rho : M \twoheadrightarrow_{\mathsf{st}} N$ whenever the reduction ρ is standard.

REMARK 1.62. (i) A residual of a leftmost-outermost redex is leftmost-outermost itself. As a consequence, the leftmost-outermost redex has at most one residual.

(ii) A reduction $\rho = R_1\,;\cdots;R_n$ contracting, at each step i, the leftmost-outermost redex R_i must be standard. We say that such a ρ is obtained following the *leftmost-outermost reduction strategy*.

The Standardization Theorem for β-reduction is due to Curry and Feys (1958).

THEOREM 1.63 (STANDARDIZATION FOR β). *Let $M, N \in \Lambda$. If $M \twoheadrightarrow_\beta N$, then there exists a standard reduction $M \twoheadrightarrow_{\mathsf{st}} N$.*

PROOF. B[1984], Theorem 11.4.7. See also Theorem 5.55, in this book. □

In the proof of the Standardization Theorem one actually shows that, for every reduction ρ, there is a unique standard reduction ν such that $\rho \sim \nu$ (B[1984], Theorem 12.3.14). So, a $\sim$-equivalence class of reductions is characterized by its unique standard reduction.

We conclude the section by presenting a Standardization Theorem for $\beta\eta$-reduction, together with its corollary, that will be used in Chapter 9. This theorem has been shown in Folkert's PhD thesis (Folkerts (1995)) making an essential use of the so-called 'actual' η-reduction. Indeed, η and β redexes may overlap, in which case the η-step becomes inessential in the sense that it can be simulated by a β-step.

EXAMPLES 1.64. (i) These η-reductions are also β-steps: $\lambda x.\mathsf{I}x \to_\beta \mathsf{I}$, $(\lambda x.y\mathsf{K}x)\mathsf{I} \to_\beta y\mathsf{KI}$.
(ii) If $x \notin \mathrm{FV}(M)$ then $\lambda x.Mx \to_\beta M$ whenever M is a λ-abstraction, i.e. $M = \lambda y.N$.

NOTATION 1.65. Let $M, N \in \Lambda$. Given a step of η-reduction $M \to_\eta N$, we write:
(i) $M \to_{\eta_\mathsf{b}} N$ whenever $M \to_\beta N$, i.e., if the η-step can be simulated by contracting a β-redex as in $(\lambda x.Px)Q \to_{\eta_\mathsf{b}} PQ$ or $\lambda x.(\lambda y.P)x \to_{\eta_\mathsf{b}} \lambda y.P$, with $x \notin \mathrm{FV}(P)$;
(ii) $M \to_{\eta_\mathsf{a}} N$, otherwise.
We say that $M \to_{\eta_\mathsf{a}} N$ is a step of *actual η-reduction.*

For a more rigorous definition of $\twoheadrightarrow_{\eta_\mathsf{a}}$, see Definition 12.38 in Chapter 12.

LEMMA 1.66. *For $M \in \Lambda$, we have:*
(i) $M \twoheadrightarrow_{\beta\eta} N \;\Rightarrow\; \exists M' \in \Lambda\,.\, M \twoheadrightarrow_\beta M' \twoheadrightarrow_\eta N.$
(ii) $M \twoheadrightarrow_{\eta} N \;\Rightarrow\; \exists M' \in \Lambda\,.\, M \twoheadrightarrow_{\eta_\mathsf{b}} M' \twoheadrightarrow_{\eta_\mathsf{a}} N.$
(iii) $M \twoheadrightarrow_{\beta\eta} N \;\Rightarrow\; \exists M' \in \Lambda\,.\, M \twoheadrightarrow_\beta M' \twoheadrightarrow_{\eta_\mathsf{a}} N.$

PROOF. (i) This is the usual postponement of η over β. See B[1984], Corollary 15.1.6.
(ii) Analogous. See Folkerts (1995).
(iii) By composing (i) and (ii). □

THEOREM 1.67 (STANDARDIZATION FOR $\beta\eta$).
Let $M \in \Lambda$. If $M \twoheadrightarrow_{\beta\eta} \lambda\vec{x}.M_0 \cdots M_n$, then $M \twoheadrightarrow_h \lambda\vec{x}z_1 \ldots z_m.M'_0 \cdots M'_n Z_1 \cdots Z_m$ for some λ-terms $\vec{M}', \vec{Z}$ such that:

- *if $M_0 = y$ then $M'_0 = y$,*
- *if $M_0 = (\lambda y.P)Q$, then $M'_0 = (\lambda y.P')Q'$ with $P' \twoheadrightarrow_{\beta\eta} P$ and $Q' \twoheadrightarrow_{\beta\eta} Q$,*
- *$M'_i \twoheadrightarrow_{\beta\eta} M_i$ for all $i \in \{1, \ldots, n\}$ and*
- *$z_j \notin \mathrm{FV}(\vec{M})$ and $Z_j \twoheadrightarrow_{\beta\eta} z_j$ for all $j \in \{1, \ldots, m\}$.*

In diagrammatic form:

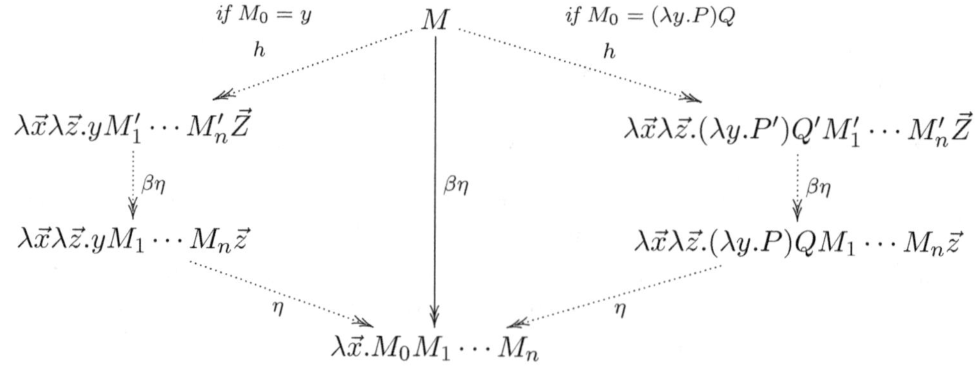

PROOF. Basically, by postponement of η_a-reduction (Lemma 1.66(iii)) and standardization for β-reduction (Theorem 1.63). For more details, see Folkerts (1995). □

In the following $\to^{\iota}_{\beta\eta}$ represents internal (namely, non-head) $\beta\eta$-reduction.

COROLLARY 1.68. *A reduction* $M \twoheadrightarrow_{\beta\eta} M_0 \cdots M_n$ *can be factorized as follows:*

$$M \twoheadrightarrow_h \lambda z_1 \ldots z_k.M'_0 \cdots M'_n Z_1 \cdots Z_k \twoheadrightarrow^{\iota}_{\beta\eta} \lambda z_1 \ldots z_k.M_0 \cdots M_n z_1 \cdots z_k \twoheadrightarrow_\eta M_0 \cdots M_n$$

for some $M'_0, \ldots, M'_n, Z_1, \ldots, Z_k$ *with* $n, k \geq 0$ *and*

- $M'_i \twoheadrightarrow_{\beta\eta} M_i$, *for all* $i\,(0 \leq i \leq n)$;
- $Z_j \twoheadrightarrow_{\beta\eta} z_j$ *and* $z_j \notin \mathrm{FV}(\vec{M})$, *for all* $j\,(1 \leq j \leq k)$.

PROOF. One has $M_0 = N_0 \cdots N_m$, with $m \geq 0$ and N_0 not an application.

Case $N_0 = \lambda y.P$ and $m + n \geq 1$. Now factorize

$$M \twoheadrightarrow_{\beta\eta} (\lambda y.P)N_1 \cdots N_m M_1 \cdots M_n$$

into

$$\begin{array}{rcl} M & \twoheadrightarrow_h & \lambda\vec{z}.(\lambda y.P')N'_1 \cdots N'_m M'_1 \cdots M'_n \vec{Z} \\ & \twoheadrightarrow^{\iota}_{\beta\eta} & \lambda\vec{z}.(\lambda y.P)N_1 \cdots N_m M_1 \cdots M_n \vec{z} \\ & \twoheadrightarrow_\eta & M_0 \cdots M_n. \end{array}$$

Case $N_0 = y$. In this case factorize $M \twoheadrightarrow_{\beta\eta} yN_0 \cdots N_m M_1 \cdots M_n$, obtaining the same conclusion. □

1.3 Reduction under substitution and consequences

In this section we present a result, reduction under substitution (RuS), of Diederik van Daalen (1980), in a slightly different style. The result is a strengthening of Proposition 1.74 below, also known as Barendregt's Lemma (BL), that was used to refute—as a challenge of Wadsworth—that the set $\{\mathsf{I}, \mathsf{KI}, \mathsf{K}^2(\mathsf{I}), \ldots\}$ is an adequate numeral system.

We introduce underlining, often used in Barendregt (1971), to motivate BL and RuS and give their proofs. We then follow the presentation in Endrullis and de Vrijer (2008), where it is shown that RuS has many known and previously unknown consequences.

DEFINITION 1.69 (LAMBDA TERMS WITH UNDERLINING).

(i) Define the set of *underlined* λ*-terms*, in notation $\underline{\Lambda}$, as follows:

$$(\underline{\Lambda}) \qquad M ::= x \mid \lambda x.M \mid MM \mid \underline{M}$$

(ii) Define the *forgetting map* $|-|: \underline{\Lambda} \to \Lambda$, erasing underlinings, as follows:

$$\begin{array}{rcl} |x| & = & x; \\ |\lambda x.P| & = & \lambda x.\,|P|; \\ |PQ| & = & |P|\,|Q|; \\ |\underline{P}| & = & |P|. \end{array}$$

(iii) On $\underline{\Lambda}$ generalize the *notion of reduction* β. For $P, Q \in \underline{\Lambda}$ and $x \in \mathrm{Var}$, define:

$$(\lambda x.P)Q \to P[x:=Q] \qquad (\beta)$$

where the substitution is extended to $\underline{\Lambda}$ by setting $\underline{M}[x:=Q] \triangleq \underline{M}$.

(iv) The reduction relation $\to_\beta$ on $\underline{\Lambda}$ is obtained as the closure of (β) under single-hole contexts whose (unique) hole does not occur within an underlined subterm.

(v) On $\underline{\Lambda}$ define the *notion of reduction* $\underline{\beta}$.

$$P \to_\beta Q \quad \Rightarrow \quad \underline{P} \to_{\underline{\beta}} \underline{Q}, \quad \text{for } P, Q \in \Lambda; \qquad (\underline{\beta})$$
$$\underline{PQ} \to_{\underline{\beta}} \underline{P\,|Q|}, \quad \text{for } P, Q \in \underline{\Lambda}.$$

EXAMPLES 1.70. (i) We have $\mathsf{I}\underline{x} \to_\beta \underline{x}$ and $(\lambda z.\underline{z})x \to_\beta \underline{z}$, but do not have $\mathsf{I}\underline{x} \to_\beta x$.

(ii) We have $\underline{\mathsf{I}x} \to_{\underline{\beta}} \underline{x}$ and $\underline{\mathsf{I}}x \to_{\underline{\beta}} \underline{\mathsf{I}x}$, but do not have $\mathsf{I}\underline{x} \to_{\underline{\beta}} \underline{x}$.

The following lemmas are useful for applications.

LEMMA 1.71. *Let $M, \vec{P} \in \Lambda$, $x, \vec{y} \in \mathrm{FV}(M)$ and a reduction*

$$\rho\colon M[x:=\underline{w}][\vec{y}:=\underline{\vec{P}}] \twoheadrightarrow_{\beta \cup \underline{\beta}} N.$$

Suppose the following.

1. *One occurrence of x in M is leftmost to all occurrences of $\vec{y}$.*

2. *ρ is a $\twoheadrightarrow_{\underline{\beta}}$ reduction.*

Then $w \in \mathrm{FV}(|N|)$.

PROOF. To get a taste of what happens, we show that the two conditions are necessary.

As to (omitting) 1. Let $M = yx$. Then $M[x:=\underline{w}][y:=\underline{\lambda xy.y}] = (\underline{\lambda xy.y})\underline{w} \twoheadrightarrow_{\underline{\beta}} \underline{\lambda y.y}$, which does not contain $\underline{w}$.

As to 2. Let $M = (\lambda xy.y)xy$. Then $M[x:=\underline{w}][y:=\underline{\mathsf{I}}] = (\lambda xy.y)\underline{w}\,\underline{\mathsf{I}} \twoheadrightarrow_\beta \underline{\mathsf{I}}$ is not an underlined reduction. Its last term does not contain $\underline{w}$.

This should suffice for an understanding. Condition 2 ensures that no erasing subterm of M is contracted to the left of x (respectively $\underline{w}$). Condition 1 ensures that this holds also after the substitution $[\vec{y} := \underline{\vec{P}}]$. □

LEMMA 1.72. *Consider $M, U \in \Lambda$, $x \in \mathrm{FV}(M)$ and a reduction*

$$\rho\colon M[x:=\underline{U}] \twoheadrightarrow_{\underline{\beta}} N.$$

If U is unsolvable, then $|N|$ must contain unsolvable subterms.

PROOF. Consider $P \in \underline{\Lambda} - \Lambda$ such that all underlined subterms of P are of the form $\underline{U}$, for some $U \in \mathrm{UNS}$. Note that $M[x:=\underline{U}]$ satisfies this property and, if N satisfies this property, then $|N|$ contains unsolvable subterms. Therefore, it is sufficient to show that this property is preserved under $\underline{\beta}$-reduction. Assume $P \to_{\underline{\beta}} P'$, there are two cases:

Case 1. A subterm of P of the form $\underline{UQ}$ $\underline{\beta}$-reduces to $\underline{U\,|Q|}$. This case follows since $U \in \mathrm{UNS}$ entails $UV \in \mathrm{UNS}$ for all $V \in \Lambda$.

Case 2. P' results from P by β-reducing an underlined subterm. This case follows since UNS is closed under β-reduction. □

DEFINITION 1.73. Let $M, L, P \in \Lambda$ define $M \leadsto_{x,L} P$ if for some $P' \in \underline{\Lambda}$ one has

$$M[x:=\underline{L}] \twoheadrightarrow_{\underline{\beta}} P' \;\&\; |P'| = P$$

The next lemma states (intuitively) that in a reduction an argument can only be used by applying it. This will follow as an immediate consequence of Theorem 1.77.

PROPOSITION 1.74 (BARENDREGT'S LEMMA). *Let $F, L, N \in \Lambda$ and $x \notin \mathrm{FV}(F)$. Then*

$$FL \twoheadrightarrow_\beta N \quad \Rightarrow \quad \exists G \in \Lambda\,.\,[Fx \twoheadrightarrow_\beta G \leadsto_{x,L} N]$$

Reduction under Substitution (RuS), Theorem 1.77, states that a reduct of a substitution $M[x:=L]$ can be reached by first reducing M sufficiently to M_1, then making the substitution $M_1[x:=L]$, after which one needs to reduce only the various occurrences of $L\vec{P}_i$, where the arguments $\vec{P}_i$ are subterms of M_1. (More precisely, the arguments are of shape $\vec{P}_i = \vec{Q}_i[x:=L]$ with the $\vec{Q}_i$ subterms of M_1.) The way L obtains power over its arguments is through the rule

$$\underline{L}P_1P_2 \to_{\underline{\beta}} \underline{LP_1}P_2 \to_{\underline{\beta}} \underline{LP_1P_2}.$$

Before going further, let us prove the following Substitution Lemma for RuS.

LEMMA 1.75. *Let $P, P', Q, Q', L \in \Lambda$ and $x, y \in \mathrm{Var}$ be such that $y \notin \mathrm{FV}(L)$. Then*

$$P \leadsto_{x,L} P' \;\&\; Q \leadsto_{x,L} Q' \quad \Rightarrow \quad P[y:=Q] \leadsto_{x,L} P'[y:=Q']$$

PROOF. Proceed by structural induction on P.

Case $P = z$, for $z \neq x, y$. Trivial.

Case $P = x$. Then $P \leadsto_{x,L} P'$ implies $P[x:=\underline{L}] = \underline{L} \twoheadrightarrow_{\underline{\beta}} P''$ with $|P''| = P'$. In fact, $L \twoheadrightarrow_\beta P'$ where $y \notin \mathrm{FV}(P')$ since $y \notin \mathrm{FV}(L)$. Thus, $P[y:=Q] = P \leadsto_{x,L} P' = P'[y:=Q']$.

Case $P = y$, for $y \neq x$. Then also $P' = y$, so $P[y:=Q] = Q \leadsto_{x,L} Q' = P'[y:=Q']$.

All other cases follow easily from the induction hypothesis. □

The RuS property is an immediate consequence of the next lemma.

LEMMA 1.76. *For all $M, L \in \Lambda$ and all variables x one has*

(i) $M \leadsto_{x,L} N \to_\beta N_1 \quad \Rightarrow \quad \exists M_1 \in \Lambda\,.\,[M \twoheadrightarrow_\beta M_1 \leadsto_{x,L} N_1]$

(ii) $M_1 \;{}_\beta\!\leftarrow M \leadsto_{x,L} N \quad \Rightarrow \quad \exists N_1 \in \Lambda\,.\,[M_1 \leadsto_{x,L} N_1 \;\&\; N \to_\beta N_1]$

In diagrammatic form

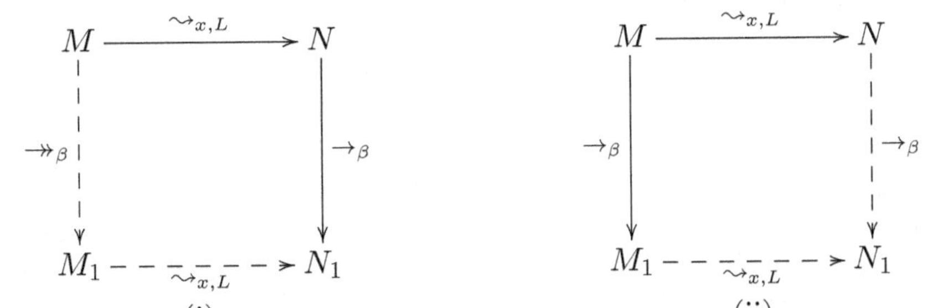

(i) (ii)

PROOF. (i) This is done by induction on the structure of M.

Case 1. $M = y$, for $x \neq y$. Trivial.

Case 2. $M = x$. Then $N = L'$ and $N_1 = L''$, with $L \twoheadrightarrow_\beta L' \to_\beta L''$. Now we can take $M_1 \triangleq M \rightsquigarrow_{x,L} L''$.

Case 3. $M = \lambda z.P$. Then $N = \lambda z.Q$, $N_1 = \lambda z.Q_1$ and $P \rightsquigarrow_{x,L} Q \to_\beta Q_1$. By the IH there exists a P_1 such that $P \twoheadrightarrow_\beta P_1 \rightsquigarrow_{x,L} Q_1$. Then take $M_1 \triangleq \lambda z.P_1$.

Case 4. $M = PQ$. There are three subcases, each with $n \geq 1$.

Subcase 4.1 $M = (\lambda z.P_0)P_1 \cdots P_n$. Similar to case 2.

Subcase 4.2 $M = zP_1 \cdots P_n$, with $z \neq x$. Similar to case 2.

Subcase 4.3 $M = xP_1 \cdots P_n$. Similar to case 1.

(ii) By induction on a derivation of $M \to_\beta M_1$.

Base case. $M = (\lambda y.P)Q$ and $M_1 = P[y{:=}Q]$. Wlog, we have $y \neq x$ and $y \notin \mathrm{FV}(L)$. If $M \rightsquigarrow_{x,L} N$ then $M[x{:=}\underline{L}] = (\lambda y.P[x{:=}\underline{L}])(Q[x{:=}\underline{L}]) \twoheadrightarrow_{\underline{\beta}} N'$, with $|N'| = N$. Since $\underline{\beta}$-reduction cannot contract the outer redex, this is only possible if $N = (\lambda y.P')Q'$ where $P \rightsquigarrow_{x,L} P'$ and $Q \rightsquigarrow_{x,L} Q'$. Conclude that $N_1 \triangleq P'[y{:=}Q']$ by applying Lemma 1.75.

All other cases follow easily from the induction hypothesis. □

THEOREM 1.77 (REDUCTION UNDER SUBSTITUTION, VAN DAALEN (1980)).

$$M[x{:=}L] \twoheadrightarrow_\beta N_1 \quad \Rightarrow \quad \exists M_1 \in \Lambda\,.\,[M \twoheadrightarrow_\beta M_1 \rightsquigarrow_{x,L} N_1].$$

PROOF. Take $N \triangleq M[x{:=}L]$ in item (i) of the previous lemma. □

Proposition 1.74 follows by taking $M \triangleq Fx$, $N_1 \triangleq G$. The required G' will be M_1.

Lemma 1.76, and whence Theorem 1.77, can be generalized in the obvious way to simultaneous substitution and multistep reduction.

COROLLARY 1.78. (i) $M \rightsquigarrow_{\vec{x},\vec{L}} N \twoheadrightarrow_\beta N_1 \quad \Rightarrow \quad \exists M_1 \in \Lambda\,.\,[M \twoheadrightarrow_\beta M_1 \rightsquigarrow_{\vec{x},\vec{L}} N_1]$.

(ii) $M_1 \;{}_\beta\!\twoheadleftarrow M \rightsquigarrow_{\vec{x},\vec{L}} N \quad \Rightarrow \quad \exists N_1 \in \Lambda\,.\,[\,M_1 \rightsquigarrow_{\vec{x},\vec{L}} N_1 \;\&\; N \twoheadrightarrow_\beta N_1\,]$.

Known consequences

We conclude the chapter by presenting some basic consequences of RuS. More advanced applications concern the Perpendicular Lines Property and are presented in Chapter 8.

DEFINITION 1.79. A *surjective pairing* is a triple (D, D_1, D_2) of λ-terms satisfying, for all $M \in \Lambda$, the following equations (where $i \in \{1, 2\}$):

$$\begin{aligned} D_i(DM_1M_2) &=_\beta M_i; \\ D(D_1M)(D_2M) &=_\beta M. \end{aligned}$$

The first equation states that we can form pairs from which the components can be retrieved. The second equation states that all terms are convertible to a pair.

The following result from Barendregt (1974) states that no surjective pairing exists. It was proved in de Vrijer (1987) using Theorem 1.77.

PROPOSITION 1.80. *There is no surjective pairing in the λ-calculus.*

PROOF. Let (D, D_1, D_2) be a surjective pairing. By definition $D(D_1\Omega)(D_2\Omega) =_\beta \Omega$, hence by confluence $D(D_1\Omega)(D_2\Omega) \twoheadrightarrow_\beta \Omega$. Write $M \triangleq D(D_1\Omega)(D_2x)$, so $M \rightsquigarrow_{x,\Omega} \Omega \twoheadrightarrow_\beta \Omega$. Then, by Theorem 1.77, there exists an M_1 such that $M \twoheadrightarrow_\beta M_1 \rightsquigarrow_{x,\Omega} \Omega$. Therefore either $M_1 = x$ or $M_1 = \Omega$ holds. In the first case $D(D_1\Omega)(D_2x) =_\beta x$, so by taking from both sides D_1 it follows that $D_1\Omega =_\beta D_1x$. Hence, by taking $x = DPQ$, it follows that $D_1\Omega =_\beta P$, for all P, thus contradicting consistency. In the second case $D(D_1\Omega)(D_2x) =_\beta \Omega$, implying similarly $Q =_\beta D_2\Omega$, for all Q, again a contradiction. □

As observed by Endrullis and de Vrijer (2008) the genericity lemma from Barendregt (1971) can be proved using Theorem 1.77. This is in fact a reconstruction of the proof in Barendregt (1971) and more intuitive than the topological proof in B[1984].

PROPOSITION 1.81 (GENERICITY LEMMA, B[1984], PROPOSITION 14.3.24). *Let U be unsolvable. Suppose $FU =_\beta \mathsf{I}$. Then $FP =_\beta \mathsf{I}$, for all $P \in \Lambda$.*

PROOF. By assumption and confluence we have $(Fx)[x := U] = FU \twoheadrightarrow_\beta \mathsf{I}$, for x fresh. By Theorem 1.77, for some M_1, one has $Fx \twoheadrightarrow_\beta M_1 \rightsquigarrow_{x,U} \mathsf{I}$. By definition of $\rightsquigarrow_{x,U}$ we get $M_1[x := U] \twoheadrightarrow_\beta \mathsf{I}$, then we can apply Lemma 1.72 and obtain $x \notin \mathrm{FV}(M_1)$ since I has no unsolvable subterms. We conclude $Fx \twoheadrightarrow_\beta M_1 = \mathsf{I}$, whence $FP =_\beta \mathsf{I}$ for any P. □

COROLLARY 1.82. *The Genericity Lemma also holds for $=_{\beta\eta}$.*

PROOF. Suppose $FU =_{\beta\eta} N$. By confluence of $\beta\eta$-reduction, they have a common reduct $FU \twoheadrightarrow_{\beta\eta} Z\ {}_{\beta\eta}\!\twoheadleftarrow N$. By applying η-postponement, we get $FU \twoheadrightarrow_\beta Z' \twoheadrightarrow_\eta Z\ {}_{\beta\eta}\!\twoheadleftarrow N$. By Genericity Lemma for $=_\beta$ (Proposition 1.81), we conclude for all $M \in \Lambda$:

$$FM =_\beta Z' =_{\beta\eta} N.$$

□

Chapter 2

Böhm trees and variations

We define three kinds of (labelled) trees representing the operational behavior of λ-terms. We start from the classical notion of Böhm(-like) trees, naturally arising from the study of solvability. We then briefly discuss Lévy-Longo trees, originated from desiderata in the practice of functional programming languages. Finally, we present Berarducci trees that have arisen from consistency studies. We will see in Chapter 6 that these trees can be seen as the normal forms of some infinitary higher-order term rewriting systems. For Böhm trees we also present an approximation theorem as well as some of its consequences.

2.1 About coinduction

In B[1984], Chapter 10, much effort is made to present possibly infinite labelled trees, like Böhm trees, as partial functions from finite sequences of natural numbers to labels. That kind of formalization is still useful to indicate specific branches in a tree, or specify subtrees rooted at some positions but, otherwise, feels obsolete. Nowadays, it is well-established that infinite trees are best described as coinductive data-types, and the coinduction principle is generally better understood. In the present book, not only we employ coinductive techniques, but we embrace the philosophy behind the article by Kozen and Silva (2017) aiming at simplifying the presentation of coinductive proofs by adopting a more familiar terminology, without reasserting the principle every time. Concretely, we will simply say that "we apply the coinductive hypothesis" whenever the coinduction principle is applied. This improves the readability. The technical explanation of why this can be done safely is that structural coinduction is inherently *productive*.

The next subsection is devoted to a comparison between inductive and coinductive techniques, but will also be the occasion to explain our terminology.

A gentle introduction

Trying to understand what is coinduction one gets the feeling that it is simply a form of ordinary induction. At the end of the story ordinary induction is used, but on a more complex statement. So there is some gain. First an observation about using induction.

PROPOSITION 2.1. (i) *Consider a binary predicate* P*. Sometimes a statement*

$$\forall k \forall n\,.\,\mathrm{P}kn$$

cannot be proved by induction on n *(applied to the subformula), but the logically equivalent statement:*

$$\forall n \forall k\,.\,\mathrm{P}kn$$

can be proved this way.

(ii) *Given a unary predicate* Z *and a binary one* P*, the following may not be provable using induction on* n

$$\forall t\,.\,[\,\mathrm{Z}t \Rightarrow \forall n\,.\,\mathrm{P}tn\,],$$

whereas the logically equivalent statement

$$\forall n \forall t\,.\,[\,\mathrm{Z}t \Rightarrow \mathrm{P}tn\,]$$

can be proved this way.

PROOF. (i) To prove the first, one has to take a number k and show by induction $\forall n.\mathrm{P}kn$. For this one needs to show $\mathrm{P}k0$ and $\forall n.(\mathrm{P}kn \Rightarrow \mathrm{P}k(n+1))$. But taking an arbitrary n and assuming $\mathrm{P}kn$, in order to conclude $\mathrm{P}k(n+1)$, one sometimes needs $\mathrm{P}kn$ and $\mathrm{P}k'n$ for different k and k'. So the attempt to prove $\forall n.\mathrm{P}kn$ fails.

If this is the case, then one may prove $\forall n \forall k.\mathrm{P}kn$, by induction on n. Indeed, $\forall k.\mathrm{P}k0$ is essentially the same as before. But now the induction step is

$$(\forall k.\mathrm{P}kn) \Rightarrow (\forall k.\mathrm{P}k(n+1))$$

Then from the induction hypothesis $(\forall k.\mathrm{P}kn)$ one may have a better chance to derive that $\mathrm{P}kn$ and $\mathrm{P}k'n$ hold; this was needed to show $\mathrm{P}k(n+1)$.

(ii) To replace $\forall t.(\mathrm{Z}t \Rightarrow \forall n.\mathrm{P}tn)$ by $\forall t \forall n.(\mathrm{Z}t \Rightarrow \mathrm{P}tn)$ seems like a trivial step. But if n ranges over the natural numbers and one wants to apply (ordinary) induction, then sometimes such a step becomes essential. Suppose that we work in, say, Peano arithmetic and want to show

$$\forall t\,.\,(\mathrm{Z}t \Rightarrow \forall n.\mathrm{P}tn)$$

Then one can consider an arbitrary t and assume $\mathrm{Z}t$ towards proving $\forall n.\mathrm{P}tn$. A natural method is applying induction on n. Now it may be the case that $\mathrm{P}t0$ is readily proved. But trying to use the induction hypothesis $\mathrm{P}tn$ towards $\mathrm{P}t(n+1)$ is impossible. One might need the induction hypothesis $\mathrm{P}tn$ not just for t, but for other values t_1, t_2 as well, in order to establish $\mathrm{P}t(n+1)$. Proving by induction the modified—but equivalent—statement, the task becomes $\forall n \forall t(\mathrm{Z}t \Rightarrow \mathrm{P}tn)$. The base case $n = 0$ goes essentially as before, but now the induction step becomes the goal to show for arbitrary n that $(\mathrm{Z}t \Rightarrow \mathrm{P}t(n+1))$ holds, assuming $\forall t(\mathrm{Z}t \Rightarrow \mathrm{P}tn)$. Now the other values t_1, t_2 of t are available, because we are proving a universal statement involving n. During a proof development in Coq we became aware of this need for reformulation of a statement to prove. In informal mathematics, in which the two equivalent forms are felt as 'the same' one may overlook this, as morally one is right to assume $\mathrm{P}t_1n$, $\mathrm{P}t_2n$ when trying to prove $\mathrm{P}t(n+1)$. But then one in fact is not proving

$$\mathrm{Z}t \;\&\; \mathrm{P}tn \Rightarrow \mathrm{P}t(n+1),$$

rather

$$\forall t\,.\,(\mathrm{Z}t \Rightarrow \mathrm{P}tn) \Rightarrow \forall t\,.\,(\mathrm{Z}t \Rightarrow \mathrm{P}t(n+1)). \qquad \square$$

Using coinduction, the subtle transition in Proposition 2.1(ii) will play a comparable role, but we do not need to realize it.

Lists as inductive objects

Lists are built up inductively (from smaller to bigger ones). Here $[B]$ denotes the type of finite (possibly empty) lists of elements of B and can be defined as follows:

$$[B] ::= \mathtt{nil} + \mathtt{cons}(b, [B]).$$

For elements b_1, b_2 of B, we have that

$$\begin{array}{rcl} & & \mathtt{nil} \\ [b_1] & \triangleq & \mathtt{cons}(b_1, \mathtt{nil}) \\ [b_1, b_2] & \triangleq & \mathtt{cons}(b_1, \mathtt{cons}(b_2, \mathtt{nil})) \end{array}$$

are all in $[B]$. So for lists one has the constructors

$$\mathtt{nil} : [B] \quad | \quad \mathtt{cons} : B \to [B] \to [B].$$

One calls $[B]$ an *inductive type*.

On lists one can define maps $f : [B] \to [B]$ inductively. An example is the following.

$$\begin{array}{rcl} f(\mathtt{nil}) & \triangleq & \mathtt{nil} \\ f(\mathtt{cons}(a, L)) & \triangleq & \mathtt{cons}(a, \mathtt{cons}(a, f(L))) \end{array}$$

Having as effect for example

$$f[a_1, a_2, a_3] = [a_1, a_1, a_2, a_2, a_3, a_3]$$

Trees as coinductive objects

Given a type A as a collection of some kind of data, we can define the type of (finitely branching, possibly infinite depth) trees over A, denoted by T_A:

$$T_A ::=_{\text{coind.}} A \times [T_A].$$

Let us write $(a; t_1, \ldots, t_k)$ for the element $(a, [t_1, \ldots, t_k])$ of T_A.

Trees are coinductive data types and cannot be built up by the functions provided to us, they can only be broken down:

$$\begin{aligned} \mathtt{head}(a; t_1, \ldots, t_k) &\triangleq a \\ \mathtt{children}(a; t_1, \ldots, t_k) &\triangleq [t_1, \ldots, t_k]. \end{aligned}$$

For trees over A one has destructors

$$\mathtt{head} : T_A \to A \quad | \quad \mathtt{children} : T_A \to [T_A]$$

One calls T_A a *coinductive type*.

Defining predicates on trees

Every predicate Q on A can be coinductively lifted to a predicate P^{Q} on T_A as follows.

DEFINITION 2.2. (i) For $\mathrm{Q} \subseteq A$, let $\mathrm{P}^{\mathrm{Q}} \subseteq T_A$ be coinductively defined by

$$\mathrm{P}^{\mathrm{Q}}(a; t_1, \ldots, t_k) \iff \mathrm{Q}(a) \;\&\; \mathrm{P}^{\mathrm{Q}}(t_1) \;\&\; \cdots \;\&\; \mathrm{P}^{\mathrm{Q}}(t_k).$$

That is, $\mathrm{P}^{\mathrm{Q}}(t)$ holds if and only if, for all labels a in t, one has $\mathrm{Q}(a)$.

(ii) An alternative inductive definition of P^{Q} is:

$$\mathrm{P}^{\mathrm{Q}}(t) \iff \forall n \in \mathbb{N} \,.\, \mathrm{P}^{\mathrm{Q}}_n(t),$$

where

$$\begin{aligned} \mathrm{P}^{\mathrm{Q}}_0(a; t_1, \ldots, t_k) &\iff \mathrm{Q}(a), \\ \mathrm{P}^{\mathrm{Q}}_{n+1}(a; t_1, \ldots, t_k) &\iff \mathrm{Q}(a) \;\&\; \mathrm{P}^{\mathrm{Q}}_n(t_1) \;\&\; \cdots \;\&\; \mathrm{P}^{\mathrm{Q}}_n(t_k). \end{aligned}$$

Intuitively, $\mathrm{P}^{\mathrm{Q}}(t)$ states that the labels at all nodes satisfy the property Q. As an example, take $A \triangleq \mathbb{N}$ and the predicate $\mathrm{P} \triangleq \textsc{Odd}$ where $\textsc{Odd}(n) \iff n \bmod 2 = 1$. Then, given $t \in T_{\mathbb{N}}$, $\mathrm{P}^{\textsc{Odd}}(t)$ holds exactly when all labels in t are odd natural numbers.

PROPOSITION 2.3 (SIMPLE COINDUCTION). *Let P^{Q} be the property of trees over A defined above. Let $\mathrm{Q}_1, \mathrm{Q}_2$ be properties of the elements of A such that*

$$\forall a \in A \,.\, \big(\mathrm{Q}_1(a) \Rightarrow \mathrm{Q}_2(a)\big). \tag{2.1}$$

Suppose that, for $t = (a; t_1, \ldots, t_k)$, we have:

$$\frac{\big(\mathrm{P}^{\mathrm{Q}_1}(t_1) \Rightarrow \mathrm{P}^{\mathrm{Q}_2}(t_1)\big) \;\&\; \cdots \;\&\; \big(\mathrm{P}^{\mathrm{Q}_1}(t_k) \Rightarrow \mathrm{P}^{\mathrm{Q}_2}(t_k)\big) \quad \text{co-IH}}{\mathrm{P}^{\mathrm{Q}_1}(t) \Rightarrow \mathrm{P}^{\mathrm{Q}_2}(t)}$$

Then

$$\forall t \in T_A \,.\, \big(\mathrm{P}^{\mathrm{Q}_1}(t) \Rightarrow \mathrm{P}^{\mathrm{Q}_2}(t)\big).$$

In words. In the situation of the proposition one likes to show a property for all trees. Assuming that the property holds for all children of a tree (coinduction hypothesis), then it holds also for the tree itself. From this implication it follows that the property holds for all trees.

PROOF. We have to show

$$\forall t \in T_A \,.\, \big((\forall m \in \mathbb{N} \,.\, \mathrm{P}_m^{\mathrm{Q}_1}(t)) \Rightarrow \forall n \in \mathbb{N} \,.\, \mathrm{P}_n^{\mathrm{Q}_2}(t)\big).$$

This is logically equivalent to

$$\forall n \in \mathbb{N}, \forall t \in T_A \,.\, \big((\forall m \in \mathbb{N} \,.\, \mathrm{P}_m^{\mathrm{Q}_1}(t)) \Rightarrow \mathrm{P}_n^{\mathrm{Q}_2}(t)\big).$$

We show this by (ordinary) induction on n.

Case $n = 0$. We must show

$$\forall t \in T_A \,.\, \big(\forall m \in \mathbb{N} \,.\, \mathrm{P}_m^{\mathrm{Q}_1}(t) \Rightarrow \mathrm{P}_0^{\mathrm{Q}_2}(t)\big).$$

So let $t = (a; t_1, \ldots, t_k) \in T_A$ be given and assume $\forall m \in \mathbb{N} \,.\, \mathrm{P}_m^{\mathrm{Q}_1}(t)$. Then $\mathrm{P}_0^{\mathrm{Q}_1}(t)$ holds, so $\mathrm{Q}_1(a)$ does, whence $\mathrm{Q}_2(a)$, which is $\mathrm{P}_0^{\mathrm{Q}_2}(t)$.

Case $n + 1$. We assume the induction hypothesis

$$\forall t \in T_A \,.\, \big((\forall m \in \mathbb{N} \,.\, \mathrm{P}_m^{\mathrm{Q}_1}(t)) \Rightarrow \mathrm{P}_n^{\mathrm{Q}_2}(t)\big)$$

in order to show

$$\forall t \in T_A \,.\, \big((\forall m \in \mathbb{N} \,.\, \mathrm{P}_m^{\mathrm{Q}_1}(t)) \Rightarrow \mathrm{P}_{n+1}^{\mathrm{Q}_2}(t)\big).$$

So let $t = (a; t_1, \ldots, t_k) \in T_A$ be given and assume

$$\forall m \in \mathbb{N} \,.\, \mathrm{P}_m^{\mathrm{Q}_1}(t), \tag{2.2}$$

towards $\mathrm{P}_{n+1}^{\mathrm{Q}_2}(t)$, which is $\mathrm{Q}_2(a) \;\&\; \mathrm{P}_n^{\mathrm{Q}_2}(t_1) \;\&\; \cdots \;\&\; \mathrm{P}_n^{\mathrm{Q}_2}(t_k)$. By assumption (2.2) we have $\mathrm{P}_0^{\mathrm{Q}_1}(t)$, so $\mathrm{Q}_1(a)$, and therefore, by (2.1), we obtain first $\mathrm{Q}_2(a)$. Now apply the induction hypothesis to $t_1, \ldots, t_k$, respectively, obtaining

$$\big((\forall m \in \mathbb{N} \,.\, \mathrm{P}_m^{\mathrm{Q}_1}(t_1)) \Rightarrow \mathrm{P}_n^{\mathrm{Q}_2}(t_1)\big) \;\&\; \cdots \;\&\; \big((\forall m \in \mathbb{N} \,.\, \mathrm{P}_m^{\mathrm{Q}_1}(t_k)) \Rightarrow \mathrm{P}_n^{\mathrm{Q}_2}(t_k)\big). \tag{2.3}$$

By assumption (2.2), we get $\mathrm{P}_n^{\mathrm{Q}_2}(t_1) \;\&\; \cdots \;\&\; \mathrm{P}_n^{\mathrm{Q}_2}(t_k)$, what remained to be proved. □

Defining functions on trees

On trees one can define maps coinductively. An example for $T_{\mathbb{N}}$ is:

$$g(n; t_1, \ldots, t_k) = (n^2; g(t_1), \ldots, g(t_k)) \tag{2.4}$$

The map g squares all elements occurring as labels in t. The coinductive definition of g can be given precisely using the coinduction principle for T_A presented below.

PROPOSITION 2.4. *The set T_A enjoys the following universal properties:*

1. *(Corecursion) Suppose a set X is given, together with two functions $h : X \to A$, $c : X \to [X]$. Then there exists a unique function $f : X \to T_A$ satisfying*

$$\begin{array}{lcl} \texttt{head}(f(x)) & = & h(x); \\ \texttt{children}(f(x)) & = & [f(x_1), \ldots, f(x_k)], \quad \textit{where } c(x) = [x_1, \ldots, x_k]. \end{array}$$

This function f is called corecursor *and denoted by* $\mathrm{corec}(X, h, c)$.

2. *(Coinduction) Let* Q *be any predicate on* A *and* (X, h, c) *be as above. Suppose a predicate* R *on* X *is given, satisfying*

$$(H_1) \quad \forall x \in X \,.\, \mathrm{R}(x) \Rightarrow \mathrm{Q}(h(x));$$

$$(H_2) \quad \forall x \in X \,.\, \mathrm{R}(x) \Rightarrow \mathrm{P}^{\mathrm{Q}}(x_1) \;\&\; \cdots \;\&\; \mathrm{P}^{\mathrm{Q}}(x_k), \quad \textit{where } c(x) = [x_1, \ldots, x_k].$$

Then for all $x \in X$, $\mathrm{R}(x) \Rightarrow \mathrm{P}^{\mathrm{Q}}(f(x))$, *where* $f = \mathrm{corec}(X, h, c)$ *is defined as above.*

DEFINITION 2.5. (i) For $f\colon A \to A$ define its *coinductive lifting* $\hat{f}\colon T_A \to T_A$ by

$$\hat{f}(a; t_1, \ldots, t_k) \triangleq (f(a); \hat{f}(t_1), \ldots, \hat{f}(t_k)).$$

That is, all labels a in t are replaced by $f(a)$.

(ii) An alternative inductive definition of $\hat{f}$ is:

$$\hat{f} \triangleq \mathrm{corec}(T_A, \lambda\!\!\lambda t. f(\texttt{head}\ t), \texttt{children})$$

For example the function $g : T_{\mathbb{N}} \to T_{\mathbb{N}}$ squaring of all elements of a tree, as in (2.4), can be defined via the corecursor as follows:

$$g = \mathrm{corec}(T_A, \lambda\!\!\lambda t. (\texttt{head}\ t)^2, \texttt{children})$$

since this yields

$$\begin{aligned} g(n; t_1, \ldots, t_k) &= (\texttt{head}(g(n; t_1, \ldots, t_k))^2; \texttt{children}(g(n; t_1, \ldots, t_k))) \\ &= (n^2; g(t_1), \ldots, g(t_k)). \end{aligned}$$

LEMMA 2.6. *Let* $\mathrm{Q} \subseteq A$ *and* $f\colon A \to A$. *Write* $\mathrm{Q} \circ f$ *for the relation*

$$(\mathrm{Q} \circ f)(a) \iff \mathrm{Q}(f(a)).$$

(i) *For all* $t \in T_A$, *we have* $\mathrm{P}^{\mathrm{Q}}(\hat{f}(t)) \iff \mathrm{P}^{\mathrm{Q} \circ f}(t)$.

(ii) *Suppose* $\forall a \in A \,.\, \mathrm{Q}(a) \Rightarrow \mathrm{Q}(f(a))$. *Then, for all* $t \in T_A$

$$\mathrm{P}^{\mathrm{Q}}(t) \Rightarrow \mathrm{P}^{\mathrm{Q}}(\hat{f}(t))$$

PROOF. (i) By coinduction. Let $t = (a; t_1, \ldots, t_k)$. Then

$$\begin{aligned} \mathrm{P}^{\mathrm{Q}}(\hat{f}(t)) &\iff \mathrm{P}^{\mathrm{Q}}((f(a); \hat{f}(t_1), \ldots, \hat{f}(t_k))) \\ &\iff \mathrm{Q}(f(a)) \;\&\; \mathrm{P}^{\mathrm{Q}}(\hat{f}(t_1)) \;\&\; \cdots \;\&\; \mathrm{P}^{\mathrm{Q}}(\hat{f}(t_k)) \\ &\iff (\mathrm{Q} \circ f)(a) \;\&\; \mathrm{P}^{\mathrm{Q}\circ f}(t_1) \;\&\; \cdots \;\&\; \mathrm{P}^{\mathrm{Q}\circ f}(t_k), \qquad \text{by co-IH}, \\ &\iff \mathrm{P}^{\mathrm{Q}\circ f}(t). \end{aligned}$$

(ii) For all $t \in T_A$ one has

$$\begin{aligned} \mathrm{P}^{\mathrm{Q}}(t) &\Rightarrow \mathrm{P}^{\mathrm{Q}\circ f}(t), \qquad \text{by assumption and Proposition 2.3}, \\ &\Rightarrow \mathrm{P}^{\mathrm{Q}}(\hat{f}(t)). \end{aligned}$$

□

So we see that the statement is clear intuitively and that the coinductive proof is a good representation of it. Moreover, the coinductive proof is certified by the possibility of turning it into an ordinary inductive proof. As the latter proof is more involved, it is useful to regularly employ coinduction when dealing with coinductive structures, like possibly infinite trees.

2.2 Numerical sequences and trees

We start by fixing some notations concerning sequences of natural numbers.

NOTATION. (i) *Let* $\mathbb{N}^*$ *(resp.* 2^**) be the set of* finite sequences *over* $\mathbb{N}$ *(resp. over* $\{0,1\}$*).*
(ii) *A sequence* α *is represented as an ordered list* $\langle n_1, \dots, n_k \rangle$ *of natural numbers.*
(iii) *In particular,* $\langle\rangle$ *represents the empty sequence.*
(iv) *The* length *of* α *is denoted* $\ell(\alpha)$*.*
(v) *Given* $\alpha = \langle n_1, \dots, n_k \rangle \in \mathbb{N}^*$ *and* $n \in \mathbb{N}$ *we write* $\alpha\,;\,n$ *for the sequence* $\langle n_1, \dots, n_k, n \rangle$*.*
(vi) *Given two sequences* $\alpha, \beta \in \mathbb{N}^*$ *we write:*
  (1) $\alpha \star \beta$ *for their concatenation,*
  (2) $\alpha \leq \beta$ *if* α *is a* prefix *of* β*,*
  (3) $\alpha < \beta$ *if* $\alpha \leq \beta$ *and* $\alpha \neq \beta$*.*

We consider fixed an effective bijective encoding of all finite sequences of natural numbers $\# : \mathbb{N}^* \to \mathbb{N}$. In particular, we assume that for all $\alpha \in \mathbb{N}^*$ and $n \in \mathbb{N}$ the code $\#(\alpha\,;\,n)$ is computable from the code $\#\alpha$ and the natural number n.

NOTATION. *The quote from Definition 1.29*(iii) *is extended to sequences* $\alpha \in \mathbb{N}^*$ *by setting*

$$\ulcorner \alpha \urcorner \triangleq \mathsf{c}_{\#\alpha}$$

DEFINITION 2.7. (i) A *tree* is a partial function $t : \mathbb{N}^* \to \mathbb{N}$ such that:

$$\forall \alpha \in \mathbb{N}^*, \forall n \in \mathbb{N}\,.\,\big[\ \alpha\,;\,n \in \mathrm{dom}(t) \iff [\alpha \in \mathrm{dom}(t)\ \&\ n < t(\alpha)]\ \big]$$

(ii) The elements of $\mathrm{dom}(t)$ are called *positions* and each position identifies a *node*.
(iii) The node of t at position $\langle\rangle$, if it exists, is called the *root* of t.
(iv) A *branch* of a tree t is a maximal subset of $\mathrm{dom}(t)$ linearly ordered w.r.t. $\leq$.
(v) A branch is called *finite* if it is finite as a set, and *infinite* otherwise.

In this definition of a tree t, $t(\alpha)$ provides the number of children of the node at position α. Therefore we have $t(\alpha) = 0$ whenever the position α corresponds to a *leaf*.

REMARK 2.8. The set $\mathrm{dom}(t)$ is closed under prefixes and $\alpha\,;\,(n+1) \in \mathrm{dom}(t)$ entails $\alpha\,;\,n \in \mathrm{dom}(t)$, whence a tree is uniquely determined by the set of its positions.

It will be sometimes convenient to identify a tree t with the set $\mathrm{dom}(t)$ itself.

DEFINITION 2.9. (i) The *subtree of a tree* t *rooted at* α is the tree $t\restriction_\alpha$ defined by

$$t\restriction_\alpha(\beta) \triangleq t(\alpha \star \beta), \quad \text{for all } \beta \in \mathbb{N}^*.$$

(ii) A tree t is called:
  (1) *empty* if $\mathrm{dom}(t) = \emptyset$;
  (2) *finite* if $\mathrm{dom}(t)$ is finite and *infinite* otherwise;
  (3) *well-founded* if it has no infinite branch;
  (4) *recursive* if $\mathrm{dom}(t)$ is a recursive set (after coding $\mathbb{N}^*$ using $\#$).
(iii) We let $\mathscr{T}_{\mathsf{rec}}$ (resp. $\mathscr{T}_{\mathsf{rec}}^{\infty}$) be the set of all (resp. infinite) recursive trees.

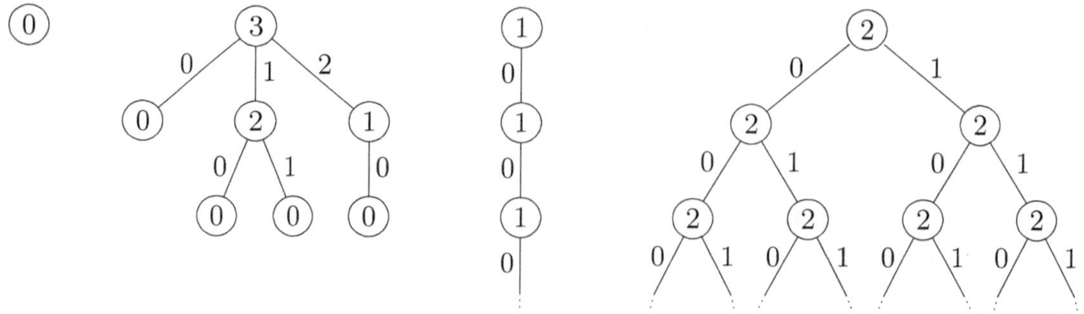

Figure 2.1: Some examples of trees. We label each node of t with the number of its children, and write the sequences $\alpha \in \mathrm{dom}(t)$ along the branches.

EXAMPLES 2.10. Figure 2.1 shows some examples of trees. From left to right:
- (i) the tree having one node (which is different from the empty tree);
- (ii) a finite tree;
- (iii) the infinite complete unary tree;
- (iv) the infinite complete binary tree.

An infinite branch of a tree can be though of as a function $f : \mathbb{N} \to \mathbb{N}$ as follows.

DEFINITION 2.11. Let t be a tree.

(i) Given a function $f : \mathbb{N} \to \mathbb{N}$ and a natural number n, the *prefix of length* n of f is the finite sequence $\hat{f}(n) = \langle f(0), \dots, f(n-1) \rangle$.

(ii) We say that a function $f : \mathbb{N} \to \mathbb{N}$ *is an infinite branch through* t if $\hat{f}(n) \in \mathrm{dom}(t)$ for all $n \in \mathbb{N}$. (Equivalently, if the set $\{\hat{f}(n) \mid n \in \mathbb{N}\}$ is a branch of t.)

(iii) We denote by $[t]_\infty$ the set of all functions being infinite branches through t.

(iv) An infinite branch through t is *recursive* if it is a recursive function.

By König's Lemma (for countable trees) a tree t is infinite if and only if $[t]_\infty \neq \emptyset$.

EXERCISE 2.12. Show that there exists an infinite recursive binary tree without infinite recursive branches.

2.3 Böhm(-like) trees

Böhm trees were introduced by Barendregt (1977), who named them after Corrado Böhm. The underlying idea is to construct a possibly infinite labelled tree by collecting the stable amounts of information coming out of the computation process.

DEFINITION 2.13. The *Böhm tree* of a λ-term M, written $\mathrm{BT}(M)$, is coinductively defined as follows:

$$
\begin{array}{rcll}
\mathrm{BT}(M) & \triangleq & \begin{array}{c} \lambda x_1 \dots x_n.y \\ \swarrow \quad \searrow \\ \mathrm{BT}(M_1) \quad \cdots \quad \mathrm{BT}(M_k) \end{array}, & \text{if } M \twoheadrightarrow_h \lambda x_1 \dots x_n.y M_1 \cdots M_k, \\
& \triangleq & \bot, & \text{otherwise.}
\end{array}
$$

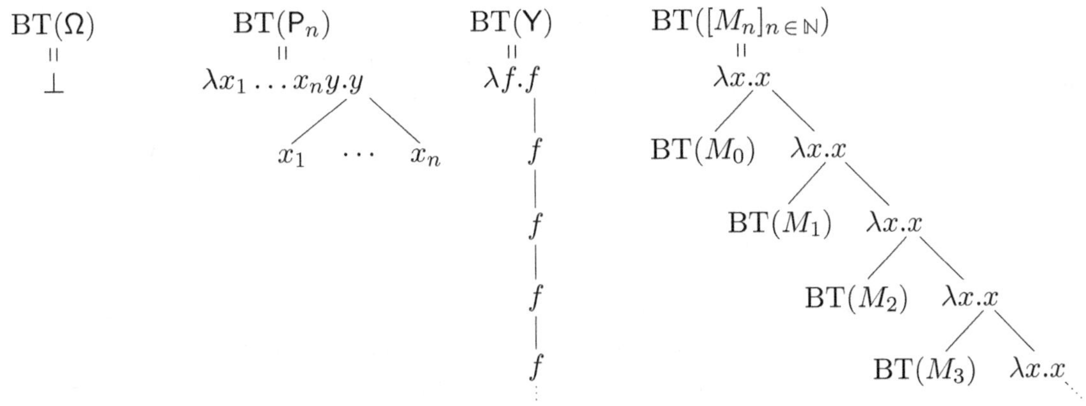

Figure 2.2: Some Böhm trees of λ-terms.

As pointed out by Jacobs and Rutten (1997), Böhm trees constitute one of the first examples of a coinductive definition. The interested reader can check the article Lassen (1999) for a more detailed discussion of the coinduction principles behind this definition.

EXAMPLES 2.14. Some Böhm-trees of λ-terms are drawn in Figure 2.2.

Notice that a variable x occurring free in a λ-term M might not occur in $\mathrm{BT}(M)$. For instance, x might occur in an unsolvable subterm as in $\mathrm{BT}(\Omega x) = \bot$ or be *pushed into infinity* along the reduction (this situation is described in Exercise 2.15(iii) below).

EXERCISE 2.15. (i) Show that $\mathrm{BT}(\Theta\mathsf{I}) = \bot$.
(ii) Show that $\mathrm{BT}(\mathsf{Y}) = \mathrm{BT}(\Theta)$ while $\mathsf{Y} \neq_\beta \Theta$.
(iii) Define a λ-term $M \in \Lambda$ satisfying, for some fresh variable x, that $Mx \twoheadrightarrow_\beta N$ entails $x \in \mathrm{FV}(N)$, while x does not occur free in $\mathrm{BT}(Mx)$.

PROPOSITION 2.16. *For* $M, N \in \Lambda$, $M =_\beta N$ *entails* $\mathrm{BT}(M) = \mathrm{BT}(N)$.

PROOF. An easy coinduction using Proposition 1.11 (confluence of β-reduction). □

A simple inspection of Figure 2.2 should convince the reader of the following.

LEMMA 2.17 (B[1984], EXAMPLE 10.1.5(V)).
Let $(M_n)_{n\in\mathbb{N}}$ *and* $(N_n)_{n\in\mathbb{N}}$ *be two effective enumerations of λ-terms. Then we have:*

$$\forall i \in \mathbb{N} \;.\; \mathrm{BT}(M_i) = \mathrm{BT}(N_i) \iff \mathrm{BT}([M_n]_{n\in\mathbb{N}}) = \mathrm{BT}([N_n]_{n\in\mathbb{N}}).$$

DEFINITION 2.18. Given $M \in \Lambda$ and $\alpha \in \mathbb{N}^*$, we define the *subterm of* M *at* α *relative to its Böhm tree*, written M_α, as follows:

$$M_\alpha \triangleq \begin{cases} M, & \text{if } \alpha = \langle\rangle, \\ (M_{i+1})_\beta, & \text{if } \alpha = \langle i\rangle \star \beta,\ M^{\mathsf{phnf}} = \lambda x_1 \dots x_n.yM_1 \cdots M_k \text{ and } i < k, \\ \Omega, & \text{otherwise.} \end{cases}$$

Notice that $M_{\langle 0\rangle}$, when different from Ω, corresponds to the first argument of the principal head normal form of M—this explains the apparent mismatch between the indices of $M_{\langle i\rangle\star\beta}$ and $(M_{i+1})_\beta$ in the second case of the definition above.

DEFINITION 2.19. (i) The set $\mathscr{B}$ of *Böhm-like trees* is defined by the following coinductive grammar:

$$(\mathscr{B}) \qquad T ::=_{\text{coind.}} \perp \mid \lambda x_1 \ldots x_n.yT\cdots T$$

(ii) The set $\mathrm{FV}(T)$ of *free variables* of a Böhm-like tree T is defined as expected and can be infinite. Böhm-like trees will be considered modulo renaming of bound variables.
(iii) We let $|T|$ denote the *underlying naked tree* of T, which is a tree in the sense of Definition 2.9(i).

REMARK 2.20. The notion of underlying naked tree given above is slightly different from the one in B[1984], Definition 10.1.7(ii). In fact, using the current version, $\mathrm{dom}(|T|)$ also includes the positions corresponding to nodes in T labelled by $\perp$.

As the two definitions are equivalent, we sometimes switch between the presentation of Böhm-like trees as coinductively generated abstract terms and 'labelled' trees.

NOTATION. *Böhm-like trees inherit the following notions from Definition 2.7:*
(i) $\mathrm{dom}(T)$ *is the set of positions in the tree (namely,* $\mathrm{dom}(|T|)$*);*
(ii) $\alpha \in T$ *is an abbreviation for* $\alpha \in \mathrm{dom}(T)$*;*
(iii) *for all* $\alpha \in T$, $T(\alpha)$ *represents the node of* T *at* α*;*
(iv) $T\restriction_\alpha$ *denotes the subtree rooted at* α*.*

DEFINITION 2.21. A Böhm-like tree T is called:
(i) *finite* if its underlying tree $|T|$ is finite; *infinite* otherwise;
(ii) $\perp$*-free* if none of its nodes has $\perp$ as label;
(iii) λ*-definable* whenever there exists $M \in \Lambda$ such that

$$T = \mathrm{BT}(M)$$

(iv) *recursively enumerable* (r.e., for short) if the partial function

$$f : \mathbb{N}^* \to \Sigma_1 \times \mathbb{N},$$

representing[1] T as a partially Σ_1-labelled tree, where

$$\Sigma_1 \triangleq \{\lambda x_1 \ldots x_n.y \mid n \geq 0, x_1, \ldots, x_n, y \in \mathrm{Var}\},$$

is partial recursive (after some coding of Σ_1);
(v) *recursive* if T is r.e. and $\mathrm{dom}(T)$ is a recursive set (after coding).

CONVENTION 2.22. *When* $T \in \mathscr{B}$ *is finite and* $\perp$*-free, we often confuse it with the corresponding* λ*-term. In other words, we treat* T *as an element* $T \in \Lambda$.

[1] In the sense of Definition 10.1.12 in B[1984].

EXAMPLES 2.23. (i) All the trees in Figure 2.2 are Böhm-like trees as well.

(ii) Let $(x_i)_{i \in \mathbb{N}}$ be an enumeration of all variables. Then the following is a Böhm-like tree T such that $\mathrm{FV}(T) = \mathrm{Var}$:

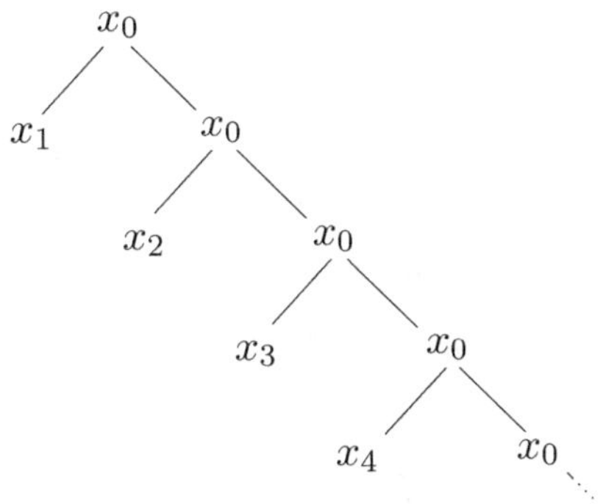

(iii) Let $(M_i)_{i \in \mathbb{N}}$ be an enumeration of all combinators having an infinite Böhm tree. As this property is undecidable, the following is a non r.e. Böhm-like tree:

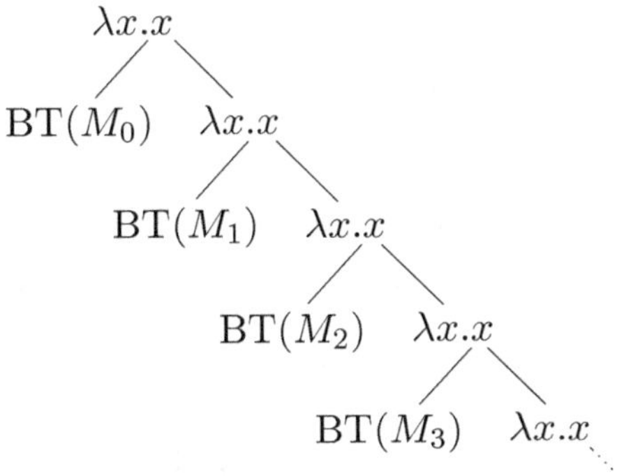

THEOREM 2.24 (B[1984], THEOREM 10.1.23).
A Böhm-like tree T is λ-definable if and only if T is r.e. and $\mathrm{FV}(T)$ is finite.

As a consequence, the Böhm-like trees from Examples 2.23(ii)-(iii) are not λ-definable.

Böhm-like tree inclusions and approximations

The constant $\perp$ in the definition of Böhm-like trees represents the absolute lack of information, therefore it makes sense to assume that $\perp$ is 'less defined' than any other Böhm-like tree T. This induces the following ordering on $\mathscr{B}$.

DEFINITION 2.25. Given $T_1, T_2 \in \mathscr{B}$, we say that T_1 *is an approximant of* T_2, in symbols $T_1 \leq_\perp T_2$, whenever T_1 results from T_2 by replacing some subtrees by $\perp$.

Böhm-like trees enjoy an approximation theorem stating that every $T \in \mathscr{B}$ is equal to the supremum of its finite approximants, namely the normal forms of a λ-calculus extended with a constant $\perp$ and a reduction relation specifying its operational behavior.

DEFINITION 2.26. (i) The set $\Lambda_\perp$ of *$\lambda\perp$-terms* is defined by the grammar:

$$(\Lambda_\perp) \qquad M ::= \perp \mid x \mid \lambda x.M \mid MM$$

(ii) The notion of *$\perp$-reduction* is given by the axioms (for $M \in \Lambda_\perp$ and $x \in \mathrm{Var}$):

$$\perp M \to \perp, \qquad \lambda x.\perp \to \perp, \qquad (\perp)$$

and generates reduction relations $\to_\perp$, $\to_{\beta\perp}$, $\twoheadrightarrow_\perp$, $\twoheadrightarrow_{\beta\perp}$ and conversions $=_\perp$, $=_{\beta\perp}$.

(iii) We define

$$\mathscr{A} \triangleq \{M \in \Lambda_\perp \mid M \text{ is in } \beta\perp\text{-normal form}\}$$

The elements of $\mathscr{A}$ are generally denoted by the metavariables P, Q and L possibly with indices.

LEMMA 2.27. *For $M \in \Lambda_\perp$, the following are equivalent:*

1. *$M \in \mathscr{A}$;*
2. *M is generated by the following grammar:*

$$\begin{array}{lcl} M & ::= & \perp \mid Q \\ Q & ::= & \lambda x.Q \mid P \\ P & ::= & x \mid PM \end{array}$$

3. *either $M = \perp$ or $M = \lambda x_1 \dots x_n.yP_1 \cdots P_k$ for some $n, k \geq 0$ and $P_i \in \mathscr{A}$.*

It follows that every $P \in \mathscr{A}$ can be seen as a finite Böhm-like tree.

DEFINITION 2.28. (i) Let $\leq_\perp$ be the least compatible preorder on $\Lambda_\perp$ satisfying $\perp \leq_\perp M$ for all $M \in \Lambda_\perp$.

(ii) For $M, N \in \Lambda_\perp$, we say that *M is an approximant of N* whenever $M \leq_\perp N$ holds.

(iii) The *direct approximant* of a $\lambda\perp$-term M is defined by induction as follows:

$$\mathsf{da}(M) \triangleq \begin{cases} \lambda x_1 \dots x_n.y\,\mathsf{da}(M_1) \cdots \mathsf{da}(M_k), & \text{if } M = \lambda x_1 \dots x_n.yM_1 \cdots M_k, \\ \perp, & \text{otherwise.} \end{cases}$$

It is easy to check that, for all $M \in \Lambda_\perp$, one has $\mathsf{da}(M) \leq_\perp M$ and $\mathsf{da}(M)$ is in $\beta\perp$-nf.

(iv) Let $M \in \Lambda_\perp$. The set $\mathscr{A}(M)$ of *finite approximants of M* is defined as follows:

$$\mathscr{A}(M) \triangleq \{\mathsf{da}(N) \mid M =_\beta N\}$$

REMARK 2.29. (i) For $M, M' \in \Lambda_\perp$, $M \to_\beta M'$ entails $\mathsf{da}(M) \leq_\perp \mathsf{da}(M')$.

(ii) For all $M \in \Lambda_\perp$ and $P \in \mathscr{A}$, we have $P \leq_\perp M$ if and only if $P \leq_\perp \mathsf{da}(M)$. In particular, as already noticed, we have $\mathsf{da}(M) \leq_\perp M$.

(iii) A $\lambda\perp$-term P belongs to $\mathscr{A}(M)$ exactly when $P \leq_\perp \mathrm{BT}(M)$.

LEMMA 2.30. *For all $M \in \Lambda$,*

$$\mathscr{A}(M) = \{P \in \mathscr{A} \mid M \twoheadrightarrow_\beta N \;\&\; P \leq_\perp N\}.$$

PROOF. ($\subseteq$) Take $P \in \mathscr{A}(M)$. By definition, we have $M =_\beta N$ for some $N \in \Lambda$ such that $\mathsf{da}(N) = P$. By confluence (Proposition 1.11), M and N have a common reduct $Z \in \Lambda$, i.e., $M \twoheadrightarrow_\beta Z \ {}_\beta\!\twoheadleftarrow N$. By Remark 2.29(i), $P = \mathsf{da}(N) \leq_\bot \mathsf{da}(Z)$ and we are done.

($\supseteq$) Take $P \in \mathscr{A}$ such that $M \twoheadrightarrow_\beta N$ and $P \leq_\bot N$. By structural induction on P.

Case $P = \bot$. Easy, since $M =_\beta \mathsf{I}N$ and $\mathsf{da}(\mathsf{I}N) = \bot$.

Case $P = \lambda\vec{x}.yP_1 \cdots P_k$. Straightforward from the IH. □

LEMMA 2.31. *For all $M \in \Lambda$, the set $\mathscr{A}(M)$ is an ideal with respect to $\leq_\bot$, i.e.:*

(i) *$\mathscr{A}(M)$ is non-empty;*
(ii) *$\mathscr{A}(M)$ is downward closed: $P \in \mathscr{A}(M)$ and $Q \leq_\bot P$ imply $Q \in \mathscr{A}(M)$;*
(iii) *for all $P_1, P_2 \in \mathscr{A}(M)$, there is $Q \in \mathscr{A}(M)$ such that $P_1 \leq_\bot Q$ and $P_2 \leq_\bot Q$.*

PROOF. Easy, using Lemma 2.30. □

The following results show that the map $\mathrm{BT}(\cdot)$ is Scott continuous in the tree topology.

THEOREM 2.32 (SYNTACTIC APPROXIMATION THEOREM).
Let $M, N \in \Lambda$. The supremum of all finite approximants in $\mathscr{A}(M)$ exists in $\mathscr{B}$ and:

$$\mathrm{BT}(M) = \bigsqcup \mathscr{A}(M)$$

Moreover $\mathrm{BT}(M) = \mathrm{BT}(N)$ *if and only if* $\mathscr{A}(M) = \mathscr{A}(N)$.

LEMMA 2.33. *Let $M, N \in \Lambda_\bot$ and $C[\,]$ be a context.*

(i) *If $M \leq_\bot N$ then $\mathscr{A}(M) \subseteq \mathscr{A}(N)$.*
(ii) *For all $P \in \mathscr{A}(M)$ we have $\mathscr{A}(C[P]) \subseteq \mathscr{A}(C[M])$.*

PROOF. See B[1984], Lemma 14.3.12 and Corollary 14.3.13 (respectively). □

PROPOSITION 2.34 (SYNTACTIC CONTINUITY, B[1984], PROPOSITION 14.3.19).
Let $C[\,]$ be a context and $M \in \Lambda$.

$$\forall P \in \mathscr{A}(C[M]), \exists Q \in \mathscr{A}(M) \,.\, P \in \mathscr{A}(C[Q])$$

PROOF SKETCH. Using Lévy's inside-out technique (1978a), the reduction of $C[M]$ generating P can be rearranged as $(\rho_1\,;\rho_2)\colon C[M] \twoheadrightarrow_\beta C[N] \twoheadrightarrow_\beta X$, where ρ_2 does not contract redexes of N, and $P \leq_\bot \mathsf{da}(X) \in \mathscr{A}(C[\mathsf{da}(N)])$. Conclude by taking $Q \triangleq \mathsf{da}(N)$. □

COROLLARY 2.35. *Let $M \in \Lambda$ and $C[\,]$ be a context.*

(i) *$\mathscr{A}(C[M]) = \bigcup\{\mathscr{A}(C[P]) \mid P \in \mathscr{A}(M)\}$.*
(ii) *$\mathscr{A}(M) \subseteq \mathscr{A}(N) \quad\Rightarrow\quad \mathscr{A}(C[M]) \subseteq \mathscr{A}(C[N])$.*

PROOF. (i) By syntactic continuity and Lemma 2.33(ii).

(ii) By (i) and Lemma 2.33(ii). □

THEOREM 2.36 (BÖHM TREES CONTEXTUALITY).
For all $M, N \in \Lambda$ and contexts $C[\,]$, we have:

$$\mathrm{BT}(M) = \mathrm{BT}(N) \quad\Rightarrow\quad \mathrm{BT}(C[M]) = \mathrm{BT}(C[N])$$

PROOF. By the Syntactic Approximation Theorem 2.32 and Corollary 2.35(ii). □

2.4 Variations of Böhm trees

Böhm trees represent the asymptotic behavior of a λ-term, but completely ignore the dynamics happening within unsolvable subterms. The assumption that all unsolvables must be equated because they never exhibit any stable information is not as innocent as it might appear. As reported in Barendregt (1992), Statman proved that any (non-trivial) co-r.e. set of λ-terms closed under β-conversion can be chosen to represent the undefined value of a partial recursive function. In this section we show how, by modifying the class of λ-terms that are considered 'meaningless', one obtains different notions of tree, equating the elements of this class.

Lévy-Longo trees

Let us consider the λ-terms Ω and $\mathsf{K}^\star$. They are both unsolvable so their Böhm tree is $\bot$, but they possess a different applicative behavior. The former not only diverges but, when plugged in a context, as in $\Omega\mathsf{I}$, is unable to interact with its environment. The latter is also called an *ogre* because it is able to 'eat' any finite amount of arguments:

$$\mathsf{K}^\star P_1 \cdots P_k \twoheadrightarrow_\beta \mathsf{K}^\star P_2 \cdots P_k \twoheadrightarrow_\beta \mathsf{K}^\star \twoheadrightarrow_\beta \lambda x_1 \dots x_n.\mathsf{K}^\star \twoheadrightarrow_\beta \cdots$$

In this respect, one says that Ω *has order zero*, while $\mathsf{K}^\star$ *has infinite order*. More generally:

Definition 2.37. The *order of a λ-term M*, in symbols $\mathrm{ord}_\beta(M) \in \mathbb{N} \cup \{\infty\}$, is given by

$$\mathrm{ord}_\beta(M) \triangleq \sup\{n \mid \exists N \in \Lambda\,.\, M \twoheadrightarrow_\beta \lambda x_1 \dots x_n.N\}$$

Examples 2.38. (i) The λ-terms x, $x\mathsf{I}$, Ω, $\Omega\mathsf{I}$, $\Omega\Omega$ and Ω_3 have order zero.
(ii) $\mathrm{ord}_\beta(\mathsf{I}) = \mathrm{ord}_\beta(\mathsf{Y}) = 1$.
(iii) $\mathrm{ord}_\beta(\mathsf{1}) = \mathrm{ord}_\beta(\mathsf{K}) = \mathrm{ord}_\beta(\mathsf{F}) = 2$.
(iv) As already mentioned, $\mathrm{ord}_\beta(\mathsf{K}^\star) = \infty$.

Inspired by the work of Lévy (1975), Longo (1983) proposed to represent the behavior of a λ-term in such a way that unsolvables of different orders have distinguished trees. These trees are nowadays called Lévy-Longo trees, and are based on the notion of 'weak' head normal form described below.

Definition 2.39. Let $M \in \Lambda$.
(i) M is in *weak head normal form* (*whnf*) if it has one of the following shapes:

$M = \lambda x.N$,

$M = xM_1 \cdots M_k$, for some $k \geq 0$.

(ii) M *has a weak head normal form* if $M \twoheadrightarrow_\beta N$ for some N in whnf.

Examples 2.40. (i) The λ-terms x, $x\mathsf{I}$, $\lambda x.\Omega$ are in whnf.
(ii) Y and $\mathsf{K}^\star$ have a whnf.
(iii) $\Omega, \Omega\Omega$, and Ω_3 have no whnf.

DEFINITION 2.41. The *Lévy-Longo tree* of M, in notation $\mathrm{LLT}(M)$, is defined coinductively by

$$
\begin{array}{llll}
\mathrm{LLT}(M) & \triangleq & \begin{array}{c} \lambda x \\ | \\ \mathrm{LLT}(N) \end{array}, & \text{if } M \twoheadrightarrow_h \lambda x.N; \\[2ex]
 & \triangleq & \begin{array}{ccc} & x & \\ \swarrow & & \searrow \\ \mathrm{LLT}(M_1) & \cdots & \mathrm{LLT}(M_k) \end{array}, & \text{if } M \twoheadrightarrow_h x M_1 \cdots M_k; \\[2ex]
 & \triangleq & \bot, & \text{otherwise, i.e. } M \text{ has no whnf.}
\end{array}
$$

The last case shows that we take as meaningless terms those having no whnf.

NOTATION 2.42. For Lévy-Longo trees, it is customary to introduce the abbreviations:

$$
\bot_n \triangleq \begin{array}{c} \lambda x_1 \\ | \\ \lambda x_2 \\ | \\ \lambda x_3 \\ \vdots \\ \lambda x_n \\ | \\ \bot \end{array}
\qquad\qquad
\mathsf{O} \triangleq \begin{array}{c} \lambda x_1 \\ | \\ \lambda x_2 \\ | \\ \lambda x_3 \\ | \\ \lambda x_4 \\ | \\ \lambda x_5 \\ | \\ \vdots \end{array}
$$

REMARK 2.43. (i) Every head normal form is a whnf too. The converse does not hold: $\lambda x.\Omega$ is a counterexample.

(ii) Our definition of LLT is taken from Endrullis et al. (2012) and it is slightly unconventional, but better crafter for infinite rewriting. In Longo's original definition $(\bot_n)_{n \in \mathbb{N}}$ and O, denoted $\top$, are actual constants and there is a case distinction (for $n, k \in \mathbb{N}$):

$$
\begin{array}{llll}
\mathrm{LLT}(M) & = & \top, & \text{if } \mathrm{ord}_\beta(M) = \infty, \\[1ex]
 & = & \bot_n, & \text{if } M \in \mathrm{UNS} \;\&\; \mathrm{ord}_\beta(M) = n, \\[1ex]
 & = & \begin{array}{ccc} & \lambda x_1 \ldots x_n.y & \\ \swarrow & & \searrow \\ \mathrm{LLT}(M_1) & \cdots & \mathrm{LLT}(M_k) \end{array}, & \text{if } M^{\mathsf{phnf}} = \lambda x_1 \ldots x_n.y M_1 \cdots M_k.
\end{array}
$$

Using (i), one readily checks that the two definitions of $\mathrm{LLT}(M)$ are equivalent, in the sense that they can be easily transformed into each other and induce the same equalities on λ-terms.

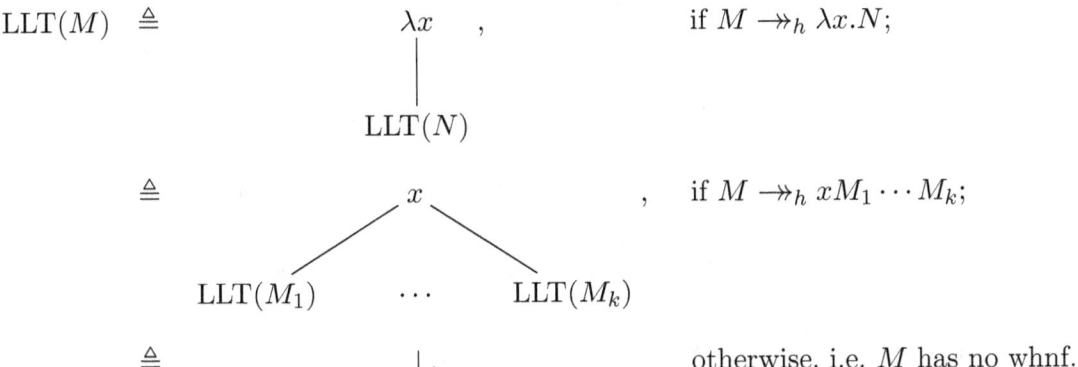

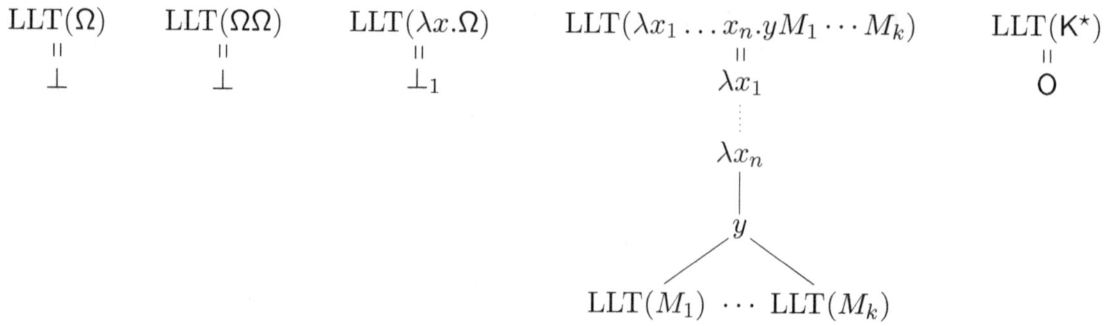

Figure 2.3: Examples of Lévy-Longo trees.

EXAMPLES 2.44. Examples of Lévy-Longo trees are provided in Figure 2.3, using the abbreviations introduced in Notation 2.42.

Berarducci trees

Berarducci trees were introduced in Berarducci (1996) and constitute a further refinement of Böhm trees. Like Lévy-Longo trees they distinguish all λ-terms having different orders, but also display more information about the rewriting process of certain unsolvables. Consider, for instance, the unsolvable Ω_3 having order zero. Recall that $\Omega_3 = \Delta_3\Delta_3$, where $\Delta_3 \triangleq \lambda x.xxx$. Therefore Ω_3 generates an infinite sequence of β-reductions:

$$\Omega_3 \to_\beta \Omega_3\Delta_3 \to_\beta \Omega_3\Delta_3\Delta_3 \to_\beta \Omega_3\Delta_3\Delta_3\Delta_3 \to_\beta \cdots$$

Note that after the first step, $\Omega_3 \to_\beta \Omega_3\Delta_3$, the argument Δ_3 becomes stable because the two components of the application cannot interact with each other. Based on this observation, Berarducci proposed this tree-representation for the behavior of Ω_3:

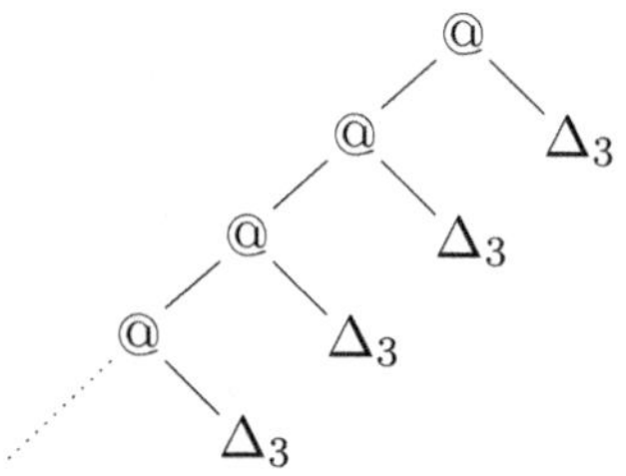

This approach is generalizable because λ-terms of order zero are known to satisfy the following properties.

LEMMA 2.45 (B[1984], LEMMA 17.3.3). *Let* $Z \in \Lambda$ *be such that* $\mathrm{ord}_\beta(Z) = 0$.

(i) *For no* $Z' \in \Lambda$ *we have* $Z \twoheadrightarrow_{\beta\eta} \lambda x.Z'$.

(ii) $Z \twoheadrightarrow_{\beta\eta} Z'$ *entails* $\mathrm{ord}_\beta(Z') = 0$.

(iii) *If* $ZM \twoheadrightarrow_{\beta\eta} N$ *then there exist* $Z', M' \in \Lambda$ *such that:*

$$N = Z'M' \ \& \ M \twoheadrightarrow_{\beta\eta} M' \ \& \ Z \twoheadrightarrow_{\beta\eta} Z'$$

(iv) $\mathrm{ord}_\beta(ZM) = 0$, *for all* $M \in \Lambda$.

More generally, λ-terms can be classified as follows.

DEFINITION 2.46. Let $M \in \Lambda$.

(i) M is in *top normal form*, abbreviated as *tnf*, if it has one of the following shapes:

$M = x$,

$M = \lambda x.N$,

$M = PQ$ with $\mathrm{ord}_\beta(P) = 0$.

(ii) M *has a top normal form* if $M \twoheadrightarrow_\beta N$ for some N in tnf.

(iii) M is called *root-active* if it has no tnf.

In the literature, root-active terms are sometimes called *mute*. See, e.g., Berarducci (1996), Carraro and Salibra (2012) and Bucciarelli et al. (2014a).

EXAMPLES 2.47. (i) $x\mathsf{I}, \mathsf{F}, \mathsf{K}$ and $\Omega_3\Delta_3$ are in tnf.

(ii) Y, K^* and Ω_3 have a tnf.

(iii) Ω is root-active.

REMARK 2.48. (i) Every λ-term in head normal form is also in top normal form. The converse does not hold, $\Omega_3\Delta_3$ being a counterexample.

(ii) If a λ-term is root-active then it must be unsolvable.

DEFINITION 2.49. The *Berarducci tree* of M, in symbols $\mathrm{BeT}(M)$, is defined coinductively by

$$\begin{array}{rcll}
\mathrm{BeT}(M) & \triangleq & x, & \text{if } M \twoheadrightarrow_h x; \\[1em]
& \triangleq & \begin{array}{c} \lambda x \\ | \\ \mathrm{BeT}(N) \end{array}, & \text{if } M \twoheadrightarrow_h \lambda x.N; \\[2em]
& \triangleq & \begin{array}{ccc} & @ & \\ \diagup & & \diagdown \\ \mathrm{BeT}(P) & & \mathrm{BeT}(Q) \end{array}, & \begin{array}{l} \text{if } M \twoheadrightarrow_h PQ \\ \text{and } \mathrm{ord}_\beta(P) = 0; \end{array} \\[2em]
& \triangleq & \bot, & \text{otherwise, i.e. } M \text{ has no tnf.}
\end{array}$$

The last case corresponds to root-active λ-terms, now considered as meaningless.

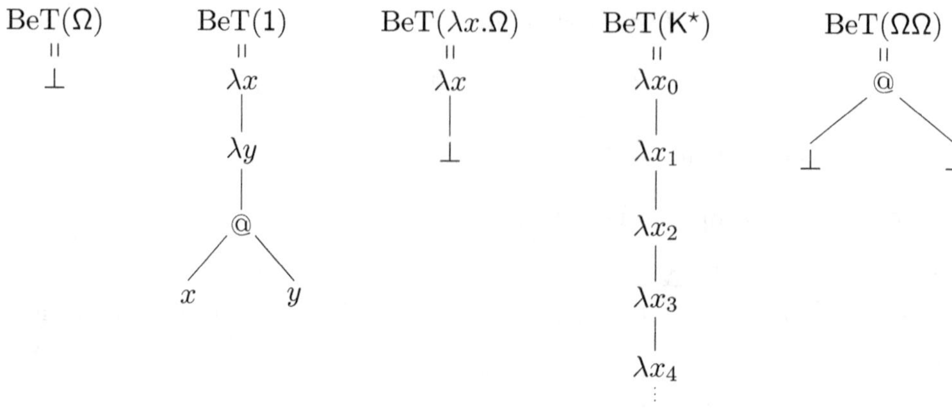

Figure 2.4: Some Berarducci trees of λ-terms.

EXAMPLES 2.50. Figure 2.4 displays some interesting examples of Berarducci trees.

EXERCISE 2.51. (i) Compare $\mathrm{BeT}(\mathsf{Y})$ with $\mathrm{BT}(\mathsf{Y})$.
(ii) Draw $\mathrm{BeT}(\Theta\Omega_3)$, where Θ is Turing's fixed point combinator.
(iii) What is the difference between $\mathrm{BeT}((\lambda x.x(xx))(\lambda x.x(xx)))$ and $\mathrm{BeT}(\mathsf{Y}(\lambda x.xxx))$?

As an example, we compare the Böhm tree, Lévy-Longo tree and Berarducci tree of some λ-terms. In the next subsection we provide a more high-level comparison.

EXAMPLES 2.52.
(i) $\mathrm{BeT}(\Omega) = \mathrm{LLT}(\Omega) = \mathrm{BT}(\Omega) = \bot$.
(ii) $\mathrm{BeT}(1) = \mathrm{LLT}(1)$ (essentially $= \mathrm{BT}(1)$).
(iii) $\mathrm{BeT}(\lambda x.\Omega) = \mathrm{LLT}(\lambda x.\Omega) =$  , but $\mathrm{BT}(\lambda x.\Omega) = \bot$.

(iv) $\mathrm{BeT}(\mathsf{K}^\star) = \mathrm{LLT}(\mathsf{K}^\star) = \mathsf{O}$. But $\mathrm{BT}(\mathsf{K}^\star) = \bot$.
(v) $\mathrm{BeT}(\Omega\Omega) =$ 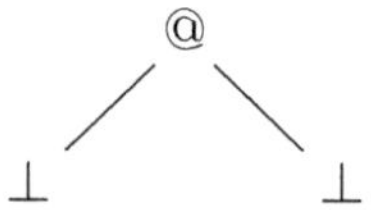 $\neq \mathrm{LLT}(\Omega\Omega) = \mathrm{BT}(\Omega\Omega) = \bot$.

(vi) A more dramatic difference is the following.

$\mathrm{BeT}(\Omega_3) =$

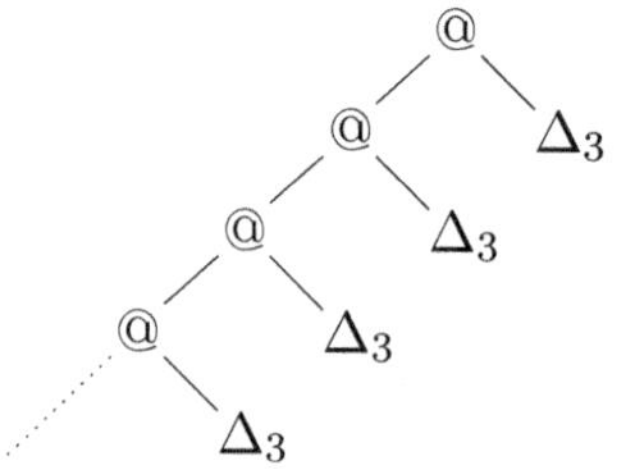

while $\mathrm{LLT}(\Omega_3) = \mathrm{BT}(\Omega_3) = \bot$.

Comparing different notions of tree

To carry out a fair comparison among the notions of labelled trees introduced so far, one should not be misled by the fact that '$\lambda\vec{x}.y$' appears as a single node in Böhm trees, while Lévy-Longo and Berarducci have a separate node for each λ-abstraction. Similarly, the fact of having a separate node '@' representing the application should be irrelevant—we need a uniform way to compare the syntactic structure of the infinitary λ-terms represented by such trees. In Figure 2.5, it is shown how to expand the compact representation of Böhm trees (called here, *hnf view*) using the general applicative notation (*applicative view*), which costs explicit application and abstraction nodes.

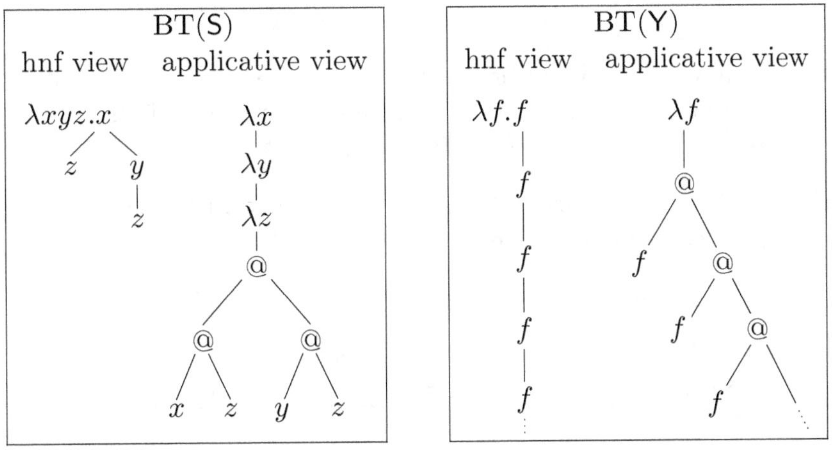

Figure 2.5: Applicative vs hnf views of Böhm trees.

The same 'applicative view' can be defined for Lévy-Longo trees in the obvious way. By uniformly adopting the applicative view to compare the three aforementioned notions of trees, it should become clear that the crucial difference lies in the class of terms that are considered 'meaningless', and therefore equated to $\perp$. The following results are taken from Barendregt (1992), Ariola and Klop (1994) and Kennaway et al. (1999), where a detailed analysis of several notions of undefinedness is performed.

THEOREM 2.53. *The following notions of 'undefined' are strictly ordered by implication:*

no top normal form (aka mute, root-active)
$\Downarrow$ (i)
no weak head normal form
$\Downarrow$ (ii)
no head normal form (aka unsolvable)
$\Downarrow$ (iii)
no normal form (aka not weakly normalizing)
$\Downarrow$ (iv)
not strongly normalizing

PROOF. (i) By Definition 2.39(i), a whnf has either the form $xM_1 \cdots M_n$ or $\lambda x.M$. The implication follows since these are tnf's as well (see Definition 2.46(i)). The converse does not hold since Ω_3 is in tnf, but does not have a whnf.

(ii) By definition, every hnf is a whnf. On the contrary, $\lambda x.\Omega$ is an unsolvable whnf. Therefore, the converse does not hold by Wadsworth's Theorem 1.23.

(iii) and (iv) are standard. □

The theorem is the contrapositive of the implications valid for properties of terms

$$\begin{array}{rcl} \text{strongly normalizing} & \Rightarrow & \text{having a nf} \\ & \Rightarrow & \text{having a hnf} \\ & \Rightarrow & \text{having a whnf} \\ & \Rightarrow & \text{having a tnf.} \end{array}$$

The theorem also shows that the notion of undefinedness embodied by root-active terms is the strongest among those considered. For more properties of this class of terms, we refer to the discussion about easiness on Page 81.

REMARK 2.54. (i) Like head normal forms, also whnf's, and tnf's can always be reached by performing head reductions. So we have, respectively:

$$\begin{array}{lcll} M \text{ has a hnf} & \Longleftrightarrow & \exists N \text{ in hnf} & .\ M \twoheadrightarrow_h N, \\ M \text{ has a whnf} & \Longleftrightarrow & \exists N \text{ in whnf} & .\ M \twoheadrightarrow_h N, \\ M \text{ has a tnf} & \Longleftrightarrow & \exists N \text{ in tnf} & .\ M \twoheadrightarrow_h N. \end{array}$$

Therefore a λ-term M has a non-trivial[2] Böhm/Lévy-Longo/Berarducci tree if and only if M has a hnf/whnf/tnf. In particular, M is root-active if and only if $\mathrm{BeT}(M) = \bot$.

(ii) By (i) and confluence, these notions of tree are all invariant under $=_\beta$.

The following result summarizes the discussion above.

COROLLARY 2.55. *For all $M \in \Lambda$, we have:*

$$\mathrm{BT}(M) \leq_\bot \mathrm{LLT}(M) \leq_\bot \mathrm{BeT}(M)$$

where $\leq_\bot$ must be defined as in Definition 2.25, but on the set of all infinitary λ-terms.

Böhm trees and extensionality

Two extensional versions of Böhm trees have been studied in the literature, under different names. We call them extensional Böhm trees and Nakajima trees, to avoid any confusion. As noticed by van Bakel et al. (2002), both can be obtained from a λ-term M by first computing its Böhm tree and then coinductively calculating its η-normal form—the difference between the two stands in the process of η-normalization which is considered. In extensional Böhm trees, introduced by Hyland (1975a) and subsequently refined by Lévy (1978a), one computes the η-normal form simply by eliminating all η-redexes.

[2] That is, different from $\bot$.

Finite λ-term	Böhm tree	Extensional Böhm tree	Nakajima tree
$\mathsf{I} = \lambda x.x$	$\mathrm{BT}(\mathsf{I}) = \mathsf{I}$	$\mathrm{nf}_\eta(\mathrm{BT}(\mathsf{I})) = \mathsf{I}$	$\mathrm{nf}_{\eta!}(\mathrm{BT}(\mathsf{I})) = \mathsf{I}$
$1 = \lambda xy.xy$	$\mathrm{BT}(1) = 1$	$\mathrm{nf}_\eta(\mathrm{BT}(1)) = \mathsf{I}$	$\mathrm{nf}_{\eta!}(\mathrm{BT}(1)) = \mathsf{I}$
$\mathsf{J} = \mathsf{Y}(\lambda jxz.x(jz))$	$\mathrm{BT}(\mathsf{J}) = \lambda xz_0.x$ $\mid$ $\lambda z_1.z_0$ $\mid$ $\lambda z_2.z_1$ $\mid$ $\lambda z_3.z_2$ $\vdots$	$\mathrm{nf}_\eta(\mathrm{BT}(\mathsf{J})) = \lambda xz_0.x$ $\mid$ $\lambda z_1.z_0$ $\mid$ $\lambda z_2.z_1$ $\mid$ $\lambda z_3.z_2$ $\vdots$	$\mathrm{nf}_{\eta!}(\mathrm{BT}(\mathsf{J})) = \mathsf{I}$

Figure 2.6: Comparison between Böhm trees and their extensional versions.

DEFINITION 2.56. The *extensional Böhm tree* of a λ-term M is given by

$$\mathrm{nf}_\eta(\mathrm{BT}(M))$$

where $\mathrm{nf}_\eta : \mathscr{B} \to \mathscr{B}$ is coinductively defined by setting $\mathrm{nf}_\eta(\bot) \triangleq \bot$ and

$$\mathrm{nf}_\eta\left(\begin{array}{c}\lambda x_1 \dots x_n.y \\ \diagup \quad \diagdown \\ T_1 \;\cdots\; T_k\end{array}\right) \triangleq \begin{cases} \mathrm{nf}_\eta\left(\begin{array}{c}\lambda x_1 \dots x_{n-1}.y \\ \diagup \quad \diagdown \\ T_1 \;\cdots\; T_{k-1}\end{array}\right), & \begin{array}{l}\text{if } n,k > 0,\ T_k \text{ is finite,} \\ T_k \twoheadrightarrow_\eta x_n \text{ and} \\ x_n \notin \mathrm{FV}(yT_1 \cdots T_{k-1}),\end{array} \\[2ex] \begin{array}{c}\lambda x_1 \dots x_n.y \\ \diagup \quad \diagdown \\ \mathrm{nf}_\eta(T_1) \;\cdots\; \mathrm{nf}_\eta(T_k)\end{array}, & \text{otherwise.}\end{cases}$$

These trees are called η-Böhm trees in van Bakel et al. (2002).

REMARK 2.57. When writing $T_k \twoheadrightarrow_\eta x_n$ above, we are employing the equivalence between finite trees and finite approximants. In fact, T_k is even $\bot$-free and $\mathrm{FV}(T_k) = \{x_n\}$.

EXERCISE 2.58. (i) Compute $\mathrm{nf}_\eta(\mathrm{BT}(1))$.

(ii) Show that, for all $M \in \mathrm{NF}_{\beta\eta}$, we have $\mathrm{nf}_\eta(\mathrm{BT}(M)) = M^{\beta\eta\text{-}\mathtt{nf}}$.

(iii) First, define a λ-term M whose Böhm tree contains infinitely many η-redexes. Second, compute $\mathrm{nf}_\eta(\mathrm{BT}(M))$.

Morally, Nakajima trees have been introduced in Nakajima (1975) and Hyland (1975b). Even if sometimes they are used only implicitly, they are widespread in the literature because they arise naturally in the characterization of the theory induced by Scott's $\mathcal{D}_\infty$. See, e.g., Chapter 10 in B[1984], Di Gianantonio et al. (1999), Ker et al. (1999), Bucciarelli and Salibra (2004), Manzonetto (2009), Severi and de Vries (2005b), Breuvart (2016), Severi and de Vries (2017) and Breuvart et al. (2018).

Intuitively, one calculates the Nakajima tree of a λ-term M by eliminating from $\mathrm{BT}(M)$ not only all η-redexes, but also η-expansions having infinite depth—the classical example being BT(J) (see Figure 2.6). Notice that BT(J) contains no finite η-redexes but, in its entirety, it can be considered as an infinite one. The following auxiliary relation specifies when a Böhm-like tree T is an infinite η-expansion of a variable x.

DEFINITION 2.59. For $T \in \mathscr{B}$ and $x \in \mathrm{Var}$, define $T \Downarrow_{\eta!} x$ by coinduction on T:

$$x \Downarrow_{\eta!} x \qquad\qquad \frac{T_1 \Downarrow_{\eta!} z_1 \quad \cdots \quad T_n \Downarrow_{\eta!} z_n \quad n > 0}{\lambda z_1 \ldots z_n.xT_1 \cdots T_n \Downarrow_{\eta!} x}$$

It is easy to check that $\mathrm{BT}(\mathsf{J}x) \Downarrow_{\eta!} x$.

DEFINITION 2.60. The *Nakajima tree* of a λ-term M is given by

$$\mathrm{nf}_{\eta!}(\mathrm{BT}(M))$$

where $\mathrm{nf}_{\eta!} : \mathscr{B} \to \mathscr{B}$ is coinductively defined by setting $\mathrm{nf}_{\eta!}(\bot) \triangleq \bot$ and

$$\mathrm{nf}_{\eta!}\left(\begin{array}{c} \lambda x_1 \ldots x_n.y \\ \diagup \quad \diagdown \\ T_1 \ \cdots \ T_k \end{array}\right) \triangleq \begin{cases} \mathrm{nf}_{\eta!}\left(\begin{array}{c} \lambda x_1 \ldots x_{n-1}.y \\ \diagup \quad \diagdown \\ T_1 \ \cdots \ T_{k-1} \end{array}\right), & \begin{array}{l} \text{if } n, k > 0 \ \& \\ T_k \Downarrow_{\eta!} x_n \ \& \\ x_n \notin \mathrm{FV}(yT_1 \cdots T_{k-1}), \end{array} \\[2ex] \begin{array}{c} \lambda x_1 \ldots x_n.y \\ \diagup \quad \diagdown \\ \mathrm{nf}_{\eta!}(T_1) \ \cdots \ \mathrm{nf}_{\eta!}(T_k) \end{array}, & \text{otherwise.} \end{cases}$$

REMARK 2.61. (i) These trees are called $\eta\infty$-Böhm trees in Severi and de Vries (2017). The definition given in Nakajima (1975) rather aims at constructing infinitely η-expanded Böhm trees—the Nakajima trees we defined are however in one-to-one correspondence with the original ones.

(ii) For $T \in \mathscr{B}$ and $x \in \mathrm{Var}$, it is easy to check that $T \Downarrow_{\eta!} x$ entails that T is $\bot$-free and $\mathrm{FV}(T) = \{x\}$. In this case T can be an infinite tree.

(iii) It is not difficult to define an $N \in \Lambda$ such that $\mathrm{nf}_{\eta!}(\mathrm{BT}(N))$ is not λ-definable, i.e., for all $M \in \Lambda$, $\mathrm{BT}(M) \neq \mathrm{nf}_{\eta!}(\mathrm{BT}(N))$ (see B[1984], Exercise 10.6.6).

EXAMPLES 2.62. Figure 2.6 displays the Böhm tree, extensional Böhm tree and Nakajima tree of some simple combinators. For more elaborated examples see Section 11.1.

Extensional Böhm trees and Nakajima trees often appear in the rest of the book:

1. In Section 6.4, we show that they can be characterized as normal forms in appropriate infinitary λ-calculi.

2. In Chapter 11, they are used to characterize observational equivalences (see Definition 3.10).

3. In Chapter 12, we provide alternative definitions, based on finite approximants, and show they satisfy suitable approximation theorems.

4. In Chapters 13 and 14, we show that they can be fruitfully used for characterizing theories of denotational models.

Chapter 3

Theories and models of λ-calculus

The λ-terms can be studied as they are when one is interested in intensional processes of computation, like reduction or conversion. On the other hand congruence relations on λ-terms, yielding equational *λ-theories* $\mathcal{T}$, become the main object of study, when one is more interested in equivalences between λ-terms. Then there are the λ-models $\mathcal{M}$ that provide a different ontology to λ-terms, coming from other areas of mathematics. They introduce new notions applicable to λ-terms, like partial orders, limits, and topologies. Every model $\mathcal{M}$, generates a λ-theory equating two λ-terms whenever they have the same interpretation in the model:

$$\mathrm{Th}(\mathcal{M}) = \{M = N \mid \mathcal{M} \models M = N\}.$$

3.1 The lattice of λ-theories

Recall that compatible binary relations $\mathrm{R} \subseteq \Lambda^2$ have been defined in Definition 1.5(iii). We say that $\mathrm{R} \subseteq \Lambda^2$ is a *congruence* if it is a compatible equivalence relation.

DEFINITION 3.1. A *λ-theory* $\mathcal{T}$ is any[1] congruence on Λ containing the β-conversion.

The set $\lambda\mathcal{T}$ of all λ-theories, ordered by set-theoretical inclusion, is a complete lattice of cardinality $2^{\aleph_0}$ (B[1984], Theorem 16.3.12). The lattice $\lambda\mathcal{T}$ constitutes a quite rich mathematical structure, as shown by Salibra and his coauthors (see Theorem 3.5 below). Manzonetto and Salibra (2010) give a gentle introduction to this topic for the algebraists.

REMARK 3.2. A λ-theory $\mathcal{T}$ is uniquely determined by its restriction to closed λ-terms. In other words, for $\mathcal{T}, \mathcal{T}' \in \lambda\mathcal{T}$, we have:

$$\mathcal{T} \cap (\Lambda^o \times \Lambda^o) = \mathcal{T}' \cap (\Lambda^o \times \Lambda^o) \iff \mathcal{T} = \mathcal{T}'.$$

For this reason, when studying λ-theories, we often focus on closed λ-terms.

[1]In B[1984] a λ-theory was required to be consistent (see Definition 3.3(i)), but it is more natural to drop this requirement. In this way the set of λ-theories forms a complete lattice.

DEFINITION 3.3. Let $\mathcal{T} \in \lambda\mathcal{T}$. We say that the λ-theory $\mathcal{T}$ is

(i) *consistent* if it does not equate all λ-terms;
(ii) *inconsistent* if it is not consistent;
(iii) *generated by* a set $X \subseteq \Lambda \times \Lambda$ of equations if it is the least λ-theory including X;
(iv) *extensional* if it contains the η-conversion;
(v) *recursively enumerable*, or simply *r.e.*, if $\mathcal{T}$ is an r.e. set (after coding);
(vi) *semi-sensible* if it does not equate a solvable and an unsolvable;
(vii) *sensible* if it is consistent and equates all unsolvable terms.

It is easy to check that all sensible λ-theories are semi-sensible.

NOTATION. (i) *Given* $\mathcal{T} \in \lambda\mathcal{T}$, *write* $\mathcal{T} \vdash M = N$, *or simply* $M =_{\mathcal{T}} N$, *for* $(M, N) \in \mathcal{T}$.

(ii) *For* $\mathcal{T} \in \lambda\mathcal{T}$ *and* $M \in \Lambda$, *write* $[M]_{\mathcal{T}}$ *for the* $\mathcal{T}$-equivalence class *of* M.

(iii) *For* $\mathcal{T} \in \lambda\mathcal{T}$ *and* $\mathcal{O} \subseteq \Lambda$,

$$\mathcal{O}/\mathcal{T} \triangleq \{[M]_{\mathcal{T}} \mid M \in \mathcal{O}\}$$

represents the set $\mathcal{O}$ *modulo* $\mathcal{T}$.

(iv) *The bottom element of* $\lambda\mathcal{T}$ *is denoted by* $\boldsymbol{\lambda}$. *Note that*

$$\boldsymbol{\lambda} \vdash M = N \iff M =_\beta N$$

(v) *The top element of* $\lambda\mathcal{T}$, *denoted by* ∇, *is the (unique) inconsistent λ-theory:*

$$\mathcal{T} = \nabla \iff \mathcal{T} \vdash \mathsf{K} = \mathsf{F}$$

(vi) *Given* $\mathcal{T}_1, \mathcal{T}_2 \in \lambda\mathcal{T}$, *we write* $\mathcal{T}_1 \wedge \mathcal{T}_2$ *for their* meet *and* $\mathcal{T}_1 \vee \mathcal{T}_2$ *for their* join.

(vii) *Given* $\mathcal{T}_1, \mathcal{T}_2 \in \lambda\mathcal{T}$, *we define the following (closed)* lattice interval

$$[\mathcal{T}_1, \mathcal{T}_2] \triangleq \{\mathcal{T} \in \lambda\mathcal{T} \mid \mathcal{T}_1 \subseteq \mathcal{T} \subseteq \mathcal{T}_2\}.$$

(viii) *Given* $\mathcal{T} \in \lambda\mathcal{T}$ *and a set* $X \subseteq \Lambda \times \Lambda$ *of equations, we write* $\mathcal{T}(X)$ *for the least λ-theory containing* $\mathcal{T} \cup X$. *When* $X = \{(M, N)\}$ *we simply write* $\mathcal{T}(M = N)$ *for* $\mathcal{T}(X)$.

(ix) *Given a λ-theory* $\mathcal{T}$, *we denote by* $\mathcal{T}\eta$ *the least extensional λ-theory including* $\mathcal{T}$. *In particular,* $\boldsymbol{\lambda}\eta$ *stands for the least extensional λ-theory, so*

$$\boldsymbol{\lambda}\eta \vdash M = N \iff M =_{\beta\eta} N$$

(x) *We denote by* $\mathcal{H}$ *the least sensible λ-theory and by* $\mathcal{B}$ *the λ-theory equating two λ-terms when they have the same Böhm tree, i.e.*

$$\mathcal{B} \vdash M = N \iff \mathrm{BT}(M) = \mathrm{BT}(N)$$

Clearly, the λ-theory $\mathcal{B}$ *is sensible and* $\mathcal{H} \subsetneq \mathcal{B}$ *holds.*

The next lemma follows from a property of equivalence relations that is well known in universal algebra (cf., Theorem 4.6 in Burris and Sankappanavar (2012)).

LEMMA 3.4. *Let* $\mathcal{T}_1, \mathcal{T}_2 \in \lambda\mathcal{T}$ *and* $M, N \in \Lambda$. *The following are equivalent:*

1. $\mathcal{T}_1 \vee \mathcal{T}_2 \vdash M = N$;
2. $\exists k \geq 0, \exists P_1, \ldots, P_k \in \Lambda \,.\, M =_{\mathcal{T}_1} P_1 =_{\mathcal{T}_2} P_2 =_{\mathcal{T}_1} \cdots =_{\mathcal{T}_1} P_{k-1} =_{\mathcal{T}_2} P_k =_{\mathcal{T}_1} N$.

Moreover, if $M, N \in \Lambda^o$ *then* $P_1, \ldots, P_k$ *can be chosen closed as well.*

Below, we summarize several interesting properties of the lattice of λ-theories.

THEOREM 3.5. *The lattice* $\lambda\mathcal{T}$ *is complete and enjoys the following properties.*

(i) *Every countable partially ordered set embeds into* $\lambda\mathcal{T}$ *by an order-preserving map. Visser (1980).*

(ii) $\lambda\mathcal{T}$ *is not modular. Salibra (2001a).*

(iii) *The meet of all coatoms is different from* $\boldsymbol{\lambda}$. *Statman (2001).*

(iv) $\lambda\mathcal{T}$ *satisfies the* Zipper condition*: for all* $\mathcal{T}_1, \mathcal{T}_2 \in \lambda\mathcal{T}$,

$$\bigvee\{\mathcal{T} \mid \mathcal{T}_1 \wedge \mathcal{T} = \mathcal{T}_2\} = \nabla \quad \Rightarrow \quad \mathcal{T}_1 = \mathcal{T}_2.$$

Lusin and Salibra (2004), using Lampe (1986).

(v) $\lambda\mathcal{T}$ *satisfies the* ET condition*: given a family of λ-theories* $(\mathcal{T}_k)_{k \in K}$, *if*

$$\bigvee_{k \in K} \mathcal{T}_k = \nabla,$$

then there is a finite sequence $\mathcal{T}_0, \ldots, \mathcal{T}_n$ *with* $\mathcal{T}_j \in \{\mathcal{T}_k \mid k \in K\}$ *for* $j \leq n$ *such that*

$$(((\cdots(((\mathcal{T}_0 \wedge \mathcal{T}) \vee \mathcal{T}_1) \wedge \mathcal{T}) \vee \mathcal{T}_2) \wedge \mathcal{T}) \vee \cdots \vee \mathcal{T}_n) \wedge \mathcal{T} = \mathcal{T}$$

for every λ-theory $\mathcal{T}$. *Lusin and Salibra (2004), again it follows from Lampe (1986).*

(vi) *For any non-trivial lattice identity* E, *there exists a natural number* n *such that the identity* E *fails in the lattice of the equational theories arising from a λ-calculus extended with* n *constants. Lusin and Salibra (2004).*

(vii) *If* $\mathcal{T}_1$ *and* $\mathcal{T}_2$ *are distinguished r.e. λ-theories, then the lattice interval* $[\mathcal{T}_1, \mathcal{T}_2]$ *has cardinality* $2^{\aleph_0}$. *Visser (1980).*

(viii) *There exists an infinite distributive sublattice of* $\lambda\mathcal{T}$. *Berline and Salibra (2006).*

As noticed by Diercks et al. (1994), condition (v) is stronger than (iv). The fact that $\lambda\mathcal{T}$ satisfies these conditions, shows that it does satisfy some non-trivial quasi-identities. Nevertheless this conjecture formulated by Salibra at the end of the nineties still stands.

CONJECTURE 3.6. The lattice $\lambda\mathcal{T}$ satisfies no non-trivial lattice identity.

Easiness

Easy terms are closed λ-terms enjoying the property that they can be consistently equated with any other combinator. For this reason, they can be considered computational processes of a completely non-informative kind.

DEFINITION 3.7. (i) A λ-term $M \in \Lambda^o$ is called *easy* if, for every $N \in \Lambda^o$, the λ-theory $\boldsymbol{\lambda}(M = N)$ generated by the equation $M = N$ is consistent.

(ii) A set $X \subseteq \Lambda^o$ is *uniformly easy* if, for every $N \in \Lambda^o$, the λ-theory generated by the set of equations $\{M = N \mid M \in X\}$ is consistent.

EXAMPLES 3.8. (i) The term Ω was shown easy by Jacopini (1975) (cf. B[1984], p. 402) via a syntactic proof. A semantic proof was proposed by Baeten and Boerboom (1979).

(ii) The term $\Omega_3\mathsf{I}$ was proved easy by Jacopini and Venturini Zilli (1985) via syntactic methods, and by Alessi et al. (2001) using filter models.

(iii) The set $\mathrm{UNS} \cap \Lambda^o$ is not uniformly easy. In fact, Jacopini (1975) has shown that even the single equation $\Omega_3 = \mathsf{I}$ generates inconsistency when added to the λ-calculus.

Many other proofs of easiness may be found in the literature. For syntactical methods, see Jacopini and Venturini Zilli (1985), Intrigila (1991), Berarducci and Intrigila (1993) and Kuper (1997). For semantic proofs, we refer to Zylberajch (1991), Berarducci (1996), Alessi and Lusin (2002) and Berline and Salibra (2006).

THEOREM 3.9 (BERARDUCCI (1996)). (i) *Every mute term $M \in \Lambda^o$ is easy.*

(ii) *The set of mute terms is uniformly easy.*

Observational preorders and equivalences

Some λ-theories are particularly interesting for computer scientists because they specify when two programs are observationally equivalent. This means that two λ-terms M and N are considered equivalent whenever one can plug either M or N into any context without noticing any difference in the global behavior. Observational equivalences therefore depend on the behavior one is interested in observing—a parametricity that will be represented by a set $\mathcal{O} \subseteq \Lambda$ of *observables*.

A subset $\mathcal{O} \subseteq \Lambda$ of observables is called *trivial* if either $\mathcal{O} = \emptyset$ or $\mathcal{O} = \Lambda$.

DEFINITION 3.10 (MORRIS (1968)).
Let $\mathcal{O} \subseteq \Lambda$ be a non-trivial set of λ-terms closed under $=_\beta$.

(i) The *observational preorder* with respect to $\mathcal{O}$, written $\sqsubseteq_{\mathcal{O}}$, is given by:

$$M \sqsubseteq_{\mathcal{O}} N \iff \forall C[\,]\,.\,[C[M] \in \mathcal{O} \;\Rightarrow\; C[N] \in \mathcal{O}].$$

(ii) The *observational equivalence* with respect to $\mathcal{O}$, written $\equiv_{\mathcal{O}}$, is given by:

$$M \equiv_{\mathcal{O}} N \iff M \sqsubseteq_{\mathcal{O}} N \;\&\; N \sqsubseteq_{\mathcal{O}} M$$

(iii) *Morris' λ-theory with observables* $\mathcal{O}$ is defined as follows:

$$\mathcal{T}_{\mathcal{O}} \triangleq \{(M, N) \mid \forall C[\,]\,.\,[C[M] \in \mathcal{O} \iff C[N] \in \mathcal{O}]\}.$$

It is not difficult to check that $\mathcal{T}_{\mathcal{O}}$ is actually a λ-theory and that $\mathcal{T}_{\mathcal{O}} \vdash M = N$ holds if and only if $M \equiv_{\mathcal{O}} N$ does. We mainly consider SOL or NF as sets of observables, and study the associated observational preorders $\sqsubseteq_{\mathrm{SOL}}$ and $\sqsubseteq_{\mathrm{NF}}$ (and equivalences).

DEFINITION 3.11. (i) The λ-theory $\mathcal{H}^*$ is defined by setting $\mathcal{H}^* \triangleq \mathcal{T}_{\mathrm{SOL}}$.
(ii) Similarly, the λ-theory $\mathcal{H}^+$ is defined by setting $\mathcal{H}^+ \triangleq \mathcal{T}_{\mathrm{NF}}$.

REMARK 3.12. In B[1984] the λ-theory $\mathcal{H}^+$ was solely denoted $\mathcal{T}_{\mathrm{NF}}$. The notation $\mathcal{H}^+$ was introduced in Manzonetto and Ruoppolo (2014) several years later. The motivating idea underlying this notational choice is that the symbol $+$ is a proper sub-symbol of $\ast$ just like $\mathcal{H}^+$ is strictly included in $\mathcal{H}^*$, namely $\mathcal{H}^+ \subsetneq \mathcal{H}^*$.

As proved by Hyland (1975b), $\mathcal{H}$ admits a unique maximal consistent extension, which is actually $\mathcal{H}^*$. See also B[1984], Theorem 16.2.6.

PROPOSITION 3.13. (i) *$\mathcal{H}^*$ is the maximal sensible λ-theory, hence a coatom of $\lambda\mathcal{T}$.*
(ii) *$\mathcal{H}^*$ is maximal among semi-sensible λ-theories.*
(iii) *Let $\mathcal{T} \in \lambda\mathcal{T}$ be sensible. Then $\mathcal{T} \vdash M = N$ entails that $M_\alpha \sim N_\alpha$, for all $\alpha \in \mathbb{N}^*$.*

As shown by Paolini (2008), to verify that two λ-terms M, N are equal in $\mathcal{H}^+$ or $\mathcal{H}^*$ it is sufficient to consider closed head contexts $H[\,]$ that are defined as follows.

DEFINITION 3.14. (i) A *head context* is a context of the form

$$H[\,] = (\lambda x_1 \ldots x_n.[\,])P_1 \cdots P_k$$

for some $n, k \in \mathbb{N}$.
(ii) A head context $H[\,]$ as above is called *applicative* when $n = 0$ and *closed* when $P_1, \ldots, P_k \in \Lambda^o$.

LEMMA 3.15 (CONTEXT LEMMA).
Let $\mathcal{O} \in \{\mathrm{NF}, \mathrm{SOL}\}$. For $M, N \in \Lambda$, the following are equivalent:

1. *$M \sqsubseteq_{\mathcal{O}} N$,*
2. *For every closed head context $H[\,]$, if $H[M] \in \mathcal{O} \cap \Lambda^o$ then $H[N] \in \mathcal{O} \cap \Lambda^o$.*

As every $M \in \Lambda^o$ having hnf $\lambda x_1 \ldots x_n.x_i M_1 \cdots M_k$ can be sent to I via the applicative context $H[\,] \triangleq [\,]\Omega^{\sim i-1}\mathsf{U}_{k+1}^{k+1}\Omega^{\sim n-i}$ satisfying $H[\Omega] =_{\mathcal{H}} \Omega$ we obtain the next result.

PROPOSITION 3.16 (SEMI-SEPARABILITY THEOREM).
For $M, N \in \Lambda$, the following are equivalent:

1. *$M \not\sqsubseteq_{\mathrm{SOL}} N$,*
2. *there is a closed head context $H[\,]$ such that $H[M] =_\beta \mathsf{I}$ and $H[N] =_{\mathcal{H}} \Omega$.*

Figure[2] 3.1 provides a graphical representation of some basic facts concerning the lattice $\lambda\mathcal{T}$ of λ-theories. In the picture $\mathcal{T}_1$ is below $\mathcal{T}_2$ whenever $\mathcal{T}_1 \subsetneq \mathcal{T}_2$.

[2]This picture is known among students in the field by the nickname of 'Nino's potato', because it was omnipresent in Salibra's talks and looks like a potato.

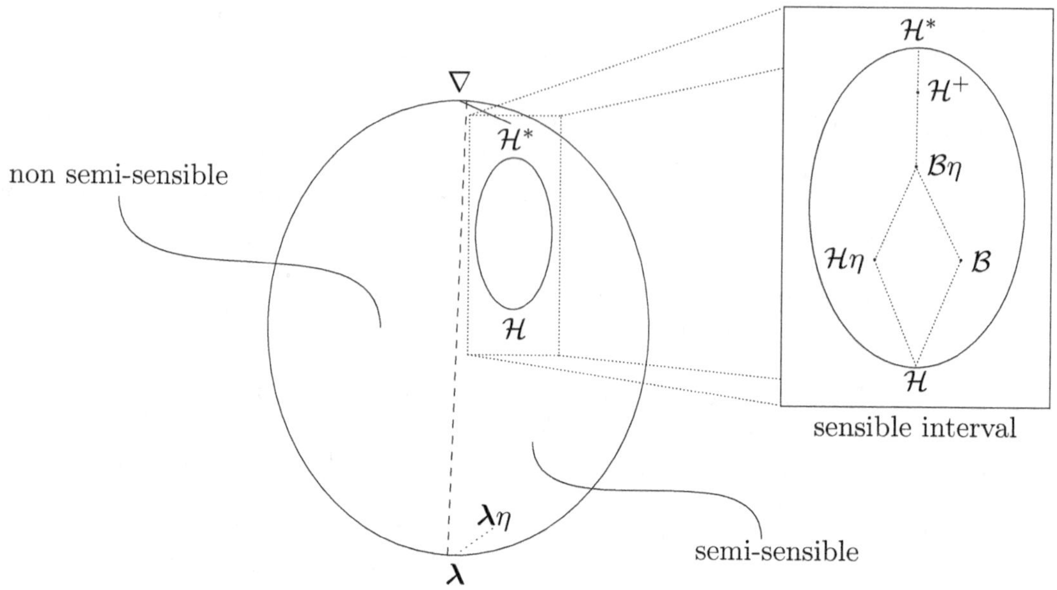

Figure 3.1: The lattice of λ-theories, with a zoom on sensible λ-theories.

Lambda theories in the arithmetical and analytical hierarchies

We have mentioned in Definition 3.3(v) that certain λ-theories are r.e., this means that they are Σ^0_1-complete sets (after coding). More generally, it is interesting to wonder where the main λ-theories $\boldsymbol{\lambda}, \boldsymbol{\lambda}\eta, \mathcal{H}, \mathcal{B}, \mathcal{H}^+, \mathcal{H}^*$ are placed in the arithmetical hierarchy. We present some well-established results and a sketch of their proofs to get the reader familiar with the techniques that are employed. For more details, see B[1984], Chapter 16.

PROPOSITION 3.17. *$\boldsymbol{\lambda}$ and $\boldsymbol{\lambda}\eta$ are r.e. λ-theories (equivalently, Σ^0_1-complete).*

PROOF. By confluence $M =_{\beta(\eta)} N$ if and only if M and N have a common $\beta(\eta)$-reduct. The latter is a semi-decidable property. □

THEOREM 3.18 (B[1984], THEOREM 16.1.11). *$\mathcal{H}$ is Σ^0_2-complete.*

PROOF. The set $\{(M,N) \mid M,N \in \mathrm{UNS}\}$ is Π^0_1, therefore it axiomatizes a Σ^0_2-predicate. We show that any Σ^0_2-predicate can be reduced to provability in $\mathcal{H}$. Take a Σ^0_2-predicate $\mathrm{P}(k) \triangleq \exists n\, \forall m\,.\, \mathrm{Q}(n,m,k)$, with Q recursive. It is sufficient to exhibit an $H \in \Lambda^o$ satisfying

$$\mathrm{P}(k) \iff \mathcal{H} \vdash H\, \mathsf{c}_k\, x\, \mathsf{c}_0 = H\, \mathsf{c}_k\, y\, \mathsf{c}_0, \text{ for } x \neq y.$$

We show that this is the case for every H satisfying the following properties:

1. $H\mathsf{c}_k z \mathsf{c}_0 \twoheadrightarrow_{\beta\Omega} M \quad \Rightarrow \quad z \in \mathrm{FV}(M)$; (for the reduction $\twoheadrightarrow_{\beta\Omega}$, see Definition 10.7)
2. $Hkzn \twoheadrightarrow_\beta [\mathsf{I}, Fkn(Hkz(\mathsf{S}^+ n))]$ where $F \in \Lambda^o$ such that

$$F\mathsf{c}_k\mathsf{c}_n =_{\mathcal{H}} \begin{cases} \Omega, & \text{if } \forall m\,.\, \mathrm{Q}(n,m,k), \\ \mathsf{I}, & \text{otherwise,} \end{cases}$$

exists by Lemma 16.1.10 in B[1984].

If there exists an $n \in \mathbb{N}$ such that $\forall m . \mathrm{Q}(n,m,k)$ holds, then $H\,\mathsf{c}_k\,z\,\mathsf{c}_n =_{\mathcal{H}} [\mathsf{I},\Omega]$, whence

$$H\,\mathsf{c}_k\,x\,\mathsf{c}_0 =_{\mathcal{H}} [\underbrace{\mathsf{I},\ldots,\mathsf{I}}_{n \text{ copies}},\Omega] =_{\mathcal{H}} H\,\mathsf{c}_k\,y\,\mathsf{c}_0.$$

Conversely, if there is no such n, then $F\mathsf{c}_k\mathsf{c}_n =_{\mathcal{H}} \mathsf{I}$, for all $n \in \mathbb{N}$. By property (1) of H, $H\mathsf{c}_k x\mathsf{c}_0$ and $H\mathsf{c}_k y\mathsf{c}_0$ cannot have a common $\beta\Omega$-reduct. Since the reduction $(\beta\Omega)$ is Church-Rosser (see Proposition 10.9), we conclude $H\mathsf{c}_k x\mathsf{c}_0 \neq_{\mathcal{H}} H\mathsf{c}_k y\mathsf{c}_0$. □

The argument used to prove the following theorem is due to Wadsworth.

THEOREM 3.19. *$\mathcal{B}$ is Π^0_2-complete.*

PROOF. First, note that $=_{\mathcal{B}}$ is a Π^0_2-predicate. Write φ_n for the partial recursive function with index n and $\simeq$ for Kleene's equality. By Proposition 1.28, every φ_n is λ-definable by some $G_n \in \Lambda^o$. Note that $\varphi_n \simeq \varphi_m$ if and only if, for all $k \in \mathbb{N}$, $G_n\mathsf{c}_k =_{\mathcal{B}} G_m\mathsf{c}_k$ holds. Let P be such that, for all $n \in \mathbb{N}$, $P\mathsf{c}_n =_{\beta} \pi_n$ and define $F_n =_{\mathcal{B}} \lambda x.Px[G_n\mathsf{c}_0, G_n\mathsf{c}_1,\ldots]$. Clearly, F_n still λ-defines φ_n and, by Lemma 2.17, we have

$$\varphi_n \simeq \varphi_m \iff \mathcal{B} \vdash F_n = F_m$$

We conclude since the F_n's can be chosen effectively from n and the relation $\varphi_n \simeq \varphi_m$ is Π^0_2-complete. □

EXERCISE 3.20. (i) Show that all λ-theories in the interval $[\mathcal{B},\mathcal{H}^*]$ are Π^0_2-hard.
(ii) Show that $\mathcal{B}\eta$ and $\mathcal{H}^+$ are Π^0_2-complete, like $\mathcal{H}^*$. [cf. B[1984], Corollary 16.2.10.]

In the examples above, adding the (η)-rule does not modify the position of the λ-theory in the hierarchy: e.g., $\boldsymbol{\lambda}$ and $\boldsymbol{\lambda}\eta$ are both r.e.; $\mathcal{H}$ and $\mathcal{H}\eta$ are Σ^0_2 and $\mathcal{B},\mathcal{B}\eta$ are Π^0_2. However, the λ-calculus possesses another notion of extensionality, known as the ω-rule, which is strictly stronger than (η) and can radically change the complexity of a λ-theory.

DEFINITION 3.21. The *ω-rule* is defined as follows (for all $M,N \in \Lambda$):

$$\forall P \in \Lambda^o \,.\, MP = NP \quad\Rightarrow\quad M = N. \qquad (\omega)$$

We write (ω^0) for the ω-rule restricted to $M,N \in \Lambda^o$.

The rule (ω) and its restriction (ω^0) can be relativized to an arbitrary λ-theory $\mathcal{T}$.

DEFINITION 3.22. Let $\mathcal{T} \in \lambda\mathcal{T}$.
(i) *$\mathcal{T}$ satisfies the ω-rule*, written $\mathcal{T} \vdash \omega$, if the following holds (for all $M,N \in \Lambda$):

$$\forall P \in \Lambda^o \,.\, MP =_{\mathcal{T}} NP \quad\Rightarrow\quad M =_{\mathcal{T}} N$$

(ii) We write $\mathcal{T} \vdash \omega^0$ if the implication above holds for all $M,N \in \Lambda^o$.
(iii) We denote by $\mathcal{T}\omega$ the closure of $\mathcal{T}$ under (ω).

The question of whether $\boldsymbol{\lambda}\eta$ satisfy the ω-rule was investigated by Barendregt (1971), during his PhD, but a negative answer has been finally given by Plotkin (1974).

The lemma below collects some general results from Section 4.1 in B[1984].

LEMMA 3.23. *Let* $\mathcal{T} \in \lambda\mathcal{T}$.

(i) $\mathcal{T}\eta \subseteq \mathcal{T}\omega$,
(ii) $\mathcal{T} \vdash \omega \iff \mathcal{T} \vdash \omega^0$,
(iii) $\mathcal{T} \subseteq \mathcal{T}'$ *entails* $\mathcal{T}\eta \subseteq \mathcal{T}'\eta$,
(iv) $\mathcal{T} \subseteq \mathcal{T}'$ *entails* $\mathcal{T}\omega \subseteq \mathcal{T}'\omega$.

The next lemma establishes the consistency of $\mathcal{H}\omega$, whence of $\boldsymbol{\lambda}\omega$ by Lemma 3.23(iv).

LEMMA 3.24 (B[1984], THEOREM 17.2.17).

$$\mathcal{H}\omega \vdash M = \mathsf{I} \iff \boldsymbol{\lambda}\eta \vdash M = \mathsf{I}.$$

The arithmetical hierarchy is known to include only a small portion of all possible sets. A typical example of non-arithmetic set is given by the set of all codes of first-order sentences that are true in arithmetic. In order to position some λ-theories, we need to consider the analytical hierarchy (see Odifreddi (1989), Section IV.2) and in particular Π^1_1-complete sets. The following characterization is due to Kleene (1955).

THEOREM 3.25. *A set* A *is* Π^1_1-complete *if and only if, for some recursive predicate* R,

$$n \in A \iff \forall f \in \mathbb{N}^{\mathbb{N}}, \exists m \in \mathbb{N} \,.\, \mathrm{R}(\#\hat{f}(m), n)$$

In the seventies Barendregt conjectured that both $\boldsymbol{\lambda}\omega$ and $\mathcal{H}\omega$ are Π^1_1-complete. We discuss in Section 16.3 some proof attempts proposed by Intrigila and Statman to confirm his conjectures.

3.2 Denotational models

The model theory of λ-calculus has been developed following mainly three approaches. The first is algebraic and based on a variety of algebras called combinatory algebras. The second, based on syntactic λ-models, is closely related to the algebraic one, but more set-theoretical. The third is category theoretic and focuses on the notion of reflexive object living in Cartesian closed categories. For more details, we refer to B[1984], Chapter 5.

The relationship among these notions has been initially explored by Koymans (1982), then revisited by Selinger (2002) and subsequently refined by Bucciarelli et al. (2007). For a different algebraic approach based on lambda abstraction algebras, see Pigozzi and Salibra (1998). A modern algebraic/categorical perspective on models and theories of λ-calculus has been recently proposed by Hyland (2017).

The algebraic models

The algebraic approach was first developed by Barendregt (1977), Scott (1980) and Meyer (1982) building on Curry and Schönfinkel's notion of combinatory algebras. We take for granted some familiarity with combinatory logic, otherwise see Hindley and Seldin (2008).

DEFINITION 3.26. (i) An *applicative structure* $\mathcal{A} = (A, \cdot)$ is an algebra where $\cdot$ is a binary operation on A called *application*.

(ii) An applicative structure $\mathcal{A}$ is *extensional* if it satisfies the following axiom:

$$\forall x \forall y (\forall z (x \cdot z = y \cdot z) \Rightarrow x = y) \qquad \text{(ext)}$$

We may write the application like an infix operator, as in $c \cdot d$, or even omit it entirely and write cd. As it is customary, we assume that application associates to the left.

DEFINITION 3.27. (i) A *combinatory algebra* $\mathcal{C} = (C, \cdot, \mathbf{k}, \mathbf{s})$ is an applicative structure for a signature with two nullary terms $\mathbf{k}$ and $\mathbf{s}$ satisfying for all $x, y, z \in \mathrm{Var}$:

$$\mathbf{k}xy = x \qquad\qquad \mathbf{s}xyz = xz(yz)$$

(ii) A λ-*algebra* is a combinatory algebra satisfying the five *Curry's axioms*:[3]

(1) $\mathbf{k} = \mathbf{s}(\mathbf{s}(\mathbf{ks})(\mathbf{s}(\mathbf{kk})\mathbf{k}))(\mathbf{k}(\mathbf{skk}))$

(2) $\mathbf{s} = \mathbf{s}(\mathbf{s}(\mathbf{ks})(\mathbf{s}(\mathbf{k}(\mathbf{s}(\mathbf{ks})))(\mathbf{s}(\mathbf{k}(\mathbf{s}(\mathbf{kk})))\mathbf{s})))(\mathbf{k}(\mathbf{k}(\mathbf{skk})))$

(3) $\mathbf{s}(\mathbf{kk}) = \mathbf{s}(\mathbf{s}(\mathbf{ks})(\mathbf{s}(\mathbf{kk})(\mathbf{s}(\mathbf{ks})\mathbf{k})))(\mathbf{kk})$

(4) $\mathbf{s}(\mathbf{ks})(\mathbf{s}(\mathbf{kk})) = \mathbf{s}(\mathbf{kk})(\mathbf{s}(\mathbf{s}(\mathbf{ks})(\mathbf{s}(\mathbf{kk})(\mathbf{skk})))(\mathbf{k}(\mathbf{skk})))$

(5) $\mathbf{s}(\mathbf{k}(\mathbf{s}(\mathbf{ks})))(\mathbf{s}(\mathbf{ks})(\mathbf{s}(\mathbf{ks}))) = \mathbf{s}(\mathbf{s}(\mathbf{ks})(\mathbf{s}(\mathbf{kk})(\mathbf{s}(\mathbf{ks})(\mathbf{s}(\mathbf{k}(\mathbf{s}(\mathbf{ks})))\mathbf{s}))))(\mathbf{ks})$

(iii) A λ-*model* is a λ-algebra satisfying the *Meyer-Scott axiom* (for $\boldsymbol{\varepsilon} \triangleq \mathbf{s}(\mathbf{k}(\mathbf{skk}))$):

$$\forall x \forall y (\forall z (xz = yz) \Rightarrow \boldsymbol{\varepsilon} x = \boldsymbol{\varepsilon} y).$$

(iv) A λ-model is *extensional* if it is extensional as an applicative structure.

The simplest way of constructing a λ-model is to start from a λ-theory and consider its term model.

DEFINITION 3.28. Let $\mathcal{T} \in \lambda\mathcal{T}$.

(i) The *(open) term model* of $\mathcal{T}$ is denoted by $\mathcal{M}(\mathcal{T})$ and defined as follows:

$$\begin{aligned} \mathcal{M}(\mathcal{T}) &\triangleq (\Lambda/\mathcal{T}, \cdot, [\mathsf{K}]_{\mathcal{T}}, [\mathsf{S}]_{\mathcal{T}}), \text{ where} \\ [M]_{\mathcal{T}} \cdot [N]_{\mathcal{T}} &\triangleq [MN]_{\mathcal{T}} \end{aligned}$$

(ii) The *closed term model* $\mathcal{M}^o(\mathcal{T})$ of a λ-theory $\mathcal{T}$ is defined analogously on $\Lambda^o/\mathcal{T}$.

The underlying set of $\mathcal{M}(\mathcal{T})$ is populated by the $\mathcal{T}$-equivalence classes of λ-terms. It is immediate to check that the application is well-defined, in the sense that it respects the equivalence classes. Corollary 5.2.13 of B[1984], gives the following examples.

[3] In B[1984], this is rather a characterization of λ-algebras. See Proposition 3.36(i) below.

EXAMPLES 3.29. (i) For every $\mathcal{T} \in \lambda\mathcal{T}$, the term model $\mathcal{M}(\mathcal{T})$ is a λ-model.

(ii) The closed term model $\mathcal{M}^o(\mathcal{T})$ is a λ-algebra, not necessarily a λ-model as first shown by Jacopini (1975) for $\mathcal{T} \triangleq \boldsymbol{\lambda}(\{\Omega\mathsf{K}Z = \Omega\mathsf{S}Z \mid Z \in \Lambda^o\})$.

(iii) From Plotkin (1974), it follows that $\mathcal{M}^o(\boldsymbol{\lambda})$ and $\mathcal{M}^o(\boldsymbol{\lambda}\eta)$ are not λ-models.

DEFINITION 3.30. A *homomorphism of combinatory algebras* $\mathcal{C}_1$ *and* $\mathcal{C}_2$ is a function $\varphi : C_1 \to C_2$ such that

$$\begin{array}{lcl} \varphi(\mathbf{k}_1) & = & \mathbf{k}_2, \\ \varphi(\mathbf{s}_1) & = & \mathbf{s}_2, \\ \varphi(x \cdot_1 y) & = & \varphi(x) \cdot_2 \varphi(y), \text{ for all } x, y \in C_1. \end{array}$$

Homomorphisms of λ-algebras (resp. *λ-models*) are defined in the same way.

A class of algebras over the same signature is called a *variety* whenever it is closed under subalgebras, homomorphic images and direct products.

LEMMA 3.31. (i) *The class* $\mathbb{CA}$ *of combinatory algebras forms a variety.*

(ii) *The class* $\mathbb{LA}$ *of λ-algebras forms a variety.*

(iii) *The class of λ-models is not a variety.*

PROOF. (i) and (ii). As proved by Birkhoff (1935), a class of algebras is a variety if and only if it is an equational class, namely if it is axiomatizable by a set of equations.

(iii) Since every $\mathcal{M}^o(\mathcal{T})$ is a homomorphic image of $\mathcal{M}(\mathcal{T})$, the discussion in Example 3.29(ii) shows that the class of λ-models is not closed under homomorphic images. □

Despite the above lemma, combinatory algebras do not satisfy the properties below, that are often considered as desirable in algebra.

PROPOSITION 3.32 (B[1984], PROPOSITION 5.1.15). *A (non-trivial) combinatory algebra cannot be commutative, associative, finite or recursive.*

Combinatory algebras have been introduced to model Schönfinkel and Curry's combinatory logic. See Curry and Feys (1958).

DEFINITION 3.33. (i) The set CL of *combinatory terms* is defined by induction:

$$P, Q \ ::= \ x \mid \mathbf{K} \mid \mathbf{S} \mid PQ \qquad \text{(CL)}$$

(ii) The set $\mathrm{FV}(P)$ of free variables of P is defined as usual.

(iii) Given $P \in \mathrm{CL}$ and $x \in \mathrm{Var}$, define the term $\lambda^* x.P \in \mathrm{CL}$ as follows:

$$\begin{array}{lcll} \lambda^* x.x & \triangleq & \mathbf{I}, & \text{where } \mathbf{I} \triangleq \mathbf{SKK}, \\ \lambda^* x.P & \triangleq & \mathbf{K}P, & \text{if } x \notin \mathrm{FV}(P), \\ \lambda^* x.(PQ) & \triangleq & \mathbf{S}(\lambda^* x.P)(\lambda^* x.Q), & \text{otherwise.} \end{array}$$

(iv) Define two maps

$$(\cdot)_{\mathrm{CL}} : \ \mathrm{CL} \to \Lambda$$
$$(\cdot)_{\lambda} : \ \Lambda \to \mathrm{CL}$$

by structural induction as follows:

$$\begin{aligned} x_{\mathrm{CL}} &\triangleq x, & x_\lambda &\triangleq x,\\ (MN)_{\mathrm{CL}} &\triangleq M_{\mathrm{CL}}\,N_{\mathrm{CL}}, & (PQ)_\lambda &\triangleq P_\lambda\,Q_\lambda,\\ (\lambda x.M)_{\mathrm{CL}} &\triangleq \lambda^* x.M_{\mathrm{CL}}, & \mathbf{K}_\lambda &\triangleq \lambda xy.x,\\ & & \mathbf{S}_\lambda &\triangleq \lambda xyz.xz(yz). \end{aligned}$$

The terms of combinatory logic can be naturally interpreted in a combinatory algebra. As a consequence, λ-terms can also be interpreted in any combinatory algebra by applying the translation $(\cdot)_{\mathrm{CL}}$, and then interpreting the resulting combinatory terms.

DEFINITION 3.34 (ALGEBRAIC INTERPRETATION).
Let $\mathcal{C}$ be a combinatory algebra.

(i) Given a set A, a *valuation in* A is any function $\rho : \mathrm{Var} \to A$.
(ii) We write Val_A for the set of all valuations in A.
(iii) Given a valuation $\rho \in \mathrm{Val}_A$, $x \in \mathrm{Var}$ and $a \in A$, define

$$\rho[x:=a](y) \triangleq \begin{cases} \rho(y), & \text{if } x \neq y,\\ a, & \text{otherwise.} \end{cases}$$

(iv) The *interpretation of a combinatory term P under a valuation* $\rho \in \mathrm{Val}_C$ is the element $[\![P]\!]^{\mathcal{C}}_\rho \in C$ inductively defined as follows:

$$[\![x]\!]^{\mathcal{C}}_\rho \triangleq \rho(x), \qquad [\![\mathbf{K}]\!]^{\mathcal{C}}_\rho \triangleq \mathbf{k}, \qquad [\![\mathbf{S}]\!]^{\mathcal{C}}_\rho \triangleq \mathbf{s}, \qquad [\![PQ]\!]^{\mathcal{C}}_\rho \triangleq [\![P]\!]^{\mathcal{C}}_\rho \cdot [\![Q]\!]^{\mathcal{C}}_\rho.$$

(v) The *interpretation of a λ-term M under a valuation* $\rho \in \mathrm{Val}_C$ is defined by setting

$$[\![M]\!]^{\mathcal{C}}_\rho \triangleq [\![M_{\mathrm{CL}}]\!]^{\mathcal{C}}_\rho$$

(vi) For $M, N \in \Lambda$, we write $\mathcal{C} \models M = N$ whenever $\forall \rho \in \mathrm{Val}_C \,.\, [\![M]\!]^{\mathcal{C}}_\rho = [\![N]\!]^{\mathcal{C}}_\rho$ holds.
(vii) The *(equational) theory induced by* $\mathcal{C}$ is defined by

$$\mathrm{Th}(\mathcal{C}) \triangleq \{(M,N) \in \Lambda^2 \mid \mathcal{C} \models M = N\}.$$

CONVENTION 3.35. *When $\mathcal{C}$ is clear from the context, we simply write $[\![M]\!]_\rho$ for $[\![M]\!]^{\mathcal{C}}_\rho$. If the λ-term M is closed, we may write $[\![M]\!]^{\mathcal{C}}$ or even $[\![M]\!]$.*

PROPOSITION 3.36. (i) *In a λ-algebra $\mathcal{C}$, the following holds (for all $M, N \in \Lambda$):*

$$\boldsymbol{\lambda} \vdash M = N \quad \Rightarrow \quad \mathcal{C} \models M = N$$

(ii) *In a λ-model $\mathcal{C}$, the following holds (for all $M, N \in \Lambda$ and $x \in \mathrm{Var}$):*

$$\mathcal{C} \models M = N \quad \Rightarrow \quad \mathcal{C} \models \lambda x.M = \lambda x.N$$

Then $\mathrm{Th}(\mathcal{C})$ is a λ-theory.
(iii) *A λ-model $\mathcal{C}$ is extensional if and only if $\mathcal{C} \models \mathsf{I} = 1$.*

PROOF. This is shown in B[1984]. Respectively: (i) Theorem 5.2.5. (ii) Proposition 5.2.9. (iii) Cf. Proposition 5.2.10. □

The set-theoretic models

The first-order definition of λ-models possesses the advantage of conferring a model-theoretic status to the model theory of λ-calculus, but has the disadvantage that the interpretation of λ-terms is difficult to handle in practice. For this reason, it is often convenient to use the notion of a syntactic λ-model introduced by Hindley and Longo (1980) and subsequently simplified in B[1984], Section 5.3.

DEFINITION 3.37. (i) A *syntactic λ-algebra* is a pair $\mathcal{S} = (\mathcal{A}, [\![-]\!]_{(\cdot)})$ such that $\mathcal{A} = (S, \cdot)$ is an applicative structure and the *interpretation function*

$$[\![-]\!]_{(\cdot)} : \Lambda \times \mathrm{Val}_S \to S$$

satisfies the following axioms:

(a) $[\![x]\!]_\rho = \rho(x)$, for all $x \in \mathrm{Var}$,

(b) $[\![MN]\!]_\rho = [\![M]\!]_\rho \cdot [\![N]\!]_\rho$,

(c) $[\![\lambda x.M]\!]_\rho \cdot s = [\![M]\!]_{\rho[x:=s]}$, for all $s \in S$,

(d) $[\![M]\!]_\rho = [\![N]\!]_{\rho'}$, whenever $\rho(x) = \rho'(x)$ for all $x \in \mathrm{FV}(M)$.

(ii) A *syntactic λ-model* is a syntactic λ-algebra satisfying moreover the axiom

(e) $\forall s \in S\,.\,[\![M]\!]_{\rho[x:=s]} = [\![N]\!]_{\rho[x:=s]} \quad \Rightarrow \quad [\![\lambda x.M]\!]_\rho = [\![\lambda x.N]\!]_\rho$.

(iii) A syntactic λ-model $\mathcal{S} = (\mathcal{A}, \cdot)$ is called *extensional* whenever $\mathcal{A}$ is extensional.

(iv) The notions $\mathcal{S} \models M = N$ and $\mathrm{Th}(\mathcal{S})$ are defined as for λ-models.

REMARK 3.38. (i) The above definition is very natural since it basically requires that the interpretation function is contextual. In practice, this notion works well for models arising from set-theoretical considerations, whilst the valuations ρ become redundant in the abstract categorical setting. See Remark 3.65, for a more detailed discussion.

(ii) Some authors impose as an additional axiom that if $y \notin \mathrm{FV}(M)$ then $[\![\lambda x.M]\!]_\rho = [\![\lambda y.M[x:=y]]\!]_\rho$. In our case this follows from the Variable Convention 1.4.

For a more thoughtful study of syntactic λ-models, see Hindley and Seldin (2008). The relationship with the notion of λ-model will be discussed at the end of the chapter.

The categorical models

Categorical models of λ-calculus have been around since 1979. For category theory we refer to Asperti and Longo (1991). We recall here some basic notions and notations.

NOTATION. *Let* $\mathbf{C}$ *be a (locally small) category and* A, B, C *be objects of* $\mathbf{C}$.

(i) *We let* $\mathbf{C}(A, B)$ *denote the* hom-set *of morphisms from* A *to* B *and* $\mathrm{Id}_A \in \mathbf{C}(A, A)$ *the* identity morphism on A.

(ii) *When* $\mathbf{C}$ *is clear from the context, we sometimes write* $f : A \to B$ *for* $f \in \mathbf{C}(A, B)$.

(iii) *Given* $f : A \to B$ *and* $g : B \to C$*, we denote their composition as* $g \circ f : A \to C$.

(iv) *We write* $A \cong B$ *to indicate that* A, B *are* isomorphic objects. *This means that there exist morphisms* $\iota_1 : A \to B$ *and* $\iota_2 : B \to A$ *such that* $\iota_2 \circ \iota_1 = \mathrm{Id}_A$ *and* $\iota_1 \circ \iota_2 = \mathrm{Id}_B$.

NOTATION. *Let* $\mathbf{C}$ *be a* Cartesian category *and* A, B, C *be objects of* $\mathbf{C}$.

(i) *We denote by* $A \times B$ *the* product *of* A *and* B, *by* $\pi_1 : A \times B \to A$ *and* $\pi_2 : A \times B \to B$ *the associated* projections.

(ii) *Given two morphisms* $f : C \to A$ *and* $g : C \to B$, *we let* $\langle f, g\rangle : C \to A \times B$ *be the unique morphism such that* $\pi_1 \circ \langle f, g\rangle = f$ *and* $\pi_2 \circ \langle f, g\rangle = g$.

(iii) *We write* $\mathbb{1}$ *for the* terminal object *and* $!_A$ *for the unique morphism in* $\mathbf{C}(A, \mathbb{1})$.

(iv) *Given* $n \in \mathbb{N}$, *we denote by* A^n *the categorical product of* n *copies of* A, *and by* $\pi_i^n : A^n \to A$ *the* i*-th projection.*

(v) *Given a set* I *and a collection of objects* $(A_i)_{i \in I}$, *we let*

$$\prod_{i \in I} A_i$$

be the I-indexed product *(when it exists)*

$$\pi_j^I : \prod_{i \in I} A_i \to A_j$$

be the j*-th projection (for* $j \in I$*) and* $\langle f_i \rangle_{i \in I}$ *denote the* I*-indexed pairing.*

(vi) *Given a set* I, A^I *denotes the* I*-indexed product of an adequate number of copies of* A *(when it exists). For all* $J \subseteq I$, *define* $\Pi_J^I \triangleq \langle \pi_j^I \rangle_{j \in J} : A^I \to A^J$.

(vii) *A Cartesian category has* all small products *(resp.* countable products*) when* $\prod_{i \in I} A_i$ *exists for any (countable) set* I.

Set-indexed products will be conveniently employed to improve the readability, as they allow to work with product objects up to permutations of their components.

REMARK 3.39. Let $\mathbf{C}$ be a Cartesian category having all small products. For all sets I, J, X such that $I \subseteq J \subseteq X$ we have:[4]

(i) $\pi_i^J = \pi_i^I \circ \Pi_I^J$, for all $i \in I$;

(ii) $\Pi_{I \cup \{i\}}^{J \cup \{i\}} = \Pi_I^J \times \mathrm{Id}$, for all $i \notin I \cup J$;

(iii) $\Pi_I^X = \Pi_I^J \circ \Pi_J^X$.

NOTATION. *Let* $\mathbf{C}$ *be a* Cartesian closed category *(CCC, for short).*

(i) *We write* $A \Rightarrow B$ *for the* exponential object *and*

$$\mathrm{Eval}_{AB} : (A \Rightarrow B) \times A \to B$$

for the corresponding evaluation morphism.

(ii) *Given* $f : C \times A \to B$, *we denote by* $\mathrm{Cur}(f) : C \to (A \Rightarrow B)$ *the* currying of f, *namely the unique morphism such that*

$$\mathrm{Eval}_{AB} \circ \langle \mathrm{Cur}(f) \circ \pi_1, \pi_2 \rangle = f$$

(iii) *The elements of* $\mathbf{C}(\mathbb{1}, A)$ *are called the* points *of* A.

[4]Item (ii) is slightly imprecise as one needs to select a precise order for the i-th component.

For the sake of readability, we often write Eval for Eval_{AB} and Id for Id_A. We recall below the basic properties of Cartesian closed categories.

THEOREM 3.40. *In every CCC* **C**, *the following equalities hold:*

$$\begin{array}{llcl} \text{(pair)} & \langle f,g\rangle\circ h & = & \langle f\circ h, g\circ h\rangle \\ \text{(beta)} & \mathrm{Eval}\circ\langle \mathrm{Cur}(f), g\rangle & = & f\circ\langle \mathrm{Id}, g\rangle \\ \text{(Curry)} & \mathrm{Cur}(f)\circ g & = & \mathrm{Cur}(f\circ(g\times\mathrm{Id})) \\ \text{(Id-Curry)} & \mathrm{Cur}(\mathrm{Eval}) & = & \mathrm{Id} \end{array}$$

where $f_1 \times f_2$ *denotes the* product map *defined by* $\langle f_1\circ\pi_1, f_2\circ\pi_2\rangle$.

DEFINITION 3.41. (i) A CCC **C** is called *well-pointed* if, for all morphisms $f, g : A \to B$,

$$f = g \iff \forall h : \mathbb{1} \to A\ .\ f\circ h = g\circ h.$$

(ii) An object A is *well-pointed* if the above property holds for all $f, g : A \to A$.

REMARK 3.42. In the 80s, a category satisfying the above condition was rather known as a category *having enough points*. Similarly, for objects. For instance, this is the terminology adopted in B[1984], Asperti and Longo (1991) and Amadio and Curien (1998). Nowadays the terminology 'well-pointed' is more common among category theorists.

Categorical models of λ-calculus are defined as follows.

DEFINITION 3.43. (i) A *reflexive object living in a CCC* **C** is a triple

$$\mathcal{U} = (U, \mathrm{App}, \mathrm{Lam})$$

where U is an object of **C** satisfying $(U \Rightarrow U) \lhd U$ via the retraction pair (App, Lam). This means that $\mathrm{App} : U \to (U \Rightarrow U)$ and $\mathrm{Lam} : (U \Rightarrow U) \to U$ are such that

$$\mathrm{App}\circ\mathrm{Lam} = \mathrm{Id}_{U\Rightarrow U}.$$

(ii) A *(categorical) model* of λ-calculus is given by any reflexive object $\mathcal{U}$ living in a Cartesian closed category **C**.

(iii) A model $\mathcal{U}$ is called *extensional* whenever $(U \Rightarrow U) \cong U$ holds via (App, Lam). In other words, we also have

$$\mathrm{Lam}\circ\mathrm{App} = \mathrm{Id}_U.$$

REMARK 3.44. The definition of categorical model presented above is definitely the most famous, and widespread in the literature, but it is not the most general. As shown by Martini (1992), to obtain a model of λ-calculus, it is sufficient to consider a reflexive objet living in a *semi-Cartesian closed category*. Semi-CCCs, also called *weak CCCs*, have been independently introduced by Wiweger (1984) and Hayashi (1985) to model typed non-extensional λ-calculi. Intuitively, in a semi-CCC **C**, one does not have an isomorphism $\mathbf{C}(A\times B, C) \cong \mathbf{C}(A, B\Rightarrow C)$, but simply a retraction

$$\mathbf{C}(A\times B, C) \lhd \mathbf{C}(A, B\Rightarrow C)$$

From this $\mathbf{C}(B,C) \triangleleft \mathbf{C}(\mathbb{1}, B \Rightarrow C)$ follows, whence $A \Rightarrow B$ is no longer an object representing exactly $\mathbf{C}(A, B)$—there may be many points that represent the same morphism. Similarly, Eval_{AB} cannot be interpreted as the usual functional application.

The level of abstraction provided by regular CCCs is sufficient for our purposes, therefore we do not need to consider semi-CCCs.

The interpretation of a λ-term M in a model $\mathcal{U}$ living in a CCC $\mathbf{C}$ is defined with respect to a sequence of variables $x_1, \ldots, x_n$ containing the free variables of M.

DEFINITION 3.45 (CATEGORICAL INTERPRETATION). Let $\mathcal{U}$ be a categorical model.
   (i) We say that a sequence $\vec{x}$ is *adequate* for M whenever $M \in \Lambda^o(\vec{x})$.
   (ii) The *(categorical) interpretation* of a λ-term M in $\mathcal{U}$ with respect to an adequate sequence $x_1, \ldots, x_n$ is the morphism
$$|M|_{\vec{x}}^{\mathcal{U}} : U^n \to U$$
defined by structural induction on M as follows:

$$
\begin{aligned}
|x_i|_{\vec{x}}^{\mathcal{U}} &= \pi_i^n, & &\text{where } i \in \{1, \ldots, n\}, \\
|PQ|_{\vec{x}}^{\mathcal{U}} &= \text{Eval} \circ \langle \text{App} \circ |P|_{\vec{x}}^{\mathcal{U}}, |Q|_{\vec{x}}^{\mathcal{U}} \rangle \\
|\lambda y.N|_{\vec{x}}^{\mathcal{U}} &= \text{Lam} \circ \text{Cur}(|N|_{\vec{x},y}^{\mathcal{U}}), & &\text{where we assume wlog } y \notin \vec{x}.
\end{aligned}
$$

   (iii) We write $\mathcal{U} \models M = N$ whenever $|M|_{\vec{x}}^{\mathcal{U}} = |N|_{\vec{x}}^{\mathcal{U}}$ holds for all $\vec{x}$ adequate for M, N.
   (iv) The *(equational) theory induced by $\mathcal{U}$* is defined by
$$\text{Th}(\mathcal{U}) \triangleq \{(M, N) \in \Lambda^2 \mid \mathcal{U} \models M = N\}.$$

CONVENTION 3.46. Whenever we write $|M|_{\vec{x}}^{\mathcal{U}}$, we assume that $\vec{x} = x_1, \ldots, x_n$ is adequate for M. If $M \in \Lambda^o$ then we can take $n = 0$ and write $|M|^{\mathcal{U}}$ for its categorical interpretation.

LEMMA 3.47. *Let $\mathcal{U}$ be a categorical model.*
   (i) *(Substitution) For $M, N \in \Lambda$ and $y \notin \vec{x}$, we have*
$$|M[y := N]|_{\vec{x}}^{\mathcal{U}} = |M|_{\vec{x},y}^{\mathcal{U}} \circ \langle \text{Id}, |N|_{\vec{x}}^{\mathcal{U}} \rangle.$$

   (ii) *(Soundness) For $M, N \in \Lambda$, we have:*
$$\lambda \vdash M = N \quad \Rightarrow \quad |M|_{\vec{x}}^{\mathcal{U}} = |N|_{\vec{x}}^{\mathcal{U}}.$$

   (iii) *(Extensionality) $\mathcal{U}$ is extensional if and only if $|\mathsf{I}|^{\mathcal{U}} = |\mathbf{1}|^{\mathcal{U}}$.*

PROOF. This is shown in B[1984], Section 5.5. Respectively:
   (i) Cf. Lemma 5.5.4(iii).
   (ii) Proposition 5.5.5.
   (iii) Proposition 5.5.7(iii). □

Comparing the definitions

The main properties one desires from a definition of a model of λ-calculus are: 1) it is as general as possible; 2) the interpretation of a λ-term is easy to compute in practice; 3) the model induces a λ-theory through the kernel of its interpretation function. Clearly, the more abstract the definition, the more difficult calculating the interpretation can be. For this reason it is convenient to have different definitions at hand, provided that they are equivalent in some technical sense. The notions of λ-models, syntactic λ-models and categorical models have been defined in order to ensure that the following holds.

PROPOSITION 3.48. (i) *If* $\mathcal{C}$ *is a λ-model, then* $\mathrm{Th}(\mathcal{C})$ *is a λ-theory.*
(ii) *If* $\mathcal{S}$ *is a syntactic λ-model, then* $\mathrm{Th}(\mathcal{S})$ *is a λ-theory.*
(iii) *If* $\mathcal{U}$ *is a categorical model, then* $\mathrm{Th}(\mathcal{U})$ *is a λ-theory.*

PROOF. It follows from the contextuality of the respective interpretation functions. The only delicate point is to show that $M = N$ entails $\lambda y.M = \lambda y.N$ in the model under consideration. (This fails in some λ-algebras for non β-convertible M and N.)
(i) By Proposition 3.36(ii).
(ii) By axiom (e) of Definition 3.37.
(iii) Assume $\mathcal{U} \models M = N$. For adequate $\vec{x}$ and $y \notin \vec{x}$, we have:

$$|\lambda y.M|_{\vec{x}}^{\mathcal{U}} = \mathrm{Lam} \circ \mathrm{Cur}(|M|_{\vec{x},y}^{\mathcal{U}}) = \mathrm{Lam} \circ \mathrm{Cur}(|N|_{\vec{x},y}^{\mathcal{U}}) = |\lambda y.N|_{\vec{x}}^{\mathcal{U}}. \qquad \square$$

We often call $\mathrm{Th}(\mathcal{C})$ *the theory of* $\mathcal{C}$. Similarly for $\mathrm{Th}(\mathcal{S})$ and $\mathrm{Th}(\mathcal{U})$.

REMARK 3.49. In a (syntactic) λ-model $\mathcal{C}$, one can also interpret λ-terms possibly containing constants $\underline{c}$, for $c \in C$, by extending the interpretation as follows: $[\![\underline{c}]\!] = c$. This is used in Theorem 3.50(iii), below. Analogously, in a categorical model $\mathcal{U}$, one can add to the syntax of λ-terms a constant $\underline{f}$ for each point $f \in \mathbf{C}(\mathbb{1}, U)$ and define $|\underline{f}|_{\vec{x}} = f \circ !_{U^n}$.

The next result clarifies the relationship among the different definitions of a model.

THEOREM 3.50. (i) *Every λ-model* $\mathcal{C}$ *corresponds to a syntactic λ-model* $((C, \cdot), [\![-]\!]^{\mathcal{C}})$.
(ii) *Every syntactic λ-model* $\mathcal{S}$ *corresponds to a λ-model* $(S, \cdot, [\![\mathsf{K}]\!]^{\mathcal{S}}, [\![\mathsf{S}]\!]^{\mathcal{S}})$.
(iii) *Every λ-model* $\mathcal{C}$ *induces a category* $\mathbf{K}_{\mathcal{C}}$ *(the* Karoubi envelope *of* $\mathcal{C}$*) defined by:*

Objects:	$c \in C$	*such that* $c \cdot c = c$,
Morphisms:	$\mathbf{K}_{\mathcal{C}}(c, d)$	$\triangleq \{f \in C \mid f = d \circ f \circ c\}$,
Composition:	$c \circ d$	$\triangleq \lambda x.\underline{c}(\underline{d}x)$,
Identity:	Id_c	$\triangleq c$.

Moreover, $\mathbf{K}_{\mathcal{C}}$ *is a CCC and* $\mathsf{I} = \lambda x.x$ *is a reflexive object via* $\mathrm{App} = \mathrm{Lam} = \mathsf{I}$.
(iv) *Every categorical model* $\mathcal{U}$ *induces a λ-algebra* $(\mathbf{C}(\mathbb{1}, U), \bullet, |\mathsf{K}|^{\mathcal{U}}, |\mathsf{S}|^{\mathcal{U}})$, *where*

$$f \bullet g \triangleq \mathrm{Eval} \circ \langle \mathrm{App} \circ f, g \rangle$$

The λ-algebra $(\mathbf{C}(\mathbb{1}, U), \bullet, |\mathsf{K}|^{\mathcal{U}}, |\mathsf{S}|^{\mathcal{U}})$ *is a λ-model if and only if the object* U *is well-pointed. This is always the case when the underlying CCC* $\mathbf{C}$ *is well-pointed.*

PROOF. (i)–(ii). By Theorem 5.3.6 in B[1984], the category of syntactic λ-algebras and the category of λ-algebras, with the respective homomorphisms, are isomorphic. Moreover, syntactic λ-models correspond exactly to λ-models under this isomorphism. As a consequence, the corresponding notions of interpretation coincide.

(iii) See B[1984], Proposition 5.5.12. Originally due to Scott (1980).

(iv) See B[1984], Theorem 5.5.6 and Proposition 5.5.7(ii). First shown by Koymans (1982). □

Working without enough points

In denotational semantics non-well-pointed CCCs arise naturally when morphisms are not functions. The literature is teeming with examples, like sequential algorithms (see Berry and Curien (1982)), strategies in various categories of games (Di Gianantonio et al. (1999); Abramsky et al. (2000); Hyland and Ong (2000)), relations (Hyland et al. (2006); Bucciarelli et al. (2007)) or profunctors (Fiore et al. (2007); Olimpieri (2021)). The fact that Koymans' construction of a λ-algebra from a categorical model $\mathcal{U}$ (Theorem 3.50(iv)) gives rise to a λ-model exactly when U is well-pointed led to some confusion concerning the model–theoretic status of reflexive objects living in non-well-pointed CCCs.

Following Bucciarelli et al. (2007), we show that it is sufficient to slightly modify Koymans' construction for turning any categorical model into a λ-model.

CONVENTION 3.51. (i) *For the sake of simplicity, we consider CCCs with countable products and discuss the general case at the end of the chapter.*

(ii) *We employ set-indexed products and consider the categorical interpretation of a λ-term M as a morphism $|M|_I : U^I \to U$, where $I = \{x_1, \dots, x_n\}$ is* adequate *in the sense that* $\mathrm{FV}(M) \subseteq I$.

DEFINITION 3.52. Let $\mathcal{U}$ be a model living in a CCC $\mathbf{C}$ with countable products.

(i) A morphism $f : U^{\mathrm{Var}} \to U$ is *finitary* if there exist $n \in \mathbb{N}$, $J = \{x_1, \dots, x_n\} \subsetneq \mathrm{Var}$, and a morphism $f_J : U^J \to U$ such that $f = f_J \circ \Pi^{\mathrm{Var}}_J$. In diagrammatic form:

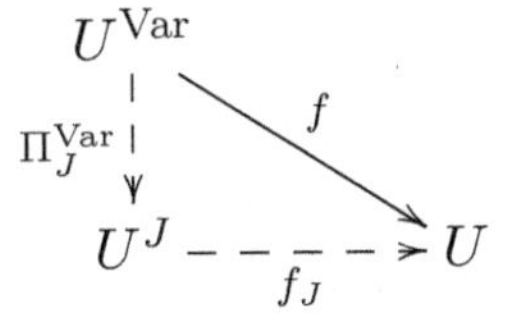

In this case, we write $(J, f_J) \Vdash f$. Notice that J must be a finite set (written $J \in \mathscr{P}_{\mathrm{f}}(\mathrm{Var})$).

(ii) Denote by $\mathbf{C}_{\mathrm{f}}(U^{\mathrm{Var}}, U)$ the set of all finitary morphisms $f : U^{\mathrm{Var}} \to U$.

(iii) The structure $\mathcal{A}_{\mathcal{U}} \triangleq (A_{\mathcal{U}}, \bullet)$ associated with $\mathcal{U}$, is given by:

$$\begin{array}{rcl} A_{\mathcal{U}} & \triangleq & \mathbf{C}_{\mathrm{f}}(U^{\mathrm{Var}}, U), \\ f \bullet g & \triangleq & \mathrm{Eval} \circ \langle \mathrm{App} \circ f, g \rangle. \end{array}$$

(iv) For $I \subseteq \mathrm{Var}$ and $\rho \in \mathrm{Val}_{\mathcal{A}_{\mathcal{U}}}$, write ρ^I for the morphism $\langle \rho(x) \rangle_{x \in I} : U^{\mathrm{Var}} \to U^I$.

REMARK 3.53. For all $x \in \mathrm{Var}$, we have $\pi_x^{\mathrm{Var}} \in \mathbf{C}_{\mathrm{f}}(U^{\mathrm{Var}}, U)$. In fact, for all $J \in \mathscr{P}_{\mathrm{f}}(\mathrm{Var})$, $x \in J$ entails $(J, \Pi^J_{\{x\}}) \Vdash \pi_x^{\mathrm{Var}}$.

LEMMA 3.54. *Let $f, g \in \mathbf{C}_{\mathrm{f}}(U^{\mathrm{Var}}, U)$. For all sets $I, J \in \mathscr{P}_{\mathrm{f}}(\mathrm{Var})$ and morphisms*

$$f_I : U^I \to U,$$
$$g_J : U^J \to U,$$

such that $(I, f_I) \Vdash f$ and $(J, g_J) \Vdash g$, we have:

$$\left(I \cup J, ((f_I \circ \Pi_I^{I\cup J}) \bullet (g_J \circ \Pi_J^{I\cup J}))\right) \Vdash f \bullet g.$$

PROOF. From Definition 3.52(i) and Remark 3.39(iii), we derive

$$f = f_I \circ \Pi_I^{\mathrm{Var}} = f_I \circ \Pi_I^{I\cup J} \circ \Pi_{I\cup J}^{\mathrm{Var}}$$

Similarly, $g = g_I \circ \Pi_J^{I\cup J} \circ \Pi_{I\cup J}^{\mathrm{Var}}$. The result follows. □

COROLLARY 3.55. *If $\mathcal{U}$ is a categorical model, then $\mathcal{A}_{\mathcal{U}}$ is an applicative structure.*

PROOF. As shown in Lemma 3.54, the set $A_{\mathcal{U}}$ is closed under $\bullet$. □

LEMMA 3.56. *Let $M \in \Lambda$ and I be adequate for M. For all $J \in \mathscr{P}_{\mathrm{f}}(\mathrm{Var})$ such that $I \subseteq J$, we have:*

(i) $\Pi_I^J \circ \rho^J = \rho^I$, *for every* $\rho \in \mathrm{Val}_{A_{\mathcal{U}}}$;

(ii) $|M|_J = |M|_I \circ \Pi_I^J$.

PROOF. (i) Easy calculations show

$$\begin{aligned}
\Pi_I^J \circ \rho^J &= \langle \pi_x^J \rangle_{x \in I} \circ \rho^J, && \text{by definition of } \Pi_I^J,\\
&= \langle \pi_x^J \circ \rho^J \rangle_{x \in I}, && \text{by (pair)},\\
&= \langle \rho(x) \rangle_{x \in I}, && \text{by definition of } \pi_x^J,\\
&= \rho^I.
\end{aligned}$$

(ii) By induction on the structure of M.

Case $M = x$. We have $|x|_J = \pi_x^J = \pi_x^I \circ \Pi_I^J = |x|_I \circ \Pi_I^J$, by Remark 3.39(i).

Case $M = PQ$. In this case, we have:

$$\begin{aligned}
|PQ|_J &= \mathrm{Eval} \circ \langle \mathrm{App} \circ |P|_J, |Q|_J \rangle, && \text{by definition of } |-|_J,\\
&= \mathrm{Eval} \circ \langle \mathrm{App} \circ |P|_I \circ \Pi_I^J, |Q|_I \circ \Pi_I^J \rangle, && \text{by IH},\\
&= \mathrm{Eval} \circ \langle \mathrm{App} \circ |P|_I, |Q|_I \rangle \circ \Pi_I^J, && \text{by (pair)},\\
&= |PQ|_I \circ \Pi_I^J, && \text{by definition of } |-|_I.
\end{aligned}$$

Case $M = \lambda x.N$. In this case, we have (for $x \notin J$):

$$\begin{aligned}
|\lambda x.N|_J &= \mathrm{Lam} \circ \mathrm{Cur}(|N|_{J\cup\{x\}}), && \text{by definition of } |-|_J,\\
&= \mathrm{Lam} \circ \mathrm{Cur}(|N|_{I\cup\{x\}} \circ \Pi_{I\cup\{x\}}^{J\cup\{x\}}), && \text{by IH},\\
&= \mathrm{Lam} \circ \mathrm{Cur}(|N|_{I\cup\{x\}} \circ (\Pi_I^J \times \mathrm{Id})), && \text{by Remark 3.39(ii)},\\
&= \mathrm{Lam} \circ \mathrm{Cur}(|N|_{I\cup\{x\}}) \circ \Pi_I^J, && \text{by (Curry)},\\
&= |\lambda x.N|_I \circ \Pi_I^J, && \text{by definition of } |-|_I.
\end{aligned}$$

□

LEMMA 3.57. *Let $M \in \Lambda$ and $\rho \in \mathrm{Val}_{A_{\mathcal{U}}}$. Then, for all $J_1, J_2 \in \mathscr{P}_{\mathrm{f}}(\mathrm{Var})$ adequate for M, we have*

$$|M|_{J_1} \circ \rho^{J_1} = |M|_{J_2} \circ \rho^{J_2}.$$

PROOF. Consider $I \in \mathscr{P}_{\mathrm{f}}(\mathrm{Var})$ such that $J_1 \subseteq I$ and $J_2 \subseteq I$.

$$\begin{array}{rcll} |M|_{J_1} \circ \rho^{J_1} & = & |M|_{J_1} \circ \Pi^I_{J_1} \circ \rho^I, & \text{by Lemma 3.56(i)}, \\ & = & |M|_I \circ \rho^I, & \text{by Lemma 3.56(ii)}, \\ & = & |M|_{J_2} \circ \Pi^I_{J_2} \circ \rho^I, & \text{by Lemma 3.56(ii)}, \\ & = & |M|_{J_2} \circ \rho^{J_2}, & \text{by Lemma 3.56(i)}. \end{array}$$

□

DEFINITION 3.58. (i) The structure $\mathcal{S}_{\mathcal{U}} \triangleq (\mathcal{A}_{\mathcal{U}}, [\![-]\!]_{(\cdot)})$ associated with $\mathcal{U}$ is defined by:

- $\mathcal{A}_{\mathcal{U}} = (A_{\mathcal{U}}, \bullet)$ is the applicative structure from Definition 3.52(iii),
- $[\![-]\!]_{(\cdot)} : \Lambda \times \mathrm{Val}_{A_{\mathcal{U}}} \to A_{\mathcal{U}}$ is defined by setting $[\![M]\!]_\rho \triangleq |M|_I \circ \rho^I$, for some adequate I.

(ii) Given $z \in \mathrm{Var}$, we define a morphism $\upsilon_z : U^{\mathrm{Var}} \times U \to U^{\mathrm{Var}}$ by specifying its behavior on each component[5] $x \in \mathrm{Var}$:

$$\upsilon_z^x \triangleq \begin{cases} \pi_2, & \text{if } x = z, \\ \pi_x^{\mathrm{Var}} \circ \pi_1, & \text{otherwise.} \end{cases}$$

(iii) Similarly, for every $x \in \mathrm{Var}$ and $J \in \mathscr{P}_{\mathrm{f}}(\mathrm{Var})$, we define componentwise a morphism $\iota_{J,x} : U^{J\cup\{x\}} \to U^{\mathrm{Var}}$ by setting:

$$\iota_{J,x}^z \triangleq \begin{cases} \pi_z^{J\cup\{x\}}, & \text{if } z \in J \cup \{x\}, \\ \mathrm{Lam} \circ \mathrm{Cur}(\mathrm{Id}_U) \circ !_{U^{J\cup\{x\}}}, & \text{otherwise.} \end{cases}$$

REMARK 3.59. (i) Intuitively, υ_z is an 'updating' morphism that allows to replace the z-th component of U^{Var} by a new value which is obtained by applying π_2.

(ii) Similarly, the reader can think of $\iota_{J,x}^z$ as a canonical injection from $U^{J\cup\{x\}}$ to U^{Var}. We claim that the choice of $\mathrm{Lam} \circ \mathrm{Cur}(\mathrm{Id}_U) \circ !_{U^{J\cup\{x\}}}$ is canonical since $\mathrm{Lam} \circ \mathrm{Cur}(\mathrm{Id}_U)$ is nothing else than the point of U corresponding to Id_U under the morphism Lam.

FACT 3.60. For all $J \in \mathscr{P}_{\mathrm{f}}(\mathrm{Var})$ and $x \notin J$, we have $\Pi^{\mathrm{Var}}_{J\cup\{x\}} \circ \iota_{J,x} = \mathrm{Id}_{U^{J\cup\{x\}}}$.

The following technical lemmas show the main properties satisfied by υ_z and $\iota_{J,x}$.

LEMMA 3.61. *Assume $f_1, \dots, f_n \in \mathbf{C}_{\mathrm{f}}(U^{\mathrm{Var}}, U)$ and $(J_k, f'_k) \Vdash f_k$, for all $k\,(1 \le k \le n)$. For every variable $z \notin \bigcup_{k=1}^n J_k$, we have:*

$$(\langle f_1, \dots, f_n\rangle \times \mathrm{Id}) = \langle\langle f_1, \dots, f_n\rangle, \pi_z^{\mathrm{Var}}\rangle \circ \upsilon_z, \tag{3.4}$$

i.e., the following diagram commutes:

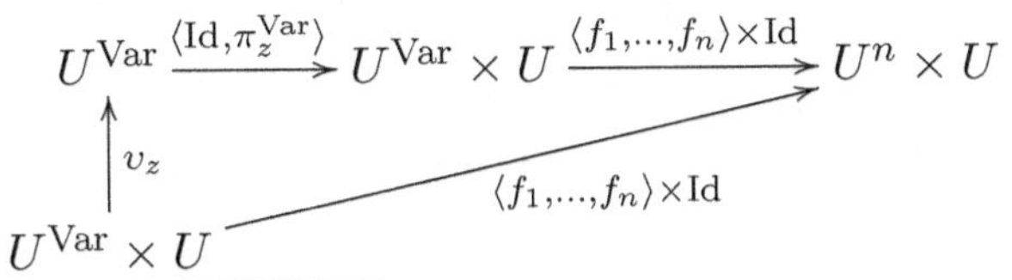

[5] In other words, υ_z^x is an abbreviation for $\pi_x^{\mathrm{Var}} \circ \upsilon_z$. Analogously, in (iii), $\iota_{J,x}^z$ stands for $\pi_z^{\mathrm{Var}} \circ \iota_{J,x}$.

PROOF. Let $f_1, \dots, f_n \in \mathbf{C}_{\mathrm{f}}(U^{\mathrm{Var}}, U)$. By (pair) and Definition 3.58(ii), one obtains:

$$\langle\langle f_1, \dots, f_n\rangle, \pi_z^{\mathrm{Var}}\rangle \circ \upsilon_z = \langle\langle f_1, \dots, f_n\rangle \circ \upsilon_z, \pi_2\rangle.$$

Hence, it is sufficient to prove that $\langle f_1, \dots, f_n\rangle \circ \upsilon_z = \langle f_1, \dots, f_n\rangle \circ \pi_1$. Consider now sets $J_k \in \mathscr{P}_{\mathrm{f}}(\mathrm{Var})$ and morphisms $f'_k : U^{J_k} \to U$ such that $(J_k, f'_k) \Vdash f_k$, for all k $(1 \leq k \leq n)$. Take $z \notin \bigcup_{k=1}^n J_k$. Conclude the proof as follows.

$$\begin{array}{rcll}
\langle f_1, \dots, f_n\rangle \circ \upsilon_z & = & \langle f'_1 \circ \Pi_{J_1}^{\mathrm{Var}}, \dots, f'_n \circ \Pi_{J_n}^{\mathrm{Var}}\rangle \circ \upsilon_z, & \text{since } f_k = f'_k \circ \Pi_{J_k}^{\mathrm{Var}}, \\
& = & \langle f'_1 \circ \Pi_{J_1}^{\mathrm{Var}} \circ \upsilon_z, \dots, f'_n \circ \Pi_{J_n}^{\mathrm{Var}} \circ \upsilon_z\rangle, & \text{by (pair)}, \\
& = & \langle f'_1 \circ \Pi_{J_1}^{\mathrm{Var}} \circ \pi_1, \dots, f'_n \circ \Pi_{J_n}^{\mathrm{Var}} \circ \pi_1\rangle, & \text{by definition of } \upsilon_z, \\
& & & \text{since } z \notin \bigcup_{k=1}^n J_k, \\
& = & \langle f_1 \circ \pi_1, \dots, f_n \circ \pi_1\rangle, & \text{since } f_k = f'_k \circ \Pi_{J_k}^{\mathrm{Var}}, \\
& = & \langle f_1, \dots, f_n\rangle \circ \pi_1, & \text{by (pair)}.
\end{array}$$

□

LEMMA 3.62. *Let $f \in \mathbf{C}_{\mathrm{f}}(U^{\mathrm{Var}}, U)$ and $(J, f_J) \Vdash f$. Then, for all $x \notin J$, we have:*

$$f \times \mathrm{Id} = \langle f, \pi_x^{\mathrm{Var}}\rangle \circ \iota_{J,x} \circ (\Pi_J^{\mathrm{Var}} \times \mathrm{Id})$$

i.e., the following diagram commutes:

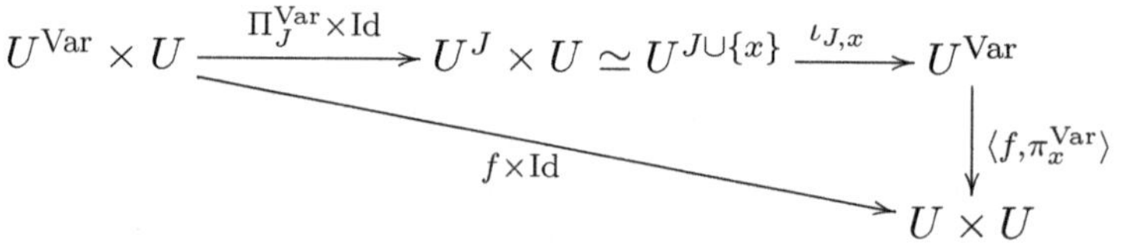

PROOF. Since by hypothesis $f = f_J \circ \Pi_J^{\mathrm{Var}}$, this is equivalent to ask that the following diagram commutes, and this is obvious by Fact 3.60 since $\langle \Pi_J^{\mathrm{Var}}, \pi_x^{\mathrm{Var}}\rangle = \Pi_{J \cup \{x\}}^{\mathrm{Var}}$.

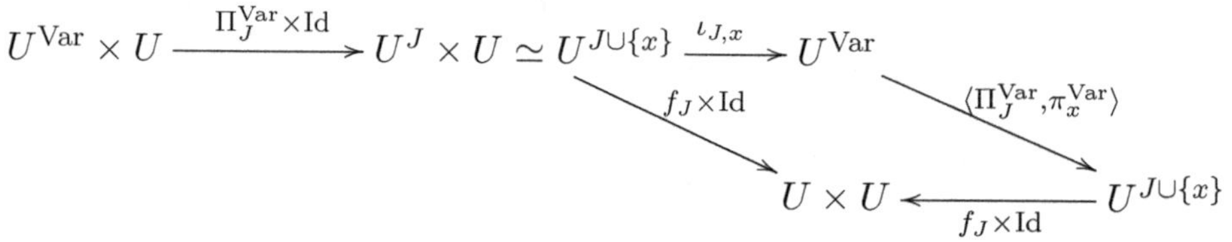

□

The next result shows that a syntactic λ-model can be constructed from any reflexive object $\mathcal{U}$, by taking as underlying set the finitary morphisms from U^{Var} to U.

THEOREM 3.63 (BUCCIARELLI ET AL. (2007)).
Let $\mathcal{U}$ be a categorical model living in a CCC $\mathbf{C}$ with countable product. Then:

(i) *$\mathcal{S}_{\mathcal{U}}$ is a syntactic λ-model,*

(ii) *$\mathcal{S}_{\mathcal{U}}$ is extensional $\iff$ $\mathcal{U}$ is extensional.*

PROOF. (i) We prove that $\mathcal{S}_{\mathcal{U}}$ satisfies the conditions (a)–(e) of Definition 3.37. In each item below, we assume that $I \subsetneq \mathrm{Var}$ is any finite set adequate for the λ-term on the LHS of the equality.

(a) $[\![z]\!]_\rho = |z|_I \circ \rho^I = \pi^I_z \circ \rho^I = \rho(z)$.

(b)
$$\begin{array}{rlr} [\![PQ]\!]_\rho = & |PQ|_I \circ \rho^I, & \text{by definition of } [\![-]\!],\\ = & (|P|_I \bullet |Q|_I) \circ \rho^I, & \text{by definition of } |-|_I,\\ = & \mathrm{Eval} \circ \langle \mathrm{App} \circ |P|_I, |Q|_I\rangle \circ \rho^I, & \text{by definition of } \bullet,\\ = & \mathrm{Eval} \circ \langle \mathrm{App} \circ |P|_I \circ \rho^I, |Q|_I \circ \rho^I\rangle, & \text{by (pair)},\\ = & [\![P]\!]_\rho \bullet [\![Q]\!]_\rho, & \text{by definition of } \bullet \text{ and } [\![-]\!]. \end{array}$$

(c)
$$\begin{array}{rlr} [\![\lambda x.P]\!]_\rho \bullet f = & (|\lambda x.P|_I \circ \rho^I) \bullet f, & \text{by def. of } [\![-]\!],\\ = & \mathrm{Eval} \circ \langle \mathrm{App} \circ (|\lambda x.P|_I \circ \rho^I), f\rangle, & \text{by definition of } \bullet,\\ = & \mathrm{Eval} \circ \langle \mathrm{App} \circ \mathrm{Lam} \circ \mathrm{Cur}(|P|_{I\cup\{x\}}) \circ \rho^I, f\rangle, & \text{by def. of } |-|_I,\\ = & \mathrm{Eval} \circ \langle \mathrm{Cur}(|P|_{I\cup\{x\}}) \circ \rho^I, f\rangle, & \text{as } \mathrm{App} \circ \mathrm{Lam} = \mathrm{Id},\\ = & \mathrm{Eval} \circ \langle \mathrm{Cur}(|P|_{I\cup\{x\}} \circ (\rho^I \times \mathrm{Id})), f\rangle, & \text{by (Curry)},\\ = & |P|_{I\cup\{x\}} \circ (\rho^I \times \mathrm{Id}) \circ \langle \mathrm{Id}, f\rangle, & \text{by (beta)},\\ = & |P|_{I\cup\{x\}} \circ \langle \rho^I, f\rangle, & \text{by (pair)},\\ = & [\![P]\!]_{\rho[x:=f]}, & \text{by def. of } [\![-]\!]. \end{array}$$

(d) Immediate since, by Lemma 3.57, $[\![M]\!]_\rho = |M|_{\mathrm{FV}(M)} \circ \rho^{\mathrm{FV}(M)}$.

(e)
$$\begin{array}{rlr} [\![\lambda z.M]\!]_\rho = & |\lambda z.M|_I \circ \rho^I, & \text{by definition of } [\![-]\!],\\ = & \mathrm{Lam} \circ \mathrm{Cur}(|M|_{I\cup\{z\}}) \circ \rho^I, & \text{by def. of } |-|_I,\\ = & \mathrm{Lam} \circ \mathrm{Cur}(|M|_{I\cup\{z\}} \circ (\rho^I \times \mathrm{Id})), & \text{by (Curry)},\\ = & \mathrm{Lam} \circ \mathrm{Cur}(|M|_{I\cup\{z\}} \circ \langle \rho^I, \pi^{\mathrm{Var}}_z\rangle \circ v_z), & \text{by Lemma 3.61},\\ = & \mathrm{Lam} \circ \mathrm{Cur}([\![M]\!]_{\rho[z:=\pi^{\mathrm{Var}}_z]} \circ v_z), & \text{by definition of } [\![-]\!]. \end{array}$$

Note that Lemma 3.61 is applicable because $\rho(x)$ is finitary for every $x \in I$, thus by definition there exists a decomposition $(J_x, f_x) \Vdash \rho(x)$. Therefore, by α-conversion, we can assume wlog that $z \notin \bigcup_{x\in I} J_x$. Now, by assumption, the equality

$$[\![M]\!]_{\rho[z:=f]} = [\![N]\!]_{\rho[z:=f]}$$

holds for every $f \in \mathbf{C}_{\mathsf{f}}(U^{\mathrm{Var}}, U)$. In particular, it holds for $f \triangleq \pi^{\mathrm{Var}}_z$. It follows that

$$\mathrm{Lam} \circ \mathrm{Cur}([\![M]\!]_{\rho[z:=\pi^{\mathrm{Var}}_z]} \circ v_z) = \mathrm{Lam} \circ \mathrm{Cur}([\![N]\!]_{\rho[z:=\pi^{\mathrm{Var}}_z]} \circ v_z)$$

Analogous calculations, show that $\mathrm{Lam} \circ \mathrm{Cur}\big([\![N]\!]_{\rho[z:=\pi^{\mathrm{Var}}_z]} \circ v_z\big) = [\![\lambda z.N]\!]_\rho$.

(ii) (⇒) By Remark 3.53, π^{Var}_x is finitary. Now, for all $f \in \mathbf{C}_{\mathsf{f}}(U^{\mathrm{Var}}, U)$ we have:

$$\begin{array}{rlr} (\mathrm{Lam} \circ \mathrm{App} \circ \pi^{\mathrm{Var}}_x) \bullet f = & \mathrm{Eval} \circ \langle \mathrm{App} \circ \mathrm{Lam} \circ \mathrm{App} \circ \pi^{\mathrm{Var}}_x, f\rangle, & \text{by definition of } \bullet,\\ = & \mathrm{Eval} \circ \langle \mathrm{App} \circ \pi^{\mathrm{Var}}_x, f\rangle, & \text{by } \mathrm{App} \circ \mathrm{Lam} = \mathrm{Id},\\ = & \pi^{\mathrm{Var}}_x \bullet f, & \text{by definition of } \bullet. \end{array}$$

If $\mathcal{S}_{\mathcal{U}}$ is extensional, this implies $\mathrm{Lam} \circ \mathrm{App} \circ \pi^{\mathrm{Var}}_x = \pi^{\mathrm{Var}}_x$. Since π^{Var}_x is an epimorphism, we get $\mathrm{Lam} \circ \mathrm{App} = \mathrm{Id}_U$.

($\Leftarrow$) Take $f, g \in \mathbf{C}_{\mathsf{f}}(U^{\mathrm{Var}}, U)$. By definition, there exist decompositions $(I, f_I) \Vdash f$ and $(J, g_J) \Vdash g$. Setting $X \triangleq I \cup J$, $f_X \triangleq f_I \circ \Pi^X_I$ and $g_X \triangleq g_J \circ \Pi^X_J$ we get $(X, f_X) \Vdash f$ and $(X, g_X) \Vdash g$. By assumption, for all $h \in \mathbf{C}_{\mathsf{f}}(U^{\mathrm{Var}}, U)$, we have that $f \bullet h = g \bullet h$. Taking $h \triangleq \pi^{\mathrm{Var}}_x$, for any $x \notin X$, we obtain $f \bullet \pi^{\mathrm{Var}}_x = g \bullet \pi^{\mathrm{Var}}_x$. By precomposing with

$$\varphi \triangleq \iota_{X,x} \circ (\Pi^{\mathrm{Var}}_X \times \mathrm{Id})$$

we obtain:

$$\begin{array}{lll}
 & \langle \mathrm{App} \circ f, \pi^{\mathrm{Var}}_x \rangle \circ \varphi = \langle \mathrm{App} \circ g, \pi^{\mathrm{Var}}_x \rangle \circ \varphi & \\
\Rightarrow & \langle \mathrm{App} \circ f, \pi^{\mathrm{Var}}_x \rangle \circ \varphi = (\mathrm{App} \circ f) \times \mathrm{Id} & \text{and} \\
 & \langle \mathrm{App} \circ g, \pi^{\mathrm{Var}}_x \rangle \circ \varphi = (\mathrm{App} \circ g) \times \mathrm{Id}, & \text{by Lemma 3.62,} \\
\Rightarrow & \mathrm{App} \circ f = \mathrm{App} \circ g & \\
\Rightarrow & \mathrm{Lam} \circ \mathrm{App} \circ f = \mathrm{Lam} \circ \mathrm{App} \circ g & \\
\Rightarrow & f = g, & \text{since } \mathrm{Lam} \circ \mathrm{App} = \mathrm{Id}_U.
\end{array}$$

This conclues the proof. □

COROLLARY 3.64 (BUCCIARELLI ET AL. (2007)). *If $\mathcal{U}$ is a categorical model living in a CCC $\mathbf{C}$ with countable products, then*

$$\mathcal{C}_{\mathcal{U}} = (\mathbf{C}_{\mathsf{f}}(U^{\mathrm{Var}}, U), \bullet, |\mathsf{K}|^{\mathcal{U}} \circ !_{U^{\mathrm{Var}}}, |\mathsf{S}|^{\mathcal{U}} \circ !_{U^{\mathrm{Var}}})$$

is a λ-model.

PROOF. By Theorem 3.50(ii), since $[\![M]\!]^{\mathcal{S}_{\mathcal{U}}}_{\rho} = |M|^{\mathcal{U}} \circ !_{U^{\mathrm{Var}}}$ for all $M \in \Lambda^o$. □

REMARK 3.65. Although some researchers did adopted this approach to describe models of λ-calculus living in non-well-pointed categories, e.g. Paolini et al. (2017), we would like to stress that the quantification over all valuations $\rho : \mathrm{Var} \to \mathbf{C}_{\mathsf{f}}(U^{\mathrm{Var}}, U)$ is irrelevant. Indeed, setting $\hat{\rho}(x) \triangleq \pi^{\mathrm{Var}}_x$ for all $x \in \mathrm{Var}$, one obtains:

$$\begin{array}{rcl}
\mathcal{S}_{\mathcal{U}} \models M = N & \iff & [\![M]\!]^{\mathcal{S}_{\mathcal{U}}}_{\hat{\rho}} = [\![N]\!]^{\mathcal{S}_{\mathcal{U}}}_{\hat{\rho}} \\
 & \iff & |M|_{\mathrm{FV}(MN)} = |N|_{\mathrm{FV}(MN)},
\end{array}$$

where the second equivalence follows from the fact that projections are epimorphisms.

Working without countable products

The construction in the previous section has been worked out under the assumption that the underlying CCC $\mathbf{C}$ has countable products. This is usually a reasonable assumption, nevertheless we explain a method for getting rid of this hypothesis. As proved by Meyer (1982) in his Lambda Algebra Theorem, every λ-algebra $\mathcal{A}$ can be embedded into the λ-model $\mathcal{A}[\mathrm{Var}]$, obtained from $\mathcal{A}$ by freely adjoining the variables as indeterminates. See also the discussion on C-monoids in the book by Lambek and Scott (1986).

More precisely, Meyer shows that the implication

$$\mathcal{A}[x_1, \ldots, x_n] \models M = N \quad \Rightarrow \quad \mathcal{A}[x_1, \ldots, x_n] \models \lambda x.M = \lambda x.N$$

holds for all λ-terms M, N whose free variables are among $x_1, \ldots, x_n$.

Moreover, Selinger (2002) has shown that if $\mathcal{A}$ is the λ-algebra associated with a categorical model $\mathcal{U}$ by Koymans' construction (Theorem 3.50(iv)), then for all $I \in \mathscr{P}_{\mathrm{f}}(\mathrm{Var})$ the free extension $\mathcal{A}[I]$ is isomorphic—in the category of combinatory algebras and homomorphisms—to $\mathbf{C}(U^I, U)$ endowed with the natural structure of combinatory algebra. Since there exist canonical homomorphisms

$$\begin{array}{lcl} \mathcal{A}[I] & \mapsto & \mathcal{A}[J], \\ \mathbf{C}(U^I, U) & \mapsto & \mathbf{C}(U^J, U), \end{array}$$

that are one-to-one whenever $I \subseteq J$ are finite subsets of Var, one can construct the inductive limit of both $\mathscr{P}_{\mathrm{f}}(\mathrm{Var})$-indexed diagrams. On the one side, this gives a λ-model isomorphic to $\mathcal{A}[\mathrm{Var}]$. On the other side, one obtains

$$X = \bigcup_{I \in \mathscr{P}_{\mathrm{f}}(\mathrm{Var})} \mathbf{C}(U^I, U)/_{\sim_{\mathrm{f}}},$$

where $\sim_{\mathrm{f}}$ is the equivalence relation defined by setting, for all $f \in \mathbf{C}(U^J, U), g \in \mathbf{C}(U^I, U)$,

$$f \sim_{\mathrm{f}} g \iff f \circ \Pi_J^{I \cup J} = g \circ \Pi_I^{I \cup J}.$$

Clearly the aforementioned isomorphism is preserved at the limit, whence the combinatory algebra $\mathcal{X}$, having X as underlying set and the natural application on the equivalence classes as binary operator, is a λ-model as well. The approach described here, although more convoluted, works even when the underlying CCC $\mathbf{C}$ does not have countable products. In case $\mathbf{C}$ does have countable products, it is easy to check the λ-model $\mathcal{X}$ so-defined is isomorphic to the λ-model $\mathcal{C}_{\mathcal{U}}$ from Corollary 3.64.

Part II

Reduction

Jean-Jacques Lévy

Jan Willem Klop

Stefano Guerrini

Andrea Asperti

Łukasz Czajka

Marc-Antoine Padovani-Dellatana
& Vincent Padovani

Vincent van Oostrom

Richard Kennaway

Paula Severi

Fer-Jan de Vries

Alessandro Berarducci

Sachio Hirokawa
(with Joly)

Chapter 4

Leaving a β-reduction plane

The reduction graph $\mathcal{G}_\beta(M)$ of a λ-term M is a directed graph having a node N for each reduct of M, and an arc from a node P to a node Q whenever $P \to_\beta Q$ holds. In λ-calculus it is not difficult to construct λ-terms whose reduction graph is finite and contains cycles. The easiest example is given by the paradigmatic looping combinator:

Another example, is the λ-term XXX where $X \triangleq \lambda yx.xyy$. Its reduction graph is a cycle of order 2, because one needs two reductions to come back to the original point:

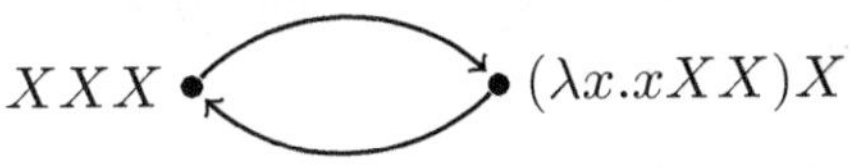

By slightly modifying the X, as in $W \triangleq \lambda xy.xyy$, we obtain a different reduction graph:

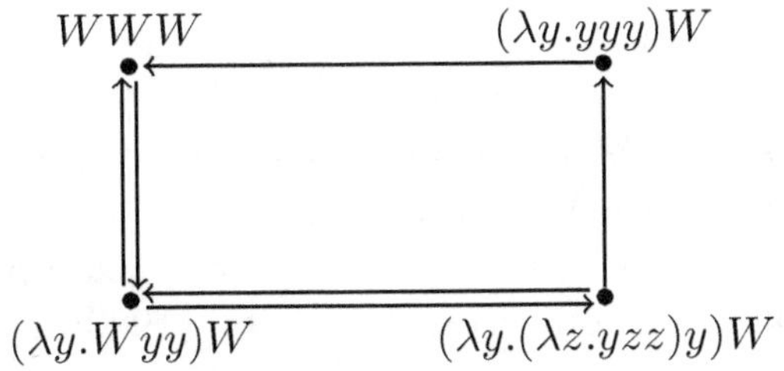

It is worth noticing that the first two examples are similar as they only differ by the length of the cycle. The third graph has a different nature. It contains several cycles, none of which is 'pure'—in a pure cycle the successor of every node is uniquely determined.

The next example is taken from Curry and Feys (1958), page 109, and is useful to explain the concept of a plane. Let $W = \lambda xy.xyy$ as above, $\Delta = \lambda x.xx$ and I be the identity. Consider the λ-term:

$$\Delta(W\mathsf{I}) \tag{4.1}$$

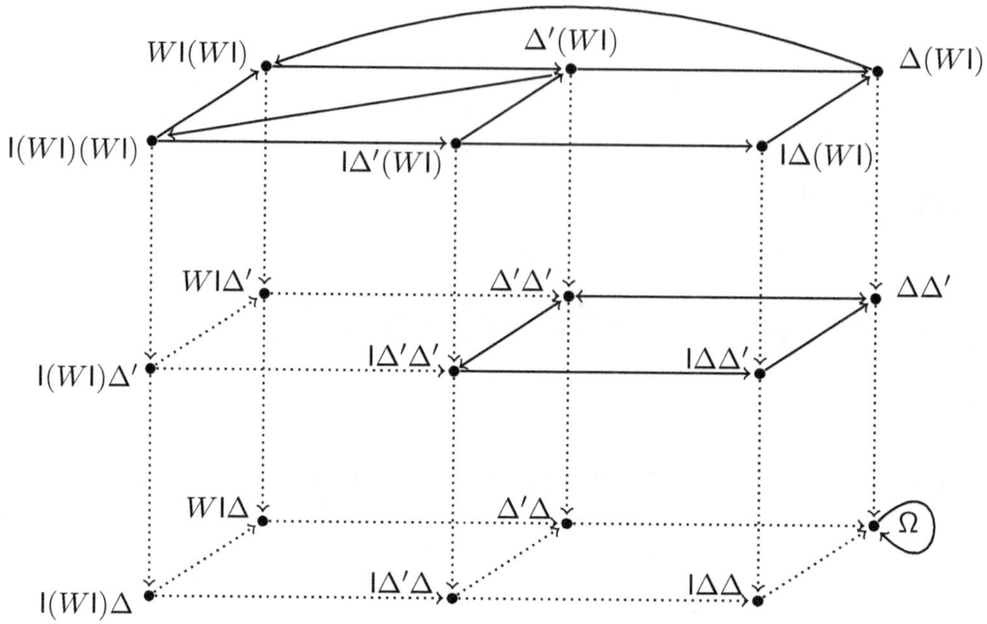

Figure 4.1: The reduction graph of $\Delta(W\mathsf{I})$.

Its reduction graph is depicted in Figure 4.1, using the auxiliary term $\Delta' \triangleq \lambda y.\mathsf{I}yy$. In the picture, solid arrows are used to connect λ-terms lying on a reduction cycle. Calling 'plane' a maximal set of inter-reducible λ-terms and 'points' its elements, one can see that this graph contains 3 planes:

- $\mathcal{P}_1 = \{W\mathsf{I}(W\mathsf{I}), \Delta'(W\mathsf{I}), \Delta(W\mathsf{I}), \mathsf{I}(W\mathsf{I})(W\mathsf{I}), \mathsf{I}\Delta'(W\mathsf{I}), \mathsf{I}\Delta(W\mathsf{I})\}$;
- $\mathcal{P}_2 = \{\Delta\Delta', \Delta'\Delta', \mathsf{I}\Delta'\Delta', \mathsf{I}\Delta\Delta'\}$;
- $\mathcal{P}_3 = \{\Omega\}$.

Now, the planes $\mathcal{P}_1$ and $\mathcal{P}_2$ can be left at any point by moving along a vertical arc, while $\mathcal{P}_3$ is a *cul de sac*. Adopting the standard terminology, one says that all points in $\mathcal{P}_1$ (and $\mathcal{P}_2$) are 'exit points', while $\mathcal{P}_3$ has no exit points. Examples of this sort, led J. W. Klop to formulate the following conjecture. See Klop (1980b), Conjecture 3.6.1.

CONJECTURE 4.1 (THE PLANE PROPERTY).
If a plane has an exit point, then every point of the plane is an exit.

This conjecture was refuted by Mulder (1986) and by Sekimoto and Hirokawa (1988). Both counterexamples were found in 1986, independently, and their complexity is comparable. In this chapter we present Mulder's counterexample, but we also show some results by Sekimoto and Hirokawa and follow their reformulation of the problem in terms of recurrent terms. We conclude by recalling some related open problems, that still stand.

4.1 Planes and cyclic reductions

Cycles inspired many combinatorial challenges in term rewriting systems (see, e.g. Terese (2003)). Moreover, cycles in λ-calculus play an important role in the study of Church–Rosser reduction strategies (see Bergstra and Klop (1979); Klop et al. (2007)), and in the analysis of unsolvability (see Jacopini and Venturini Zilli (1985), but also Chapter 9 of this book). We now present the formal definitions behind the notions discussed above.

DEFINITION 4.2. Let $\rho : M_0 \to_\beta M_1 \to_\beta \cdots \to_\beta M_n$ be a non-empty reduction sequence, that is, $n > 0$.

(i) We say that ρ is a *(reduction) cycle* if $M_0 = M_n$ and ρ contains no proper sub-cycles, i.e.:

$$\forall i, j\,(0 \le i < j \le n)\,.\,[\,M_i = M_j \iff i = 0 \;\&\; j = n\,]$$

(ii) A cycle ρ is *pure* if every M_i contains exactly one redex (for $0 \le i \le n$);
(iii) If $\rho : M_0 \to_\beta M_1 \to_\beta \cdots \to_\beta M_n$ is a pure cycle, then n is called its *order*.
(iv) Pure cycles of order 1 are called *loops*.

The terminology introduced above is not completely standard. Other authors call 'cycle' any reduction of the form $M \twoheadrightarrow_\beta M$, while our definition of a cycle requires an extra 'minimality' condition.[1] Note that the order of a cycle is a property of the reduction, and should not be confused with the order of a λ-term (Definition 2.37).

EXAMPLES 4.3. (i) Let $X \triangleq \lambda xy.x(yy)x$. The following is a reduction cycle:

$$\begin{array}{ccccc}
\mathsf{I}(XX)\mathsf{I} & \xrightarrow[\beta]{} & XX\mathsf{I} & \xrightarrow[\beta]{} & (\lambda y.X(yy)X)\mathsf{I} \\
\uparrow{\scriptstyle\beta} & & & & \downarrow{\scriptstyle\beta} \\
(\lambda y.\mathsf{I}(yy)\mathsf{I})X & \xleftarrow[\beta]{} & X\mathsf{I}X & \xleftarrow[\beta]{} & X(\mathsf{I}\mathsf{I})X
\end{array}$$

which is not pure. For instance $\mathsf{I}(XX)\mathsf{I}$ has 2 redexes.

(ii) $\Omega \to_\beta \Omega$ and $\lambda x.x\Omega \to_\beta \lambda x.x\Omega$ are loops.
(iii) Let $Y \triangleq \lambda xy.yxy$. Then $Y\mathsf{I}Y \to_\beta (\lambda y.y\mathsf{I}y)Y \to_\beta Y\mathsf{I}Y$ is a pure cycle of order 2.

By a counting argument one can show that there is exactly one loop in λ-calculus having the shape of a β-redex.

LEMMA 4.4 (LERCHER (1976)). *Let R be a β-redex such that $R \to_\beta R$. Then*

$$R = \Omega = (\lambda x.xx)(\lambda x.xx).$$

PROOF. For $M, N \in \Lambda$ and $x \in \mathrm{Var}$, write

$$\begin{array}{rcl}
\#_x(M) & \triangleq & \text{number of free occurrences of } x \text{ in } M; \\
\#_\lambda(M) & \triangleq & \text{number of } \lambda\text{-abstraction in } M; \\
\#_N(M) & \triangleq & \text{number of subterm occurrences of } N \text{ in } M.
\end{array}$$

[1] In fact, our cycles are called 'minimal cycles' in Endrullis et al. (2016). In the same paper, the authors use the terminology 'deterministic' cycle for our notion of a pure cycles.

Let $R = (\lambda x.M)N$ be such that $R \to_\beta R$, that is

$$(\lambda x.M)N = M[x{:=}N], \tag{4.2}$$

where, by the Variable Convention, we can assume $x \notin \mathrm{FV}(N)$. We start by counting the number of abstractions on the two sides of (4.2). On the LHS we have

$$\#_\lambda((\lambda x.M)N) = \#_\lambda(M) + \#_\lambda(N) + 1.$$

The reduction removes a 'λx' and copies $\#_x(M)$ times the abstractions occurring in N, so the RHS is $\#_\lambda(M[x:=N]) = \#_\lambda(M) + \#_\lambda(N) \times \#_x(M)$. By equality (4.2) we get the Diophantine equation

$$\#_\lambda(M) + \#_\lambda(N) + 1 = \#_\lambda(M) + \#_\lambda(N) \times \#_x(M)$$

having as unique solution $\#_\lambda(N) = 1$ and $\#_x(M) = 2$. As M must contain 2 occurrences of x, we can determine the number of subterms equal to N on both sides of (4.2). On the LHS we have $\#_N((\lambda x.M)N) = \#_N(\lambda x.M)+1$, while $x \notin \mathrm{FV}(N)$ implies on the RHS

$$\#_N(M[x{:=}N]) = \#_N(M) + \#_x(M) = \#_N(M) + 2,$$

giving

$$\#_N(\lambda x.M) + 1 = \#_N(M) + 2.$$

This is only possible if $\lambda x.M = N$ and $\#_N(M) = 0$. Since $M[x:=N]$ is an application and $M \neq x$, we must have $M = M_1M_2$, where $M_1[x{:=}N] = \lambda x.M$ and $M_2[x{:=}N] = N$. Gathering all pieces together, we obtain

$$M_1[x{:=}N] = \lambda x.M = N = M_2[x{:=}N]$$

from which we conclude that $M_1 = M_2 = x$, so that $\lambda x.M = \lambda x.xx = N$, i.e. $R = \Omega$. □

EXERCISE 4.5. (i) [Visser] Show that the only λ-terms reducing in one step to themselves have shape $C[\Omega]$, for some context $C[\,]$. That is, Ω is the only minimal loop in λ-calculus.
(ii) Show that Lemma 4.4 and (i) hold for $\beta\eta$ as well.
(iii) Give an alternative proof of Lemma 4.4 by case analysis on the stucture of the terms involved.

REMARK 4.6. (i) For a fixed n, it is decidable to check whether a λ-term has a pure cycle of order n. However, it is not decidable whether a λ-term admits a cyclic reduction.
(ii) Endrullis et al. (2016) have characterized all pure cycles of order 2, *bicycles* using their terminology. It turns out that there exist infinitely many different bicycles.

The following class of combinators provides examples of pure cycles of any order (see Lemma 4.9). They deserve to stand alone as they will play a crucial role in Section 9.1.

DEFINITION 4.7. Define an infinite sequence of combinators $(\mathsf{X}_n)_{n\in\mathbb{N}}$ by setting:

$$\begin{aligned} \mathsf{X}_n &\triangleq \mathsf{W}_n\mathsf{W}_n^{\sim n+1}, \text{ where} \\ \mathsf{W}_n &\triangleq \lambda y_1 \ldots y_n x.xxy_1 \cdots y_n. \end{aligned}$$

REMARK 4.8. (i) For all $n \geq 0$, X_n is a λ-term of order 0, i.e. $\mathrm{ord}_\beta(\mathsf{X}_n) = 0$.
(ii) Note that $\mathsf{X}_0 = \Omega$.
(iii) For X_1, one gets a bicycle: $\mathsf{X}_1 = \mathsf{W}_1\mathsf{W}_1\mathsf{W}_1 \to_\beta (\lambda x.xx\mathsf{W}_1)\mathsf{W}_1 \to_\beta \mathsf{X}_1$.
(iv) For X_2, one gets a pure cycle of order 3:

$$\mathsf{X}_2 = \mathsf{W}_2\mathsf{W}_2\mathsf{W}_2\mathsf{W}_2 \to_\beta (\lambda y_2 x.xx\mathsf{W}_2 y_2)\mathsf{W}_2\mathsf{W}_2 \to_\beta (\lambda x.xx\mathsf{W}_2\mathsf{W}_2)\mathsf{W}_2 \to_\beta \mathsf{X}_2$$

More generally, the following holds.

LEMMA 4.9. *For all $n \in \mathbb{N}$, X_n has a pure reduction cycle of order $n+1$.*

PROOF. The cases $n < 3$ are discussed in Remark 4.8(ii)–(iv). For $n \geq 3$, we have:

$$\begin{array}{ccc} \mathsf{X}_n = \quad \mathsf{W}_n\mathsf{W}_n^{\sim n+1} & \xrightarrow[\beta]{} & (\lambda y_2\ldots y_n x.xx\mathsf{W}_n y_2\cdots y_n)\mathsf{W}_n^{\sim n} \\ \beta \uparrow & & \downarrow \beta\ (n-2\text{ steps}) \\ (\lambda x.xx\mathsf{W}_n^{\sim n})\mathsf{W}_n & \xleftarrow[\beta]{} & (\lambda y_n x.xx\mathsf{W}_n^{\sim n-1}y_n)\mathsf{W}_n^{\sim 2} \end{array}$$

Moreover, every term in the cycle has only one redex, hence the cycle is pure. □

DEFINITION 4.10. (i) For $M, N \in \Lambda$, define the relation $\circlearrowleft_\beta$ by setting:

$$M \circlearrowleft_\beta N \iff M \twoheadrightarrow_\beta N \ \&\ N \twoheadrightarrow_\beta M$$

It is easy to check that $\circlearrowleft_\beta$ is an equivalence relation.
(ii) Therefore, $[M]_{\circlearrowleft_\beta} = \{N \in \Lambda \mid M \circlearrowleft_\beta N\}$ represents the *$\circlearrowleft_\beta$-equivalence class* of M.
(iii) Whenever $M \circlearrowleft_\beta N$ holds, we say that M and N are *cyclically equivalent*.
(iv) A subset $\mathcal{P} \subseteq \Lambda$ is called a *plane* if it is an equivalence class of $\circlearrowleft_\beta$. Equivalently:

$$\begin{array}{ll} (1) & \forall M, N \in \mathcal{P}\,.\, M \twoheadrightarrow_\beta N; \\ (2) & \forall M \in \mathcal{P}, N \in \Lambda\,.\,[M \circlearrowleft_\beta N \quad \Rightarrow \quad N \in \mathcal{P}]. \end{array}$$

(v) The elements of a plane $\mathcal{P}$ are called *points*.
(vi) A point $M \in \mathcal{P}$ is called an *exit* if $M \to_\beta N$ for some $N \notin \mathcal{P}$.
(vii) Given a plane $\mathcal{P}$, we denote by $\mathcal{E}_\mathcal{P}$ the set of its exit points.

EXAMPLES 4.11. (i) We have $\Omega \circlearrowleft_\beta \Omega$ and $\mathsf{X}_2 \circlearrowleft_\beta (\lambda x.xx\mathsf{W}_2\mathsf{W}_2)\mathsf{W}_2$.
(ii) On the contrary, $\mathsf{I}\Omega \not\circlearrowleft_\beta \Omega$ and $\mathsf{F}\Omega \not\circlearrowleft_\beta \mathsf{I}$.
(iii) The following are planes:
(1) $\mathcal{P}_1 = \{\Omega\}$;
(2) $\mathcal{P}_2 = \{\mathsf{X}_2, (\lambda y_2 x.xx\mathsf{W}_2 y_2)\mathsf{W}_2\mathsf{W}_2, (\lambda x.xx\mathsf{W}_2\mathsf{W}_2)\mathsf{W}_2\}$;
(3) $\mathcal{P}_3 = \{\mathsf{F}\mathsf{X}_2, \mathsf{F}((\lambda y_2 x.xx\mathsf{W}_2 y_2)\mathsf{W}_2\mathsf{W}_2), \mathsf{F}((\lambda x.xx\mathsf{W}_2\mathsf{W}_2)\mathsf{W}_2)\}$.

For $i \in \{1, 2\}$ we have $\mathcal{E}_{\mathcal{P}_i} = \emptyset$, while $\mathcal{E}_{\mathcal{P}_3} = \mathcal{P}_3$ because $\mathsf{F}M \to_\beta \mathsf{I} \notin \mathcal{P}_3$ for all $M \in \Lambda$.
(iv) $\mathcal{P}_4 = \{\mathsf{X}_2, (\lambda y_2 x.xx\mathsf{W}_2 y_2)\mathsf{W}_2\mathsf{W}_2\}$ is not a plane as $\mathsf{X}_2 \circlearrowleft_\beta (\lambda x.xx\mathsf{W}_2\mathsf{W}_2)\mathsf{W}_2 \notin \mathcal{P}_4$.

4.2 Recurrent terms and the Plane Property

The notion of recurrent terms is due to Venturini Zilli (1978). See also Böhm and Micali (1980), where recurrent terms are called 'minimal'. One-step recurrent terms were introduced by Hirokawa (1984), and root-recurrent terms by Polonsky in 2014.

DEFINITION 4.12. Let $M \in \Lambda$.
(i) M is *recurrent* if $M \twoheadrightarrow_\beta N$ entails $N \twoheadrightarrow_\beta M$.
(ii) M is *one-step recurrent* if $M \to_\beta N$ entails $N \twoheadrightarrow_\beta M$.
(iii) M is *root-recurrent* if it is recurrent and reduces to a redex, i.e. $M \twoheadrightarrow_\beta (\lambda x.P)Q$.

EXAMPLES 4.13. (i) $\mathsf{I}, \mathsf{K}, \mathsf{F}$, and $\lambda x.x\Omega$ are (one-step) recurrent, but not root-recurrent.
(ii) $\Omega, \lambda x.\Omega, \Omega\mathsf{I}$ and $\Omega\Omega$ are one-step recurrent; Ω and the X_n's are root-recurrent.
(iii) $M = (\lambda x.xx)((\lambda xy.xyy)\mathsf{I})$, namely the λ-term $\Delta(W\mathsf{I})$ from (4.1), is not recurrent because $M \twoheadrightarrow_\beta \Omega$, and $\Omega \twoheadrightarrow_\beta N$ entails $N = \Omega$. (See Figure 4.1.)

REMARK 4.14. (i) Every β-normal form is recurrent. (Even if this can be considered as a degenerate form of recurrence.)
(ii) Every recurrent term is one-step recurrent.
(iii) Every root-recurrent term is a (possibly open) recurrent unsolvable of order 0.
(iv) Every closed recurrent term of order 0 is root-recurrent. The former class of combinators was studied in Jacopini and Venturini Zilli (1985).

LEMMA 4.15. *Let $\mathcal{P}$ be a plane and $M \in \mathcal{P}$. Then*

$$M \notin \mathcal{E}_{\mathcal{P}} \quad \Rightarrow \quad M \textit{ is one-step recurrent.}$$

PROOF. By hypothesis, $M \to_\beta N$ entails $N \in \mathcal{P}$. Since $\mathcal{P}$ is a plane, we get $N \circlearrowleft_\beta M$. We conclude because N is an arbitrary one-step reduct of M. □

We show some interesting properties of recurrent terms. The reader who is only interested in the refutation of Klop's conjecture can safely skip to the next subsection.

LEMMA 4.16. *Let $M \in \Lambda$ be recurrent.*
(i) *$M = \lambda x.N$ implies that N is recurrent.*
(ii) *$M = PQ$ and $\mathrm{ord}_\beta(P) = 0$ imply that P, Q are recurrent.*

PROOF. (i) Immediate. Use the fact that $M \to_\beta M' \iff M' = \lambda x.N' \ \&\ N \to_\beta N'$.
(ii) Easy. By applying Lemma 2.45(iii). □

The results presented below are taken from Endrullis et al. (2016).

LEMMA 4.17. *Every root-recurrent term is easy.*

PROOF. Let M be root-recurrent. Then every reduct of M reduces to a redex via M, and this entails that $\mathrm{BeT}(M) = \bot$. By Theorem 3.9, we conclude that M is easy. □

The next proposition shows that root-recurrent terms constitute the 'building blocks' of the recurrent ones.

PROPOSITION 4.18. *Let $M \in \Lambda$. If M is recurrent, then there exist $n \geq 0$, an n-ary context $C[-_1, \ldots, -_n]$ and root-recurrent terms $M_1, \ldots, M_n \in \Lambda$ such that:*
 (i) *$M = C[M_1, \ldots, M_n]$;*
 (ii) *$C[x_1, \ldots, x_n]$, for $\vec{x}$ fresh, is a β-normal form.*

PROOF. As an induction loading, also verify that

$$\text{(iii)}\ \exists C'[\,] \,.\, C[-_1, \ldots, -_n] = \lambda y.C'[-_1, \ldots, -_n] \iff \exists N \in \Lambda, y \in \mathrm{Var} \,.\, M = \lambda y.N$$

Proceed by structural induction on M.

Case $M = x$. Take for $C[\,]$ the constant context $C[\,] \triangleq x$.

Case $M = \lambda y.N$. By Lemma 4.16(i), we know that N is also recurrent. So, we take $C[x_1, \ldots, x_n] \triangleq \lambda y.C'[x_1, \ldots, x_n]$ where $C'[\,]$ is obtained by induction hypothesis.

Case $M = PQ$. We distinguish two cases.

Subcase $P \twoheadrightarrow_\beta \lambda x.N$. Then $M \twoheadrightarrow_\beta (\lambda x.N)Q$, hence M is root-recurrent. We can take $C[x_1] \triangleq x_1$, which is normal and satisfies $M = C[M]$.

Subcase $\mathrm{ord}_\beta(P) = 0$. By Lemma 4.16(ii), P and Q are both recurrent. By induction hypothesis, there are (normal) contexts $C_1[x_1, \ldots, x_k]$, $C_2[y_1, \ldots, y_m]$ and root-recurrent terms $M_1, \ldots, M_k$, $N_1, \ldots, N_m$ such that

$$P = C_1[M_1, \ldots, M_k] \ \&\ Q = C_2[N_1, \ldots, N_m]$$

Setting

$$C[\vec{x}, \vec{y}] \triangleq C_1[\vec{x}]\, C_2[\vec{y}]$$

we have

$$M = PQ = C_1[\vec{M}]\, C_2[\vec{N}] = C[\vec{M}, \vec{N}].$$

Since P has order 0, by (iii), $C_1[\vec{x}]$ does not start with an abstraction, therefore $C_1[\vec{x}]\, C_2[\vec{y}]$ is a β-normal form. □

Refuting Klop's Conjecture

In Example 4.13 we did not give any example of a one-step recurrent term which is not recurrent—this was intentional. Indeed, the existence of such a term constitutes a counterexample for the Plane Property.

PROPOSITION 4.19. *The following are equivalent:*

1. *Klop's Conjecture 4.1: for all planes $\mathcal{P}$, either $\mathcal{E}_\mathcal{P} = \emptyset$ or $\mathcal{E}_\mathcal{P} = \mathcal{P}$.*

2. *Every one-step recurrent term is recurrent.*

PROOF. ($1 \Rightarrow 2$) Let M be a one-step recurrent term. Consider the plane $\mathcal{P} \triangleq [M]_{\circlearrowleft_\beta}$. Since M is one-step recurrent it cannot be an exit point, so we have $\mathcal{E}_\mathcal{P} \neq \mathcal{P}$. By (1) we infer $\mathcal{E}_\mathcal{P} = \emptyset$, therefore M must be recurrent.

($2 \Rightarrow 1$) Take a plane $\mathcal{P}$ and $M, N \in \mathcal{P}$. Assume that $M \in \mathcal{E}_\mathcal{P}$, i.e. $M \to_\beta M' \notin \mathcal{P}$, while $N \notin \mathcal{E}_\mathcal{P}$. By Lemma 4.15, N is one-step recurrent and, by (2), is also recurrent. Therefore $N \twoheadrightarrow_\beta M \to_\beta M' \twoheadrightarrow_\beta N$, which contradicts the fact that $M' \notin \mathcal{P}$. □

THEOREM 4.20. *There exists a one-step recurrent term M which is not recurrent.*

PROOF. Such a λ-term M was defined by Hans Mulder as follows:

$$\begin{array}{rcll} M & \triangleq & HH(\lambda x.P(Px)), & \text{where:} \\ H & \triangleq & \lambda fg.ff(\lambda y.g(gy)), & \\ P & \triangleq & (\lambda y.\mathsf{I})z, & \text{for a fresh } z \in \mathrm{Var}. \end{array}$$

For the rest of the proof we fix the notation

$$P^\eta \triangleq \lambda x.Px, \text{ where } x \notin \mathrm{FV}(P).$$

Note that $P^\eta x \to_\beta Px$, while $P^\eta \not\twoheadrightarrow_\beta P$. First, we show that M is a one-step recurrent term. There are three cases to consider, one for each redex occurrence in M:

$$M = \underline{(\lambda fg.ff(\lambda y.g(gy)))H}_1(\lambda x.\underline{P}_2(\underline{P}_3 x))$$

Case 1, the head redex is contracted. Underlying the contracted redexes, we have:

$$\begin{array}{rcll} M & \to_\beta & (\lambda g.HH(\lambda y.g(gy)))(\lambda x.\underline{P}(Px)) & \\ & \to_\beta & (\lambda g.HH(\lambda y.g(gy)))(\lambda x.\underline{\mathsf{I}(Px)}), & \text{as } P \to_\beta \mathsf{I}, \\ & \to_\beta & \underline{(\lambda g.HH(\lambda y.g(gy)))P^\eta}, & \text{by def. of } P^\eta, \\ & \to_\beta & HH(\lambda y.\underline{P^\eta}(\underline{P^\eta y})) & \\ & \twoheadrightarrow_\beta & HH(\lambda y.P(Py)), & \text{as } P^\eta x \to_\beta Px, \\ & = & M & \end{array}$$

Cases 2 and 3 can be treated together since we have:

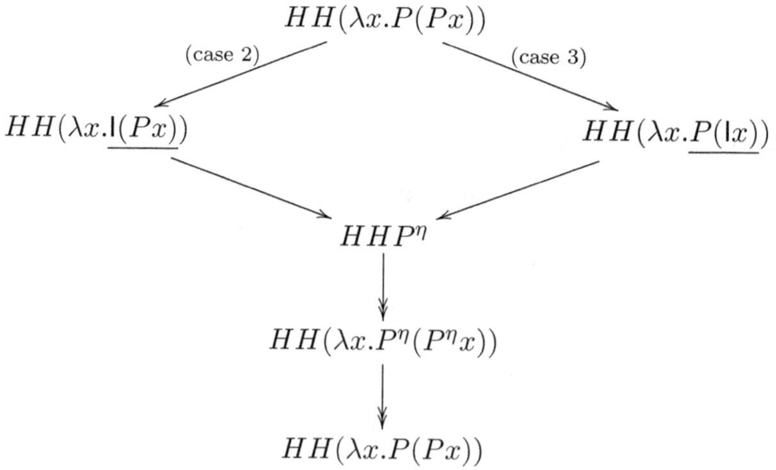

This proves that M is one-step recurrent. To show that M is not recurrent, it is enough to verify that

$$M = HH(\lambda x.\underline{P}(Px)) \to_\beta HH(\lambda x.\mathsf{I}(\underline{P}x)) \to_\beta HH(\lambda x.\mathsf{I}(\mathsf{I}x)).$$

Indeed $z \in \mathrm{FV}(M)$, while $\mathrm{FV}(HH(\lambda x.\mathsf{I}(\mathsf{I}x))) = \emptyset$ whence $HH(\lambda x.\mathsf{I}(\mathsf{I}x)) \not\twoheadrightarrow_\beta M$. □

By Proposition 4.19, this refutes Klop's conjecture.

REMARK 4.21. (i) The use of a free variable z in the proof of Theorem 4.20 is useful to prove that a certain reduction sequence is impossible, as λ-terms do not produce free variables along the reduction. However, it is not essential to obtain the result—the example constructed by Sekimoto and Hirokawa (1988) is a closed λ-term.

(ii) Hirokawa (1984) has proved that every one-step recurrent term having 2 redexes is also recurrent. Therefore, the example constructed by Mulder is minimal, in the sense that it contains a minimal amount of redexes. The same can be said for the example by Sekimoto and Hirokawa.

The question whether combinatory logic does satisfy the Plane Property is still open. As remarked by Mulder, a simple adaptation of his counterexample does not work because combinatory logic does not satisfy the ξ-rule. Indeed, from $P(P^n x) \to P(Px)$ one cannot infer $\lambda^* x.P(P^n x) \to \lambda^* x.P(Px)$, the former being a normal form.

CONJECTURE 4.22 (KLOP).
Combinatory logic satisfies the Plane Property.

A digression on hyper-recurrent terms

Statman proposed to push the notion of recurrent terms even further, by introducing a notion of 'hyper-recurrence' (Statman (1993a)). He proposed a similar strengthening for the concept of universal generator (U.G.), by adding a uniformity condition (see below). In both cases, the conditions imposed are so strong that it becomes unclear whether the resulting definitions are actually populated by some λ-term.

Recall from B[1984], Definition 8.2.7(ii), that M is a U.G. whenever, for every $N \in \Lambda^o$, there exists a context $C[\,]$ such that $M \twoheadrightarrow_\beta C[N]$.

DEFINITION 4.23. Let $M \in \Lambda^o$.

(i) M is *hyper-recurrent* if, for every $N \in \Lambda^o$, $N =_\beta M$ implies that N is recurrent.

(ii) M is a *uniform universal generator* if there exists a context $C[\,]$ such that

$$\forall N \in \Lambda^o \,.\, M \twoheadrightarrow_\beta C[N]$$

REMARK 4.24. (i) Clearly, every hyper-recurrent term must be unsolvable.

(ii) On the contrary, if M is a uniform U.G. then so is $\lambda x.xM$ which is solvable.

(iii) By (i) and (ii) above, the two notions do not coincide.

LEMMA 4.25. *Every hyper-recurrent λ-term M is a uniform universal generator.*

PROOF. For every $N \in \Lambda^o$, we have $\mathsf{K}MN \twoheadrightarrow_\beta M$ so, assuming that $\mathsf{K}MN$ is recurrent, we conclude $M \twoheadrightarrow_\beta C[N]$ for $C[\,] \triangleq \mathsf{K}M[\,]$. □

A similar argument shows that the condition 'N closed' in Definition 4.23(i) is crucial. Indeed, $\mathsf{K}Mx \twoheadrightarrow_\beta M$ but $M \twoheadrightarrow_\beta \mathsf{K}Mx$ is impossible since M cannot produce a free variable along reduction. The following characterization of hyper-recurrent terms has been given by Statman (1993a).

LEMMA 4.26. *For $M \in \Lambda^o$, the following are equivalent:*

1. *M is hyper-recurrent.*

2. *For all $P, Q \in \Lambda^o$, $P =_\beta M =_\beta Q \quad \Rightarrow \quad P \twoheadrightarrow_\beta Q$.*

3. *For all $N \in \Lambda^o$, $N \twoheadrightarrow_\beta M \quad \Rightarrow \quad M \twoheadrightarrow_\beta N$.*

The question whether hyper-recurrent terms actually exist, arises naturally when applying the Ershov-Visser theory to $\twoheadrightarrow_\beta$, rather than $=_\beta$ as done in Visser (1980). In his 1993 paper, Statman has proposed the following conjectures that still stand.

CONJECTURE 4.27 (STATMAN).
(i) There are no hyper-recurrent terms.
(ii) There are no uniform universal generators.

By Lemma 4.25, if no uniform U.G. exist then no hyper-recurrent terms can exist either. Statman (1991) proved that hyper-recurrent combinators do not exist in combinatory logic, however the two calculi are known to satisfy different properties as term rewriting systems. For instance, in combinatory logic based on $\{\mathsf{S}, \mathsf{K}, \mathsf{I}\}$ there are no pure cycles, as shown by Endrullis et al. (2016).

Chapter 5

Optimal lambda reduction

Given a λ-term $(\lambda x.M)N$ a reduction strategy may choose to evaluate the argument N first, and subsequently contract the head redex, as in

$$(\lambda x.M)N \twoheadrightarrow_\beta (\lambda x.M)N' \rightarrow_h M[x:=N'],$$

or to contract the head redex and then continue the reduction within the copies of N:

$$(\lambda x.M)N \rightarrow_h M[x:=N] \twoheadrightarrow_\beta M[x:=N'].$$

On the one hand, if there are many occurrences of x in M, the former strategy seems more efficient because the contraction of the head redex produces several copies of N that need to be reduced individually. On the other hand, the reduction strategy evaluating the argument N first cannot guarantee that no unnecessary work is performed, in general. As a worst case scenario, consider the case where N is unsolvable and $x \notin \mathrm{FV}(M)$: the contraction of the head redex erases the argument N, while the strategy reducing the argument first leads to a non-terminating reduction even when the term has a β-nf.

Duplication of work can be avoided in the strategy that contracts the head redex first by adopting a shared representation of the argument N. Indeed, the two strategies above become equally efficient in an implementation of β-reduction that maintains shared all the occurrences of the subterm N, and of its reducts. From a theoretical point of view, this corresponds to introduce a parallel reduction where each step simultaneously contracts the same redex in all the copies of the subterm. To define such a parallel reduction, we need a way to characterize those redexes arising from the duplication of a common redex. The main difficulty is that it is not sufficient to check whether two redexes are copies of a same redex to conclude that they are shareable. In order to characterize all the redexes that are potentially shareable, it is necessary to identify also those redexes having a term that will eventually be instantiated into a redex as their common origin.

In 1978, Lévy introduced a notion of *redex family* in order to formalize the intuitive idea of *optimal sharing* between 'copies' of the same redex. Consider a redex occurrence R in a λ-term N. There may be other redex occurrences R' in N, even some with $R' = R$. Such an R' may be related to the original R, or not, depending on how R and R' did appear in N, i.e. depending on the 'past' of N. We say that a reduction $\rho : M \twoheadrightarrow_\beta N$ is a 'history' of N, from the perspective of the initial term M. Within this history, it makes sense to define when two redex occurrences R and R' belong to the same family, and are then potentially shareable. This is the case if some Z exists along the history ρ having a redex occurrence R_Z (created or not) of which both R and R' are residuals. E.g., in Figure 5.1, the two occurrences of $\mathsf{I}y$ in $N \triangleq \mathsf{I}y(\mathsf{I}y)$ are related in the history of N from the perspective of $M \triangleq \Delta(\langle y\rangle\mathsf{I})$ via either $M; R_2; R_4$ or $M; R_1; R_3$ since they are both copies of R; but not from the perspective of $\langle y\rangle\mathsf{I}(\langle y\rangle\mathsf{I})$, neither via $R_2; R_4$ or $R_1; R_3$.

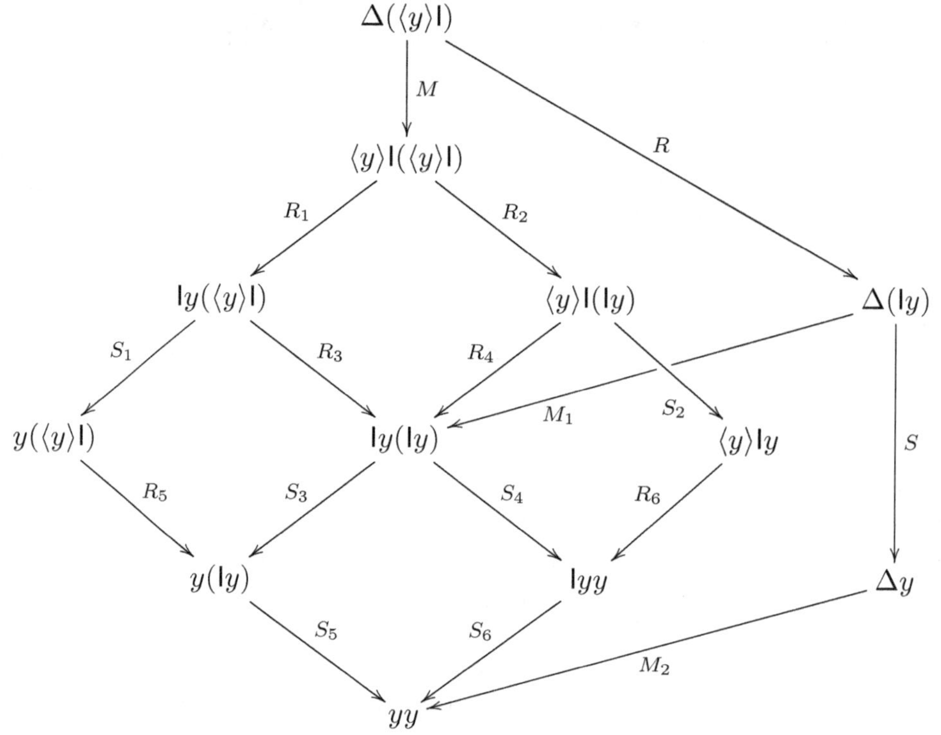

Figure 5.1: Define $M \triangleq \underline{\Delta(\langle y\rangle\mathsf{I})}$, $N \triangleq \mathsf{I}y(\mathsf{I}y)$, $R \triangleq \Delta(\underline{\langle y\rangle\mathsf{I}})$ and $S \triangleq \Delta(\underline{\mathsf{I}y})$.

In general, determining the kinship of redex occurrences is a complex matter that requires a fine-grained analysis of the dependencies in their creation histories. To fully explore this notion of family, Lévy proposed alternative (equivalent) definitions.

1. THE ZIG-ZAG RELATION. The most abstract approach to the notion of family is based on the so-called *zig-zag* relation. In this axiomatization, which is presented in Section 5.1, the duplication of redexes is formalized as residuals modulo permutation equivalence. More precisely, a redex R with history ρ represents a *copy* of a redex S with history ν when ν is a prefix of ρ, i.e. $\nu \precsim \rho$, and R is a residual of S across ρ/ν. Intuitively, this means that R and S are instances of a single redex at different stages of the evaluation started from the initial term of ρ and ν. The family relation $\simeq$ is then given by the symmetric and transitive closure of the copy-relation. Pictorially, this gives rise to a 'zig-zag' from which it takes its name.

2. THE EXTRACTION RELATION. Another approach, which is presented in Section 5.2, consists in considering the *causal history* of the redex. Intuitively, two redexes can be shared whenever they have been created 'in the same way' along the reduction. It is possible to capture this idea by defining a simplification process, called *extraction*, that eliminates from the history ρ of a redex R those redexes that are not relevant to its creation. The canonical form obtained at the end of this (terminating) process essentially expresses the causal dependencies of R along the reduction ρ. We actually obtain causal *chains* since only standard reductions are considered.

3. THE LABELED λ-CALCULUS. The most operational approach to capture the family relation is based on a labeled variant of the λ-calculus introduced in Lévy (1975), and presented in Section 5.3. Before the discovery of genome sequencing, finding out whether two humans are relatives did depend on knowing their ancestral history. With the possibility of genome sequencing one also can determine kinship, without having to know this ancestral history. Similarly for λ-terms, we can annotate their subterms with labels, to be suitably modified along a reduction, from which it can be read whether two redex occurrences R and R' are related. Unlike the case with the human genome, where some genes go back to our uni-cellular ancestors, for $R, R' \in N$ 'being related' according to their labels is relative to the history of N.

The equivalence between zig-zag and extraction is not particularly difficult (Lévy (1980)). The equivalence between extraction (or zig-zag) and labeling is much more convoluted and constitutes the technical core of Lévy's PhD thesis (1978b). In this chapter we only need, and prove, one side of this equivalence (Theorem 5.63).

Let $M \twoheadrightarrow_\beta N$. The following will be shown.

(i) Each equivalence class of the family relation has a unique canonical representative and extraction gives an effective procedure to find it (Theorem 5.24 and Corollaries 5.26 and 5.27). Therefore, kinship of redex occurrences from the perspective of M is decidable.

(ii) In the labeled λ-calculus, two redexes reduced by ρ are in the same family only if they have the same label (Theorem 5.63). We also prove that the labeled λ-calculus enjoys confluence (Theorem 5.54) and standardization (Theorems 5.55 and 5.57). Such results give an alternative proof of the corresponding properties for the unlabeled case.

(iii) A family reduction $\rho\colon M \twoheadrightarrow_\beta N$ is a reduction of the form

$$\rho\colon M = M_0 \twoheadrightarrow_\beta M_1 \twoheadrightarrow_\beta \cdots \twoheadrightarrow_\beta M_n = N,$$

where, for all $i\,(0 \leq i < n)$, $M_i \twoheadrightarrow_\beta M_{i+1}$ can be seen as a single step contracting in parallel a collection of all the redex occurrences in M_i in the same family from the perspective of M. The number n is called the *cost* of this family reduction.

(iv) Family reductions satisfy many nice properties, including a one-step diamond property implying that permutationally equivalent family reductions have the same cost. For call-by-need reductions, in which every reduced family contains at least a 'needed' redex, the cost of a normalizing reduction (if any) is independent from the strategy.

For the presentation of the material in this chapter, we mainly follow the reworking of Lévy (1980) made by Asperti and Guerrini (1998) in their book.

5.1 Families of redexes

Adopting Lévy's terminology, the set of redexes in the reducts of a λ-term M for which we can find a common origin form a *family*, from the perspective of the initial term M. Accordingly, in the example depicted in Figure 5.1, we identify three families constituted by those redexes having as origin M, R, and S, respectively. In this section families are described via a zig-zag relation, defined starting from the notion of a *copy* of a redex.

Redexes with history and the copy relation

As already mentioned, taking redex histories into account allows to distinguish among redexes corresponding to the same subterm, but that shall not belong to the same family. For instance $\mathsf{I}(\mathsf{I}x)$ has two redexes: the whole λ-term $R \triangleq \mathsf{I}(\mathsf{I}x)$ and the subterm $S \triangleq \mathsf{I}x$. Assuming that the initial term does not contain any shared redexes,[1] these two redexes are distinct, and therefore unrelated from the perspective of $\mathsf{I}(\mathsf{I}x)$, even if their contraction leads to the same λ-term $\mathsf{I}x$. After contracting S the redex $\mathsf{I}x$ becomes a residual R' of R, while after contracting R it is a residual S' of S. Since R and S are unrelated in $\mathsf{I}(\mathsf{I}x)$, one concludes that the resulting R' and S' are equal only coincidentally. In other words, we can establish that the redex $\mathsf{I}x$ with history R and the redex $\mathsf{I}x$ with history S are unrelated only once their histories have been adequately analyzed.

We also expect that the contractions of two independent redexes have no impact in the creation history of a redex and that, more generally, redex histories are consistent with permutation equivalence $\sim$. For instance, the two histories

$$\begin{array}{llllll} \rho_1: & \langle y\rangle(\mathsf{II}) & \to_\beta & \mathsf{II}y & \to_\beta & \mathsf{I}y \\ \rho_2: & \langle y\rangle(\mathsf{II}) & \to_\beta & \langle y\rangle\mathsf{I} & \to_\beta & \mathsf{I}y \end{array}$$

only differ by the order of the contraction of two distinct redexes, thus we have $\rho_1 \sim \rho_2$. In this respect, it is natural to assume that $\mathsf{I}y$ with history ρ_1 and $\mathsf{I}y$ with history ρ_2 indeed represent the 'same' redex.

The simplest situation where two redexes with history belong to the same family, is when one is a 'copy' of the other one. Intuitively, this means that both redexes are instances of a same redex at different stages of a reduction starting from the initial term.

DEFINITION 5.1. (i) A *redex with history* is a pair $\langle \rho, R\rangle$ where $\rho : M \twoheadrightarrow_\beta N$ and R is a redex of N. In this case, we also say that *R is a redex with history ρ*.

(ii) A redex with history $\langle \rho, R\rangle$ is a *copy* of a redex with history $\langle \nu, S\rangle$, in symbols $\langle \nu, S\rangle \preceq \langle \rho, R\rangle$, whenever

$$\nu \precsim \rho \ \ \& \ \ R \in S/(\rho/\nu)$$

Lemma 1.59 and Corollary 1.56 imply that a redex R with history ρ is a copy of a redex S with history ν if and only if there exists a reduction ν_0 such that $\nu\,;\nu_0 \sim \rho$ and R is a residual of S across ν_0 (written $R \in S/\nu_0$). This situation is depicted in Figure 5.2(a).

[1] This is an important assumption of the whole theory.

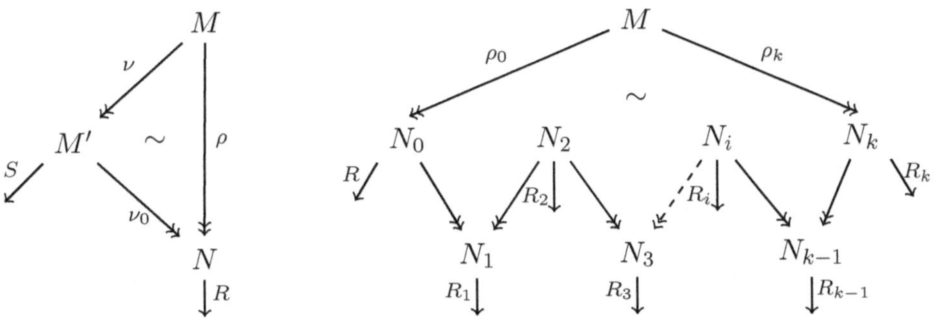

(a) Copy (b) Family (zig-zag)

Figure 5.2: Copy and family relations

EXAMPLES 5.2. (i) The λ-term $M = \mathsf{I}((\lambda y.\mathsf{I}x)z)$ has three redexes (that are underlined). Consider now the reductions

$$\begin{aligned} \rho: &\quad \mathsf{I}((\lambda y.\mathsf{I}x)z) \to_h (\lambda y.\mathsf{I}x)z \to_h \mathsf{I}x; \\ \nu: &\quad \mathsf{I}((\lambda y.\mathsf{I}x)z) \to_\beta \mathsf{I}((\lambda y.x)z) \to_\beta \mathsf{I}x. \end{aligned}$$

We notice that $\langle o, \mathsf{I}x\rangle \preceq \langle \rho, \mathsf{I}x\rangle$ and $\langle o, \mathsf{I}((\lambda y.\mathsf{I}x)z)\rangle \preceq \langle \nu, \mathsf{I}x\rangle$. This is an example where the redex histories allow disambiguation of the residual relation (Definition 1.40(iii)). Ignoring histories, we would have that $\mathsf{I}x$ is a residual of both the redex $\mathsf{I}x \in M$ and of the redex M. Considering histories, we see instead that these two residuals are unrelated.

(ii) The redexes with history generated by Ω are unrelated even if they all correspond to the same term. Indeed, although Ω reduces to itself by contracting its only redex, the redex in its contractum is newly created, not a residual of the redex in the initial term. E.g., for $o : \Omega$, $\rho : \Omega \to_\beta \Omega$ and $\nu : \Omega \to_\beta \Omega \to_\beta \Omega$, we have $o \not\preceq \rho \not\preceq \nu$, so the 'status' of each of these threes redexes Ω is different with respect to the corresponding histories.

EXAMPLES 5.3. Consider the following redexes with history from Figure 5.1. As already remarked, these three redexes are related and must belong to the same family.

history									*redex*
$o:$	$\Delta(\langle y\rangle\mathsf{I})$								R
$\nu:$	$\Delta(\langle y\rangle\mathsf{I})$	$\xrightarrow{M}_\beta$	$\langle y\rangle\mathsf{I}(\langle y\rangle\mathsf{I})$	$\xrightarrow{R_1}_\beta$	$\mathsf{I}y(\langle y\rangle\mathsf{I})$				R_3
$\rho:$	$\Delta(\langle y\rangle\mathsf{I})$	$\xrightarrow{M}_\beta$	$\langle y\rangle\mathsf{I}(\langle y\rangle\mathsf{I})$	$\xrightarrow{R_1}_\beta$	$\mathsf{I}y(\langle y\rangle\mathsf{I})$	$\xrightarrow{S_1}_\beta$	$\mathsf{I}y(\langle y\rangle\mathsf{I})$		R_5

(i) The redex R_3 with history ν is a residual of R with (the empty) history o across the reduction ν, and clearly $o; \nu \sim \nu$. Thus, we have $\langle o, R\rangle \preceq \langle \nu, R_3\rangle$.

(ii) Analogously, the redex R_5 with history ρ is a residual of R_3 across S_1 and $\nu; S_1 \sim \rho$. It follows that $\langle \nu, R_3\rangle \preceq \langle \rho, R_5\rangle$.

(iii) One can easily combine (i) and (ii), to conclude that $\langle o, R\rangle \preceq \langle \rho, R_5\rangle$.

Symbol	Relation	Definition
$\rho \sim \nu$	permutation equivalence	1.52
$\rho \precsim \nu$	prefix of	1.58
$\langle \rho, R \rangle \simeq \langle \nu, S \rangle$	family relation	5.8(i)
$\langle \rho, R \rangle \preceq \langle \nu, S \rangle$	copy of	5.1(ii)

Table 5.1: Relations on reductions and on redexes with history.

Example 5.3(iii) provides a first hint of the fact that the copy relation $\preceq$ is actually transitive. In general, the following holds.

LEMMA 5.4. *The relation $\preceq$ is a preorder.*

PROOF. (Reflexivity.) By definition of $\precsim$ and reflexivity of $\sim$, we obtain $\rho \precsim \rho$. Easy calculations give $R \in R/(\rho/\rho) = R/o = \{R\}$. Therefore, we conclude $\langle \rho, R \rangle \preceq \langle \rho, R \rangle$.

(Transitivity.) Assume that $\langle \rho_1, R_1 \rangle \preceq \langle \nu, S \rangle$ and $\langle \nu, S \rangle \preceq \langle \rho_2, R_2 \rangle$. From the former, we get $\rho_1 \precsim \nu$ and $S \in R_1/(\nu/\rho_1)$. From the latter, we derive $\nu \precsim \rho_2$ and $R_2 \in S/(\rho_2/\nu)$. By transitivity of $\precsim$, we obtain then $\rho_1 \precsim \rho_2$. Since $\rho_2 \sim \nu; (\rho_2/\nu) \sim \rho_1; (\nu/\rho_1); (\rho_2/\nu)$ holds, by the definition of residual and Lemmas 1.59 and 1.55, we conclude

$$R_2 \in (R_1/(\nu/\rho_1))/(\rho_2/\nu) = R_1/((\nu/\rho_1); (\rho_2/\nu)) = R_1/(\rho_2/\rho_1). \qquad \square$$

The copy relation $\preceq$ becomes even a partial order if one equates all the redexes with history $\langle \rho, R \rangle$ modulo permutation equivalence of ρ. In other words, as proved by the following lemma, $\preceq$ is antisymmetric on redexes with history equated modulo $\sim$.

LEMMA 5.5. *For all reductions ρ, ν and redexes R, S, the following are equivalent:*

1. $\langle \rho, R \rangle \preceq \langle \nu, S \rangle \;\; \& \;\; \langle \nu, S \rangle \preceq \langle \rho, R \rangle$;

2. $\rho \sim \nu \;\; \& \;\; R = S$.

PROOF. $(1 \Rightarrow 2)$ By definition, we have $\rho \precsim \nu$ and $\nu \precsim \rho$. By Lemma 1.59(vi), we obtain $\rho \sim \nu$. Therefore, R and S are redexes of the same term, and $R \in S/(\rho/\nu) = S/o = \{S\}$.

$(2 \Rightarrow 1)$ Trivial. $\square$

An immediate consequence of the previous lemma is that the copy relation $\preceq$ behaves well with respect to permutation equivalence. In fact, the copy relation is invariant under permutation equivalence of redex histories.

COROLLARY 5.6. *Assume $\rho \sim \rho'$ and $\nu \sim \nu'$. Then:*

$$\langle \rho, R \rangle \preceq \langle \nu, S \rangle \quad \Rightarrow \quad \langle \rho', R \rangle \preceq \langle \nu', S \rangle$$

PROOF. Assume $\langle \rho, R \rangle \preceq \langle \nu, S \rangle$. Then both $\langle \rho', R \rangle \preceq \langle \rho, R \rangle$ and $\langle \nu, S \rangle \preceq \langle \nu', S \rangle$ hold by Lemma 5.5. Conclude by transitivity. $\square$

The next lemma proves that, when $\langle\rho_1, R_1\rangle \preceq \langle\rho_2, R_2\rangle$, a sequence of residuals leading from R_1 to R_2 can be traced in a unique way in the reduction ν such that $\rho_1;\nu \precsim \rho_2$.

LEMMA 5.7. (i) (interpolation) *Let* $\langle\rho_1, R_1\rangle \preceq \langle\rho_2, R_2\rangle$. *For any reduction* ρ *satisfying* $\rho_1 \precsim \rho \precsim \rho_2$ *there exists a redex* R *such that*

$$\langle\rho_1, R_1\rangle \preceq \langle\rho, R\rangle \preceq \langle\rho_2, R_2\rangle$$

(ii) (uniqueness) $\langle\rho, R\rangle \preceq \langle\nu, S\rangle \;\&\; \langle\rho, R'\rangle \preceq \langle\nu, S\rangle \quad \Rightarrow \quad R = R'$.

PROOF. (i) By assumption, there exist reductions ν_1 and ν_2 such that $\rho_1;\nu_1 \sim \rho$ and $\rho_2 \sim \rho\,;\nu_2 \sim \rho_1\,;\nu_1\,;\nu_2$ and $R_2 \in R_1/(\rho_2/\rho_1)$. By Lemma 1.59(iv), we get $\nu_1\,;\nu_2 \sim \rho_2/\rho_1$, and $\nu_1 \sim \rho/\rho_1$, and $\nu_2 \sim \rho_2/\rho$. By Corollary 1.56, we obtain $R_2 \in R_1/(\nu_1\,;\nu_2)$. Then, since R_1 has a residual R_2 across $\nu_1\,;\nu_2$, there exists a residual R of R_1 across ν_1 such that R_2 is a residual of R across ν_2. In other words, there is $R \in R_1/\nu_1$ such that $R_2 \in R/\nu_2$. From this, we conclude that $\langle\rho_1, R_1\rangle \preceq \langle\rho, R\rangle \preceq \langle\rho_2, R_2\rangle$ holds.

(ii) Since a redex in the final term of a reduction has at most one ancestor in any intermediate term of the reduction. □

The family relation

We have seen that two redexes with history $\langle\rho, R\rangle \preceq \langle\nu, S\rangle$ belong to the same family since they actually represent instances of the same redex at different stages of a computation. More generally, it is natural to require that two redexes $\langle\nu, S'\rangle$ and $\langle\nu, S''\rangle$ of the same term are family-related, also when they are copies of a same ancestor, that is, when both $\langle\rho, R\rangle \preceq \langle\nu, S'\rangle$ and $\langle\rho, R\rangle \preceq \langle\nu, S''\rangle$ hold, for some $\langle\rho, R\rangle$.

Reexamining the example in Figure 5.1:

- from the perspective of M, the redexes R and $R_1, \ldots, R_6$ with the corresponding histories are in the same family, since they are all copies of $\langle o, R\rangle$.
- Analogously, the redexes M, M_1, and M_2 with the corresponding histories are in the same family since they are all residuals of $\langle o, M\rangle$.
- The case of the redexes S and $S_1, \ldots, S_6$ is subtler. Intuitively, all of these redexes have the same origin, since they come from the reduction of the subterm $\langle y\rangle \mathsf{I}$ of the initial term M, however they are not all residuals of a common ancestor. More precisely, the redexes S_3, S_4, S_5, and S_6 with the corresponding histories are copies of $\langle R, S\rangle$, while S_1 and S_2 arise in two distinct terms and have no common ancestor. A relation between S_1, S_2 and the other redexes S_i can be found if we look at the reducts of the terms in which S_1 and S_2 appear. In fact, S_3 is a residual of S_1 across R_3, while S_4 is a residual of S_2 across R_4. Therefore, even if they are not residuals of S, we can say that both S_1 and S_2 are indirectly related to it through their residuals, which are residuals of S as well. In other words, we need to assume that two redexes belong to the same families not only when they have a common family-related ancestor, but also when two of their residuals

have a common ancestor. Under this hypothesis, we can establish that the redexes $\langle M; R_1, S_1\rangle$ and $\langle M; R_2, S_2\rangle$ belong to the same family of the other redexes S_i and S with the corresponding histories.

The previous example leads to define the family relation as follows.

DEFINITION 5.8. (i) The *family* relation $\simeq$ is defined as the equivalence relation obtained from the symmetric and transitive closure of the copy relation. The equivalence class

$$\langle \rho, R\rangle_{\simeq} \triangleq \{\langle \nu, S\rangle \mid \langle \nu, S\rangle \simeq \langle \rho, R\rangle\}$$

is called the *family class* of $\langle \rho, R\rangle$.

(ii) We say that $\langle \rho, R\rangle$ and $\langle \nu, S\rangle$ *belong to the same family* if $\langle \rho, R\rangle \simeq \langle \nu, S\rangle$ holds.

REMARK 5.9. (i) By definition $\preceq \subseteq \simeq$, i.e. $\langle \nu, R\rangle \preceq \langle \nu, S\rangle$ entails $\langle \nu, R\rangle \simeq \langle \nu, S\rangle$.

(ii) A summary of the relations hitherto defined on reductions and on redexes with history is given in Table 5.1 (Page 122).

DEFINITION 5.10. Let $\langle \rho, R\rangle$ and $\langle \rho', R'\rangle$ be redexes with history and $k \in \mathbb{N}$. We say that the redexes with history $\langle \nu_i, S_i\rangle_{i \in \{0,\ldots,k\}}$ are *zig-zag witnesses of* $\langle \rho, R\rangle \simeq \langle \rho', R'\rangle$ when

$$\begin{aligned}
&\langle \rho, R\rangle = \langle \nu_0, S_0\rangle,\\
&\forall i\,(0 \le i < k)\,.\,\big[\big(\langle \nu_i, S_i\rangle \preceq \langle \nu_{i+1}, S_{i+1}\rangle\big) \vee \big(\langle \nu_{i+1}, S_{i+1}\rangle \preceq \langle \nu_i, S_i\rangle\big)\big],\\
&\langle \nu_k, S_k\rangle = \langle \rho', R'\rangle.
\end{aligned}$$

The family relation is sometimes called 'zig-zag' relation because of the following characterization (see also Figure 5.2(b)).

PROPOSITION 5.11. *For $\langle \rho, R\rangle$ and $\langle \nu, S\rangle$, the following are equivalent:*

1. $\langle \rho, R\rangle \simeq \langle \nu, S\rangle$;
2. *there exist $k \in \mathbb{N}$ and zig-zag witnesses $\langle \nu_i, S_i\rangle_{i \in \{0,\ldots,k\}}$ of $\langle \rho, R\rangle \simeq \langle \nu, S\rangle$.*

PROOF. $(1 \Rightarrow 2)$ Easy induction on a derivation of $\langle \rho, R\rangle \simeq \langle \nu, S\rangle$.

$(2 \Rightarrow 1)$ By transitivity and reflexivity of $\simeq$, using Remark 5.9(i). □

As the copy relation, also the family relation behaves well with respect to $\sim$.

COROLLARY 5.12. *Assume $\rho \sim \rho'$ and $\nu \sim \nu'$. Then*

$$\langle \rho, R\rangle \simeq \langle \nu, S\rangle \iff \langle \rho', R\rangle \simeq \langle \nu', S\rangle$$

PROOF. An immediate consequence of Corollary 5.6. □

5.2 Extraction

We now present another approach to the family relation, formalizing the intuition that redexes inessential to the creation of a redex can be safely eliminated from its history. More precisely, we shall define a simplification process of a redex with history $\langle \rho, R \rangle$, called *extraction*, erasing from its history ρ all the redexes that are not relevant to the creation of R. The result of this process is a new redex with history $\langle \nu, S \rangle$ in the same family of $\langle \rho, R \rangle$ and in which all the redexes contribute to the creation of S in some sense, thus $\langle \nu, S \rangle$ is a sort of *canonical representative* of R with respect to ρ.

Ideally, we would like that $\langle \rho_1, R_1 \rangle$ and $\langle \rho_2, R_2 \rangle$ belong to the same family whenever they have the same canonical representative. However, when several redexes participate in the creation of R_1 and R_2, their canonical representatives might differ by the order in which such redexes are contracted. In order to avoid working modulo the permutation equivalence, that would make it very difficult to compare histories in inductive reasoning, when studying canonical representatives we shall restrict to redexes whose histories are standard. Since every reduction has a unique permutationally equivalent standard reduction and the copy and family relations are compatible with permutation equivalence (Corollaries 5.6 and 5.12), this can be assumed without loss of generality.

We shall then obtain the following main results:

1. If ρ_1^s and ρ_2^s are the standard reductions $\sim$-equivalent to ρ_1 and ρ_2, respectively, then $\langle \rho_1, R_1 \rangle$ and $\langle \rho_2, R_2 \rangle$ are in the same family whenever $\langle \rho_1^s, R_1 \rangle$ and $\langle \rho_2^s, R_2 \rangle$ have the same canonical representative (Corollary 5.26).

2. Every family has a unique canonical representative, that can be effectively computed from any member of the family (Corollary 5.28).

Before entering into the technical details of extraction, let us remark again that all this machinery is required to handle those redexes that are created along the reduction. The case where the redexes already occur in the initial term M is particularly easy, as being in the same family of a redex R in M is the same as being a residual of R or, equivalently, a copy of $\langle o, R \rangle$.

LEMMA 5.13. $\langle o, R \rangle \simeq \langle \nu, S \rangle \iff S \in R/\nu$.

PROOF. ($\Rightarrow$) Assume $\langle o, R \rangle \simeq \langle \nu, S \rangle$. By Proposition 5.11, there exist $k \in \mathbb{N}$ and zig-zag witnesses $\langle \nu_i, S_i \rangle_{i \in \{0,\ldots,k\}}$ of $\langle o, R \rangle \simeq \langle \nu, S \rangle$. That is, $\langle o, R \rangle = \langle \nu_0, S_0 \rangle$, $\langle \nu_k, S_k \rangle = \langle \nu, S \rangle$ and either $\langle \nu_i, S_i \rangle \preceq \langle \nu_{i+1}, S_{i+1} \rangle$ or $\langle \nu_{i+1}, S_{i+1} \rangle \preceq \langle \nu_i, S_i \rangle$ holds, for every i $(0 \leq i < k)$. Therefore, we may proceed by induction on k.

Base case: $k = 0$. Then $\nu = o$ and $R = S$, so we trivially have $S \in \{R\} = R/o = R/\nu$.

Induction case: $k > 0$. By IH we have $S_{k-1} \in R/\nu_{k-1}$, whence $\langle o, R \rangle \preceq \langle \nu_{k-1}, S_{k-1} \rangle$. There are two possibilities:

1. $\langle \nu_{k-1}, S_{k-1} \rangle \preceq \langle \nu, S \rangle$. By Definition 5.1(ii), we have $S \in S_{k-1}/(\nu/\nu_{k-1})$ and $\nu \sim \nu_{k-1}\,; (\nu/\nu_{k-1})$. Therefore, $S \in (R/\nu_{k-1})/(\nu/\nu_{k-1}) = R/(\nu_{k-1}\,; (\nu/\nu_{k-1})) = R/\nu$.

2. $\langle \nu, S\rangle \preceq \langle \nu_{k-1}, S_{k-1}\rangle$. From Definition 5.1(ii), we get $\nu \precsim \nu_{k-1}$. Therefore, since $o \precsim \nu \precsim \nu_{k-1}$, we can apply the interpolation property (Lemma 5.7(i)) and obtain an S' such that $\langle o, R\rangle \preceq \langle \nu, S'\rangle \preceq \langle \nu_{k-1}, S_{k-1}\rangle$. From the uniqueness property (Lemma 5.7(ii)), we get $S = S'$. We conclude that $S \in R/(\nu/o) = R/\nu$.

($\Leftarrow$) Assume $S \in R/\nu$. By definition, we have $\nu \sim o;\nu$, whence $\langle o, R\rangle \preceq \langle \nu, S\rangle$ holds. Conclude since $\preceq \subseteq \simeq$. □

Parallelization

The key case for the extraction relation that we will define in the next subsection is the one of the redexes S_1 and S_2 in Figure 5.1, already discussed on page 123. These two redexes, with the corresponding histories, are not copies of a common ancestor, nevertheless the intuition leads to find in the redex S a common representative. In fact, by analyzing $\langle M; R_1, S_1\rangle$ and $\langle M; R_2, S_2\rangle$, one can see that the redex M duplicates the subterm $\langle y\rangle \mathsf{I}$ before the creation of the redex $\mathsf{I}y$, which will then appear independently into the two copies of $\langle y\rangle \mathsf{I}$. So, we might have created instances of S_1 and S_2 by directly contracting the redex R in the initial term. In some sense, M is not relevant to the creation of S_1 and S_2, which depends instead on the contraction of R_1 and R_2, and should not be included in their casual history. By extracting this redex from their histories, both $\langle M; R_1, S_1\rangle$ and $\langle M; R_2, S_2\rangle$ should then become $\langle R, S\rangle$. Since $\langle M; R_1, S_1\rangle \preceq \langle R; M_1, S_3\rangle$ and $\langle M; R_2, S_2\rangle \preceq \langle R; M_1, S_4\rangle$, such a simplification process is compatible with the notion of family, as $\langle R, S\rangle \preceq \langle R; M_1, S_3\rangle$ and $\langle R, S\rangle \preceq \langle R; M_1, S_4\rangle$ also hold.

This particular case[2] corresponds to the more general situation where we reduce inside a copy of the argument of a β-redex. We notice that, under the assumption that the copies of the argument N of a contracted β-redex $R = (\lambda x.M)N$ are kept shared, contracting R corresponds to reducing all instances of N in parallel. Indeed, in this shared setting, the redexes S_1 and S_2 of the previous example are indistinguishable, since they have the same representation. The operation of parallelization that we are going to introduce (Definition 5.16) formally captures this phenomenon. In order to define it properly, we first need to fix some notations to indicate the i-th copy of N in $M[x:=N]$, and some terminology to express the fact that a reduction takes place within this subterm.

DEFINITION 5.14. Let $M \in \Lambda$ and $n \geq 0$ be the number of free occurrences of x in M. For $i \in \{1, \ldots, n\}$, call the *i-th occurrence of x in M* the one identified by the index i, counting from left to right.

(i) The *i-th occurrence* of N in $M[x:=N]$ is the one replacing the i-th occurrence of x in M.

(ii) We denote by $M^x[i := N' \mid N]$ the λ-term obtained by replacing the i-th occurrence of x in M by $N' \in \Lambda$ and all other occurrences by N.

(iii) Consider a subterm N of M, i.e. $M = C[N]$, for some single-hole context $C[\,]$. A reduction $\rho \in \mathcal{G}_\beta(M)$ is *internal to N* when $\rho : C[N] \twoheadrightarrow_\beta C[N']$ by contracting redexes in N and in its reducts only. The empty reduction o is trivially internal to N, and every redex R such that $\rho : C[N] \twoheadrightarrow_\beta C[N'''] \xrightarrow{R}_\beta C[N''] \twoheadrightarrow_\beta C[N']$ is a redex of N'''.

[2]In the example: the reductions $R_1; S_1$ and $R_2; S_2$ in the argument $\langle y\rangle \mathsf{I}$ of the β-redex M.

(iv) Two coinitial reductions $\rho_1, \rho_2 \in \mathcal{G}_\beta(M)$ are called *disjoint* if they are internal to two disjoint subterms N_1 and N_2 of M (equivalently, if there exists a binary context $C[-_1, -_2]$ such that $M = C[N_1, N_2]$, where ρ_1 is internal to N_1 and ρ_2 is internal to N_2).

It is readily seen that, when $\rho : C[N] \xrightarrow{\rho_1}\!\!\!\twoheadrightarrow_\beta C[N''] \xrightarrow{\rho_2}\!\!\!\twoheadrightarrow_\beta C[N']$ is internal to N, then ρ_1 is internal to N, and ρ_2 is internal to N''.

REMARK 5.15. Consider a redex $R = (\lambda x.M)N$ and a finite reduction ρ internal to the i-th occurrence of N in $M[x:=N]$. That is,

$$R = (\lambda x.M)N \xrightarrow{R}_\beta M^x[i := N \mid N] \xrightarrow{\rho}\!\!\!\twoheadrightarrow_\beta M^x[i := N' \mid N]$$

The following facts are noteworthy.

(i) There exists a reduction $\rho' : N \twoheadrightarrow_\beta N'$ isomorphic to ρ, obtained by applying the contractions in ρ to the λ-term N itself.

(ii) There is a reduction $\rho \Uparrow R : (\lambda x.M)N \twoheadrightarrow_\beta (\lambda x.M)N'$ isomorphic to ρ (and ρ'), which applies the contractions in ρ to the argument N of the redex R.

(iii) For every j $(1 \le j \le n)$, there exists a reduction

$$\rho_j : M^x[j := N \mid N] \twoheadrightarrow_\beta M^x[j := N' \mid N]$$

which applies the contractions in ρ to the j-th occurrence of N in $M[x:=N]$.

(iv) Permuting the reduction of $\rho \Uparrow R$ and the contraction of the redex R leads to a reduction

$$R \sqcup (\rho \Uparrow R) \sim (\rho \Uparrow R) \sqcup R : (\lambda x.M)N \twoheadrightarrow_\beta M[x:=N'],$$

in which all the occurrences of the argument N of R have been reduced to N', as in $R; (\rho_1 \sqcup \cdots \sqcup \rho_n)$.

(v) The reduction $(\rho \Uparrow R)/R$ will be denoted by $\rho \parallel R$ and called the *parallelization of ρ with respect to R*, since it corresponds to applying in parallel the isomorphic reductions $\rho_1 \ldots, \rho_n$ of the n occurrences of N in $M[x:=N]$, that is, $\rho \parallel R \sim \rho_1 \sqcup \cdots \sqcup \rho_n$.

Since the reduction $\rho \Uparrow R$ (and parallelization $\rho \parallel R$) plays a key role in the definition of the extraction relation and in the proofs of its properties, let us give an inductive definition of it.

DEFINITION 5.16 (PARALLELIZATION). Consider a redex $R = (\lambda x.M)N$.

(i) Let $S : M[x:=N] \to_\beta M^x[i := N'' \mid N]$ be a redex in the i-th occurrence of N in $M[x:=N]$. Denote by $S \Uparrow R$ the unique redex inside the subterm N of R such that

$$S \Uparrow R : R \to_\beta (\lambda x.M)N'' \ \&\ S \in (S \Uparrow R)/R$$

(ii) If $\rho : M[x:=N] \twoheadrightarrow_\beta M^x[i := N' \mid N]$ is a reduction internal to the i-th occurrence of N in $M[x:=N]$, then the reduction

$$\rho \Uparrow R : (\lambda x.M)N \twoheadrightarrow_\beta (\lambda x.M)N'$$

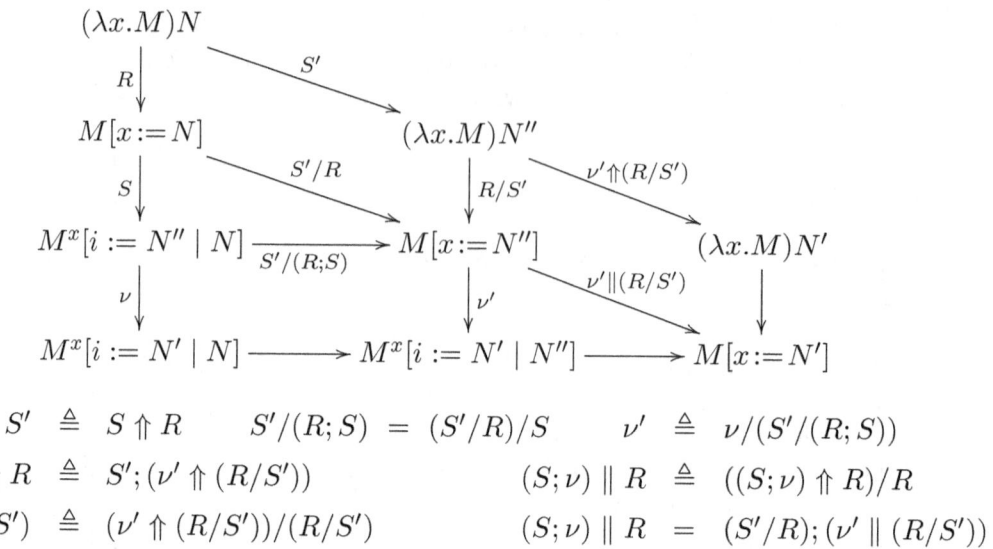

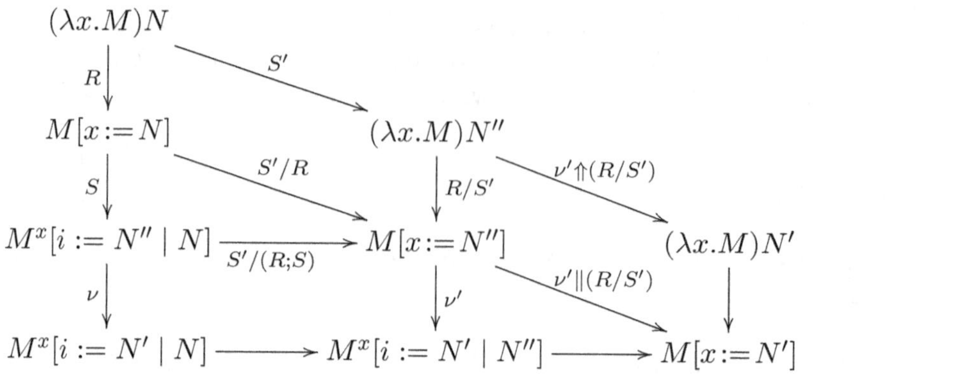

$$
\begin{array}{rclrclrcl}
S' & \triangleq & S \Uparrow R & S'/(R;S) & = & (S'/R)/S & \nu' & \triangleq & \nu/(S'/(R;S)) \\
(S;\nu) \Uparrow R & \triangleq & S';(\nu' \Uparrow (R/S')) & & & (S;\nu) \parallel R & & \triangleq & ((S;\nu) \Uparrow R)/R \\
\nu' \parallel (R/S') & \triangleq & (\nu' \Uparrow (R/S'))/(R/S') & & & (S;\nu) \parallel R & & = & (S'/R);(\nu' \parallel (R/S'))
\end{array}
$$

Figure 5.3: Inductive case of parallelization.

is defined by induction on ρ as follows:

$$
\begin{array}{rcl}
o \Uparrow R & \triangleq & o; \\
(S;\nu) \Uparrow R & \triangleq & S';(\nu' \Uparrow (R/S')), \qquad \text{with } S' = S \Uparrow R \ \& \ \nu' = \nu/(S'/(R;S)).
\end{array}
$$

and the reduction ρ *parallelized by* R, in symbols $\rho \parallel R$, is defined by:

$$\rho \parallel R \triangleq (\rho \Uparrow R)/R$$

The inductive case ($\rho = S;\nu$) of the previous definition is depicted in Figure 5.3. From the figure, it is readily seen that $\rho \Uparrow R$ is well defined. Moreover, we also see that

$$\rho \parallel R : M[x{:=}N] \twoheadrightarrow_\beta M[x{:=}N']$$

Indeed, by definition, we have:

$$R;(\rho \parallel R) = R;((\rho \Uparrow R)/R) = R \sqcup (\rho \Uparrow R) \sim (\rho \Uparrow R) \sqcup R$$

In Lévy (1978b; 1980), Jean-Jacques Lévy provides directly an inductive definition of $\rho \parallel R$ without explicitly introducing the reduction $\rho \Uparrow R$. It is readily seen that the two definitions of parallelization are equivalent. In fact, for the base case, $(o \parallel R) = o$ trivially holds, while for the induction case, from Figure 5.3, we see that

$$(S;\nu) \parallel R = (S'/R);(\nu' \parallel (R/S')).$$

LEMMA 5.17. *Let* $(\nu;S)$ *be a reduction internal to the* i*-th occurrence of* N *in the contractum* $M[x{:=}N]$ *of* $R = (\lambda x.M)N$*. The set of residuals of* R *after* $R/(\nu \Uparrow R)$ *and the set of residuals of* S *after* $(\nu \Uparrow R)/(R;\nu)$ *are both singletons, and*

$$(\nu;S) \Uparrow R = (\nu \Uparrow R);(S' \Uparrow R'),$$

with $R' \triangleq R/(\nu \Uparrow R)$ *and* $S' \triangleq S/((\nu \Uparrow R)/(R;\nu))$.

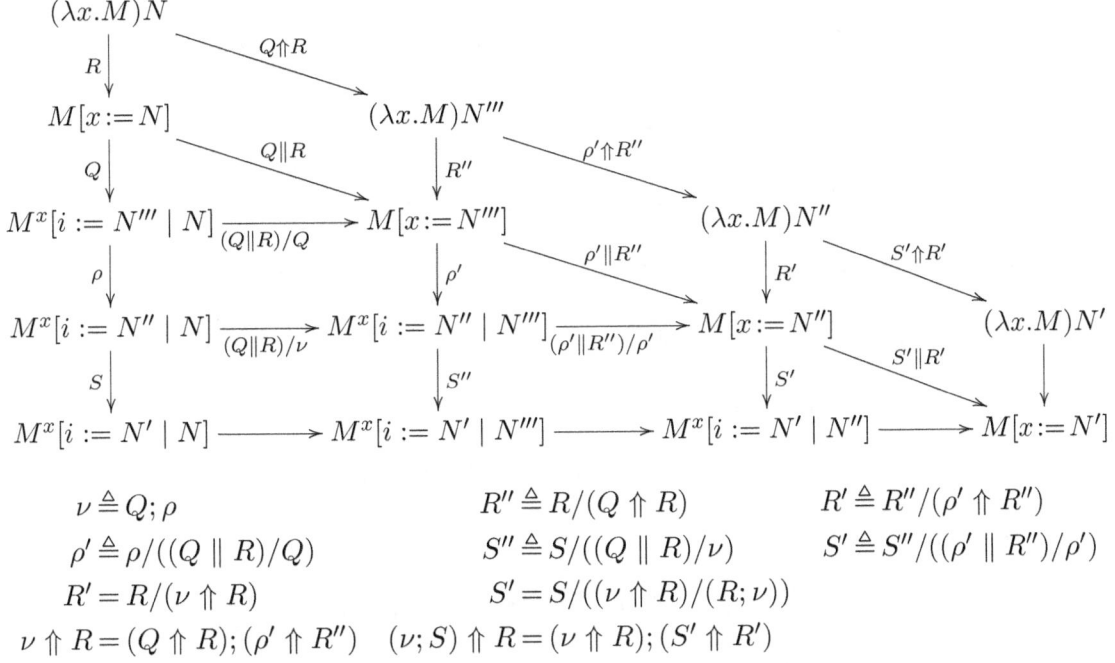

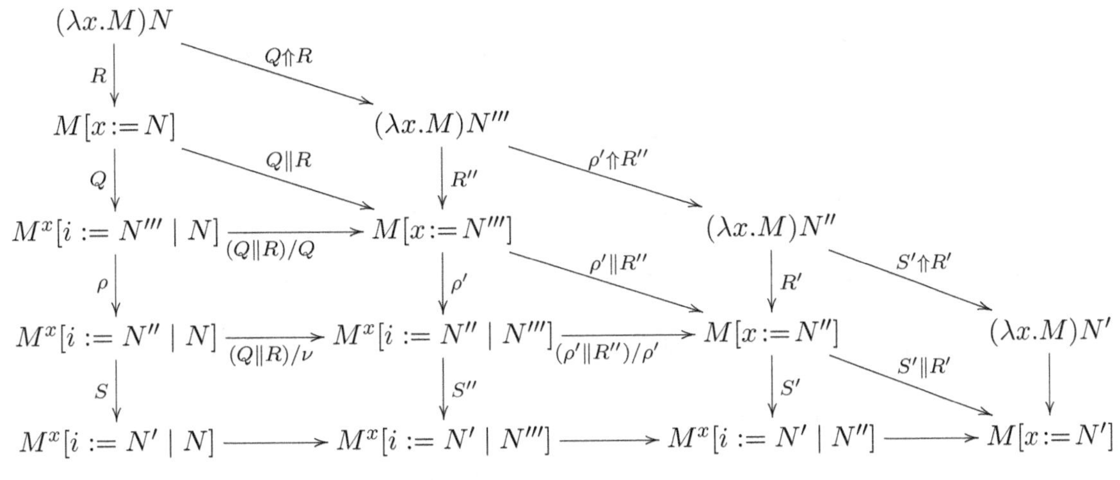

$$\nu \triangleq Q;\rho \qquad R'' \triangleq R/(Q \Uparrow R) \qquad R' \triangleq R''/(\rho' \Uparrow R'')$$

$$\rho' \triangleq \rho/((Q \parallel R)/Q) \qquad S'' \triangleq S/((Q \parallel R)/\nu) \qquad S' \triangleq S''/((\rho' \parallel R'')/\rho')$$

$$R' = R/(\nu \Uparrow R) \qquad S' = S/((\nu \Uparrow R)/(R;\nu))$$

$$\nu \Uparrow R = (Q \Uparrow R);(\rho' \Uparrow R'') \quad (\nu;S) \Uparrow R = (\nu \Uparrow R);(S' \Uparrow R')$$

Figure 5.4: Induction case of Lemma 5.17.

PROOF. Proceed by induction on ν.

The case $\nu = o$ is trivial, since this implies $\nu \Uparrow R = o$, and $(\nu \Uparrow R)/(R;\nu) = o$. Thus, $R' = R$ and $S' = S$, and the main equivalence reduces to the identity $R \Uparrow S = R \Uparrow S$.

For the induction case $\nu = Q;\rho$, we refer to Figure 5.4. By Definition 5.16, we have

$$(\nu;S) \Uparrow R = (Q;\rho;S) \Uparrow R = (Q \Uparrow R);(\rho'' \Uparrow R''),$$

with

$$R'' \triangleq R/(Q \Uparrow R) \qquad \& \qquad \rho'' \triangleq (\rho;S)/((Q \parallel R)/Q) = \rho';S'',$$

where

$$\rho' \triangleq \rho/((Q \parallel R)/Q) \qquad \& \qquad S'' \triangleq S/((Q \parallel R)/\nu) = S/(((Q \parallel R)/Q)/\rho).$$

Moreover, S'' is a singleton, since S is internal to a reduct of the i-th occurrence of N in $M[x:=N]$, while$(\nu \Uparrow R)/(R;\nu)$ is internal to the other occurrences of N. By induction hypothesis (noting that ρ' has the same length as ρ), we then have:

$$\rho'' \Uparrow R'' = (\rho' \Uparrow R'');(S' \Uparrow R'),$$

with the singletons (again by induction hypothesis)

$$R' \triangleq R''/(\rho' \Uparrow R'') = (R/(Q \Uparrow R))/(\rho' \Uparrow R'') = R/((Q \Uparrow R);(\rho' \Uparrow R'')),$$

and

$$\begin{aligned} S' \triangleq S''/((\rho' \parallel R'')/\rho') = & \ (S/((Q \parallel R)/\nu))/((\rho' \parallel R'')/\rho') \\ = & \ S/(((Q \parallel R)/\nu);((\rho' \parallel R'')/\rho')). \end{aligned}$$

After some calculations (or by analyzing the diagram in Figure 5.4), we can see that:

$$((Q \parallel R)/\nu);((\rho' \parallel R'')/\rho') \sim ((Q \Uparrow R);(\rho' \Uparrow R''))/(R;\nu)$$

and S' is a singleton (again by induction hypothesis). Then, by Lemma 1.55,

$$S' = S/(((Q \parallel R)/\nu);((\rho' \parallel R'')/\rho')) = S/((Q \Uparrow R);(\rho' \Uparrow R''))/(R;\nu).$$

Finally, since by Definition 5.16,

$$\nu \Uparrow R = (Q;\rho) \Uparrow R = (Q \Uparrow R);(\rho' \Uparrow R'')$$

we can conclude that:

$$\begin{aligned} R' &= R/((Q \Uparrow R);(\rho' \Uparrow R'')) = R/(\nu \Uparrow R), \\ S' &= S/(((Q \Uparrow R);(\rho' \Uparrow R''))/(R;\nu)) = S/((\nu \Uparrow R)/(R;\nu)), \\ (\nu;S) \Uparrow R &= (Q \Uparrow R);(\rho'' \Uparrow R'') = (Q \Uparrow R);(\rho' \Uparrow R'');(S' \Uparrow R') \\ &= (\nu \Uparrow R);(S' \Uparrow R'). \end{aligned}$$

□

The extraction relation

We now have all the ingredients needed to define the extraction relation.

DEFINITION 5.18 (EXTRACTION). (i) The *contraction by extraction* is the relation $\rhd$ obtained by the union of the following four contraction relations:

(1) $\langle \rho; R, S \rangle \rhd_1 \langle \rho, S' \rangle$, if $S \in S'/R$;

(2) $\langle \rho; (R \sqcup \nu), S \rangle \rhd_2 \langle \rho; \nu, S' \rangle$, if ν is not empty, R and $\nu; S'$ are disjoint reductions, and $S \in S'/(R/\nu)$. In a diagram:

$$\begin{array}{ccccccc} \xrightarrow{\rho} C[M,N] & \xrightarrow{\nu} & C[M,N'] & \xrightarrow{S'} & C[M,N''] \\ \downarrow R & & \downarrow R/\nu & & \\ C[M',N] & \xrightarrow{\nu/R} & C[M',N'] & \xrightarrow{S} & C[M',N''] \end{array}$$

(3) $\langle \rho; (R \sqcup \nu), S \rangle \rhd_3 \langle \rho; \nu, S' \rangle$, if ν is not empty and $\nu; S'$ is internal to the function part M of $R = (\lambda x.M)N$, and $S \in S'/(R/\nu)$. In a diagram:

$$\begin{array}{ccccc} \xrightarrow{\rho} C[(\lambda x.M)N] & \xrightarrow{\nu} & C[(\lambda x.M')N] & \xrightarrow{S'} & C[(\lambda x.M'')N] \\ \downarrow R & & \downarrow R/\nu & & \\ C[M[x:=N]] & \xrightarrow{\nu/R} & C[M'[x:=N]] & \xrightarrow{S} & C[M''[x:=N]] \end{array}$$

(4) $\langle \rho; R; \nu, S \rangle \triangleright_4 \langle \rho; (\nu \Uparrow R), (S/\nu') \Uparrow R' \rangle$, where $\nu' = (\nu \Uparrow R)/(R;\nu)$ and $R' = R/(\nu \Uparrow R)$, if ν is not empty and $\nu; S$ is internal to the i-th occurrence of the argument of $R = (\lambda x.M)N$ in its contractum $M[x:=N]$. In a diagram:

$$
\begin{array}{ccccc}
\xrightarrow{\rho} C[(\lambda x.M)N] & \xrightarrow{\nu \Uparrow R} & C[(\lambda x.M)N'] & \xrightarrow{(S/\nu') \Uparrow R'} & C[(\lambda x.M)N''] \\
\downarrow{\scriptstyle R} & & \downarrow{\scriptstyle R'=R/(\nu \Uparrow R)} & & \\
C[M[x:=N]] & \xrightarrow{\nu \| R = (\nu \Uparrow R)/R} & C[M[x:=N']] & \xrightarrow{S/\nu'} & C[M^x[i := N'' \mid N']] \\
\downarrow{\scriptstyle \nu} & \nearrow {\scriptstyle \nu'=(\nu \Uparrow R)/(R;\nu)} & & & \\
C[M^x[i := N' \mid N]] & & & & \\
\downarrow{\scriptstyle S} & & & & \\
C[M^x[i := N'' \mid N]] & & & &
\end{array}
$$

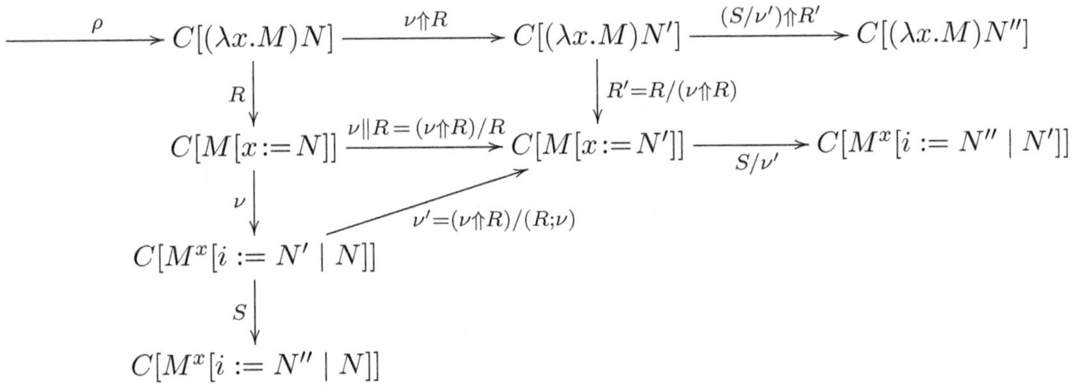

(ii) For $i \in \{1, 2, 3, 4\}$, the relation $\trianglerighteq_i$ is the transitive and reflexive closure of the corresponding relation $\triangleright_i$.

(iii) The *extraction* relation $\trianglerighteq$ is the transitive and reflexive closure of $\triangleright$.

Note that Lemma 5.17 is needed to ensure that the case $\triangleright_4$ of the previous definition is well defined. In particular, it ensures also that, when $\langle \rho; R; \nu, S \rangle \trianglerighteq_4 \langle \rho; \nu'', S' \rangle$, then $\nu''; S' = (\nu; S) \Uparrow R$.

THEOREM 5.19. *The extraction $\trianglerighteq$ is confluent and strongly normalizing.*

PROOF. See Lévy (1980). □

Extraction and families

In Theorem 5.24 we establish the equivalence, for redexes with standard history, between the approaches based on zig-zag and on extraction. We also prove that the extraction process is effectively given and terminating, thus kinship of redexes is decidable.

In most of the theorems that follow we have to restrict to redexes with standard histories. We start then by showing that contraction by extraction preserves the standardness of reductions—a property that will be silently used in the proof of Lemma 5.23.

LEMMA 5.20. *If $\langle \rho, R \rangle$ is a redex with standard history and $\langle \rho, R \rangle \trianglerighteq \langle \rho', R' \rangle$, then the history ρ' is standard too.*

PROOF. By analysis of the contraction rules in Definition 5.18. □

The goal is to prove that two redexes with standard history are in the same family if and only if the extraction reduces them to some common reduct (Theorem 5.24). Since we have just seen that extraction is strongly normalizing and confluent, every redex with history has a unique $\trianglerighteq$-normal form. This implies the decidability of the family relation for redexes with standard history, which, by standardization, can then be extended to all the redexes with history (Corollary 5.27).

The first and easy step is to prove that extraction reduces a redex with history to another redex with history in the same family.

LEMMA 5.21. *Let ρ_1 and ρ_2 be standard reductions.*

$$\langle \rho_1, R_1 \rangle \trianglerighteq \langle \rho_2, R_2 \rangle \quad \Rightarrow \quad \langle \rho_1, R_1 \rangle \simeq \langle \rho_2, R_2 \rangle$$

PROOF. We proceed by induction on a derivation of $\langle \rho_1, R_1 \rangle \trianglerighteq \langle \rho_2, R_2 \rangle$.

Case $\langle \rho_1, R_1 \rangle \triangleright \langle \rho_2, R_2 \rangle$. In the first three cases of Definition 5.18(i), we obviously have $\langle \rho_2, R_2 \rangle \preceq \langle \rho_1, R_1 \rangle$, which implies $\langle \rho_1, R_1 \rangle \simeq \langle \rho_2, R_2 \rangle$. In the last case, there exists a redex S satisfying $\rho_1 = \rho_1'; S; \rho_1''$ and $\rho_2 = \rho_1'; \nu$, for some ρ_1' and ρ_1'' such that $\rho_1''; R_1$ is internal to the i-th occurrence of N, for some i, and $\nu = \rho_1'' \Uparrow S$. In diagrammatic form:

$$\begin{array}{ccccc}
\xrightarrow{\rho_1'} C[(\lambda x.M)N] & \xrightarrow{\nu=\rho_1''\Uparrow S} & C[(\lambda x.M)N'] & \xrightarrow{R_2} & C[(\lambda x.M)N''] \\
S \downarrow & & \downarrow S/\nu & & \\
C[M[x{:=}N]] & \xrightarrow{\nu/S \,=\, \rho_1''\|S} & C[M[x{:=}N']] & & \\
\rho_1'' \downarrow & & \| & & \\
C[M^x[i := N' \mid N]] & \xrightarrow{\nu'=\nu/(S;\rho_1'')} & C[M^x[i := N' \mid N']] & & \\
R_1 \downarrow & & \downarrow R' & & \\
C[M^x[i := N'' \mid N]] & & C[M^x[i := N'' \mid N']] & &
\end{array}$$

Calling $\nu' \triangleq \nu/(S;\rho_1'')$ we obtain $\rho_1''; \nu' \sim \nu/S$, and thus $S;\, \rho_1'';\, \nu' \sim (S \sqcup \nu) \sim (\nu \sqcup S)$. Remark now that R_1 has a unique residual R' after ν', namely $R' \triangleq R_1/\nu'$, which is indeed a residual of R_2 after S/ν too, that is, $R' \in R_2/(S/\nu)$. In fact, the redex R_2 is internal to N' in $C[(\lambda x.M)N']$, and R' is the image of R_2 in the i-th copy of N' in $C[M[x{:=}N']]$, after reducing S/ν. Therefore, we have that $\langle \rho_1, R_1 \rangle \preceq \langle \rho_1; \nu', R' \rangle = \langle \rho_1'; S; \rho_1''; \nu', R' \rangle \preceq \langle \rho_1'; (\nu \sqcup S), R' \rangle$, since $S; \rho_1''; \nu' \sim (\nu \sqcup S)$, and that $\langle \rho_2, R_2 \rangle = \langle \rho_1'; \nu, R_2 \rangle \preceq \langle \rho_1'; (\nu \sqcup S), R' \rangle$. Hence, $\langle \rho_1, R_1 \rangle \simeq \langle \rho_1'; (\nu \sqcup S), R' \rangle \simeq \langle \rho_2, R_2 \rangle$. Conclude by transitivity.

Case $\rho_1 = \rho_2$ and $R_1 = R_2$ (reflexivity). Immediate.

Case $\langle \rho_1, R_1 \rangle \trianglerighteq \langle \nu, S \rangle$ and $\langle \nu, S \rangle \trianglerighteq \langle \rho_2, R_2 \rangle$ (transitivity). It follows from the IH and the transitivity of $\simeq$. □

The proof of the converse property, namely that every pair of redexes in the same family can be reduced by extraction to the same redex with history, is more involved. Moreover, it is an argument that requires to restrict to redexes with *standard* history. But first, let us see the base case in which the family contains a redex in the initial term.

LEMMA 5.22. *Let ρ be a standard reduction.*

$$\langle o, R_1 \rangle \simeq \langle \rho, R_2 \rangle \quad \Rightarrow \quad \langle \rho, R_2 \rangle \trianglerighteq \langle o, R_1 \rangle$$

PROOF. By Lemma 5.13, we have $R_2 \in R_1/\rho$. An easy induction on the length of ρ using interpolation (Lemma 5.7(i)) establishes that $\langle \rho, R_2 \rangle \trianglerighteq_1 \langle o, R_1 \rangle$. □

Let us then prove the case in which $\langle\rho_2, R_2\rangle$ is a copy of $\langle\rho_1, R_1\rangle$.

LEMMA 5.23. *Let $\langle\rho_1, R_1\rangle$ and $\langle\rho_2, R_2\rangle$ be two redexes with standard history such that $\langle\rho_1, R_1\rangle \preceq \langle\rho_2, R_2\rangle$. There exists a redex $\langle\rho, R\rangle$ with standard history such that $\langle\rho_1, R_1\rangle \trianglerighteq \langle\rho, R\rangle$ and $\langle\rho_2, R_2\rangle \trianglerighteq \langle\rho, R\rangle$.*

PROOF. First, let us remark that if $\rho_1 = o$ then, by applying Lemma 5.22, the statement holds with $\langle\rho, R\rangle = \langle o, R_1\rangle$. We proceed then by induction on the length of ρ_2.

Base case: $\rho_2 = o$. Trivial, since it reduces to the already seen case $\rho_1 = o$. In fact, by hypothesis, $\rho_1 \precsim \rho_2$ and then $\rho_1 = o$ too.

Induction case. Because of the initial remark, it suffices to analyze the case in which ρ_1 and ρ_2 are both non-empty. So, let us assume $\rho_1 = S_1; \nu_1$ and $\rho_2 = S_2; \nu_2$.

Subcase $S_1 = S_2$. By hypothesis $\rho_2 \sim \rho_1; (\rho_2/\rho_1)$, therefore $\nu_2 \sim \nu_1; (\rho_2/\rho_1)$ by left cancellation, i.e. $\nu_1 \precsim \nu_2$. Since $R_2 \in R_1/(\rho_2/\rho_1)$ by hypothesis, we get $\langle\nu_1, R_1\rangle \preceq \langle\nu_2, R_2\rangle$. The reductions ν_1 and ν_2 are standard, thus, by IH, there is $\langle\nu, R\rangle$ such that $\langle\nu_1, R_1\rangle \trianglerighteq \langle\nu, R\rangle$ and $\langle\nu_2, R_2\rangle \trianglerighteq \langle\nu, R\rangle$. We conclude by taking $\rho \triangleq S_2; \nu$.

Subcase $S_1 \neq S_2$. Remark that S_1 cannot be external or to the left of S_2, otherwise the redex S_1 would have a residual across ρ_2, thus contradicting the hypothesis $\langle\rho_1, R_1\rangle \preceq \langle\rho_2, R_2\rangle$, which implies that $\rho_1 \precsim \rho_2$ and then $\rho_1/\rho_2 \sim o$. As a consequence, S_2 must be external or to the left of S_1. To better describe and analyze the situation, we introduce some additional notation.

For $i = 1, 2$, consider $\rho_i : U_0 \twoheadrightarrow_\beta U_i$ and $R_i : U_i \rightarrow_\beta V_i$. Assuming $S_2 = (\lambda x.M)N$, there is a context $C_0[\,]$ such that $U_0 = C_0[S_2] = C_0[(\lambda x.M)N]$ and, by the assumption on S_2, there exist a unique residual $S_2/\rho_1 = (\lambda x.M')N'$ and a context $C_1[\,]$ such that $U_1 = C_1[(\lambda x.M')N']$. Moreover, $\rho_1 = \rho_f \sqcup \rho_a \sqcup \rho_r$ with:

- $\rho_f : C_0[(\lambda x.M)N] \twoheadrightarrow_\beta C_0[(\lambda x.M')N]$ internal to the function part of S_2;
- $\rho_a : C_0[(\lambda x.M)N] \twoheadrightarrow_\beta C_0[(\lambda x.M)N']$ internal to the argument part of S_2;
- $\rho_r : C_0[(\lambda x.M)N] \twoheadrightarrow_\beta C_1[(\lambda x.M)N]$ to the right of, and then external to S_2.

We then proceed by analyzing the position of R_1 with respect to $S_2/\rho_1 = (\lambda x.M')N'$ in $U_1 = C_1[(\lambda x.M')N']$. We distinguish four cases.

1. $R_1 : C_1[(\lambda x.M')N'] \rightarrow_\beta C_2[(\lambda x.M')N']$ is external, or to left of S_2/ρ_1. By induction on ρ_1, one can see that R_1 is the unique residual of a redex $R_1' : C_0[(\lambda x.M)N] \rightarrow_\beta C_0'[(\lambda x.M)N]$ of U_0. Then $\langle o, R_1'\rangle \preceq \langle\rho_1, R_1\rangle$ holds and, by transitivity, we obtain $\langle o, R_1'\rangle \preceq \langle\rho_2, R_2\rangle$, since $\langle\rho_1, R_1\rangle \preceq \langle\rho_2, R_2\rangle$ holds by hypothesis. We can then take $\langle\rho, R\rangle \triangleq \langle o, R_1'\rangle$, since $\langle\rho_1, R_1\rangle \trianglerighteq \langle o, R_1'\rangle$ and $\langle\rho_2, R_2\rangle \trianglerighteq \langle o, R_1'\rangle$, by Lemma 5.22.

2. $R_1 : C_1[(\lambda x.M')N'] \rightarrow_\beta C_1[(\lambda x.M'')N']$ is internal to the function part (the term M') of $R_2/\rho_1 = (\lambda x.M')N'$. For a diagram with the reductions in the proof see Figure 5.6.

   The proof follows the sketch in Figure 5.5. First of all, in subitem (a) below (see the upper left triangle in Figure 5.5 also), we shall prove that we can extract ρ_a

$$
\begin{array}{ccccccc}
\langle \rho_1, R_1 \rangle & \text{(a)} & \langle S_2 \sqcup \rho_f, R_1'' \rangle & \text{(b)IH} & \langle \rho_2, R_2 \rangle \\
& \trianglerighteq & & \trianglerighteq & \\
& \langle \rho_f, R_1' \rangle & \text{(c)CR} & \langle S_2; \rho', R' \rangle & \\
& & \trianglerighteq & \trianglerighteq & \\
& & \langle \rho, R \rangle & &
\end{array}
$$

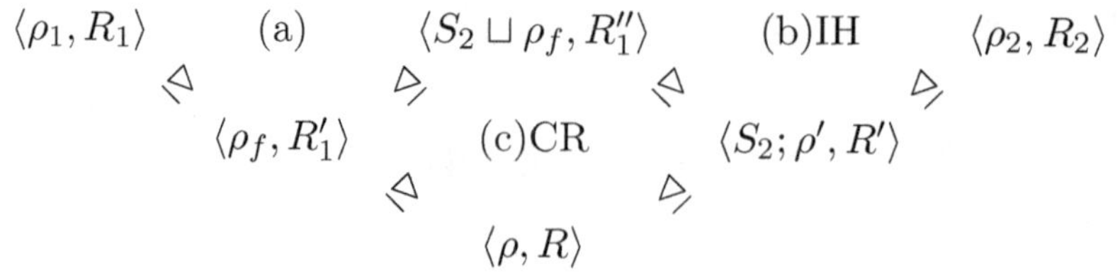

Figure 5.5: Lemma 5.23(2). R_1 is internal to the function part of R_2/ρ_1: extractions.

and ρ_r from $\langle \rho_1, R_1 \rangle$ obtaining then $\langle \rho_1, R_1 \rangle \trianglerighteq \langle \rho_f, R_1' \rangle$ for some R_1'. We shall then consider $\rho_2 = S_2; \nu_2$ and the residual R_1'' of R_1' along S_2/ρ_f and we will see that $\langle S_2 \sqcup \rho_f, R_1'' \rangle \trianglerighteq \langle \rho_f, R_1' \rangle$. In subitem (b) (see the upper right triangle in Figure 5.5 also), by applying the induction hypothesis to $\langle \rho_f/S_2, R_1'' \rangle$ and $\langle \nu_2, R_2 \rangle$, we shall then see that $\langle S_2 \sqcup \rho_f, R_1'' \rangle \trianglerighteq \langle S_2; \rho', R' \rangle$ and that $\langle \rho_2, R_2 \rangle \trianglerighteq \langle S_2; \rho', R' \rangle$. Finally, in subitem (c) (see the central diamond in Figure 5.5), by Church-Rosser applied to $\langle S_2 \sqcup \rho_f, R_1'' \rangle \trianglerighteq \langle \rho_f, R_1' \rangle$ and $\langle S_2 \sqcup \rho_f, R_1'' \rangle \trianglerighteq \langle S_2; \rho', R' \rangle$, we will get $\langle \rho_f, R_1' \rangle \trianglerighteq \langle \rho, R \rangle$ and $\langle S_2; \rho', R' \rangle \trianglerighteq \langle \rho, R \rangle$, for some $\langle \rho, R \rangle$. From which, we conclude.

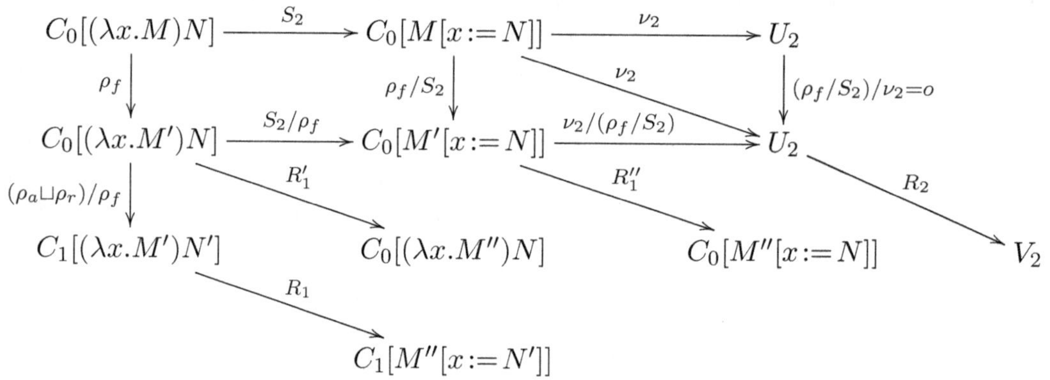

$\rho_1 \triangleq \rho_f \sqcup \rho_a \sqcup \rho_r,\quad S_2 \triangleq (\lambda x.M)N,\quad \langle \rho_f, R_1' \rangle \preceq \langle \rho_1, R_1 \rangle \preceq \langle \rho_2, R_1 \rangle,$
$\rho_2 \triangleq S_2; \nu_2,\quad R_1 \triangleq R_1'/(\rho_a \sqcup \rho_r)/\rho_f,\quad \langle \rho_f/S_2, R_1'' \rangle \preceq \langle \nu_2, R_2 \rangle,$
$\langle \rho_1, R_1 \rangle \preceq \langle \rho_2, R_2 \rangle,\quad R_1'' \triangleq R_1'/(S_2/\rho_f).$

Figure 5.6: Lemma 5.23(2). R_1 is internal to the function part of R_2/ρ_1: reductions.

Let us then see the proof in detail.

(a) The redex R_1 is the unique residual of a redex $R_1' : C_0[(\lambda x.M')N] \to_\beta C_0[(\lambda x.M'')N]$; that is, $R_1 = R_1'/((\rho_a \sqcup \rho_r)/\rho_f)$. Then, $R_1/\rho_1 \trianglerighteq R_1'/\rho_f$. Now, let us take $R_1'' \triangleq R_1'/(S_2/\rho_f)$. Since the reduction $\rho_f; R_1'$ is internal to the function part of S_2, we have that $\langle S_2 \sqcup \rho_f, R_1'' \rangle \trianglerighteq_3 \langle \rho_f, R_1' \rangle$.

(b) We show now that $\langle \rho_f/S_2, R_1'' \rangle \preceq \langle \nu_2, R_2 \rangle$ holds (recall that $\rho_2 = S_2; \nu_2$). Since $\langle \rho_f, R_1' \rangle \preceq \langle \rho_f, R_2 \rangle$, we have

$$S_2; \nu_2 \sim \rho_f; (\rho_2/\rho_f) = \rho_f; (S_2/\rho_f); (\nu_2/(\rho_f/S_2)) \sim S_2; (\rho_f/S_2); (\nu_2/(\rho_f/S_2)).$$

By left cancellation, $\nu_2 \sim (\rho_f/S_2);(\nu_2/(\rho_f/S_2))$, which implies $\rho_f/S_2 \precsim \nu_2$, and

$$\begin{aligned} R_2 \in R_1'/(\rho_2/\rho_f) = {} & R_1'/((S_2/\rho_f);(\nu_2/(\rho_f/S_2))) \\ = {} & (R_1'/(S_2/\rho_f))/(\nu_2/(\rho_f/S_2)) = R_1''/(\nu_2/(\rho_f/S_2)). \end{aligned}$$

Both ρ_f/S_2 and ν_2 are standard so, by IH, there exists $\langle \rho', R' \rangle$ such that

$$\langle \rho_f/S_2, R_1'' \rangle \trianglerighteq \langle \rho', R' \rangle \qquad \text{and} \qquad \langle \nu_2, R_2 \rangle \trianglerighteq \langle \rho', R' \rangle$$

By adding S_2 at the beginning of the histories, this implies also (see Figure 5.5)

$$\langle S_2 \sqcup \rho_f, R_1'' \rangle \trianglerighteq \langle S_2;\rho', R' \rangle \qquad \text{and} \qquad \langle \rho_2, R_2 \rangle \trianglerighteq \langle S_2;\rho', R' \rangle$$

(c) By Church-Rosser for $\trianglerighteq$, there is $\langle \rho, R \rangle$ such that (see Figure 5.5)

$$\langle S_2 \sqcup \rho_f, R_1'' \rangle \trianglerighteq_3 \langle \rho_f, R_1' \rangle \trianglerighteq \langle \rho, R \rangle \qquad \text{and} \qquad \langle S_2 \sqcup \rho_f, R_1'' \rangle \trianglerighteq \langle S_2;\rho', R' \rangle \trianglerighteq \langle \rho, R \rangle$$

From which, we can finally conclude that

$$\langle \rho_1, R_1 \rangle \trianglerighteq \langle \rho_f, R_1' \rangle \trianglerighteq \langle \rho, R \rangle \qquad \text{and} \qquad \langle \rho_2, R_2 \rangle \trianglerighteq \langle S_2;\rho', R' \rangle \trianglerighteq \langle \rho, R \rangle$$

3. $R_1 : C_1[(\lambda x.M')N'] \to_\beta C_2[(\lambda x.M')N']$ is to the right of, and then external to $R_2/\rho_1 = (\lambda x.M')N'$. For a diagram with the reductions in the proof see Figure 5.8.

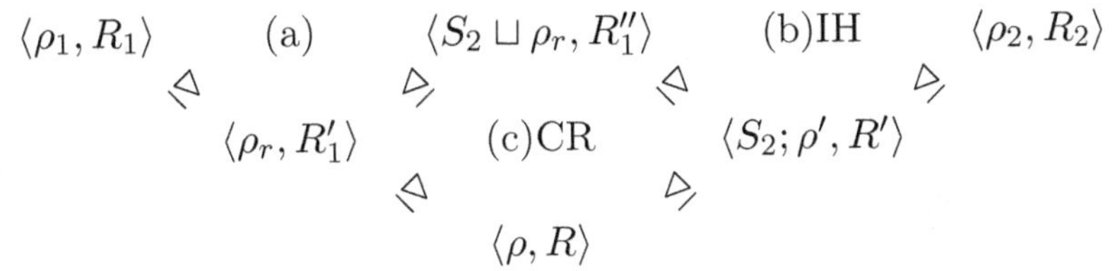
⟨ρ1, R1⟩
(a)
⟨S2 ⊔ ρr, R1''⟩
(b)IH
⟨ρ2, R2⟩
⟨ρr, R1'⟩
(c)CR
⟨S2; ρ', R'⟩
⟨ρ, R⟩

Figure 5.7: Lemma 5.23(3). R_1 is to the right of R_2/ρ_1: extractions

The proof follows the same steps of the previous case with ρ_r replacing ρ_f and its sketch can be seen in Figure 5.7. First of all, in subitem (a) (see the left upper triangle in Figure 5.7), we get $\langle \rho_1, R_1 \rangle \trianglerighteq \langle \rho_a, R' \rangle$ for some R' by extracting ρ_a and ρ_f from $\langle \rho_1, R_1 \rangle$, and we see that $\langle S_2 \sqcup \rho_r, R_1'' \rangle \trianglerighteq \langle \rho_r, R_1' \rangle$ also, where S_2 is the first redex of $\rho_2 = S_2;\nu_2$ and R_1'' is the residual of R_1' along S_2/ρ_r. In subitem (b) (see the upper right triangle in Figure 5.7), we exploit the induction hypothesis on $\langle \rho_r/S_2, R_1'' \rangle$ and $\langle \nu_2, R_2 \rangle$ to prove that $\langle \rho_2, R_2 \rangle \trianglerighteq \langle S_2;\rho', R' \rangle$ and $\langle \rho_r/S_2, R_1'' \rangle \trianglerighteq \langle S_2;\rho', R' \rangle$. Finally, in subitem (c), we conclude by Church-Rosser for $\trianglerighteq$ (see the central diamond in Figure 5.7).

Let us see the proof in detail.

(a) By induction on $\rho_f \sqcup \rho_a$, one can see that there is a redex with history $\langle \rho_r, R_1' \rangle$, with $R_1' : C_0[(\lambda x.M)N] \to_\beta C_1[(\lambda x.M)N]$ (see Figure 5.8), such that R_1 is the unique residual of R_1' across $(\rho_f \sqcup \rho_a)/\rho_r$, that is, $R_1 = R_1'/((\rho_f \sqcup \rho_a)/\rho_r)$. Then, $\langle \rho_r, R_1' \rangle \preceq \langle \rho_1, R_1 \rangle$. Moreover, since ρ_r and $\rho_f \sqcup \rho_a$ are disjoint, we have $\langle \rho_1, R_1 \rangle = \langle (\rho_f \sqcup \rho_a \sqcup \rho_r), R_1 \rangle \trianglerighteq_2 \langle \rho_r, R_1' \rangle$. By taking $R_1'' \triangleq R_1'/(S_2/\rho_f)$, we observe also that $\langle S_2 \sqcup \rho_r, R_1'' \rangle \trianglerighteq_2 \langle \rho_r, R_1' \rangle$.

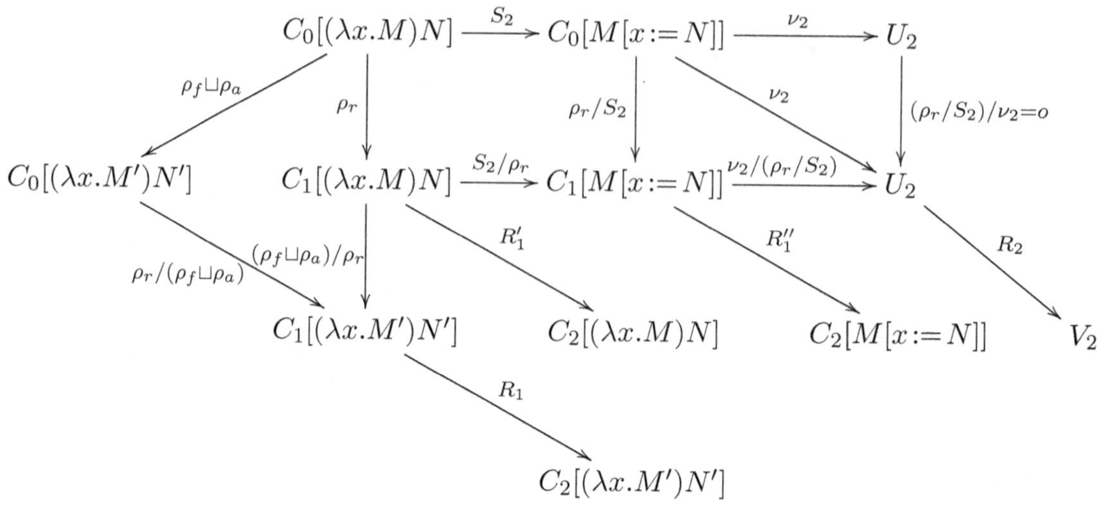

$$\rho_1 \triangleq \rho_f \sqcup \rho_a \sqcup \rho_r, \quad S_2 \triangleq (\lambda x.M)N, \quad \langle \rho_r, R_1' \rangle \preceq \langle \rho_1, R_1 \rangle \preceq \langle \rho_2, R_2 \rangle,$$
$$\rho_2 \triangleq S_2; \nu_2, \quad R_1 \triangleq R_1'/((\rho_f \sqcup \rho_a)/\rho_r), \quad \langle \rho_r/S_2, R_1'' \rangle \preceq \langle \nu_2, R_2 \rangle,$$
$$\langle \rho_1, R_1 \rangle \preceq \langle \rho_2, R_2 \rangle, \quad R_1'' \triangleq R_1'/(S_2/\rho_r).$$

Figure 5.8: Lemma 5.23(3). R_1 is to the right of R_2/ρ_1: reductions.

(b) As in case 2, we see that $\langle \rho_r/S_2, R_1'' \rangle \preceq \langle \nu_2, R_2 \rangle$ holds. Then, by IH, there exists $\langle \nu, R' \rangle$ such that

$$\langle \rho_f/S_2, R_1'' \rangle \trianglerighteq \langle \rho', R' \rangle \qquad \& \qquad \langle \nu_2, R_2 \rangle \trianglerighteq \langle \rho', R' \rangle$$

which implies also (see Figure 5.7)

$$\langle S_2 \sqcup \rho_r, R_1'' \rangle \trianglerighteq \langle S_2; \rho', R' \rangle \qquad \& \qquad \langle \rho_2, R_2 \rangle \trianglerighteq \langle S_2; \rho', R' \rangle$$

(c) By Church-Rosser for $\trianglerighteq$, there is $\langle \rho, R \rangle$ such that (see Figure 5.7 again)

$$\langle S_2 \sqcup \rho_r, R_1'' \rangle \trianglerighteq_3 \langle \rho_r, R_1' \rangle \trianglerighteq \langle \rho, R \rangle \qquad \& \qquad \langle S_2 \sqcup \rho_r, R_1'' \rangle \trianglerighteq \langle S_2; \rho', R' \rangle \trianglerighteq \langle \rho, R \rangle$$

From which, we can finally conclude that

$$\langle \rho_1, R_1 \rangle \trianglerighteq \langle \rho_r, R_1' \rangle \trianglerighteq \langle \rho, R \rangle \qquad \& \qquad \langle \rho_2, R_2 \rangle \trianglerighteq \langle S_2; \rho', R' \rangle \trianglerighteq \langle \rho, R \rangle$$

4. $R_1 : C_1[(\lambda x.M')N'] \to_\beta C_1[(\lambda x.M')N'']$ is internal to the argument part N' of $R_2/\rho_1 = (\lambda x.M')N'$. For a diagram of the reductions in the proof see Figure 5.10.

   The sketch of the proof, see the diagram in Figure 5.9, is similar to those of the previous cases (2) and (3), in which we may recognize three main steps consisting in proving the upper-left triangle of the diagram (subitem (a)); using the induction hypothesis to prove the upper-right triangle (subitem (b)); showing the central diamond (subitem (c)) by applying the Church-Rosser property.

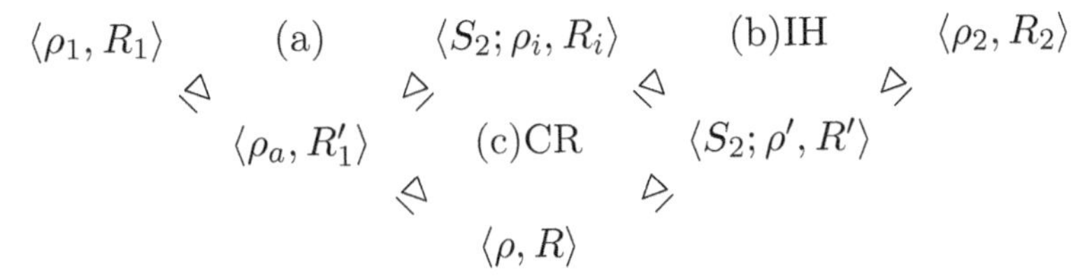

Figure 5.9: Lemma 5.23(4). R_1 is internal to the argument part of R_2/ρ_1: extractions.

The main difference with the previous cases is that we cannot apply the IH to $\langle \rho_a/S_2, R_1''\rangle$ with $R_1'' = R_1'/(S_2/\rho_a)$ to find the central vertex $\langle S_2 \sqcup \rho_a, R_1''\rangle$ of a zig-zag between $\langle \rho_1, R_1\rangle$ and $\langle \rho_2, R_2\rangle$, as done with $\langle \rho_f/S_2, R_1''\rangle$ and $\langle S_2 \sqcup \rho_f, R_1''\rangle$ in case (2) (see Figure 5.5) and with ρ_r/S_2 and $\langle S_2 \sqcup \rho_r, R_1''\rangle$ in case (3) (see Figure 5.7). In fact, when the bound variable x of S_2 has $n > 1$ occurrences in M, the reduction ρ_a/S_2 is not standard, and the residual $R_1'/(S_2/\rho_a)$ is not a singleton, but a set with n redexes.

However, despite this, it is still true that $S_2; (\rho_a/S_2) \precsim S_2; \nu_2 = \rho_2$, and then that $\rho_a/S_2 \precsim \nu_2$ (see Figure 5.10). Thus, since $\langle \rho_a, R_1'\rangle \preceq \langle \rho_1, R_1\rangle \preceq \langle \rho_2, R_2\rangle$, by Lemma 5.7, there is a redex R_1'' such that $\langle \rho_a, R_1'\rangle \preceq \langle S_2; (\rho_a/S_2), R_1''\rangle \preceq \langle \rho_2, R_2\rangle$, which implies $\langle (\rho_a/S_2), R_1''\rangle \preceq \langle \nu_2, R_2\rangle$ too. It is readily seen that such a redex R_1'' is internal to the reduced argument N' in $M[x:=N']$, and that there is $\langle \rho_i, R_i\rangle$ such that

- ρ_i and R_i are internal to an instance of N in $M[x := N]$, and $\rho_a = \rho_i \Uparrow S_2$. Therefore, we have $\langle S_2; \rho_i, R_i\rangle \trianglerighteq_3 \langle \rho_a, R_1'\rangle$.
- $\rho_a/S_2 = \rho_i \parallel S_2$. Then, we have $\langle \rho_i, R_i\rangle \preceq \langle \rho_a/S_2, R_1''\rangle \preceq \langle \nu_2, R_2\rangle$. Moreover, since ρ_a is standard, ρ_i is standard too.

We now have all the ingredients to apply the three steps of the proof.

(a) By induction on $\rho_f \sqcup \rho_r$, we can prove that there is a redex with history $\langle \rho_a, R_1'\rangle$ such that $R_1 = R_1'/((\rho_f \sqcup \rho_r)/\rho_a)$, and such that $\langle \rho_1, R_1\rangle \trianglerighteq \langle \rho_a, R_1'\rangle$. In fact, $R_1 = R_1'/((\rho_f \sqcup \rho_r)/\rho_a) = R_1'/((\rho_f/\rho_a); ((\rho_r/\rho_f)/(\rho_a/\rho_f))) = (R_1'/(\rho_f/\rho_a))/(\rho_r/(\rho_f \sqcup \rho_a))$. It follows that

$$\langle \rho_1, R_1\rangle = \langle \rho_f \sqcup \rho_a \sqcup \rho_r, R_1\rangle \trianglerighteq_1 \langle \rho_f \sqcup \rho_a, R_1'/(\rho_f/\rho_a)\rangle \trianglerighteq_2 \langle \rho_a, R_1'\rangle$$

(by induction on ρ_f, since ρ_f and ρ_a are disjoint). Moreover, we have already seen that $\langle S_2; \rho_i, R_i\rangle \trianglerighteq_3 \langle \rho_a, R_1'\rangle$.

(b) Since we have seen that $\langle \rho_i, R_i\rangle \preceq \langle \nu_2, R_2\rangle$ with ρ_i standard, we can apply the induction hypothesis to $\langle \rho_i, R_i\rangle$ and $\langle \nu_2, R_2\rangle$ to find a $\langle \rho', R'\rangle$ such that

$$\langle \rho_i, R_i\rangle \trianglerighteq \langle \rho', R'\rangle \qquad \& \qquad \langle \nu_2, R_2\rangle \trianglerighteq \langle \rho', R'\rangle$$

from which, we also have that (see Figure 5.9)

$$\langle S_2; \rho_i, R_i\rangle \trianglerighteq \langle S_2; \rho', R'\rangle \qquad \& \qquad \langle \rho_2, R_2\rangle = \langle S_2; \nu_2, R_2\rangle \trianglerighteq \langle S_2; \rho', R'\rangle$$

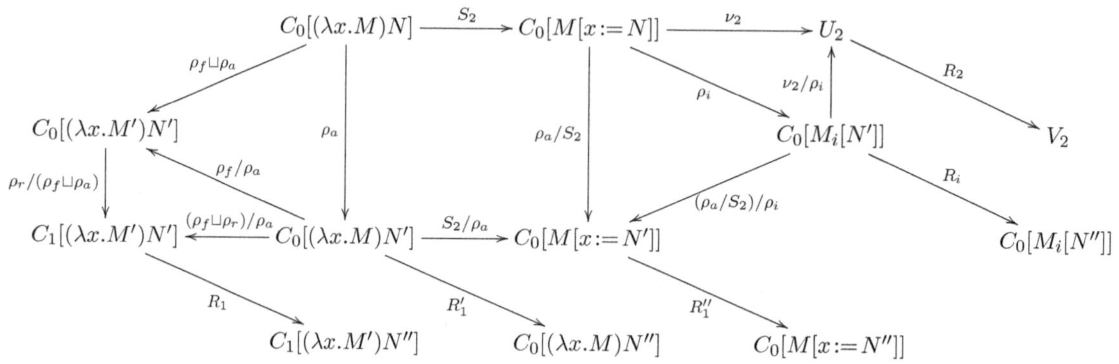

$$\begin{array}{lll}
\rho_1 \triangleq \rho_f \sqcup \rho_a \sqcup \rho_r, & S_2 \triangleq (\lambda x.M)N, & M_i[] \triangleq M^x[i := [] \mid N], \\
\rho_2 \triangleq S_2; \nu_2, & R_1 \triangleq R_1'/((\rho_f \sqcup \rho_a)/\rho_r), & \langle \rho_r, R_1' \rangle \preceq \langle \rho_1, R_1 \rangle \preceq \langle \rho_2, R_2 \rangle, \\
\langle \rho_1, R_1 \rangle \preceq \langle \rho_2, R_2 \rangle, & R_1'' \triangleq R_1'/(S_2/\rho_r), & \langle \rho_r/S_2, R_1'' \rangle \preceq \langle \nu_2, R_2 \rangle.
\end{array}$$

Figure 5.10: Lemma 5.23(4). R_1 is in the argument part of R_2/ρ_1: reductions.

(c) By Church-Rosser, there exists $\langle \rho, R \rangle$ such that (see Figure 5.9)

$$\langle S_2; \rho_i, R_i \rangle \trianglerighteq \langle \rho_a, R_1' \rangle \trianglerighteq \langle \rho, R \rangle \qquad \& \qquad \langle S_2; \rho_i, R_i \rangle \trianglerighteq \langle S_2; \rho', R' \rangle \trianglerighteq \langle \rho, R \rangle$$

From which, we can finally conclude that

$$\langle \rho_1, R_1 \rangle \trianglerighteq \langle \rho_a, R_1' \rangle \trianglerighteq \langle \rho, R \rangle \qquad \& \qquad \langle \rho_2, R_2 \rangle \trianglerighteq \langle S_2; \rho', R' \rangle \trianglerighteq \langle \rho, R \rangle$$

as needed. □

We now have all the ingredients to prove the main property of extraction.

Theorem 5.24 (Decidability of extraction). *Let ρ_1 and ρ_2 be standard reductions and R_1, R_2 be redexes. The following are equivalent:*

1. $\exists \nu, S \,.\, \langle \rho_1, R_1 \rangle \trianglerighteq \langle \nu, S \rangle \;\&\; \langle \rho_2, R_2 \rangle \trianglerighteq \langle \nu, S \rangle$,
2. $\langle \rho_1, R_1 \rangle \simeq \langle \rho_2, R_2 \rangle$.

Proof. ($1 \Rightarrow 2$) By Lemma 5.21, we obtain $\langle \rho_1, R_1 \rangle \simeq \langle \nu, S \rangle \simeq \langle \rho_2, R_2 \rangle$. Conclude by transitivity of $\simeq$.

($2 \Rightarrow 1$) Assume $\langle \rho_1, R_1 \rangle \simeq \langle \rho_2, R_2 \rangle$. By Proposition 5.11, there exist zig-zag witnesses $\langle \nu_i, S_i \rangle_{i \in \{0,\ldots,k\}}$ of $\langle \rho_1, R_1 \rangle \simeq \langle \rho_2, R_2 \rangle$. Wlog, we may assume that all such ν_i are standard reductions. By Lemma 5.23, there exist redexes with history $\langle \nu_i', S_i' \rangle_{i \in \{0,\ldots,k\}}$ such that

$$\langle \nu_i, S_i \rangle \trianglerighteq \langle \nu_i', S_i' \rangle \;\&\; \langle \nu_{i+1}, S_{i+1} \rangle \trianglerighteq \langle \nu_i', S_i' \rangle,$$

for all $i\,(0 \leq i < k)$. By Theorem 5.19 (confluence of $\trianglerighteq$), we conclude that there exists $\langle \nu, S \rangle$ such that both $\langle \rho_1, R_1 \rangle \trianglerighteq \langle \nu, S \rangle$ and $\langle \rho_2, R_2 \rangle \trianglerighteq \langle \nu, S \rangle$ hold. □

EXAMPLES 5.25. Consider the example in Figure 5.1 and the redexes S_1 and S_4 with their standard histories, i.e., $\langle M; R_1, S_1\rangle$ and $\langle M; R_1; R_3, S_4\rangle$. By applying extraction:

(i) $\langle M; R_1, S_1\rangle \rhd_4 \langle R, S\rangle$ and

(ii) $\langle M; R_2, S_2\rangle \rhd_2 \langle M; R_2, S_2\rangle \rhd_4 \langle R, S\rangle$.

Moreover, if $\rho_i, \rho_i', \rho_i''$ are the standard histories of the corresponding redexes:

(i) $\langle \rho_i, S_i\rangle \unrhd \langle R, S\rangle$, with $1 \leq i \leq 6$,

(ii) $\langle \rho_i', R_i\rangle \unrhd \langle o, R\rangle$, with $1 \leq i \leq 6$,

(iii) $\langle \rho_i'', M_i\rangle \unrhd \langle o, M\rangle$, with $1 \leq i \leq 2$.

The above theorem not only establishes a precise correspondence between zig-zag and extraction, but also provides an *effective* procedure for deciding the family relation. Indeed, we now prove that two redexes with history are in the same family exactly when their permutationally equivalent standard redexes with history have the same $\unrhd$-nf's.

COROLLARY 5.26. *Given the redexes with history $\langle \rho, R\rangle$ and $\langle \rho', R'\rangle$, let ρ_s and ρ_s' be the standard reductions such that $\rho \sim \rho_s$ and $\rho' \sim \rho_s'$. Assume that $\langle \rho_s, R\rangle \unrhd \langle \rho_c, R_c\rangle$ and $\langle \rho_s', R'\rangle \unrhd \langle \rho_c', R_c'\rangle$, where $\langle \rho_c, R_c\rangle$ and $\langle \rho_c', R_c'\rangle$ are $\unrhd$-normal forms. Then,*

$$\langle \rho, R\rangle \simeq \langle \rho', R'\rangle \iff \langle \rho_c, R_c\rangle = \langle \rho_c', R_c'\rangle$$

PROOF. By Theorem 5.24 and the fact that, since extraction is strong normalizing and confluent, every redex with history has a unique $\unrhd$-normal form. □

Since standardization and $\unrhd$-normalization are effective and terminating processes, the problem of determining whether two redexes with history belong to the same family is decidable. Another corollary is that every family class has a canonical representative.

COROLLARY 5.27. *Let $\mathscr{F}$ be a family class.*

(i) There is a unique $\unrhd$-normal form $\langle \rho_c, R_c\rangle \in \mathscr{F}$ such that ρ_c is standard.

(ii) $\langle \rho, R\rangle \in \mathscr{F}$ if and only if $\langle \rho_s, R\rangle \unrhd \langle \rho_c, R_c\rangle$, where ρ_s is a standard reduction such that $\rho_s \sim \rho$.

PROOF. Since every redex with history has a normal formal, every family $\mathscr{F}$ contains a $\unrhd$-normal form $\langle \rho_c, R_c\rangle$ (just take the $\unrhd$-normal form of any member of the family).

(i) By Corollary 5.26.

(ii) By Corollary 5.12, any standard reduction $\rho_s \sim \rho$ is such that $\langle \rho_s, R\rangle \in \mathscr{F}$. So, its unique $\unrhd$-normal form is equal to the unique $\unrhd$-normal form in the family. □

DEFINITION 5.28 (CANONICAL REPRESENTATIVE). The *canonical representative of a family class $\mathscr{F}$* is the unique $\langle \rho_c, R_c\rangle \in \mathscr{F}$ such that:

(i) $\langle \rho_c, R_c\rangle$ is in $\unrhd$-normal form;

(ii) ρ_c is standard.

By Corollary 5.27, the canonical representative of a family can be effectively constructed from each member of the family.

5.3 The labeled λ-calculus

We now define and study the labeled λ-calculus introduced by Lévy (1975) in order to assign names to redexes and subterms and therefore track their origin in a reduction. This technique generalizes an idea of Vuillemin (1973) for recursive program schemes.

DEFINITION 5.29. (i) The set $\mathcal{L}$ of *labels* over a countable alphabet $\{a, b, c, \dots\}$ of *atomic labels* is inductively defined as follows (where a is an atomic label):

$$\alpha, \beta, \gamma, \delta, \varepsilon ::= a \mid \lceil \alpha \rceil \mid \lfloor \alpha \rfloor \mid \alpha\beta \qquad (\mathcal{L})$$

$\lceil \alpha \rceil$ is called an *overlining of* α, while $\lfloor \alpha \rfloor$ is called an *underlining*. The operation of juxtaposition $\alpha\beta$ is assumed to be associative.

(ii) The *size of a label* $\alpha \in \mathcal{L}$, written $\mathrm{size}(\alpha)$, is defined by induction on α:

$$\begin{array}{lcll} \mathrm{size}(a) & \triangleq & 1, & \text{if } a \text{ is atomic;} \\ \mathrm{size}(\beta\gamma) & \triangleq & \mathrm{size}(\beta) + \mathrm{size}(\gamma); & \\ \mathrm{size}(\lceil \beta \rceil) & \triangleq & \mathrm{size}(\beta) + 1; & \\ \mathrm{size}(\lfloor \beta \rfloor) & \triangleq & \mathrm{size}(\beta) + 1. & \end{array}$$

(iii) The *depth of a label* $\alpha \in \mathcal{L}$, in symbols $\mathrm{depth}(\alpha)$, is defined by induction on α:

$$\begin{array}{lcll} \mathrm{depth}(a) & = & 0, & \text{if } a \text{ is atomic;} \\ \mathrm{depth}(\beta\gamma) & = & \max\{\mathrm{depth}(\beta), \mathrm{depth}(\gamma)\}; & \\ \mathrm{depth}(\lfloor \beta \rfloor) & = & \mathrm{depth}(\beta) + 1; & \\ \mathrm{depth}(\lceil \beta \rceil) & = & \mathrm{depth}(\beta) + 1. & \end{array}$$

Intuitively, $\mathrm{depth}(\alpha)$ is the maximal amount of nested overlinings or underlinings in α.

EXAMPLES 5.30. (i) For $\alpha \triangleq abc$, we have: $\alpha = (ab)c = a(bc)$; $\mathrm{size}(\alpha) = 3$; $\mathrm{depth}(\alpha) = 0$.

(ii) For $\alpha \triangleq \lfloor \lceil \lfloor a \rfloor \rceil \rfloor$, we obtain $\mathrm{size}(\alpha) = 4$ and $\mathrm{depth}(\alpha) = 3$.

(iii) For $\alpha \triangleq a \lceil \lfloor b \rfloor \lceil c \rceil \rceil$, we get $\mathrm{size}(\alpha) = 6$ and $\mathrm{depth}(\alpha) = 2$.

DEFINITION 5.31. (i) The set $\Lambda^{\mathcal{L}}$ of *labeled λ-terms* is generated by the following grammar (for $\alpha \in \mathcal{L}$):

$$M ::= x^\alpha \mid (MM)^\alpha \mid (\lambda x.M)^\alpha \qquad (\Lambda^{\mathcal{L}})$$

Labeled λ-terms are considered up to renaming of bound variables (not labels).

(ii) Given $M \in \Lambda^{\mathcal{L}}$, we denote by $\ell_{\mathrm{e}}(M)$ its *external label*. Formally:

$$\begin{array}{lcl} \ell_{\mathrm{e}}(x^\alpha) & \triangleq & \alpha; \\ \ell_{\mathrm{e}}((PQ)^\alpha) & \triangleq & \alpha; \\ \ell_{\mathrm{e}}((\lambda x.P)^\alpha) & \triangleq & \alpha. \end{array}$$

We sometimes write $M^\alpha \in \Lambda^{\mathcal{L}}$ to denote an $M \in \Lambda^{\mathcal{L}}$ with external label $\ell_{\mathrm{e}}(M) = \alpha$.

(iii) The *size of a labeled term* $M^\alpha \in \Lambda^{\mathcal{L}}$, written $\text{size}(M^\alpha)$, is defined by structural induction on M^α:

$$\begin{array}{lcl} \text{size}(x^\alpha) & \triangleq & 1; \\ \text{size}((PQ)^\alpha) & \triangleq & \text{size}(P) + \text{size}(Q) + 1; \\ \text{size}((\lambda x.P)^\alpha) & \triangleq & \text{size}(P) + 1. \end{array}$$

(iv) The *concatenation* of $\alpha \in \mathcal{L}$ and $M^\beta \in \Lambda^{\mathcal{L}}$ is the labeled λ-term defined by

$$\alpha \cdot M^\beta \triangleq M^{\alpha\beta}.$$

(v) Given $M \in \Lambda^{\mathcal{L}}$ and $\alpha_1, \ldots, \alpha_k \in \mathcal{L}$, we write $\alpha_1 \cdots \alpha_k \cdot M$ for $(\alpha_1 \cdots \alpha_k) \cdot M$.

REMARK 5.32. (i) For all $\alpha \in \mathcal{L}$, we have $\text{size}(\alpha) > 0$.
(ii) Similarly, for all $M \in \Lambda^{\mathcal{L}}$, we have $\text{size}(M) > 0$.
(iii) The size of a labeled λ-term is independent from its labels.

DEFINITION 5.33. (i) Given $M^\alpha, N \in \Lambda^{\mathcal{L}}$ and $x \in \text{Var}$, the *capture-free substitution of N for all free occurrences of x in M^α*, written $M^\alpha[x{:=}N]$, is defined by structural induction on M^α:

$$\begin{array}{lcl} y^\alpha[x{:=}N] & \triangleq & \begin{cases} \alpha \cdot N, & \text{if } x = y, \\ y^\alpha, & \text{otherwise;} \end{cases} \\ (PQ)^\alpha[x{:=}N] & \triangleq & (P[x{:=}N](Q[x{:=}N]))^\alpha; \\ (\lambda y.P)^\alpha[x{:=}N] & \triangleq & (\lambda y.P[x{:=}N])^\alpha; \end{array}$$

where, in the abstraction case, we assume (wlog) $y \neq x$.

(ii) The *labeled λ-calculus* is endowed with a notion of *labeled β-reduction*:

$$((\lambda x.M)^\alpha N)^\beta \to \beta\lceil\alpha\rceil \cdot M[x{:=}\lfloor\alpha\rfloor \cdot N] \qquad (\beta_{\mathcal{L}})$$

This generates reduction relations $\to_{\beta_{\mathcal{L}}}$ and $\twoheadrightarrow_{\beta_{\mathcal{L}}}$ in the usual way.

(iii) The *name of a (labeled) redex* $R = ((\lambda x.M)^\alpha N)^\beta$ is the label α of its functional part, in symbols, $\mathsf{name}(R) = \alpha$.

REMARK 5.34. (i) Note that $\beta\lceil\alpha\rceil$ is *not* the external label of $P \triangleq \beta\lceil\alpha\rceil \cdot M[x{:=}\lfloor\alpha\rfloor \cdot N]$. Indeed $M[x{:=}\lfloor\alpha\rfloor \cdot N]$ is a labeled λ-term, whence it has its own external label, say, γ. As a consequence the external label of P is rather $\ell_e(P) = \beta\lceil\alpha\rceil\gamma$.

(ii) In his PhD thesis Klop (1980a) has shown that it is unnecessary to distinguish underlinings and overlinings. Subsequently, Gonthier et al. (1992b) pointed out that both over- and under- linings could be removed entirely: in fact, they can be reconstructed as their language is LR(1).

When contracting a $\beta_{\mathcal{L}}$-redex R, its label $\mathsf{name}(R)$ gets surrounded by the labels of the other edges incident to the nodes in R. More precisely, in the contractum of $R = ((\lambda x.M^\gamma)^\alpha N^\varepsilon)^\beta$, the name α of R is captured:

- between the label β of the application and the label γ of the body of the abstraction,

- and between the label of any occurrence of the variable x in M and the label ε of the argument part of R.

In the contractum, the pairs of edges corresponding to the previous pairs of labels are replaced by new connections. If one forgets the linings, the labels of the new edges are obtained by composing the labels of the corresponding edges in the natural way:

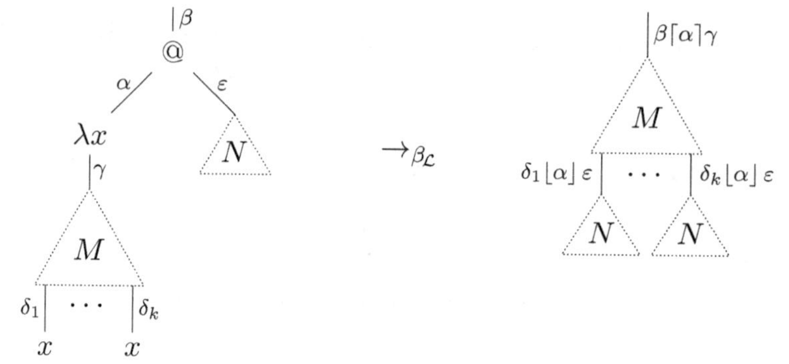

Overlining $\lceil\cdot\rceil$ and underlining $\lfloor\cdot\rfloor$ represent the two ways in which these new connections are created:

- upwards: from the surrounding context to the body M, or
- downwards: from the occurrences of the variable x in M to the instances of the argument N.

EXAMPLES 5.35. (i) The labeled λ-term $M = ((\lambda x.x^c)^b((\lambda x.x^f)^e y^g)^d)^a$ has two possible normalizing $\beta_{\mathcal{L}}$-reduction sequences:

$$M \to_{\beta_{\mathcal{L}}} ((\lambda x.x^f)^e y^g)^{a\lceil b\rceil c\lfloor b\rfloor d} \to_{\beta_{\mathcal{L}}} y^{a\lceil b\rceil c\lfloor b\rfloor d\lceil e\rceil f\lfloor e\rfloor g};$$

$$M \to_{\beta_{\mathcal{L}}} ((\lambda x.x^c)^b y^{d\lceil e\rceil f\lfloor e\rfloor g})^a \to_{\beta_{\mathcal{L}}} y^{a\lceil b\rceil c\lfloor b\rfloor d\lceil e\rceil f\lfloor e\rfloor g}.$$

This is a first hint of the fact that $\to_{\beta_{\mathcal{L}}}$ is actually confluent.

(ii) Consider the following labeled version of Ω: $((\lambda x.(x^d x^e)^c)^b(\lambda x.(x^h x^i)^g)^f)^a$. The labeled β-reduction proceeds as follows:

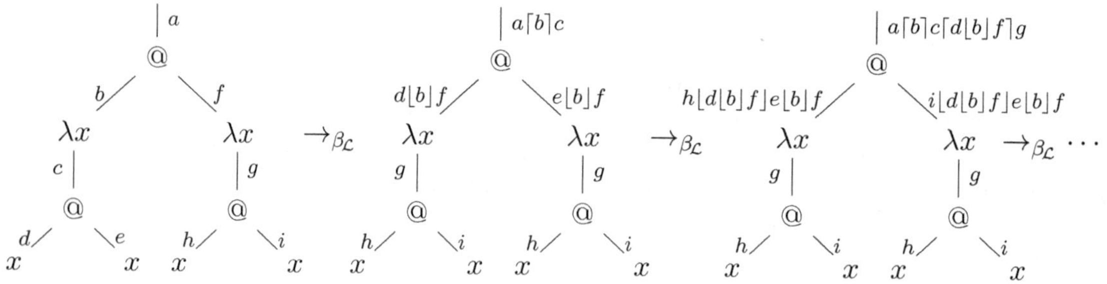

LEMMA 5.36. *Let $M, N, P \in \Lambda^{\mathcal{L}}$ and $\alpha \in \mathcal{L}$.*

(i) $(\alpha \cdot M)[x{:=}N] = \alpha \cdot (M[x{:=}N])$.

(ii) $M[x{:=}N][y{:=}P] = M[y{:=}P][x{:=}(N[y{:=}P])]$, *when* $x \notin \mathrm{FV}(P)$

PROOF. Both items follow by a straightforward induction on the structure of M. □

The strongly normalizing setting

The original labeled λ-calculus, introduced by Hyland (1975b) and Wadsworth (1976) as a tool for examining $\mathcal{D}_\infty$-models, is strongly normalizing. Lévy's calculus $\Lambda^{\mathcal{L}}$ is more general and does not satisfy strong normalization, as shown in Example 5.35(ii) above. To harness its computational power, we are going to parametrize the reduction by a predicate $\mathcal{P}$ on labels that specifies whether a redex is contractible.

DEFINITION 5.37. Let $\mathcal{P}$ be any unary predicate on $\mathcal{L}$ (equivalently, $\mathcal{P} \subseteq \mathcal{L}$).

(i) The *$\beta_{\mathcal{L}}$-reduction restricted to $\mathcal{P}$*, written $\rightarrow_{\beta_{\mathcal{P}}}$, is defined by allowing the contraction of a redex R only if $\mathcal{P}(\mathsf{name}(R))$ holds (equivalently, $\mathsf{name}(R) \in \mathcal{P}$).

(ii) The corresponding notions of *head reduction* $\twoheadrightarrow_{h_{\mathcal{P}}}$ and *standard reduction* $\twoheadrightarrow_{\mathsf{st}_{\mathcal{P}}}$ are defined as in the $\rightarrow_\beta$ case.

When $\mathcal{P} = \mathcal{L}$, the reduction relation $\rightarrow_{\beta_{\mathcal{P}}}$ coincides with the labeled β-reduction $\rightarrow_{\beta_{\mathcal{L}}}$ from Definition 5.33(ii). The statements of the following results are formulated for an arbitrary $\mathcal{P} \subseteq \mathcal{L}$ but, in practice, we are going to use predicates that force the calculus to enjoy strong normalization.

LEMMA 5.38. *Let $M, N \in \Lambda^{\mathcal{L}}$, $\alpha, \beta \in \mathcal{L}$ and $\mathcal{P} \subseteq \mathcal{L}$. Then*

(i) $\alpha \cdot (\beta \cdot M) \quad = \quad \alpha\beta \cdot M.$

(ii) $M \rightarrow_{\beta_{\mathcal{P}}} N \quad \Rightarrow \quad \alpha \cdot M \rightarrow_{\beta_{\mathcal{P}}} \alpha \cdot N.$

(iii) $M \twoheadrightarrow_{\beta_{\mathcal{P}}} N \quad \Rightarrow \quad \alpha \cdot M \twoheadrightarrow_{\beta_{\mathcal{P}}} \alpha \cdot N.$

PROOF. (i) Easy. By structural induction on M, using the associativity of juxtaposition.

(ii) By induction on a derivation of $M \rightarrow_{\beta_{\mathcal{P}}} N$. The only interesting case is $M = ((\lambda x.P^\gamma)Q)^\beta$, $N = \beta\lceil\gamma\rceil \cdot P[x:=\lfloor\gamma\rfloor \cdot Q]$ and $\mathcal{P}(\gamma)$ holds. Then

$$\begin{array}{rlll} \alpha \cdot M & = & ((\lambda x.P^\gamma)Q)^{\alpha\beta} & \\ & \rightarrow_{\beta_{\mathcal{P}}} & \alpha\beta\lceil\gamma\rceil \cdot P[x:=\lfloor\gamma\rfloor \cdot Q], & \text{as } \mathcal{P}(\gamma) \text{ holds,} \\ & = & \alpha \cdot (\lceil\beta\rceil\gamma \cdot P[x:=\lfloor\gamma\rfloor \cdot Q]), & \text{by (i),} \\ & = & \alpha \cdot N. & \end{array}$$

(iii) By induction on the length of the reduction $M \twoheadrightarrow_{\beta_{\mathcal{P}}} N$, using (ii). □

The next lemmas show how the size and the depth of some labels evolve during the reduction of a term M. In particular, on the external label of M, both measures may only increase along its reduction.

LEMMA 5.39. *Let $M^\alpha, N^\beta \in \Lambda^{\mathcal{L}}$ and $\mathcal{P} \subseteq \mathcal{L}$. If $M^\alpha \twoheadrightarrow_{\beta_{\mathcal{P}}} N^\beta$ then* $\mathsf{size}(\alpha) \leq \mathsf{size}(\beta)$.

PROOF. Immediate. □

LEMMA 5.40. (i) *If $M^\alpha \rightarrow_{\beta_{\mathcal{P}}} N^\beta$ then* $\mathsf{depth}(\alpha) \leq \mathsf{depth}(\beta)$.

(ii) *If $M^\alpha \twoheadrightarrow_{\beta_{\mathcal{P}}} N^\beta$ then* $\mathsf{depth}(\alpha) \leq \mathsf{depth}(\beta)$.

(iii) *Let $(((\cdots(M_0^{\alpha_0} M_1)^{\alpha_1} M_2)^{\alpha_2}) \cdots M_k)^{\alpha_k} \twoheadrightarrow_{\beta_{\mathcal{P}}} (\lambda x.N)^\beta$, for some $k \geq 0$. Then*

$$\mathsf{depth}(\alpha_0) \leq \mathsf{depth}(\beta).$$

PROOF. (i) If the contracted redex is a proper subterm of M, then the external label does not change, i.e. $\alpha = \beta$. Otherwise, $M = ((\lambda x.P)^{\gamma}Q)^{\alpha}$ and $N = \alpha\lceil\gamma\rceil \cdot P[x := \lfloor\gamma\rfloor \cdot Q]$. Conclude since $\text{depth}(\alpha) \leq \max\{\text{depth}(\alpha), \text{depth}(\gamma)+1\} = \text{depth}(\alpha\lceil\gamma\rceil) \leq \text{depth}(\beta)$.

(ii) By induction on the length of the reduction, using (i).

(iii) By induction on k.

Case $k = 0$. By (ii).

Case $k > 0$. In this case, we must have:

$$\begin{array}{rcl} (((\cdots(M_0^{\alpha_0}M_1)^{\alpha_1}M_2)^{\alpha_2})\cdots M_{k-1})^{\alpha_{k-1}} & \twoheadrightarrow_{\beta_{\mathcal{P}}} & (\lambda y.P)^{\gamma}, \text{ for some } P \text{ and } \gamma; \\ M_k & \twoheadrightarrow_{\beta_{\mathcal{P}}} & M'_k; \\ ((\lambda y.P)^{\gamma}M'_k)^{\alpha_k} & \twoheadrightarrow_{\beta_{\mathcal{P}}} & \alpha_k\lceil\gamma\rceil \cdot P[y := \lfloor\gamma\rfloor \cdot M'_k]; \end{array}$$

where $\text{depth}(\alpha_0) \leq \text{depth}(\gamma)$ holds by IH. Calling $\varepsilon \triangleq \ell_e(P[y := \lfloor\gamma\rfloor \cdot M'_k])$, we have $\beta = \alpha_k\lceil\gamma\rceil\varepsilon$. We conclude as follows:

$$\begin{array}{rcl} \text{depth}(\alpha_0) & < & \text{depth}(\gamma)+1 \\ & = & \text{depth}(\lceil\gamma\rceil) \\ & \leq & \max\{\text{depth}(\alpha_k), \text{depth}(\lceil\gamma\rceil), \text{depth}(\varepsilon)\} \\ & = & \text{depth}(\alpha_k\lceil\gamma\rceil\varepsilon) \\ & = & \text{depth}(\beta). \end{array}$$

□

The following properties satisfied by β-reduction are also enjoyed by $\rightarrow_{\beta_{\mathcal{P}}}$.

LEMMA 5.41. *Let* $M, M', N, N' \in \Lambda^{\mathcal{L}}$, $x \in \text{Var}$ *and* $\mathcal{P} \subseteq \mathcal{L}$. *Then*

(i) $M \rightarrow_{\beta_{\mathcal{P}}} M' \Rightarrow M[x := N] \rightarrow_{\beta_{\mathcal{P}}} M'[x := N]$;

(ii) $M \twoheadrightarrow_{\beta_{\mathcal{P}}} M' \Rightarrow M[x := N] \twoheadrightarrow_{\beta_{\mathcal{P}}} M'[x := N]$;

(iii) $N \twoheadrightarrow_{\beta_{\mathcal{P}}} N' \Rightarrow M[x := N] \twoheadrightarrow_{\beta_{\mathcal{P}}} M[x := N']$.

PROOF. (i) Proceed by structural induction on M. The only interesting case is when $M = ((\lambda y.M_1)^{\alpha}M_2)^{\beta}$, $M' = \beta\lceil\alpha\rceil \cdot M_1[y := \lfloor\alpha\rfloor \cdot M_2]$ and $\mathcal{P}(\alpha)$ holds. Wlog, we may assume $x \neq y$ and $y \notin \text{FV}(N)$. On the one hand, by Definition 5.33(i), we have

$$M[x := N] = ((\lambda y.M_1[x := N])^{\alpha}(M_2[x := N]))^{\beta}$$

On the other hand, we obtain

$$\begin{array}{rcll} M'[x := N] & = & (\beta\lceil\alpha\rceil \cdot M_1[y := \lfloor\alpha\rfloor \cdot M_2])[x := N] & \\ & = & \beta\lceil\alpha\rceil \cdot M_1[y := \lfloor\alpha\rfloor \cdot M_2][x := N], & \text{by Lemma 5.36(i)}, \\ & = & \beta\lceil\alpha\rceil \cdot M_1[x := N][y := (\lfloor\alpha\rfloor \cdot M_2)[x := N]], & \text{by Lemma 5.36(ii)}, \\ & = & \beta\lceil\alpha\rceil \cdot M_1[x := N][y := \lfloor\alpha\rfloor \cdot M_2[x := N]], & \text{by Lemma 5.36(i)}. \end{array}$$

Since $\mathcal{P}(\alpha)$ holds, we conclude $M[x := N] \rightarrow_{\beta_{\mathcal{P}}} M'[x := N]$.

(ii) By induction on the length of $M \twoheadrightarrow_{\beta_{\mathcal{P}}} M'$, using (i).

(iii) By structural induction on M, using Definition 5.33(i).

Case $M = x^{\alpha}$. By Lemma 5.38(iii), $x^{\alpha}[x := N] = \alpha \cdot N \twoheadrightarrow_{\beta_{\mathcal{P}}} \alpha \cdot N' = x^{\alpha}[x := N']$.

Case $M = y^{\alpha}$ with $x \neq y$. Trivial.

Case $M = (\lambda y.M_1)^\alpha$ where, wlog, $x \neq y \notin \mathrm{FV}(N)$. Then

$$\begin{array}{rcll} (\lambda y.M_1)^\alpha[x:=N] & = & (\lambda y.M_1[x:=N])^\alpha, & \text{by Definition 5.33(i)}, \\ & \twoheadrightarrow_{\beta_\mathcal{P}} & (\lambda y.M_1[x:=N'])^\alpha, & \\ & & \text{since } M_1[x:=N] \twoheadrightarrow_{\beta_\mathcal{P}} M_1[x:=N'] \text{ by IH}, & \\ & = & (\lambda y.M_1)^\alpha[x:=N']. & \end{array}$$

Case $M = (M_1M_2)^\alpha$. It follows analogously from the IH. □

COROLLARY 5.42. *For all* $M, M', N, N' \in \Lambda^\mathcal{L}$, $x \in \mathrm{Var}$ *and* $\mathcal{P} \subseteq \mathcal{L}$, *we have:*

$$M \twoheadrightarrow_{\beta_\mathcal{P}} M' \ \&\ N \twoheadrightarrow_{\beta_\mathcal{P}} N' \quad \Rightarrow \quad M[x:=N] \twoheadrightarrow_{\beta_\mathcal{P}} M'[x:=N']$$

PROOF. By Lemma 5.41(ii)-(iii). □

PROPOSITION 5.43 (LOCAL CONFLUENCE). *Let* $M \in \Lambda^\mathcal{L}$, $R_1, R_2 \in M$ *and* $\mathcal{P} \subseteq \mathcal{L}$. *If* $M \xrightarrow{R_1}_{\beta_\mathcal{P}} N_1$ *and* $M \xrightarrow{R_2}_{\beta_\mathcal{P}} N_2$ *then there is* $Z \in \Lambda^\mathcal{L}$ *and two reductions* ρ_1, ρ_2 *such that*

$$\begin{array}{rl} \rho_1: & N_1 \twoheadrightarrow_{\beta_\mathcal{P}} Z \\ \rho_2: & N_2 \twoheadrightarrow_{\beta_\mathcal{P}} Z \end{array}$$

Moreover, $\rho_1 = R_2/R_1$ *and* $\rho_2 = R_1/R_2$. *In diagrammatic form:*

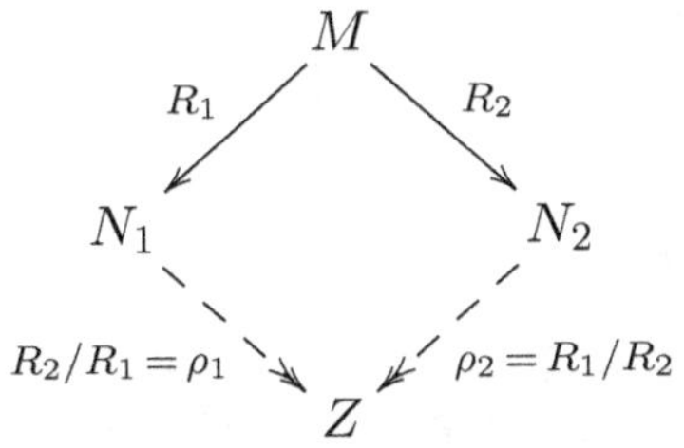

PROOF. By structural induction on M. Assume $R_1 \neq R_2$, otherwise it is trivial.

Case $M = x^\alpha$. Vacuous.

Case $M = (\lambda x.M_1)^\alpha$. Then, we must have $R_1, R_2 \in M_1$. By contracting these redexes, we obtain $M_1 \xrightarrow{R_1}_{\beta_\mathcal{P}} N_1'$ and $M_2 \xrightarrow{R_2}_{\beta_\mathcal{P}} N_2'$, for some $N_1', N_2' \in \Lambda^\mathcal{L}$. From the IH, we obtain two reductions $\rho_1' : N_1' \twoheadrightarrow_{\beta_\mathcal{P}} Z'$ and $\rho_2' : N_2' \twoheadrightarrow_{\beta_\mathcal{P}} Z'$, for some $Z' \in \Lambda^\mathcal{L}$. We conclude by setting $Z \triangleq \lambda x.Z'$, $\rho_1 \triangleq \lambda x.\rho_1'$ and $\rho_2 \triangleq \lambda x.\rho_2'$.

Case $M = (M_1M_2)^\alpha$. There are two subcases.

1. Both redexes occur in the same component, say, $R_1, R_2 \in M_1$. This case is similar to the previous one: simply apply the IH and proceed as above.

2. The two redexes occur in different components, say, $R_1 \in M_1$ and $R_2 \in M_2$. Then $M_1 \xrightarrow{R_1}_{\beta_\mathcal{P}} N_1'$ and $M_2 \xrightarrow{R_2}_{\beta_\mathcal{P}} N_2'$. We close the diagram in one step, by contracting the unique residual of R_1 in M_2, and of R_2 in M_1.

Case $M = R_1 = ((\lambda x.M_1)^\alpha M_2)^\beta$, whence $N_1 = \beta\lceil\alpha\rceil \cdot M_1[x := \lfloor\alpha\rfloor \cdot M_2]$. We split into subcases, depending on the position of R_2 in M.

1. Subcase $R_2 \in M_1$. Let N_1' be such that $M_1 \xrightarrow{R_2}_{\beta_\mathcal{P}} N_1'$. Then $((\lambda x.N_1')^\alpha M_2)^\beta \to_{\beta_\mathcal{P}} \beta\lceil\alpha\rceil \cdot N_1'[x := \lfloor\alpha\rfloor \cdot M_2]$. Since $M_1 \to_{\beta_\mathcal{P}} N_1'$, by Lemmas 5.38(iii) and 5.41(i), we conclude that $N_1 \to_{\beta_\mathcal{P}} \beta\lceil\alpha\rceil \cdot N_1'[x:=\lfloor\alpha\rfloor \cdot M_2]$.

2. Subcase $R_2 \in M_2$. Let N_2' be such that $M_2 \xrightarrow{R_2}_{\beta_\mathcal{P}} N_2'$. Then $((\lambda x.M_1)^\alpha N_2')^\beta \to_{\beta_\mathcal{P}} \beta\lceil\alpha\rceil \cdot M_1[x := \lfloor\alpha\rfloor \cdot N_2']$. Since $M_2 \to_{\beta_\mathcal{P}} N_2'$, by Lemmas 5.38(iii) and 5.41(iii) we conclude that $N_1 \twoheadrightarrow_{\beta_\mathcal{P}} \beta\lceil\alpha\rceil \cdot M_1[x:=\lfloor\alpha\rfloor \cdot N_2']$.

Case $M = R_2 = ((\lambda x.M_1)^\alpha M_2)^\beta$. Symmetric to the previous case.

Note that, in each of the above cases, we took $\rho_1 = R_2/R_1$ and $\rho_2 = R_1/R_2$. □

Recall that $\mathcal{G}_{\beta_\mathcal{L}}(M)$ denotes the $\beta_\mathcal{L}$-reduction graph of $M \in \Lambda^\mathcal{L}$.

DEFINITION 5.44. Consider a predicate $\mathcal{P} \subseteq \mathcal{L}$.

(i) An $M \in \Lambda^\mathcal{L}$ is called *strongly normalizing* with respect to $\to_{\beta_\mathcal{P}}$ if all reductions in $\mathcal{G}_{\beta_\mathcal{P}}(M)$ are finite.

(ii) Let $\mathrm{SN}_{\beta_\mathcal{P}}$ be the set of all labeled λ-terms strongly normalizing w.r.t. $\to_{\beta_\mathcal{P}}$.

(iii) For $M \in \mathrm{SN}_{\beta_\mathcal{P}}$, max-red$(M)$ denotes the maximal length of reductions in $\mathcal{G}_{\beta_\mathcal{P}}(M)$.

A celebrated result in the theory of rewriting systems – known as Newman's Lemma – states that, in a strongly normalizing setting, local confluence implies confluence. The reader may consult the original paper by Newman (1942) or Terese (2003), Theorem 1.2.1. The next proposition is an instance of this property, but the statement is more precise since it specifies the form of the two reductions needed for closing the confluence diagram.

PROPOSITION 5.45. *Let $\mathcal{P} \subseteq \mathcal{L}$ and $M \in \mathrm{SN}_{\beta_\mathcal{P}}$. If $\rho : M \twoheadrightarrow_{\beta_\mathcal{P}} N_1$ and $\nu : M \twoheadrightarrow_{\beta_\mathcal{P}} N_2$, then there exists $Z \in \Lambda^\mathcal{L}$ such that $\nu/\rho : N_1 \twoheadrightarrow_{\beta_\mathcal{P}} Z$ and $\rho/\nu : N_2 \twoheadrightarrow_{\beta_\mathcal{P}} Z$.*

PROOF. Since $M \in \mathrm{SN}_{\beta_\mathcal{P}}$, we may proceed by induction on max-red$(M) = n$.

Case $n = 0$. Then M is in $\beta_\mathcal{P}$-normal form. Hence, $\rho = \nu = o$.

Case $n > 0$. If $\rho = o$ then $M = N_1$, $\rho/\nu = o$ and $\nu/\rho = \nu \neq o$. Take $Z \triangleq N_2$.

If $\nu = o$ then the reasoning is symmetric.

Assume now that $\rho, \nu \neq o$. In this case, ρ and ν are of the form

$$\begin{array}{rcl} \rho & = & M \xrightarrow{R}_{\beta_\mathcal{P}} P_1 \xrightarrow{\rho_1}\!\!\!\!\twoheadrightarrow_{\beta_\mathcal{P}} N_1, \\ \nu & = & M \xrightarrow{S}_{\beta_\mathcal{P}} P_2 \xrightarrow{\nu_2}\!\!\!\!\twoheadrightarrow_{\beta_\mathcal{P}} N_2. \end{array}$$

By Proposition 5.43, there is $Q \in \Lambda^\mathcal{L}$ such that $S/R : P_1 \twoheadrightarrow_{\beta_\mathcal{P}} Q$ and $R/S : P_2 \twoheadrightarrow_{\beta_\mathcal{P}} Q$. By construction, max-red(P_1) and max-red(P_2) are strictly smaller than n. By applying the IH to $N_1 \,{}_{\beta_\mathcal{P}}\!\!\overset{\rho_1}{\twoheadleftarrow} P_1 \overset{S/R}{\twoheadrightarrow}_{\beta_\mathcal{P}} Q$ we obtain a $Q_1 \in \Lambda^\mathcal{L}$ such that

$$\begin{array}{rl} (S/R)/\rho_1 : & N_1 \twoheadrightarrow_{\beta_\mathcal{P}} Q_1, \\ \rho_1/(S/R) : & Q \twoheadrightarrow_{\beta_\mathcal{P}} Q_1. \end{array}$$

Analogously, the IH on $Q \,{}_{\beta_\mathcal{P}}\!\!\overset{R/S}{\twoheadleftarrow} P_2 \overset{\nu_2}{\twoheadrightarrow}_{\beta_\mathcal{P}} N_2$ gives a $Q_2 \in \Lambda^\mathcal{L}$ satisfying

$$\begin{array}{rl} \nu_2/(R/S) : & Q \twoheadrightarrow_{\beta_\mathcal{P}} Q_2, \\ (R/S)/\nu_2 : & N_2 \twoheadrightarrow_{\beta_\mathcal{P}} Q_2. \end{array}$$

Since $Q \in \mathcal{G}_{\beta_\mathcal{P}}(P_1) \cap \mathcal{G}_{\beta_\mathcal{P}}(P_2)$, the longest reduction in $\mathcal{G}_{\beta_\mathcal{P}}(Q)$ has length smaller than n. We can therefore apply the IH on $\rho_1/(S/R) : Q \twoheadrightarrow_{\beta_\mathcal{P}} Q_1$ and $\nu_2/(R/S) : Q \twoheadrightarrow_{\beta_\mathcal{P}} Q_2$, and obtain a $Z \in \Lambda^\mathcal{L}$ and two reductions

$$\begin{array}{rl} (\nu_2/(R/S))/(\rho_1/(S/R)) : & Q_1 \twoheadrightarrow_{\beta_\mathcal{P}} Z, \\ (\rho_1/(S/R))/(\nu_2/(R/S)) : & Q_2 \twoheadrightarrow_{\beta_\mathcal{P}} Z. \end{array}$$

It remains to be checked that:

$$\begin{array}{rcll} \nu/\rho & = & (S\,;\nu_2)/(R\,;\rho_1) & \\ & = & (S/(R;\rho_1)); (\nu_2/((R;\rho_1)/S)), & \text{by Lemma 1.51(i)}, \\ & = & (S/(R;\rho_1)); (\nu_2/(R/S); (\rho_1/(S/R))), & \text{by Lemma 1.51(i)}, \\ & = & ((S/R)/\rho_1)\,; (\nu_2/(R/S))/(\rho_1/(S/R)), & \text{by Lemma 1.51(ii)}. \end{array}$$

Analogous calculations give $((R/S)/\rho_2)\,; (\rho_1/(S/R))/(\nu_2/(R/S)) = \rho/\nu$. □

Recall that the depth of a label has been introduced in Definition 5.29(iii).

DEFINITION 5.46. (i) The *depth of a predicate* $\mathcal{P} \subseteq \mathcal{L}$ is defined as follows:

$$\text{depth}(\mathcal{P}) \triangleq \max\{\text{depth}(\alpha) \mid \alpha \in \mathcal{P}\} \in \mathbb{N} \cup \{\infty\}$$

(ii) A predicate $\mathcal{P}$ is called *bounded* if it has a finite depth, i.e., $\text{depth}(\mathcal{P}) \in \mathbb{N}$.

REMARK 5.47. If a predicate $\mathcal{P}$ is finite, then it is necessarily bounded.

LEMMA 5.48. *Let $\mathcal{P} \subseteq \mathcal{L}$. Every standard reduction $\rho : M[x:=N] \twoheadrightarrow_{\mathsf{st}_\mathcal{P}} (\lambda y.P)^\alpha$ can be factorized in one of the following ways:*

1. *$M \twoheadrightarrow_{\beta_\mathcal{P}} (\lambda y.M')^\alpha$ and $M'[x:=N] \twoheadrightarrow_{\beta_\mathcal{P}} P$;*
2. *$M \twoheadrightarrow_{\beta_\mathcal{P}} M' \triangleq ((\cdots((x^{\beta_0}M_1)^{\beta_1}M_2)^{\beta_2}\cdots M_k)^{\beta_k}$ and $M'[x:=N] \twoheadrightarrow_{\beta_\mathcal{P}} (\lambda y.P)^\alpha$.*

PROOF. By induction on the length n of ρ.

Case $n = 0$. In this case $M[x:=N] = (\lambda y.P)^\alpha$. There are two possibilities. Either $M = (\lambda y.M')^\alpha$ and $M'[x:=N] = P$, or $M = x^\beta$ and $x^\beta[x:=N] = (\lambda y.P)^\alpha$.

Case $n > 0$. We distinguish subcases depending on the structure of M.

- Subcase $M = (\lambda y.M')^\beta$. In this case $M[x:=N] = (\lambda y.M'[x:=N])^\beta$. Therefore, we get $\alpha = \beta$ and $M'[x:=N] \twoheadrightarrow_{\beta_\mathcal{P}} P$.

- Subcase $M = (\cdots((y^{\beta_0}M_1)^{\beta_1}M_2)^{\beta_2}\cdots M_k)^{\beta_k}$. We must have $x = y$, for otherwise $M[x:=N]$ could not reduce to an abstraction. Therefore, taking we can take $M' \triangleq M$ and obtain $M'[x:=N] \twoheadrightarrow_{\beta_\mathcal{P}} (\lambda y.P)^\alpha$.

- Subcase $M = (\cdots((((\lambda y.X)^\gamma Y)^{\beta_0}M_1)^{\beta_1}M_2)^{\beta_2}\cdots M_k)^{\beta_k}$. In this case, the redex $((\lambda y.X)^\gamma Y)^{\beta_0}$ must be contracted along the standard reduction ρ, otherwise we could not get as a result an abstraction. Moreover, since $((\lambda y.X)^\gamma Y)^{\beta_0}$ is the head redex in M, it is the first one contracted in $M[x:=N] \twoheadrightarrow_{\mathsf{st}_\mathcal{P}} (\lambda y.P)^\alpha$. Setting

  $$Q \triangleq (\cdots(((\beta_0 \cdot \lceil\gamma\rceil \cdot X[y:=\lfloor\gamma\rfloor \cdot Y])M_1)^{\beta_1}M_2)^{\beta_2}\cdots M_k)^{\beta_k},$$

  we obtain a reduction $\nu : Q[x:=N] \twoheadrightarrow_{\mathsf{st}_\mathcal{P}} (\lambda y.P)^\alpha$ strictly shorter than ρ. Since $M \twoheadrightarrow_{\beta_\mathcal{P}} Q$, we conclude by applying the IH to ν. □

PROPOSITION 5.49. *Let $\mathcal{P} \subseteq \mathcal{L}$ and $M \in \mathrm{SN}_{\beta_\mathcal{P}}$. For every $\rho : M \twoheadrightarrow_{\beta_\mathcal{P}} N$, there exists a standard reduction $\nu : M \twoheadrightarrow_{\mathsf{st}_\mathcal{P}} N$ such that $\rho \sim \nu$.*

PROOF. By induction on max-red(M).

Case max-red$(M) = 0$. Trivial.

Case max-red$(M) > 0$. We proceed by structural induction on M and call IH2 the corresponding induction hypothesis.

- $M = x^\alpha$. This case is vacuous.
- $M = (\lambda x.M_1)^\alpha$. In this case, $N = (\lambda x.N_1)^\alpha$ and $M_1 \twoheadrightarrow_{\beta_\mathcal{P}} N_1$. Since M_1 is a strict subterm of M, we have $M_1 \twoheadrightarrow_{\mathsf{st}_\mathcal{P}} N_1$ by IH2, so we obtain $\nu : M \twoheadrightarrow_{\mathsf{st}_\mathcal{P}} N$.
- $M = (M_1 M_2)^\alpha$. Let us distinguish two subcases:

  Subcase 1. $N = (N_1 N_2)^\alpha$, and the reduction $\rho : M \twoheadrightarrow_{\beta_\mathcal{P}} N$ is composed of two separate reductions $\rho_1 : M_1 \twoheadrightarrow_{\beta_\mathcal{P}} N_1$ and $\rho_2 : M_2 \twoheadrightarrow_{\beta_\mathcal{P}} N_2$. By IH2, we get $M_1 \twoheadrightarrow_{\mathsf{st}_\mathcal{P}} N_1$, $M_2 \twoheadrightarrow_{\mathsf{st}_\mathcal{P}} N_2$, whence the standard reduction ν proceeds as follows:

  $$M = (M_1 M_2)^\alpha \twoheadrightarrow_{\mathsf{st}_\mathcal{P}} (N_1 M_2)^\alpha \twoheadrightarrow_{\mathsf{st}_\mathcal{P}} (N_1 N_2)^\alpha = N.$$

  Subcase 2. The reduction ρ can be decomposed as follows:

  $$(M_1 M_2)^\alpha \twoheadrightarrow_{\beta_\mathcal{P}} ((\lambda x.N_1)^\beta N_2)^\alpha \to_{\beta_\mathcal{P}} \alpha\lceil\beta\rceil \cdot N_1[x := \lfloor\beta\rfloor \cdot N_2] \twoheadrightarrow_{\beta_\mathcal{P}} N$$

  with $M_1 \twoheadrightarrow_{\beta_\mathcal{P}} (\lambda x.N_1)^\beta$ and $M_2 \twoheadrightarrow_{\beta_\mathcal{P}} N_2$. By IH2, there exists a standard reduction $\nu_1 : M_1 \twoheadrightarrow_{\mathsf{st}_\mathcal{P}} (\lambda x.N_1)^\beta$. Now, let us call $\lambda x.M_3$ the first appearance of an external abstraction in the head reduction sequence of M_1, namely:

  $$\nu_1 : M_1 \twoheadrightarrow_{h_\mathcal{P}} (\lambda x.M_3)^\beta \twoheadrightarrow_{\mathsf{st}_\mathcal{P}} (\lambda x.N_1)^\beta,$$

  where $M_3 \twoheadrightarrow_{\beta_\mathcal{P}} N_1$. From this and $M_2 \twoheadrightarrow_{\beta_\mathcal{P}} N_2$, we obtain the reduction

  $$\begin{array}{lll} (M_1 M_2)^\alpha & \twoheadrightarrow_{h_\mathcal{P}} & ((\lambda x.M_3)^\beta M_2)^\alpha \\ & \to_{h_\mathcal{P}} & \alpha\lceil\beta\rceil \cdot M_3[x := \lfloor\beta\rfloor \cdot M_2] \\ & \twoheadrightarrow_{\beta_\mathcal{P}} & \alpha\lceil\beta\rceil \cdot N_1[x := \lfloor\beta\rfloor \cdot N_2], \quad \text{by Corollary 5.42} \\ & & \qquad\qquad \text{and Lemma 5.38(iii),} \\ & \twoheadrightarrow_{\beta_\mathcal{P}} & N \end{array}$$

  Because of the head step $\to_{h_\mathcal{P}}$, we must have max-red$(\alpha\lceil\beta\rceil \cdot M_3[x := \lfloor\beta\rfloor \cdot M_2]) <$ max-red(M). By IH we get a standard $\nu_2 : \alpha\lceil\beta\rceil \cdot M_3[x := \lfloor\beta\rfloor \cdot M_2] \twoheadrightarrow_{\mathsf{st}_\mathcal{P}} N$. Gathering all the pieces together, the standard reduction ν is given by:

  $$\begin{array}{lll} M = (M_1 M_2)^\alpha & \twoheadrightarrow_{h_\mathcal{P}} & ((\lambda x.M_3)^\beta M_2)^\alpha \\ & \to_{h_\mathcal{P}} & \alpha\lceil\beta\rceil \cdot M_3[x := \lfloor\beta\rfloor \cdot M_2] \twoheadrightarrow_{\mathsf{st}_\mathcal{P}} N. \end{array}$$

We leave as an exercise to check that $\rho \sim \nu$ holds. □

LEMMA 5.50. *Let $\mathcal{P}$ be bounded. Then*

$$M, N \in \mathrm{SN}_{\beta_\mathcal{P}} \quad \Rightarrow \quad M[x := N] \in \mathrm{SN}_{\beta_\mathcal{P}}$$

PROOF. We prove the statement by induction on the lexicographically ordered triple:

$$(\mathrm{depth}(\mathcal{P}) \dot{-} \mathrm{depth}(\ell_\mathrm{e}(N)), \text{max-red}(M), \mathrm{size}(M))$$

where $n \dot{-} m \triangleq \max\{0, n - m\}$. Note that $\mathrm{depth}(\mathcal{P}) \neq \infty$ since $\mathcal{P}$ is bounded. Moreover, if $\mathcal{P}(\alpha)$ holds, then $\mathrm{depth}(\mathcal{P}) - \mathrm{depth}(\alpha) \geq 0$.

Case $(0, 0, 0)$. Vacuous, since necessarily $\mathrm{size}(M) > 0$.

Base case $(0, 0, 1)$. Since $\mathrm{size}(M) = 1$, then $M = y^\alpha$. There are two possibilities.
Subcase $y \neq x$. Trivial, since $M[x := N] = y^\alpha$.
Subcase $y = x$. Then $M[x := N] = N$, which is strongly normalizing by hypothesis.

Induction case. We distinguish several cases, depending on the structure of M.

1. $M = y^\alpha$. We can proceed like in the base case, since the reasoning did not rely on the hypothesis $\mathrm{depth}(\mathcal{P}) \dot{-} \mathrm{depth}(\ell_\mathrm{e}(N)) = 0$.

2. $M = (\lambda y.M_1)^\alpha$. Consider the triple

$$(\mathrm{depth}(\mathcal{P}) \dot{-} \mathrm{depth}(\ell_\mathrm{e}(N)), \text{max-red}(M_1), \mathrm{size}(M_1)).$$

   Since M is strongly normalizing, then so is M_1, and $\text{max-red}(M_1) \leq \text{max-red}(M)$. Moreover, $\mathrm{size}(M_1) < \mathrm{size}(M)$ so the IH gives that $M_1[x := N] \in \mathrm{SN}_{\beta_\mathcal{P}}$.

3. $M = (M_1 M_2)^\alpha$. By Definition 5.33(i), $M[x := N] = (M_1[x := N](M_2[x := N]))^\alpha$. For every $i \in \{1, 2\}$, we have $\text{max-red}(M_i) \leq \text{max-red}(M)$ and $\mathrm{size}(M_i) < \mathrm{size}(M)$. Therefore, each $M_i[x := N]$ is strongly normalizing by IH. Two cases are possible:

   (a) $M_1[x := N]$ never reduces to a λ-abstraction. In this case any reduction $\rho \in \mathcal{G}_{\beta_\mathcal{P}}(M[x := N])$ must factorize as the composition of two independent reductions $\rho_1 \in \mathcal{G}_{\beta_\mathcal{P}}(M_1[x := N])$ and $\rho_2 \in \mathcal{G}_{\beta_\mathcal{P}}(M_2[x := N])$. Since $M_1[x := N]$ and $M_2[x := N]$ are strongly normalizing, we conclude $M[x := N] \in \mathrm{SN}_{\beta_\mathcal{P}}$.

   (b) $M_1[x := N] \twoheadrightarrow_{\beta_\mathcal{P}} (\lambda y.P)^\beta \in \mathrm{SN}_{\beta_\mathcal{P}}$. The case in which $\mathcal{P}(\beta)$ is false is similar to the previous one. Assume now that $\mathcal{P}(\beta)$ holds. Recall that in this case $\mathrm{depth}(\mathcal{P}) \geq \mathrm{depth}(\beta)$. Easy calculations give:

$$M[x := N] = (M_1[x := N](M_2[x := N]))^\alpha \twoheadrightarrow_{\beta_\mathcal{P}} ((\lambda y.P)^\beta (M_2[x := N]))^\alpha.$$

   We need to show that its reduct $\alpha \cdot \lceil \beta \rceil \cdot P[y := \lfloor \beta \rfloor \cdot M_2[x := N]]$ belongs to $\mathrm{SN}_{\beta_\mathcal{P}}$. Since $M_1[x := N] \in \mathrm{SN}_{\beta_\mathcal{P}}$, by Proposition 5.49, there is a standard reduction $M_1[x := N] \twoheadrightarrow_{\mathsf{st}_\mathcal{P}} (\lambda y.P)^\beta$. By Lemma 5.48, there are two possibilities:

(i) $M_1 \twoheadrightarrow_{\beta_{\mathcal{P}}} (\lambda y.M_3)^\beta$ and $M_3[x:=N] \twoheadrightarrow_{\beta_{\mathcal{P}}} P$. From $M_1[x:=N] \in \mathrm{SN}_{\beta_{\mathcal{P}}}$, it follows that $M' \triangleq \alpha\lceil\beta\rceil \cdot M_3[y:=\lfloor\beta\rfloor \cdot M_2] \in \mathrm{SN}_{\beta_{\mathcal{P}}}$. By Lemma 5.36(i)-(ii), we obtain

$$M'[x:=N] = \alpha\lceil\beta\rceil \cdot M_3[x:=N][y:=\lfloor\beta\rfloor \cdot M_2[x:=N]]$$

By Lemmas 5.38(iii) and 5.41(ii), we obtain:

$$M'[x:=N] \twoheadrightarrow_{\beta_{\mathcal{P}}} \alpha\lceil\beta\rceil \cdot P[y:=\lfloor\beta\rfloor \cdot M_2[x:=N]]$$

Note that $M \twoheadrightarrow_{\beta_{\mathcal{P}}} M'$, therefore $\mathrm{depth}(\ell_e(M)) \leq \mathrm{depth}(\ell_e(M'))$ holds by Lemma 5.40(i), and $\text{max-red}(M') < \text{max-red}(M)$.

$$\begin{aligned}&(\mathrm{depth}(\mathcal{P}) - \mathrm{depth}(\ell_e(N)), \text{max-red}(M'), \mathrm{size}(M'))\\ &\quad < (\mathrm{depth}(\mathcal{P}) - \mathrm{depth}(\ell_e(N)), \text{max-red}(M), \mathrm{size}(M)).\end{aligned}$$

By applying the IH we conclude that $M'[x:=N] \in \mathrm{SN}_{\beta_{\mathcal{P}}}$.

(ii) $M_1 \twoheadrightarrow_{\beta_{\mathcal{P}}} M_1' \triangleq (\cdots((x^\gamma X_1)^{\alpha_1} X_2)^{\alpha_2}) \cdots X_n)^{\alpha_n}$ for some M_1' such that $M_1'[x:=N] \twoheadrightarrow_{\beta_{\mathcal{P}}} (\lambda y.P)^\beta$. Since $M_1[x:=N] \in \mathrm{SN}_{\beta_{\mathcal{P}}}$, then also $P \in \mathrm{SN}_{\beta_{\mathcal{P}}}$. Moreover:

$$M_1'[x:=N] = (\cdots((\gamma \cdot N)(X_1[x:=N])) \cdots (X_n[x:=N]))^{\alpha_n}$$

By Lemma 5.40(iii), $\mathrm{depth}(\ell_e(\gamma \cdot N)) \leq \mathrm{depth}(\beta)$. Therefore, we have:

$$\begin{aligned}\mathrm{depth}(\ell_e(N)) &\leq \mathrm{depth}(\ell_e(\gamma \cdot N)) &&\leq \mathrm{depth}(\beta)\\ &< \mathrm{depth}(\lfloor\beta\rfloor) &&\leq \mathrm{depth}(\ell_e(\lfloor\beta\rfloor \cdot M_2[x:=N]))\end{aligned}$$

Then, by HI, we conclude $\alpha\lceil\beta\rceil \cdot P[y:=\lfloor\beta\rfloor \cdot M_2[x:=N]] \in \mathrm{SN}_{\beta_{\mathcal{P}}}$. □

The following proposition captures the intuition that a bounded predicate $\mathcal{P}$ only allows finitely many $\beta_{\mathcal{P}}$-reduction steps.

PROPOSITION 5.51. *If $\mathcal{P}$ is bounded, every $M \in \Lambda^{\mathcal{L}}$ is strongly normalizing w.r.t. $\rightarrow_{\beta_{\mathcal{P}}}$.*

PROOF. By structural induction on M.

Case $M = x^\alpha$. Trivial.

Case $M = (\lambda x.N)^\alpha$. Easy, since $N \in \mathrm{SN}_{\beta_{\mathcal{P}}}$ by IH.

Case $M = (N_1 N_2)^\alpha$. By IH, we get $N_1, N_2 \in \mathrm{SN}_{\beta_{\mathcal{P}}}$. Now, if N_1 never reduces to an abstraction, then the result is obvious. Otherwise, there is a reduction $N_1 \twoheadrightarrow_{\beta_{\mathcal{P}}} (\lambda x.N_1')^\beta$. Then, we have

$$M \twoheadrightarrow_{\beta_{\mathcal{P}}} ((\lambda x.N_1')^\beta N_2)^\alpha \rightarrow_{\beta_{\mathcal{P}}} \alpha\lceil\beta\rceil \cdot N_1'[x:=\lfloor\beta\rfloor \cdot N_2],$$

where $\lfloor\beta\rfloor \cdot N_2 \in \mathrm{SN}_{\beta_{\mathcal{P}}}$ because $N_2 \in \mathrm{SN}_{\beta_{\mathcal{P}}}$. By applying Lemma 5.50 we obtain that $N_1'[x:=\lfloor\beta\rfloor \cdot N_2] \in \mathrm{SN}_{\beta_{\mathcal{P}}}$. This allows to conclude. □

Confluence and standardization

We have seen that the labeled λ-calculus endowed with the reduction $\to_{\beta_{\mathcal{P}}}$ satisfies confluence (Proposition 5.45) and standardization (Proposition 5.49), provided that the predicate $\mathcal{P}$ is bounded. We employed proof techniques relying on the fact that, under this hypothesis, the reduction $\to_{\beta_{\mathcal{P}}}$ enjoys strong normalization (Proposition 5.51). We now show that confluence and standardization can be easily transferred to the unrestricted labeled reduction $\to_{\beta_{\mathcal{L}}}$. Indeed, starting from any finite $\beta_{\mathcal{L}}$-reduction ρ, one can construct an isomorphic $\beta_{\mathcal{P}}$-reduction ρ', by constructing a suitable predicate $\mathcal{P}$. Since this predicate can always be taken finite, and therefore bounded, it is possible to apply the previous results holding in the strongly normalizing setting.

DEFINITION 5.52. Let $M, N \in \Lambda^{\mathcal{L}}$, $\mathcal{P} \subseteq \mathcal{L}$ and $\rho : M \twoheadrightarrow_{\beta_{\mathcal{L}}} N$ be a reduction of length n. Call R_i the $\beta_{\mathcal{L}}$-redex contracted at the i-th step of ρ. In symbols: $\rho = R_1 ; \cdots ; R_n$.

(i) We say that $\mathcal{P}$ *is adequate for* ρ whenever $\mathcal{P}(\mathsf{name}(R_i))$ holds, for all i $(1 \leq i \leq n)$.

(ii) Define the finite predicate $\mathcal{P}_\rho \subseteq \mathcal{L}$ as follows:

$$\mathcal{P}_\rho(\alpha) \iff \exists i\,(1 \leq i \leq n)\,.\,\mathsf{name}(R_i) = \alpha$$

LEMMA 5.53. *Let $M \in \Lambda^{\mathcal{L}}$ and $\rho \in \mathcal{G}_{\beta_{\mathcal{P}}}(M)$. Then:*

(i) *$\mathcal{P}_\rho$ is adequate for ρ;*

(ii) *If $\mathcal{P}$ is adequate for ρ and $\mathcal{P} \subseteq \mathcal{P}'$, then $\mathcal{P}'$ is adequate for ρ.*

PROOF. (i) For any redex R contracted in ρ, $\mathcal{P}_\rho(\mathsf{name}(R))$ holds by construction of $\mathcal{P}_\rho$.

(ii) Immediate. □

This allows to prove confluence and standardization for $\Lambda^{\mathcal{L}}$.

THEOREM 5.54 (CONFLUENCE FOR $\Lambda^{\mathcal{L}}$). *The notion of reduction $(\beta_{\mathcal{L}})$ is confluent.*

PROOF. Let $M \in \Lambda^{\mathcal{L}}$. Consider $\rho : M \twoheadrightarrow_{\beta_{\mathcal{L}}} N_1$ and $\nu : M \twoheadrightarrow_{\beta_{\mathcal{L}}} N_2$. By Lemma 5.53(i), we know that $\mathcal{P}_\rho$ and $\mathcal{P}_\nu$ are adequate for ρ and ν, respectively. By Lemma 5.53(ii), $\mathcal{P} \triangleq \mathcal{P}_\rho \cup \mathcal{P}_\nu$ is adequate for both ρ and ν. Moreover, $\mathcal{P}$ is finite, and therefore bounded (see Remark 5.47). By Proposition 5.51, we obtain that $M \in \mathrm{SN}_{\beta_{\mathcal{P}}}$. By Proposition 5.45, the reductions $\nu/\rho : N_1 \twoheadrightarrow_{\beta_{\mathcal{P}}} Z$ and $\rho/\nu : N_2 \twoheadrightarrow_{\beta_{\mathcal{P}}} Z$ give a common reduct Z. □

THEOREM 5.55 (STANDARDIZATION FOR $\Lambda^{\mathcal{L}}$). *Let $M, N \in \Lambda^{\mathcal{L}}$. If $\rho : M \twoheadrightarrow_{\beta_{\mathcal{L}}} N$, then there exists a unique standard reduction $\nu : M \twoheadrightarrow_{\mathsf{st}_{\mathcal{L}}} N$.*

PROOF. The predicate $\mathcal{P}_\rho$ is finite, bounded (see Remark 5.47), and adequate for ρ by Lemma 5.53(i). Therefore, we have $M \in \mathrm{SN}_{\beta_{\mathcal{P}}}$ by Proposition 5.51. By Proposition 5.49, there exists a standard reduction $\nu : M \twoheadrightarrow_{\mathsf{st}_{\mathcal{L}}} N$. We need to check that it is unique.

The proof of uniqueness proceeds by induction on the length n of ν.

Base case: $n = 0$. Then $M = N = x^\alpha$. The only standard reduction is $\nu = o$.

Induction case: $n > 0$. We proceed by structural induction on M and call IH2 the corresponding induction hypothesis.

Case $M = x^\alpha$. Vacuous, since $n > 0$.

Case $M = (\lambda x.M_1)^\alpha$. Then $N = (\lambda x.N_1)^\alpha$ and $M_1 \twoheadrightarrow_{\mathsf{st}_\mathcal{L}} N_1$, with a standard reduction ν' of length n. By IH2 we have that ν' is unique, therefore so is ν.

Case $M = (M_1 M_2)^\alpha$. Then, any standard reduction $M \twoheadrightarrow_{\mathsf{st}_\mathcal{L}} N$ must have one of the following shapes:

1. $(M_1 M_2)^\alpha \twoheadrightarrow_{\mathsf{st}_\mathcal{L}} (N_1 M_2)^\alpha \twoheadrightarrow_{\mathsf{st}_\mathcal{L}} (N_1 N_2)^\alpha = N$;

2. $(M_1 M_2)^\alpha \twoheadrightarrow_{h_\mathcal{L}} ((\lambda x.M_3)^\beta M_2)^\alpha \rightarrow_{h_\mathcal{L}} \alpha\lceil\beta\rceil \cdot M_3[x{:=}\lfloor\beta\rfloor \cdot M_2] \twoheadrightarrow_{\mathsf{st}_\mathcal{L}} N$.

Let us show that if there are two standard reductions $\nu, \nu' : M \twoheadrightarrow_{\mathsf{st}_\mathcal{L}} N$ then they must have the same shape. Suppose, by the way of contradiction, that ν is of shape (1) and ν' is of shape (2). On the one hand, following ν, we must have that $\ell_\mathrm{e}(M) = \ell_\mathrm{e}(N) = \alpha$. On the other hand, following ν', we obtain

$$\begin{array}{rcll} \mathsf{size}(\ell_\mathrm{e}(M)) & = & \mathsf{size}(\alpha) & \\ & < & \mathsf{size}(\alpha) + \mathsf{size}(\lceil\beta\rceil) + \mathsf{size}(\ell_\mathrm{e}(M_3[x{:=}\lfloor\beta\rfloor \cdot M_2])) & \\ & = & \mathsf{size}(\ell_\mathrm{e}(\alpha\lceil\beta\rceil \cdot M_3[x{:=}\lfloor\beta\rfloor \cdot M_2])) & \\ & \leq & \mathsf{size}(\ell_\mathrm{e}(N)), & \text{by Lemma 5.39.} \end{array}$$

Therefore $\ell_\mathrm{e}(M) \neq \ell_\mathrm{e}(N)$, and we derive a contradiction. As a consequence both ν and ν' must have the same shape. It remains to be shown that $\nu = \nu'$.

We distinguish two cases, depending on the shape of ν.

Subcase: ν has shape (1). Then $M_1 \twoheadrightarrow_{\mathsf{st}_\mathcal{L}} N_1$ and $M_2 \twoheadrightarrow_{\mathsf{st}_\mathcal{L}} N_2$, so the result follows from the IH2.

Subcase: ν has shape (2). Then also ν' must have shape (2). Since the initial part of the reduction is composed by head steps, it must be common to every standard reduction. This part is not empty, therefore the length of $\alpha\lceil\beta\rceil \cdot M_3[x{:=}\lfloor\beta\rfloor \cdot M_2] \twoheadrightarrow_{\mathsf{st}_\mathcal{L}} N$ is strictly smaller than n. We can therefore apply the IH, and conclude $\nu = \nu'$. □

Clearly, every $M \in \Lambda$ can be turned into a labeled λ-term $P \in \Lambda^\mathcal{L}$ by annotating its subterms with distinguished atomic labels. Therefore, every reduction $M \twoheadrightarrow_\beta N$ induces a unique isomorphic reduction $M' \twoheadrightarrow_{\beta_\mathcal{L}} Q$, where Q is some labeling of N. As a consequence, the two theorems above imply the respective properties for the (unlabeled) λ-calculus, namely the confluence of $\rightarrow_\beta$ (part of Proposition 1.10) and the Standardization Theorem 1.63.

Theorem 1.63 can be strengthened by showing that for every $M \in \Lambda$ and every $M \twoheadrightarrow_\beta N$ there exists a unique standard reduction $\rho_s : M \twoheadrightarrow_{\mathsf{st}} N$ such that $\rho_s \sim \rho$ (B[1984], Theorem 12.3.14). Which does not exclude that, in general, we may have more than one standard reduction reducing a term M to a term N. This can be seen a sort of 'syntactic accident' that cannot arise in the labeled case, in which there is a unique standard labeled reduction between a labeled term and any of its reducts. So, if $\rho_s : M \twoheadrightarrow_\beta N$ and $\nu_s : M \twoheadrightarrow_\beta N$ are two distinct standard unlabeled reductions, given a labeling P of M, the corresponding standard labeled reductions $\rho_s^\ell : P \twoheadrightarrow_{\beta_\mathcal{L}} Q_1$ and $\nu_s^\ell : P \twoheadrightarrow_{\beta_\mathcal{L}} Q_2$ end with two distinct labelings of N, the labeled terms Q_1 and Q_2. Even if, by confluence, we have $Q_1 \twoheadrightarrow_{\beta_\mathcal{L}} Q$ and $Q_2 \twoheadrightarrow_{\beta_\mathcal{L}} Q$, for some Q.

EXAMPLES 5.56. Let $M = \mathsf{I}(\mathsf{I}x)$ and $N = \mathsf{I}x$. There are two standard reductions $\rho_s, \nu_s : M \twoheadrightarrow_{\mathsf{st}} N$, respectively contracting the outer and the inner redex of M. However, if we consider the labeled term $P = ((\lambda z.z^{\alpha})^{\beta}((\lambda y.y^{\gamma})^{\delta}x^{\epsilon})^{\eta})^{\theta}$, contracting the outer redex we obtain the term $Q_1 = ((\lambda y.y^{\gamma})^{\delta}x^{\epsilon})^{\theta\lceil\beta\rceil\alpha\lfloor\beta\rfloor\eta}$, while contracting the inner redex we obtain $Q_2 = ((\lambda y.y^{\alpha})^{\beta}x^{\eta\lceil\delta\rceil\gamma\lfloor\delta\rfloor\epsilon})^{\theta}$. Both of them eventually reduce to $Q = x^{\theta\lceil\beta\rceil\alpha\lfloor\beta\rfloor\eta\lceil\delta\rceil\gamma\lfloor\delta\rfloor\epsilon}$.

In a sense, the labeled λ-calculus introduces a higher level of precision that avoids identifying λ-terms coming out from 'different' reductions. In fact, given $P, Q \in \Lambda^{\mathcal{L}}$, if $P \twoheadrightarrow_{\beta_{\mathcal{L}}} Q$, all the reductions between P and Q are permutationally equivalent.

THEOREM 5.57. *Let $P \in \Lambda^{\mathcal{L}}$ be a labeling of M. Consider two reductions*

$$\begin{aligned}\rho &: M \twoheadrightarrow_{\beta} N,\\ \nu &: M \twoheadrightarrow_{\beta} L,\end{aligned}$$

as well as the corresponding labeled reductions

$$\begin{aligned}\rho^{\ell} &: P \twoheadrightarrow_{\beta_{\mathcal{L}}} X,\\ \nu^{\ell} &: P \twoheadrightarrow_{\beta_{\mathcal{L}}} Y.\end{aligned}$$

Then:

(i) $\rho \sim \nu \iff X = Y$;

(ii) $\rho \precsim \nu \iff X \twoheadrightarrow_{\beta_{\mathcal{L}}} Y$.

PROOF. (i) By the Standardization Theorem 5.55, there are unique standard labeled reductions ρ^{ℓ}_s and ν^{ℓ}_s satisfying $\rho^{\ell}_s \sim \rho^{\ell}$ and $\nu^{\ell}_s \sim \nu^{\ell}$. Clearly, we have the following chain of equivalences:

$$\rho \sim \nu \iff \rho^{\ell} \sim \nu^{\ell} \iff \rho^{\ell}_s = \nu^{\ell}_s.$$

By Theorem 5.55, we also have $\rho^{\ell}_s = \nu^{\ell}_s \iff X = Y$. From which, we conclude.

(ii) By Definition 1.58, $\rho \precsim \nu$ holds if and only if there exists ρ_1 such that $\rho; \rho_1 \sim \nu$. Take the labeled reduction $\rho^{\ell}_1 : X \twoheadrightarrow_{\beta_{\mathcal{L}}} Z$ isomorphic to ρ_1. By (i), we have that $\rho; \rho_1 \sim \nu$ holds if and only if $Y = Z$ does. We conclude that $\rho \precsim \rho_1$ if and only if $X \twoheadrightarrow_{\beta_{\mathcal{L}}} Y$. □

5.4 Optimal reductions

In this section we show that the properties of the labeled λ-calculus can be fruitfully applied to study the family relation $\simeq$ of the (unlabeled) λ-calculus. More precisely, we prove that two redexes (with history) belong to the same family exactly when their labels are identical. The second part of the section is devoted to prove that there exists an *optimal* (in Lévy's sense) reduction strategy for the λ-calculus.

Since we need to work with both Λ and $\Lambda^{\mathcal{L}}$, it is convenient to fix some notations.

NOTATION (IN THIS SECTION ONLY). *We denote (unlabeled) λ-terms by $L, M, N \in \Lambda$, and labeled λ-terms by $P, Q, X, Y, W, Z \in \Lambda^{\mathcal{L}}$.*

DEFINITION 5.58. (i) Given a labeled λ-term $P^\alpha \in \Lambda^\mathcal{L}$, we denote by $|P^\alpha| \in \Lambda$ the λ-term obtained by erasing its labels. Formally, $|P^\alpha|$ is defined by structural induction:

$$\begin{array}{lcl} |x^\alpha| & \triangleq & x; \\ |(XY)^\alpha| & \triangleq & |X|\ |Y|; \\ |(\lambda x.X)^\alpha| & \triangleq & \lambda x.|X|; \end{array}$$

(ii) Given $M \in \Lambda$ and $P \in \Lambda^\mathcal{L}$, we say that P *is a labeling of* M whenever $M = |P|$.
(iii) Given a reduction $\rho : M \twoheadrightarrow_\beta N$ and a labeling P of M, write ρ^ℓ for the isomorphic labeled reduction $\rho^\ell : P \twoheadrightarrow_{\beta_\mathcal{L}} Q$, where $N = |Q|$. Similarly, given a redex $R \in M$, we write R^ℓ for the corresponding labeled redex in P.

Labeling and families

We start with some preliminary results connecting the labeled reduction with the notions of permutation equivalence and of residual.

PROPOSITION 5.59. *Let $P, Q \in \Lambda^\mathcal{L}$ and $S \in P$ be a redex. If $P \twoheadrightarrow_{\beta_\mathcal{L}} Q$ then all residuals of S in Q have the same name as S.*

PROOF. We show the case $P \xrightarrow{R}_{\beta_\mathcal{L}} Q$. Let $R = ((\lambda x.X_1)^\alpha Y_1)^\beta$ and $S = ((\lambda y.X_2)^\gamma Y_2)^\delta$. We distinguish several subcases, depending on the mutual positions of R and S in P.

Subcase 1. R and S are disjoint. Trivial.

Subcase 2. If S contains $R \in X_2$, then the unique residual of S in Q is of the form $((\lambda y.X_2')^\gamma Y_2)^\delta$, which has the same name as S.

Subcase 3. If S contains $R \in Y_2$, then the unique residual of S in Q is of the form $((\lambda y.X_2)^\gamma Y_2')^\delta$, which has the same name as S.

Subcase 4. If R contains $S \in X_1$, then S has a unique residual in Q of the form $((\lambda y.X_2[x := \lfloor\alpha\rfloor \cdot Y_1])^\gamma (Y_2[x := \lfloor\alpha\rfloor \cdot Y_1]))^\varepsilon$, where only the external label δ can be modified into ε.

Subcase 5. If R contains S in Y_1, then all residuals of the redex S in Q are of the form $((\lambda y.X_2)^\gamma Y_2)^\varepsilon$, where only the external label δ can be modified into ε.

The general statement follows by induction on the length of $P \twoheadrightarrow_{\beta_\mathcal{L}} Q$. □

PROPOSITION 5.60. *Let $P, Q \in \Lambda^\mathcal{L}$. If $P \xrightarrow{R}_{\beta_\mathcal{L}} Q$ and $S \in Q$ is created by R, then*

$$\text{size}(\mathsf{name}(R)) < \text{size}(\mathsf{name}(S))$$

PROOF. Let $R = ((\lambda x.X)^\alpha Y)^\beta$. Therefore $\mathsf{name}(R) = \alpha$. By Lévy's Lemma 1.43 about the creation of redexes, only three cases are possible:

Case 1. There is a subterm of P having shape $(((\lambda x.X)^\alpha Y)^\beta Z)^\delta$ where $X = (\lambda y.W)^\gamma$. In this case, $S = ((\beta\lceil\alpha\rceil \cdot X[x := \lfloor\alpha\rfloor \cdot Y])Z)^\delta = ((\lambda y.W[x := \lfloor\alpha\rfloor \cdot Y])^{\beta\lceil\alpha\rceil\gamma} Z)^\delta$ and $\text{size}(\alpha) < \text{size}(\beta\lceil\alpha\rceil\gamma) = \text{size}(\mathsf{name}(S))$.

Case 2. There is a subterm of X having shape $(x^\varepsilon Z)^\delta$ and $Y = (\lambda y.W)^\gamma$. In this case we have $S = (x^\varepsilon Z)^\delta[x := \lceil\alpha\rceil \cdot (\lambda y.W)^\gamma] = ((\lambda y.W)^{\varepsilon\lceil\alpha\rceil\gamma} Z)^\delta$ and $\text{size}(\alpha) < \text{size}(\varepsilon\lceil\alpha\rceil\gamma) = \text{size}(\mathsf{name}(S))$.

Case 3. There is a subterm of P of the form $(((\lambda x.X)^\alpha Y)^\beta Z)^\delta$ where $X = x^\varepsilon$ and $Y = (\lambda y.W)^\gamma$. Then, we have

$$\begin{array}{rcll} S & = & ((\beta\lceil\alpha\rceil \cdot X[x{:=}\lfloor\alpha\rfloor \cdot Y])Z)^\delta & \\ & = & ((\beta\lceil\alpha\rceil \cdot \varepsilon \cdot \lfloor\alpha\rfloor \cdot Y)Z)^\delta & \\ & = & ((\beta\lceil\alpha\rceil\varepsilon\lfloor\alpha\rfloor \cdot Y)Z)^\delta, & \text{by Lemma 5.38(i),} \\ & = & ((\lambda y.W)^{\beta\lceil\alpha\rceil\varepsilon\lfloor\alpha\rfloor\gamma} Z)^\delta & \end{array}$$

We conclude since $\text{size}(\alpha) < \text{size}(\beta\lceil\alpha\rceil\varepsilon\lfloor\alpha\rfloor\gamma) = \text{size}(\mathsf{name}(S))$. □

DEFINITION 5.61. We say that $P \in \Lambda^{\mathcal{L}}$ is *well initialized* whenever the labels of all its subterms are atomic and pairwise distinct.

The initial terms introduced in Example 5.35 are examples of well-initialized labeled λ-terms, but their reducts are not. We now show that the converse of Proposition 5.59 holds under the hypothesis that the initial term is well initialized.

PROPOSITION 5.62. *Let $P \in \Lambda^{\mathcal{L}}$ be well initialized. For any reduction $\rho : P \twoheadrightarrow_{\beta_{\mathcal{L}}} Q$ and redexes $R \in P$, $S \in Q$, the following are equivalent:*

1. *$S \in R/\rho$;*

2. *$\mathsf{name}(S) = \mathsf{name}(R)$.*

PROOF. ($\Rightarrow$) By Proposition 5.59.

($\Leftarrow$) By induction on the length n of the reduction ρ.

Case $n = 0$. Then $P = Q$ and there are no residuals of R nor redexes sharing the same name, since P is well initialized.

Case $n > 0$. In this case ρ can be written as $P \twoheadrightarrow_{\beta_{\mathcal{L}}} Q' \xrightarrow{R'}_{\beta_{\mathcal{L}}} Q$, where $R \in P$, $S \in Q$ and $\mathsf{name}(R) = \mathsf{name}(S)$. Since P is well initialized, all its redexes have atomic names, whence $\mathsf{name}(S)$ must be atomic as well. This means that S cannot be created by the contraction of R', for otherwise $\text{size}(\mathsf{name}(S)) > 1$. Therefore, S must be the residual of an $S' \in Q'$ having the same name as R. By IH, S' is a residual of R and, by transitivity, so must be S. □

Let us show that all redexes belonging to the same family share the same name.

THEOREM 5.63. *Let $\langle\rho, R\rangle$ and $\langle\nu, S\rangle$ be two redexes with their associated histories*

$$\begin{array}{l} \rho : M \twoheadrightarrow_\beta N_1; \\ \nu : M \twoheadrightarrow_\beta N_2. \end{array}$$

Consider a labeling P of M and the isomorphic reductions in $\Lambda^{\mathcal{L}}$

$$\begin{array}{l} \rho^\ell : P \twoheadrightarrow_\beta Q_1; \\ \nu^\ell : P \twoheadrightarrow_\beta Q_2; \end{array}$$

where $N_i = |Q_i|$, for $i \in \{1,2\}$. Then

$$\langle\rho, R\rangle \simeq \langle\nu, S\rangle \quad\Rightarrow\quad \mathsf{name}(R^\ell) = \mathsf{name}(S^\ell)$$

where R^ℓ and S^ℓ denote the labeled redexes corresponding to R and S, respectively.

PROOF. By Theorem 5.57 and Proposition 5.59. □

Notice that the previous theorem holds for any labeling P of the initial term M. According to the intended interpretation of labeling as the name associated to all the edges with the same origin, the previous theorem states that this interpretation is sound. The converse implication holds under the additional hypothesis that P is well initialized. The original proof of this fact occupies over 40 pages of Lévy's thesis (Lévy (1978b), pp. 68–113), but a shorter proof was subsequently given by Asperti and Laneve (1995).

Reduction by families and optimality

Our next goal is to find the syntactic counterpart of an evaluator that never duplicates redexes, according to the notion of copy induced by the family relation $\simeq$ (Definition 5.8(i)). We introduce a reduction strategy 'by families', i.e. a parallel reduction where, at each step, several redexes belonging to the same family are contracted. By the interpolation property (Lemma 5.7(i)), if one reduces in parallel all the redexes belonging to a given family then no member of that family may appear subsequently in the reduction.

DEFINITION 5.64. (i) Let ρ be a finite reduction. The set of *family classes contained in* ρ is defined by

$$\mathsf{FAM}(\rho) \triangleq \{\langle \mathcal{F}_1; \cdots ; \mathcal{F}_i, R\rangle_{\simeq} \mid \rho = \mathcal{F}_1; \cdots ; \mathcal{F}_n \ \&\ R \in \mathcal{F}_{i+1} \ \&\ i < n\}$$

(ii) Given a set $\mathscr{F}$ of family classes, we say that a reduction ρ is:
(1) *relative to* $\mathscr{F}$ whenever $\mathsf{FAM}(\rho) \subseteq \mathscr{F}$ holds;
(2) a *development of* $\mathscr{F}$ if there is no redex R such that $\langle \rho, R\rangle_{\simeq} \in \mathscr{F}$.

The following is a generalization of the Finite Developments$^+$ Theorem 1.45.

THEOREM 5.65 (GENERALIZED FINITE DEVELOPMENTS).
Let $\mathscr{F}$ be a finite set of family classes.
(i) *There is no infinite reduction relative to $\mathscr{F}$.*
(ii) *If ρ and ν are two developments of $\mathscr{F}$, then $\rho \sim \nu$.*

PROOF. (i) Let $M \in \Lambda$ and $\mathscr{F}$ be a finite set of family classes of reductions having M as initial term. Take a well-initialized labeling P of M. Recall that every $\rho; R \in \mathcal{G}_\beta(M)$ has a corresponding labeled reduction sequence $\rho^\ell; R^\ell \in \mathcal{G}_{\beta_\mathcal{L}}(P)$. Define

$$\begin{array}{rcl} \mathcal{N} & \triangleq & \{\mathsf{name}(R^\ell) \mid \langle \rho, R\rangle_{\simeq} \in \mathscr{F}\}; \\ \mathcal{P} & \triangleq & \{\alpha \mid \mathrm{depth}(\alpha) \leq \max\{\mathrm{depth}(\beta) \mid \beta \in \mathcal{N}\}\}. \end{array}$$

By Theorem 5.63 and the fact that P is well-initialized, the redexes with history in a family class have the same name. Thus, the predicate $\mathcal{N}$ is finite and this entails that $\mathcal{P}$ is bounded. By Proposition 5.51, if $\mathcal{P}$ is adequate for a labeled reduction ν, then such a ν must be finite. We conclude since, for any reduction ρ relative to $\mathscr{F}$, the predicate $\mathcal{P}$ is adequate for ρ^ℓ.

(ii) If ρ and ν are relative to $\mathscr{F}$, then ν/ρ, ρ/ν, $\rho \sqcup \nu$ and $\nu \sqcup \rho$ are also relative to $\mathscr{F}$. By definition, when ρ and ν are developments of $\mathscr{F}$, the final terms of ρ and ν do not contain redexes in the family classes in $\mathscr{F}$. Therefore, since ν/ρ and ρ/ν are relative to $\mathscr{F}$, they are just a sequence of empty steps. We conclude then by $\rho \sim \rho \sqcup \nu \sim \nu \sqcup \rho \sim \nu$. □

DEFINITION 5.66. Let $M \in \Lambda$, $n > 0$ and $\rho \in \mathcal{G}_\beta(M)$.

(i) We say that ρ is *normalizing* if it is finite and its final term is in β-normal form.

(ii) The set $\mathcal{R}(\rho)$ of *redexes in M having some residual which is contracted along* ρ, is defined as follows:

$$\mathcal{R}(\rho) \triangleq \{R \mid \rho = \mathcal{F}_1; \cdots; \mathcal{F}_n \;\&\; \exists i > 0\,.\,(R/\mathcal{F}_1; \cdots; \mathcal{F}_{i-1}) \cap \mathcal{F}_i \neq \emptyset\}$$

where $\mathcal{F}_1, \ldots, \mathcal{F}_n$ represent sets of redexes in M.

(iii) Say that a redex $R \in M$ is *needed* if, for every normalizing reduction $\nu \in \mathcal{G}_\beta(M)$, we have $R \in \mathcal{R}(\nu)$.

(iv) A reduction $\rho : M = M_1 \xrightarrow{\mathcal{F}_1} M_2 \xrightarrow{\mathcal{F}_2} \cdots \xrightarrow{\mathcal{F}_{n-1}} M_n \xrightarrow{\mathcal{F}_n} M_{n+1}$ is called:

(1) *call-by-need* if every $\mathcal{F}_i$ contains at least one needed redex of M_i;

(2) *complete* if, for every $i\,(1 \leq i \leq n)$, $\mathcal{F}_i \neq \emptyset$ and $\mathcal{F}_i$ is a maximal set of redexes satisfying

$$\forall R, S \in \mathcal{F}_i\,.\,\langle \mathcal{F}_1; \cdots; \mathcal{F}_{i-1}, R\rangle \simeq \langle \mathcal{F}_1; \cdots; \mathcal{F}_{i-1}, S\rangle$$

To prove that, after the development of a family class $\mathscr{F}$, no redex belonging to a family contained in $\mathscr{F}$ can be created along the reduction, we need an auxiliary lemma. Recall that the canonical representative of a family has been introduced in Definition 5.28.

LEMMA 5.67. *Let $\langle \rho_c, R_c\rangle$ be the canonical representative of $\langle \rho, R\rangle_\simeq$. Then*

$$\rho_c \precsim \rho \iff \langle \rho_c, R_c\rangle \preceq \langle \rho, R\rangle$$

PROOF. ($\Leftarrow$) This implication follows directly from Definition 5.1(ii).

($\Rightarrow$) Let ρ_s be the standard reduction such that $\rho_s \sim \rho$. Let us start by remarking that $\langle \rho, R\rangle \simeq \langle \rho_s, R\rangle \trianglerighteq \langle \rho_c, R_c\rangle$ and $\rho_c \precsim \rho \sim \rho_s$. Then, since $\rho \sim \rho_s$ implies $\langle \rho_s, R\rangle \preceq \langle \rho, R\rangle$ also, we see that to complete the proof it suffices to show that $\langle \rho_c, R_c\rangle \preceq \langle \rho_s, R\rangle$. We proceed by induction on the length n of ρ_s.

Case $n = 0$, i.e. $\rho_s = o$. As $\langle o, R\rangle \trianglerighteq \langle \rho_c, R_c\rangle$ implies $\rho_c = o$ and $R_c = R$, we are done.

Case $n > 0$, i.e. $\rho_s = S; \nu$. Since both ρ_c and ρ_s are standard, there are two cases:

Subcase 1. $\rho_c = S; \nu_c$, where $\langle \nu_c, R_c\rangle$ is the canonical representative of $\langle \nu, R\rangle_\simeq$. In this case, $\rho_c \precsim \rho_s$ implies $\nu_c \precsim \nu$ by left deletion (Lemma 1.59(iii)). By IH we obtain $\langle \nu_c, R_c\rangle \preceq \langle \nu, R\rangle$, from which we conclude $\langle \rho_c, R_c\rangle \preceq \langle \rho_s, R\rangle$.

Subcase 2. $\langle S; \nu_c, R_c'\rangle \trianglerighteq \langle \rho_c, R_c\rangle$, where $\langle \nu_c, R_c'\rangle$ is the canonical representative of $\langle \nu, R\rangle_\simeq$. Then $\nu_c \precsim \rho_c/S$, by definition of $\trianglerighteq$. This fact and $\rho_c \precsim \rho_s$ imply $\rho_c/S \precsim \rho_s/S = \nu$. Therefore, we obtain $\nu_c \precsim \rho_c/S \precsim \nu$ and, by applying the IH, also $\langle \nu_c, R_c'\rangle \preceq \langle \nu, R\rangle$. By the interpolation property (Lemma 5.7(i)), there exists a redex S' such that

$$\langle \nu_c, R_c'\rangle \preceq \langle \rho_c/S, S'\rangle \preceq \langle \nu, R\rangle.$$

Thus, we have $\langle S; (\rho_c/S), S'\rangle \preceq \langle S; \nu, R\rangle = \langle \rho_s, R\rangle$. Moreover, $\langle \rho_c; (S/\rho_c), S'\rangle \preceq \langle \rho_s, R\rangle$, where $S; (\rho_c/S) \sim \rho_c; (S/\rho_c)$. Finally, since $\langle \rho_c, R_c\rangle$ is the canonical representative of the family class and $\langle \rho_c; (S/\rho_c), S'\rangle \simeq \langle \rho_c, R_c\rangle$, we obtain $R \in R_c/(S/\rho_c)$. We conclude that $\langle \rho_c, R_c\rangle \preceq \langle \rho_s, R\rangle$. □

PROPOSITION 5.68. *Let ρ be a development of a set of family classes $\mathscr{F}$.*
(i) *If $\langle \rho_c, R_c \rangle$ is the canonical representative of $\langle \rho, R \rangle_{\simeq}$, then $\langle \rho_c, R_c \rangle \preceq \langle \rho, R \rangle$.*
(ii) *For every $\langle \nu, S \rangle$ such that $\rho \precsim \nu$, we have $\langle \nu, S \rangle_{\simeq} \notin \mathsf{FAM}(\rho)$.*

PROOF. (i) We can see that the reduction ρ_c is relative to $\mathscr{F}$. Then $\rho_c \precsim \rho$, since ρ_c can always be extended to a development $\rho_c; \rho'$ of $\mathscr{F}$ and $\rho_c; \rho' \sim \rho$, by the Generalized Finite Developments Theorem 5.65. Thus $\langle \rho_c, R_c \rangle \preceq \langle \rho, R \rangle$, by Lemma 5.67.

(ii) Assume that $\langle \nu, S \rangle \in \mathsf{FAM}(\rho)$ holds, towards a contradiction. Let $\rho = \rho_1; \mathcal{F}; \rho_2$ and $R \in \mathcal{F}$ be such that $\langle \rho_1, R \rangle \simeq \langle \nu, S \rangle$. Let $\langle \rho_c, R_c \rangle$ be the canonical representative of $\langle \rho_1, R \rangle_{\simeq}$ (equivalently, of $\langle \nu, S \rangle_{\simeq}$). By definition the reduction ρ_c is relative to $\mathsf{FAM}(\rho)$. As a consequence, we have $\rho_c \precsim \rho$ and, by transitivity, also $\rho_c \precsim \nu$. By Lemma 5.67, we obtain $\langle \rho_c, R_c \rangle \preceq \langle \nu, S \rangle$. By the interpolation property (Lemma 5.7(i)), there exists a redex S' such that $\langle \rho_c, R_c \rangle \preceq \langle \rho, S' \rangle \preceq \langle \nu, S \rangle$. This contradicts the hypothesis that ρ is a development of $\mathscr{F}$, since in this case ρ should also be a development of $\mathsf{FAM}(\rho)$. □

We now show that complete reductions ρ are particular developments whose length is exactly the cardinality of $\mathsf{FAM}(\rho)$.

LEMMA 5.69. *Let ρ be a complete reduction. Then*
(i) *ρ is a development of $\mathsf{FAM}(\rho)$;*
(ii) *the length of ρ is equal to $|\mathsf{FAM}(\rho)|$.*

PROOF. (i) By induction on the length n of ρ.

Case $n = 0$. Obvious, since $\rho = o$.

Case $n > 0$. Then $\rho = \nu; \mathcal{F}$, for some maximal set $\mathcal{F}$ of redexes in the same family. From the IH, we obtain that ν is a development of $\mathsf{FAM}(\nu)$. Now, by Proposition 5.68(ii), there exists no $\langle \nu', S \rangle$ such that $\nu \precsim \nu'$ and $\langle \nu', S \rangle_{\simeq} \in \mathsf{FAM}(\nu)$. By contradiction, assume that $\nu; \mathcal{F}$ is not a development of $\mathsf{FAM}(\nu; \mathcal{F})$, i.e. that there exists a redex R such that $\langle \nu; \mathcal{F}, R \rangle \simeq \langle \nu, S \rangle$. Let $\nu_c; S_c$ be the canonical reduction associated with $\nu; S$. Then $\nu_c \precsim \nu \precsim \nu; \mathcal{F}$. Now, by Proposition 5.68(i), we get $\langle \nu_c, S_c \rangle \preceq \langle \nu; \mathcal{F}, R \rangle$ and, by the interpolation property (Lemma 5.7(i)), there exists a redex R' such that $\langle \nu, R' \rangle \preceq \langle \nu; \mathcal{F}, R \rangle$. This means that $\langle \nu, R' \rangle \simeq \langle \nu, S \rangle$ with $R \in R'/\mathscr{F}$, thus contradicting the hypothesis that $\mathscr{F}$ is a maximal set of redexes in the same family.

(ii) By (i), and the fact that the steps of a complete reduction are non-empty. □

Since complete reductions must contract maximal set of copies of a single redex, the above lemma entails that—when considering complete derivations—the process for deciding whether two redexes are in the same family may be safely reduced to checking the copy-relation.

LEMMA 5.70. *For a reduction $\rho = \mathcal{F}_1; \cdots; \mathcal{F}_n$, the following are equivalent:*

1. *ρ is complete;*

2. *every $\mathcal{F}_i$ $(1 \leq i \leq n)$ is a maximal set of copies, i.e.*

$$\forall i \in \{1, \ldots, n\}, \exists \nu_i, \nu_i', S_i \,.\, [\ \ \nu_i; \nu_i' \sim \rho_i \ \&\ \mathcal{F}_i = S_i/\nu_i'\ \ ].$$

PROOF. (1 ⇒ 2) Let ρ be a complete reduction and R be a redex with history ρ. Let $\mathcal{F} \triangleq \{S \mid \langle\rho, R\rangle \simeq \langle\rho, S\rangle\}$ and $\mathcal{F}'$ be a set of redexes containing R and such that there exist ν, ν', S with $\nu; \nu' \sim \rho$ and $\mathcal{F}' = S/\nu'$. To conclude, it suffices to prove that $\mathcal{F} = \mathcal{F}'$.

We show the two inclusions.

($\subseteq$) By definition of $\simeq$.

($\supseteq$) By the completeness of ρ and Lemma 5.69(i), ρ is a development of $\mathsf{FAM}(\rho)$. Let $\langle\rho_c, R_c\rangle$ be the canonical representative of $\langle\rho, R\rangle_{\simeq}$. By Proposition 5.68(i), we get $\langle\rho_c, R_c\rangle \preceq \langle\rho, R\rangle$. So, $\langle\rho_c, R_c\rangle \preceq \langle\rho, S\rangle$ for any $S \in \mathcal{F}$, whence $\mathcal{F} \subseteq \mathcal{F}'$, for $\mathcal{F}'$ is maximal.

(2 ⇒ 1) By the way of contradiction, assume that ρ is a complete reduction but there exists an index i such that $\mathcal{F}_i$ is not a maximal set of copies. Then, for every $R \in \mathcal{F}_i$, there exists a redex $R' \notin \mathcal{F}_i$ such that

$$\langle\nu, S\rangle \preceq \langle\mathcal{F}_1; \cdots; \mathcal{F}_{i-1}, R\rangle \ \&\ \langle\nu, S\rangle \preceq \langle\mathcal{F}_1; \cdots; \mathcal{F}_{i-1}, R'\rangle,$$

for some ν, S. We obtain $\langle\mathcal{F}_1; \cdots; \mathcal{F}_{i-1}, R\rangle \simeq \langle\mathcal{F}_1; \cdots; \mathcal{F}_{i-1}, R'\rangle$, thus contradicting the hypothesis that $\mathcal{F}_i$ is a maximal set of redexes in the same family. □

Recall that an optimal evaluation strategy must satisfy two constraints: (i) it must elude the duplication of redexes; (ii) it needs to avoid contracting redexes that are subsequently erased. The former objective is achieved through completeness, the latter is fulfilled by exclusively contracting needed redexes.

LEMMA 5.71. *Every λ-term is either β-normal or contains at least a needed redex.*

PROOF. Assume that M is not in β-nf, then it contains some redexes. Take its leftmost-outermost redex R and consider a reduction $\rho : M \twoheadrightarrow_\beta N$ such that $R \notin \mathcal{R}(\rho)$. By definition, there is a redex $S \in N$ such that $R/\rho = \{S\}$. We conclude that R is needed. □

LEMMA 5.72. *Let ρ, ν be coinitial reductions such that ρ is complete and call-by-need. If ν is normalizing then*

$$\mathsf{FAM}(\rho) \subseteq \mathsf{FAM}(\nu).$$

PROOF. By induction on the length of ρ.

Case $\rho = o$. Obvious.

Case $\rho = \rho'; \mathcal{F}$. By definition we have $\mathsf{FAM}(\rho) = \mathsf{FAM}(\rho') \cup \{\langle\rho', S\rangle_{\simeq}\}$, where $S \in \mathcal{F}$. From the induction hypothesis, we obtain $\mathsf{FAM}(\rho') \subseteq \mathsf{FAM}(\nu)$. Since ρ is call-by-need, there exists a needed redex $R \in \mathcal{F}$. As ν is normalizing, we have that $R \in \mathcal{R}(\nu/\rho')$, which means that $\langle\rho', R\rangle \in \mathsf{FAM}(\nu)$ and $\mathsf{FAM}(\rho) \subseteq \mathsf{FAM}(\nu)$. □

We can finally define a cost measure for reductions, recalling that an evaluator keeping shared copies of a redex performs a reduction of all the copies in one step.

DEFINITION 5.73. (i) The *unitary cost* of a redex $\mathcal{F}$ with history ρ, written $\mathsf{cost}(\langle\rho, \mathcal{F}\rangle)$, is defined as follows:

$$\mathsf{cost}(\langle\rho, \mathcal{F}\rangle) \triangleq |\{\langle\rho, R\rangle \mid R \in \mathcal{F}\}/_{\simeq}|$$

(ii) The *cost of a reduction* $\rho = \mathcal{F}_1; \cdots ; \mathcal{F}_n$, still denoted by $\mathsf{cost}(\rho)$, is defined as

$$\mathsf{cost}(\rho) \triangleq \sum_{i=1}^{n} \mathsf{cost}(\langle \mathcal{F}_1; \cdots ; \mathcal{F}_{i-1}, \mathcal{F}_i \rangle)$$

REMARK 5.74. By Lemma 5.70, if ρ is a complete reduction of length n then $\mathsf{cost}(\rho) = n$. Moreover, by Lemma 5.69(ii), we know that $n = |\mathsf{FAM}(\rho)|$. Now, it is easy to check that every reduction sequence ν satisfies the inequality

$$\mathsf{cost}(\nu) \geq |\mathsf{FAM}(\nu)|.$$

We conclude that the cost of reducing copies is minimal for complete reductions. However, in order to be optimal, a reduction must avoid useless computations as well.

PROPOSITION 5.75. *Let $M \in \mathrm{NF}$ and ρ be a call-by-need reduction starting from M.*
(i) *If ρ is long enough, then it is normalizing.*
(ii) *If ρ is complete, then it computes the β-normal form of M with optimal cost.*

PROOF. (i) By the Standardization Theorem 1.63, the leftmost-outermost reduction ν_{st} starting from M reaches its β-normal form, i.e., we have $\nu_{\mathsf{st}} : M \twoheadrightarrow_{\mathsf{st}} M^{\beta\text{-}\mathbf{nf}}$. Let

$$\rho = M \xrightarrow{\mathcal{F}_1} M_1 \xrightarrow{\mathcal{F}_2} \cdots M_{n-1} \xrightarrow{\mathcal{F}_n} M_n \xrightarrow{\mathcal{F}_{n+1}} \cdots$$

We show that the length of the leftmost-outermost reduction starting from M, written $\mathsf{len}_{\mathsf{st}}(M)$, decreases along the above reduction. In other words, we prove that:

$$\mathsf{len}_{\mathsf{st}}(M) > \mathsf{len}_{\mathsf{st}}(M_1) > \cdots > \mathsf{len}_{\mathsf{st}}(M_{n-1}) > \mathsf{len}_{\mathsf{st}}(M_n) > \cdots$$

Call R_i the redex contracted at the i-th step of ν_{st}, in other words, assume $\nu_{\mathsf{st}} = R_1; \cdots ; R_k$. Since the residuals of a leftmost-outermost redex are leftmost-outermost as well, we have $\nu_{\mathsf{st}}/\mathcal{F}_1 = \mathcal{S}_1; \cdots ; \mathcal{S}_k$ where every $\mathcal{S}_i$ is either empty or a singleton. Since ρ is call-by-need, there is a needed redex $S \in \mathcal{F}_1$. Therefore, we have $S \in \mathcal{R}(\nu_{\mathsf{st}})$, from which it follows $R_j \in S/R_1 ; \cdots ; R_{j-1}$, for some j. So, we get $\mathcal{S}_j = \emptyset$ and $\mathsf{len}_{\mathsf{st}}(M) > \mathsf{len}_{\mathsf{st}}(M_1)$. By iterating this argument, we obtain $\mathsf{len}_{\mathsf{st}}(M_{i-1}) > \mathsf{len}_{\mathsf{st}}(M_i)$, for any index $i > 1$. By Lemma 5.71, if ρ is long enough, then there exists $k \in \mathbb{N}$ such that $\mathsf{len}_{\mathsf{st}}(M_k) = 0$. We conclude that $M_k = M^{\beta\text{-}\mathbf{nf}}$.

(ii) We show that the cost of every normalizing reduction ν starting from M is greater than the cost of ρ. By Lemma 5.72, we have $\mathsf{FAM}(\rho) \subseteq \mathsf{FAM}(\nu)$. Therefore, we obtain

$$\begin{array}{rcll} \mathsf{cost}(\nu) & \geq & |\mathsf{FAM}(\nu)| & \\ & \geq & |\mathsf{FAM}(\rho)| & \\ & = & \mathsf{cost}(\rho), & \text{by Lemmas 5.70 and 5.69(ii).} \end{array}$$ □

As a consequence, we obtain the main result of this chapter.

THEOREM 5.76. *Every call-by-need reduction strategy for complete reductions is optimal and normalizing.*

This theorem states that the length of a normalizing reduction (if it exists) does not depend on the adopted strategy, provided that only needed redexes are contracted. For a long time, this supported the conjecture that family reduction could provide a suitable measure of the 'intrinsic complexity' of λ-terms, i.e. of the cost required to compute the normal form of a λ-term independently from the reduction technique employed. In 2001, Asperti and Mairson (2001) refuted this conjecture by proving that the cost of parallel β-reduction is not bounded by any elementary recursive function.

5.5 Sharing graphs: Lévy's optimal implementations

Lévy developed his theory of optimal sharing long before an implementation for family reduction was available. His labeled λ-calculus showed that such an implementation was possible, since in the labeled reduction two redexes belong to the same family exactly when they have the same label and, to completely reduce a family, it suffices to contract in parallel all the redexes which share the same label. Nevertheless, such an implementation is not reasonable, since the sizes of labels can grow more than exponentially during reduction.

The problem of finding an implementation remained open for quite a long time. It was finally solved in Lamping (1989) by introducing the so-called 'sharing graphs'. The core of Lamping's sharing graphs approach rests on a representation of λ-terms in which each syntactic construct is mapped into a node of the graph, with a rewriting rule between the nodes representing application and λ-abstraction, corresponding to β-contraction, plus some sharing nodes allowing to share, and unshare, parts of the λ-term being reduced. In particular, the task of the sharing nodes is to keep shared the copies of the argument of a contracting β-redex, and to implement a lazy duplication of their nodes only when strictly needed.

The key of Lamping's breakthrough was finding a smart way to manage this sharing and unsharing part of the algorithm by means of an indexing of the nodes in the graphs and some auxiliary nodes called 'brackets' which assured a correct matching between the dual nodes for sharing and unsharing. Remarkably, Gonthier et al. (1992a,b) showed that node indexes and the operators associated to the rewriting of brackets were strictly related to the Linear Logic (Girard (1987)) interpretation of λ-calculus and, in particular, to the so-called Geometry of Interaction (Girard (1989)). The relation with Linear Logic allows us to distinguish three kinds of rewriting rules:

1. the one corresponding to the actual contraction of β-redexes,
2. the lazy duplication rules implemented by the sharing and unsharing nodes, and
3. the bracket rewriting rules, usually referred to as 'bookkeeping' to mean that their role is to record the additional information required to properly unfold the shared representation of a term.

The graph β-rule and the duplication rules form the core of the sharing graph implementations and are unavoidable in any implementation based on sharing. The bookkeeping part of the algorithm differ instead in the various implementations proposed in the literature: in addition to Lamping's original algorithm (1989) and Gonthier et al. (1992b), we mention the Bologna Optimal Higher Order Machine (BOHM) by Asperti et al. (1996) and Lambdascope by van Oostrom et al. (2004).

Lévy defined the cost of family reduction as the number of families contracted by any call-by-need strategy. In some sense this is the potential optimum that one can get, since at every step we reduce a redex whose contraction is unavoidable in any reduction sequence and since family reduction contracts at every step a maximal set of shareable redexes, hence the name *optimal reductions.*

We have already remarked that Asperti and Mairson (2001), firstly presented in Asperti and Mairson (1998), showed that Lévy's cost estimation was too optimistic and that the corresponding potential optimum cannot be reached by any implementation, even if it remains a lower bound for β-reduction. We would like to stress however that the important result proved in Asperti and Mairson (1998) should not be seen as a negative result about Lévy's family reduction and sharing implementations—it just shows that in λ-terms, due to the presence of higher-order functions, one can have much more sharing than expected. In particular, one can reduce a simply typed λ-term in a number of family reductions that is approximately linear in its size. Technically, this implies that (on a sequential machine) the cost of sharing a single redex cannot be bound by any elementary function, but this is merely due to the enormous amount of sharing that is inherent in λ-terms. As a consequence, we can clearly state that:

- the computational cost of every parallel step of family reduction may be huge;
- the length of family reduction is not a good measure of the intrinsic computational complexity of reducing λ-terms.

However, this does not mean that another type of implementation can do better. In fact, the negative result about optimal reduction cost estimation is proved by showing that otherwise one would violate some constraint arising from the intrinsic computational complexity of β-reduction—but such a constraint cannot be possibly violated by any implementation.

Even the result in Accattoli and Dal Lago (2016) can be understood in this sense. In this paper the authors showed that the length of a leftmost-outermost derivation to normal form is an invariant cost model. In fact, by a suitable shared modification of leftmost-outermost derivation (which exploits some of the ideas introduced by sharing optimal reductions) they construct a reduction machine that can simulate, and can be simulated by Turing machines, with a polynomial time overhead only. The point is that in order to simulate a (bounded) Turing machine one just needs to encode the transition function between configurations, that is a linear function, and have the possibility to iterate it. On these trivial λ-terms even a simple strategy like leftmost-outermost reduction turns out to be effective. Of course, this tells nothing about the best way of evaluating λ-terms in general. If one really wants to extrapolate a general lesson from this result it should be that, in order to encode Turing machines, one does not really need the full expressive power of λ-terms, and in particular one does not need higher-order terms.

Chapter 6

Infinitary lambda calculus

Infinitary λ-calculus is a generalization of λ-calculus that allows to consider infinite terms and transfinite reductions. Already in the finite setting one observes that λ-terms like $F \triangleq (\lambda mx.mm)(\lambda mx.mm)$ have an infinite rewriting sequence which is rather regular:

$$F \to_\beta \lambda x_0.F \to_\beta \lambda x_0 x_1.F \to_\beta \lambda x_0 x_1 x_2.F \to_\beta \cdots$$

Intuitively, this sequence converges towards the infinite λ-term O satisfying $\mathsf{O} = \lambda x.\mathsf{O}$, which is only reached at the 'limit'. Infinitary rewriting, as defined in Kennaway et al. (1995), Berarducci (1996) and Kennaway et al. (1997), allows to make this statement precise by constructing the set of infinitary λ-terms as the metric completion of Λ w.r.t.

$$\mathrm{d}(M, N) = \inf\{2^{-n} \mid M \text{ and } N \text{ have the same symbols up to depth } n\}$$

and considering limits of infinite reduction sequences that are converging in the corresponding topology. Unfortunately the infinitary rewriting system that arises does not satisfy confluence, a problem that can be overcome by introducing $\bot$-rules for rewriting all pathological terms—traditionally called *meaningless*—to a constant $\bot$. By modifying the notion of 'depth' in the distance above, as well as the kind of terms that are considered meaningless, one obtains several confluent infinitary λ-calculi. As shown in Kennaway and de Vries (2003), standard choices allow to retrieve Böhm trees, Lévy-Longo trees and Berarducci trees as normal forms (see Section 2.4). Building on previous work by Endrullis and Polonsky (2011), Czajka (2020) has shown that a coinductive definition of infinitary λ-terms and infinitary rewriting opens the way for a uniform presentation of these results. The only parameter of the resulting system remains the notion of meaningless terms. We conclude the chapter by discussing two extensional cases, that allow to retrieve extensional Böhm trees and Nakajima trees while maintaining some peculiarities.

6.1 The infinitary λ-calculus

The first infinitary version of λ-calculus we discuss is called λ^∞-*calculus* and consists in infinitary λ-terms equipped with an infinitary notion of β-reduction denoted by $\twoheadrightarrow\!\!\!\!\!\!\rightarrow_\beta$.

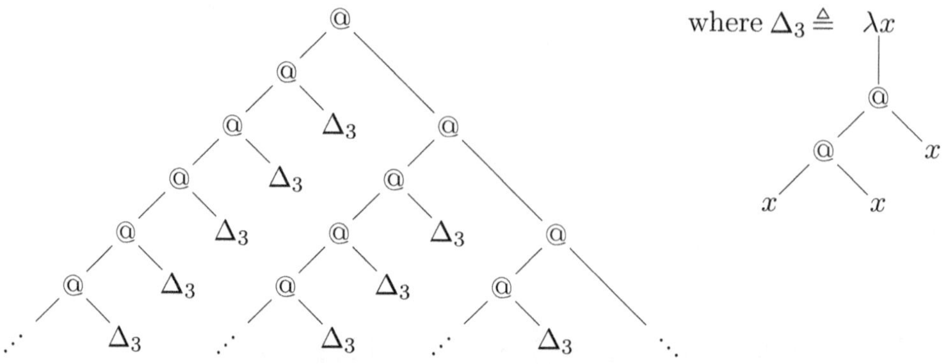

Figure 6.1: A labelled tree representing an infinitary λ-term.

Its syntax

We consider fixed an infinite set Var of variables denoted by $x, y, z, \ldots$

DEFINITION 6.1. (i) The set Λ^∞ of *infinitary λ-terms* is coinductively defined by:

$$(\Lambda^\infty) \qquad T ::=_{\text{coind.}} x \mid \lambda x.T \mid TT$$

Uppercase letters like T, U, V are used as metavariables for infinitary λ-terms. As for Λ, we assume that application associates to the left and has higher priority than abstraction.

(ii) Given $T, U \in \Lambda^\infty$, we write $T = U$ whenever T and U are *bisimilar*, i.e., the binary relation $=$ on Λ^∞ is coinductively defined by:

$$\frac{}{x = x} \qquad \frac{T = U}{\lambda x.T = \lambda x.U} \qquad \frac{T = T' \quad U = U'}{TU = T'U'}$$

The double inference lines are used to emphasize the coinductive nature of a rule. Intuitively, this means that a derivation tree of $T = U$ can be infinite.

As discussed in Section 2.4, infinitary λ-terms will be often represented via the so-called 'applicative view' as unary-binary branching labelled trees where:

- applications nodes @ are binary branching,
- abstraction nodes λx are unary branching,
- variable nodes x are the leaves.

An example of such a tree representation is given in Figure 6.1. We say that $T \in \Lambda^\infty$ is *infinite* if its representation as a tree is infinite, otherwise we say that T is *finite*.

NOTATION. *Given $T \in \Lambda^\infty$, write:*

(i) T^ω *for the infinite λ-term satisfying* $T^\omega = T(T^\omega)$;

(ii) ${}^\omega T$ *for the infinite λ-term satisfying* ${}^\omega T = ({}^\omega T)T$.

DEFINITION 6.2. Infinitary λ-terms in Λ^∞ inherit the following notions from the finite ones, in the usual way.

Consider $T, U \in \Lambda^\infty$.

(i) The set $\mathrm{FV}(T)$ of free variables of T, whose cardinality can be $\aleph_0$.
(ii) The renaming of the bound variables of T (aka, α-conversion).
(iii) The capture-free substitution $T[x{:=}U]$ of U for all free occurrences of x in T.
(iv) The notions of reduction on Λ^∞ (also called *infinitary notions of reduction*):

$$\begin{aligned} \beta &= \{(T_1, T_2) \mid T_1 = (\lambda x.T)U \ \&\ T_2 = T[x{:=}U]\}; \\ \eta &= \{(T, U) \mid T = \lambda x.Ux \quad \& \quad x \notin \mathrm{FV}(U)\}. \end{aligned}$$

CONVENTION 6.3. *Infinitary λ-terms are considered up to α-conversion. Moreover, we assume that a fresh variable $x \notin \mathrm{FV}(T)$ may always be chosen.*

REMARK 6.4. The above convention is more subtle than Convention 1.4, nevertheless there are several solutions for working modulo α-equivalence. E.g., one can define Λ^∞ as

- the space of α-equivalence classes with a certain metric (Terese (2003), Section 12.4);
- the final coalgebra of an appropriate functor in the category of nominal sets (Kurz et al. (2013)).

Postulating the existence of a fresh $x \notin \mathrm{FV}(T)$ is crucial to provide a proper definition of substitution by guarded coinduction. However, if Var is countable, then there exists $T \in \Lambda^\infty$ such that $\mathrm{FV}(T) = \mathrm{Var}$. (Note that Hilbert's Hotel trick is not applicable here, as free variables cannot be renamed.) This issue can be solved by assuming that Var is uncountable, or that $\mathrm{FV}(T)$ is always finite, or by working with de Bruijn indices.

DEFINITION 6.5. Let $\mathrm{R} \subseteq \Lambda^\infty \times \Lambda^\infty$ be a notion of reduction on infinitary λ-terms.

(i) The *one-step* R*-reduction*, written $\to_\mathrm{R}$, is defined as the *compatible closure* of R:

$$\frac{(T,U) \in \mathrm{R}}{T \to_\mathrm{R} U} \qquad \frac{T \to_\mathrm{R} T'}{TU \to_\mathrm{R} T'U} \qquad \frac{T \to_\mathrm{R} T'}{UT \to_\mathrm{R} UT'} \qquad \frac{T \to_\mathrm{R} T'}{\lambda x.T \to_\mathrm{R} \lambda x.T'}$$

(ii) The *multi-step* R*-reduction* $\twoheadrightarrow_\mathrm{R}$ is defined starting from $\to_\mathrm{R}$ as follows:

$$\frac{}{T \twoheadrightarrow_\mathrm{R} T} \qquad \frac{T \to_\mathrm{R} U \quad U \twoheadrightarrow_\mathrm{R} V}{T \twoheadrightarrow_\mathrm{R} V}$$

(iii) The *infinitary closure* of $\to_\mathrm{R}$, in symbols $\twoheadrightarrow\!\!\!\!\!\rightarrow_\mathrm{R}$, is defined as follows:

$$\frac{T \twoheadrightarrow_\mathrm{R} x}{T \twoheadrightarrow\!\!\!\!\!\rightarrow_\mathrm{R} x} \qquad \frac{T \twoheadrightarrow_\mathrm{R} \lambda x.U' \quad U' \twoheadrightarrow\!\!\!\!\!\rightarrow_\mathrm{R} U}{T \twoheadrightarrow\!\!\!\!\!\rightarrow_\mathrm{R} \lambda x.U}$$

$$\frac{T \twoheadrightarrow_\mathrm{R} U'V' \quad U' \twoheadrightarrow\!\!\!\!\!\rightarrow_\mathrm{R} U \quad V' \twoheadrightarrow\!\!\!\!\!\rightarrow_\mathrm{R} V}{T \twoheadrightarrow\!\!\!\!\!\rightarrow_\mathrm{R} UV}$$

(iv) The *infinitary conversion* $\overset{\infty}{=}_\mathrm{R}$ is the transitive closure of $({}_\mathrm{R}\!\leftarrow\!\!\!\!\!\twoheadleftarrow \circ \twoheadrightarrow\!\!\!\!\!\rightarrow_\mathrm{R})$, where $\circ$ stands for relational composition.

Items (i)–(ii) above are completely standard, and are presented simply to stress the fact that multi-steps reductions have finite length also in this context. The real novelties in Λ^∞ are infinitary reduction relations like $\mathrel{\rightarrow\mkern-14mu\twoheadrightarrow}_\beta$.

EXAMPLES 6.6. (i) Write Curry's fixed point combinator as $\mathsf{Y} = \lambda f.WW$ with $W = \lambda x.f(xx)$. The infinitary reduction $\mathsf{Y}f \mathrel{\rightarrow\mkern-14mu\twoheadrightarrow}_\beta f^\omega$ can be derived like this:

$$\dfrac{\mathsf{Y}f \twoheadrightarrow_\beta f(WW) \quad f \mathrel{\rightarrow\mkern-14mu\twoheadrightarrow}_\beta f \quad \dfrac{WW \twoheadrightarrow_\beta f(WW) \quad f \mathrel{\rightarrow\mkern-14mu\twoheadrightarrow}_\beta f \quad \overline{WW \mathrel{\rightarrow\mkern-14mu\twoheadrightarrow}_\beta f^\omega}}{WW \mathrel{\rightarrow\mkern-14mu\twoheadrightarrow}_\beta f^\omega}}{\mathsf{Y}f \mathrel{\rightarrow\mkern-14mu\twoheadrightarrow}_\beta f^\omega}$$

The figure above displays a compact representation of an infinite derivation tree through the loop $\dashrightarrow$.

(ii) Similarly, writing $\Omega_3 = \Delta_3\Delta_3$ with $\Delta_3 = \lambda x.xxx$, we may derive $\Omega_3 \mathrel{\rightarrow\mkern-14mu\twoheadrightarrow}_\beta {}^\omega\Delta_3$.

$$\dfrac{\Omega_3 \twoheadrightarrow_\beta \Omega_3\Delta_3 \quad \overline{\Omega_3 \mathrel{\rightarrow\mkern-14mu\twoheadrightarrow}_\beta {}^\omega\Delta_3} \quad \Delta_3 \mathrel{\rightarrow\mkern-14mu\twoheadrightarrow}_\beta \Delta_3}{\Omega_3 \mathrel{\rightarrow\mkern-14mu\twoheadrightarrow}_\beta {}^\omega\Delta_3}$$

EXERCISE 6.7. (i) Show that $T \twoheadrightarrow_{\mathrm{R}} U \mathrel{\rightarrow\mkern-14mu\twoheadrightarrow}_{\mathrm{R}} V$ implies $T \mathrel{\rightarrow\mkern-14mu\twoheadrightarrow}_{\mathrm{R}} V$.
(ii) Show that $T \mathrel{\rightarrow\mkern-14mu\twoheadrightarrow}_{\mathrm{R}} U \twoheadrightarrow_{\mathrm{R}} V$ implies $T \mathrel{\rightarrow\mkern-14mu\twoheadrightarrow}_{\mathrm{R}} V$.
(iii) Show that $T \twoheadrightarrow_{\mathrm{R}} V$ implies $T \mathrel{\rightarrow\mkern-14mu\twoheadrightarrow}_{\mathrm{R}} V$.
(iv) Show that $\mathrel{\rightarrow\mkern-14mu\twoheadrightarrow}_{\mathrm{R}}$ is reflexive and transitive.
(v) Show that $[\mathsf{Y}x, \mathsf{Y}z] \mathrel{\rightarrow\mkern-14mu\twoheadrightarrow}_\beta [x^\omega, z^\omega]$.
(vi) Show that $\mathsf{Y}\Omega_3 \mathrel{\rightarrow\mkern-14mu\twoheadrightarrow}_\beta ({}^\omega\Delta_3)^\omega$, namely, the infinitary λ-term from Figure 6.1.

As a brief excursion in the infinitary λ-calculus, we present an interesting construction giving rise to infinite fixed point combinators (fpc's). Consider first the finite case. Let

$$\boldsymbol{\delta} \triangleq \lambda yx.x(yx) =_\beta \mathsf{SI}$$

This term also attracted Smullyan's attention, in his beautiful fable about fixed point combinators figuring as birds in an enchanted forest:

> "An extremely interesting bird is the owl ʘ defined by the following condition: ʘ$xy = y(xy)$." pp. 133–134, Smullyan (1985).

The notation $\boldsymbol{\delta}$ is more common in the literature. Recall that an fpc Y is 'reducing' if $Yx \twoheadrightarrow_\beta x(Yx)$ for x fresh. Böhm and van der Mey made the following remarks.

FACT 6.8. (i) If Y is a fixed point combinator then both $\boldsymbol{\delta}Y$ and $Y\boldsymbol{\delta}$ are fpc's.
(ii) If the fpc Y is moreover reducing then so is $Y\boldsymbol{\delta}$ as it is shown by

$$Y\boldsymbol{\delta}x \twoheadrightarrow_\beta \boldsymbol{\delta}(Y\boldsymbol{\delta})x \twoheadrightarrow_\beta x(Y\boldsymbol{\delta}x) \twoheadrightarrow_\beta x^n(Y\boldsymbol{\delta}x). \tag{6.2}$$

These observations suggest a recipe for constructing new fpc's from old ones.

DEFINITION 6.9. The *Böhm-van der Mey sequence* $(Y_n)_{n \in \mathbb{N}}$ is defined, starting from Curry's fixed point combinator Y, by induction on n:

$$Y_0 \triangleq \mathsf{Y}, \qquad\qquad Y_{n+1} \triangleq Y_n \boldsymbol{\delta}.$$

Notice that Turing's fixed point combinator occurs as Y_1 in the sequence: $\Theta =_\beta Y_1$.

PROPOSITION 6.10. *Consider the above sequence* $(Y_n)_{n \in \mathbb{N}}$.
 (i) *All* Y_n *are fixed point combinators.*
 (ii) *The sequence* $(Y_n)_{n \in \mathbb{N}}$ *contains no repetition (modulo* $=_\beta$*).*

PROOF. (i) By Fact 6.8.
 (ii) B[1984], Exercise 6.8.9. See also Klop (2007). □

By performing the reduction (6.2) in an infinitary way, we obtain:

$$Y\boldsymbol{\delta}x \twoheadrightarrow\!\!\!\!\rightarrow_\beta (\lambda f.f^\omega)\boldsymbol{\delta}x \to_\beta \boldsymbol{\delta}^\omega x = \boldsymbol{\delta}(\boldsymbol{\delta}^\omega)x \twoheadrightarrow_\beta x(\boldsymbol{\delta}^\omega x) \twoheadrightarrow\!\!\!\!\rightarrow_\beta x^\omega.$$

Therefore $Y\boldsymbol{\delta}$ is indeed behaving as an fpc in the λ^∞-calculus: $Y\boldsymbol{\delta} \stackrel{\infty}{=}_\beta \lambda x.x^\omega \stackrel{\infty}{=}_\beta Y$. Moreover, the reduction above shows that $\boldsymbol{\delta}^\omega$ is an infinite reducing fpc. We mention the existence of countably many infinite fpc's. It is indeed easy to verify that, for every $n \in \mathbb{N}$, the infinite λ-term $(\mathsf{SS})^\omega\mathsf{S}^{\sim n}\mathsf{I}$ enjoys this property. This is shown in Klop (2007).

Basic properties of λ^∞-calculus

We now discuss some basic properties of the λ^∞-calculus endowed with infinitary β-reduction. It is well-known that this higher-order infinitary rewriting system is neither confluent nor weakly normalizing, nevertheless its β-normal forms turn out to be unique.

DEFINITION 6.11. Consider Λ^∞ endowed with an infinitary notion of reduction R.
 (i) We say that $T \in \Lambda^\infty$ is an R-*normal form* if it does not contain any R-redex.
 (ii) We denote by $\Lambda^\infty_{\text{R-nf}}$ the set of all infinitary λ-terms in R-normal form.
 (iii) Define the following properties (for all $T \in \Lambda^\infty$):

$$\mathrm{CR}^\infty: \quad T \twoheadrightarrow\!\!\!\!\rightarrow_{\mathrm{R}} T_1 \;\&\; T \twoheadrightarrow\!\!\!\!\rightarrow_{\mathrm{R}} T_2 \quad\Rightarrow\quad \exists V \in \Lambda^\infty.\, T_1 \twoheadrightarrow\!\!\!\!\rightarrow_{\mathrm{R}} V \;\&\; T_2 \twoheadrightarrow\!\!\!\!\rightarrow_{\mathrm{R}} V$$

$$\mathrm{UN}^\infty: \quad T \twoheadrightarrow\!\!\!\!\rightarrow_{\mathrm{R}} T_1 \;\&\; T \twoheadrightarrow\!\!\!\!\rightarrow_{\mathrm{R}} T_2 \;\&\; T_1, T_2 \in \Lambda^\infty_{\text{R-nf}} \quad\Rightarrow\quad T_1 = T_2$$

$$\mathrm{WN}^\infty: \quad \exists V \in \Lambda^\infty_{\text{R-nf}}.\, T \twoheadrightarrow\!\!\!\!\rightarrow_{\mathrm{R}} V$$

(Intuitively CR = Church-Rosser; UN = uniqueness of nf's; WN = weak normalization.)

For coinductively defined infinitary reductions there exists no general notion of strong normalization (SN^∞). The counterexample to CR^∞ presented below was found independently by Ariola and Klop (1994) and Berarducci (1996). Such a counterexample is quite powerful as it shows that even the Parallel Moves Lemma (PML^∞), i.e. the infinitary analogue of Lemma 1.48(i), fails in the λ^∞-calculus. For examples of infinitary term rewriting systems satisfying PML^∞—but not CR^∞—see Endrullis et al. (2012). Perhaps unexpectedly, even if the λ^∞-calculus does not satisfy CR^∞, it does enjoy UN^∞.

PROPOSITION 6.12. (i) *The properties* CR^∞ *and* WN^∞ *fail in the* λ^∞*-calculus.*
(ii) *This restricted form of* PML^∞ *does hold. For all* $T \in \Lambda^\infty$*, we have:*

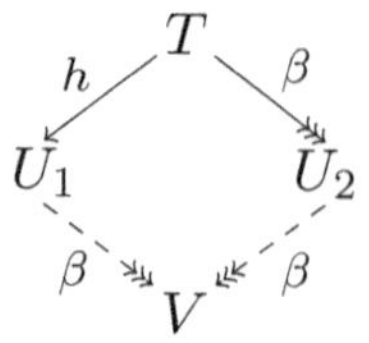

where $T \to_h U_1$ *denotes one head-reduction step.*
(iii) *The property* UN^∞ *holds in the* λ^∞*-calculus.*

PROOF. (i) The λ-term YI gives a counterexample to CR^∞. On the one hand we have:

$$\mathsf{YI} \to_\beta (\lambda x.\mathsf{I}(xx))(\lambda x.\mathsf{I}(xx)) \twoheadrightarrow_\beta \mathsf{I}^\omega,$$

and on the other hand we obtain:

$$\mathsf{YI} \to_\beta (\lambda x.\mathsf{I}(xx))(\lambda x.\mathsf{I}(xx)) \to_\beta (\lambda x.xx)(\lambda x.\mathsf{I}(xx)) \to_\beta \Omega.$$

It follows that $\mathsf{YI} \twoheadrightarrow_\beta \mathsf{I}^\omega$ and $\mathsf{YI} \twoheadrightarrow_\beta \Omega$. Now, both I^ω and Ω only reduce to themselves, so they cannot have a common reduct. They also constitute conterexamples to the property WN^∞.
(ii) See Theorem 12.6.3(ii) in Terese (2003).
(iii) It follows from Lemma 5.4 in Ketema and Simonsen (2009), as the λ^∞-calculus is confluent modulo so-called 'hypercollapsing'[1] subterms. □

The normal forms of λ^∞-calculus admit the following characterization.

THEOREM 6.13. *All normal forms of* λ^∞*-calculus are built (coinductively) by composing three kinds of building blocks, namely hnf-contexts, the ogre and infinite 'left-spines':*

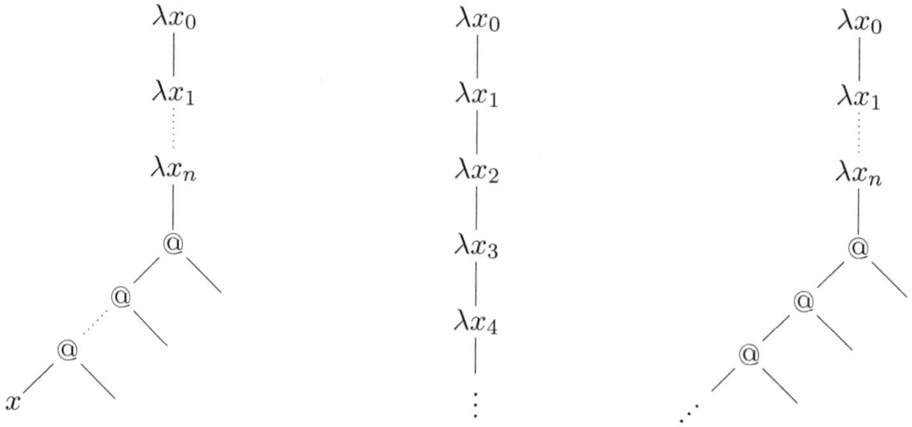

Notice that the variable case 'x' is an instance of the first scenario (hnf-context).

[1]In the λ^∞-calculus, hypercollapsing terms coincide with the root-active ones (Definition 6.43, below).

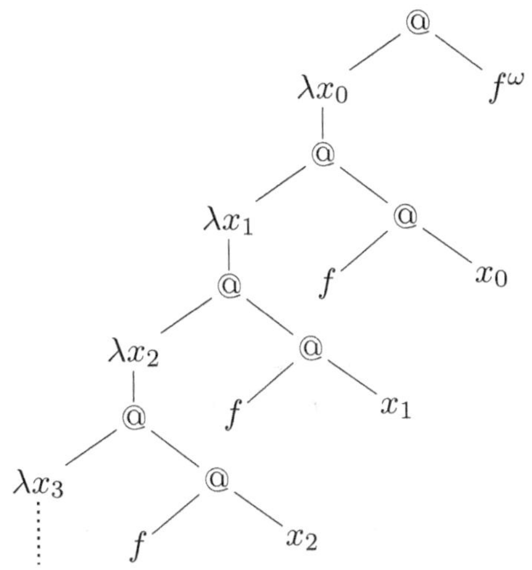

Figure 6.2: An example of a cascade.

PROOF. It is clear that any possibly infinite composition of these building blocks gives rise to a β-normal form, as it contains no β-redex. *Vice versa* any infinitary β-normal form can be decomposed into building blocks by coloring all the edges in the tree that are oriented down or left. It is easy to check that the maximal connected colored components in the tree, together with variables occurrences at the end of some branches, constitute the desired decomposition. □

For the finite λ-calculus it is not difficult to characterize looping terms as those λ-terms having Ω as a subterm, see Exercice 4.5. The infinitary case is more complicated.

DEFINITION 6.14. Let $T \in \Lambda^\infty$. We say that T is

(i) a *looping term* if $T \to_\beta T$;
(ii) a *root-looping term* if $T \to_\beta T$ by contracting a redex at the top of the tree;
(iii) a *cascade* if it has shape

$$T = (\lambda x_0.(\lambda x_1.(\lambda x_2.(\lambda x_3.\cdots)T_3)T_2)T_1)T_0 \tag{6.3}$$

where, every T_{i+1} is obtained from T_i by:

(1) renaming all variables x_j to x_{j+1} (for $j \in \mathbb{N}$), then
(2) replacing an arbitrary (possibly infinite) number of occurrences of T_0 by x_0.

EXAMPLES 6.15. (i) The λ-terms $\lambda x.\Omega$, $\lambda x.x\Omega$ are looping, but not root-looping.

(ii) The infinitary λ-terms Ω and I^ω are root-looping.

(iii) The example in Figure 6.2 is root-looping and a cascade. For the latter property, notice that it has the same structure as (6.3), instantiated with $T_0 = f^\omega$ and $T_{j+1} = fx_j$. In particular, the condition $T_0 = f^\omega = f(f^\omega) = (fx_0)[x_0 := T_0] = T_1[x_0 := T_0]$ is satisfied.

It is not difficult to check that $T \in \Lambda^\infty$ is looping if and only if $T = C[U]$ for some infinitary context $C[]$ and root-looping term U. In order to characterize looping terms, it is therefore enough to consider root-looping ones.

LEMMA 6.16 (ENDRULLIS AND POLONSKY (2011)).
Given $T \in \Lambda^\infty$, the following are equivalent:

1. *T is root-looping,*

2. *T has one of the following forms:*

   (i) Ω,
   (ii) I^ω,
   (iii) BB, *where B is the infinite solution of* $B = \lambda x.xB$,
   (iv) $(\lambda x_0.(\lambda x_1.(\lambda x_2.\cdots)T_2)T_1)T_0$, *satisfying the conditions of a cascade.*

PROOF. ($1 \Rightarrow 2$) By an exhaustive case analysis.

($2 \Rightarrow 1$) The only interesting case is when T is a cascade. Then, by contracting the redex at the root, we obtain:

$$\begin{array}{lll} T & \to_\beta & (\lambda x_1.(\lambda x_2.(\lambda x_3.(\lambda x_4.\cdots)T_4^*)T_3^*)T_2^*)T_1^*, \quad \text{where } T_i^* \triangleq T_i[x_0 := T_0], \\ & =_\alpha & (\lambda x_0.(\lambda x_1.(\lambda x_2.(\lambda x_3.\cdots)T_3)T_2)T_1)T_0, \quad \text{by definition of cascade}, \\ & = & T. \end{array}$$

□

Standardization for the λ^∞-calculus

We now show a more advanced property of λ^∞-calculus, namely the Standardization Theorem 6.21 (below). For its proof, which is a generalization of the proof of standardization for finite rewriting by Plotkin (1975), we follow Endrullis and Polonsky (2011).

DEFINITION 6.17. (i) The notion of (one-step) *weak head reduction* $\to_w$ is defined by:

$$\frac{}{(\lambda x.T)U \to_w T[x := U]} \qquad \frac{T \to_w T'}{TU \to_w T'U}$$

(ii) The *infinitary standard reduction* $\twoheadrightarrow\!\!\!\!\twoheadrightarrow_{\mathsf{st}}$ is given by the same rules defining $\twoheadrightarrow\!\!\!\!\twoheadrightarrow_\beta$, except for the fact that finite prefixes are now required to be weak head reductions:

$$\frac{T \twoheadrightarrow_w x}{T \twoheadrightarrow\!\!\!\!\twoheadrightarrow_{\mathsf{st}} x} \qquad \frac{T \twoheadrightarrow_w \lambda x.U' \quad U' \twoheadrightarrow\!\!\!\!\twoheadrightarrow_{\mathsf{st}} U}{T \twoheadrightarrow\!\!\!\!\twoheadrightarrow_{\mathsf{st}} \lambda x.U}$$

$$\frac{T \twoheadrightarrow_w U'V' \quad U' \twoheadrightarrow\!\!\!\!\twoheadrightarrow_{\mathsf{st}} U \quad V' \twoheadrightarrow\!\!\!\!\twoheadrightarrow_{\mathsf{st}} V}{T \twoheadrightarrow\!\!\!\!\twoheadrightarrow_{\mathsf{st}} UV}$$

In other words, $\twoheadrightarrow\!\!\!\!\twoheadrightarrow_{\mathsf{st}}$ is the infinitary closure of $\to_w$.

(iii) Define an auxiliary infinitary β-reduction $\twoheadrightarrow\!\!\!\!\twoheadrightarrow_{\mathsf{aux}}$ taking as prefixes infinitary standard reductions:

$$\frac{T \twoheadrightarrow\!\!\!\!\twoheadrightarrow_{\mathsf{st}} x}{T \twoheadrightarrow\!\!\!\!\twoheadrightarrow_{\mathsf{aux}} x} \qquad \frac{T \twoheadrightarrow\!\!\!\!\twoheadrightarrow_{\mathsf{st}} \lambda x.U' \quad U' \twoheadrightarrow\!\!\!\!\twoheadrightarrow_{\mathsf{aux}} U}{T \twoheadrightarrow\!\!\!\!\twoheadrightarrow_{\mathsf{aux}} \lambda x.U}$$

$$\frac{T \twoheadrightarrow\!\!\!\!\twoheadrightarrow_{\mathsf{st}} U'V' \quad U' \twoheadrightarrow\!\!\!\!\twoheadrightarrow_{\mathsf{aux}} U \quad V' \twoheadrightarrow\!\!\!\!\twoheadrightarrow_{\mathsf{aux}} V}{T \twoheadrightarrow\!\!\!\!\twoheadrightarrow_{\mathsf{aux}} UV}$$

We need a couple of technical lemmas.

LEMMA 6.18. *Let* $T, U, V \in \Lambda^\infty$.

(i) *The relation* $\twoheadrightarrow_w$ *is transitive.*

(ii) $T \twoheadrightarrow_w U \;\&\; U \twoheadrightarrow\!\!\!\!\twoheadrightarrow_{\mathsf{st}} V \quad\Rightarrow\quad T \twoheadrightarrow\!\!\!\!\twoheadrightarrow_{\mathsf{st}} V$.

(iii) $T \twoheadrightarrow\!\!\!\!\twoheadrightarrow_{\mathsf{st}} T' \;\&\; U \twoheadrightarrow\!\!\!\!\twoheadrightarrow_{\mathsf{st}} U' \quad\Rightarrow\quad T[x:=U] \twoheadrightarrow\!\!\!\!\twoheadrightarrow_{\mathsf{st}} T'[x:=U']$.

(iv) *For* $\to \,\in \{\to_\beta, \twoheadrightarrow_\beta, \twoheadrightarrow_w\}$, *we have:*

$$T \twoheadrightarrow\!\!\!\!\twoheadrightarrow_{\mathsf{st}} U \;\&\; U \to V \quad\Rightarrow\quad T \twoheadrightarrow\!\!\!\!\twoheadrightarrow_{\mathsf{st}} V$$

PROOF. (i) By induction.

(ii) By case distinction, using (i) to concatenate the prefixes.

(iii) By coinduction, using property (2) below (which follows from (1) by induction):

$$\begin{array}{llll} 1. & T \to_w U & \Rightarrow & T[x:=V] \to_w U[x:=V] \\ 2. & T \twoheadrightarrow_w U & \Rightarrow & T[x:=V] \twoheadrightarrow_w U[x:=V] \end{array}$$

(iv) By induction on $U \to V$, using (iii) for the redex base case. □

LEMMA 6.19. *Let* $T, U, V \in \Lambda^\infty$.

(i) *The relation* $\twoheadrightarrow\!\!\!\!\twoheadrightarrow_{\mathsf{st}}$ *is transitive.*

(ii) $T \twoheadrightarrow\!\!\!\!\twoheadrightarrow_{\mathsf{st}} U \;\&\; U \twoheadrightarrow\!\!\!\!\twoheadrightarrow_{\mathsf{aux}} V \quad\Rightarrow\quad T \twoheadrightarrow\!\!\!\!\twoheadrightarrow_{\mathsf{aux}} U$.

(iii) $T \twoheadrightarrow\!\!\!\!\twoheadrightarrow_{\mathsf{aux}} T' \;\&\; U \twoheadrightarrow\!\!\!\!\twoheadrightarrow_{\mathsf{aux}} U' \quad\Rightarrow\quad T[x:=U] \twoheadrightarrow\!\!\!\!\twoheadrightarrow_{\mathsf{aux}} T'[x:=U']$.

(iv) *For* $\to \,\in \{\to_\beta, \twoheadrightarrow_\beta, \twoheadrightarrow_w, \twoheadrightarrow\!\!\!\!\twoheadrightarrow_{\mathsf{st}}\}$, *we have:*

$$T \twoheadrightarrow\!\!\!\!\twoheadrightarrow_{\mathsf{aux}} U \;\&\; U \to V \quad\Rightarrow\quad T \twoheadrightarrow\!\!\!\!\twoheadrightarrow_{\mathsf{aux}} V$$

(v) $T \twoheadrightarrow\!\!\!\!\twoheadrightarrow_{\mathsf{aux}} U \;\&\; U \twoheadrightarrow\!\!\!\!\twoheadrightarrow_{\mathsf{aux}} V \quad\Rightarrow\quad T \twoheadrightarrow\!\!\!\!\twoheadrightarrow_{\mathsf{aux}} U$.

PROOF. (i) By coinduction, with case analysis on $U \twoheadrightarrow\!\!\!\!\twoheadrightarrow_{\mathsf{st}} V$, we prove that

$$T \twoheadrightarrow\!\!\!\!\twoheadrightarrow_{\mathsf{st}} U \;\&\; U \twoheadrightarrow\!\!\!\!\twoheadrightarrow_{\mathsf{st}} V \quad\Rightarrow\quad T \twoheadrightarrow\!\!\!\!\twoheadrightarrow_{\mathsf{st}} V$$

Case $U \twoheadrightarrow_w x = V$. Then we have $T \twoheadrightarrow\!\!\!\!\twoheadrightarrow_{\mathsf{st}} x$, by Lemma 6.18(iv).

Case $V = V_1V_2$, $U \twoheadrightarrow_w U_1U_2$ and $U_i \twoheadrightarrow\!\!\!\!\twoheadrightarrow_{\mathsf{st}} V_i$, for every $i \in \{1,2\}$. By Lemma 6.18(iv), we have $T \twoheadrightarrow\!\!\!\!\twoheadrightarrow_{\mathsf{st}} U_1U_2$. Hence we obtain $T \twoheadrightarrow_w U_1'U_2'$, with $U_i' \twoheadrightarrow\!\!\!\!\twoheadrightarrow_{\mathsf{st}} U_i$ for all $i \in \{1,2\}$. By co-IH, we know that $U_i' \twoheadrightarrow\!\!\!\!\twoheadrightarrow_{\mathsf{st}} V_i$, whence, using $T \twoheadrightarrow_w U_1'U_2'$, we derive $T \twoheadrightarrow\!\!\!\!\twoheadrightarrow_{\mathsf{st}} V_1V_2$.

Case $V = \lambda x.V_1$, $U \twoheadrightarrow_w \lambda x.U_1$ and $U_1 \twoheadrightarrow\!\!\!\!\twoheadrightarrow_{\mathsf{st}} V_1$. Therefore, we have $T \twoheadrightarrow_w \lambda x.U_1'$ with $U_1' \twoheadrightarrow\!\!\!\!\twoheadrightarrow_{\mathsf{st}} U_1$. By co-IH, we have $U_1' \twoheadrightarrow\!\!\!\!\twoheadrightarrow_{\mathsf{st}} V_1$ and, using the fact that $T \twoheadrightarrow_w \lambda x.U_1'$, we conclude $T \twoheadrightarrow\!\!\!\!\twoheadrightarrow_{\mathsf{st}} \lambda x.V_1$.

(ii)–(v) Analogous to the proof of Lemma 6.18. □

LEMMA 6.20. *Let* $T, U \in \Lambda^\infty$.

(i) $T \twoheadrightarrow\!\!\!\!\twoheadrightarrow_{\mathsf{st}} U \quad\Rightarrow\quad T \twoheadrightarrow\!\!\!\!\twoheadrightarrow_\beta U$.

(ii) $T \twoheadrightarrow_\beta U \quad\Rightarrow\quad T \twoheadrightarrow\!\!\!\!\twoheadrightarrow_{\mathsf{st}} U$.

(iii) $T \twoheadrightarrow\!\!\!\!\twoheadrightarrow_\beta U \quad\Rightarrow\quad T \twoheadrightarrow\!\!\!\!\twoheadrightarrow_{\mathsf{aux}} U$.

(iv) $T \twoheadrightarrow\!\!\!\!\twoheadrightarrow_{\mathsf{st}} U \iff T \twoheadrightarrow\!\!\!\!\twoheadrightarrow_{\mathsf{aux}} U$.

PROOF. (i) Immediate, since every weak head prefix is also a β-prefix.

(ii) By induction on $T \twoheadrightarrow_\beta U$, using Lemma 6.18(iv) and the reflexivity of $\twoheadrightarrow\!\!\!\!\!\rightarrow_{\mathsf{st}}$.

(iii) By (ii).

(iv) ($\Rightarrow$) By composing (i) and (iii).

($\Leftarrow$) By coinduction on $T \twoheadrightarrow\!\!\!\!\!\rightarrow_{\mathsf{aux}} U$.

Case $T \twoheadrightarrow\!\!\!\!\!\rightarrow_{\mathsf{st}} x = U$. Trivial.

Case $U = U_1U_2$, $T \twoheadrightarrow\!\!\!\!\!\rightarrow_{\mathsf{st}} T_1T_2$, and $T_i \twoheadrightarrow\!\!\!\!\!\rightarrow_{\mathsf{aux}} U_i$ for every $i \in \{1,2\}$. Therefore, we have $T \twoheadrightarrow_w T_1'T_2'$, with $T_i' \twoheadrightarrow\!\!\!\!\!\rightarrow_{\mathsf{st}} T_i$ for every i. By (iv)($\Rightarrow$), we obtain $T_i' \twoheadrightarrow\!\!\!\!\!\rightarrow_{\mathsf{aux}} T_i$. By Lemma 6.19(v), we have $T_i' \twoheadrightarrow\!\!\!\!\!\rightarrow_{\mathsf{aux}} U_i$ and, by co-IH, we know that $T_i' \twoheadrightarrow\!\!\!\!\!\rightarrow_{\mathsf{st}} U_i$. Using the fact that $T \twoheadrightarrow_w T_1'T_2'$, we derive $T \twoheadrightarrow\!\!\!\!\!\rightarrow_{\mathsf{st}} U_1U_2$.

Case $U = \lambda x.U'$, $T \twoheadrightarrow\!\!\!\!\!\rightarrow_{\mathsf{st}} \lambda x.T_1$ and $T_1 \twoheadrightarrow\!\!\!\!\!\rightarrow_{\mathsf{aux}} U'$. In this case, $T \twoheadrightarrow_w \lambda x.T_1'$ with $T_1' \twoheadrightarrow\!\!\!\!\!\rightarrow_{\mathsf{st}} T_1$. Now, by (iv)($\Rightarrow$) we have $T_1' \twoheadrightarrow\!\!\!\!\!\rightarrow_{\mathsf{aux}} T_1$ and by Lemma 6.19(v) we obtain $T_1' \twoheadrightarrow\!\!\!\!\!\rightarrow_{\mathsf{aux}} U'$. By co-IH, we know that $T_1' \twoheadrightarrow\!\!\!\!\!\rightarrow_{\mathsf{st}} U'$ holds, therefore, using $T \twoheadrightarrow_w \lambda x.T_1'$, we conclude $T \twoheadrightarrow\!\!\!\!\!\rightarrow_{\mathsf{st}} \lambda x.U'$. □

The following appears as Theorem 7 in Endrullis and Polonsky (2011).

THEOREM 6.21 (STANDARDIZATION FOR λ^∞). *For $T, U \in \Lambda^\infty$, we have:*

$$T \twoheadrightarrow\!\!\!\!\!\rightarrow_\beta U \quad \Rightarrow \quad T \twoheadrightarrow\!\!\!\!\!\rightarrow_{\mathsf{st}} U$$

PROOF. By Lemma 6.20(iii) and (iv). □

Old fashioned infinitary rewriting

The coinductive treatment of infinitary reductions is relatively recent. In the pioneering paper by Kennaway et al. (1997) the idea was to consider reduction sequences of transfinite length α (an ordinal), as in

$$T_0 \to_\beta T_1 \to_\beta T_2 \to_\beta \cdots T_\omega \to_\beta T_{\omega+1} \to_\beta \cdots T_{\omega+\omega} \to_\beta \cdots T_\alpha$$

The crucial point is how to control what happens at limit ordinals γ. In other words, to have a meaningful rewriting system, one needs to specify correctly the relationship between T_γ and its predecessors for, otherwise, they might remain completely unrelated. The first attempt is to require that T_γ is for a limit ordinal γ the Cauchy limit of the previous T_δ, for $\delta < \gamma$, with the usual distance metric $\mathrm{d}(-,-)$ on infinitary λ-terms:

$$\mathrm{d}(T,V) = \begin{cases} 0, & \text{if } T = V; \\ 2^{-n}, & \text{if } T, V \text{ coincide only up to depth } n. \end{cases}$$

Cauchy convergence has been studied by Dershowitz et al. (1991) but, without additional requirements, the resulting rewriting theory does not enjoy fundamental properties like Standardization and Compression (see Proposition 6.25, below). Intuitively, the problem is that the topology does not respect the 'rewriting structure' of the terms, whence some information concerning the shape of subterms might be lost when passing to a Cauchy convergent limit. As a result, there is no meaningful notion of descendants and one cannot perform projections of reductions. To recover these properties, it is sufficient to impose a further requirement on the limit behavior of reduction sequences, namely that the depth of the redexes contracted in the successive steps tends to infinity when approaching a limit ordinal from below. This rules out the possibility that the action of redex contraction remains confined at the top, or stagnates at some finite depth level.

Hereafter α, γ, δ denote ordinals, while we keep β for the usual notion of reduction.

DEFINITION 6.22. Let $\mathrm{R} \subseteq \Lambda^\infty \times \Lambda^\infty$.

(i) An R-*reduction sequence of length* α is given by a map $T : (\alpha+1) \to \Lambda^\infty$ together with a sequence $\rho \triangleq (T_\delta \to_{\mathrm{R}} T_{\delta+1})_{\delta<\alpha}$. Such a reduction sequence is called:

Cauchy convergent: if, for every limit ordinal $\gamma \leq \alpha$, the term T_γ is the limit of the sequence $(T_\delta)_{\delta<\gamma}$ in the metric topology on infinitary λ-terms.

Strongly convergent: if ρ is Cauchy convergent and, for every limit ordinal $\gamma \leq \alpha$, $\lim_{\delta\to\gamma} d_\gamma = \infty$, where d_γ denotes the depth of the redex contracted in the step ρ_γ. This means that for every $k \in \mathbb{N}$ there exists $\delta < \gamma$ such that, for all δ' with $\delta \leq \delta' < \gamma$, the redex contracted in the step $\rho_{\delta'}$ occurs at depth greater than k.

(ii) If ρ is a strongly convergent R-reduction sequence from U to V, then we write

$$\begin{array}{rl} \rho : U \xrightarrow{\alpha}_{\mathrm{R}} V, & \text{in case the length of } \rho \text{ is } \alpha; \\ \rho : U \xrightarrow{\leq\alpha}_{\mathrm{R}} V, & \text{in case the length of } \rho \text{ is at most } \alpha; \\ \rho : U \xrightarrow{\infty}_{\mathrm{R}} V, & \text{in case } \rho : U \xrightarrow{\alpha}_{\mathrm{R}} V \text{ holds for some ordinal } \alpha; \\ U \xrightarrow{X}_{\mathrm{R}} V, & \text{with } X \in \{\alpha, \leq\alpha, \infty\}, \text{ if } \rho : U \xrightarrow{X}_{\mathrm{R}} V \text{ for some } \rho. \end{array}$$

The notations introduced in Definition 6.22(ii) reflect the fact that we are mainly interested in strongly convergent reductions.

EXAMPLES 6.23. (i) Recall that Θ stands for Turing's fixed point combinator. As Θ is reducing, we get $\Theta f \twoheadrightarrow_\beta f^n(\Theta f)$ for every $n \in \mathbb{N}$, from which it follows that $\Theta f \xrightarrow{\omega}_\beta f^\omega$.

(ii) From the previous example, we have

$$\rho_1 : [\Theta x, \Theta z] \xrightarrow{\omega}_\beta [x^\omega, \Theta z] \qquad\qquad \rho_2 : [x^\omega, \Theta z] \xrightarrow{\omega}_\beta [x^\omega, z^\omega]$$

By composing the reduction sequences ρ_1 and ρ_2, we obtain a $\rho : [\Theta x, \Theta z] \xrightarrow{\omega+\omega}_\beta [x^\omega, z^\omega]$.

EXERCISE 6.24. (i) Show that all finite reductions are strongly converging.

(ii) Show that there exists a reduction $\rho : [\Theta x, \Theta z] \xrightarrow{\omega}_\beta [x^\omega, z^\omega]$.

[Hint: Compress a reduction of length $\omega+\omega$ by alternating the contraction of a redex to the left and to the right.]

(iii) Show that $\Theta\Omega_3 \xrightarrow{\infty}_\beta ({}^\omega\Delta_3)^\omega$ and determine the length of the chosen reduction.

(iv) Show that also the reduction sequence in (iii) can be compressed to a reduction sequence of length ω, by permuting the steps of the original sequence (if needed).

At this point, it is natural to wonder whether every strongly convergent reduction of length α (countable) can be compressed to one with the same initial and final terms, but having finite length, or length ω. We show that the λ^∞-calculus satisfies this property.

Proposition 6.25 (Compression Lemma). *Let $\rho : M \xrightarrow{\alpha}_\beta N$, for some countable ordinal α. Then there exists an infinitary reduction ρ' of length at most ω, i.e.,*

$$\rho' : M \xrightarrow{\leq\omega}_\beta N.$$

Proof. It is possible to obtain ρ' from ρ by compression, as a straightforward application of 'dove-tailing'. See Terese (2003), Chapter 12 (p. 690). □

Remark 6.26. (i) The Compression Lemma does not hold in general for infinitary reductions that are merely Cauchy convergent, without the depth-to-infinity requirement.

(ii) Not all infinitary term rewriting systems enjoy the Compression Lemma for strongly convergent reductions. For counterexamples, see Section 6.4 or Endrullis et al. (2012).

(iii) We can now define infinitary strong normalization SN^∞ by requiring that all reduction sequences are strongly convergent. The λ^∞-calculus does not satisfy SN^∞.

In fact, there is no general notion of SN^∞ for the coinductive definition of reduction since $\twoheadrightarrow\!\!\!\!\rightarrow_\beta$ actually captures the notion of strongly convergent reduction. This result is shown below and first appeared as Theorem 3 in Endrullis and Polonsky (2011).

Theorem 6.27. *For $T, V \in \Lambda^\infty$, the following are equivalent:*

1. *$T \twoheadrightarrow\!\!\!\!\rightarrow_\beta V$;*

2. *there exists a strongly convergent reduction $\rho : T \xrightarrow{\infty}_\beta V$.*

Proof. ($1 \Rightarrow 2$) By traversing the (possibly infinite) derivation tree of $T \twoheadrightarrow\!\!\!\!\rightarrow_\beta V$ in breadth-first order, and concatenating the finite β-prefixes, one obtains a reduction sequence of length ω satisfying the depth-to-infinity requirement by construction.

($2 \Rightarrow 1$) Consider a strongly convergent reduction $\rho : T \xrightarrow{\alpha}_\beta V$, for some ordinal α. First, by induction on α, we show that $T \twoheadrightarrow\!\!\!\!\rightarrow_{\mathsf{aux}} V$ is derivable.

Case $\alpha = 0$. Then $T \xrightarrow{0}_\beta V$ which means $T = V$, whence $T \twoheadrightarrow\!\!\!\!\rightarrow_{\mathsf{aux}} V$ holds.

Case $\alpha = \gamma + 1$. In this case $T \xrightarrow{\gamma+1}_\beta V$, i.e., $T \xrightarrow{\gamma}_\beta T' \to_\beta V$. Then $T' \twoheadrightarrow\!\!\!\!\rightarrow_{\mathsf{st}} V$, $T' \twoheadrightarrow\!\!\!\!\rightarrow_{\mathsf{aux}} V$ and, by IH, $T \twoheadrightarrow\!\!\!\!\rightarrow_{\mathsf{aux}} T'$. We obtain $T \twoheadrightarrow\!\!\!\!\rightarrow_{\mathsf{aux}} V$ by Lemma 6.19(v).

Limit case $T \xrightarrow{\alpha}_\beta V$, for α a limit ordinal. We construct an infinite derivation of $T \twoheadrightarrow\!\!\!\!\rightarrow_\beta V$, coinductively. By the depth condition, there exists a $\delta < \alpha$ such that, for every $\gamma \geq \delta$, the redex contracted in ρ at γ occurs at depth greater than 0. Call V_δ the infinitary λ-term having index δ in ρ. By IH, we have $T \twoheadrightarrow\!\!\!\!\rightarrow_{\mathsf{aux}} V_\delta$, and $T \twoheadrightarrow\!\!\!\!\rightarrow_{\mathsf{st}} V_\delta$ by Lemma 6.20(iv). We split into cases, depending on the shape of V_δ.
Case $V_\delta = x$. Impossible, since then V_δ cannot reduce to anything, while we assumed that $\delta < \gamma$.

Case $V_\delta = \lambda x.V'$. In this case, we have $V = \lambda x.U$ and $V' \xrightarrow{\leq\alpha}_\beta U$. Therefore, $V' \twoheadrightarrow\!\!\!\!\rightarrow_{\mathsf{aux}} U$ by coinduction. We get $T \twoheadrightarrow\!\!\!\!\rightarrow_\beta \lambda x.U$ by the abstraction rule of $\twoheadrightarrow\!\!\!\!\rightarrow_{\mathsf{aux}}$.
Case $V_\delta = V_1 V_2$. Then $V = U_1 U_2$ and the tail of the reduction ρ past δ can be split into two parts $V_i \xrightarrow{\leq\alpha}_\beta U_i$ ($i \in \{0,1\}$) of length at most α. Then $V_0 \twoheadrightarrow\!\!\!\!\rightarrow_{\mathsf{aux}} U_0$ and $V_1 \twoheadrightarrow\!\!\!\!\rightarrow_{\mathsf{aux}} U_1$ by coinduction. Thus, $T \twoheadrightarrow\!\!\!\!\rightarrow_\beta U_1 U_2$ by the application rule of $\twoheadrightarrow\!\!\!\!\rightarrow_{\mathsf{aux}}$.

From $T \twoheadrightarrow\!\!\!\!\rightarrow_{\mathsf{aux}} V$ we conclude $T \twoheadrightarrow\!\!\!\!\rightarrow_\beta V$, by applying (iv) and (i) of Lemma 6.20. □

6.2 Relative computability

The ideas of this section are taken from Barendregt and Klop (2009). A fundamental result in λ-calculus establishes that all recursive numerical functions f are λ-definable (cf. Proposition 1.28). Thanks to this property, it is not difficult to construct a combinator F whose Böhm tree represents f, that is, employing the notations introduced in Notation 1.35:

$$\mathrm{BT}(F) = [\mathsf{c}_{f(0)}, \mathsf{c}_{f(1)}, \mathsf{c}_{f(2)}, \dots]$$

In the λ^∞-calculus, because of the coinductive definition of Λ^∞, we are able to represent all total numerical functions f, including non-recursive ones, as infinite normal forms T^∞_f. We show that it is possible to use T^∞_f itself as an oracle in the computation of another function $g : \mathbb{N} \to \mathbb{N}$, where the actual computation is performed by a finite λ-term through infinitary β-reduction. That is, we can capture the notion of relative computability, $f \leadsto_T g$, meaning that g can be computed using f as an oracle, entirely within the infinitary λ-calculus. As an intermediate step we consider the finite calculus λf, namely λ-calculus enriched with a constant f representing an arbitrary numerical function f. As shown by Kleene (1963), we have that $f \leadsto_T g$ holds exactly when g can be λ-defined in λf. Finally, in Theorem 6.42, we connect the finite λf-calculus with the λ^∞-calculus.

DEFINITION 6.28. (i) Let $\Lambda^\infty_{\mathsf{f}}$ be the set of *infinitary λ-terms possibly containing a constant* f, intended to represent a function $f : \mathbb{N} \to \mathbb{N}$.

(ii) Define the following notion of reduction on $\Lambda^\infty_{\mathsf{f}}$:

$$\mathsf{f}\mathsf{c}_n \to \mathsf{c}_{f(n)} \qquad (f)$$

(iii) The *$\lambda^\infty f$-calculus* is given by $\Lambda^\infty_{\mathsf{f}}$ endowed with reduction relations $\to_{\beta f}, \twoheadrightarrow\!\!\!\!\rightarrow_{\beta f}$. Its finite restriction is called *λf-calculus* and its underlying set is denoted Λ_{f}.

(iv) Similarly, we write Λ^o_{f} for the subset of combinators in Λ_{f}.

The $\lambda^\infty f$-calculus enjoys the same fundamental properties as the λ^∞-calculus, and for the same reasons.

LEMMA 6.29. (i) *On Λ_{f}, the notions of reduction (f) and (βf) are Church-Rosser.*

(ii) *The properties* CR^∞ *and* WN^∞ *fail in the $\lambda^\infty f$-calculus.*

(iii) *The property* UN^∞ *holds in the $\lambda^\infty f$-calculus.*

(iv) *For $T \in \Lambda^\infty_{(\mathsf{f})}$ and $M \in \Lambda$, we have*

$$T \twoheadrightarrow\!\!\!\!\rightarrow_{\beta(f)} M \quad \Rightarrow \quad T \twoheadrightarrow_{\beta(f)} M$$

(v) *Suppose that $T \in \Lambda^\infty_{(\mathsf{f})}$ and $N \in \Lambda_{(\mathsf{f})}$ is a finite β-normal form. Then*

$$T \overset{\infty}{=}_{\beta(f)} N \quad \Rightarrow \quad T =_{\beta(f)} N.$$

PROOF. (i) Similar to the proof of Mitschke's Theorem 15.3.3 in B[1984].

(ii) The same counterexamples given in the proof of Proposition 6.12 apply.

(iii) Since the $\lambda^\infty f$-calculus is also confluent modulo hypercollapsing subterms.
(iv) By structural induction on M.
(v) By (iv) and the UN^∞ property for $\lambda^\infty f$, which is (iii). □

DEFINITION 6.30. (i) Let $\mathbb{B} \triangleq \bigcup_{k\in\mathbb{N}} \mathbb{B}_k$, where $\mathbb{B}_k \triangleq \mathbb{N}^k \to \mathbb{N}$ and we identify $\mathbb{B}_0$ with $\mathbb{N}$.
(ii) Let $[n_0, \ldots, n_{k-1}]$ be some coding of sequences of numbers such that:
(1) for all $k \in \mathbb{N}$, the function $\lambda\!\!\lambda x_0 \ldots x_{k-1}.[x_0, \ldots, x_{k-1}] \in \mathbb{B}_k$ is recursive;
(2) there exists a recursive function $\lambda\!\!\lambda px.(x)_p \in \mathbb{B}_2$ projecting a sequence onto its components, i.e., satisfying:

$$\forall k, p, n_0, \ldots, n_{k-1} \in \mathbb{N}\,.\,[p < k \quad \Rightarrow \quad ([n_0, \ldots, n_{k-1}])_p = n_p\,].$$

(iii) For all $k \in \mathbb{N}$ and $g \in \mathbb{B}_k$, define $[g] \in \mathbb{B}_1$ by setting

$$[g](n) \triangleq \begin{cases} g, & \text{if } k = 0, \\ g((n)_0, \ldots, (n)_{k-1}), & \text{if } k > 0. \end{cases}$$

(iv) For $A \subseteq \mathbb{B}_0$, its *characteristic function* χ_A is defined by

$$\chi_A(x) \triangleq \begin{cases} 1, & \text{if } x \in A, \\ 0, & \text{else.} \end{cases}$$

We recall some notions concerning relative computability. The interested reader is invited to consult Odifreddi (1989) or Longley and Normann (2015).

DEFINITION 6.31. Let $f \in \mathbb{B}$.
(i) The set $\mathcal{C}_f$ of *functions recursive in* f is the least class of total functions that contains as initial functions:

- f, as an *oracle*,
- the successor,
- the constant zero function,
- the projections $\lambda\!\!\lambda x_1 \ldots x_n.x_i$,

and is closed under composition, primitive recursion and minimization.
(ii) Let P be an n-ary relation on $\mathbb{N}$. Then
(1) P is *recursive in* f if $\chi_{\mathrm{P}} \in \mathbb{B}_n$ is recursive in f.
(2) P is *r.e. in* f if there exists an $n+1$-ary relation R, recursive in f, satisfying

$$\mathrm{P}(\vec{x}) \iff \exists y\,.\,\mathrm{R}(\vec{x}, y)$$

DEFINITION 6.32. Let $f, g \in \mathbb{B}$.
(i) We say that f *computes* g, written $f \rightsquigarrow_T g$ if g is recursive in f.
(ii) We say that f *and* g *compute each other* (or are *computably equivalent*), written $f =_T g$, if $f \rightsquigarrow_T g$ and $g \rightsquigarrow_T f$.

(iii) In computability theory $f \rightsquigarrow_T g$ is usually called *g is Turing-computable in f* and is written as $g \leq_T f$. If $f =_T g$, then f and g are called *Turing-equivalent*.

REMARK 6.33. Relative computability inherits this result from the classical one: a function $g : \mathbb{N} \to \mathbb{N}$ is recursive in f if and only if its graph $\{[n,m] \mid m = g(n)\}$ is r.e. in f. (By the usual dove-tailing argument, one shows that its complement is also r.e. in f.)

LEMMA 6.34. *Let $k \in \mathbb{N}$ and $g \in \mathbb{B}_k$. Then $g =_T [g]$.*

PROOF. Let $k > 0$ and $g \in \mathbb{B}_k$. It is enough to notice that, for all $n_0, \ldots, n_{k-1}$, we have:

$$\begin{array}{rcl} [g](n) & = & g((n)_0, \ldots, (n)_{k-1}); \\ g(n_0, \ldots, n_{k-1}) & = & [g]([n_0, \ldots, n_{k-1}]). \end{array} \qquad \square$$

DEFINITION 6.35. (i) Let $g \in \mathbb{B}_1$. We associate to g an infinite λ-term $\mathsf{T}^\infty_g \in \Lambda^\infty$ as follows, depicted in hnf view.

$\mathsf{T}^\infty_g \triangleq$

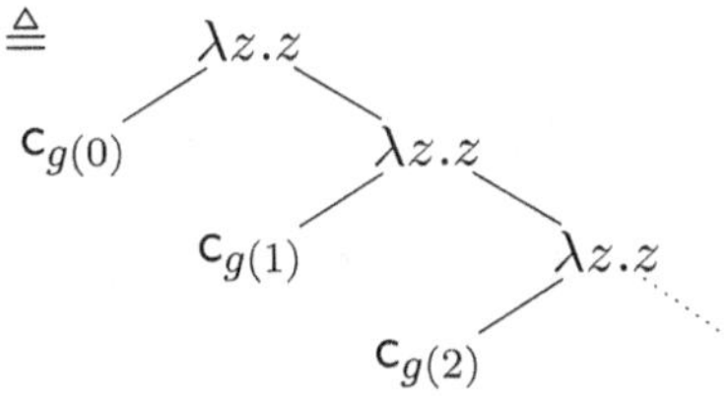

(ii) For $g \in \mathbb{B}_k$ with $k > 1$, we define $\mathsf{T}^\infty_g \triangleq \mathsf{T}^\infty_{[g]}$.

DEFINITION 6.36. Let $f \in \mathbb{B}_1$, $G, U, V \in \Lambda^\infty_{(\mathsf{f})}$.
Define the following notions and notations.

Definiendum	Notation	Definiens
G $\lambda(f)$-defines V	$G\,\mathrm{D}(f)\,V$	$G \overset{\infty}{=}_{\beta(f)} V$
V is $\lambda(f)$-definable	$\mathrm{D}(f)\ V$	$\exists G \in \Lambda^o_{(\mathsf{f})} \,.\, G\,\mathrm{D}(f)\ V$
G $\lambda(f)$-defines V from U	$G\colon U \rightsquigarrow_{\lambda(f)} V$	$GU\,\mathrm{D}(f)\,V$
V is $\lambda(f)$-definable from U	$U \rightsquigarrow_{\lambda(f)} V$	$\exists G \in \Lambda^o_{(\mathsf{f})} \,.\, G\colon U \rightsquigarrow_{\lambda(f)} V$
U, V are $\lambda(f)$-equivalent	$U \sim_{\lambda(f)} V$	$U \rightsquigarrow_{\lambda(f)} V \;\&\; V \rightsquigarrow_{\lambda(f)} U$

EXERCISE 6.37. (i) Let $A \subseteq \mathbb{N}$ and $A^c \triangleq \mathbb{N} - A$. Show that $\mathsf{T}^\infty_{\chi_A} \sim_\lambda \mathsf{T}^\infty_{\chi_{A^c}}$.
(ii) Show that $\mathsf{T}^\infty_f \sim_\lambda \mathsf{T}^\infty_{\chi_{\mathcal{G}_f}}$, where $f \in \mathbb{B}_1$ and $\mathcal{G}_f = \{[n,m] \mid n, m \in \mathbb{N} \;\&\; f(n) = m\}$.

DEFINITION 6.38. Let $f \in \mathbb{B}_1, g \in \mathbb{B}_k$, $G, U \in \Lambda^\infty_{(\mathsf{f})}$.
Define the following notions and notations.

Definiendum	Notation	Definiens
G $\lambda(f)$-defines g	$G\,\mathrm{D}(f)\,g$	$\forall n_1, \ldots, n_k . G\mathsf{c}_{n_1} \cdots \mathsf{c}_{n_k} \overset{\infty}{=}_{\beta(f)} \mathsf{c}_{g(n_1,\ldots,n_k)}$
g is $\lambda(f)$-definable	$\mathrm{D}\ g$	$\exists G \in \Lambda^o_{(\mathsf{f})} \,.\, G\,\mathrm{D}(f)\ g$
G $\lambda(f)$-defines g from U	$G\colon U \rightsquigarrow_{\lambda(f)} g$	$GU\,\mathrm{D}(f)\ g$
g is $\lambda(f)$-definable from U	$U \rightsquigarrow_{\lambda(f)} g$	$\exists G \in \Lambda^o \,.\, G\colon U \rightsquigarrow_{\lambda(f)} g$
g is λ-definable from f	$\mathsf{f} \rightsquigarrow_{\lambda f} g$	$\exists G \in \Lambda^o \,.\, G\colon \mathsf{f} \rightsquigarrow_{\lambda f} g$

LEMMA 6.39. *Let* $f \in \mathbb{B}_1$, $g \in \mathbb{B}_k$, *for* $k > 0$, *and* $U \in \Lambda^{\infty}_{(\mathsf{f})}$. *Then*
(i) $\mathsf{T}^{\infty}_g \rightsquigarrow_{\lambda} g$.
(ii) $G\,\mathrm{D}(f)\,g \quad \Rightarrow \quad G \rightsquigarrow_{\lambda(f)} \mathsf{T}^{\infty}_g$.
(iii) $\mathsf{f} \rightsquigarrow_{\lambda f} \mathsf{T}^{\infty}_f$.
(iv) g *is* λ*-definable* $\iff$ T^{∞}_g *is* λ*-definable.*

PROOF. (i) Wlog assume $k = 1$, as the case $k > 1$ is obtained similarly, using the λ-definability of the functions showing $g =_T [g]$, Lemma 6.34. Define

$$P_1 \triangleq \Theta(\lambda ftn.\mathsf{ifz}(n, (t\mathsf{K}), f(t\mathsf{F})(\mathsf{S}^- n))),$$

where the combinators $\Theta, \mathsf{ifz}, \mathsf{K}, \mathsf{F}$ and S^- have been defined in Notation 1.12 (see also Table 1.1). For all $T \in \Lambda^{\infty}$, P_1 satisfies the following primitive recursion scheme

$$\begin{array}{lcl} P_1 T \mathsf{c}_0 & =_{\beta} & T\mathsf{K}; \\ P_1 T \mathsf{c}_{n+1} & =_{\beta} & P_1(T\mathsf{F})\mathsf{c}_n. \end{array}$$

Towards $\mathsf{T}^{\infty}_g \rightsquigarrow_{\lambda} g$ one verifies by induction on n that $P_1 \mathsf{T}^{\infty}_g \mathsf{c}_n = \mathsf{c}_{g(n)}$.

$$\begin{array}{lclclcl} P_1 \mathsf{T}^{\infty}_g \mathsf{c}_0 & =_{\beta} & \mathsf{T}^{\infty}_g \mathsf{K} & = & [\mathsf{c}_{g(0)}, \mathsf{c}_{g(1)}, \mathsf{c}_{g(2)}, \ldots]\mathsf{K} & =_{\beta} & \mathsf{c}_{g(0)}, \\ P_1 \mathsf{T}^{\infty}_g \mathsf{c}_{n+1} & =_{\beta} & P_1(\mathsf{T}^{\infty}_g \mathsf{F})\mathsf{c}_n & = & P_1[\mathsf{c}_{g(1)}, \mathsf{c}_{g(2)}, \mathsf{c}_{g(3)}, \ldots]\mathsf{c}_n & =_{\beta} & \mathsf{c}_{g(n+1)}. \end{array}$$

(ii) Define $H \triangleq \Theta(\lambda hgn.[gn, hg(\mathsf{S}^+ n)])$, where S^+ is the successor. For all $n \in \mathbb{N}$:

$$HG\mathsf{c}_n \twoheadrightarrow_{\beta} [G\mathsf{c}_n, HG\mathsf{c}_{n+1}].$$

Take $F_1 \triangleq \lambda f.Hf\mathsf{c}_0$, towards $F_1 \colon G \rightsquigarrow_{\lambda(f)} \mathsf{T}^{\infty}_g$. Suppose $G\,\mathrm{D}(f)\,g$. Then

$$\begin{array}{lllll} F_1 G & \to_{\beta} & HG\mathsf{c}_0 & \twoheadrightarrow_{\beta} & [G\mathsf{c}_0, HG\mathsf{c}_1] \\ & & & \twoheadrightarrow_{\beta} & [G\mathsf{c}_0, G\mathsf{c}_1, HG\mathsf{c}_2] \\ & & & \twoheadrightarrow_{\beta} & [G\mathsf{c}_0, G\mathsf{c}_1, \ldots, HG\mathsf{c}_k] \\ & & & \twoheadrightarrow\!\!\!\!\twoheadrightarrow_{\beta} & [G\mathsf{c}_0, G\mathsf{c}_1, G\mathsf{c}_2, \ldots] \\ & & & \twoheadrightarrow\!\!\!\!\twoheadrightarrow_{\beta(f)} & [\mathsf{c}_{g(0)}, \mathsf{c}_{g(1)}, \mathsf{c}_{g(2)}, \ldots] \\ & & & = & \mathsf{T}^{\infty}_g. \end{array}$$

(iii) Since $\mathsf{f}\mathsf{c}_n \to_f \mathsf{c}_{f(n)}$ one has $\mathsf{f}\;\mathrm{D}(f)\;f$. Hence (ii) applies to $G = \mathsf{f}$.
(iv) ($\Leftarrow$) By (i). ($\Rightarrow$) By (ii). □

EXERCISE 6.40. Let $U, V \in \Lambda^{\infty}$. Check that the following are equivalent.
(i) $\exists H \in \Lambda^o \,.\, H\mathsf{f} \colon U \rightsquigarrow_{\lambda f} V$.
(ii) $\exists H \in \Lambda^o \,.\, H\mathsf{T}^{\infty}_f \colon U \rightsquigarrow_{\lambda} V$.

PROPOSITION 6.41. *Let* g *be defined from* $g_1, \ldots, g_k \in \mathbb{B}$ *by composition, primitive recursion or minimization. Then for some* $F \in \Lambda^o$

$$F\mathsf{T}^{\infty}_{g_1} \cdots \mathsf{T}^{\infty}_{g_k} \overset{\infty}{=}_{\beta} \mathsf{T}^{\infty}_g.$$

PROOF. Wlog we do this for composition: $g(x,y) = g_1(g_2(x,y), g_3(x))$ where $g_1, g_2 \in \mathbb{B}_2$ and $g_3 \in \mathbb{B}_1$. By Lemma 6.39(i), for every $k = 1, 2, 3$, there exists $P_k \in \Lambda^o$ satisfying

$$P_k \mathsf{T}^{\infty}_{g_k} \mathsf{c}_{n_1} \cdots \mathsf{c}_{n_k} \overset{\infty}{=}_\beta \mathsf{c}_{g_k(n_1,\ldots,n_k)}.$$

Setting $G \triangleq \lambda xy.P_1 \mathsf{T}^{\infty}_{g_1}(P_2 \mathsf{T}^{\infty}_{g_2} xy)(P_3 \mathsf{T}^{\infty}_{g_3} x)$, we have:

$$\begin{aligned}
G\mathsf{c}_n\mathsf{c}_m &=_\beta P_1 \mathsf{T}^{\infty}_{g_1}(P_2 \mathsf{T}^{\infty}_{g_2} \mathsf{c}_n \mathsf{c}_m)(P_3 \mathsf{T}^{\infty}_{g_3} \mathsf{c}_n), && \text{by definition of } G,\\
&\overset{\infty}{=}_\beta P_1 \mathsf{T}^{\infty}_{g_1} \mathsf{c}_{g_2(n,m)} \mathsf{c}_{g_3(n)}, && \text{by the choice of } P_2, P_3,\\
&\overset{\infty}{=}_\beta \mathsf{c}_{g_1(g_2(n,m),g_3(n))}, && \text{by the choice of } P_1,\\
&= \mathsf{c}_{g(n,m)}, && \text{by definition of } g.
\end{aligned}$$

Then G λ-defines g in Λ^∞. By Lemma 6.39(ii) one has $HG \overset{\infty}{=}_\beta \mathsf{T}^{\infty}_g$, for some $H \in \Lambda^o$. Now for an appropriate $F \in \Lambda^o$, we have:

$$F \mathsf{T}^{\infty}_{g_1} \mathsf{T}^{\infty}_{g_2} \mathsf{T}^{\infty}_{g_3} \overset{\infty}{=}_\beta H(\lambda xy.P_1 \mathsf{T}^{\infty}_{g_1}(P_2 \mathsf{T}^{\infty}_{g_2} xy)(P_3 \mathsf{T}^{\infty}_{g_3} x)) = HG \overset{\infty}{=}_\beta \mathsf{T}^{\infty}_g. \qquad \square$$

THEOREM 6.42 (BARENDREGT AND KLOP (2009)). *Let $f \in \mathbb{B}_1, g \in \mathbb{B}_k$, with $k > 0$. Then the following are equivalent.*

1. $f \rightsquigarrow_T g$ *(g is Turing computable in f).*
2. $\mathsf{f} \rightsquigarrow_{\lambda f} g$ *(g is λf-definable from* f*).*
3. $\mathsf{T}^{\infty}_f \rightsquigarrow_\lambda \mathsf{T}^{\infty}_g$ *(T^{∞}_g is λ-definable from T^{∞}_f).*

PROOF. ($3 \Rightarrow 2$) Suppose that $\mathsf{T}^{\infty}_f \rightsquigarrow_\lambda \mathsf{T}^{\infty}_g$, i.e., that $H_1 \mathsf{T}^{\infty}_f \overset{\infty}{=}_\beta \mathsf{T}^{\infty}_g$ holds for some $H_1 \in \Lambda^o$. By Lemma 6.39(iii) one has $H_2\mathsf{f} \overset{\infty}{=}_{\beta f} \mathsf{T}^{\infty}_f$, for some $H_2 \in \Lambda^o$. Thus $H_1(H_2\mathsf{f}) \overset{\infty}{=}_{\beta f} \mathsf{T}^{\infty}_g$. By Lemma 6.39(i) one has $P\mathsf{T}^{\infty}_g \mathsf{c}_n =_\beta \mathsf{c}_{g(n)}$, for some $P \in \Lambda^o$. So $P(H_1(H_2\mathsf{f}))\mathsf{c}_n \overset{\infty}{=}_{\beta f} P\mathsf{T}^{\infty}_g \mathsf{c}_n \overset{\infty}{=}_{\beta f} \mathsf{c}_{g(n)}$, for all $n \in \mathbb{N}$. Hence, by Lemma 6.29(v), also $P(H_1(H_2\mathsf{f}))\mathsf{c}_n =_{\beta f} \mathsf{c}_{g(n)}$. We conclude that $\mathsf{f} \rightsquigarrow_{\lambda f} g$ holds.

($2 \Rightarrow 1$) Suppose $H\mathsf{f} \rightsquigarrow_{\lambda f} g$. Then for all $n \in \mathbb{N}$ one has $H\mathsf{f}\mathsf{c}_n =_{\beta f} \mathsf{c}_{g(n)}$. We will show that the graph of g is r.e. in f. Note that

$$g(n) = m \iff \mathsf{c}_{g(n)} =_{\beta f} \mathsf{c}_m \iff H\mathsf{f}\mathsf{c}_n =_{\beta f} \mathsf{c}_m.$$

The latter is clearly r.e. in f.

($1 \Rightarrow 3$) Assume $f \rightsquigarrow_T g$. Then g can be obtained from the initial functions extended by f using composition, primitive recursion and minimization. By induction on this generation we show that $\mathsf{T}^{\infty}_f \rightsquigarrow_\lambda \mathsf{T}^{\infty}_g$, that is $H\mathsf{T}^{\infty}_f \overset{\infty}{=}_\beta \mathsf{T}^{\infty}_g$, for some $H \in \Lambda^o$.

Case 1(a): $g = f$. Then we can take $H \triangleq \mathsf{I}$.

Case 1(b): g is an initial function. Then g is recursive, hence $G \overset{\infty}{=}_\beta \mathsf{T}^{\infty}_g$ for some $G \in \Lambda^o$, by Lemma 6.39(iv), and we can take $H \triangleq \mathsf{K}G$.

Case 2: g is constructed from $g_1, \ldots, g_k$ using composition, primitive recursion or minimization. By Proposition 6.41 one has $\mathsf{T}^{\infty}_g \overset{\infty}{=}_\beta Q\mathsf{T}^{\infty}_{g_1} \cdots \mathsf{T}^{\infty}_{g_k}$, for some $Q \in \Lambda^o$. By the IH there exist $H_1, \ldots, H_k$ such that $H_i \mathsf{T}^{\infty}_f \overset{\infty}{=}_\beta \mathsf{T}^{\infty}_{g_i}$. Then $H \triangleq \lambda x.Q(H_1 x) \cdots (H_k x)$ does the job. $\square$

6.3 Restoring confluence via Böhm reduction

Besides showing that CR^∞ fails in the λ^∞-calculus, Berarducci proposed a method to restore this confluence property by collapsing a class of problematic terms. The class of terms that he has identified as 'meaningless', and therefore worth being equated, contains those terms that are root-active in the sense that their root never stabilizes.

The notion of root-activeness was introduced in Definition 2.46(iii) for finite λ-terms, and generalizes to infinitary λ-terms in the following way.

DEFINITION 6.43. A term $T \in \Lambda^\infty$ is called *root-active* if every reduct of T can be reduced $\twoheadrightarrow_\beta$ to a β-redex $(\lambda x.U)V$. We write $\mathcal{R}$ for the set of all root-active infinitary λ-terms.

It is interesting to note that although root-active terms may not have common reducts, they are all pairwise interconvertible.

LEMMA 6.44. *Let $T \in \Lambda^\infty$. If T is root-active, then there exists $V \in \Lambda^\infty$ such that*

$$V \twoheadrightarrow\!\!\!\!\rightarrow_\beta \mathsf{I}^\omega \;\&\; V \twoheadrightarrow\!\!\!\!\rightarrow_\beta T$$

PROOF. For any T, write T_I for the term obtained by simultaneously replacing every application UV in T by $\mathsf{I}(UV)$. Clearly, we have $T_\mathsf{I} \twoheadrightarrow\!\!\!\!\rightarrow_\beta T$. An easy coinduction, with case analysis on T, shows the bisimilarity $(T[x:=V])_\mathsf{I} = T_\mathsf{I}[x:=V_\mathsf{I}]$. Therefore, one gets:

$$((\lambda x.U)V)_\mathsf{I} = \mathsf{I}((\lambda x.U_\mathsf{I})V_\mathsf{I}) \to_\beta \mathsf{I}(U_\mathsf{I}[x:=V_\mathsf{I}]) = \mathsf{I}(U[x:=V])_\mathsf{I} \to_\beta (U[x:=V])_\mathsf{I}.$$

This allows T_I to mimic any reduction sequence starting from T. Thus, any reduction ρ having T as initial term and infinitely many root steps can be converted into a reduction ρ_I having T_I as initial term and infinitely many root steps. Thus, ρ_I can be turned into a reduction $T_\mathsf{I} \twoheadrightarrow\!\!\!\!\rightarrow_\beta \mathsf{I}^\omega$ by omitting its root steps contracting redexes of the form $\mathsf{I}U$. □

As shown in Berarducci (1996), by equating all root-active terms to $\bot$ along the reduction, one obtains as infinitary normal form of a λ-term M its Berarducci tree $\mathrm{BeT}(M)$, as introduced in Definition 2.49. It turns out that the class of root-active terms is not the only one giving rise to a confluent infinitary λ-calculus once its elements are collapsed. Kennaway et al. (1995) have shown that, by considering as meaningless terms those without head normal form (resp. weak hnf), one retrieves as infinitary normal forms the Böhm trees (resp. the Lévy-Longo trees). At the end of the section, we will see that there are uncountably many meaningful notions of meaningless terms.

Collapsing meaningless terms

The first general axiomatic description of sets of meaningless terms was proposed by Kennaway et al. (1999). Subsequently, Severi and de Vries (2011b) have refined this notion by identifying necessary and sufficient conditions for a set of terms in order to be considered as meaningless. (Intuitively, a class of terms can be considered as meaningless if the corresponding infinitary λ-calculus is guaranteed to be confluent.) All results in this section have been originally proved by Kennaway, Klop, Severi, Sleep, van Oostrom and de Vries, or a subset of them. When it is not the case, we explicitly mention it.

For the presentation of the material, we mainly follow Kennaway et al. (2005).

DEFINITION 6.45. Consider a subset $\mathcal{U} \subseteq \Lambda^\infty$.

(i) Define a binary relation $\overset{\mathcal{U}}{\longleftrightarrow}$ on Λ^∞ by coinduction as follows:

$$\frac{}{x \overset{\mathcal{U}}{\longleftrightarrow} x} \qquad \frac{U, V \in \mathcal{U}}{U \overset{\mathcal{U}}{\longleftrightarrow} V} \qquad \frac{T \overset{\mathcal{U}}{\longleftrightarrow} T' \quad U \overset{\mathcal{U}}{\longleftrightarrow} U'}{TU \overset{\mathcal{U}}{\longleftrightarrow} T'U'} \qquad \frac{T \overset{\mathcal{U}}{\longleftrightarrow} V}{\lambda x.T \overset{\mathcal{U}}{\longleftrightarrow} \lambda x.V}$$

When $T \overset{\mathcal{U}}{\longleftrightarrow} U$ holds, we say that T and U are *indiscernible modulo* $\mathcal{U}$.

(ii) We say that $\mathcal{U}$ is a *set of meaningless terms*, or simply a *meaningless set*, if it satisfies the following axioms:

Root-activeness: $\mathcal{U}$ contains all root-active terms;

Closure under substitution: if $U \in \mathcal{U}$ then $U^\sigma \in \mathcal{U}$, for all substitutions $\sigma : \mathrm{Var} \to \Lambda^\infty$;

Closure under β-reduction: if $U \in \mathcal{U}$ and $U \twoheadrightarrow\!\!\!\rightarrow_\beta V$ then $V \in \mathcal{U}$;

Closure under β-anti-reduction: if $U \in \mathcal{U}$ and $V \twoheadrightarrow\!\!\!\rightarrow_\beta U$ then $V \in \mathcal{U}$;

Overlap: if $\lambda x.U \in \mathcal{U}$ then $(\lambda x.U)T \in \mathcal{U}$, for all $T \in \Lambda^\infty$;

Indiscernibility: For all $U, V \in \Lambda^\infty$ such that $U \overset{\mathcal{U}}{\longleftrightarrow} V$, we have $U \in \mathcal{U} \iff V \in \mathcal{U}$.

REMARK 6.46. (i) The only recursive set of meaningless terms is Λ^∞.

(ii) Since Ω is root-active, it belongs to any set $\mathcal{U}$ of meaningless terms.

(iii) Intuitively, $U \overset{\mathcal{U}}{\longleftrightarrow} V$ means that U can be obtained from V by replacing some (possibly, infinitely many) subterms in U by other terms from $\mathcal{U}$.

(iv) The closure under β-anti-reductions first appeared in Kennaway et al. (2005). This axiom is unnecessary to obtain a confluent term rewriting system, but the coinductive proof of confluence by Czajka (2020) does rely on it (see Theorem 6.48, below).

(v) Even without requiring the closure under β-anti-reduction, by root-activeness and indiscernibility, one has that $M \in \mathcal{U}$ entails, e.g., $\mathsf{I}(\mathsf{I}M), \mathsf{K}MN \in \mathcal{U}$.

DEFINITION 6.47. (i) The set $\Lambda^\infty_\bot$ of *infinitary $\lambda\bot$-terms* is coinductively defined by:

$$(\Lambda^\infty_\bot) \qquad T ::=_{\text{coind.}} \bot \mid x \mid \lambda x.T \mid TT$$

The substitution $T[x := U]$ of $U \in \Lambda^\infty_\bot$ for all free occurrences of x in $T \in \Lambda^\infty_\bot$ is defined as usual, namely by adding the case $\bot[x := U] \triangleq \bot$.

(ii) The definition of the infinitary closure $\twoheadrightarrow\!\!\!\rightarrow_{\mathrm{R}}$ is given as in Definition 6.5(ii), but with an additional axiom. Indeed, the constant $\bot$ is treated as a variable:

$$\frac{T \twoheadrightarrow_{\mathrm{R}} \bot}{T \twoheadrightarrow\!\!\!\rightarrow_{\mathrm{R}} \bot}$$

(iii) Given a set $\mathcal{U}$ of meaningless terms, define the infinitary notion of reduction $(\bot_\mathcal{U})$ on $\Lambda^\infty_\bot$ as follows:

$$\frac{T[\bot := \Omega] \in \mathcal{U} \quad T \neq \bot}{T \to \bot} \qquad (\bot_\mathcal{U})$$

where $T[\bot := \Omega]$ denotes the infinitary λ-term obtained from T by substituting Ω for all occurrences of $\bot$. The choice of Ω is due to the fact that it always belongs to $\mathcal{U}$.

(iv) The notion of reduction $(\beta\bot_{\mathcal{U}}) \triangleq (\beta) \cup (\bot_{\mathcal{U}})$ is called *Böhm reduction* relative to $\mathcal{U}$.
(v) The $\lambda^{\infty}_{\mathcal{U}}$*-calculus* is given by $\Lambda^{\infty}_{\bot}$ endowed with Böhm reduction relative to $\mathcal{U}$.

By Remark 6.46(i), the reduction $\to_{\bot_{\mathcal{U}}}$ is not effective (unless $\mathcal{U} = \Lambda^{\infty}$) but it provides a mathematically convenient way of talking about confluence modulo undefinedness.

THEOREM 6.48. *Let $\mathcal{U}$ be a set of meaningless terms. Then,*
(i) *the $\lambda^{\infty}_{\mathcal{U}}$-calculus enjoys both* CR^{∞} *and* WN^{∞};
(ii) *the $\bot_{\mathcal{U}}$-reduction can be postponed after β-reduction, i.e., for all $T, U \in \Lambda^{\infty}_{\bot}$:*

$$T \twoheadrightarrow\!\!\!\!\rightarrow_{\beta\bot_{\mathcal{U}}} U \quad \Rightarrow \quad \exists V \in \Lambda^{\infty}_{\bot}\,.\, T \twoheadrightarrow\!\!\!\!\rightarrow_{\beta} V \twoheadrightarrow\!\!\!\!\rightarrow_{\bot_{\mathcal{U}}} U$$

PROOF. (i) For these properties on strongly convergent reductions $\xrightarrow{\infty}_{\beta\bot_{\mathcal{U}}}$, cf. Terese (2003), Theorems 12.9.6 (CR^{∞}) and 12.9.11 (WN^{∞}). In fact, by Corollary 12.9.15 (*ibidem*), this confluence property holds for Cauchy-convergent reduction sequences as well. For the coinductively defined infinitary reduction relation $\twoheadrightarrow\!\!\!\!\rightarrow_{\beta\bot_{\mathcal{U}}}$, see Czajka (2020).
(ii) By Lemma 5.14 in Czajka (2020). □

Therefore, the $\lambda^{\infty}_{\mathcal{U}}$-calculus is a weakly normalizing and confluent extension of the finite λ-calculus. Moreover, since this calculus enjoys CR^{∞}, it also satisfies UN^{∞}. The postponement property is useful to connect Böhm reductions with plain β-reductions.

NOTATION. *Let $\mathcal{U}$ be a set of meaningless terms.*
(i) *Given $T \in \Lambda^{\infty}$, we denote by $T^{\mathcal{U}\text{-nf}}$ its $\beta\bot_{\mathcal{U}}$-normal form.*
(ii) *Given $\mathcal{X} \subseteq \Lambda^{\infty}$, we let* $\mathrm{NF}_{\mathcal{U}}(\mathcal{X}) \triangleq \{T^{\mathcal{U}\text{-nf}} \mid T \in \mathcal{X}\}$.

REMARK 6.49. (i) By Remark 6.46(ii), $\Omega^{\mathcal{U}\text{-nf}} = \bot$ for every set $\mathcal{U}$ of meaningless terms.
(ii) Although we mainly work with the coinductively defined infinitary notion of reduction $\twoheadrightarrow\!\!\!\!\rightarrow_{\beta\bot_{\mathcal{U}}}$, it is worth mentioning that the Compression Lemma 6.25 does hold for strongly converging reduction in $\lambda^{\infty}_{\mathcal{U}}$. Also, using the 'metric' definition of reduction, SN^{∞} keep failing because of divergent reduction sequences like

$$\Omega \to_{\beta} \Omega \to_{\beta} \Omega \to_{\beta} \cdots$$

One might obtain SN^{∞} by imposing that $\bot_{\mathcal{U}}$-reductions take precedence over β-reductions.

LEMMA 6.50. *Let $\mathcal{U}$ be a set of meaningless terms. For $T \in \Lambda^{\infty}_{\bot}$, we have:*

$$T^{\mathcal{U}\text{-nf}} = \bot \iff T[\bot := \Omega] \in \mathcal{U}$$

PROOF. ($\Rightarrow$) Assume that $T \twoheadrightarrow\!\!\!\!\rightarrow_{\beta\bot_{\mathcal{U}}} \bot$. By Theorem 6.48(ii), this reduction factorizes as $T \twoheadrightarrow\!\!\!\!\rightarrow_{\beta} V \twoheadrightarrow\!\!\!\!\rightarrow_{\bot_{\mathcal{U}}} \bot$. By applying the substitution $[\bot := \Omega]$, we obtain the reduction

$$T[\bot := \Omega] \twoheadrightarrow\!\!\!\!\rightarrow_{\beta} V[\bot := \Omega] \twoheadrightarrow\!\!\!\!\rightarrow_{\bot_{\mathcal{U}}} V \twoheadrightarrow\!\!\!\!\rightarrow_{\bot_{\mathcal{U}}} \bot.$$

From the definition of $\bot_{\mathcal{U}}$-reduction and the indiscernability property of $\mathcal{U}$, we get that $V[\bot := \Omega] \twoheadrightarrow\!\!\!\!\rightarrow_{\bot_{\mathcal{U}}} \bot$ entails $V[\bot := \Omega] \to_{\bot_{\mathcal{U}}} \bot$, from which it follows $V[\bot := \Omega] \in \mathcal{U}$. Finally, from the closure of $\mathcal{U}$ under β-anti-reduction, we derive $T[\bot := \Omega] \in \mathcal{U}$.
($\Leftarrow$) Easy. □

As we did in Definition 6.43 for root-activeness, we need to transfer the notions of hnf, whnf and top nf from the finite to the infinitary λ-calculus.

DEFINITION 6.51. Let $T \in \Lambda^\infty$. We say that T is a:
(i) *head normal form* (hnf) if it has shape $T = \lambda x_1 \ldots x_n.yT_1 \cdots T_k$, for some $n, k \in \mathbb{N}$;
(ii) *weak head normal form* (whnf) if either it is a hnf or it has shape $T = \lambda x.V$;
(iii) *top normal form* (top nf) if either it is a whnf or it has shape $T = UV$ for some U that does not reduce to a λ-abstraction, i.e. $U \not\twoheadrightarrow_\beta \lambda x.U'$.

For a while the following three sets were the only known sets of meaningless terms.

NOTATION. *Let us fix the following subsets of* Λ^∞*:*

$$\begin{aligned} \mathcal{HN} &\triangleq \{T \in \Lambda^\infty \mid T \twoheadrightarrow_\beta V \textit{ in head normal form } \}; \\ \mathcal{WN} &\triangleq \{T \in \Lambda^\infty \mid T \twoheadrightarrow_\beta V \textit{ in weak head normal form } \}; \\ \mathcal{TN} &\triangleq \{T \in \Lambda^\infty \mid T \twoheadrightarrow_\beta V \textit{ in top normal form } \}. \end{aligned}$$

We denote by $\overline{\mathcal{HN}}$, $\overline{\mathcal{WN}}$ *and* $\overline{\mathcal{TN}}$ *their respective complements in* Λ^∞. *E.g.,*

$$\overline{\mathcal{HN}} = \Lambda^\infty - \mathcal{HN}$$

EXERCISE 6.52. (i) Show that $\mathcal{HN} \subseteq \mathcal{WN} \subseteq \mathcal{TN}$.
(ii) Show that UNS $\subsetneq \overline{\mathcal{HN}}$.
(iii) Show that $\overline{\mathcal{TN}} = \mathcal{R}$.

LEMMA 6.53. *The following are sets of meaningless terms:*
(i) $\overline{\mathcal{HN}}$;
(ii) $\overline{\mathcal{WN}}$;
(iii) $\overline{\mathcal{TN}} = \mathcal{R}$.

PROOF. Let us check the closure under β-anti-reduction, all other axioms being proved in Kennaway et al. (1999).

Case (i). Assume that $U \in \overline{\mathcal{HN}}$ and $V \twoheadrightarrow\!\!\!\!\!\twoheadrightarrow_\beta U$. By the way of contradiction, assume that V has a hnf. Then $V \twoheadrightarrow_h \lambda\vec{x}.yV_1 \cdots V_k$, for some $V_i \in \Lambda^\infty$ with $i\,(1 \le i \le k \ge 0)$. By iterated applications of the restricted PML$^\infty$ property (Proposition 6.12(ii)) we obtain that U and $\lambda\vec{x}.yV_1 \cdots V_k$ have a common reduct T. But $\lambda\vec{x}.yV_1 \cdots V_k \twoheadrightarrow\!\!\!\!\!\twoheadrightarrow_\beta T$ entails $T = \lambda\vec{x}.yT_1 \cdots T_k$ for some T_i such that $V_i \twoheadrightarrow\!\!\!\!\!\twoheadrightarrow_\beta T_i$, whence U has a hnf. Contradiction.

Cases (ii) and (iii) are similar. □

As promised, Böhm trees, Lévy-Longo trees and Berarducci trees can be reconstructed as normal forms in the appropriate $\lambda^\infty_\mathcal{U}$-calculus.

PROPOSITION 6.54. *Let* $M \in \Lambda$.
(i) *If* $\mathcal{U} = \overline{\mathcal{HN}}$ *then* $M^{\mathcal{U}\text{-nf}} = \mathrm{BT}(M)$.
(ii) *If* $\mathcal{U} = \overline{\mathcal{WN}}$ *then* $M^{\mathcal{U}\text{-nf}} = \mathrm{LLT}(M)$.
(iii) *If* $\mathcal{U} = \mathcal{R}$ *then* $M^{\mathcal{U}\text{-nf}} = \mathrm{BeT}(M)$.

PROOF. (i) Note that $\mathrm{BT}(M)$ is a $\beta\bot_{\mathcal{U}}$-normal form since it does not contain any β- nor $\bot_{\mathcal{U}}$-redex. We show that $M \mathrel{\rightarrow\!\!\!\!\twoheadrightarrow}_{\beta\bot_{\mathcal{U}}} \mathrm{BT}(M)$ holds by coinduction, splitting into cases depending on the solvability of M.

Case $M \in \mathrm{UNS} \subseteq \overline{\mathcal{HN}}$. Then $M \rightarrow_{\bot_{\mathcal{U}}} \bot$, whence $M \mathrel{\rightarrow\!\!\!\!\twoheadrightarrow}_{\bot_{\mathcal{U}}} \bot$ (by Exercise 6.7(iii)).

Case $M \in \mathrm{SOL}$. Then $M \twoheadrightarrow_{\beta} \lambda\vec{x}.yM_1 \cdots M_k$. By co-IH, we have $M_i \mathrel{\rightarrow\!\!\!\!\twoheadrightarrow}_{\beta\bot_{\mathcal{U}}} \mathrm{BT}(M_i)$ for every $i\,(1 \le i \le k)$. So, writing $\twoheadrightarrow$ and $\mathrel{\rightarrow\!\!\!\!\twoheadrightarrow}$ for the corresponding Böhm reductions, we construct the (possibly infinite) derivation

$$\dfrac{M \twoheadrightarrow \lambda\vec{x}.N \qquad \dfrac{N \twoheadrightarrow T_k M_k \qquad \dfrac{\dfrac{T_2 \twoheadrightarrow T_1 M_1 \quad T_1 \mathrel{\rightarrow\!\!\!\!\twoheadrightarrow} y \quad M_1 \mathrel{\rightarrow\!\!\!\!\twoheadrightarrow} \mathrm{BT}(M_1)}{\vdots}}{T_k \mathrel{\rightarrow\!\!\!\!\twoheadrightarrow} y\mathrm{BT}(M_1)\cdots\mathrm{BT}(M_{k-1})} \qquad M_k \mathrel{\rightarrow\!\!\!\!\twoheadrightarrow} \mathrm{BT}(M_k)}{N \mathrel{\rightarrow\!\!\!\!\twoheadrightarrow} y\mathrm{BT}(M_1)\cdots\mathrm{BT}(M_k)}}{M \mathrel{\rightarrow\!\!\!\!\twoheadrightarrow} \lambda\vec{x}.y\mathrm{BT}(M_1)\cdots\mathrm{BT}(M_k)}$$

For compactness, the rules generating the outer abstractions $\lambda\vec{x}$ have been gathered.

(ii) and (iii). Analogous. □

The following construction allows to build the so-called *normal form models* of the finite λ-calculus.

THEOREM 6.55. *Given a set $\mathcal{U}$ of meaningless terms, the algebra*

$$\mathcal{M}_{\mathcal{U}} = (\mathrm{NF}_{\mathcal{U}}(\Lambda), \bullet, \mathsf{K}, \mathsf{S})$$

where the application $\bullet$ is defined by

$$M^{\mathcal{U}\text{-}\mathbf{nf}} \bullet N^{\mathcal{U}\text{-}\mathbf{nf}} \triangleq (MN)^{\mathcal{U}\text{-}\mathbf{nf}}$$

is a λ-model. Moreover, for all $M \in \Lambda^o$, we have $[\![M]\!]^{\mathcal{M}_{\mathcal{U}}} = M^{\mathcal{U}\text{-}\mathbf{nf}}$.

PROOF. Using CR^∞ and WN^∞ (Theorem 6.48), one promptly verifies the following

1. $M^{\mathcal{U}\text{-}\mathbf{nf}}\, N^{\mathcal{U}\text{-}\mathbf{nf}} \mathrel{\rightarrow\!\!\!\!\twoheadrightarrow}_{\beta\bot_{\mathcal{U}}} (MN)^{\mathcal{U}\text{-}\mathbf{nf}}$;
2. $(M^{\mathcal{U}\text{-}\mathbf{nf}})^{\mathcal{U}\text{-}\mathbf{nf}} = M^{\mathcal{U}\text{-}\mathbf{nf}}$;
3. $(MN)^{\mathcal{U}\text{-}\mathbf{nf}} = (M^{\mathcal{U}\text{-}\mathbf{nf}}\, N^{\mathcal{U}\text{-}\mathbf{nf}})^{\mathcal{U}\text{-}\mathbf{nf}}$;
4. $[\![M]\!]^{\mathcal{M}_{\mathcal{U}}}_{\mathrm{id}} = M^{\mathcal{U}\text{-}\mathbf{nf}}$, where $\mathrm{id}(x) \triangleq x$ for all $x \in \mathrm{Var}$.

As a consequence, we obtain the chain of equivalences

$$\mathcal{M}_{\mathcal{U}} \models M = N \iff [\![M]\!]^{\mathcal{M}_{\mathcal{U}}}_{\mathrm{id}} = [\![N]\!]^{\mathcal{M}_{\mathcal{U}}}_{\mathrm{id}} \iff M^{\mathcal{U}\text{-}\mathbf{nf}} = N^{\mathcal{U}\text{-}\mathbf{nf}}$$

It is now easy to check that the axioms of a λ-model are satisfied. □

COROLLARY 6.56. *Taking as $\mathcal{U}$ the sets $\overline{\mathcal{HN}}, \overline{\mathcal{WN}}, \mathcal{R}$ of meaningless terms, one retrieves the usual Böhm, Lévy-Longo and Berarducci models, respectively.*

PROOF. By the above theorem and Proposition 6.54. □

The lattice of meaningless sets

Although in the initial literature on infinitary λ-calculus only $\overline{\mathcal{HN}}$, $\overline{\mathcal{WN}}$ and $\overline{\mathcal{TN}}$ were presented as set of meaningless terms, these sets are not the only examples. More recently Severi and de Vries explored the whole lattice of meaningless sets, whose cardinality is the continuum, and exposed some of its properties. See, e.g., Severi and de Vries (2011a), Severi and de Vries (2005a) and Severi and de Vries (2005b).

The rest of the section is devoted to present some of their results.

The next lemma generalizes Wadsworth's remark concerning the possible shapes of a λ-term (see page 34) to infinitary $\lambda\bot$-terms.

LEMMA 6.57. *Every $T \in \Lambda_\bot^\infty$ has one of the following shapes:*

1. $\lambda\vec{x}.yT_1 \cdots T_k$;
2. $\lambda\vec{x}.(\lambda y.V)UT_1 \cdots T_k$;
3. $\lambda\vec{x}.\bot T_1 \cdots T_k$;
4. $\lambda\vec{x}.(((\cdots)T_3)T_2)T_1$;
5. $\mathsf{O} = \lambda x_0.\lambda x_1.\lambda x_2.\ldots$

PROOF. By case analysis on the coinductive grammar in Definition 6.47(i). □

DEFINITION 6.58. For $T \in \Lambda_\bot^\infty$, we introduce the following terminology.

T is called:		T has shape:
head active	$\Longleftrightarrow$	$T = \lambda\vec{x}.UT_1 \cdots T_k \ \&\ U \in \mathcal{R}$
strong active	$\Longleftrightarrow$	$T = UT_1 \cdots T_k \ \&\ U \in \mathcal{R}$
infinite left spine	$\Longleftrightarrow$	$T = \lambda\vec{x}.(((\cdots)T_3)T_2)T_1$
strong infinite left spine	$\Longleftrightarrow$	$T = (((\cdots)T_3)T_2)T_1$

NOTATION. *Let us fix the following subsets of Λ^∞:*

$$\begin{array}{lcl}
\mathcal{SA} & \triangleq & \{T \in \Lambda^\infty \mid T \twoheadrightarrow_\beta V \ \&\ V \textit{ is strong active }\}; \\
\mathcal{HA} & \triangleq & \{T \in \Lambda^\infty \mid T \twoheadrightarrow_\beta V \ \&\ V \textit{ is head active }\}; \\
\mathcal{IL} & \triangleq & \{T \in \Lambda^\infty \mid T \twoheadrightarrow\!\!\!\!\rightarrow_\beta V \ \&\ V \textit{ is an infinite left spine }\}; \\
\mathcal{SIL} & \triangleq & \{T \in \Lambda^\infty \mid T \twoheadrightarrow\!\!\!\!\rightarrow_\beta V \ \&\ V \textit{ is a strong infinite left spine }\}; \\
\mathcal{O} & \triangleq & \{T \in \Lambda^\infty \mid T \twoheadrightarrow\!\!\!\!\rightarrow_\beta \mathsf{O}\}.
\end{array}$$

EXERCISE 6.59. (i) Verify that $\mathcal{SA} \subseteq \mathcal{HA}$ and $\mathcal{SIL} \subseteq \mathcal{IL}$.

(ii) Show that $\overline{\mathcal{WN}} = \mathcal{SA} \cup \mathcal{SIL}$.

(iii) Show that $\overline{\mathcal{HN}} = \mathcal{HA} \cup \mathcal{IL} \cup \mathcal{O}$.

Exercise 6.59(ii) reveals that the infinitary λ-terms without weak hnf are precisely those terms of which no instance can reduce to an abstraction.

PROPOSITION 6.60. *The collection* $\{\mathcal{HN}, \mathcal{HA}, \mathcal{IL}, \mathcal{O}\}$ *forms a partition of* Λ^∞.

PROOF. Using infinitary confluence modulo root-active subterms (Czajka (2020)), one verifies that these sets are disjoint. By Lemma 6.57, their union covers the whole set Λ^∞. Note that item (3) of that lemma does not apply to terms in Λ^∞. □

LEMMA 6.61. *Let* $\mathcal{U}$ *be a meaningless set.*

(i) *If* $\lambda x.T \in \mathcal{U}$ *then* $T \in \mathcal{U}$.
(ii) *If* $\lambda x.T \in \mathcal{U}$ *then* $\mathcal{HA} \subseteq \mathcal{U}$. *If moreover* $\mathsf{O} \in \mathcal{U}$, *then* $\mathcal{HA} \cup \mathcal{O} \subseteq \mathcal{U}$.
(iii) *If* $\lambda x.T \in \mathcal{U}$ *then* $\lambda x.U \in \mathcal{U}$, *for all* $U \in \mathcal{U}$.
(iv) *If* $\lambda x.T \in \mathcal{U}$ *and* $\mathcal{SA} \cup \mathcal{SIL} \subseteq \mathcal{U}$ *then* $\mathcal{HA} \cup \mathcal{IL} \subseteq \mathcal{U}$.
(v) *If* $\mathcal{SIL} \subseteq \mathcal{U}$ *then* $\mathcal{SA} \subseteq \mathcal{U}$.
(vi) *If* $\mathcal{IL} \subseteq \mathcal{U}$ *then* $\mathcal{HA} \subseteq \mathcal{U}$.
(vii) *If there exists* $U \in \mathcal{U} \cap \mathcal{HN}$, *then* $\mathcal{U} = \Lambda^\infty$.

PROOF. (i) By the overlap axiom, $\lambda x.T \in \mathcal{U}$ implies $(\lambda x.T)x \in \mathcal{U}$. Therefore, we obtain $T \in \mathcal{U}$ by closure under β-reduction.

(ii) By the overlap axiom, $(\lambda x.T)V \in \mathcal{U}$ for all $V \in \Lambda^\infty$. By indiscernibility, we obtain $UV \in \mathcal{U}$ for all $U \in \mathcal{R}$. By repeating this reasoning, $U \in \mathcal{R}$ entails $UV_1 \cdots V_k \in \mathcal{U}$ for all $V_i \in \Lambda^\infty$. Now, by indiscernibility, we get $\lambda x.U \in \mathcal{U}$ whence $\lambda x_1 \ldots x_n.UV_1 \cdots V_k \in \mathcal{U}$. Finally, $\mathsf{O} \in \mathcal{U}$ entails $\mathcal{O} \subseteq \mathcal{U}$ by closure under β-anti-reduction.

(iii) Assume $\lambda x.T \in \mathcal{U}$. By (i), we obtain $T \in \mathcal{U}$. Conclude by indiscernibility.

(iv) By (iii).

(v) Let $({}^\omega T) \triangleq (((\cdots)T)T)$. We have that $({}^\omega T) = ({}^\omega T)T \in \mathcal{U}$. By indiscernibility, we get $UT \in \mathcal{U}$ for any $U \in \mathcal{R}$. It follows that $UV_1 \cdots V_k \in \mathcal{U}$ for all $V_i \in \Lambda^\infty$.

(vi) By (ii), since $\lambda x.({}^\omega \mathsf{I}) \in \mathcal{IL} \subseteq \mathcal{U}$.

(vii) From $U \in \mathcal{HN}$ we get $U \twoheadrightarrow\!\!\!\!\rightarrow_\beta \lambda x_1 \ldots x_n.yT_1 \cdots T_k$. For simplicity, consider the case $n > 0$ and $y = x_1$. Since $U \in \mathcal{U}$ and $\mathcal{U}$ is closed under β-reduction, we obtain $\lambda x_1 \ldots x_n.x_1T_1 \cdots T_k \in \mathcal{U}$. By overlap, we get $(\lambda x_1 \ldots x_n.x_1T_1 \cdots T_k)\mathsf{U}^{k+1}_{k+1}x_2 \cdots x_nV \in \mathcal{U}$ where $\mathsf{U}^{k+1}_{k+1} = \lambda x_1 \cdots x_{k+1}.x_{k+1}$ while $V \in \Lambda^\infty$ is arbitrary. Easy calculations give:

$$\begin{array}{llll}(\lambda x_1 \ldots x_n.x_1T_1 \cdots T_k)\mathsf{U}^{k+1}_{k+1}x_2 \cdots x_nV & \twoheadrightarrow_\beta & \mathsf{U}^{k+1}_{k+1}T'_1 \cdots T'_kV, & \text{for } T'_i \triangleq T_i[x_1 := \mathsf{U}^{k+1}_{k+1}], \\ & \twoheadrightarrow_\beta & V \in \mathcal{U}, & \text{by closure under } \twoheadrightarrow\!\!\!\!\rightarrow_\beta.\end{array}$$

Since V is arbitrary, we conclude $\mathcal{U} = \Lambda^\infty$. The case $y \notin \vec{x}$ is analogous, but one needs to apply the closure under substitution. □

LEMMA 6.62. *The following are meaningless sets:*

(i) Λ^∞,
(ii) $\mathcal{SA}$,
(iii) $\mathcal{HA}$,
(iv) $\mathcal{HA} \cup \mathcal{O}$,
(v) $\mathcal{SA} \cup \mathcal{SIL}$,
(vi) $\mathcal{HA} \cup \mathcal{IL}$,
(vii) $\mathcal{HA} \cup \mathcal{IL} \cup \mathcal{O}$.

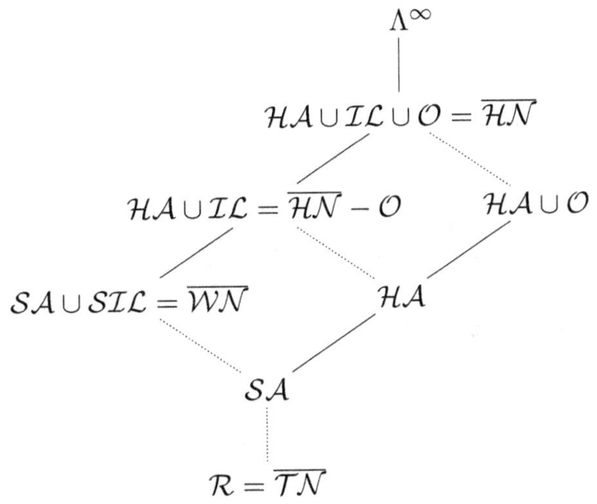

Figure 6.3: The lattice of meaningless sets. Dotted lines represent lattice intervals of cardinality $2^{\aleph_1}$. The other intervals have cardinality 2.

PROOF. (i) Trivial.
(ii) - (vii) See Severi and de Vries (2005b). □

The sets of meaningless terms, ordered by set-theoretical inclusion, form a lattice. As usual, the operation *meet* is given by the intersection and the *join* of two elements is the least meaningless set encompassing both of them. This lattice is depicted in Figure 6.3, where a node $\mathcal{U}_1$ is below a node $\mathcal{U}_2$ whenever $\mathcal{U}_1 \subsetneq \mathcal{U}_2$ holds.

REMARK 6.63. Not every combination of basic sets $\overline{\mathcal{WN}}, \overline{\mathcal{HN}}, \mathcal{R}, \mathcal{SA}, \mathcal{HA}, \mathcal{IL}, \mathcal{SIL}, \mathcal{O}$ gives rise to a set of meaningless terms.

(i) The sets $\mathcal{SIL}$, $\mathcal{IL}$ and $\mathcal{O}$ do not satisfy root-activeness.

(ii) The sets $\mathcal{R}\cup\mathcal{SIL}$ and $\mathcal{R}\cup\mathcal{IL}$ do not satisfy indiscernibility. Indeed, the term ${}^\omega\mathsf{I} = (((\cdots)\mathsf{I})\mathsf{I})\mathsf{I}$ belongs to $\mathcal{SIL}\cap\mathcal{IL}$. Now, since ${}^\omega\mathsf{I} = ({}^\omega\mathsf{I})\mathsf{I}$, any set satisfying indiscernibility should contain $\Omega\mathsf{I}$ as well. However, it is easy to check that $\Omega\mathsf{I} \notin \mathcal{SIL}\cup\mathcal{IL}$.

(iii) The set $\mathcal{R}\cup\mathcal{O}$ contains O but does not contain $\lambda x.\Omega$. Since $\mathsf{O} = \lambda x.\mathsf{O}$, this contradicts indiscernibility.

Let us characterize the cardinalities of the intervals displayed in Figure 6.3.

THEOREM 6.64. *Consider the lattice of meaningless sets.*

(i) *The following intervals have cardinality* 2*:*

$[\mathcal{HA}\cup\mathcal{IL}\cup\mathcal{O};\ \Lambda^\infty]$

$[\mathcal{HA}\cup\mathcal{IL};\ \mathcal{HA}\cup\mathcal{IL}\cup\mathcal{O}]$

$[\mathcal{SA}\cup\mathcal{SIL};\ \mathcal{HA}\cup\mathcal{IL}]$

$[\mathcal{HA};\ \mathcal{HA}\cup\mathcal{O}]$

$[\mathcal{SA};\ \mathcal{HA}]$

(ii) *The following intervals have cardinality* $2^{\aleph_1}$:

$[\mathcal{HA}\cup\mathcal{O};\ \mathcal{HA}\cup\mathcal{IL}\cup\mathcal{O}]$
$[\mathcal{HA}\cup\mathcal{IL};\ \mathcal{HA}\cup\mathcal{IL}]$
$[\mathcal{SA};\ \mathcal{SA}\cup\mathcal{SIL}]$
$[\mathcal{R};\ \mathcal{SA}]$

PROOF. (i) By Lemma 6.61(vii), $\overline{\mathcal{HN}} = \mathcal{HA}\cup\mathcal{IL}\cup\mathcal{O}$ is the largest meaningless set strictly included in Λ^∞. By Lemma 6.61(ii), $\mathcal{SA}\subsetneq\mathcal{U}\subsetneq\mathcal{HA}\cup\mathcal{O}$ entails $\mathcal{U}=\mathcal{HA}$. Finally, assume that $\mathcal{SA}\cup\mathcal{SIL}\subsetneq\mathcal{U}\subsetneq\mathcal{HA}\cup\mathcal{IL}\cup\mathcal{O}$. Then there exists $U\in\mathcal{U}$ such that $U \twoheadrightarrow\!\!\!\!\to_\beta \lambda x.T$. By closure under β-reduction, we get $\lambda x.T\in\mathcal{U}$. Since $\mathcal{SA}\cup\mathcal{SIL}\subsetneq\mathcal{U}$, Lemma 6.61(iv) allows to conclude $\mathcal{U}=\mathcal{HA}\cup\mathcal{SIL}$.

(ii) As the whole powerset of Λ^∞ has cardinality $2^{\aleph_1}$, it is enough to check that these intervals contain uncountably many elements. Let $\mathcal{X}\subseteq\Lambda^\infty$ be a set of closed normal forms without infinite left spines. In Severi and de Vries (2005b), the authors show that the following are meaningless sets:

$$\begin{aligned}\mathcal{SA}_\mathcal{X} &\triangleq \{T\in\Lambda^\infty \mid T\twoheadrightarrow_\beta UV_1\cdots V_n \;\&\; U\in\mathcal{R} \;\&\; \vec{V}\in\mathcal{X}\};\\ \mathcal{IL}_\mathcal{X} &\triangleq \{T\in\Lambda^\infty \mid T\twoheadrightarrow\!\!\!\!\to_\beta \lambda\vec{x}.((\cdots V_3)V_2)V_1 \;\&\; (V_i)_{i\in\mathbb{N}}\in\mathcal{X}\};\\ \mathcal{SIL}_\mathcal{X} &\triangleq \{T\in\Lambda^\infty \mid T\twoheadrightarrow\!\!\!\!\to_\beta ((\cdots V_3)V_2)V_1 \;\&\; (V_i)_{i\in\mathbb{N}}\in\mathcal{X}\}.\end{aligned}$$

Moreover, so are the sets $\mathcal{HA}\cup\mathcal{IL}_\mathcal{X}\cup\mathcal{O}, \mathcal{HA}\cup\mathcal{IL}_\mathcal{X}, \mathcal{SA}\cup\mathcal{SIL}_\mathcal{X}, \mathcal{SA}_\mathcal{X}$ that belong to the corresponding intervals listed in item (ii) of the theorem. We conclude since there are $2^{\aleph_1}$ sets $\mathcal{X}$ satisfying the above conditions. □

We draw some conclusions concerning the normal form models from Theorem 6.55.

COROLLARY 6.65. *There exist* $2^{\aleph_1}$ *distinguished normal form models* $\mathcal{M}_\mathcal{U}$.

PROOF. Immediate from the previous theorem. □

COROLLARY 6.66. *Let* $\mathcal{U}$ *be a meaningless set.*

(i) $\mathcal{M}_\mathcal{U}$ *is consistent* $\iff \mathcal{R}\subseteq\mathcal{U}\subseteq\overline{\mathcal{HN}}$.

(ii) *If* $\mathcal{U}\neq\Lambda^\infty$ *then, for all* $M,N\in\Lambda$, *we have the following chain of implications:*

$$\mathrm{BeT}(M)=\mathrm{BeT}(N) \quad\Rightarrow\quad \mathcal{M}_\mathcal{U}\models M=N \quad\Rightarrow\quad \mathrm{BT}(M)=\mathrm{BT}(N).$$

PROOF. (i) ($\Rightarrow$) Since $\mathcal{U}$ enjoys root-activeness, we know that $\mathcal{R}\subseteq\mathcal{U}$. Assume, by the way of contradiction, that $\mathcal{U}\not\subseteq\overline{\mathcal{HN}}$. By Lemma 6.61(vii), we get $\mathcal{U}=\Lambda^\infty$. It follows that $M^{\mathcal{U}\text{-nf}}=\bot$, for all $M\in\Lambda$. By Theorem 6.55, we conclude that $\mathrm{Th}(\mathcal{M}_\mathcal{U})=\nabla$.

($\Leftarrow$) If $\mathcal{U}\subseteq\overline{\mathcal{HN}}$ then $\mathsf{K}^{\mathcal{U}\text{-nf}}\neq\mathsf{F}^{\mathcal{U}\text{-nf}}$. By Theorem 6.55, $\mathrm{Th}(\mathcal{M}_\mathcal{U})$ is consistent.

(ii) Using Theorem 6.55, it is easy to check that $\mathcal{U}_1\subseteq\mathcal{U}_2$ entails $\mathrm{Th}(\mathcal{M}_{\mathcal{U}_2})\subseteq\mathrm{Th}(\mathcal{M}_{\mathcal{U}_2})$. We conclude since all meaningless terms $\mathcal{U}\neq\Lambda^\infty$ belong to the interval $[\mathcal{R},\overline{\mathcal{HN}}]$. □

Since the cardinality of the lattice of meaningless sets is $2^{\aleph_1}$, while $\lambda\mathcal{T}$ is $2^{\aleph_0}$, there must exist uncountably many normal form models inducing the same λ-theory.

EXERCISE 6.67. Construct two distinguished sets $\mathcal{U}_1,\mathcal{U}_2$ of meaningless terms such that

$$\mathrm{Th}(\mathcal{M}_{\mathcal{U}_1})=\mathrm{Th}(\mathcal{M}_{\mathcal{U}_2})$$

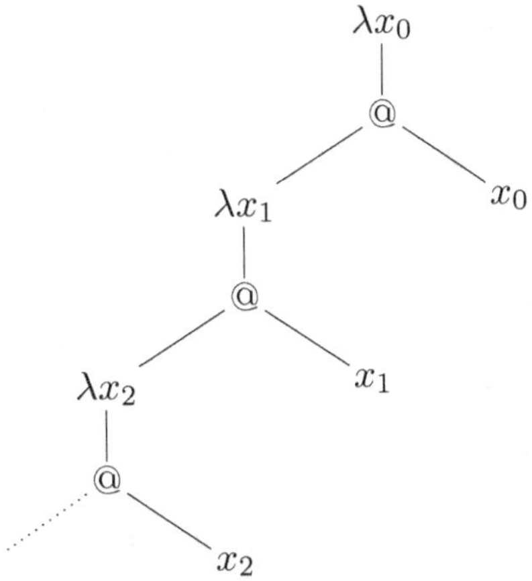

Figure 6.4: The infinite λ-term Ω_η.

6.4 Extensional infinitary λ-calculi

We have mentioned in Definition 6.2(iv) that the λ^∞-calculus inherits from Λ an infinitary notion of η-reduction. It is therefore tempting to extend the $\lambda^\infty_\mathcal{U}$-calculus with η-reduction in order to obtain extensional versions of Böhm trees, Lévy-Longo trees and Berarducci trees. Unfortunately, in the last two cases, η-reduction does not behave well and one looses the property of confluence. This was shown by V. van Oostrom via the λ-term Ω_η depicted in Figure 6.4 and satisfying $\Omega_\eta = \lambda x.(\Omega_\eta)x$, for x fresh. Similar to Ω, which β-reduces to itself in only one step, this term η-reduces to itself in one step.

THEOREM 6.68. *Let* $\mathcal{U} = \overline{\mathcal{WN}}$ *or* $\mathcal{U} = \overline{\mathcal{TN}}$. *Then* CR^∞ *fails in the* $\lambda^\infty_\mathcal{U}$*-calculus with* η.

PROOF. For $M \in \Lambda$, define $\Omega_M \triangleq (\lambda x.M(xx))(\lambda x.M(xx))$. Now, consider Ω_1 where $1 = \lambda xy.xy$ is an η-expansion of the identity I. Notice that the body of the outer abstraction in Ω_η has no weak (top) normal form. A similar remark holds for Ω_I. Therefore, we get:

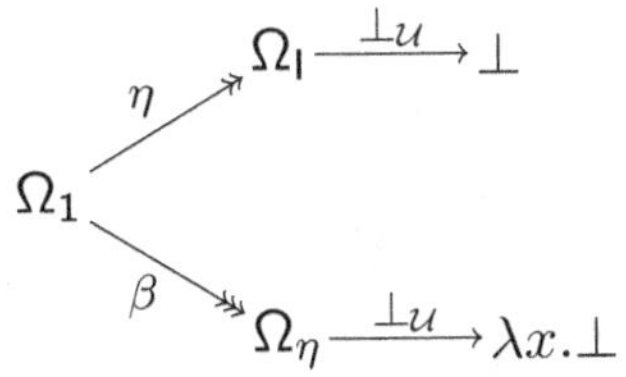

We conclude as $\bot$ and $\lambda x.\bot$ cannot be joined because $\lambda x.\Omega \notin \overline{\mathcal{WN}} \cup \overline{\mathcal{TN}}$. □

For Böhm trees, the situation is different. Indeed, when $\mathcal{U} = \overline{\mathcal{HN}}$ we have $\lambda x.\bot \to_{\bot_\mathcal{U}} \bot$, whence the configuration above no longer constitutes a counterexample to confluence. In fact, it turns out that Böhm trees admit two extensional versions (as mentioned at the end of Section 2.4) and that each version arises from a confluent infinitary λ-calculus.

Characterizing extensional Böhm trees

From now on, and until the end of the chapter, we consider fixed $\mathcal{U} = \overline{\mathcal{HN}}$ as a set of meaningless terms. We can therefore adopt the following lighter notation.

NOTATION. *We simply denote* $\bot_{\overline{\mathcal{HN}}}$*-reduction by* $\to_\bot$.

The results in this section are due to Severi and de Vries (2002) and we follow their presentation. In particular, in order to get rid of terms like Ω_η that are pathological from the rewriting viewpoint, we consider infinitary $\lambda\bot$-terms living in a proper subset of $\Lambda_\bot^\infty$. Moreover, we work with the classical notion of infinitary reduction $\xrightarrow{\infty}$ defined in terms of strongly convergent reduction sequences. We will however discuss the issues arising with coinductively defined infinitary reduction at the end of the chapter.

DEFINITION 6.69. (i) Let $M \in \Lambda_\bot$ and $n \in \mathbb{N}$. The *truncation of M at depth n*, in symbols $(M)^n$, is defined by induction on M as follows:

$$
\begin{aligned}
(M)^0 = (\bot)^n &\triangleq \bot,\\
(x)^{n+1} &\triangleq x,\\
(\lambda x.N)^{n+1} &\triangleq \lambda x.(N)^{n+1},\\
(PQ)^{n+1} &\triangleq (P)^{n+1}(Q)^n.
\end{aligned}
$$

I.e., the depth n of a truncation only decreases on the right-hand side of an application.

(ii) For $M, N \in \Lambda_\bot$, define a distance metric $\mathrm{h}(M,N)$ by setting:

$$
\mathrm{h}(M,N) = \begin{cases} 0, & \text{if } M = N;\\ 2^{-m}, & \text{where } m \triangleq \max\{n \in \mathbb{N} \mid (M)^n = (N)^n\}, \text{ otherwise.}\end{cases}
$$

(iii) The set $\Lambda_\bot^{\mathrm{h}\infty}$ is defined as the metric completion of the set $\Lambda_\bot$ of finite $\lambda\bot$-terms with respect to the metric $\mathrm{h}(-,-)$.

(iv) The $\lambda_{\mathrm{R}}^{\mathrm{h}\infty}$*-calculus* is given by $\Lambda_\bot^{\mathrm{h}\infty}$ together with the strongly convergent (w.r.t. h) reduction relation $\xrightarrow{\infty}_{\mathrm{R}}$. For instance, the $\lambda_{\beta\bot\eta}^{\mathrm{h}\infty}$*-calculus* stands for $\Lambda_\bot^{\mathrm{h}\infty}$ with $\xrightarrow{\infty}_{\beta\bot\eta}$. Similarly, the $\lambda_\eta^{\mathrm{h}\infty}$*-calculus* is given by the set $\Lambda_\bot^{\mathrm{h}\infty}$ endowed with $\xrightarrow{\infty}_\eta$ alone.

EXAMPLES 6.70. (i) For $M = \lambda y.\Omega$, we have $(M)^0 = \bot$, $(M)^1 = \lambda y.(\lambda x.x\bot)\bot$, $(M)^2 = \lambda y.(\lambda x.xx)(\lambda x.x\bot)$ and $(M)^n = M$ for $n > 2$.

(ii) For $M = x(y(zu))$ and $N = x(y(zv))$ we have $\mathrm{h}(M,N) = 2^{-3}$.

(iii) For all $M \in \Lambda$, we have $\mathrm{BT}(M) \in \Lambda_\bot^{\mathrm{h}\infty}$. E.g., $\mathrm{BT}(\mathsf{Y}f) = f(f(f(\cdots))) \in \Lambda_\bot^{\mathrm{h}\infty}$.

(iv) Infinite λ-terms like Ω_η, $\mathsf{O} = \lambda x_0x_1x_2\ldots$ or $(((\cdots)f)f)f$ do not belong to $\Lambda_\bot^{\mathrm{h}\infty}$.

(v) More generally, we have $\Lambda_\bot \subsetneq \Lambda_\bot^{\mathrm{h}\infty} \subsetneq \Lambda_\bot^\infty$.

REMARK 6.71. (i) The set $\Lambda_\bot^{\mathrm{h}\infty}$ can be equivalently described as a grammar where inductive and coinductive non-terminal symbols coexist:

$$
T ::= \bot \mid x \mid \lambda x.T \mid T\,T^{\text{co-ind}} \qquad (\Lambda_\bot^{\mathrm{h}\infty})
$$

Intuitively, this means that the rule $T, U \in \Lambda_\bot^{\mathrm{h}\infty} \Rightarrow TU \in \Lambda_\bot^{\mathrm{h}\infty}$ can be repeated finitely many times in the first argument and infinitely many times in the second argument.

(ii) As shown in Terese (2003), Chapter 12, the $\lambda^{\mathrm{h}\infty}_{\beta\perp}$-calculus satisfies CR^∞, WN^∞ the Compression Lemma and $\beta\perp$-postponement. In the following, we will keep referring to Theorem 6.48 for these properties even if, technically, it is stated for the $\lambda^{\infty}_{\mathcal{U}}$-calculus.

(iii) The $\xrightarrow{\infty}_{\beta\perp\eta}$ reduction is not compressible. As a counterexample, consider the infinite λ-term

$$T = \lambda fz.\mathsf{K}fz(\mathsf{K}fz(\cdots))z$$

Note that $T \in \Lambda^{\mathrm{h}\infty}_{\perp}$, and all its finite β-reducts are in η-normal form. However, the term T β-reduces in ω steps to $\lambda fz.f(f(f(\cdots)))z$, which can η-reduce further to $\lambda f.f(f(f(\cdots)))$. Clearly, this $\beta\eta$-reduction sequence cannot be compressed to a shorter one.

We start by presenting some properties of η-reduction.

LEMMA 6.72. *Any Cauchy-converging η-reduction sequence starting from a term in $\Lambda^{\mathrm{h}\infty}_{\perp}$ is also* h*-strongly convergent.*

PROOF. For $T \in \Lambda^{\mathrm{h}\infty}_{\perp}$, write $|T|_n$ for the number of nodes at h-depth n. Note that $|T|_n$ decreases whenever an η-redex in T at h-depth n is contracted. Assume, by the way of contradiction, that there exists a transfinite Cauchy-converging η-reduction sequence $T_0 \to_\eta T_1 \to_\eta T_2 \to_\eta \cdots$ that is not h-strongly convergent. This means that the contracted redexes occur infinitely often at some h-depth n. Then infinitely many inequalities in the sequence $|T_0|_n \geq |T_1|_n \geq |T_2|_n \geq \cdots$ must be strict, which is impossible. □

LEMMA 6.73. *The $\lambda^{\mathrm{h}\infty}_{\eta}$-calculus is compressible.*

PROOF. By transfinite induction on α, we prove that every reduction sequence $T \xrightarrow{\alpha}_\eta U$ can be compressed to a reduction $T \xrightarrow{\omega}_\eta U$. The argument at the limit case is standard. Let α be a successor ordinal, wlog $\alpha = \omega + 1$. Setting $T_0 \triangleq T$ and $T_\omega \triangleq U$, we have

$$\begin{array}{ccccccc}
T_0 \twoheadrightarrow_\eta \lambda z.T_k z & \xrightarrow{\eta} & \lambda z.T_{k+1} z & \xrightarrow{\eta} & \lambda z.T_{k+2} z & \cdots\cdots & \lambda z.T_\omega z \\
\downarrow \eta & & \downarrow \eta & & \downarrow \eta & & \downarrow \eta \\
T_k & \xrightarrow{\eta} & T_{k+1} & \xrightarrow{\eta} & T_{k+2} & \cdots\cdots & T_\omega
\end{array}$$

By constructing the dashed squares, we find a reduction of length ω starting from T_0 and, after $k+1$ steps, continuing as $T_k \to_\eta T_{k+1} \to_\eta \cdots$. Its limit is clearly T_ω. □

THEOREM 6.74. (i) *The $\lambda^{\mathrm{h}\infty}_{\eta}$-calculus enjoys* CR^∞.

(ii) *The $\lambda^{\mathrm{h}\infty}_{\eta}$-calculus enjoys* WN^∞.

PROOF. (i) By simultaneous induction on the length of two coinitial η-reductions, show that they can be joined with a tiling diagram construction in which all horizontal and vertical reductions are strongly convergent. By Lemma 6.73, one can assume that their length is at most ω. For more details see Severi and de Vries (2002), Theorem 3.

(ii) Let $T_0 \in \Lambda^{\mathrm{h}\infty}_{\perp}$. Consider the reduction $T_0 \to_\eta T_1 \to_\eta T_2 \to_\eta \cdots$ where each T_{i+1} is obtained from its predecessor T_i by contracting the leftmost η-redex of least h-depth in T_i. By Lemma 6.72, this reduction is h-strongly convergent. It is easy to check that its final term is in η-nf since the reduction strategy does not overlook any redex. □

We now study the interactions between $\xrightarrow{\infty}_\eta$ and the Böhm reduction $\xrightarrow{\infty}_{\beta\bot}$.

REMARK 6.75. (i) In general, the commutation of η- and $\bot$-reduction does not hold.

(ii) Already the local commutation of $\to_\eta$ and $\to_\bot$ fails when the contracted $\bot$-redex is not outermost: e.g., $\Omega\ {}_\eta\!\!\leftarrow \lambda x.\Omega x \to_\bot \lambda x.\bot$.

Despite the discussion above, for proving confluence of the $\lambda^{\mathrm{h}\infty}_{\beta\bot\eta}$-calculus it is sufficient to check that η-reduction commutes with outermost $\bot$-reduction.

DEFINITION 6.76. Let $T \in \Lambda^{\mathrm{h}\infty}_\bot$.

(i) A subterm U of T is an *outermost* $\bot$*-redex of* T if $U[\bot:=\Omega]$ is a maximal subterm of $T[\bot:=\Omega]$ without head normal form.

(ii) We say that $T \to_\bot T'$ is a step of *outermost* $\bot$*-reduction*, written $T \to_{\bot\text{-out}} T'$, if T' is obtained from T by contracting an outermost $\bot$-redex of T.

EXAMPLES 6.77. (i) The λ-term $M \triangleq x((\lambda y.\Omega y)z)$ has four $\bot$-redexes, namely Ω, Ωy, $\lambda y.\Omega y$ and $(\lambda y.\Omega y)z$ but only the latter is an outermost $\bot$-redex. Thus, $M \to_{\bot\text{-out}} x\bot$.

(ii) The two occurrences of Ω in $x\Omega\Omega$ are both maximal, therefore $x\Omega\Omega \to_{\bot\text{-out}} x\bot\Omega \to_{\bot\text{-out}} x\bot\bot$, but also $x\Omega\Omega \to_{\bot\text{-out}} x\Omega\bot \to_{\bot\text{-out}} x\bot\bot$.

The following commutation and postponement properties constitute the key ingredient in the proof of CR^∞ for the $\lambda^{\mathrm{h}\infty}_{\beta\bot\eta}$-calculus.

PROPOSITION 6.78 (COMMUTATIONS). *Let* $T, T_1, T_2 \in \Lambda^{\mathrm{h}\infty}_\bot$.

(i) $T \xrightarrow{\infty}_\beta T_1 \quad \& \; T \xrightarrow{\infty}_\eta T_2 \quad \Rightarrow \quad \exists T_3 \in \Lambda^{\mathrm{h}\infty}_\bot \,.\, T_1 \xrightarrow{\infty}_\eta T_3 \;\&\; T_2 \xrightarrow{\infty}_\beta T_3.$

(ii) $T \xrightarrow{\infty}_{\bot\text{-out}} T_1 \;\&\; T \xrightarrow{\infty}_\eta T_2 \quad \Rightarrow \quad \exists T_3 \in \Lambda^{\mathrm{h}\infty}_\bot \,.\, T_1 \xrightarrow{\infty}_\eta T_3 \;\&\; T_2 \xrightarrow{\infty}_{\bot\text{-out}} T_3.$

PROOF. See Severi and de Vries (2002), Theorems 5 and 6, respectively. □

THEOREM 6.79 (POSTPONEMENT OF $\xrightarrow{\infty}_\eta$ OVER $\xrightarrow{\infty}_{\beta\bot}$).
For $T, T_1, T_2 \in \Lambda^{\mathrm{h}\infty}_\bot$, *we have:*

$$T_1 \xrightarrow{\infty}_\eta T_2 \xrightarrow{\infty}_{\beta\bot} T_3 \quad \Rightarrow \quad \exists T_2' \in \Lambda^{\mathrm{h}\infty}_\bot \,.\, T_1 \xrightarrow{\infty}_{\beta\bot} T_2' \xrightarrow{\infty}_\eta T_3.$$

Moreover, if $T_2 \xrightarrow{\infty}_{\beta\bot} T_3$ *is finite then* $T_1 \xrightarrow{\infty}_{\beta\bot} T_2'$ *is finite as well.*

PROOF. See Severi and de Vries (2002), Corollary 1. □

LEMMA 6.80. *Let* $T, T' \in \Lambda^{\mathrm{h}\infty}_\bot$.

$$T \xrightarrow{\infty}_{\beta\bot\eta} T' \quad \Rightarrow \quad \exists T_0 \in \Lambda^{\mathrm{h}\infty}_\bot \,.\, T \xrightarrow{\infty}_{\beta\bot} T_0 \xrightarrow{\infty}_\eta T'$$

PROOF. (i) We proceed by transfinite induction on the number α of maximal subsequences of the form

$$V_0 \xrightarrow{\infty}_\eta V_1 \xrightarrow{\infty}_{\beta\bot} V_2$$

in $T \xrightarrow{\infty}_{\beta\bot\eta} T'$, the case $\alpha = 0$ being trivial.

Case $\alpha = \gamma + 1$. This case follows directly from Theorem 6.79.

Case α is a limit ordinal, say, $\alpha = \omega$. In this case, we have

$$T = V_{10} \xrightarrow{\infty}_{\eta} V_{11} \xrightarrow{\infty}_{\beta\perp} V_{21} \xrightarrow{\infty}_{\eta} V_{22} \xrightarrow{\infty}_{\beta\perp} V_{32} \xrightarrow{\infty}_{\beta\perp\eta} V_{\omega\omega} = T'$$

for the appropriate intermediate terms. By applying the induction hypothesis, we construct the following diagram (where $\xrightarrow{\infty}$ is drawn as $\twoheadrightarrow$) row by row:

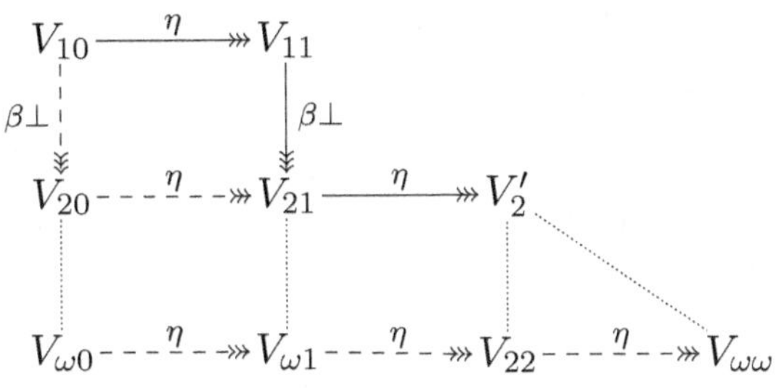

As the diagonal is strongly converging and the horizontal η-reductions do not modify the depth of the vertical $\beta\perp$-reductions, all vertical reductions are strongly convergent and have limits. By IH, they are connected by η-reduction sequences. By Lemma 6.72, the combined reduction at the bottom row is strongly converging. By the uniform nature of the strong convergence of all vertical reductions the limit of reductions at the bottom row and at the diagonal are the same.

The case for arbitrary limits is analogous. □

PROPOSITION 6.81. *Let* $T, V \in \Lambda_{\perp}^{\mathrm{h}\infty}$.

(i) $T \xrightarrow{\infty}_{\eta} V \;\Rightarrow\; T^{\beta\perp\text{-nf}} \xrightarrow{\infty}_{\eta} V^{\beta\perp\text{-nf}}$.

(ii) $T \xrightarrow{\infty}_{\beta\perp\eta} V \;\Rightarrow\; T^{\beta\perp\text{-nf}} \xrightarrow{\infty}_{\eta} V^{\beta\perp\text{-nf}}$.

In diagrammatic form:

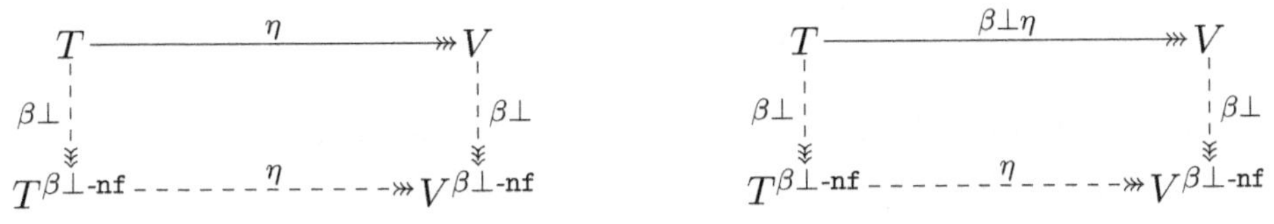

(i) (ii)

PROOF. (i) As the $\lambda_{\beta\perp}^{\mathrm{h}\infty}$-calculus enjoys WN^{∞} (Theorem 6.48(i)), we get $T \xrightarrow{\infty}_{\beta\perp} T^{\beta\perp\text{-nf}}$. By Theorem 6.48(ii), $\perp$-reductions can be postponed. Moreover, since the $\perp$-reduction sequence ends in a $\beta\perp$-normal form, we can remove all non-outermost $\perp$-steps. Thus, we can construct the following diagram using the two commutation properties.

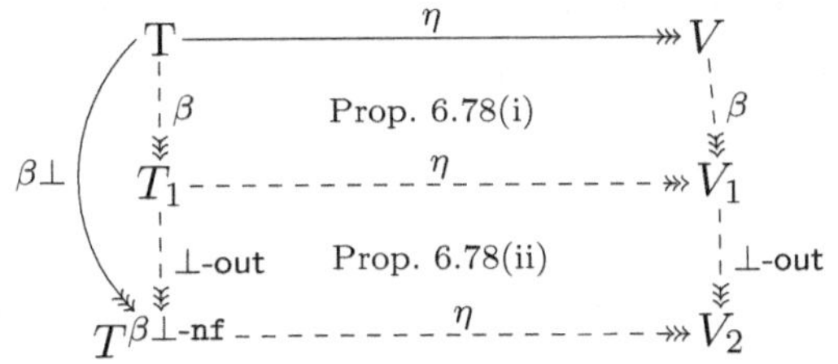

The term V_2 is in $\beta\perp$-nf, whence $V_2 = V^{\beta\perp\text{-nf}}$ by CR^{∞} for $\lambda_{\beta\perp}^{\mathrm{h}\infty}$-calculus (Theorem 6.48(i)).

(ii) Assume $T \xrightarrow{\infty}_{\beta\bot\eta} V$ and proceed as in the diagram:

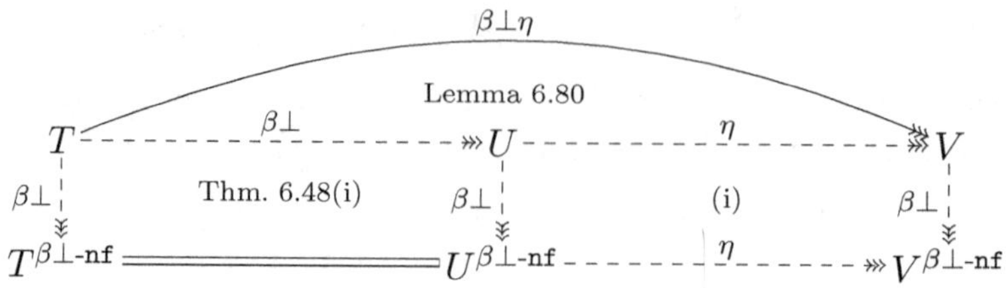

□

We present the main result of the section, stating that the $\lambda^{\mathrm{h}\infty}_{\beta\bot\eta}$-calculus is confluent and weakly normalizing. As a consequence, the extensional Böhm tree of a λ-term can be obtained by first performing leftmost-outermost $\beta\bot$-reduction and then η-normalization.

THEOREM 6.82 (SEVERI AND DE VRIES (2002)).
(i) *The* $\lambda^{\mathrm{h}\infty}_{\beta\bot\eta}$*-calculus enjoys* CR^∞ *and* WN^∞.
(ii) *For all* $M \in \Lambda$, *we have*

$$M^{\beta\bot\eta\text{-}\mathtt{nf}} = \mathrm{nf}_\eta(\mathrm{BT}(M))$$

PROOF. (i) To prove CR^∞, we proceed as follows:

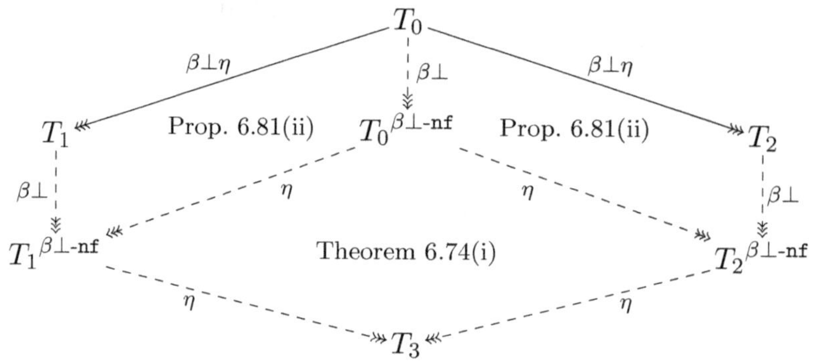

WN^∞ for $\lambda^{\infty}_{\beta\bot\eta}$ is obtained by combining the same property for $\lambda^{\mathrm{h}\infty}_{\beta\bot}$ (Theorem 6.48(i)) and for $\lambda^{\mathrm{h}\infty}_{\eta}$ (Theorem 6.74(ii)).

(ii) The $\lambda^{\mathrm{h}\infty}_{\beta\bot}$-calculus enjoys CR^∞, WN^∞ (Theorem 6.48(i)), so $M \xrightarrow{\infty}_{\beta\bot} M^{\beta\bot\text{-}\mathtt{nf}}$ and $M^{\beta\bot\text{-}\mathtt{nf}} = \mathrm{BT}(M)$, by Corollary 6.56 with $\mathcal{U} = \overline{\mathcal{HN}}$. By coinduction on $\mathrm{BT}(M)$, we show $\mathrm{BT}(M)^{\eta\text{-}\mathtt{nf}} = \mathrm{nf}_\eta(\mathrm{BT}(M))$, splitting into cases depending on the solvability of M.

Case $M \in \mathrm{UNS}$. Then $\mathrm{BT}(M) = \bot$ and $\bot = \bot^{\eta\text{-}\mathtt{nf}} = \mathrm{nf}_\eta(\mathrm{BT}(M))$.

Case $M \in \mathrm{SOL}$. Then we can write its Böhm tree as follows (for $m, k \geq 0$):

$$\mathrm{BT}(M) = \lambda\vec{x}z_1 \ldots z_m.y\mathrm{BT}(M_1)\cdots\mathrm{BT}(M_k)Z_1^{\beta\text{-}\mathtt{nf}}\cdots Z_m^{\beta\text{-}\mathtt{nf}},$$

with $\vec{z} \notin \mathrm{FV}(\mathrm{BT}(yM_1\cdots M_k))$ and $Z_j^{\beta\text{-}\mathtt{nf}} \twoheadrightarrow_\eta z_j$ for every $j\,(1 \leq j \leq m)$. We obtain:

$$\begin{array}{rcll}
\mathrm{BT}(M)^{\eta\text{-}\mathtt{nf}} & = & \lambda\vec{x}.y\mathrm{BT}(M_1)^{\eta\text{-}\mathtt{nf}}\cdots\mathrm{BT}(M_k)^{\eta\text{-}\mathtt{nf}}, & \text{by } \mathrm{WN}^\infty,\\
& = & \lambda\vec{x}.y\,\mathrm{nf}_\eta(\mathrm{BT}(M_1))\cdots\mathrm{nf}_\eta(\mathrm{BT}(M_k)), & \text{by co-IH},\\
& = & \mathrm{nf}_\eta(\lambda\vec{x}\vec{z}.y\,\mathrm{BT}(M_1)\cdots\mathrm{BT}(M_k)Z_1^{\beta\text{-}\mathtt{nf}}\cdots Z_m^{\beta\text{-}\mathtt{nf}}), & \text{since}\\
& & & \vec{z} \notin \mathrm{FV}(\mathrm{BT}(y\vec{M})),\\
& = & \mathrm{nf}_\eta\mathrm{BT}(M), & \text{by definition.}
\end{array}$$

□

About Nakajima trees

We have seen that extensional Böhm trees can be obtained as normal forms of the $\lambda^{h\infty}_{\beta\bot\eta}$-calculus. A question that naturally arises is whether there exists a notion of infinitary η-reduction allowing to retrieve Nakajima trees as well. A positive answer was given by Severi and de Vries (2017) through the notion of $\eta!$-reduction, which allows to get rid of infinitely many η-expansions in one step.

DEFINITION 6.83. (i) On $\Lambda^{h\infty}_{\bot}$, define the notion of *η-expansion* as follows (for $T \in \Lambda^{h\infty}_{\bot}$):

$$\frac{x \notin \mathrm{FV}(T)}{T \to \lambda x.Tx} \qquad (\eta^{-1})$$

(ii) The infinitary notion of *$\eta!$-reduction* is defined on $\Lambda^{h\infty}_{\bot}$ by the rule (for $T, U \in \Lambda^{h\infty}_{\bot}$):

$$\frac{x \twoheadrightarrow\!\!\!\!\!\twoheadrightarrow_{\eta^{-1}} U \quad x \notin \mathrm{FV}(T)}{\lambda x.TU \to T} \qquad (\eta!)$$

(iii) The *$\lambda^{h\infty}_{\beta\bot\eta!}$-calculus* is given by the set $\Lambda^{h\infty}_{\bot}$ endowed with $\xrightarrow{\infty}_{\beta\bot\eta!}$.

EXAMPLES 6.84. We explain the behavior of $\to_{\eta!}$ on the combinator J from Table 1.1. The Böhm tree of J is easy to compute (see Figure 2.6). By Proposition 6.54(i), we have that $\mathsf{J} \xrightarrow{\infty}_{\beta} \lambda x z_0.x(\lambda z_1.z_0(\lambda z_2.z_1(\cdots))) = \mathrm{BT}(\mathsf{J})$. In one $\eta!$-step, we get $\mathrm{BT}(\mathsf{J}) \to_{\eta!} \mathsf{I}$ since

$\eta!$

$$\lambda x.x \to_{\eta^{-1}} \lambda x z_0.x z_0 \to_{\eta^{-1}} \lambda x z_0.x(\lambda z_1.z_0 z_1) \twoheadrightarrow\!\!\!\!\!\twoheadrightarrow_{\eta^{-1}} \mathrm{BT}(\mathsf{J})$$

THEOREM 6.85 (SEVERI AND DE VRIES (2017)).
(i) *The $\lambda^{h\infty}_{\beta\bot\eta!}$-calculus satisfies CR^{∞} and WN^{∞}.*
(ii) *For all $M \in \Lambda$, we have*

$$M^{\beta\bot\eta!\text{-}\mathtt{nf}} = \mathrm{nf}_{\eta!}(\mathrm{BT}(M))$$

PROOF. The proof structure is similar to the one for proving Theorem 6.82. □

REMARK 6.86. (i) Although $\xrightarrow{\infty}_{\beta\bot}$, $\xrightarrow{\infty}_{\eta}$ and $\xrightarrow{\infty}_{\eta!}$ satisfy the Compression Lemma, recall that neither $\xrightarrow{\infty}_{\beta\bot\eta}$ nor $\xrightarrow{\infty}_{\beta\bot\eta!}$ are compressible by Remark 6.71(iii).

(ii) For $\xrightarrow{\infty}_{\beta\bot\eta!}$ one can simply take the combinator J as a counterexample, since all its finite reducts are in η-nf and the reduction $\mathsf{J} \xrightarrow{\omega}_{\beta} \mathrm{BT}(\mathsf{J}) \to_{\eta!} \mathsf{I}$ is not compressible.

Both $\xrightarrow{\infty}_{\beta\bot\eta}$ and $\xrightarrow{\infty}_{\beta\bot\eta!}$ satisfy the weaker form of compression below.

LEMMA 6.87 (WEAK COMPRESSION LEMMA). *Let $T \in \Lambda^{h\infty}_{\bot}$. Every reduction sequence $T \xrightarrow{\infty}_{\beta\bot\eta(!)} U$, with U in $\beta\bot\eta(!)$-nf, can be compressed into a reduction of length $2.\omega$.*

PROOF. It follows from Theorem 6.82 for $\xrightarrow{\infty}_{\beta\bot\eta}$ and from Theorem 6.85 for $\xrightarrow{\infty}_{\beta\bot\eta!}$. Intuitively, the $\beta\bot\eta(!)$-nf of T can be obtained by first calculating its $\beta\bot$-nf and then the $\eta(!)$-nf. By postponement, which is also satisfied by $\xrightarrow{\infty}_{\eta!}$, we obtain a reduction sequence of length $\omega + \omega$ since both reductions are individually compressible. □

Set	Meaningless set $\mathcal{U}$	Rules	Reductions	Normal forms
$\Lambda_{\bot}^{\infty}$	$\mathcal{R}$	$\beta, \bot_{\mathcal{U}}$	$\twoheadrightarrow\!\!\!\!\!\!\twoheadrightarrow$	Berarducci trees
$\Lambda_{\bot}^{\infty}$	$\overline{\mathcal{WN}}$	$\beta, \bot_{\mathcal{U}}$	$\twoheadrightarrow\!\!\!\!\!\!\twoheadrightarrow$	Lévy-Longo trees
$\Lambda_{\bot}^{\infty}$	$\overline{\mathcal{HN}}$	$\beta, \bot_{\mathcal{U}}$	$\twoheadrightarrow\!\!\!\!\!\!\twoheadrightarrow$	Böhm trees
$\Lambda_{\bot}^{\mathrm{h}\infty}$	$\Lambda_{\bot}^{\mathrm{h}\infty} - \mathcal{HN}$	$\beta, \bot_{\mathcal{U}}, \eta$	$\xrightarrow{\infty}$	extensional Böhm trees
$\Lambda_{\bot}^{\mathrm{h}\infty}$	$\Lambda_{\bot}^{\mathrm{h}\infty} - \mathcal{HN}$	$\beta, \bot_{\mathcal{U}}, \eta!$	$\xrightarrow{\infty}$	Nakajima trees

Table 6.1: Trees as normal forms.

Table 6.1 proposes a summary of the characterizations provided in this chapter. Recall that these are partial characterizations. For instance, not all $\beta\bot$-normal forms in $\Lambda_{\bot}^{\mathrm{h}\infty}$ (and *a fortiori* in $\Lambda_{\bot}^{\infty}$) arise as Böhm trees of finite λ-terms (by Theorem 2.24).

Coinductive extensional rewriting and related issues

An advantage of describing infinitary rewriting coinductively, is that such an approach is more prone to be formalized in proof assistants.[2] For instance, both the Standardization Theorem in Endrullis and Polonsky (2011) and the confluence proof in Czajka (2020) have been written in Coq, by the respective authors. The reader might wonder whether it is possible to improve the last two lines of Table 6.1 by:

1. considering the whole set $\Lambda_{\bot}^{\infty}$, rather than its restriction $\Lambda_{\bot}^{\mathrm{h}\infty}$;

2. describing $\xrightarrow{\infty}_{\beta\bot\eta}$ and $\xrightarrow{\infty}_{\beta\bot\eta!}$ coinductively.

The problem is that when working on $\Lambda_{\bot}^{\infty}$ the commutation of β with η (namely, Proposition 6.78(i)) fails. Despite that, we believe that CR^{∞} and WN^{∞} are still in place as all pathological terms should rewrite to $\bot$. One should elaborate a different proof-technique, though. It is easy to check that in $\Lambda_{\bot}^{\infty}$ the reductions $\xrightarrow{\infty}_{\beta\bot\eta}$ and $\twoheadrightarrow\!\!\!\!\!\!\twoheadrightarrow_{\beta\bot\eta}$ are equivalent, so we formulate the following conjecture.

Conjecture 6.88. The $\lambda_{\beta\eta\bot_{\overline{\mathcal{HN}}}}^{\infty}$-calculus $(\Lambda_{\bot}^{\infty}, \twoheadrightarrow\!\!\!\!\!\!\twoheadrightarrow_{\beta\bot\eta})$ is CR^{∞} and WN^{∞}.

For $\xrightarrow{\infty}_{\beta\bot\eta!}$, the answer might still be positive, but considering the infinitary closure $\twoheadrightarrow\!\!\!\!\!\!\twoheadrightarrow_{\beta\bot\eta!}$ in the sense of Definition 6.5(iii) does not give rise to an equivalent infinitary reduction relation. Indeed, in general, $T \xrightarrow{\infty}_{\beta\bot\eta!} x$ does not imply $T \twoheadrightarrow_{\beta\bot\eta!} x$ as one can easily verify by taking $T = \mathsf{J}x$. On the contrary, $T \twoheadrightarrow\!\!\!\!\!\!\twoheadrightarrow_{\beta\bot\eta!} x$ holds exactly when $T \twoheadrightarrow_{\beta\bot\eta!} x$ does.

The problem of finding a coinductive framework for capturing strong convergent sequences of arbitrary ordinal length without mentioning ordinals or metric convergence

[2] M. Vermaat formalized infinitary rewriting in Coq using metric convergence (rather than strong convergence). While this formalization could be extended to strong convergence, it remains to be investigated to what extent it can be used for the further development of the theory of infinitary rewriting.

has been solved in Endrullis et al. (2015), for first-order infinitary term rewriting systems. The idea is to mix together induction and coinduction. Given a one-step R-reduction $\to_{\mathrm{R}}$, using the least (μ) and greatest (ν) fixed-point operators, one defines:

$$\twoheadrightarrow^{\infty}_{\mathrm{R}} \;\triangleq\; \mu \mathrm{P}.\, \nu \mathrm{S}.\, (\to_{\mathrm{R}} \cup \overline{\mathrm{P}})^* \,; \mathrm{S} \tag{6.4}$$

where semicolon ; denotes relational composition in diagrammatic order, $(\cdot)^*$ stands for the transitive-reflexive closure, and the relation $\overline{\mathrm{P}}$ is obtained from P as follows:

$$T\,\overline{\mathrm{P}}\,T \qquad \frac{V\;\mathrm{P}\;V'}{\lambda x.V\;\overline{\mathrm{P}}\;\lambda x.V'} \qquad \frac{U\;\mathrm{P}\;U' \quad V\;\mathrm{P}\;V'}{UV\;\overline{\mathrm{P}}\;U'V'}$$

In (6.4), the greatest fixed-point defined using the variable S is a coinductively defined relation. Thus only the last step in the sequence $(\to_{\mathrm{R}} \cup \overline{\mathrm{P}})^* ; \mathrm{S}$ is coinductive. This corresponds to the fact that, in a reduction $\rho : T \xrightarrow{\alpha} U$ of ordinal length α, every strict prefix of ρ must be shorter than α, while strict suffixes may have length α, just like ρ.

The relation $\twoheadrightarrow^{\infty}_{\mathrm{R}}$ in (6.4) is described via a coinductive system of inference rules, using two auxiliary relations $\xrightarrow{\infty}\!\!\!\!\!\!\!\!/\!\!/\!\!/_{\mathrm{R}}$ and $\xrightarrow{<\infty}\!\!\!\!\!\!\!\!/\!\!/\!\!/_{\mathrm{R}}$, the latter being a 'marked' version of the former. This marking is needed to forbid an infinite nesting of $\xrightarrow{<\infty}\!\!\!\!\!\!\!\!/\!\!/\!\!/_{\mathrm{R}}$ in a derivation:

$$T \xrightarrow{(<)\infty}\!\!\!\!\!\!\!\!/\!\!/\!\!/_{\mathrm{R}}\; T \qquad \frac{T(\to_{\mathrm{R}} \cup \xrightarrow{<\infty}\!\!\!\!\!\!\!\!/\!\!/\!\!/_{\mathrm{R}})^* \,; \xrightarrow{\infty}\!\!\!\!\!\!\!\!/\!\!/\!\!/_{\mathrm{R}}\; U}{T \twoheadrightarrow^{\infty}_{\mathrm{R}} U}$$

$$\frac{U \twoheadrightarrow^{\infty}_{\mathrm{R}} U'}{\lambda x.U \xrightarrow{(<)\infty}\!\!\!\!\!\!\!\!/\!\!/\!\!/_{\mathrm{R}}\; \lambda x.U'} \qquad \frac{U \twoheadrightarrow^{\infty}_{\mathrm{R}} U' \quad V \twoheadrightarrow^{\infty}_{\mathrm{R}} V'}{UV \xrightarrow{(<)\infty}\!\!\!\!\!\!\!\!/\!\!/\!\!/_{\mathrm{R}}\; U'V'}$$

It is easy to check that $\twoheadrightarrow_{\mathrm{R}} \subsetneq \twoheadrightarrow^{\infty}_{\mathrm{R}}$ holds.

PROPOSITION 6.89. *For $T, U \in \Lambda^{\infty}$, the following are equivalent:*

1. $T \xrightarrow{\infty}_{\beta\bot\eta!} U$*;*

2. $T \twoheadrightarrow^{\infty}_{\beta\bot\eta!} U$*.*

PROOF. For first-order infinitary TRSs, the equivalence between $\xrightarrow{\infty}_{\mathrm{R}}$ and $\twoheadrightarrow^{\infty}_{\mathrm{R}}$ is shown in Endrullis et al. (2018), Theorem 5.2. The proof for $\lambda^{\infty}_{\beta\bot\eta!}$ is a simple adaptation. □

The question of whether $\Lambda^{\infty}_{\bot}$ with $\twoheadrightarrow^{\infty}_{\beta\bot\eta!}$ satisfies infinitary confluence and weak normalization remains open.

Chapter 7

Starlings

A set $\mathcal{B} \subseteq \Lambda^o$ constitutes a basis of λ-calculus if every $M \in \Lambda^o$ is obtainable by applying combinators from $\mathcal{B}$ with each other. More precisely, writing $\mathcal{B}^\bullet$ for the set of all applicative combinations of elements of $\mathcal{B}$, we say that $\mathcal{B}$ is a *basis* whenever

$$\forall M \in \Lambda^o, \exists N \in \mathcal{B}^\bullet . N =_\beta M.$$

The most famous basis of λ-calculus is $\{\mathsf{K}, \mathsf{S}\}$, where $\mathsf{K} = \lambda xy.x$ and $\mathsf{S} = \lambda xyz.xz(yz)$. Perhaps surprisingly, it was shown by Meredith and Prior (1963) that there exist a basis consisting of a single element, namely $\{\lambda xywz.wz(x(\mathsf{K}z))\}$ (see B[1984], Exercise 8.5.16). Probably the simplest singleton basis is $\mathcal{B} = \{\langle \mathsf{K}, \mathsf{S}, \mathsf{K}\rangle\}$, found by Rosser (1971), by simplifying a construction due to Barendregt. In general, one may wonder what operational properties the elements of a set $\mathcal{B}$ need to satisfy to form a basis. The problem becomes more manageable when considering bases only containing *proper* combinators, i.e. closed λ-terms having all abstractions at the root as in $\lambda\vec{x}.M$. Under this additional hypothesis, Craig was able to identify some necessary conditions.

PROPOSITION 7.1. *Any basis $\mathcal{B}$ of proper combinators must contain*
(i) *a selector (see Definition 1.16, notice that also* I *is a selector);*
(ii) *a combinator N with a* cancellative effect, *i.e. for some $n > 0$ and $Q \in \Lambda$*

$$Nx_1 \cdots x_n =_\beta Q$$

and (at least) one of the variables $x_1, \dots, x_n$ does not occur in Q;
(iii) *a combinator N with a* duplicative effect, *i.e. for some $n > 0$ and $Q \in \Lambda$*

$$Nx_1 \cdots x_n =_\beta Q$$

and some variable x_i occurs at least twice in Q.

PROOF. This was essentially shown by Craig in Curry and Feys (1958). Although his original proof was incomplete, it was subsequently corrected by Bellot (1985).
(i) A selector is needed in order to construct other selectors.
(ii) Since the $\lambda\mathsf{I}$-terms are closed under application, one never can obtain K.
(iii) A duplicative effect is required to obtain $\Delta = \lambda x.xx$. □

In spite of this there are sets $\mathcal{B}_1 = \{\mathsf{I}, \mathsf{B}, \mathsf{C}, \mathsf{S}\}$ and $\mathcal{B}_2 = \{\mathsf{I}, \mathsf{J}\}$, for $\mathsf{C} \triangleq \lambda xyz.xzy$ and $\mathsf{J} \triangleq \lambda xyzw.zy(xwz)$, that do not contain a combinator with cancellative effect but are such that $\mathcal{B}_1^\bullet = \mathcal{B}_2^\bullet$ equals the set of $\lambda\mathsf{I}$-terms and therefore can represent all partial recursive functions (see B[1984], Theorem 9.2.16). In this context, in the seminar leading to Barendregt (1975), one started to study the set $\{\mathsf{S}\}^\bullet$ and asked the question whether it contains an element without normal form. Not only this turned out to be the case, but this class became subject to studies revealing an unexpectedly rich array of properties.

We need to mention a slight change of perspective: researchers mainly focused on **S**-terms, i.e. terms living in the **S**-fragment of combinatory logic, because the resulting rewriting system is cleaner. In this framework, they examined the following problems.

1. TERMINATION. The aforementioned problem concerning the existence of a non-terminating **S**-term was positively solved by Barendregt in 1975 (see Barendregt et al. (1976)). The decidability of normalization for **S**-terms has been conjectured by Zachos (1978) and proved by Waldmann (1998) during his PhD.

2. INFINITARY BEHAVIOR. As discussed in Chapter 6, both λ-calculus and CL admit infinitary reductions that generate Böhm trees and Berarducci trees, when pushed to infinity. Does the set of finite **S**-terms satisfy the infinitary strong normalization property (SN^∞)? Waldmann (1997) positively answered this question by showing that every **S**-term is top-normalizing (Definition 7.32). Waldmann's argument is based on the so-called $\mathcal{QQQ}$ *criterion* identifying sufficient conditions for **S**-terms to generate an infinite head reduction sequence.

3. WORD PROBLEMS. Is the word problem for **S**-terms decidable? Equivalently, is it decidable whether two **S**-terms are inter-convertible? This problem is still open. First raised by Barendregt in 1975, it is probably the most important open problem concerning the **S**-fragment of combinatory logic.

In this chapter we present the results mentioned above, that are nowadays well-established, but also previously unpublished results obtained by Vincent Padovani in the period 2012–2020. Among the wealth of his results, the following stand out:

- DECIDABILITY OF HEAD-NORMALIZATION. Padovani isolated a new criterion $\mathcal{Q}_{\mathrm{III}}$ which is a refinement of Waldmann's criterion and allows to provide a complete characterization of those **S**-terms having a terminating head-reduction sequence. It follows that head-normalization is decidable for **S**-terms.

- DIFFERENT TERMS, SAME BERARDUCCI TREE. To solve the word problem for **S** it is interesting to know whether checking the inter-convertibility of two **S**-terms is equivalent to verifying if they have the same Berarducci trees. Padovani has proved that this property is false, by constructing a counterexample.

For the presentation of the classical results we mainly follow Barendregt et al. (2018); all the figures in this chapter have been made by Jörg Endrullis and appear thanks to the kind permission of the authors and Springer. The presentation of Padovani's original results is due to Padovani himself, who participated actively in the writing of this chapter.

7.1 Combinatory Logic — the S-fragment

We recall some basic properties of combinatory logic and introduce its subsystem $\mathbb{S}_{\mathrm{CL}}$ generated by the combinator $\mathbf{S}$ alone. Recall that the set CL of combinatory terms has been introduced in Definition 3.33(i). Combinatory logic can be seen as a term rewriting system by endowing combinatory terms with the so-called *weak reduction* $\to_w$, generated by the rules

$$\mathbf{K}xy \to x, \qquad \mathbf{S}xyz \to xz(yz) \tag{w}$$

Working in combinatory logic, we adopt the following standard notations.

NOTATION. (i) *For every* $P \in \mathrm{CL}$, $P^{w\text{-}\mathbf{nf}}$ *stands for its* w-normal form, *if it exists.*

(ii) *As usual,* $\to_w$ *generates* multi-step weak reduction $\twoheadrightarrow_w$ *and* weak conversion $=_w$ *relations. Moreover, we denote by* $\to_w^+$ *the transitive closure of* $\to_w$.

(iii) *Given* $P, Q \in \mathrm{CL}$, *we let* $PQ^{\sim n} \triangleq PQ \cdots Q$ *and* $P^n(Q) \triangleq P(P(\cdots(PQ)))$ *(n times).*

The confluence of weak reduction $\to_w$ was been first established by Rosser (1935). In his thesis, Klop (1980a) showed that $(\mathrm{CL}, \to_w)$ is an 'orthogonal' rewriting system. Intuitively, a term rewriting system is *orthogonal* if it is left-linear[1] and redexes do not overlap. Orthogonal term rewriting systems are known to satisfy several interesting properties, the most important being confluence. The reader who wishes to explore further the rewriting properties of CL is advised to study Section 3.3 in Terese (2003).

REMARK 7.2. The terminology 'weak reduction' for $\to_w$ was introduced by Church and Rosser (1936) and is not related to λ-calculus weak β-reduction, the latter being defined by forbidding the contraction of β-redexes under the scope of a λ-abstraction. In CL 'weak' refers to the fact that $\to_w$ is—in general—not sufficiently powerful to simulate $\to_\beta$ on the combinatory terms representing λ-terms. The typical example is given by $\lambda z.(\lambda x.x)z \to_\beta \lambda x.x$, while $(\lambda z.(\lambda x.x)z)_{\mathrm{CL}} = \mathbf{S}(\mathbf{KI})\mathbf{I} \not\twoheadrightarrow_w \mathbf{I} = (\lambda x.x)_{\mathrm{CL}}$.

DEFINITION 7.3. (i) Given an applicative structure $(A, \cdot)$ and a subset $\mathcal{X} \subseteq A$, let us denote by $\mathcal{X}^\bullet$ the least subset of A containing $\mathcal{X}$ that is closed under application $\cdot$.

(ii) In $(\mathrm{CL}, \cdot)$ we define the set of $\mathbf{S}$-*terms* by setting $\mathbb{S}_{\mathrm{CL}} \triangleq \{\mathbf{S}\}^\bullet$.

(iii) The $\mathbf{S}$-*fragment of combinatory logic* is given by $(\mathbb{S}_{\mathrm{CL}}, \to_w)$, where we consider weak reduction $\to_w$ restricted to $\mathbf{S}$-terms.

(iv) $\mathbf{S}$-*contexts* $C[\,]$ are $\mathbf{S}$-terms possibly containing occurrences of a hole denoted by $[\,]$. Given $P \in \mathbb{S}_{\mathrm{CL}}$, write $C[P]$ for the $\mathbf{S}$-term obtained by substituting P for the hole in $C[\,]$.

(v) The *size of an* $\mathbf{S}$-*term* P, in symbols $\mathrm{size}(P)$, is defined as follows:

$$\begin{aligned} \mathrm{size}(\mathbf{S}) &\triangleq 1; \\ \mathrm{size}(P_1P_2) &\triangleq \mathrm{size}(P_1) + \mathrm{size}(P_2) + 1. \end{aligned}$$

(vi) The *weight of an* $\mathbf{S}$-*term* P, in symbols $\mathrm{weight}(P)$, is defined as follows:

$$\begin{aligned} \mathrm{weight}(\mathbf{S}) &\triangleq 1; \\ \mathrm{weight}(P_1P_2) &\triangleq 2 \cdot \mathrm{weight}(P_1) + \mathrm{weight}(P_2). \end{aligned}$$

[1]This means that each variable occurs only once on the left-hand side of each reduction rule.

EXAMPLES 7.4. The following **S**-terms will be used to construct examples.

Notation		Definition	Size	Weight
T	$\triangleq$	**SS**	3	3
A	$\triangleq$	**SSS**	5	7
D	$\triangleq$	**SAA**	13	25
E	$\triangleq$	**SAD**	21	43

Table 7.1: Notations for specific **S**-terms.

REMARK 7.5. (i) For $P, Q, W \in \mathbb{S}_{\text{CL}}$, we have $\text{size}(\mathbf{S}PQW) < \text{size}(PW(QW))$ whenever $W \neq \mathbf{S}$. Otherwise, $\text{size}(\mathbf{S}PQ\mathbf{S}) = \text{size}(P\mathbf{S}(Q\mathbf{S}))$ and $\text{weight}(\mathbf{S}PQ\mathbf{S}) > \text{weight}(P\mathbf{S}(Q\mathbf{S}))$.
(ii) By monotonicity of contexts, it follows that $P \to_w Q$ entails $\text{size}(P) \leq \text{size}(Q)$.

Basic rewriting properties of S

Redex occurrences in an **S**-term P are classified accordingly to their position in P, in particular redexes occurring in head position and at the root will play a special role.

DEFINITION 7.6. Let $P \in \mathbb{S}_{\text{CL}}$ and R be a w-redex occurrence in P. Then R is called:
(i) the *head redex* of P if it occurs in head position, i.e.

$$P = RX_1 \cdots X_n,$$

for some $n \in \mathbb{N}$ and $\vec{X} \in \mathbb{S}_{\text{CL}}$.
(ii) the *root redex* of P whenever $P = R$.

By definition every root redex is a head redex, while the converse does not hold.

EXAMPLES 7.7. We exhibit the weak reduction steps originating from the terms $\mathbf{SS}^{\sim n}$, for all $n \leq 5$. In each step we contract the head redex.
(i) The terms **S**, **A**, **T**, **D** and **E** are in weak normal form;
(ii) $\mathbf{SSSS} \to_w \mathbf{SS(SS)}$. Equivalently, using the above notations, $\mathbf{AS} \to_w \mathbf{TT}$;
(iii) $\mathbf{ASS} \to_w \mathbf{TTS} \to_w \mathbf{TA}$;
(iv) $\mathbf{ASSS} \to_w \mathbf{TTT} \to_w \mathbf{TAS} \to_w \mathbf{T(AS)} \to_w \mathbf{T(TT)}$.

EXERCISE 7.8. (i) Show that $\mathbf{SS}^{\sim 6} \twoheadrightarrow_w \mathbf{T(TA)}$.
(ii) Show that $\mathbf{SS}^{\sim 7} \twoheadrightarrow_w \mathbf{T(T(TT))}$.
(iii) Formulate a conjecture about the shape of $(\mathbf{SS}^{\sim n})^{w\text{-}\mathbf{nf}}$, for an arbitrary $n > 3$.
(iv) Show that $\mathbf{ATS} \twoheadrightarrow_w \mathbf{S(STS)(S(STS))}$.
(v) Show that $\mathbb{S}_{\text{CL}}$ does not admit pure cycles: $\nexists P \in \mathbb{S}_{\text{CL}} \,.\, P \to_w P$.

The fact that every term of the form $\mathbf{SS}^{\sim n}$, for $n \in \mathbb{N}$, has a w-normal form follows from a more general result found by Klop, stating that every flat CL term is strongly normalizing—a term being 'flat' if it can be built from **K** and **S** without the use of visible parentheses. This result first appeared in the so-called 'Blue preprint' by Barendregt et al. (1976), but the proof is also available in Barendregt et al. (2018), Theorem 18. The w-normal forms of flat **S**-terms admit the following characterization.

LEMMA 7.9 (PADOVANI 2012). *For all $n \in \mathbb{N}$, we have:*

$$\mathbf{AS}^{\sim n} =_w \begin{cases} \mathbf{T}^m(\mathbf{A}), & \textit{if } n = 2 \cdot m, \\ \mathbf{T}^m(\mathbf{T}), & \textit{if } n = 2 \cdot m + 1. \end{cases}$$

PROOF. First, one shows by induction on m that, for all $m \in \mathbb{N}$, the following holds:

$$\begin{array}{lcl} \mathbf{T}^m(\mathbf{T}) & =_w & \mathbf{T}^m(\mathbf{A}) \\ \mathbf{T}^m(\mathbf{A}) & =_w & \mathbf{T}^{m+1}(\mathbf{T}) \end{array} \tag{7.1}$$

Subsequently one proves the main statement by induction on n, using (7.1) to insert the rightmost occurrence of **S** in the term obtained from the IH. □

The next theorem summarizes some well-known properties enjoyed by the **S**-fragment of CL. The first two properties are in common with λ-calculus and CL, the third is not, but it is rather common for first-order term rewriting systems.

THEOREM 7.10. (i) $(\mathfrak{S}_{\mathrm{CL}}, \to_w)$ *is an orthogonal term rewriting system.*
(ii) *The reduction $\to_w$ on $\mathfrak{S}_{\mathrm{CL}}$ is confluent.*
(iii) *For all $P \in \mathfrak{S}_{\mathrm{CL}}$, there are finitely many **S**-terms Q such that $Q \twoheadrightarrow_w P$.*

PROOF. (i) It follows immediately from the analogous property of $(\mathrm{CL}, \to_w)$.
(ii) By (i).
(iii) Take $P \in \mathfrak{S}_{\mathrm{CL}}$ and call $k \triangleq \mathrm{size}(P) \in \mathbb{N}$. By Remark 7.5(ii), we obtain that

$$\{Q \in \mathfrak{S}_{\mathrm{CL}} \mid Q \twoheadrightarrow_w P\} \subseteq \{Q \in \mathfrak{S}_{\mathrm{CL}} \mid \mathrm{size}(Q) \leq k\}$$

We conclude since the latter set is clearly finite. □

Theorem 7.10(iii) alone does not ensure the acyclicity of a term rewriting system, nor the absence of infinite sequences of anti-reduction $\cdots \to_w P_n \to_w \cdots \to_w P_1 \to_w P_0$. Consider for instance $\mathfrak{O}_\lambda \triangleq \{\Omega\}^\bullet$ endowed with β-reduction: every $P \in \mathfrak{O}_\lambda$ has only one ancestor, namely P itself, whence it generates a 1-cycle $P \to_w P$ and the corresponding infinite anti-reduction sequence $\cdots \to_w P \to_w P$. That said, it is possible to show that $\mathfrak{S}_{\mathrm{CL}}$ actually satisfies these two properties by mixing together the properties of the size and of the weight of a term.

LEMMA 7.11. *Let $P, Q \in \mathfrak{S}_{\mathrm{CL}}$.*
(i) *If $P \to_w Q$ and $\mathrm{size}(P) = \mathrm{size}(Q)$ then the redex contracted in P has shape $\mathbf{S}VW\mathbf{S}$, for some $V, W \in \mathfrak{S}_{\mathrm{CL}}$.*
(ii) *For all **S**-contexts $C[\,]$, we have*

$$\mathrm{weight}(P) > \mathrm{weight}(Q) \quad \Rightarrow \quad \mathrm{weight}(C[P]) > \mathrm{weight}(C[Q])$$

PROOF. (i) By structural induction on P, using Remark 7.5(i).
(ii) By structural induction on $C[\,]$. □

PROPOSITION 7.12 (BERGSTRA AND KLOP (1979)).
(i) $\mathbf{S}_{\mathrm{CL}}$ *is acyclic.*
(ii) *The weak anti-reduction* ${}_w\!\leftarrow$ *is strongly normalizing.*

PROOF. (i) Assume by contradiction that a cycle exists. Then, for some $n > 0$, we have:

$$P_0 \to_w P_1 \to_w \cdots \to_w P_n = P_0$$

By Remark 7.5(ii), we have $\mathrm{size}(P_0) = \mathrm{size}(P_1) = \cdots = \mathrm{size}(P_n)$. By Lemma 7.11(i), every contracted redex is of the form $\mathbf{S}VW\mathbf{S}$. Easy calculations give:

$$\begin{aligned}
\mathrm{weight}(\mathbf{S}VW\mathbf{S}) &= 8\cdot\mathrm{weight}(\mathbf{S}) + 4\cdot\mathrm{weight}(V) + 2\cdot\mathrm{weight}(W) + \mathrm{weight}(\mathbf{S})\\
&= 4\cdot\mathrm{weight}(V) + 2\cdot\mathrm{weight}(W) + 9\\
\mathrm{weight}(V\mathbf{S}(W\mathbf{S})) &= 4\cdot\mathrm{weight}(V) + 2\cdot\mathrm{weight}(\mathbf{S}) + 2\cdot\mathrm{weight}(W) + \mathrm{weight}(\mathbf{S})\\
&= 4\cdot\mathrm{weight}(V) + 2\cdot\mathrm{weight}(W) + 3
\end{aligned}$$

By Lemma 7.11(i) we obtain $\mathrm{weight}(P_0) > \mathrm{weight}(P_1) > \cdots > \mathrm{weight}(P_n) > \mathrm{weight}(P_0)$, which constitutes a contradiction.

(ii) By Theorem 7.10(iii) every infinite anti-reduction $\cdots \to_w P_n \to_w \cdots \to_w P_0$ must contain some cycles. We conclude by (i). □

Non-definability of quasi-identities

In combinatory logic the term $(\lambda x.x)_{\mathrm{CL}} = \mathbf{SKK}$ satisfies $\mathbf{SKK}x \to_w \mathbf{K}x(\mathbf{K}x) \to_w x$. Now, in CL there exist other combinators that behave like the identity, e.g. $\mathbf{SKS}$. Thus, one might wonder whether the identity can be constructed using the combinator $\mathbf{S}$ alone.

DEFINITION 7.13. Let $k \in \mathbb{N}$, $P, P_0, \dots, P_k \in \mathbf{S}_{\mathrm{CL}}$.
(i) We say that P is an *identity* whenever $PQ =_w Q$ holds, for all $Q \in \mathbf{S}_{\mathrm{CL}}$.
(ii) Similarly, $Q \in \mathbf{S}_{\mathrm{CL}}$ admits *quasi-identities* $P_0, \dots, P_k$ if $P_0(\cdots(P_kQ)) =_w Q$.
(iii) Define the set $\mathcal{R}(P)$ of *rightmost subterms of* P by setting:

$$\mathcal{R}(P) \triangleq \{X \in \mathbf{S}_{\mathrm{CL}} \mid \exists n \in \mathbb{N}, \exists Q_1, \dots, Q_n \in \mathbf{S}_{\mathrm{CL}} \,.\, Q_1(\cdots(Q_nX)\cdots) = P\} \qquad (7.2)$$

(iv) The set of *rightmost strict subterms of* P is given by $\mathcal{R}^-(P) \triangleq \mathcal{R}(P) - \mathcal{P}$.

Clearly, we obtain $P \in \mathcal{R}(P)$ by taking $n = 0$ in equation (7.2). Also, remark that the equality used in (7.2) is syntactical, therefore X is actually a subterm of P. Finally, if an identity $\mathbf{I}$ is definable in $\mathbf{S}_{\mathrm{CL}}$ then every Q admits a quasi-identity ($\mathbf{I}$ itself).

We present a non-definability result due to Padovani stating that no $\mathbf{S}$-term admits quasi-identities. In particular, the identity is not definable in $\mathbf{S}_{\mathrm{CL}}$. The result follows easily from the lemmas below.

LEMMA 7.14 (PADOVANI 2012). *For all* $X_1, X_2, W \in \mathbf{S}_{\mathrm{CL}}$ *satisfying* $X_1X_2 \twoheadrightarrow_w W$ *there exist* $U_1U_2 \in \mathcal{R}(W)$ *and* $V \in \mathbf{S}_{\mathrm{CL}}$ *such that:*

1. $X_1 \twoheadrightarrow_w V$ *and* $U_1 \in \mathcal{R}(V)$;

2. $X_2 \twoheadrightarrow_w U_2$.

PROOF. The proof is by induction on the lexicographically ordered pair $(k, \text{size}(W))$, where k is the sum of the lengths of all reductions $X_1X_2 \twoheadrightarrow_w W$. By Proposition 7.12(ii), this induction is well-founded. Two cases need to be considered.

Case 1. No root redex is contracted in any reduction from X_1X_2 to W. Then we can simply take $U_1U_2 = W$ and $V = U_1$.

Case 2. Assume $X_1 \twoheadrightarrow_w \mathbf{S}P_1P_2$ and $X_2 \twoheadrightarrow_w Q_2$ for appropriate **S**-terms P_1, P_2, Q_2. Then the reduction $X_1X_2 \twoheadrightarrow_w W$ factorizes as follows:

$$X_1X_2 \twoheadrightarrow_w \mathbf{S}P_1P_2Q_2 \to_w P_1Q_2(P_2Q_2) \twoheadrightarrow_w W.$$

By applying the IH to $P_1Q_2(P_2Q_2) \twoheadrightarrow_w W$ we obtain a term $W_1W_2 \in \mathcal{R}(W)$ such that $P_2Q_2 \twoheadrightarrow_w W_2$. Now, notice that the sum of the lengths of all reductions from P_2Q_2 to W_2 is less than or equal to k, and the term W_2 is a strict subterm of W. We can therefore apply the IH and obtain terms $U_1U_2 \in \mathcal{R}(W_2)$ and $V_0 \in \mathbb{S}_{\text{CL}}$ satisfying

$$P_2 \twoheadrightarrow_w V_0, \qquad U_1 \in \mathcal{R}(V_0), \qquad X_2 \twoheadrightarrow_w Q_2 \twoheadrightarrow_w U_2.$$

From the fact that $W_2 \in \mathcal{R}^-(W)$, we derive $\mathcal{R}(W_2) \subsetneq \mathcal{R}(W)$ and therefore $U_1U_2 \in \mathcal{R}(W)$. Moreover $X_1 \twoheadrightarrow_w \mathbf{S}P_1P_2 \twoheadrightarrow_w \mathbf{S}P_1V_0$ and $U_1 \in \mathcal{R}(V_0) \subsetneq \mathcal{R}(\mathbf{S}P_1V_0)$, so we can conclude by taking $V \triangleq \mathbf{S}P_1V_0$. □

LEMMA 7.15 (PADOVANI 2012).
Given $P \in \mathbb{S}_{\text{CL}}$, there exist no $X \in \mathbb{S}_{\text{CL}}$, $n \in \mathbb{N}$ and $Q_0, \dots, Q_n \in \mathbb{S}_{\text{CL}}$ such that:

1. *$X \twoheadrightarrow_w P$, and*
2. *$X \twoheadrightarrow_w Q_0(\cdots(Q_nP)\cdots)$.*

PROOF. By structural induction on P.

Case $P = \mathbf{S}$. Trivial, since $X \twoheadrightarrow_w P$ entails $X = P = \mathbf{S}$.

Case $P = P_1P_2$. Assume that $X \twoheadrightarrow_w P_1P_2$ and $X \twoheadrightarrow_w Q_0(\cdots(Q_n(P_1P_2))\cdots)$ hold, towards a contradiction. There are two possible subcases.

Subcase 1. In the former case, we have:

$$\begin{array}{lll} X & \twoheadrightarrow_w & X_1X_2 \twoheadrightarrow_w P_1P_2, \\ X_1 & \twoheadrightarrow_w & Q_0, \\ X_2 & \twoheadrightarrow_w & Q_1(\cdots(Q_n(P_1P_2))\cdots). \end{array}$$

By applying Lemma 7.14 to the reduction $X_1X_2 \twoheadrightarrow_w P_1P_2$, we get $U_1U_2 \in \mathcal{R}(P_1P_2)$ such that $X_2 \twoheadrightarrow_w U_2$. By Definition 7.13(iii), there exist $V_1, \dots, V_m \in \mathbb{S}_{\text{CL}}$ such that $V_1(\cdots(V_m(U_1U_2))\cdots) = P_1P_2$. Therefore, we obtain

$$\begin{array}{lll} X_2 & \twoheadrightarrow_w & U_2; \\ X_2 & \twoheadrightarrow_w & Q_1(\cdots(Q_n(P_1P_2))\cdots) = Q_1(\cdots(Q_n(V_1(\cdots(V_m(U_1U_2))\cdots)))\cdots). \end{array}$$

Since U_2 is a strict subterm of P, we derive a contradiction from the IH.

Subcase 2. In the latter case, there exist $X_1, X_2 \in \mathbb{S}_{\mathrm{CL}}$ such that:

$$\begin{array}{lcl} X & \twoheadrightarrow_w & X_1X_2 \twoheadrightarrow_w Q_0(\cdots(Q_n(P_1P_2))\cdots); \\ X_1 & \twoheadrightarrow_w & P_1; \\ X_2 & \twoheadrightarrow_w & P_2. \end{array}$$

It follows from the IH that the term X_2 cannot reduce to $Q_1(\cdots(Q_n(P_1P_2))\cdots)$. So the reduction $X_1X_2 \twoheadrightarrow_w Q_0(\cdots(Q_n(P_1P_2))\cdots)$ must involve the contraction of (at least) a top redex. That is, there must exist $U_1, U_2, U_3 \in \mathbb{S}_{\mathrm{CL}}$ such that:

$$\begin{array}{rcl} X_1 & \to_w & \mathbf{S}U_1U_2; \\ X_2 & \twoheadrightarrow_w & U_3; \\ U_1U_3(U_2U_3) & \twoheadrightarrow_w & Q_0(\cdots(Q_n(P_1P_2))\cdots). \end{array}$$

By Lemma 7.14 there exist $W_1W_2 \in \mathcal{R}(Q_0(\cdots(Q_n(P_1P_2))\cdots))$ such that $U_2U_3 \twoheadrightarrow_w W_2$. By applying Lemma 7.14 again, there exist $V_1V_2 \in \mathcal{R}(W_2)$ and an **S**-term U_2' such that $U_2 \twoheadrightarrow_w U_2'$, $V_1 \in \mathcal{R}(U_2')$ and $X_2 \to_w U_3 \twoheadrightarrow_w V_2$. Now $V_1V_2 \in \mathcal{R}(W_2) \subset \mathcal{R}^-(Q_0(\cdots(Q_n(P_1P_2))\cdots))$.

- Suppose $V_1V_2 \in \mathcal{R}^-(P_1P_2) = \mathcal{R}(P_2)$. Then there exist $Z_1, \ldots, Z_m \in \mathbb{S}_{\mathrm{CL}}$ such that $X_2 \twoheadrightarrow_w V_2$ and $X_2 \twoheadrightarrow_w P_2 = Z_1(\cdots(Z_m(V_1V_2))\cdots) = P_2$. A contradiction follows from the IH.
- Suppose $V_1V_2 = Q_i(\cdots(Q_n(P_1P_2))\cdots)$ where $i \le n$. Then $X_2 \twoheadrightarrow_w P_2$ and $X_2 \twoheadrightarrow_w V_2 = Q_{i+1}(Q_i(\cdots(Q_n(P_1P_2))\cdots))$, thus contradicting the IH.

In other words, we have $V_1V_2 = P_1P_2$. So $X_1 \twoheadrightarrow_w P_1$ and there exists $Z_1, \ldots, Z_m$ such that $X_1 \twoheadrightarrow_w \mathbf{S}U_1U_2 \twoheadrightarrow_w \mathbf{S}U_1(Z_1(\cdots(Z_mV_1)\cdots)) = \mathbf{S}U_1(Z_1(\cdots(Z_mP_1)\cdots))$. Again, a contradiction follows from the induction hypothesis. □

PROPOSITION 7.16 (PADOVANI 2012). *No* **S**-*term admits quasi-identities, hence the identity function is not definable in* $\mathbb{S}_{\mathrm{CL}}$.

PROOF. Assume by contradiction that $P =_w Q_0(\cdots(Q_nP)\cdots)$, for some $P, \vec{Q} \in \mathbb{S}_{\mathrm{CL}}$. By confluence there is an **S**-term W such that $P \twoheadrightarrow_w W$ and $Q_0(\cdots(Q_nP)\cdots) \twoheadrightarrow_w W$. It follows that $Q_0(\cdots(Q_nP)\cdots) \twoheadrightarrow_w Q_0(\cdots(Q_nW)\cdots)$. By applying Lemma 7.15, where we take $X \triangleq Q_0(\cdots(Q_nP)\cdots)$, a contradiction follows immediately. □

7.2 Normalization

Note that in the first part of the chapter we did not provide any evidence of the existence of an **S**-term generating an infinite reduction sequence. The problem of finding a non-terminating **S**-term was first proposed by Jan Bergstra during a research seminar at Utrecht University in 1974. This problem was solved by Barendregt who constructed the **S**-term **A**, whose non-termination will be shown in Section 7.3.

It is worth mentioning that—unless one is already familiar with the system $\mathbb{S}_{\mathrm{CL}}$—it is a non-trivial exercise to construct a term that does not enjoy strong normalization (SN), or even weak normalization (WN) due to the following equivalence.

PROPOSITION 7.17. *For all* $P \in \mathbb{S}_{\mathrm{CL}}$, $P \in \mathrm{WN} \iff P \in \mathrm{SN}$.

PROOF. By Theorem 7.10(i) $\mathbb{S}_{\mathrm{CL}}$ is orthogonal. Since each variable that occurs on the left-hand side of the rule $\mathbf{S}xyz \to_w xz(yz)$ also occur on its right-hand side, it is also non-erasing. We conclude since Huet and Lévy (1991), generalyzing Church (1941), proved that this equivalence holds in every term rewriting system satisfying these properties. □

We show that non-terminating terms exist, but also that normalization is decidable.

Spiralling terms

In order to construct a non-normalizing term the usual strategy is to construct a term having a cyclic reduction, but we have already seen $\mathbb{S}_{\mathrm{CL}}$ admits no such reductions. The second best option would be to construct a term P that reduces to a bigger term containing P as a subterm—terms satisfying this property are called 'spiralling'.

Recall that $\to_w^+$ denotes the transitive closure of $\to_w$.

DEFINITION 7.18. $P \in \mathbb{S}_{\mathrm{CL}}$ is called *spiralling* if $P \to_w^+ C[P]$, for some context $C[\,]$.

It was proved by Waldmann that spiralling **S**-terms do not exist.

PROPOSITION 7.19 (WALDMANN (2000)). *There is no spiralling* $P \in \mathbb{S}_{\mathrm{CL}}$.

PROOF. We define a function $\ell(\cdot) : \mathbb{S}_{\mathrm{CL}} \to \mathbb{N}$ as follows:

$$\ell(\mathbf{S}) \triangleq 1, \qquad \ell(PQ) \triangleq 1 + \ell(Q)$$

Note that $\ell(P)$ is measuring the length of the right-spine of P. Consider the set $\overline{\mathbb{S}}$ of **S**-terms whose application symbols are labeled with indices $k \in \mathbb{N}$, as in $P \cdot^k Q$. For $P \in \mathbb{S}_{\mathrm{CL}}$, define $\overline{P} \in \overline{\mathbb{S}}$ by annotating its applications with the length of their right-spine:

$$\overline{\mathbf{S}} \triangleq \mathbf{S}, \qquad \overline{PQ} \triangleq \overline{P} \cdot^{\ell(PQ)} \overline{Q}$$

and consider the following labelled **S**-rule (for $k, m, n \in \mathbb{N}$)

$$\mathbf{S} \cdot^k P \cdot^m Q \cdot^n W \to (P \cdot^n W) \cdot^{n+1} (Q \cdot^n W) \qquad (\mathbf{S}^\ell)$$

together with the label-lifting rule (for $n \in \mathbb{N}$):

$$P \cdot^n Q \to P \cdot^{n+1} Q \qquad (\ell\ell)$$

For all **S**-terms P, Q, it is easy to check that

$$P \to_w Q \quad \Rightarrow \quad \overline{P} \to_{\mathbf{S}^\ell} P' \twoheadrightarrow_{\ell\ell} \overline{Q} \qquad (7.3)$$

where the reduction $\twoheadrightarrow_{\ell\ell}$ might be needed to increase labels in the context of the step.

Assume that there exists a spiralling reduction $P \twoheadrightarrow_w C[P]$. Then there is a spiralling labelled reduction $P \twoheadrightarrow_{\mathbf{S}^\ell\ell\ell} C[P]$, by (7.3). The term P is a subterm of $C[P]$, whence we could continue reducing forever using only a bounded set of labels. As a consequence, it suffices to show the rules $(\mathbf{S}^\ell)$ and $(\ell\ell)$ are strongly normalizing when bounding the labels. Let $k \in \mathbb{N}$ and consider the rewrite system $\overline{\mathbb{S}}_k$ consisting of all instances of rules $(\mathbf{S}^\ell)$, $(\ell\ell)$ that contain only labels smaller than or equal to k. The system $\overline{\mathbb{S}}_k$ is terminating by the recursive partial order (*rpo*, see Dershowitz (1979)): $(\cdot)^m < (\cdot)^n \iff m > n$. □

The absence of cycles and spiralling terms in $\mathbb{S}_{\mathrm{CL}}$ seems to suggest that non-terminating **S**-terms might not exist. We now present a result going in the opposite direction—there exist **S**-terms whose normal form size grows extremely fast. The precise definition of the growth factor of a term is given below. We also introduce a reduction $\hookrightarrow_w$ which turns out to be useful for proving that an **S**-term has no w-nf.

DEFINITION 7.20. (i) Define the *growth factor of* $P \in \mathbb{S}_{\mathrm{CL}}$, written $\mathrm{growth}(P) \in \mathbb{N} \cup \{\infty\}$, as follows.

$$\begin{array}{rcll} \mathrm{growth}(P) & \triangleq & \dfrac{\mathrm{size}(P^{w\text{-nf}})}{\mathrm{size}(P)}, & \text{if } P \text{ has a } w\text{-nf;} \\ & \triangleq & \infty, & \text{otherwise.} \end{array}$$

(ii) Write $P \hookrightarrow_w Q$ whenever $P \to_w C[Q]$, for some context $C[\,]$.
(iii) As usual, $\hookrightarrow\!\!\!\!\!\rightarrow_w$ denotes the transitive and reflexive closure of $\hookrightarrow_w$.
(iv) Similarly, $\hookrightarrow_w^+$ denotes the transitive closure of $\hookrightarrow_w$.

By definition we have $\to_w \subseteq \hookrightarrow_w$, whence $\to_w^+ \subseteq \hookrightarrow_w^+$ and $\twoheadrightarrow_w \subseteq \hookrightarrow\!\!\!\!\!\rightarrow_w$.

PROPOSITION 7.21. (i) *For every* $k \in \mathbb{N}$ *there exists a* $P_k \in \mathbb{S}_{\mathrm{CL}}$ *such that*

$$k < \mathrm{growth}(P_k) < \infty.$$

(ii) *There exists a* $P \in \mathbb{S}_{\mathrm{CL}}$ *with* $\mathrm{growth}(P) = \infty$, *i.e.* P *has no* w*-nf.*

PROOF. (i) Let $Q_n \triangleq \mathbf{A}^n(\mathbf{A})$, for every $n \in \mathbb{N}$. It is easy to check that $Q_n \in \mathrm{SN}$, whence its growth is finite. Now, notice that the sequence $\langle \mathrm{size}(Q_n) \rangle_{n \in \mathbb{N}}$ starts with $\mathrm{size}(Q_0) = \mathrm{size}(\mathbf{A}) = 5$ and grows linearly, since $\mathrm{size}(Q_{n+1}) = \mathrm{size}(Q_n) + 6$. By induction on n, we show that $\mathrm{size}((Q_n)^{w\text{-nf}}) > 2^n$.

Case $n = 0$. We have $\mathrm{size}(\mathbf{A}^{w\text{-nf}}) = \mathrm{size}(\mathbf{SSS}) = 5 > 2^0 = 1$.

Case $n > 0$. We have $Q_n \to_w \mathbf{S}(Q_n)(\mathbf{S}(Q_n)) \twoheadrightarrow_w \mathbf{S}((Q_n)^{w\text{-nf}})(\mathbf{S}((Q_n)^{w\text{-nf}}))$. Easy calculations give:

$$\begin{array}{rcll} \mathrm{size}(\mathbf{S}((Q_n)^{w\text{-nf}})(\mathbf{S}((Q_n)^{w\text{-nf}}))) & = & 2 \cdot \mathrm{size}((Q_n)^{w\text{-nf}}) + 5 & \\ & > & 2 \cdot 2^n + 5, & \text{by IH,} \\ & = & 2^{n+1} + 5. & \end{array}$$

So, the sequence $\langle \mathrm{size}((Q_n)^{w\text{-nf}}) \rangle_{n \in \mathbb{N}}$ has an exponential growth, from which (i) follows by taking $P_n \triangleq Q_{\log_2 n}$.

(ii) We show that $\mathbf{SAA}(\mathbf{SAA})$ has no w-nf. By straightforward induction on n, one shows that

$$\mathbf{A}^n(X)Y \hookrightarrow\!\!\!\!\!\rightarrow_w XY \tag{7.4}$$

for all $X, Y \in \mathbb{S}_{\mathrm{CL}}$. Therefore, for every $n \in \mathbb{N}$, we obtain:

$$\begin{array}{rcll} \mathbf{A}^n(\mathbf{SAA})(\mathbf{A}^n(\mathbf{SAA})) & \hookrightarrow\!\!\!\!\!\rightarrow_w & \mathbf{SAA}(\mathbf{A}^n(\mathbf{SAA})), & \text{by (7.4),} \\ & \to_w & \mathbf{A}^{n+1}(\mathbf{SAA})(\mathbf{A}^{n+1}(\mathbf{SAA})). & \end{array}$$

As a consequence, $\mathbf{SAA}(\mathbf{SAA})$ has an infinite $\hookrightarrow_w$-reduction, whence no w-nf. □

Normalization is decidable

We now turn to another result on **S**-terms by Waldmann, namely the decidability of normalization. Its proof is definitely more complicated than the proofs of the previous theorems, and it is beyond the scope of the chapter to provide all its details.

	0	1	2	3	4	5	6	7	8	9	10	11	12	13	14	15	16	17	18	19	20	21	22	23	24	25	26	27	28	29	30	31	32	33	34	35	36	37
0	1	2	3	4	5	6	6	8	5	10	5	12	5	4	5	10	5	5	5	20	6	6	6	6	6	6	6	6	6	6	6	6	21	6	6	6	21	6
1	7	9	11	13	14	5	6	15	16	15	17	11	18	13	14	15	16	17	18	19	21	21	23	23	24	25	26	27	28	29	30	31	32	33	34	35	36	37
2	19	22	24	25	25	25	25	26	25	26	25	25	25	25	25	26	25	25	25	25	25	25	25	25	25	25	25	25	25	29	30	31	31	31	29	29	31	31
3	27	28	29	29	30	30	30	29	30	29	30	29	30	29	30	29	30	30	30	29	30	30	30	30	30	30	30	30	30	30	30	30	29	30	30	30	29	30
4	31	31	29	29	30	30	30	29	30	29	30	29	30	29	30	29	30	30	30	29	30	30	30	30	30	30	30	30	30	30	30	30	29	30	30	30	29	30
5	29	30	30	30	30	30	30	30	30	30	30	30	30	30	30	30	30	30	30	30	30	30	30	30	30	30	30	30	30	30	30	30	30	30	30	30	30	30
6	30	30	30	30	30	30	30	30	30	30	30	30	30	30	30	30	30	30	30	30	30	30	30	30	30	30	30	30	30	30	30	30	30	30	30	30	30	30
7	9	24	29	30	30	30	30	29	30	29	30	30	30	30	30	29	30	30	30	30	30	30	30	30	30	30	30	30	30	30	30	30	30	30	30	30	30	30
8	32	33	29	29	29	29	29	34	29	34	29	29	29	29	29	34	29	29	29	29	29	29	29	29	29	29	29	29	29	29	30	29	29	29	29	29	29	29
9	15	26	29	29	30	30	30	34	30	35	30	29	30	29	30	35	30	30	30	30	30	30	30	30	30	30	30	30	30	30	30	30	30	30	30	30	30	30
10	36	37	29	29	29	29	29	35	29	35	29	29	29	29	29	35	29	29	29	29	29	29	29	29	29	29	29	29	29	29	30	29	29	29	29	29	29	29
11	19	25	30	30	30	30	30	29	30	29	30	30	30	30	30	29	30	30	30	30	30	30	30	30	30	30	30	30	30	30	30	30	30	30	30	30	30	30
12	37	31	29	29	30	30	30	29	30	29	30	29	30	29	30	29	30	30	30	29	30	30	30	30	30	30	30	30	30	30	30	30	29	30	30	30	29	30
13	27	25	30	30	30	30	30	29	30	29	30	30	30	30	30	29	30	30	30	30	30	30	30	30	30	30	30	30	30	30	30	30	30	30	30	30	30	30
14	31	31	30	30	30	30	30	29	30	29	30	30	30	30	30	29	30	30	30	30	30	30	30	30	30	30	30	30	30	30	30	30	30	30	30	30	30	30
15	15	25	30	30	30	30	30	29	30	29	30	30	30	30	30	29	30	30	30	30	30	30	30	30	30	30	30	30	30	30	30	30	30	30	30	30	30	30
16	32	31	30	30	30	30	30	29	30	29	30	30	30	30	30	29	30	30	30	30	30	30	30	30	30	30	30	30	30	30	30	30	30	30	30	30	30	30
17	36	31	30	30	30	30	30	29	30	29	30	30	30	30	30	29	30	30	30	30	30	30	30	30	30	30	30	30	30	30	30	30	30	30	30	30	30	30
18	37	31	30	30	30	30	30	29	30	29	30	30	30	30	30	29	30	30	30	30	30	30	30	30	30	30	30	30	30	30	30	30	30	30	30	30	30	30
19	24	29	30	30	30	30	30	30	30	30	30	30	30	30	30	30	30	30	30	30	30	30	30	30	30	30	30	30	30	∞	∞	∞	∞	∞	∞	∞	∞	∞
20	29	29	29	29	30	30	30	29	30	29	30	29	30	29	30	29	30	30	30	30	30	30	30	30	30	30	30	30	30	30	30	30	30	30	30	30	30	30
21	29	29	30	30	30	30	30	29	30	29	30	30	30	30	30	29	30	30	30	30	30	30	30	30	30	30	30	30	30	30	30	30	30	30	30	30	30	30
22	29	35	30	30	30	30	30	29	30	29	30	30	30	30	30	29	30	30	30	30	30	30	∞	∞	∞	∞	∞	30	∞	∞	∞	∞	∞	∞	∞	∞	∞	∞
23	29	29	30	30	30	30	30	29	30	29	30	30	30	30	30	29	30	30	30	30	30	30	∞	∞	∞	∞	∞	30	∞	∞	∞	∞	∞	∞	∞	∞	∞	∞
24	30	30	30	30	30	30	30	30	30	30	30	30	30	30	30	30	30	30	30	30	30	30	∞	∞	∞	∞	∞	30	∞	∞	∞	∞	∞	∞	∞	∞	∞	∞
25	30	30	30	30	30	30	30	30	30	30	30	30	30	30	30	30	30	30	30	∞	30	30	∞	∞	∞	∞	∞	∞	∞	∞	∞	∞	∞	∞	∞	∞	∞	∞
26	29	30	30	30	30	30	30	30	30	30	30	30	30	30	30	30	30	30	30	∞	30	30	∞	∞	∞	∞	∞	∞	∞	∞	∞	∞	∞	∞	∞	∞	∞	∞
27	29	30	30	30	30	30	30	30	30	30	30	30	30	30	30	30	30	30	30	30	30	30	30	30	30	30	30	30	30	∞	∞	∞	∞	∞	∞	∞	∞	∞
28	30	29	30	30	30	30	30	30	30	30	30	30	30	30	30	30	30	30	30	∞	30	30	∞	∞	∞	∞	∞	∞	∞	∞	∞	∞	∞	∞	∞	∞	∞	∞
29	30	∞	∞	∞	∞	∞	∞	∞	∞	∞	∞	∞	∞	∞	∞	∞	∞	∞	∞	∞	∞	∞	∞	∞	∞	∞	∞	∞	∞	∞	∞	∞	∞	∞	∞	∞	∞	∞
30	∞	∞	∞	∞	∞	∞	∞	∞	∞	∞	∞	∞	∞	∞	∞	∞	∞	∞	∞	∞	∞	∞	∞	∞	∞	∞	∞	∞	∞	∞	∞	∞	∞	∞	∞	∞	∞	∞
31	30	30	∞	∞	∞	∞	∞	∞	∞	∞	∞	∞	∞	∞	∞	∞	∞	∞	∞	∞	∞	∞	∞	∞	∞	∞	∞	∞	∞	∞	∞	∞	∞	∞	∞	∞	∞	∞
32	26	30	∞	∞	∞	∞	∞	∞	∞	∞	∞	∞	∞	∞	∞	∞	∞	∞	∞	∞	∞	∞	∞	∞	∞	∞	∞	∞	∞	∞	∞	∞	∞	∞	∞	∞	∞	∞
33	34	30	∞	∞	∞	∞	∞	∞	∞	∞	∞	∞	∞	∞	∞	∞	∞	∞	∞	∞	∞	∞	∞	∞	∞	∞	∞	∞	∞	∞	∞	∞	∞	∞	∞	∞	∞	∞
34	35	∞	∞	∞	∞	∞	∞	∞	∞	∞	∞	∞	∞	∞	∞	∞	∞	∞	∞	∞	∞	∞	∞	∞	∞	∞	∞	∞	∞	∞	∞	∞	∞	∞	∞	∞	∞	∞
35	29	∞	∞	∞	∞	∞	∞	∞	∞	∞	∞	∞	∞	∞	∞	∞	∞	∞	∞	∞	∞	∞	∞	∞	∞	∞	∞	∞	∞	∞	∞	∞	∞	∞	∞	∞	∞	∞
36	25	30	∞	∞	∞	∞	∞	∞	∞	∞	∞	∞	∞	∞	∞	∞	∞	∞	∞	∞	∞	∞	∞	∞	∞	∞	∞	∞	∞	∞	∞	∞	∞	∞	∞	∞	∞	∞
37	29	30	∞	∞	∞	∞	∞	∞	∞	∞	∞	∞	∞	∞	∞	∞	∞	∞	∞	∞	∞	∞	∞	∞	∞	∞	∞	∞	∞	∞	∞	∞	∞	∞	∞	∞	∞	∞

Figure 7.1: Tree automaton accepting the set of strongly normalizing terms in $\mathfrak{S}_{\mathrm{CL}}$. More precisely: every **S** is interpreted as 0, and the table above gives the interpretation for the application symbol '·'. If P has interpretation x and Q has interpretation y then the interpretation of PQ is given in the intersection of row x and column y.

THEOREM 7.22. (i) *The set of normalizing* **S**-*terms is a regular tree language accepted by the automaton shown in Figure 7.1. The accepting states* $0, \ldots, 37$ *capture strongly normalizing terms, the symbol* ∞ *designates that the term has no interpretation in the automaton and hence has an infinite reduction (it is not normalizing).*

(ii) *Strong normalization of* **S**-*terms is decidable in linear time.*

The original proof of this theorem given by Waldmann (1998) was based on an elaborate analysis of reductions involving an *ad hoc* construction of a tree grammar generating the normalizing **S**-terms (see also Waldmann (2000)). The minimal deterministic tree automaton corresponding to the grammar is shown in Figure 7.1. Also the termination proof was hand-crafted. We note that the original proof by Waldmann made use of a computer program to check that one of the complicated case distinctions was indeed exhaustive. In Cheilaris et al. (2011) a proof is given without the use of computers.

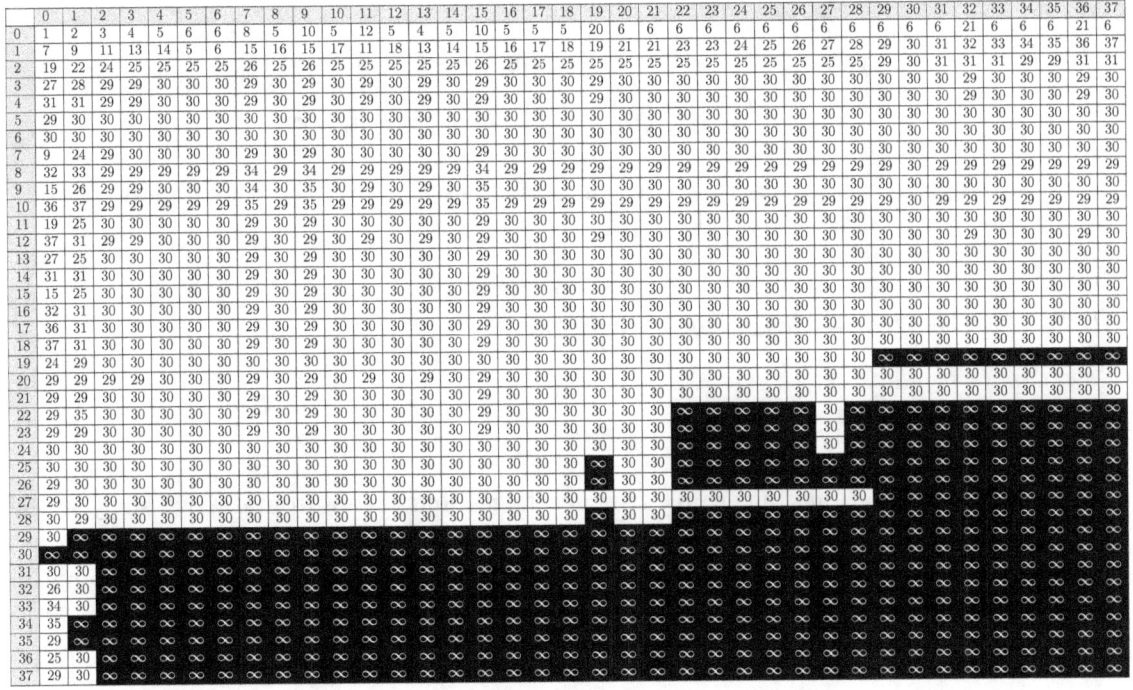

In Endrullis et al. (2009) and Endrullis et al. (2010) a method has been presented which allows to fully automatically find the tree automaton accepting the set of normalizing terms and prove termination. We will briefly sketch the ideas behind this approach.

DEFINITION 7.23. (i) A *tree language* $\mathcal{L}$ over an alphabet Σ is any subset $\mathcal{L} \subseteq \mathscr{T}_\Sigma$, where $\mathscr{T}_\Sigma$ denotes the smallest set of strings such that $a(t_1, \ldots, t_k) \in \mathscr{T}_\Sigma$ whenever $a \in \Sigma$ has arity k and $t_1, \ldots, t_k \in \mathscr{T}_\Sigma$.

(ii) The *Nerode congruence* $\sim_\mathcal{L}$ for a tree language $\mathcal{L}$ is defined by

$$s \sim_\mathcal{L} t \iff (\forall C \,.\, C[s] \in \mathcal{L} \iff C[t] \in \mathcal{L}) \qquad (7.5)$$

These languages enjoy nice structural properties captured by the famous Myhill-Nerode theorem, which is recalled below.

THEOREM 7.24 (MYHILL (1957) & NERODE (1958)).
Given a tree language $\mathcal{L}$, the following are equivalent:

1. *$\mathcal{L}$ is is regular;*

2. *the number of equivalence classes with respect to $\sim_\mathcal{L}$ is finite.*

For more information concerning tree languages and tree automata, we suggest the book by Comon et al. (1997). In the following we are interested in $\mathcal{L}$ being the set of strongly normalizing **S**-terms. However, using (7.5) to compute the tree automaton (corresponding to $\sim_\mathcal{L}$) raises the following problems:

- *a priori* we do not know how to decide membership in $\mathcal{L}$;
- we cannot test an infinite number of contexts $C[\,]$.

However, both problems can be approximated as follows. Given $p, q \in \mathbb{N}$, we approximate $\mathcal{L}$ by the set of terms $\mathcal{L}_p$ that admit reductions of length at most p, moreover, we check only contexts up to size q. With these two approximations in place, the right-hand side of (7.5) becomes decidable.

NOTATION. (i) *Write $\sim_{p,q}$ for the decidable approximation of $\sim_\mathcal{L}$ described above.*

(ii) *Let $\mathbb{S}_n$ denote the set of* **S***-terms whose syntax-tree has height at most n.*

(iii) *Write $[P]_{p,q}$ for the equivalence class of P in $\mathbb{S}_n$ with respect to $\sim_{p,q}$.*

Then we can search a natural number $n = 1, 2, 3, \ldots$ such that for every $P, Q \in \mathbb{S}_n$ there exists $O \in \mathbb{S}_n$ satisfying $PQ \sim_{p,q} O$. In case this condition is fulfilled, we obtain a tree automaton as follows.

- The states of the automaton are the equivalence classes $\mathbb{S}_n/_{\sim_{p,q}}$.

- The final states are those equivalence classes containing terms in $\mathcal{L}_n$ (admitting reductions of length at most p).

- The interpretation of **S** is $[\mathbf{S}]_{p,q}$.

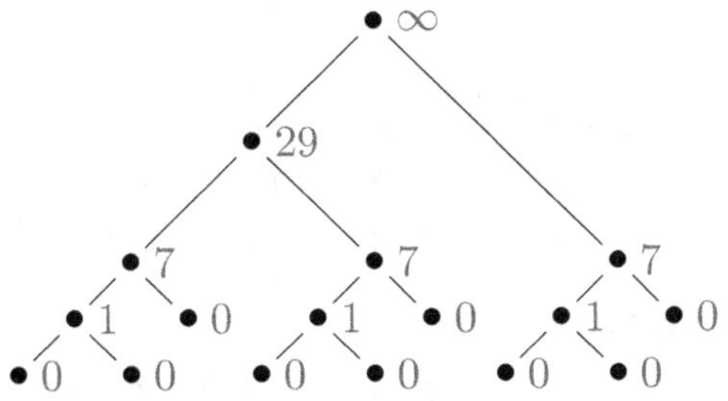

Figure 7.2: Non-termination of **AAA** as established by the automaton in Figure 7.1.

- The interpretation of application is $[\![\cdot]\!]([P]_{p,q}, [Q]_{p,q}) \triangleq [PQ]_{p,q}$.

This search for n is guaranteed to terminate for any choice of the numbers p, q. However, if p, q are chosen too small, the resulting automaton is not guaranteed to accept the correct language.

So how do we know whether the choice was good?

First, the resulting tree automaton must be a model of the rewrite system since termination is invariant under rewriting. It then can be checked easily that the set of non-accepted terms is closed under rewriting (the model property) and that they always contain a redex (subset of the regular language of terms containing redexes). Second, we can prove *local termination* on all accepted terms by following Endrullis et al. (2009). This approach employs semantic labelling of the rewrite system with the states of the tree automaton, see Zantema (1995). Afterwards *global* termination of the labelled system can be proven using automated termination tools. Thus it can be automatically verified that the resulting automaton accepts precisely the terminating **S**-terms. More details can be found in Endrullis et al. (2009).

7.3 Infinite reductions

In the proof of Proposition 7.21 we exhibited an **S**-term, first constructed by Pettorossi in 1976, without a w-nf. Other examples are given in Table 7.2, below.

P	Year found	By
AAA	1975	Barendregt
S(SS)SSSS	1976	Duboué & Baron
SAA(SAA)	1976	Pettorossi
STT(STT)	1978	Zachos

Table 7.2: Examples of $P \in \mathbb{S}_{\rm CL}$ without a w-nf.

We now show that these **S**-terms do not satisfy the same rewriting properties: some are not SN but are head normalizing, others generate an infinite head reduction sequence.

Böhm trees

In combinatory logic a term is in head normal form if it has no redex in head position. This means that it starts with either **K** or **S**, but without enough arguments to become a redex. For **S**-terms the notion of head redex has been introduced in Definition 7.6(i).

DEFINITION 7.25. Let $P, Q \in \mathbb{S}_{\mathrm{CL}}$.

(i) $P \to_w Q$ is a *head reduction step* if the redex contracted in P occurs in head position. In this case, we write $P \to_h Q$.

(ii) An **S**-term is in *head normal form* (*hnf*) if it has no head redex.

(iii) An **S**-term P *is head normalizable* if $P \twoheadrightarrow_w Q$, for some $Q \in \mathbb{S}_{\mathrm{CL}}$ in hnf.

(iv) We let $\perp\!\!\!\perp$ be the set of non head normalizable **S**-terms.

In $\mathbb{S}_{\mathrm{CL}}$ head normal forms admit the following characterization.

LEMMA 7.26. *Let* $P \in \mathbb{S}_{\mathrm{CL}}$.

(i) *P is in hnf if and only if P is of the form* $\mathbf{S}$, $\mathbf{S}X$, $\mathbf{S}XY$, *for some* $X, Y \in \mathbb{S}_{\mathrm{CL}}$.

(ii) *P is head normalizable if and only if its head reduction terminates.*

(iii) *If P is not head normalizable then it has no w-nf.*

PROOF. (i) By an easy case analysis.

(ii) ($\Rightarrow$) We follow an argument by Padovani. Assume $P \twoheadrightarrow_w P'$, for some P' in hnf. We show that there exists P'' in hnf such that $P \twoheadrightarrow_h P'' \twoheadrightarrow_w P'$. By Proposition 7.12(ii), we can proceed by induction on the sum k of the lengths of all reductions $P \twoheadrightarrow_w P'$.

Case $k = 0$. Then P is in hnf, whence $P = P'$ and we can take $P'' \triangleq P$.

Case $k > 0$. Then either P is in hnf, in which case we conclude easily, or the reduction $P \twoheadrightarrow_w P'$ must be of the form (for some $n \in \mathbb{N}$, and appropriate $\vec{U}, \vec{V} \in \mathbb{S}_{\mathrm{CL}}$):

$$P = \mathbf{S}U_1U_2U_3U_4 \cdots U_{n+3} \twoheadrightarrow_w \mathbf{S}V_1V_2V_3V_4 \cdots V_{n+3} \to_w V_1V_3(V_2V_3)V_4 \cdots V_{n+3} \twoheadrightarrow_w P'$$

where $U_i \twoheadrightarrow_w V_i$, for every $i \in \mathbb{N}$. We can rearrange the reduction above by starting with an head step:

$$\begin{aligned} P = \mathbf{S}U_1U_2U_3U_4 \cdots U_{n+3} &\to_h U_1U_3(U_2U_3)U_4 \cdots U_{n+3} \\ &\twoheadrightarrow_w V_1V_3(V_2V_3)V_4 \cdots V_{n+3} \twoheadrightarrow_w P' \end{aligned}$$

such that the sum of the lengths of all reductions $U_1U_3(U_2U_3)U_4 \cdots U_{n+3} \twoheadrightarrow_w P'$ is strictly smaller than k. By IH, $U_1U_3(U_2U_3)U_4 \cdots U_{n+3} \twoheadrightarrow_h P''$ for some hnf P'' satisfying $P'' \twoheadrightarrow_w P'$. We conclude by composing the two head reductions.

($\Leftarrow$) Easy, using (i).

(iii) Assume that P has no hnf. By (ii), the head reduction sequence starting from P is infinite. We conclude by Proposition 7.17. □

By coinductively collecting the stable information coming out from the head-reduction of a term, and equating to $\perp$ all terms without hnf, one retrieves the usual definition of a Böhm tree recasted for **S**-terms.

DEFINITION 7.27. Let $P \in \mathbb{S}_{\mathrm{CL}}$. The *Böhm tree of P*, denoted by $\mathrm{BT}(P)$, is coinductively defined by:

$\mathrm{BT}(P) \triangleq \mathbf{S}$, if $P = \mathbf{S}$;

$\triangleq$

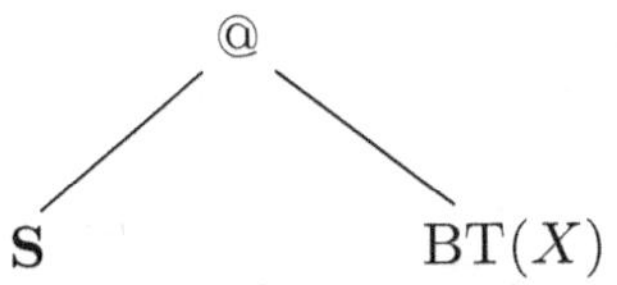

, if $P = \mathbf{S}X$;

$\triangleq$

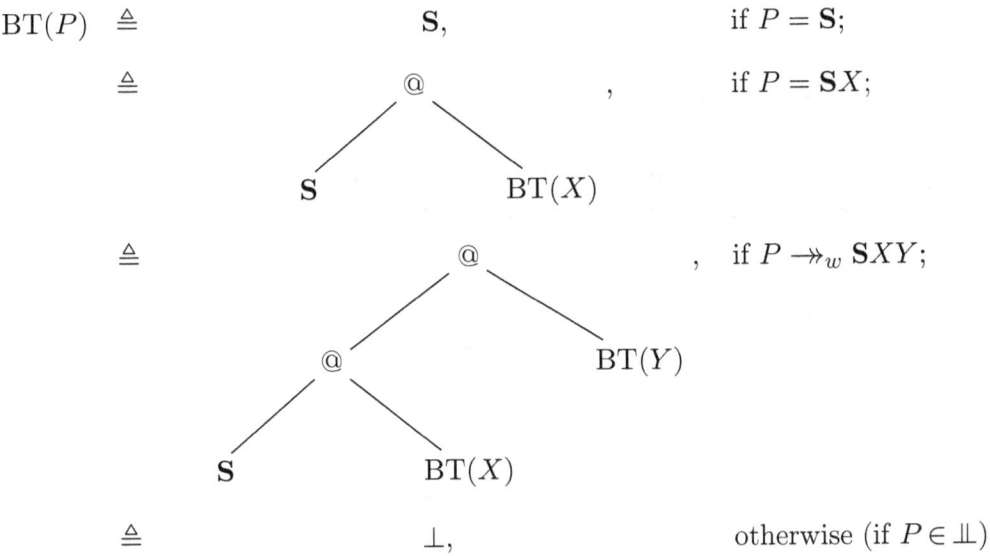

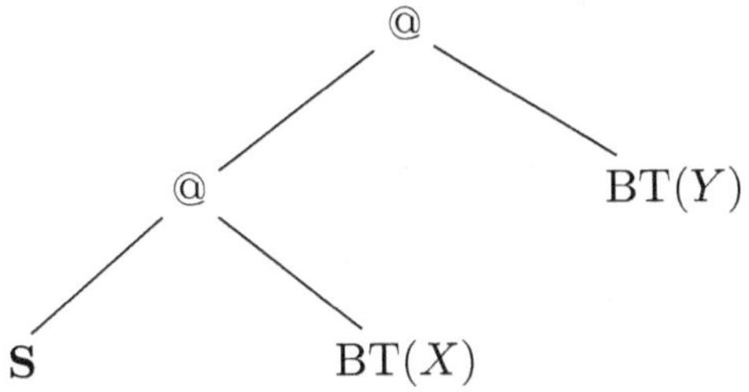

, if $P \twoheadrightarrow_w \mathbf{S}XY$;

$\triangleq \bot$, otherwise (if $P \in \perp\!\!\!\perp$).

As a warm up exercise we rewrite Zachos' term $\mathbf{STT}(\mathbf{STT})$ defined in Table 7.2. This term enjoys the property that its reduction graph is a single line—equivalently—each reduct contains exactly one redex. We place this redex in a $\boxed{\text{box}}$ whenever it is a root-redex. In order to reach its hnf, it is enough to perform two head steps:

$$\begin{array}{rl} & \boxed{\mathbf{STT}(\mathbf{STT})} \\ \to_h & \mathbf{T}(\mathbf{STT})(\mathbf{T}(\mathbf{STT})) \\ = & \boxed{\mathbf{SS}(\mathbf{STT})(\mathbf{T}(\mathbf{STT}))} \\ \to_h & \mathbf{S}(\mathbf{T}(\mathbf{STT}))(\mathbf{STT}(\mathbf{T}(\mathbf{STT}))) \end{array}$$

Note that $\mathbf{S}(\mathbf{T}(\mathbf{STT}))$ is in w-nf and applying a term does not create a head redex, whence the head position of the term has become stable. We can continue to perform reduction on the rightmost subterm:

$$\begin{array}{rl} & \mathbf{S}(\mathbf{T}(\mathbf{STT}))(\mathbf{STT}(\mathbf{T}(\mathbf{STT}))) \\ \to_w & \mathbf{S}(\mathbf{T}(\mathbf{STT}))(\mathbf{T}(\mathbf{T}(\mathbf{STT}))(\mathbf{T}(\mathbf{T}(\mathbf{STT})))) \\ \to_w & \mathbf{S}(\mathbf{T}(\mathbf{STT}))(\mathbf{S}(\mathbf{T}(\mathbf{T}(\mathbf{STT})))(\mathbf{T}(\mathbf{STT})(\mathbf{T}(\mathbf{T}(\mathbf{STT}))))) \\ \to_w & \mathbf{S}(\mathbf{T}(\mathbf{STT}))(\mathbf{S}(\mathbf{T}(\mathbf{T}(\mathbf{STT})))(\mathbf{S}(\mathbf{T}(\mathbf{T}(\mathbf{STT})))(\mathbf{STT}(\mathbf{T}(\mathbf{T}(\mathbf{STT})))))) \\ \to_w & \mathbf{S}(\mathbf{T}(\mathbf{STT}))(\mathbf{S}(\mathbf{T}(\mathbf{T}(\mathbf{STT})))(\mathbf{S}(\mathbf{T}(\mathbf{T}(\mathbf{STT})))(\mathbf{T}(\mathbf{T}(\mathbf{T}(\mathbf{STT})))(\mathbf{T}(\mathbf{T}(\mathbf{T}(\mathbf{STT}))))))) \end{array}$$

To improve the readability, it is convenient to introduce the notations $X_0 \triangleq \mathbf{STT}$ and $X_{n+1} \triangleq \mathbf{T}(X_n)$, for all $n \in \mathbb{N}$. Employing these notations, we can rewrite the last term above as $\mathbf{S}X_1(\mathbf{S}X_2(\mathbf{S}X_2(X_3X_3)))$. More generally, Zachos (1978) noticed that all redexes occurring in a reduct of $\mathbf{STT}(\mathbf{STT})$ have the form X_nX_m for some $n, m \in \mathbb{N}$.

In fact, there are two kinds of reduction steps. One going downwards:

$$X_{n+1}X_m = \mathbf{SS}X_nX_m \to_w \mathbf{S}X_m(\underline{X_nX_m})$$

and the other going upwards:

$$X_0X_m = \mathbf{STT}X_m \to_w \mathbf{T}X_m(\mathbf{T}X_m) = X_{m+1}X_{m+1}.$$

In particular this shows that Zachos' term has an infinite w-reduction sequence, whose limit is the infinite term $\mathbf{S}X_1((\mathbf{S}X_2)^2((\mathbf{S}X_3)^3((\mathbf{S}X_4)^4(\cdots))))$. Thus, its Böhm tree is

$\mathrm{BT}(\mathbf{STT}(\mathbf{STT})) =$

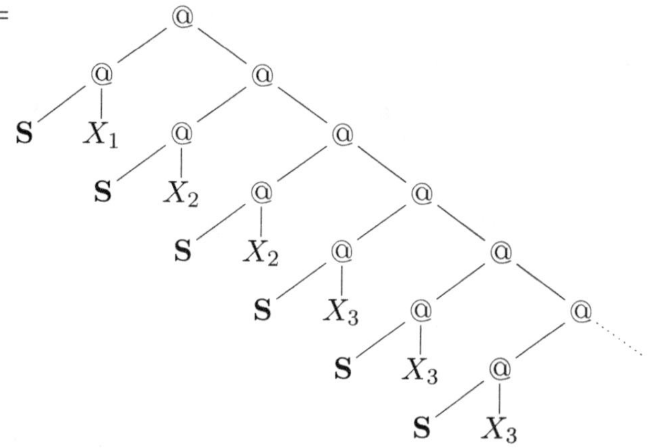

For a comparison, let us compute the head reduction starting from Barendregt's term **AAA**. Head redexes are underlined and root redexes (that are also head) are boxed:

$$
\begin{array}{ll}
 & \mathbf{AAA} \\
= & \underline{\mathbf{SSSA}}\,\mathbf{A} \\
\to_h & \boxed{\mathbf{SA(SA)A}} \\
\to_h & \mathbf{AA(SAA)} \\
= & \underline{\mathbf{SSSA}}\,\mathbf{(SAA)} \\
\to_h & \boxed{\mathbf{SA(SA)(SAA)}} \\
\to_h & \mathbf{A(SAA)(SA(SAA))} \\
= & \underline{\mathbf{SSS(SAA)}}\,\mathbf{(SA(SAA))} \\
\to_h & \boxed{\mathbf{S(SAA)(S(SAA))(SA(SAA))}} \\
\to_h & \underline{\mathbf{SAA(SA(SAA))}}\,\mathbf{(S(SAA)(SA(SAA)))} \\
\to_h & \mathbf{A(SA(SAA))(A(SA(SAA)))(S(SAA)(SA(SAA)))} \\
= & \underline{\mathbf{SSS(SA(SAA))}}\,\mathbf{(A(SA(SAA)))(S(SAA)(SA(SAA)))} \\
\to_h & \underline{\mathbf{S(SA(SAA))(S(SA(SAA)))(A(SA(SAA)))}}\,\mathbf{(S(SAA)(SA(SAA)))} \\
\to_h & \underline{\mathbf{SA(SAA)(A(SA(SAA)))}}\,\mathbf{(S(SA(SAA))(A(SA(SAA))))(S(SAA)(SA(SAA)))} \\
\to_h & \cdots
\end{array}
$$

After the three root reduction steps displayed above, no further root steps take place: the root has become stable. We show that **AAA** has an infinite head reduction sequence.

THEOREM 7.28 (HENK BARENDREGT 1974; UPGRADED[2]). *The term*

$$\mathbf{SSS(SSS)(SSS)}$$

has an infinite head reduction (thus, no w-normal form).

PROOF. We use the notations $\mathbf{D} = \mathbf{SAA}$ and $\mathbf{E} = \mathbf{SAD}$ introduced in Examples 7.4. We also set $\underline{n} \triangleq \mathbf{A}^n(\mathbf{E})$, for every $n \in \mathbb{N}$. We define

$$P \rhd Q \iff P = Q\vec{R}, \text{ for some non-empty sequence of } \mathbf{S}\text{-terms } \vec{R},$$

and $\rhd_h \triangleq (\to_h \cup \rhd)$. Note that $\rhd$ is an instance of the reduction $\hookrightarrow_w$, obtained by taking an applicative context $C[\,] \triangleq [\,]\vec{R}$. Finally, we denote by $\rhd_h^+$ the transitive closure of $\rhd_h$ and by $\rhd_h^*$ its transitive and reflexive closure. We prove the following claim.

Claim 7.28.1. (i) $\mathbf{AAA} \rhd_h^+ \underline{1}\,\underline{1}$;

(ii) $\underline{0}\,x \rhd_h^+ x(\mathbf{D}x) \rhd x$;

(iii) $\underline{n+1}\,x \rhd_h^+ \underline{n}\,x$;

(iv) $\underline{n}\,x \rhd_h^* \underline{0}\,x \rhd_h x$;

(v) $\underline{0}\,\underline{n} \rhd_h^+ \underline{n+1}\,\underline{n+1}$;

(vi) $\underline{n}\,\underline{n} \rhd_h^+ \underline{n+1}\,\underline{n+1}$.

Subproof. (i)
$$\begin{aligned}\mathbf{AAA} = \mathbf{SSSAA} &\to_h \mathbf{SA(SA)A} \to_h \mathbf{AAD} = \mathbf{SSSAD}\\ &\to_h \mathbf{SA(SA)D} \to_h \mathbf{AD(SAD)} = \mathbf{SSSDE} \to_h \mathbf{SD(SD)E}\\ &\to_h \mathbf{DE(SDE)} \rhd \mathbf{DE} = \mathbf{SAAE} \to_h \mathbf{AE(AE)} = \underline{1}\,\underline{1}.\end{aligned}$$

(ii)
$$\begin{aligned}\underline{0}\,x = \mathbf{E}x = \mathbf{SAD}x &\to_h \mathbf{A}x(\mathbf{D}x) = \mathbf{SSS}x(\mathbf{D}x)\\ &\to_h \mathbf{S}x(\mathbf{S}x)(\mathbf{D}x) \to_h x(\mathbf{D}x)(\mathbf{S}x(\mathbf{D}x)) \rhd x(\mathbf{D}x) \rhd x\end{aligned}$$

(iii)
$$\begin{aligned}\underline{n+1}\,x = \mathbf{A}^{n+1}(\mathbf{E})x = \mathbf{SSS}(\mathbf{A}^n(\mathbf{E}))x &\to_h \mathbf{S}(\mathbf{A}^n(\mathbf{E}))(\mathbf{S}(\mathbf{A}^n(\mathbf{E})))x\\ &\to_h \mathbf{A}^n(\mathbf{E})x(\mathbf{S}(\mathbf{A}^n(\mathbf{E}))x) \rhd \mathbf{A}^n(\mathbf{E})x = \underline{n}\,x\end{aligned}$$

(iv) The claim follows by induction using (iii) and (ii).

(v)
$$\begin{aligned}\underline{0}\,\underline{n} &\rhd_h^+ \underline{n}(\mathbf{D}\underline{n}) \quad \text{by (ii)}\\ &\rhd_h^* \mathbf{D}\,\underline{n} \quad \text{by clause (iv)}\\ &= \mathbf{SAA}\,\underline{n} \to_h \mathbf{A}\,\underline{n}(\mathbf{A}\,\underline{n}) = \underline{n+1}\,\underline{n+1}\end{aligned}$$

(vi) $\underline{n}\,\underline{n} \rhd_h^* \underline{0}\,\underline{n} \rhd_h^+ \underline{n+1}\,\underline{n+1}$ by (iv) and (v).

This concludes the proof of the claim. ∎

[2]The original proof did not use a head reduction strategy, and required the invocation of a theorem about non-erasing reduction (now known as Church's Theorem, see Klop (1980a), Theorem 5.9.3, or Terese (2003), Theorem 4.8.5) to conclude that the term **AAA** has no w-nf.

From these clauses it follows that **SSS**(**SSS**)(**SSS**) has an infinite $\rhd_h$-reduction:

$$\mathbf{AAA} \rhd_h^+ \underline{1}\,\underline{1} \rhd_h^+ \underline{2}\,\underline{2} \rhd_h^+ \underline{3}\,\underline{3} \rhd_h^+ \underline{4}\,\underline{4} \rhd_h^+ \cdots$$

First note that this $\rhd_h$-reduction contains an infinite number of head steps (since $\rhd$ is well-founded). We have $\rhd \circ \to_h \subseteq \to_h \circ \rhd$ and consequently this infinite $\rhd_h$-path can be transformed into an infinite head reduction sequence:

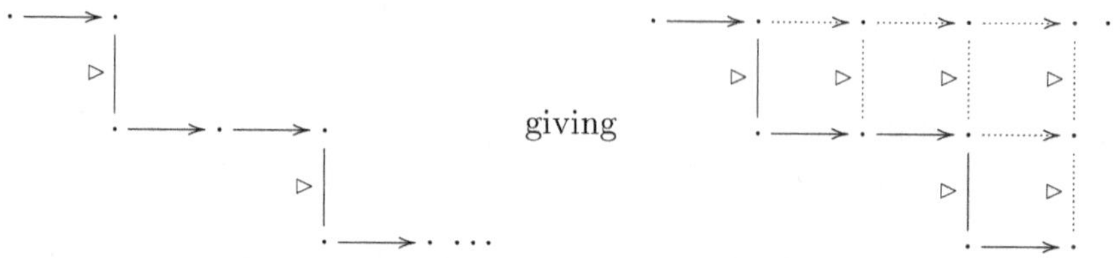

Since **S**-reduction is non-erasing, this infinite head reduction shows that **AAA** does not have a w-normal form. □

COROLLARY 7.29. $\mathrm{BT}(\mathbf{AAA}) = \bot \neq \mathrm{BT}(\mathbf{STT}(\mathbf{STT}))$.

EXAMPLES 7.30. Let us say that $P \in \mathfrak{S}_{\mathrm{CL}}$ has length n if it contains exactly n occurrences of the symbol **S**.

(i) All **S**-terms having length at most 7 are strongly normalizing. This was automatically verified by Waldmann using the tree automaton in Figure 7.1.

(ii) The shortest **S**-terms without a head normal form have length 8:

S(SSS)S(SSS)	**(SSS)(S(SSS))S**	**S(SS)S(SSS)S**
S(S(SSS))(SS)S	**(SSS)(SS)(SS)S**	**S(SSS)S(SS)S**
S(SS)SS(SS)S	**(SSS)(S(SS))SS**	**S(S(SS))(SS)SS**
S(SS)S(SS)SS	**S(S(SSS))SSS**	**S(S(SS)S)SSS**
(SSS)(SS)SSS	**S(S(SS))SSSS**	**S(SS)SSSSS**

Waldmann (2022) pointed out that the automaton in Figure 7.1 can be turned into a complete deterministic tree automaton with 35 states, accepting all head normalizing **S**-terms.

Berarducci trees

In Theorem 7.28 we have seen that **AAA** admits an infinite head reduction sequence:

$$\mathbf{AAA} = P_0 \to_h P_1 \to_h P_2 \to_h P_3 \to_h \cdots$$

The limit of this reduction is the infinite term $\lim_{n\to\infty} P_n$ in the usual metric on terms:

$$\begin{aligned} \mathrm{d}(P,Q) &= 0, && \text{if } P = Q, \\ \mathrm{d}(P,Q) &= 2^{-n}, && \text{if } P \neq Q \text{ and } n \text{ is the least depth at which } P \text{ and } Q \text{ differ.} \end{aligned}$$

We will see that any infinite reduction starting from a finite **S**-term has a limit in this sense since every finite prefix eventually becomes stable. Figure 7.3 displays the infinite limit of the head reduction of **AAA** adopting the following conventions.

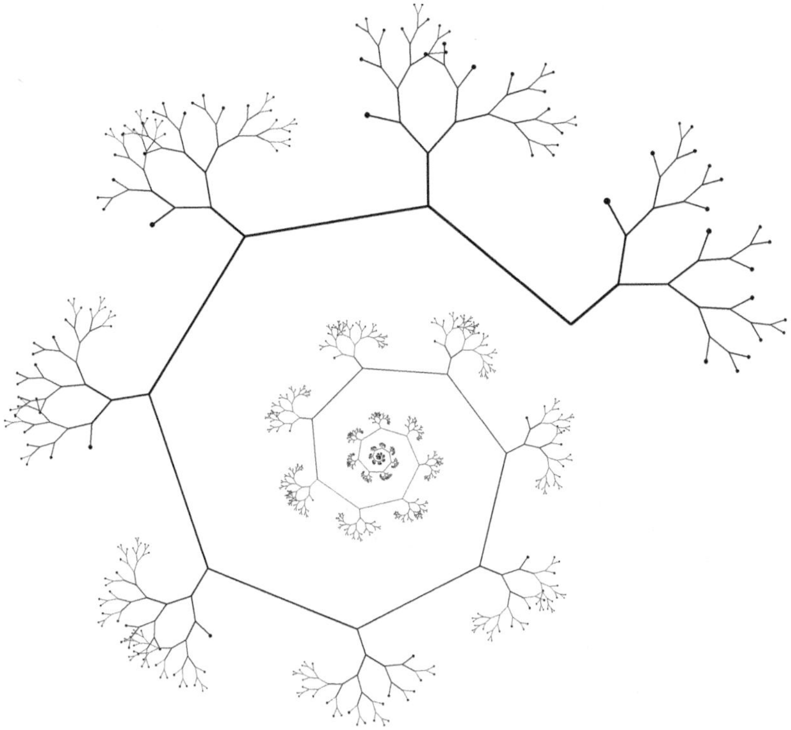

Figure 7.3: Limit of the infinite head reduction of **AAA**.

CONVENTION 7.31. *When drawing binary trees depicting infinite* **S**-*terms:*

- *the symbols for the binary application nodes (@) are suppressed and appear implicitly at every corner where line segments meet;*
- *the terminal nodes are combinators* **S**'*s;*
- *the infinite path represents the left spine of the term—this is the path starting at the root, in each step progressing to the first argument (which would be drawn on the left in the usual top-down rendering of a term tree).*

The infinite **S**-term in Figure 7.3 does not contain head redexes—but it is not an infinite normal form yet since it contains plenty of other redexes. In order to complete its evaluation, one needs to define the Berarducci tree which does not record the progressive stabilization of head normal forms, but of top normal forms. In fact, the limit of the infinite head reduction of **AAA** can be seen as a partially evaluated Berarducci tree. Recall that the notion of root redex for **S**-terms has been introduced in Definition 7.6(ii).

DEFINITION 7.32. Let $P, Q \in \mathfrak{S}_{\mathrm{CL}}$.

(i) $P \to_w Q$ is a *root step* if the redex contracted in P is a root redex.

(ii) An **S**-term P is in *top normal form* (*tnf*) if either it is a hnf or it has shape $P = XY$ and there are no $X', Y' \in \mathfrak{S}_{\mathrm{CL}}$ such that $X \twoheadrightarrow_w \mathbf{S}X'Y'$.

(iii) An **S**-term P *is top normalizable* if $P \twoheadrightarrow_w Q$ for some tnf $Q \in \mathfrak{S}_{\mathrm{CL}}$.

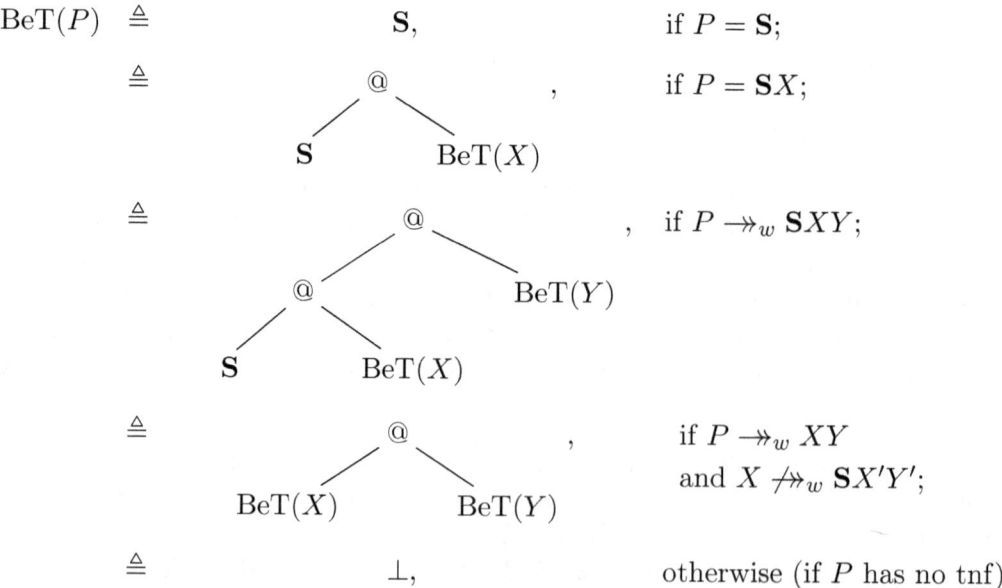

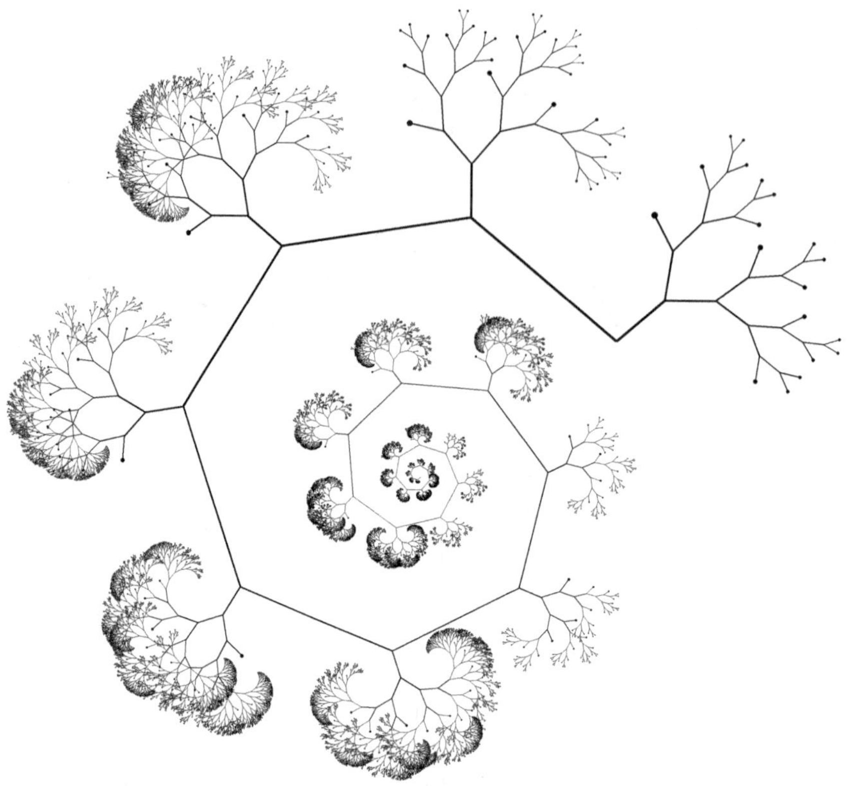

Figure 7.4: The Berarducci tree $\mathrm{BeT}(\mathbf{AAA})$.

This leads to the following definition of Berarducci trees (cf. Definition 2.49).

DEFINITION 7.33. Let $P \in \mathbb{S}_{\mathrm{CL}}$. The *Berarducci tree of* P, denoted by $\mathrm{BeT}(P)$, is coinductively defined by:

$$\mathrm{BeT}(P) \triangleq \mathbf{S}, \qquad \text{if } P = \mathbf{S};$$

$$\triangleq \begin{array}{ccc} & @ & \\ \mathbf{S} & & \mathrm{BeT}(X) \end{array}, \qquad \text{if } P = \mathbf{S}X;$$

$$\triangleq \begin{array}{ccccc} & & & @ & \\ & @ & & & \mathrm{BeT}(Y) \\ \mathbf{S} & & \mathrm{BeT}(X) & & \end{array}, \qquad \text{if } P \twoheadrightarrow_w \mathbf{S}XY;$$

$$\triangleq \begin{array}{ccc} & @ & \\ \mathrm{BeT}(X) & & \mathrm{BeT}(Y) \end{array}, \qquad \text{if } P \twoheadrightarrow_w XY \text{ and } X \not\twoheadrightarrow_w \mathbf{S}X'Y';$$

$$\triangleq \bot, \qquad \text{otherwise (if } P \text{ has no tnf).}$$

EXAMPLES 7.34. The Berarducci tree of **AAA** is depicted in Figure 7.4. The subtrees we encounter on the spine may be finite (0) or infinite (1). The sequence starts with

$$0011110011111000111\ldots$$

and is computable because head normalization (being 0 or 1) is decidable.

REMARK 7.35. (i) Contrary to head reduction—that can always be used to reach the hnf of a term (if it exists)—the top normal form of an **S**-term is not necessarily reachable by contracting root redexes exclusively. See, e.g., the reduction sequence on page 214.

(ii) An **S**-term is top normalizable if finitely many top reduction steps occur along its reduction sequences.

(iii) If P is top normalizable, then its tnf is reachable by performing head reductions.

In Definition 7.33 we included the last case for the sake of completeness, but the following result reveals that it is vacuous. We present its statement together with a proof-sketch, because the complete demonstration is not difficult, but lengthy.

THEOREM 7.36 (WALDMANN (1997)). *Every* $P \in \mathfrak{S}_{\mathrm{CL}}$ *has a tnf, whence* $\mathrm{BeT}(P) \neq \perp$.

PROOF SKETCH. We need to use the $\mathcal{QQQ}$ criterion that will be explained in Section 7.4. The idea is to show that in any head reduction of P having infinitely many root steps, there is a root redex of the form $\mathbf{S}\mathcal{QQQ}$. Thus P reaches $\mathcal{QQQ}$ in one additional head-step. By Proposition 7.40, we derive a contradiction. □

An immediate consequence of the above result is that every finite **S**-terms is infinitary strongly normalizing.

COROLLARY 7.37. $(\mathfrak{S}_{\mathrm{CL}}, \xrightarrow{\infty}_w)$ *satisfies* SN^∞.

Padovani also constructed the **S**-term XY below, whose Berarducci trees is infinite, but such that all the subtrees one encounters on its spine are finite.

LEMMA 7.38 (PADOVANI 2018). *Define:*

$$X \triangleq \mathbf{SAS}, \qquad Y \triangleq \mathbf{S}(\mathbf{S}X)(\mathbf{S}(\mathbf{S}X)).$$

Then, writing ν *for the greatest fixed point operator, we have:*

$$\mathrm{BeT}(XY) \in \nu\mathcal{X}.(\mathcal{X} \mapsto \mathcal{X}\mathcal{N})$$

where $\mathcal{N}$ *represents the set of* w*-normal forms and* $\mathcal{X}\mathcal{N} \triangleq \{PQ \mid P \in \mathcal{X} \;\&\; Q \in \mathcal{N}\}$.

PROOF. As a matter of notation, we write:

$$\begin{array}{lcll} Y_0 & \triangleq & Y, & \text{and, for all } i \in \mathbb{N}, \\ Y_{i+1} & \triangleq & \mathbf{S}Y_i(\mathbf{S}Z_i), & Z_i \triangleq \mathbf{S}(\mathbf{S}X)(\mathbf{S}Y_i). \end{array}$$

We give the set of rules allowing the computation of every finite prefix of $\mathrm{BeT}(XY_0)$. We surround by brackets some terms without hnf, to emphasize the fact that each rule leaves a w-normal term as rightmost subterm. A brace under a subterm indicates which rule is applicable.

Rule (0) For all $W \in \bigcup_{j\in\mathbb{N}}\{Y_j, Z_j\}$:

$$XW = \mathbf{SAS}W \to_w \mathbf{SSS}W(\mathbf{S}W) \to_w \mathbf{S}W(\mathbf{S}W)(\mathbf{S}W) \to_w [W(\mathbf{S}W)](\mathbf{S}W(\mathbf{S}W)).$$

Rule (1) For all $j\in\mathbb{N}$, $\underbrace{XY_j}_{(0)} \to_w^+ \underbrace{[Y_j(\mathbf{S}Y_j)]}_{(2)}(\mathbf{S}Y_j(\mathbf{S}Y_j))$.

Rule (2) We identify two subrules, depending on the index i of Y_i.

$$\begin{array}{lrcl} (2.0) & Y_0(\mathbf{S}Y_j) & \to_w & \mathbf{S}X(\mathbf{S}Y_j)(\mathbf{S}(\mathbf{S}X)(\mathbf{S}Y_j)) \\ & & = & \mathbf{S}X(\mathbf{S}Y_j)Z_j \to_w \underbrace{[XZ_j]}_{(3)}(\mathbf{S}Y_jZ_j) \end{array}$$

$$(2.1)\ \ Y_{i+1}(\mathbf{S}Y_j) = \mathbf{S}Y_i(\mathbf{S}Z_i)(\mathbf{S}Y_j) \to_w \underbrace{[Y_i(\mathbf{S}Y_j)]}_{(2)}(\mathbf{S}Z_i(\mathbf{S}Y_j)).$$

Rule (3) $\underbrace{XZ_j}_{(0)} \to_w^+ \underbrace{[Z_j(\mathbf{S}Z_j)]}_{(4)}(\mathbf{S}Z_j(\mathbf{S}Z_j))$.

Rule (4)
$$\begin{array}{rcl} Z_j(\mathbf{S}Z_j) & = & \mathbf{S}(\mathbf{S}X)(\mathbf{S}Y_j)(\mathbf{S}Z_j) \ \ \to_w \ \ \mathbf{S}X(\mathbf{S}Z_j)(\mathbf{S}Y_j(\mathbf{S}Z_j)) \\ & = & \mathbf{S}X(\mathbf{S}Z_j)Y_{j+1} \ \ \to_w \ \ \underbrace{[XY_{j+1}]}_{(1)}(\mathbf{S}Z_jY_{j+1}). \end{array}$$ □

7.4 Head-normalization is decidable

In this section we present a simple criterion due to Padovani for deciding whether an **S**-term is head normalizable. First, we introduce some auxiliary notations.

NOTATION. *Let $P \in \mathfrak{S}_{\mathrm{CL}}$ and $\mathcal{X}, \mathcal{Y} \subseteq \mathfrak{S}_{\mathrm{CL}}$.*

(i) *We write*

$$\begin{array}{rcl} \mathcal{X}^{\leftarrow} & \triangleq & \{\, P \mid \exists P' \in \mathcal{X}\,.\, P \twoheadrightarrow_w P' \,\}, \\ \mathcal{X}\mathcal{Y} & \triangleq & \{\, PQ \mid P \in \mathcal{X}\ \&\ Q \in \mathcal{Y} \,\}, \\ \mathcal{X} \to \mathcal{Y} & \triangleq & \{\, P \in \mathcal{X} \mid \exists Q \in \mathcal{Y}\ \&\ P \to_w Q \,\}. \end{array}$$

(ii) *Inside of an expression, we write $*$ for $\mathfrak{S}_{\mathrm{CL}}$. E.g., $\mathcal{X}* = \mathcal{X}\mathfrak{S}_{\mathrm{CL}}$ and $*\mathcal{Y} = \mathfrak{S}_{\mathrm{CL}}\mathcal{Y}$.*

(iii) *We confuse singletons and terms. E.g., for $\mathcal{X} = \{P\}$, we may write $P\mathcal{Y}$ for $\mathcal{X}\mathcal{Y}$.*

(iv) *We say that $\mathcal{X}$* reaches *$\mathcal{Y}$, written $\mathcal{X} \twoheadrightarrow_w \mathcal{Y}$, if for all $P \in \mathcal{X}$ there exists $Q \in \mathcal{Y}$ such that $P \twoheadrightarrow_w Q$. The notations $\mathcal{X} \to_w \mathcal{Y}$ and $\mathcal{X} \to_w^+ \mathcal{Y}$ are defined analogously.*

DEFINITION 7.39. (i) Define an infinite sequence $(\mathbf{S}_n)_{n>0}$ of **S**-terms as follows:

$$\mathbf{S}_1 \triangleq \mathbf{S}, \qquad\qquad \mathbf{S}_{n+1} \triangleq \mathbf{S}(\mathbf{S}_n), \quad \text{for } n > 1.$$

(ii) Define the following sets of **S**-terms:

$$\begin{array}{rcl} \mathcal{P} & ::= & \mathbf{S} \mid \mathbf{S}\mathcal{P} \\ \mathcal{Q} & ::= & *** \mid \mathbf{S}\mathcal{Q} \end{array}$$

Note that $\mathbf{S}_{n+1}$ is an alternative notation for $\mathbf{S}^n(\mathbf{S})$ and $\mathcal{P} = \{\mathbf{S}_n \mid n \geq 1\}$. Moreover, $\mathcal{Q} = \mathfrak{S}_{\mathrm{CL}} - \mathcal{P}$ is the set of **S**-terms having some subterm with left-spine of length > 1.

The $\mathcal{QQQ}$ criterion

The following result identifies a sufficient condition for **S**-terms to generate an infinite head reduction sequence.

PROPOSITION 7.40 (WALDMANN (2000)). *Every term $P \in \mathcal{QQQ}$ admits an infinite head reduction, i.e.* $\mathrm{BT}(P) = \bot$.

PROOF. It suffices to show that for every $P \in \mathcal{QQQ}$ there is $Q \in \mathcal{QQQ}$ such that $P \to_h Q$. Let $P \in \mathcal{QQQ}$, that is, $P = P_1 P_2 P_3$ for some $P_1, P_2, P_3 \in \mathcal{Q}$.

We perform a case distinction on the shape of P_1:

1. If $P_1 \in \mathbf{S}\mathcal{Q}$, then $P_1 P_2 P_3 \to_h \mathcal{Q}P_3(P_2 P_3) \subseteq \mathcal{QQQ}$.

2. If $P_1 \in \mathbf{S} * *$, then $P_1 P_2 P_3 \to_h * P_2(* P_2) P_3 \subseteq \mathcal{QQQ}$.

3. If $P_1 \in \mathbf{S} * * * \cdots$, then $P_1 \to_h P_1' \in \mathcal{Q}$ and $P_1' P_2 P_3 \in \mathcal{QQQ}$.

In each case, the term P has a head reduct in $\mathcal{QQQ}$. This concludes the proof. □

As a consequence, if $P \twoheadrightarrow_w C[\mathcal{QQQ}]$ for some head context $C[]$ then $\mathrm{BT}(P) = \bot$. We use this property to show non-termination of some of the terms mentioned earlier.

EXAMPLES 7.41. (i) Barendregt's term **AAA** is clearly of the form $\mathcal{QQQ}$.

(ii) Also Pettorossi's term **SAA(SAA)** is of this form:

$$\underbrace{\mathbf{SA}}_{\in \mathcal{Q}} \underbrace{\mathbf{A}}_{\in \mathcal{Q}} \underbrace{\mathbf{(SAA)}}_{\in \mathcal{Q}}$$

(iii) Zachos' term **STT(STT)** does not belong to $\mathcal{QQQ}$, and neither do its reducts. This can be seen by inspection of the complete reduction graph of this term as given in Waldmann (1998), but it is also expected since its head reduction is terminating.

(iv) Consider now $X_0 \triangleq \mathbf{S}_4 \mathbf{S}_3$. The terms in $X_0 X_0 \mathcal{P}$, first defined in Waldmann (1998), constitute examples of **S**-terms never reaching $\mathcal{QQQ}$, despite possessing an infinite head reduction. In fact, writing $X_{i+1} \triangleq \mathbf{S}_3 X_i$, we have as upward step

$$X_0 X_j \mathcal{P} = \mathbf{S(ST)(ST)} X_j \mathcal{P} \to_h \mathbf{ST} X_j (\mathbf{ST} X_j) \mathcal{P} = X_{j+1} X_{j+1} \mathcal{P},$$

and as downward step

$$X_{i+1} X_j \mathcal{P} = \mathbf{ST} X_i X_j \mathcal{P} \to_h \mathbf{T} X_j (X_i X_j) \mathcal{P} \to_h \mathbf{S}(X_i X_j) \mathcal{Q} \mathcal{P} \in \mathcal{QQP}$$
$$\to_h X_i X_j \mathcal{P} \mathcal{Q} \in \mathcal{QQP}*$$

Therefore $\mathrm{BeT}(X_0 X_0) \in \nu \mathcal{X}.(\mathcal{X} \to \mathbf{S}\mathcal{X}*)$, and $X_0 X_0 \mathcal{P}$ has an infinite head reduction whose head redexes have the form $\mathbf{S}\mathcal{QQP}$. Since the set

$$\mathbf{S}\mathcal{QQP}$$

is disjoint from $\mathcal{QQQ}$, it follows that $X_0 X_0 \mathcal{P}$ never reach $\mathcal{QQQ}$ along its head reduction.

EXERCISE 7.42. Show that $\mathbf{STSSSS} \notin \mathcal{QQQ}$ admits a rewrite sequence ending in a term of the shape $C[\mathcal{QQQ}]$, for an appropriate context $C[]$.

The $\mathcal{Q}_{\mathrm{III}}$ criterion

As discussed in Example 7.41(iv), Waldmann's criterion is a sufficient, but not necessary, condition for establishing that a term is not head normalizing. More recently, Padovani proved that the set $\mathcal{QQQ}$ is merely a portion of a more general picture.

DEFINITION 7.43. Let $\mathcal{Q}_{\mathrm{I}}, \mathcal{Q}_{\mathrm{II}}, \mathcal{Q}_{\mathrm{III}} \subseteq \mathbb{S}_{\mathrm{CL}}$ be the sets of $\mathbf{S}$-terms defined by the grammar:

$$\begin{array}{lcl} \mathcal{Q}_{\mathrm{I}} & ::= & \mathcal{Q} \\ \mathcal{Q}_{\mathrm{II}} & ::= & \mathcal{Q}_{\mathrm{I}}\mathcal{Q}_{\mathrm{I}} \mid \mathcal{Q}_{\mathrm{II}} * \\ \mathcal{Q}_{\mathrm{III}} & ::= & \mathcal{Q}_{\mathrm{II}}\mathcal{Q}_{\mathrm{I}} \mid \mathcal{Q}_{\mathrm{III}} * \end{array}$$

The following easy lemma implies that all elements of $\mathcal{Q}_{\mathrm{III}}$ have no hnf, and that all head normalizable $\mathbf{S}$-terms belonging to $\mathcal{Q}_{\mathrm{II}}$ reach $\mathbf{S}\mathcal{QQ}$.

LEMMA 7.44. *The sets $\mathcal{Q}_{\mathrm{I}}$, $\mathcal{Q}_{\mathrm{II}}$, $\mathcal{Q}_{\mathrm{III}}$ are closed under reduction.*

PROOF. By cases on the grammars defining these sets, using the property $*\,\mathcal{Q} \subsetneq \mathcal{Q}$. □

Note that $\mathcal{Q}_{\mathrm{III}}$ strictly contains $\mathcal{QQQ}$, e.g. $\mathcal{QQQ} \subsetneq \mathcal{QQPQ} \subsetneq \mathcal{Q}_{\mathrm{III}}$. Moreover, the fact that $\mathcal{Q}_{\mathrm{III}} = \mathcal{Q}_{\mathrm{III}} \to \mathcal{Q}_{\mathrm{III}}$ implies that all elements in $\mathcal{QQPQ}$ have no hnf as well.

LEMMA 7.45 (PADOVANI 2014). *There is no $P \in \mathbb{S}_{\mathrm{CL}}$ whose Berarducci tree belongs to the greatest fixed point*

$$\mathcal{F} \triangleq \nu\mathcal{X}.(\mathcal{X} \mapsto \mathcal{XP})$$

PROOF. The reader can check that $(*\,\mathcal{P})^{\leftarrow}$ is given by the union of the sets:

1. $*\,\mathcal{P}$;
2. $\mathbf{S} * \mathbf{S}\mathcal{P}$; $\qquad (*\,\mathcal{P} \supseteq *\,\mathcal{P}(\mathbf{S}\mathcal{P}) \leftarrow \mathbf{S} * \mathbf{S}\mathcal{P})$
3. $\mathbf{S}(\mathbf{S}\,*)\mathbf{SS}$. $\qquad (\mathbf{S} * \mathbf{S}\mathcal{P} \supseteq \mathbf{S} * \mathbf{SS}_2 \leftarrow \mathbf{S}(\mathbf{S}\,*)\mathbf{SS})$

So, $\mathrm{BeT}(P) \in \nu\mathcal{X}.(\mathcal{X} \mapsto \mathcal{XP})$ holds exactly when one of the following conditions does:

1. $P = XY$, for $Y \in \mathcal{P}$ and $\mathrm{BeT}(X) \in \mathcal{F}$,
2. $P = (\mathbf{S}X\mathbf{S})Y \to_h XY(\mathbf{S}Y)$ where $Y \in \mathcal{P}$ and
   - either $\mathrm{BeT}(XY) \in \mathcal{F}$,
   - or $XY = \mathbf{S}(\mathbf{S}X')\mathbf{SS} \to_h \mathbf{S}X'\mathbf{SS}_2 \to_h X'\mathbf{S}_2\mathbf{S}_3$ and $\mathrm{BeT}(X'\mathbf{S}_2) \in \mathcal{F}$,
3. $P = \mathbf{S}(\mathbf{S}X)\mathbf{SS} \to_h \mathbf{S}X\mathbf{SS}_2 \to_h X\mathbf{S}_2\mathbf{S}_3$ and $\mathrm{BeT}(X\mathbf{S}_2) \in \mathcal{F}$.

So, there is no $P \in \mathbb{S}_{\mathrm{CL}}$ having minimal size such that $\mathrm{BeT}(P) \in \mathcal{F}$, since every possible form for P yields a strictly smaller term whose Berarducci tree also belongs to $\mathcal{F}$. □

THEOREM 7.46 (PADOVANI 2014). *Every $\mathbf{S}$-term without hnf reaches $\mathcal{Q}_{\mathrm{III}}$.*

PROOF. By Theorem 7.36 every $\mathbf{S}$-term P has a tnf, hence, by Lemma 7.45, if P has no hnf then $\mathrm{BeT}(P)$ must contain an infinite number of elements of $\mathcal{Q}$ on the right branches of its left spine. *A fortiori*, P must reach $\mathcal{Q}_{\mathrm{III}}$. □

COROLLARY 7.47. *It is decidable whether an* **S**-*term* P *has a hnf.*

PROOF. By cases on the shape of P.

Cases $P = \mathbf{S}$ or $P = \mathbf{S}*$. Trivial.

Case $P \in ***$. By Lemma 7.26(ii) and Theorem 7.46 either $P \twoheadrightarrow_h \mathbf{S}**$ or $P \twoheadrightarrow_h \mathcal{Q}_{\mathrm{III}}$. We conclude since both properties are semi-decidable. □

7.5 The word problem for S

For all $P, Q \in \mathbb{S}_{\mathrm{CL}}$ the problem of determining if $P =_w Q$ holds is known as *the word problem for* **S**. The question whether the word problem for **S** is decidable has been repeatedly posed in the literature (see Barendregt (1975); Waldmann (1998); Waldmann (2000); Barendregt et al. (2018)) and is still open. Recently, Padovani has proved that showing $\mathrm{BeT}(P) = \mathrm{BeT}(Q)$ is not sufficient to ensure that P and Q are inter-convertible. Remark that checking whether the Berarducci tree of two terms P and Q are distinct and determining whether P and Q are inter-convertible, are both semi-decidable property. Thus, the decidability of the word problem for **S** would have followed from this property, which however does not hold. The proof relies on the existence of non-convertible terms that nevertheless are extensionally equivalent.

Consider terms A, B of shape $A \triangleq \mathbf{S}(\mathbf{S}U(\mathbf{S}VU))(\mathbf{S}(\mathbf{S}_2V)U)$ and $B \triangleq \mathbf{S}(\mathbf{S}_3U)(\mathbf{S}VU)$. Independently from the values of $U, V \in \mathbb{S}_{\mathrm{CL}}$ we have $A \neq_w B$, and for every x,

$$\begin{array}{rll} Ax = \mathbf{S}(\mathbf{S}U(\mathbf{S}VU))(\mathbf{S}(\mathbf{S}_2V)U)x & \twoheadrightarrow_w & Ux(Vx(Ux))(x(Ux)(Vx(Ux))) \\ & {}_w\!\twoheadleftarrow & \mathbf{S}(\mathbf{S}_3U)(\mathbf{S}VU)x \qquad = Bx \end{array}$$

By choosing the *ad hoc* values $U \triangleq V \triangleq \mathbf{S}_3$, we build two non inter-convertible terms AA and BB which leave no trace of their differences in the corresponding Berarducci trees.

LEMMA 7.48 (PADOVANI 2020). *Consider any set* $\mathcal{X}$ *of terms closed under* w-*reduction. Let* $\mathcal{Y}$ *be the set of all terms having no subterm of one of the following forms:*

1. $**\mathbf{S}$
2. $*(\mathbf{S}\mathcal{X})$.

Then, the set $\mathcal{Y}$ *is closed under* w-*reduction.*

PROOF. Let $P \in \mathcal{Y}$. We prove by induction on P that if $P \to_w P'$ then $P' \in \mathcal{Y}$. The only non-trivial case is $P = \mathbf{S}UVW \to_w P' = UW(VW)$, so we have $P' \notin **\mathbf{S}$. There are several subcases to consider.

- If $UW \in **\mathbf{S}$ or $VW \in **\mathbf{S}$ then $P = \mathbf{S}UV\mathbf{S} \in **\mathbf{S}$.
- If $\mathbf{S}UVW \in *(\mathbf{S}\mathcal{X})$ or $VW * (\mathbf{S}\mathcal{X})$ then $P = \mathbf{S}UVW \in *(\mathbf{S}\mathcal{X})$.
- Finally, if $P' = UW(VW) \in *(\mathbf{S}\mathcal{X})$ then $\mathbf{S}UV \in \mathbf{S}*\mathbf{S} \subsetneq **\mathbf{S}$.

In each of the three cases above, we get a contradiction. □

THEOREM 7.49 (PADOVANI 2020). *Let*

$$\begin{aligned} A &\triangleq \mathbf{S}(\mathbf{S}_4(\mathbf{S}_4\mathbf{S}_3))(\mathbf{S}(\mathbf{S}_2\mathbf{S}_3)\mathbf{S}_3), \\ B &\triangleq \mathbf{S}(\mathbf{S}_3\mathbf{S}_3)(\mathbf{S}_4\mathbf{S}_3). \end{aligned}$$

Then

(i) $AA \neq_w BB$, *but*
(ii) $\mathrm{BeT}(AA) = \mathrm{BeT}(BB)$.

PROOF. (i) Since A and B are w-nf's, Lemma 7.14 ensures that every reduct of AA contains the subterm A, and every reduct of BB contains the subterm B. To show $AA \neq_w BB$, it is sufficient to prove that no reduct of AA can contain the subterm B, and vice versa. This trivially holds for AA and BB themselves.

Note that AA and BB both belong to the set $\mathcal{Y}$ as defined in Lemma 7.48 for $\mathcal{X} =_w \mathcal{Q}$. Recall that $\mathcal{Q}$ is closed under w-reduction by Lemma 7.44, whence so is $\mathcal{Y}$ by Lemma 7.48. Suppose now that $AA \twoheadrightarrow_w P \rightarrow_w P'$ where B is not a subterm of P, whereas it is a subterm of P'. There is no term W such that $W \rightarrow_w B$, so the subterm B must have been created in P' by reducing a redex having one the following forms (for some $V \in \mathbb{S}_{\mathrm{CL}}$):

(1) $\mathbf{S}(\mathbf{S}(\mathbf{S}_3\mathbf{S}_3))V(\mathbf{S}_4\mathbf{S}_3) \rightarrow_w \mathbf{S}(\mathbf{S}_3\mathbf{S}_3)(\mathbf{S}_4\mathbf{S}_4)(V(\mathbf{S}_4\mathbf{S}_3)) = B(V(\mathbf{S}_4\mathbf{S}_3))$,

(2) $\mathbf{S}V(\mathbf{S}(\mathbf{S}_3\mathbf{S}_3))(\mathbf{S}_4\mathbf{S}_3) \rightarrow_w \mathbf{S}V(\mathbf{S}_4\mathbf{S}_3)(\mathbf{S}(\mathbf{S}_3\mathbf{S}_3)(\mathbf{S}_4\mathbf{S}_3)) = V(\mathbf{S}_4\mathbf{S}_3)B$.

In the first case, we have $\mathbf{S}(\mathbf{S}(\mathbf{S}_3\mathbf{S}_3)) \in *(\mathbf{S}\mathcal{Q})$. In the second case, $\mathbf{S}V(\mathbf{S}(\mathbf{S}_3\mathbf{S}_3)) \in *(\mathbf{S}\mathcal{Q})$. In both cases, a contradiction follows from Lemma 7.48.

Similarly, there exists no term W such that $W \rightarrow_w A$, so the subterm A must have been created by reducing a redex of one the following forms (for some $V \in \mathbb{S}_{\mathrm{CL}}$):

$$\begin{aligned} (1)\quad & \mathbf{S}(\mathbf{S}(\mathbf{S}_4(\mathbf{S}_4\mathbf{S}_3)))V(\mathbf{S}(\mathbf{S}_2\mathbf{S}_3)\mathbf{S}_3) &\rightarrow_w\; & \mathbf{S}(\mathbf{S}_4(\mathbf{S}_4\mathbf{S}_3))(\mathbf{S}(\mathbf{S}_2\mathbf{S}_3)\mathbf{S}_3)(V(\mathbf{S}(\mathbf{S}_2\mathbf{S}_3)\mathbf{S}_3)) \\ & &=\; & A(V(\mathbf{S}(\mathbf{S}_2\mathbf{S}_3)\mathbf{S}_3)), \\ (2)\quad & \mathbf{S}V(\mathbf{S}(\mathbf{S}_4(\mathbf{S}_4\mathbf{S}_3)))(\mathbf{S}(\mathbf{S}_2\mathbf{S}_3)\mathbf{S}_3) &\rightarrow_w\; & V(\mathbf{S}(\mathbf{S}_2\mathbf{S}_3)\mathbf{S}_3)(\mathbf{S}_4(\mathbf{S}_4\mathbf{S}_3))(\mathbf{S}(\mathbf{S}_2\mathbf{S}_3)\mathbf{S}_3) \\ & &=\; & V(\mathbf{S}(\mathbf{S}_2\mathbf{S}_3)\mathbf{S}_3)A. \end{aligned}$$

In the former case, we have $\mathbf{S}(\mathbf{S}A) \in *(\mathbf{S}\mathcal{Q})$. In the latter case, $\mathbf{S}V(\mathbf{S}A) \in *(\mathbf{S}\mathcal{Q})$. In either case, a contradiction follows from Lemma 7.48.

(ii) Given $X_0 \in \mathbb{S}_{\mathrm{CL}}$, let $(X_n)_{n \in \mathbb{N}}$ be the sequence defined by (for all $i \in \mathbb{N}$):

$$X_{i+1} \triangleq \mathbf{S}_3 X_i.$$

We consider the two sequences generated by taking $X_0 \triangleq A$ and $X_0 \triangleq B$, respectively. We present a set of rules for calculating every finite prefix of the Berarducci tree of X_0X_0. Each rule shows how to perform a finite number of reduction steps in a term of a given shape (the left-hand side of the rule) to produce another term (its right-hand side). All subterms surrounded by square brackets are element of $\mathcal{Q}_{\mathrm{III}} \subsetneq \perp\!\!\!\perp$, no matter what the values of i, j and k are. A brace under a subterm indicates which rule(s) are applicable.

(1) Rules for $\mathbf{S}_3(\mathcal{X}\mathcal{X})$ and $\mathcal{X}\mathcal{X}$, where $\mathcal{X} \triangleq \{X_i \mid i \in \mathbb{N}\}$.

$$(1.0)\quad [X_0X_j] =_w [\mathbf{S}_3X_j(\mathbf{S}_3X_j(\mathbf{S}_3X_j))[X_j(\mathbf{S}_3X_j)(\mathbf{S}_3X_j(\mathbf{S}_3X_j))]] = \underbrace{[X_{j+1}(X_{j+1}X_{j+1})[X_jX_{j+1}(X_{j+1}X_{j+1})]]}_{(4)((2)(1))}$$

$$(1.1)\quad X_{i+1}X_j = \mathbf{S}_3X_iX_j =_w \mathbf{S}_2X_j(X_iX_j) =_w \mathbf{S}\,\underbrace{(X_iX_j)}_{(1)}\,\underbrace{(X_j(X_iX_j))}_{(3)}$$

$$(1.2)\quad \mathbf{S}_3(X_iX_j) = \mathbf{S}_3\,\underbrace{(X_iX_j)}_{(1)}$$

(2)(r) Rules for $\mathcal{X}\mathcal{X}W$, where W is the left-hand side of any rule marked as (r).

$$(2.0)(r)\quad [[X_0X_j]W] = [\underbrace{[X_0X_j]}_{(1.0)}\,\underbrace{W}_{(r)}]$$

$$(2.1)(r)\quad [\underbrace{X_{i+1}X_j}_{(1.1)}W] =_w [\mathbf{S}(X_iX_j)(X_j(X_iX_j))W] =_w \underbrace{[X_iX_jW]}_{(2)(r)}\,\underbrace{[X_j(X_iX_j)W]}_{(4)(r)}$$

(3) Rules for $\mathcal{X}(\mathcal{X}\mathcal{X})$

$$(3.0)\quad [X_0W],\ \text{where } W = X_jX_k,\quad =_w \underbrace{[\mathbf{S}_3W(\mathbf{S}_3W(\mathbf{S}_3W))[[W(\mathbf{S}_3W)](\mathbf{S}_3W(\mathbf{S}_3W))]]}_{(7)}$$

$$(3.1)\quad X_{i+1}(X_jX_k) = \mathbf{S}_3X_i(X_jX_k) =_w \mathbf{S}_2(X_jX_k)(X_i(X_jX_k)) =_w \mathbf{S}\,\underbrace{(X_i(X_jX_k))}_{(3)}\,\underbrace{[X_jX_k(X_i(X_jX_k))]}_{(2)(3)}$$

(4)(r) Rules for $\mathcal{X}(\mathcal{X}\mathcal{X})W$, where W is the left-hand side of a rule marked as (r).

$$(4.0)(r)\quad [\underbrace{[X_0(X_jX_k)]}_{(3.0)}\,\underbrace{W}_{(r)}]$$

$$(4.1)(r)\quad [\underbrace{X_{i+1}(X_jX_k)}_{(3.1)}W] =_w [\mathbf{S}(X_i(X_jX_k))[X_jX_k(X_i(X_jX_k))]W] =_w \underbrace{[X_i(X_jX_k)W]}_{(4)(r)}[\underbrace{[X_jX_k(X_i(X_jX_k))]}_{(2)(3)}\,\underbrace{W}_{(r)}]$$

The last three rules are required to finish the head reduction of the right-hand side of the Rule (3.0).

(5) $\mathbf{S}_3W(\mathbf{S}_3W)$, where $W = X_iX_j$,

$$=_w \mathbf{S}_2(\mathbf{S}_3W)[W(\mathbf{S}_3W)]$$

$$= \mathbf{S}_2(\mathbf{S}_3W)U, \quad \text{where } U = \underbrace{[W(\mathbf{S}_3W)]}_{(2)(1.2)},$$

$$=_w \mathbf{S}U(\mathbf{S}_3WU)$$

$$=_w \mathbf{S}U(\mathbf{S}_2U[WU])$$

$$=_w \mathbf{S}U(\mathbf{S}_2UV), \quad \text{where } V = \underbrace{[WU]}_{(2)((2)(1.2))},$$

$$=_w \mathbf{S}U(\mathbf{S}V[UV])$$

(6) $\mathbf{S}_3W(\mathbf{S}_3W(\mathbf{S}_3W))$, where $W = X_iX_j$,

$$=_w \mathbf{S}_2(\mathbf{S}_3W(\mathbf{S}_3W))[W(\mathbf{S}_3W(\mathbf{S}_3W))]$$

$$= \mathbf{S}_2(\mathbf{S}_3W(\mathbf{S}_3W))T, \quad \text{where } T = \underbrace{[W(\mathbf{S}_3W(\mathbf{S}_3W))]}_{(2)(5)},$$

$$=_w \mathbf{S}T[\underbrace{\mathbf{S}_3W(\mathbf{S}_3W)}_{(5)}T]$$

$$=_w \mathbf{S}T[\mathbf{S}U(\mathbf{S}V[UV])T], \quad \text{where } U = \underbrace{[W(\mathbf{S}_3W)]}_{(2)(1.2)} \text{ and } V = \underbrace{[WU]}_{(2)((2)(1.2))},$$

$$=_w \mathbf{S}T[UT[\mathbf{S}V[UV]T]]$$

$$=_w \mathbf{S}T[UT[VT[UVT]]]$$

(7) $[\mathbf{S}_3W(\mathbf{S}_3W(\mathbf{S}_3W))[[W(\mathbf{S}_3W)](\mathbf{S}_3W(\mathbf{S}_3W))]]$, where $W = X_iX_j$,

$$= [\underbrace{\mathbf{S}_3W(\mathbf{S}_3W(\mathbf{S}_3W))}_{(6)}R], \quad \text{where } R = [\underbrace{[W(\mathbf{S}_3W)]}_{(2)(1.2)}\underbrace{(\mathbf{S}_3W(\mathbf{S}_3W))}_{(5)}],$$

$$=_w [\mathbf{S}T[UT[VT[UVT]]]R], \quad \text{where } T = \underbrace{[W(\mathbf{S}_3W(\mathbf{S}_3W))]}_{(2)(5)}, U = \underbrace{[W(\mathbf{S}_3W)]}_{(2)(1.2)},$$

$$\text{and } V = \underbrace{[WU]}_{(2)((2)(1.2))},$$

$$=_w [TR[UT[VT[UVT]]R]]$$

It is easy to check that this set of rules is deterministic, complete and 'top-terminating' in the following sense: in every right-hand side, each subterm having minimal depth and whose top reduction is not terminated is the left-hand side of one, and only one, rule; every left-hand side becomes root stable after the application of at most two rules. As a consequence, these rules allow the computation of all the prefixes of the Berarducci trees of their left-hand sides, in particular X_0X_0 (Rule (1.0)).

It is also easy to check that the resulting trees will not depend on the choice of X_0. Indeed, only two rules (Rule (1.0), Rule (3.0)) have a left-hand side of the form X_0W, but the following equivalence does not depend on the choice of X_0:

$$X_0W = \mathbf{S}_3W(\mathbf{S}_3W(\mathbf{S}_3W))(W(\mathbf{S}_3W)(\mathbf{S}_3W(\mathbf{S}_3W)))$$

Moreover, no right-hand side contains a top-terminated subterm of minimal depth of the form WX_i. So, a trivial induction on $n \in \mathbb{N}$ shows that the shapes of the Berarducci trees of all left-hand sides do not depend on the choice of X_0, at least up to depth n. □

The order of S-terms

Another noteworthy aspect of **S**-terms is that they can consume only a limited amount of resources. In λ-calculus the quantity of arguments a term is capable of consuming depends on the number of outer abstractions it eventually produces, i.e. it depends on its order (Definition 2.37). The order of a combinatory logic term can be defined as follows.

DEFINITION 7.50. Let $P \in \mathrm{CL}$. The *order of* P is $n \in \mathbb{N}$, in notation $\mathrm{ord}(P) = n$, if

1. $Px_0x_1 \cdots x_n \twoheadrightarrow_w Q$ implies that $Q = Q'x_n$ where x_n does not occur in Q', and
2. n is minimal with this property.

Otherwise, we say that the order of P is infinite, written $\mathrm{ord}(P) = \infty$.

We therefore need to consider **S**-terms possibly containing free variables.

DEFINITION 7.51. (i) Let Var be a countable set of variables $x, y, z \ldots$.
(ii) Let $\mathbb{S}$ be the set of all terms built from **S** and variables in Var:

$$\mathbb{S} \triangleq \mathrm{S} \mid \mathrm{Var} \mid \mathbb{S}\mathbb{S} \qquad (\mathbb{S})$$

(iii) A *substitution* σ is any finite map Var to $\mathbb{S}$. Given $P \in \mathbb{S}$, we write P^σ for the **S**-term obtained from P by simultaneously substituting every $x \in \mathrm{dom}(\sigma)$ by $\sigma(x)$.

EXAMPLES 7.52. (i) $\mathrm{ord}(\mathbf{AAA}) = 0$. By Theorem 7.28, this term has an infinite head reduction, while it is necessary to have a hnf to consume one or more arguments.
(ii) $\mathrm{ord}(\mathbf{SSS}) = 1$.
(iii) $\mathrm{ord}(\mathbf{SS}) = 2$.
(iv) $\mathrm{ord}(\mathbf{S}) = 3$.

We now prove that the order of an **S**-term is either 0, 1, 2, or 3. This property was shown by Visser during his PhD dissertation, but his proof got lost. Another proof was found by Vincent van Oostrom, but the argument we present here is due to Padovani.

LEMMA 7.53. *For all $P \in \mathbb{S}$, if $P \twoheadrightarrow_w Q$ then $P^\sigma \twoheadrightarrow_w Q^\sigma$ for all substitutions σ.*

PROOF. By an immediate induction on the length k of any reduction $P \twoheadrightarrow_w Q$.
Case $k = 0$. Immediate.
Case $k \geq 1$. It follows from an immediate induction on the size of P. □

LEMMA 7.54. *If P is a closed* **S**-*term and* $\mathrm{ord}_\beta(P) = n$, *for* $n > 2$, *then*

$$Px_0 \cdots x_{n-2} \twoheadrightarrow_w \mathbf{S}\mathbb{S}\mathbb{S}$$

PROOF. The set $\mathbb{S}\mathbb{S}\mathbb{S}$ is obviously closed under reduction and, since $n > 2$, the term $Px_0 \cdots x_{n-2}$ cannot be equal to **S**, cannot belong to $\mathbf{S}\mathbb{S}$ nor it can have an infinite head reduction. So $Px_0 \cdots x_{n-2}$ must reach $\mathbf{S}\mathbb{S}\mathbb{S}$. Then $Px_0 \cdots x_{n-2}x_{n-1} \twoheadrightarrow_w Q$ implies $Q = Q'x_{n-1}$ where $Px_0 \cdots x_{n-2} \twoheadrightarrow_w Q'$. In other words, the order of P is at most $n-1$. Contradiction. □

PROPOSITION 7.55. *(Visser 1975, van Oostrom[3] 2002) If* $P \in \mathbb{S}_{\mathrm{CL}}$ *then* $\mathrm{ord}(P) \leq 3$.

PROOF. Assume, towards a contradiction, that P is a closed **S**-term of order $n > 3$. By Lemma 7.54 we have $Px_0 \cdots x_{n-2} \twoheadrightarrow_w \mathbf{S}\mathbb{S}\mathbb{S}$. Now, let σ be any substitution mapping $x_0, \ldots x_{n-2}$ to arbitrary elements of $\mathcal{Q}$. By Lemma 7.53, we have

$$(Px_0 \cdots x_{n-1})^\sigma \twoheadrightarrow_w (\mathbf{S}\mathbb{S}\mathbb{S})^\sigma = \mathbf{S} * *$$

However $(Px_0 \cdots x_{n-1})^\sigma \in \mathcal{Q}_{\mathrm{III}}$. By Lemma 7.44 the set $\mathcal{Q}_{\mathrm{III}}$ is closed under reduction, hence no term belonging to this set can have a hnf. Contradiction. □

7.6 Non-normalizing patterns

The following terms are well-known examples of non-normalizing terms:

1. $\mathbf{S}_3\mathbf{S}_2(\mathbf{S}_3\mathbf{S}_2)$, aka Zachos' term (using the notation from Definition 7.39(i));
2. $\mathbf{A}_2\mathbf{A}_2\mathbf{A}_2$, where $\mathbf{A}_k \triangleq \mathbf{S}_k\mathbf{S}$, for all $k > 0$;
3. $\mathbf{S}\mathbf{A}_2\mathbf{A}_2(\mathbf{S}\mathbf{A}_2\mathbf{A}_2)$.

Now, it is relatively easy to check to produce hand-crafted proofs showing that these terms are not normalizing. (The term (1) is only head-normalizing; for (2)–(3) simply apply the $\mathcal{Q}_{\mathrm{III}}$-criterion). A more interesting question raised by Padovani is the following:

> is it possible to find a common property shared by these terms explaining why their reduction is maintained forever?

All we can say for now is that these three terms share some kind of common 'main thread' which makes at least parts of their infinite reductions very similar in their principle.

The closest analogue of Ω — the $\mathcal{I}$ pattern

We have seen in Section 4.1 that λ-calculus admits cycles of any length, in particular it admits pure cycles: $\Omega = \Delta\Delta \to_\beta \Delta\Delta$. Similarly, in combinatory logic we have $\mathbf{SII}(\mathbf{SII}) \to_w^+ \mathbf{SII}(\mathbf{SII})$. On the contrary—due to the acyclicity of $\mathbb{S}_{\mathrm{CL}}$ (Proposition 7.12(i))—it is impossible to define an **S**-term X satisfying $XX \to_w^+ XX$. The closest analogue of Ω that we can construct is an **S**-term X satisfying $XX \hookrightarrow_w^+ XX$.

[3] Vincent van Oostrom argument is presented in Barendregt et al. (2018).

DEFINITION 7.56. (i) For every $W \in \mathfrak{S}_{CL}$, we denote by $\langle W \rangle$ the set defined as follows:

$$\langle W \rangle \quad ::= \quad W \mid \mathbf{S}\langle W \rangle * \mid \mathbf{S} * \langle W \rangle \qquad (\langle W \rangle)$$

(ii) We let $\mathcal{I}$ be the set of terms defined by:

$$\mathcal{I} \quad ::= \quad \mathbf{S} * \mid \mathbf{S}_2 * \mid \mathbf{S}_3 \mathcal{I} \mid \mathbf{S}_4 \mathbf{S} \mid \mathbf{S}_5 \mathbf{S} \qquad (\mathcal{I})$$

The following proposition gives a characterization of those $\mathbf{S}$-terms reaching $\mathcal{I}$.

PROPOSITION 7.57 (PADOVANI 2014). $\mathcal{I}^{\leftarrow} = \{P \in \mathfrak{S}_{CL} \mid \forall W \in \mathfrak{S}_{CL} \,.\, PW \twoheadrightarrow_w \langle W \rangle\}$.

PROOF. ($\subseteq$) Easy, since $\mathcal{I}\, W \twoheadrightarrow_w \langle W \rangle$, for all $W \in \mathfrak{S}_{CL}$.

($\supseteq$) Conversely, consider an $\mathbf{S}$-term P satisfying $PW \twoheadrightarrow_w W' \in \langle W \rangle$, for all $W \in \mathfrak{S}_{CL}$. In particular, this holds for $W \in \perp\!\!\!\perp$. By Proposition 7.16, $W' \neq W$, thus W' is in hnf. We prove that $P \in \mathcal{I}^{\leftarrow}$ by induction on the sum of the lengths of all reductions $PW \twoheadrightarrow_w W'$.

Note that either P belongs to $\mathbf{S}*$ or P reaches $\mathbf{S} * *$. In the latter case, because of $\mathbf{S}\mathcal{Q} * \mathcal{Q} \to \mathcal{Q}_{\mathrm{III}} \subsetneq \perp\!\!\!\perp$, the term P must reach $\mathbf{S}\mathcal{P}*$. In other words P reaches $\mathcal{P}*$, and Lemma 7.26(ii) ensures that the head reduction of P also reaches this set. Moreover W' is a hnf, hence the same lemma ensures that the head reduction of PW reaches a term $P'W$ where $P' \in \mathcal{P}*$ is yielded by a head reduction of P and $P'W \twoheadrightarrow_w W'$.

If $P \to_w^+ P'$ then the conclusion follows immediately from the induction hypothesis. Otherwise P is of the form $\mathbf{S}_k U$. We shall examine all possible values of k in $PW = \mathbf{S}_k UW$, starting from $k = 1$, and conclude in each possible case that P belongs to $\mathcal{I}$:

Case $PW = \mathbf{S}UW$

Case $PW = \mathbf{S}_2 UW \to_w^+ \mathbf{S}W(UW)$

Case $PW = \mathbf{S}_3 UW \to_w^+ \mathbf{S}(UW)(W(UW))$
The term $W(UW) \in \perp\!\!\!\perp$, so we must have $UW \twoheadrightarrow_w \langle W \rangle$. By the IH, we get $U \in \mathcal{I}^{\leftarrow}$.

Case $PW = \mathbf{S}_4 UW \to_w^+ \mathbf{S}(W(UW))(UW(W(UW)))$
We must have $(UW(W(UW))) \twoheadrightarrow_w \langle W \rangle$. Because of the $\mathcal{Q}_{\mathrm{III}}$-criterion, the term U must belong to $\mathcal{P}$. It is easy to check that $(UW(W(UW))) \to_w^+ \mathbf{S}\perp\!\!\!\perp\perp\!\!\!\perp$, for all values of $U \in \mathcal{P}$ except $\mathbf{S}$, for which $(UW(W(UW))) \in \langle W \rangle$.

Case $PW = \mathbf{S}_5 UW \to_w^+ \mathbf{S}(UW(W(UW)))(W(UW)(UW(W(UW))))$.
Again, the only possible value for U is $\mathbf{S}$.

Case $PW = \mathbf{S}_k UW \to_w^+ \mathbf{S}\perp\!\!\!\perp\perp\!\!\!\perp$ for all $k > 5$, hence this case is impossible. □

Note that the three non-terminating $\mathbf{S}$-terms mentioned at the beginning of the section reach $\mathbf{S}\mathcal{I}\mathcal{I}(\mathbf{S}\mathcal{I}\mathcal{I})$. Indeed, easy calculations give:

1. $\mathbf{S}_3\mathbf{S}_2(\mathbf{S}_3\mathbf{S}_2) \;\in\; \mathbf{S}(\mathbf{S}*)(\mathbf{S}*)(\mathbf{S}(\mathbf{S}*)(\mathbf{S}*)) \;\subsetneq\; (\mathbf{S}\mathcal{I}\mathcal{I})(\mathbf{S}\mathcal{I}\mathcal{I})$.
2. $\mathbf{A}_2\mathbf{A}_2\mathbf{A}_2 \;\to_w\; \mathbf{S}\mathbf{A}_2(\mathbf{S}\mathbf{A}_2)\mathbf{A}_2 \to_w \mathbf{A}_2\mathbf{A}_2(\mathbf{S}\mathbf{A}_2\mathbf{A}_2) \;\to_w\; \mathbf{S}\mathbf{A}_2(\mathbf{S}\mathbf{A}_2)(\mathbf{S}\mathbf{A}_2\mathbf{A}_2)$
$\in\; \mathbf{S}(\mathbf{S}_2*)(\mathbf{S}*)(\mathbf{S}(\mathbf{S}_2*)(\mathbf{S}_2*)) \;\subsetneq\; (\mathbf{S}\mathcal{I}\mathcal{I})(\mathbf{S}\mathcal{I}\mathcal{I})$.
3. $\mathbf{S}\mathbf{A}_2\mathbf{A}_2(\mathbf{S}\mathbf{A}_2\mathbf{A}_2) \;\in\; \mathbf{S}(\mathbf{S}_2*)(\mathbf{S}_2*)(\mathbf{S}(\mathbf{S}_2*)(\mathbf{S}_2*)) \;\subsetneq\; (\mathbf{S}\mathcal{I}\mathcal{I})(\mathbf{S}\mathcal{I}\mathcal{I})$.

Therefore, we have a uniform way of describing the infinite reductions of these terms. More generally, the following (class of) **S**-term(s) can be seen as an analogue of the looping combinator Ω

$$\langle \mathbf{S}\mathcal{I}\mathcal{I}\rangle\langle \mathbf{S}\mathcal{I}\mathcal{I}\rangle$$

in the sense that it satisfies the following property.

PROPOSITION 7.58 (PADOVANI 2014). $\langle \mathbf{S}\mathcal{I}\mathcal{I}\rangle\langle \mathbf{S}\mathcal{I}\mathcal{I}\rangle \hookrightarrow_w^+ \langle \mathbf{S}\mathcal{I}\mathcal{I}\rangle\langle \mathbf{S}\mathcal{I}\mathcal{I}\rangle$.

PROOF. Indeed $\langle \mathbf{S}\mathcal{I}\mathcal{I}\rangle\langle \mathbf{S}\mathcal{I}\mathcal{I}\rangle \hookrightarrow\!\!\!\twoheadrightarrow_w \mathbf{S}\mathcal{I}\mathcal{I}\langle \mathbf{S}\mathcal{I}\mathcal{I}\rangle \rightarrow_w \mathcal{I}\langle \mathbf{S}\mathcal{I}\mathcal{I}\rangle(\mathcal{I}\langle \mathbf{S}\mathcal{I}\mathcal{I}\rangle)$
$\twoheadrightarrow_w \langle\langle \mathbf{S}\mathcal{I}\mathcal{I}\rangle\rangle\,\langle\langle \mathbf{S}\mathcal{I}\mathcal{I}\rangle\rangle = \langle \mathbf{S}\mathcal{I}\mathcal{I}\rangle\langle \mathbf{S}\mathcal{I}\mathcal{I}\rangle$. □

As an immediate consequence, we obtain that every term belonging to $\langle \mathbf{S}\mathcal{I}\mathcal{I}\rangle\langle \mathbf{S}\mathcal{I}\mathcal{I}\rangle$ has no w-nf. We can be slightly more precise in the description of the infinite reduction of our three exemplary terms. Consider an arbitrary **S**-term having shape:

$$\mathbf{S}AB(\mathbf{S}CD)$$

where $A, B, C, D \in \mathcal{I}$. Let $Y_0 \triangleq B(\mathbf{S}CD)$ and $Y_{i+1} \triangleq DY_i$, for every $i \in \mathbb{N}$. Then:

- $\mathbf{S}AB(\mathbf{S}CD) \rightarrow_w A(\mathbf{S}CD)Y_0 \hookrightarrow_w^+ (\mathbf{S}CD)Y_0 \hookrightarrow\!\!\!\twoheadrightarrow_w (CY_0)Y_1 \hookrightarrow\!\!\!\twoheadrightarrow_w Y_0Y_1$.
- $Y_{i+1}Y_j = DY_iY_j \hookrightarrow\!\!\!\twoheadrightarrow_w Y_iY_j$.
- $Y_0Y_j = B(\mathbf{S}CD)Y_j \hookrightarrow\!\!\!\twoheadrightarrow_w \mathbf{S}CDY_j \hookrightarrow\!\!\!\twoheadrightarrow_w CY_jY_{j+1} \hookrightarrow\!\!\!\twoheadrightarrow_w Y_jY_{j+1}$.

The $\mathcal{J}$ pattern

Another pattern generating infinite reduction sequences is presented below.

DEFINITION 7.59. We let $\mathcal{C}, \mathcal{D}, \mathcal{J}$ be the sets of terms defined by:

$$\mathcal{C} ::= \mathbf{S}\mathcal{I}\mathbf{S} \qquad (\mathcal{C})$$
$$\mathcal{D} ::= \mathbf{S}(\mathbf{S}\mathcal{C})\,\mathcal{J} \qquad (\mathcal{D})$$
$$\mathcal{J} ::= \mathbf{S}_3 * \mid \mathbf{S}_4 * \mid \mathbf{S}_5\mathbf{S} \mid \mathbf{S}_5\mathbf{S}_2 \mid \mathbf{S}_5\mathbf{S}_3 \mid \mathbf{S}_2\,\mathcal{J} \qquad (\mathcal{J})$$

Note that for all $C \in \mathcal{C}$ and for all $W \in \mathfrak{S}_{\mathrm{CL}}$, we have $CW \twoheadrightarrow_w \langle W\rangle(\mathbf{S}W)$.

PROPOSITION 7.60 (PADOVANI 2014). $\mathcal{J}^{\leftarrow} = \{P \in \mathfrak{S}_{\mathrm{CL}} \mid \forall W \in \mathfrak{S}_{\mathrm{CL}} \,.\, P(\mathbf{S}W) \twoheadrightarrow_w \langle W\rangle\}$.

PROOF. Analogous to the proof of Proposition 7.57. □

PROPOSITION 7.61 (PADOVANI 2014). $\langle\mathcal{D}\rangle(\mathbf{S}\langle\mathcal{D}\rangle) \rightarrow^+ \langle\mathcal{D}\rangle(\mathbf{S}\langle\mathcal{D}\rangle) *$

PROOF. We have

$$\begin{aligned}
\langle\mathcal{D}\rangle\,(\mathbf{S}\,\langle\mathcal{D}\rangle) &\hookrightarrow\!\!\!\twoheadrightarrow_w \mathcal{D}\,(\mathbf{S}\,\langle\mathcal{D}\rangle) = \mathbf{S}\,(\mathbf{S}\mathcal{C})\,\mathcal{J}\,(\mathbf{S}\langle\mathcal{D}\rangle) \rightarrow_w \mathbf{S}\mathcal{C} * (\mathcal{J}\,(\mathbf{S}\langle\mathcal{D}\rangle))\\
&\rightarrow_w \mathbf{S}\mathcal{C} * \langle\langle\mathcal{D}\rangle\rangle \rightarrow_w \mathcal{C}\,\langle\langle\mathcal{D}\rangle\rangle * \rightarrow_w^+ \langle\langle\mathcal{D}\rangle\rangle\,(\mathbf{S}\langle\langle\mathcal{D}\rangle\rangle) *\\
&\subsetneq \langle\mathcal{D}\rangle\,(\mathbf{S}\langle\mathcal{D}\rangle) *
\end{aligned}$$

□

An example of a term whose infinite reduction matches this pattern is:

$$(\mathbf{S}(\mathbf{S}\mathbf{A}_3)\mathbf{A}_3)(\mathbf{S}(\mathbf{S}(\mathbf{S}\mathbf{A}_3)\mathbf{A}_3))$$

with $\mathbf{A}_3 = \mathbf{S}_3\mathbf{S} \in \mathcal{C} \cap \mathcal{D}$.

7.7 Translating S-terms into λ-calculus

A substantial amount of work has been dedicated to the comparison between λ-calculus and combinatory logic as term rewriting systems. The main properties that captured the interest of researchers are normalization and convertibility, see B[1984]. We perform a similar investigation on the restriction of λ-calculus generated by the λ-term S.

DEFINITION 7.62. (i) In $(\Lambda, \cdot)$ we define $\mathbb{S}_\lambda \triangleq \{\mathsf{S}\}^\bullet$, where $\mathsf{S} = \lambda xyz.xz(yz)$ and the operator $(\cdot)^\bullet$ has been introduced in Definition 7.3(i).

(ii) We call $(\mathbb{S}_\lambda, \rightarrow_\beta)$ the S-*fragment of λ-calculus*.

Clearly, rewrite sequences of **S**-terms in CL can be interpreted directly in λ-calculus. However, reduction in λ-calculus is more fine-grained, thus causing a number of differences. For example, consider the following **S**-terms from Waldmann (2000):

$$\mathbf{S}(\mathbf{STS})\mathbf{A} \qquad \text{and} \qquad \mathbf{S}(\mathbf{TA})(\mathbf{SAS})$$

In CL, both terms are w-nfs, whence not inter-convertible because syntactically different. However, in λ-calculus these terms turn out to have the same β-normal form:

$$\lambda x.xG(xG(\lambda y.xy(Gy)))$$

where $G = xF(\lambda y.xy\lambda z.xz(yz))$ and $F = \lambda wy.xy(wy)$ both belong to $\Lambda^o(x)$. Therefore, we can conclude that convertibility in CL and λ-calculus differ for **S**-terms.

One may wonder whether a normalizing **S**-term also possesses a β-nf after being translated into λ-calculus. We will see in Examples 7.63 that this is not always the case, since λ-calculus may provide considerably more reduction power. The first phenomenon one observes when β-normalizing the λ-terms populating $\mathbb{S}_\lambda$ is that the growth factor of a term can be much higher than the one of the corresponding **S**-term. As an attempt to contain the size of a β-normal form $M^{\beta\text{-nf}}$ of some $M \in \mathbb{S}_\lambda$, let us count only the number of its λ-abstractions and of its variable occurrences. (Notice that this information is sufficient to reconstruct the actual size of the whole term.) Nevertheless, we obtain the following examples.

EXAMPLES 7.63. (i) The **S**-term **SS**(**SSS**)(**S**(**S**(**S**(**SS**))**SSS**)), having a relatively short w-nf, when translated into λ-calculus, has a β-nf whose length is the astronomic number

$$114866151345992335476101988845880839899949 7$$

i.e., it is in the order of 10^{42} symbols.

(ii) More extremely, the term **S**(**S**(**STT**))**S**(**S**(**STT**)) is in w-nf, but as a λ-term it is not even β-normalizing.

(iii) Other examples of exploding **S**-terms are given in Table 7.3.

$P \in$ CL	size($P^{w\text{-nf}}$)	size($(P_\lambda)^{\beta\text{-nf}}$)	source
(SS)(SS)(SS)SS	169	485	BBKV
S(SS)S(SS)S	79	963955	van der Poel
SS(SSS)(S(S(S(SS))SSS))	251	$\sim 10^{42}$	Waldmann
S(S(S(SS)(SS)))(S(S(SS)(SS)))	25	∞	Waldmann

Table 7.3: Some exploding **S**-terms.

Efficiently computing the size of a β-normalized λ-term

In order to compute the third column of Table 7.3, Waldmann has devised a method for computing efficiently the size of normal forms of λ-terms, provided that they exist. The idea is to avoid any explicit manipulation of the term structure (e.g., named variables or de Bruijn indices), and use instead higher-order abstract syntax in `Haskell`.

Recall that a λ-term in β-normal form is either an abstraction, or a series of applications where the head subterm is a variable and the other subterms are β-normal forms. The latter case is intended to include the case of a variable standing alone.

Therefore, we may use the following semantic domain

```
data Val = Fun (Val → Val) | Val Natural
```

Here, Fun f represents the value of an abstraction, and Val s represents the size of a stacked application with a head variable (such terms are 'stable' because they do not use their argument). To find the size of the normal form, we use the function

```
measure :: Val → Natural
measure (Val s) = s
measure (Fun f) = succ (measure (f (Val 1)))
```

The first case of the definition is self-explanatory. In the second case, where we need to normalize 'under the lambda', we feed the function with an argument Val 1 that represents a variable (whose size is 1). The resulting size is increased by one (using the successor **succ** :: Natural → Natural) to account for the lambda. For evaluating the application of a function f to an argument a, as in (fa), there are two cases:

```
apply :: Val → Val → Val
apply (Fun f) a = f a
apply (Val s) a = Val (s + measure a)
```

The combinator S is expressed as

```
s :: Val
s = Fun (\x → Fun (\y → Fun (\z → apply(apply x z)(apply y z))))
```

The size of the normal form of SS is obtained via 'measure (apply s s)' which indeed gives the correct value 8 (counting 3 lambda's and 5 variable occurrences). It a simple exercise to construct the terms in Table 7.3 and compute the size of their β-nf. Interestingly, the computation of size($\mathsf{SS(SSS)(S(S(S(SS))SSS))}^{\beta\text{-nf}}$) gives the output almost immediately. Since the last term of Table 7.3 has no β-nf, the corresponding computation diverges.

Efficiently computing the size of a w-normalized S-term

We have seen in Proposition 7.21 that weak reduction tends to make the size of **S**-terms grow exponentially. However, the amount of information added by reduction grows only linearly. This fact has been exploited by Padovani, who proposed a compact representation for elements of $\mathfrak{S}_{\mathrm{CL}}$, as well as a simple algorithm to perform reductions efficiently and calculate the size of an **S**-term in linear time.

DEFINITION 7.64. Let $P \in \mathfrak{S}_{\mathrm{CL}}$. Given an array T and $i \in \mathbb{N}$, we say that P *is encoded in a position* i *in* T, written $|T[i]| = P$, if either :

- P is equal to **S** and $T[i]$ contains a singular value which is not a pair,
- $P = QV$ and $T[i] = (j, k)$, for some $j, k \in \mathbb{N}$ such that $|T[j]| = Q$, $|T[k]| = V$.

Now, suppose $|T[i]| = P$ and that T is sufficiently large to have at least two unused indices. Suppose also that, for appropriate indices $i, j_1, k_1, l_1 \in \mathbb{N}$, we have:

- $T[i] = (j_1, j_2)$ and $|T[j_2]| = W$;
- $T[j_1] = (k_1, k_2)$ and $|T[k_2]| = V$;
- $T[k_1] = (l_1, l_2)$ and $|T[l_2]| = Q$;
- $|T[l_1]| = \mathbf{S}$.

In other words, we have $P = \mathbf{S}QVW$. In order to encode its reduct $P' = QW(VW)$ at position i, we simply need to perform the following operations:

1. store the pair (l_2, j_2) at some unused position n_1 in T,
2. store the pair (k_2, j_2) at some unused position n_2 in T,
3. replace $T[i]$ with (n_1, n_2).

Note that, even if W is duplicated in P', the set of indices used in T has only increased by 2. In fact, the term W is still encoded at index j_2 in T, but only at this index.

This technique allows the computation of reducts of **S**-terms in a very efficient way. It also permits to retrieve the size of a term in linear time (in $O(N)$, where N is the number of distinct indices used to encode a term and its subterms, which can be far less than its actual size, see below). Indeed, suppose that $|T[i]| = P$ and that we want to compute the size of P. Consider a new array S having the same size as T, initialized with zeros. Compute $S[i]$ as follows:

1. if $S[i] > 0$ then return,
2. otherwise, if $|T[i]| = \mathbf{S}$ then write $S[i] := 1$ and return,
3. finally, if $T[i] = (j_1, j_2)$ then

- compute (recursively) $S[j_1]$,
- compute (recursively) $S[j_2]$,
- write $S[i] := 1 + S[j_1] + S[j_2]$, then return.

Clearly this computation terminates with $S[i]$ equal to the size of P. Note that each $S[j]$ is written at most once—in the two recursive calls, the computation of $S[j_k]$ will immediately return if $S[j_k] > 0$, that is, if the size of the term encoded at position j_k has already been determined.

This method allows the computation of size of terms which would be hardly feasible by other means such as storing the full tree structure of terms in memory. For instance, the size of **AAA** after 1000 steps of head reduction (and also parallel reductions, induced by subterms sharing) can be computed almost instantaneously with a small `Ocaml` program (using an array T of length 2017, that is, 17 to encode the term, plus $2 \cdot 1000$ to compute its reduction, with an auxiliary array S of arbitrarily large integers). It is exactly equal to:

$$1374376519827782371006672158950891112994571 7$$

Part III

Conversion

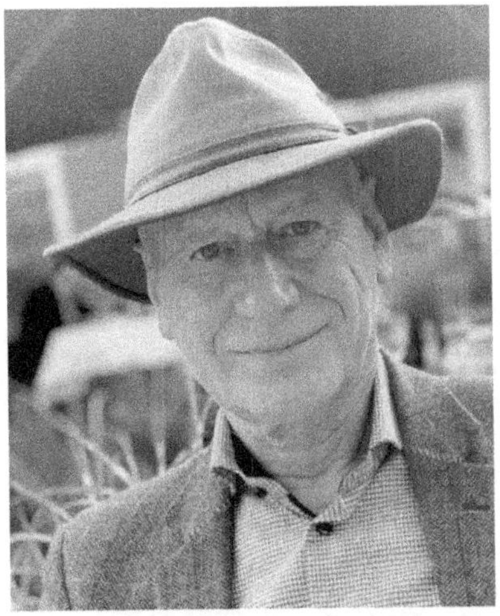

Roel de Vrijer

Richard Statman

Jörg Endrullis

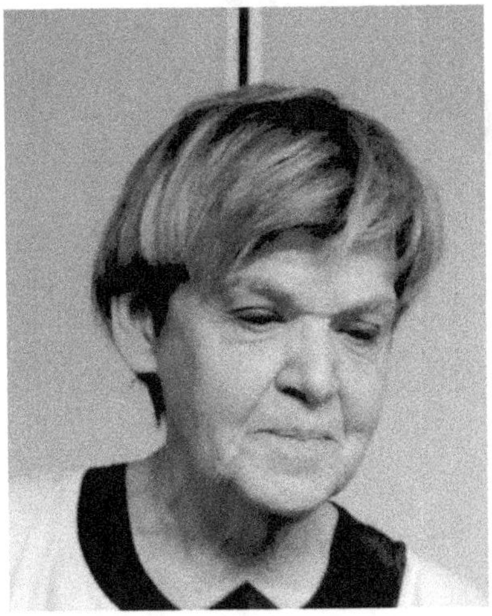

Ines Maria Margaria

Enno Folkerts

Mariangiola Dezani-Ciancaglini

Chapter 8

The Perpendicular Lines Property

In the Cartesian plane $\mathbb{R}^2$ the lines $\{(x, 2) \mid x \in \mathbb{R}\}$ and $\{(3, y) \mid y \in \mathbb{R}\}$ are perpendicular (and intersect in the point $(3, 2)$). In $\mathbb{R}^3$ the lines $\{(x, 2, 3) \mid x \in \mathbb{R}\}$, $\{(1, y, 4) \mid y \in \mathbb{R}\}$, and $\{(0, 1, z) \mid z \in \mathbb{R}\}$ are perpendicular (these lines have no point in common). Translated to λ-terms one says that $\{(X, \mathsf{I}) \mid X \in \Lambda\}$ and $\{(\mathsf{K}, Y) \mid Y \in \Lambda\}$ are 'perpendicular', and similarly in higher dimensions. The Perpendicular Lines Property (PLP) states that if a λ-definable function F of k arguments is constant on k perpendicular lines, then F is constant everywhere. The validity of this result depends on which notion of equality is considered and on whether only closed terms are considered or also open terms are allowed. The result—first formulated in B[1984], Theorem 14.4.12, for Böhm-tree equality $=_{\mathcal{B}}$ and open terms—follows from Berry's Sequentiality Theorem (1978). It was distilled from Coppo et al. (1978), who used it implicitly to derive a characterization of separability of finite sets of λ-terms. In this chapter it will be shown that the PLP holds for the λ-algebras $\mathcal{M}(\mathcal{B}), \mathcal{M}^o(\mathcal{B}), \mathcal{M}(\boldsymbol{\lambda})$, but not for $\mathcal{M}^o(\boldsymbol{\lambda})$.

DEFINITION 8.1. Let $\mathcal{M}$ be a λ-algebra and let $n \in \mathbb{N}$.

(i) An element $F \in \mathcal{M}$ is *constant on n perpendicular lines* if there exist $N_1, \ldots, N_n$ and $M_{ij} \in \mathcal{M}$, $1 \le i \ne j \le n$, such that

$$\forall Z \in \mathcal{M} \quad \left\{ \begin{array}{cccccccc} F & Z & M_{12} & \cdots & M_{1(n-1)} & M_{1n} & = & N_1 \\ F & M_{21} & Z & \cdots & M_{2(n-1)} & M_{2n} & = & N_2 \\ \vdots & \vdots & & \ddots & & \vdots & & \vdots \\ F & M_{(n-1)1} & M_{(n-1)2} & \cdots & Z & M_{(n-1)n} & = & N_{n-1} \\ F & M_{n1} & M_{n2} & \cdots & M_{n(n-1)} & Z & = & N_n \end{array} \right. \tag{8.1}$$

(ii) $\mathcal{M}$ satisfies the *perpendicular lines property* ($\mathcal{M} \models$ PLP, for short) if, for all $n \in \mathbb{N}$, every λ-definable $F \in \mathcal{M}$ that is constant on n perpendicular lines, is in fact constant everywhere. That is, assuming the system of equations (8.1), one has

$$\forall Z_1, \ldots, Z_n \in \mathcal{M} \,.\, F Z_1 \cdots Z_n = N_1 = \cdots = N_n.$$

For term models, the PLP is often formulated using n-ary contexts, namely contexts $C[\,]$ possibly containing multiple occurrences of n distinct holes $[\,]_i$ $(1 \le i \le n)$. See e.g. Chapter 14 of B[1984]; Endrullis and de Vrijer (2008); Barbarossa and Manzonetto (2020). The following theorem shows that the two formulations are equivalent.

THEOREM 8.2. *For $\mathcal{T} \in \lambda\mathcal{T}$, the following are equivalent:*

1. $\mathcal{M}(\mathcal{T}) \models \mathrm{PLP}$
2. *Let $C[-_1, \ldots, -_n]$ be an n-ary context and $N_1, \ldots, N_n, M_{ij} \in \Lambda$ for $1 \leq i \neq j \leq n$. Then*

$$\forall Z \in \Lambda \quad \begin{cases} C[Z, & M_{12}, & \ldots & M_{1(n-1)}, & M_{1n}] & =_{\mathcal{T}} & N_1 \\ C[M_{21}, & Z, & \ldots & M_{2(n-1)}, & M_{2n}] & =_{\mathcal{T}} & N_2 \\ \vdots & & \ddots & & \vdots & & \vdots \\ C[M_{n1}, & M_{n2}, & \ldots & M_{n(n-1)}, & Z\ ] & =_{\mathcal{T}} & N_n \end{cases} \tag{8.2}$$

entails

$$\forall Z_1, \ldots, Z_n \in \Lambda \,.\, C[Z_1, \ldots, Z_n] =_{\mathcal{T}} N_1 =_{\mathcal{T}} \cdots =_{\mathcal{T}} N_n.$$

PROOF. ($1 \Rightarrow 2$) Take $C[\,]$, M_{ij}, $N_1, \ldots, N_n$ satisfying (8.2). By the Variable Convention, we may assume that all the bound variables occurring in these terms are pairwise distinct. For every i ($1 \leq i \leq n$), let $\vec{y}_i$ be the list of all the abstracted variables encountered along the paths from the root of $C[\,]$ to the occurrences of its i-th hole $[\,]_i$.

For fresh variables $x_1, \ldots, x_n$, define $F \in \Lambda$ by:

$$F \triangleq \lambda x_1 \ldots x_n . C[x_1\vec{y}_1, \ldots, x_n\vec{y}_n]$$

Remark that, for arbitrary $Z_1, \ldots, Z_n \in \Lambda$, we may have $FZ_1 \cdots Z_n \neq_\beta C[Z_1\vec{y}_1, \ldots, Z_n\vec{y}_n]$. For example, taking $n = 1$, $C[\,] \triangleq \lambda y.[\,]$ and $Z \triangleq y$ one obtains $FZ =_\beta \lambda y'.yy' \neq_\beta \Delta = C[Zy]$. However, the equality $FZ_1 \cdots Z_n =_\beta C[Z_1\vec{y}_1, \ldots, Z_n\vec{y}_n]$ does hold whenever $\vec{y}_i \notin \mathrm{FV}(Z_i)$. Thus, setting $M'_{ij} \triangleq \lambda\vec{y}_j.M_{ij}$ ($1 \leq i \neq j \leq n$), we have for every $i \in \{1, \ldots, n\}$ and $Z' \in \Lambda$:

$$\begin{array}{lll} & FM'_{i1} \cdots M'_{i(i-1)} Z' M'_{i(i+1)} \cdots M'_{in} & \\ =_\beta & C[M'_{i1}\vec{y}_1, \ldots, M'_{i(i-1)}\vec{y}_{i-1}, x_i\vec{y}_i, M'_{i(i+1)}\vec{y}_{i+1}, \ldots, M'_{in}\vec{y}_n]^\sigma, & \text{for } \sigma \triangleq [x_i := Z'], \\ =_\beta & C[M_{i1}, \ldots, M_{i(i-1)}, x_i\vec{y}_i, M_{i(i+1)}, \ldots, M_{in}]^\sigma, & \text{by def. of } M'_{ij}, \\ =_{\mathcal{T}} & N_i^\sigma, & \text{taking } Z \triangleq x_i\vec{y}_i \text{ in} \\ & & \text{the } i\text{-th row of (8.2)}, \\ =_{\mathcal{T}} & N_i, & \text{since } x_i \text{ is fresh.} \end{array}$$

Therefore F is constant on n perpendicular lines thus, by (1), it is constant everywhere. In particular, given $Z_1, \ldots, Z_k \in \Lambda$, we have

$$\begin{array}{rcl} C[Z_1, \ldots, Z_k] & =_\beta & C[(\lambda\vec{y}_1.Z_1)\vec{y}_1, \ldots, (\lambda\vec{y}_n.Z_n)\vec{y}_n] \\ & =_\beta & F(\lambda\vec{y}_1.Z_1) \cdots (\lambda\vec{y}_n.Z_n) \\ & =_{\mathcal{T}} & N_1 \ =_{\mathcal{T}} \ \cdots \ =_{\mathcal{T}} \ N_n \end{array}$$

($2 \Rightarrow 1$) Given F satisfying (8.1), simply take $C[\,] \triangleq F[\,]_1 \cdots [\,]_n$. □

In the rest of the chapter we will use the formulation given in Definition 8.1(ii).

8.1 Validity of PLP in $\mathcal{M}(\boldsymbol{\lambda})$

We show that $\mathcal{M}(\boldsymbol{\lambda}) \models$ PLP following a proof by Endrullis and de Vrijer (2008), using reduction under substitution developed in Section 1.3.

PROPOSITION 8.3 (ENDRULLIS AND DE VRIJER (2008)). $\mathcal{M}(\boldsymbol{\lambda}) \models$ PLP.

PROOF. For notational convenience we prove PLP for dimension $n = 2$, but we will be careful to provide a proof that does generalize to higher dimensions.

Suppose that, for all $Z \in \Lambda$, one has

$$\begin{cases} F \quad Z \quad M_{12} =_\beta N_1 \\ F \quad M_{21} \quad Z =_\beta N_2 \end{cases}$$

in order to show that for all $Z_1, Z_2 \in \Lambda$ one has $FZ_1Z_2 =_\beta N_1 =_\beta N_2$. (In this case the fact that $N_1 =_\beta N_2$ is trivial: $N_1 =_\beta FM_{21}M_{12} =_\beta N_2$, but we will not use this shortcut, that is not valid for $n \geq 3$.) Choose a fresh variable $w \notin \mathrm{FV}(N_1N_2)$. By definition and the Church-Rosser theorem, we have

$$Fxy \rightsquigarrow_{xy,wM_{12}} FwM_{12} \twoheadrightarrow_\beta N_1' \;{}_\beta\!\twoheadleftarrow N_1,$$
$$Fxy \rightsquigarrow_{xy,M_{21}w} FM_{21}w \twoheadrightarrow_\beta N_2' \;{}_\beta\!\twoheadleftarrow N_2.$$

By Corollary 1.78(i) there are $M_1', M_2' \in \Lambda$ such that

$$Fxy \twoheadrightarrow_\beta M_1' \rightsquigarrow_{xy,wM_{12}} N_1',$$
$$Fxy \twoheadrightarrow_\beta M_2' \rightsquigarrow_{xy,M_{21}w} N_2'.$$

By confluence there exists an $M'' \in \Lambda$ such that $M_i' \twoheadrightarrow_\beta M''$, for each $i = 1, 2$. Now, by applying Corollary 1.78(ii) we obtain a situation as follows.

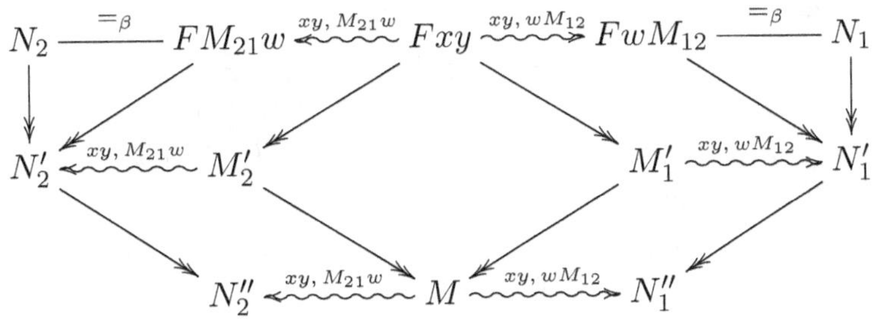

Claim x, y are not in $\mathrm{FV}(M)$. Suppose towards a contradiction that one of them did. Then the leftmost occurrence of x or y, substituted by w, would survive in one of the N_i'', by Lemma 1.71, contradicting that $\mathrm{FV}(N_i'') \subseteq \mathrm{FV}(N_1N_2) \not\ni w$, as w is chosen fresh. By the claim, for all $Z_1, Z_2 \in \Lambda$, one concludes

$$FZ_1Z_2 =_\beta M =_\beta N_1 =_\beta N_2. \qquad \square$$

8.2 $\mathcal{M}(\mathcal{B}), \mathcal{M}^o(\mathcal{B}) \models$ PLP

Using Berry's Sequentiality Theorem, B[1984], Theorem 14.4.8, it can be proved that $\mathcal{M}(\mathcal{B}) \models$ PLP, ibidem Theorem 14.4.12. Ingemarie Bethke observed that Berry's result remains valid if in the definition of the notion 'is caused by' the implication[1]

$$M_i'|\beta = z \quad \Rightarrow \quad C[\vec{M}']|\alpha \neq \bot$$

is replaced by

$$M_i'|\beta \neq \bot \quad \Rightarrow \quad C[\vec{M}']|\alpha \neq \bot$$

implying $\mathcal{M}^o(\mathcal{B}) \models$ PLP. More recently, Barbarossa and Manzonetto (2020) proposed an inductive proof exploiting the Taylor expansion technique introduced by Ehrhard and Regnier (2006). We will not follow these ideas, but present a new proof based on coinduction showing simultaneously that $\mathcal{M}^{(o)}(\mathcal{B}) \models$ PLP.

THEOREM 8.4. (i) $\mathcal{M}^o(\mathcal{B}) \models$ PLP.
(ii) $\mathcal{M}(\mathcal{B}) \models$ PLP.

PROOF. (i) Assume, wlog, $n = 3$. We split into claims.

Claim 8.4.1. Let $P \in \Lambda$, $\vec{z} = z_1, z_2, z_3$ and write $P[X_1, X_2, X_3]$ for $P[z_i := X_i]_{i \in \{1,2,3\}}$. Assume that P satisfies (for $M_{ij}, N_i \in \Lambda^o$):

$$\forall Z \in \Lambda^o \begin{cases} P[Z, & M_{12}, & M_{13}] & =_{\mathcal{B}} & N_1 \\ P[M_{21}, & Z, & M_{23}] & =_{\mathcal{B}} & N_2 \\ P[M_{31}, & M_{32}, & Z\ ] & =_{\mathcal{B}} & N_3 \end{cases} \tag{8.3}$$

Then $z_i \notin \mathrm{BT}(P)$, $1 \leq i \leq 3$.

Subproof. We prove the statement by coinduction on $\mathrm{BT}(P)$.

If P is unsolvable, then $\mathrm{BT}(P) = \bot$ and we are done.

Otherwise $P \twoheadrightarrow_h \lambda x_1 \ldots x_m.yP_1 \cdots P_k$ for some $m, k \geq 0$, $y \in \mathrm{Var}$ and $P_1, \ldots, P_k \in \Lambda$. There are two subcases.

- $y = z_i$ for some $i \leq 3$, say, $i = 2$. By instantiating Z in the second equation of (8.3) with the combinators $Z_1 \triangleq \mathsf{K}^k(\mathsf{K})$ and $Z_2 \triangleq \mathsf{K}^k(\mathsf{F})$, respectively, we would get $P[M_{21}, Z_1, M_{23}] \twoheadrightarrow_h \lambda\vec{x}.Z_1(P_1[M_{21}, Z_1, M_{23}]) \cdots (P_k[M_{21}, Z_1, M_{23}]) \twoheadrightarrow_\beta \lambda\vec{x}.\mathsf{K} =_{\mathcal{B}} N_2$ and $P[M_{21}, Z_2, M_{23}] \twoheadrightarrow_\beta \lambda\vec{x}.\mathsf{F} =_{\mathcal{B}} N_2$, which is impossible since $\lambda\vec{x}.\mathsf{K} \neq_{\mathcal{B}} \lambda\vec{x}.\mathsf{F}$.
- $y \in \mathrm{Var} - \{z_1, z_2, z_3\}$. For all $i \leq 3$, by using the i-th equation of (8.3), we derive that $N_i \twoheadrightarrow_h \lambda\vec{x}.yN_{i1} \cdots N_{ik}$ such that, for all $j = 1, \ldots, k$, we have:

$$\forall Z \in \Lambda^o \begin{cases} P_j\,[Z, & M_{12}, & M_{13}] & =_{\mathcal{B}} & N_{1j} \\ P_j\,[M_{21}, & Z, & M_{23}] & =_{\mathcal{B}} & N_{2j} \\ P_j\,[M_{31}, & M_{32}, & Z\ ] & =_{\mathcal{B}} & N_{3j} \end{cases} \tag{8.4}$$

By co-IH, we have $\vec{z} \notin \mathrm{BT}(P_j)$ for all such j, from which it follows $\vec{z} \notin \mathrm{BT}(P)$. ∎

[1] See B[1984], Definition 14.4.2(ii)(2).

Claim 8.4.2. Let $F \in \Lambda$ satisfying (8.1) for $n = 3$. Then $F \in \mathrm{SOL}$ implies $\mathrm{ord}_\beta(F) \geq 3$.

Subproof. Assume by contradiction that $F \twoheadrightarrow_h \lambda z_1 \ldots z_m.yP_1 \cdots P_k$ for $m < 3$, say $m = 2$. If $y = z_i$ we get a contradiction from the i-th equation of (8.1), as above. Otherwise, the third equation gives $FM_{31}M_{32}Z =_\beta y(P_1[M_{31}, M_{32}, z_3]) \cdots (P_k[M_{31}, M_{32}, z_3])Z =_{\mathcal{B}} N_3$ for all $Z \in \Lambda^o$, which is also impossible. ■

If $F \in \mathrm{UNS}$ then $FZ_1Z_2Z_3 =_{\mathcal{B}} \Omega$ for all Z_1, Z_2, Z_3 and the result follows. Otherwise, $F \in \mathrm{SOL}$ and, by Claim 8.4.2, we get $F \twoheadrightarrow_\beta \lambda z_1 z_2 z_3.P$, for some $P \in \Lambda$. As F satisfies (8.1), P satisfies the hypotheses of Claim 8.4.1. Therefore

$$\forall Z_1, Z_2, Z_3 \in \Lambda^o \,.\, F\vec{Z} =_{\mathcal{B}} N_1 =_{\mathcal{B}} N_2 =_{\mathcal{B}} N_3.$$

(ii) Same proof. Indeed, in the claims above we do not assume $F \in \Lambda^o$ and, by the variable convention, we are allowed to take $\vec{z}$ not occurring in M_{ij}, N_i. □

8.3 Invalidity of PLP in $\mathcal{M}^o(\boldsymbol{\lambda})$

The fact that $\mathcal{M}^o(\boldsymbol{\lambda}) \not\models$ PLP follows from a special case of the main result in Statman and Barendregt (1999), Theorem 1.5. We will not present this theorem, but do formulate and prove the following consequence.

THEOREM 8.5 (CLUSTER LEMMA). *Consider two disjoint r.e. sets $\mathcal{A}_0, \mathcal{A}_1 \subseteq \Lambda^o$ closed under $=_\beta$. Then there exist $F, M_0, M_1 \in \Lambda^o$ such that $M_0 \neq_\beta M_1$ and*

$$\forall A \in \mathcal{A}_i \,.\, FA =_\beta M_i, \; i \in \{0, 1\}.$$

To get motivated to read the involved proof, let us show first that the non-validity of PLP for the closed term model is an easy consequence.

COROLLARY 8.6. $\mathcal{M}^o(\boldsymbol{\lambda}) \not\models$ PLP.

PROOF. It is enough to show that PLP fails for some n. We show that it fails for $n = 2$. Define the disjoint r.e. sets

$$\begin{aligned} \mathcal{A}_0 &\triangleq \{\langle \mathsf{I}, Z\rangle \mid Z \in \Lambda^o\}^\beta \cup \{\langle Z, \mathsf{I}\rangle \mid Z \in \Lambda^o\}^\beta; \\ \mathcal{A}_1 &\triangleq \{\langle \mathsf{S}, \mathsf{S}\rangle\}^\beta; \end{aligned}$$

where $\mathcal{X}^\beta$ stands for the closure of $\mathcal{X}$ under $=_\beta$. By the Cluster Lemma there exist $F, M_0, M_1 \in \Lambda^o$ such that for all $A_i \in \mathcal{A}_i$, $i \in \{0, 1\}$, we have in $\mathcal{M}^o(\boldsymbol{\lambda})$:

$$FA_0 = M_0 \neq M_1 = FA_1.$$

Hence one has in $\mathcal{M}^o(\boldsymbol{\lambda})$, for all $Z \in \Lambda^o$,

$$F\langle \mathsf{I}, Z\rangle = F\langle Z, \mathsf{I}\rangle \neq F\langle \mathsf{S}, \mathsf{S}\rangle.$$

Therefore F is constant on the perpendicular lines of which $\mathcal{A}_0$ is the union, but not globally constant. This refutes the PLP for the λ-term $F' \triangleq \lambda xy.F\langle x, y\rangle$. □

We need to define some preliminary notions.

DEFINITION 8.7. Let $\mathcal{A} \subseteq \Lambda^o$.

(i) $\mathcal{A}^\beta \triangleq \{M \in \Lambda^o \mid \exists N \in \mathcal{A} \,.\, M =_\beta N\}$.

(ii) $\mathcal{A}$ is *β-closed* whenever $\mathcal{A} = \mathcal{A}^\beta$.

DEFINITION 8.8. Let $\mathcal{A}_0, \mathcal{A}_1 \subseteq \Lambda^o$. Let $\mathcal{P} = \{\mathcal{A}_0, \mathcal{A}_1\}$.

(i) The family $\mathcal{P}$ is an *Ershov-partition* if (for $i \in \{0,1\}$):

$\#\mathcal{A}_i$ is r.e.;

$\mathcal{A}_i$ is non-empty and β-closed;

$\mathcal{A}_0, \mathcal{A}_1$ are non-overlapping, i.e. $\mathcal{A}_0 \cap \mathcal{A}_1 = \emptyset$.

(ii) For $M, N \in \Lambda^o$ define

$$M \sim_{\mathcal{P}} N \iff M, N \in \mathcal{A}_0 \ \vee\ M, N \in \mathcal{A}_1 \ \vee\ M =_\beta N.$$

(iii) Let $H \in \Lambda^o$ and $\mathcal{P}$ be an Ershov-partition. Then H is a *cluster function* for $\mathcal{P}$ if (for all $M, N \in \Lambda^o$):

$$HM =_\beta HN \iff M \sim_{\mathcal{P}} N. \tag{8.5}$$

The statement of the Cluster Lemma 8.5 is trivial if (at least) one of the $\mathcal{A}_i$'s is empty. E.g., if $\mathcal{A}_0 = \emptyset$ then one may simply take $F \triangleq \mathsf{KI}$, $M_0 \triangleq \Omega$ and $M_1 \triangleq \mathsf{I}$. The interesting case where both $\mathcal{A}_0, \mathcal{A}_1$ are not empty is an immediate consequence of the next theorem.

THEOREM 8.9 (CLUSTER THEOREM). *Let $\mathcal{A}_0, \mathcal{A}_1 \subseteq \Lambda^o$ such that $\mathcal{P} = \{\mathcal{A}_0, \mathcal{A}_1\}$ is an Ershov-partition. Then there exists an $H \in \Lambda^o$ clustering $\mathcal{P}$. This implies*

$$\begin{array}{rcl} \forall M, N \in \mathcal{A}_i \,.\, HM & =_\beta & HN. \\ \forall M_0 \in \mathcal{A}_0, M_1 \in \mathcal{A}_1 \,.\, HM_0 & \neq_\beta & HM_1. \end{array}$$

The proof will occupy Steps 1-6.

Step 1. Since $\mathcal{A}_0, \mathcal{A}_1$ are r.e. and non-empty there are recursive functions f_0, f_1 such that $\#\mathcal{A}_i = \{f_i(n) \mid n \in \mathbb{N}\}$. There is a recursive function f such that

$$f(2^n 3^i) \;=\; f_i(n).$$

Now let $F \in \Lambda^o$ λ-define f and define $\mathsf{E}^{\mathcal{P}} \triangleq \mathsf{E} \circ F$. Then $\forall n \in \mathbb{N} \,.\, \mathsf{E}^{\mathcal{P}} \mathsf{c}_n =_\beta \mathsf{E}\mathsf{c}_{f(n)}$.

LEMMA 8.10. $\mathcal{A}_i = \{\mathsf{E}^{\mathcal{P}} \mathsf{c}_{2^n 3^i} \mid n \in \mathbb{N}\}^\beta$, *for $i \in \{0,1\}$. That is*

(i) $\mathcal{A}_0 = \{\mathsf{E}^{\mathcal{P}}\mathsf{c}_1, \mathsf{E}^{\mathcal{P}}\mathsf{c}_2, \mathsf{E}^{\mathcal{P}}\mathsf{c}_4, \mathsf{E}^{\mathcal{P}}\mathsf{c}_8, \mathsf{E}^{\mathcal{P}}\mathsf{c}_{16}, \ldots\}^\beta$;

(ii) $\mathcal{A}_1 = \{\mathsf{E}^{\mathcal{P}}\mathsf{c}_3, \mathsf{E}^{\mathcal{P}}\mathsf{c}_6, \mathsf{E}^{\mathcal{P}}\mathsf{c}_{12}, \mathsf{E}^{\mathcal{P}}\mathsf{c}_{24}, \mathsf{E}^{\mathcal{P}}\mathsf{c}_{48}, \ldots\}^\beta$.

PROOF. We have

$$\begin{array}{rcll} \mathcal{A}_i & = & \{M \mid M \in \mathcal{A}_i\}, & \\ & = & \{\mathsf{E}\ulcorner M \urcorner \mid M \in \mathcal{A}_i\}^\beta, & \text{by Theorem 1.31}, \\ & = & \{\mathsf{E}\mathsf{c}_{\#M} \mid M \in \mathcal{A}_i\}^\beta, & \text{as } \ulcorner M \urcorner = \mathsf{c}_{\#M}, \\ & = & \{\mathsf{E}\mathsf{c}_n \mid n \in \#\mathcal{A}_i\}^\beta, & \\ & = & \{\mathsf{E}\mathsf{c}_{f_i(n)} \mid n \in \mathbb{N}\}^\beta, & \text{as } \mathcal{A}_i = \{f_i(n) \mid n \in \mathbb{N}\}, \\ & = & \{\mathsf{E}\mathsf{c}_{f(2^n 3^i)} \mid n \in \mathbb{N}\}^\beta, & \text{as } f_i(n) = f(2^n 3^i), \\ & = & \{\mathsf{E}^{\mathcal{P}}\mathsf{c}_{2^n 3^i} \mid n \in \mathbb{N}\}^\beta. & \end{array}$$

□

Step 2. Constructing H.

DEFINITION 8.11. Let a be a fresh variable. Recall that c_n is the n-th Church numeral and S^+ the λ-representation of the successor function $S\colon \mathbb{N} \to \mathbb{N}$. Define

$$\begin{aligned} T &\triangleq \lambda xyz.xy(xyz), && \text{then } T\mathsf{c}_n \twoheadrightarrow_\beta \mathsf{c}_{2n};\\ A &\triangleq \lambda fgxyz.fx(a(\mathsf{E}^{\mathcal{P}}x))(f(\mathsf{S}^+x)y(g(\mathsf{S}^+x))z),\\ B &\triangleq \lambda fgx.f(\mathsf{S}^+x)(a(\mathsf{E}^{\mathcal{P}}(Tx)))(g(\mathsf{S}^+x))(gx). \end{aligned}$$

(ii) By the double fixed-point theorem there exist λ-terms F, G such that

$$\begin{aligned} F &\twoheadrightarrow_\beta AFG && \twoheadrightarrow_\beta \lambda xyz.Fx(a(\mathsf{E}^{\mathcal{P}}x))(F(\mathsf{S}^+x)y(G(\mathsf{S}^+x))z);\\ G &\twoheadrightarrow_\beta BFG && \twoheadrightarrow_\beta \lambda x.F(\mathsf{S}^+x)(a(\mathsf{E}^{\mathcal{P}}(Tx)))(G(\mathsf{S}^+x))(Gx). \end{aligned}$$

To be explicit, take

$$\begin{aligned} G &\triangleq \Theta(\lambda u.B(\Theta(\lambda x.Aux))u);\\ F &\triangleq \Theta(\lambda u.AuG). \end{aligned}$$

(iii) Finally, define

$$H \triangleq \lambda xa.F\mathsf{c}_1(ax)(G\mathsf{c}_1).$$

Step 3. Preliminary properties.

NOTATION. *Let a be a fresh variable and $k \in \mathbb{N}$. Write*

$$\begin{aligned} F_k &\triangleq F\mathsf{c}_k;\\ G_k &\triangleq G\mathsf{c}_k;\\ \mathsf{E}^{\mathcal{P}}_k &\triangleq \mathsf{E}^{\mathcal{P}}\mathsf{c}_k;\\ a_k &\triangleq a\mathsf{E}^{\mathcal{P}}_k;\\ H_k[\,] &\triangleq F_k[\,]G_k;\\ C_k[\,] &\triangleq F_k a_k([\,]G_k). \end{aligned}$$

Note that, by construction,

$$\begin{aligned} F_kMN &\twoheadrightarrow_\beta F_k a_k(F_{k+1}MG_{k+1}N);\\ G_k &\twoheadrightarrow_\beta F_{k+1}a_{2k}G_{k+1}G_k. \end{aligned}$$

By reducing F, G, respectively, it follows that

$$\begin{aligned} H_k[a_p] &= F_k a_p G_k \twoheadrightarrow_\beta C_k[H_{k+1}[a_p]]; && (8.6)\\ H_k[a_k] &= F_k a_k G_k \twoheadrightarrow_\beta C_k[H_{k+1}[a_{2k}]]. && (8.7) \end{aligned}$$

Step 4. Now the first half of the Cluster Theorem follows, namely the implication $(\Leftarrow)$ of (8.5).

PROPOSITION 8.12. *If $M \sim_{\mathcal{P}} N$ then $HM =_\beta HN$.*

PROOF. Suppose $M \sim_{\mathcal{P}} N$. If $M =_\beta N$ then $HM =_\beta HN$ follows immediately. Assume now $M, N \in \mathcal{A}_i$, for some $i \in \{0,1\}$. By the characterization of $\mathcal{A}_i$ in terms of $\mathsf{E}^{\mathcal{P}}$ (Lemma 8.10), it suffices to show that $H\mathsf{E}^{\mathcal{P}}_k =_\beta H\mathsf{E}^{\mathcal{P}}_{2k}$, for all $k > 0$. Indeed,

$$\begin{array}{rcll} H\mathsf{E}^{\mathcal{P}}_k & =_\beta & \lambda a.H_1[a_k] \twoheadrightarrow_\beta \lambda a.C_1[H_2[a_k]], & \\ & =_\beta & \lambda a.C_1[C_2[\cdots C_{k-1}[H_k[a_k]]\cdots]], & \text{by multiple use of (8.6)}, \\ & =_\beta & \lambda a.C_1[C_2[\cdots C_{k-1}[C_k[H_{k+1}[a_{2k}]]]\cdots]], & \text{by (8.7)}. \\ H\mathsf{E}^{\mathcal{P}}_{2k} & =_\beta & \lambda a.H_1[a_{2k}], & \\ & =_\beta & \lambda a.C_1[C_2[\cdots C_{k-1}[C_k[H_{k+1}[a_{2k}]]]\cdots]], & \text{by (8.6)}. \end{array}$$

□

Step 5. For the converse implication we need the fine structure of the reduction.

DEFINITION 8.13. Remember that $\Theta = (\lambda xy.y(xxy))(\lambda xy.y(xxy))$ is Turing's fixed point combinator. Define

$$\begin{array}{rcl} D^0_k[M] & \triangleq & F_k(aM) = \Theta(\lambda u.AuG)\mathsf{c}_k(aM) \\ D^1_k[M] & \triangleq & (\lambda y.y(\Theta y))(\lambda u.AuG)\mathsf{c}_k(aM) \\ D^2_k[M] & \triangleq & (\lambda u.AuG)F\mathsf{c}_k(aM) \\ D^3_k[M] & \triangleq & AFG\mathsf{c}_k(aM) \\ D^4_k[M] & \triangleq & (\lambda gxyz.Fx(a(\mathsf{E}^{\mathcal{P}}x))(F(\mathsf{S}^+x)y(g(\mathsf{S}^+x))z))G\mathsf{c}_k(aM) \\ D^5_k[M] & \triangleq & (\lambda xyz.Fx(a(\mathsf{E}^{\mathcal{P}}x))(F(\mathsf{S}^+x)y(G(\mathsf{S}^+x))z))\mathsf{c}_k(aM) \\ D^6_k[M] & \triangleq & (\lambda yz.F_k(a\mathsf{E}^{\mathcal{P}}_k)(F(\mathsf{S}^+\mathsf{c}_k)y(G(\mathsf{S}^+\mathsf{c}_k))z))(aM) \\ D^7_k[M] & \triangleq & \lambda z.F_k(a\mathsf{E}^{\mathcal{P}}_k)(F(\mathsf{S}^+\mathsf{c}_k)(aM)(G(\mathsf{S}^+\mathsf{c}_k))z). \end{array}$$

LEMMA 8.14. *Let $F_k(aM)N$ head-reduce to W in n steps. Then, writing $n = 8p+q$ for appropriate $p \geq 0$ and $q \leq 7$, we have*[2]

$$W = \begin{cases} D^q_k[M]N, & \text{if } p = 0; \\ D^q_k[\mathsf{E}^{\mathcal{P}}_k]((H_{k+1}[\mathsf{E}^{\mathcal{P}}_k])^{p-1}(H_{k+1}[M]N)), & \text{else.} \end{cases}$$

PROOF. Note that $F_k(aM)N = D^0_k[M]N$. Moreover,

$$\begin{array}{rcll} D^q_k[M]N & \to_h & D^{q+1}_k[M]N, & \text{for } q < 7, \\ D^7_k[M]N & \to_h & D^0_k[\mathsf{E}^{\mathcal{P}}_k](H_{k+1}[M]N). & \end{array}$$

The rest is clear. At steps 16, 24 we obtain, for example,

$$\begin{array}{rcl} D^7_k[\mathsf{E}^{\mathcal{P}}_k](H_{k+1}[M]N) & \to_h & D^0_k[\mathsf{E}^{\mathcal{P}}_k]((H_{k+1}[\mathsf{E}^{\mathcal{P}}_k])(H_{k+1}[M]G_k)). \\ D^7_k[\mathsf{E}^{\mathcal{P}}_k]((H_{k+1}[\mathsf{E}^{\mathcal{P}}_k])(H_{k+1}[M]G_k)) & \to_h & D^0_k[\mathsf{E}^{\mathcal{P}}_k]((H_{k+1}[\mathsf{E}^{\mathcal{P}}_k])^2(H_{k+1}[M]G_k)). \end{array}$$

□

[2] Note that the first case corresponds to $n \leq 7$, the second to $n > 7$.

Remember that a standard reduction $\sigma\colon M \twoheadrightarrow_{\mathsf{st}} N$ always consists of a head reduction followed by an internal reduction:

$$\sigma\colon M \twoheadrightarrow_h W \twoheadrightarrow^{\iota}_{\beta} N.$$

NOTATION 8.15. Write $M =_{\mathsf{st}\leq n} N$ if there are standard reductions of length $\leq n$ from M and N to a common reduct Z. Similarly, $M =_{\iota\leq n} N$ for internal standard reductions. Also, the notations $=_{\mathsf{st}<n}$ and $=_{\iota<n}$ will be used.

LEMMA 8.16. *For $M, N, M', N' \in \Lambda^o$ and $q, q' \leq 7$ and $k, n \in \mathbb{N}$ one has*

$$\begin{array}{rlllll} \text{(i)} & D^q_k[M]N & =_{\iota\leq n} & D^{q'}_k[M']N' & \Rightarrow & q = q' \;\&\; N =_{\mathsf{st}\leq n} N'. \\ \text{(ii)} & D^q_k[M]N & =_{\iota\leq n} & D^q_k[M']N' \;\&\; q < 7 & \Rightarrow & M =_{\mathsf{st}\leq n} M'. \\ \text{(iii)} & D^7_k[M]N & =_{\iota\leq n} & D^7_k[M']N' & \Rightarrow & H_{k+1}[M] =_{\mathsf{st}\leq n} H_{k+1}[M']. \end{array}$$

PROOF. (i) Suppose $D^q_k[M]N =_{\iota\leq n} D^{q'}_k[M']N'$. Then by observing where the free variable a occurs, one can conclude that $q = q'$. Since the reductions to a common reduct are internal, the positions of N, N' are not changed, and hence $N =_{\mathsf{st}\leq n} N'$.

(ii) Obvious from the definition of D^i_k.

(iii) In this case it follows that

$$D^0_k[\mathsf{E}^{\mathcal{P}}_k](H_{k+1}[M]z) =_{\iota\leq n} D^0_k[\mathsf{E}^{\mathcal{P}}_k](H_{k+1}[M']z).$$

The conclusion $H_{k+1}[M] =_{\mathsf{st}\leq n} H_{k+1}[M']$ depends upon the fact that there are the free variables z to mark the residuals. □

LEMMA 8.17. *Suppose $G_k =_{\mathsf{st}\leq n} (H_{k+1}[\mathsf{E}^{\mathcal{P}}_k])^d(H_{k+1}[M]G_k)$. Then*

$$H_{k+1}[\mathsf{E}^{\mathcal{P}}(T\mathsf{c}_k)] =_{\mathsf{st}<n} H_{k+1}[M].$$

PROOF. By induction on d. If $d = 0$, then we have $G_k =_{\mathsf{st}\leq n} H_{k+1}[M]G_k$. So there are standard reductions of these two terms to a common reduct. Observe that the head-reduction starting with G_k begins as follows:

$$\begin{array}{rcl} G_k & = & \Theta(\lambda u.B(\Theta(\lambda v.Avu))u)\mathsf{c}_k \\ & \to_h & (\lambda y.y(\Theta y))(\lambda u.B(\Theta(\lambda v.Avu))u)\mathsf{c}_k \\ & \to_h & (\lambda u.B(\Theta(\lambda v.Avu))u)G\mathsf{c}_k \\ & \to_h & BFG\mathsf{c}_k \\ & \to_h & (\lambda gx.F(\mathsf{S}^+x)(a(\mathsf{E}^{\mathcal{P}}(Tx)))(g(\mathsf{S}^+x))(gx))G\mathsf{c}_k \\ & \to_h & (\lambda x.F(\mathsf{S}^+x)(a(\mathsf{E}^{\mathcal{P}}(Tx)))(G(\mathsf{S}^+x))(Gx))\mathsf{c}_k \\ & \to_h & F(\mathsf{S}^+\mathsf{c}_k)(a(\mathsf{E}^{\mathcal{P}}(T\mathsf{c}_k)))(G(\mathsf{S}^+\mathsf{c}_k))(G\mathsf{c}_k). \end{array}$$

The heads of these terms are not of order 0 except the last one, but $H_{k+1}[X]$ is always of order 0. Therefore, the mentioned standard reduction of G_k goes at least to this last term $H_{k+1}[\mathsf{E}^{\mathcal{P}}(T\mathsf{c}_k)]G_k$, but then $H_{k+1}[\mathsf{E}^{\mathcal{P}}(T\mathsf{c}_k)] =_{\mathsf{st}<n} H_{k+1}[M]$.

If $d > 0$, then start the same argument as above, but at the intermediate conclusion

$$H_{k+1}[\mathsf{E}^{\mathcal{P}}(T\mathsf{c}_k)]G_k =_{\mathsf{st}<n} (H_{k+1}[\mathsf{E}^{\mathcal{P}}_k])^d(H_{k+1}[M]G_k),$$

one proceeds by concluding that

$$G_k =_{\mathsf{st}<n} (H_{k+1}[\mathsf{E}^{\mathcal{P}}_k])^{d-1}(H_{k+1}[M]G_k)$$

and uses the induction hypotheses. □

PROPOSITION 8.18. $H_k[M] =_\beta H_k[N] \Rightarrow M, N \in \mathcal{A}_0 \vee M, N \in \mathcal{A}_1 \vee M =_\beta N.$

PROOF. By the standardization theorem, it suffices to show for all n that

$$\forall k \in \mathbb{N}\,.\,[\,H_k[M] =_{\mathsf{st}\leq n} H_k[N] \quad\Rightarrow\quad M \sim_{\mathcal{P}} N\,].$$

This will be done by induction on n. From $H_k[M] =_{\mathsf{st}\leq n} H_k[N]$, it follows that

$$\begin{array}{lllll} H_k[M] & \twoheadrightarrow_h & W_M & \twoheadrightarrow^{\iota}_{\beta} & Z; \\ H_k[N] & \twoheadrightarrow_h & W_N & \twoheadrightarrow^{\iota}_{\beta} & Z, \end{array}$$

for some W_M, W_N, Z.

Case 1. W_M, W_N are both reached after $q, q' < 8$ steps, respectively. Then by Lemma 8.14, one has

$$\begin{array}{lll} W_M & = & D^q_k[M]G_k; \\ W_N & = & D^{q'}_k[N]G_k. \end{array}$$

By Lemma 8.16(i), it follows that $q = q'$. If $q < 7$, then by Lemma 8.16(ii) one has $M =_\beta N$, so $M \sim_{\mathcal{P}} N$. If $q = 7$, then by Lemma 8.16(iii) one obtains that $H_{k+1}[M] =_{\mathsf{st}<n} H_{k+1}[N]$, and by the induction hypothesis one concludes $M \sim_{\mathcal{P}} N$.

Case 2. W_M is reached after $p \geq 8$ steps and W_N after $q < 8$ steps. Then $p = 8d+q$ and, keeping in mind Lemma 8.16(i), it follows that

$$\begin{array}{lll} W_M & = & D^q_k[M]G_k; \\ W_N & = & D^q_k[\mathsf{E}^{\mathcal{P}}_k]R; \\ G_k & =_{\mathsf{st}<n} & R, \end{array}$$

where $R = (H_{k+1}[\mathsf{E}^{\mathcal{P}}_k])^{d-1}(H_{k+1}[N]G_k)$. Then as in case 1, it follows that $M \sim_{\mathcal{P}} \mathsf{E}^{\mathcal{P}}_k$. Moreover, by Lemma 8.17 $H_{k+1}[\mathsf{E}^{\mathcal{P}}_{2k}] =_{\mathsf{st}<n} H_{k+1}[N]$, so by the IH $\mathsf{E}^{\mathcal{P}}_{2k} \sim_{\mathcal{P}} N$. So $M \sim_{\mathcal{P}} \mathsf{E}^{\mathcal{P}}_k \sim_{\mathcal{P}} \mathsf{E}^{\mathcal{P}}_{2k} \sim_{\mathcal{P}} N$.

Case 3. Both W_M, W_N are reached after ≥ 8 steps. Then

$$\begin{array}{lll} W_M & = & D^j_k[\mathsf{E}^{\mathcal{P}}_k]((H_{k+1}[\mathsf{E}^{\mathcal{P}}_k])^d(H_{k+1}[M]G_k)); \\ W_N & = & D^j_k[\mathsf{E}^{\mathcal{P}}_k]((H_{k+1}[\mathsf{E}^{\mathcal{P}}_k])^{d'}(H_{k+1}[N]G_k)). \end{array}$$

If $d = d'$ then, by Lemma 8.16(i), we have

$$(H_{k+1}[\mathsf{E}_k^{\mathcal{P}}])^d(H_{k+1}[M]G_k) =_{\mathsf{st}<n} (H_{k+1}[\mathsf{E}_k^{\mathcal{P}}])^d(H_{k+1}[N]G_k),$$

so

$$H_{k+1}[M] =_{\mathsf{st}<n} H_{k+1}[N],$$

since $H_{k+1}[X]$ is always of order 0. Hence $M \sim_{\mathcal{P}} N$, by the induction hypothesis. If, on the other hand, say, $d < d'$, then writing $d' = d + e$ it follows that

$$\begin{array}{rcl} W_M & = & D_k^j[\mathsf{E}_k^{\mathcal{P}}]((H_{k+1}[\mathsf{E}_k^{\mathcal{P}}])^d(H_{k+1}[M] \qquad\qquad G_k \qquad\qquad)); \\ W_N & = & D_k^j[\mathsf{E}_k^{\mathcal{P}}]((H_{k+1}[\mathsf{E}_k^{\mathcal{P}}])^d(H_{k+1}[\mathsf{E}_k^{\mathcal{P}}]\underline{((H_{k+1}[\mathsf{E}_k^{\mathcal{P}}])^{e-1}(H_{k+1}[N]G_k))})); \end{array}$$

so

$$\begin{array}{rcl} H_{k+1}[M] & =_{\mathsf{st}<n} & H_{k+1}[\mathsf{E}_k^{\mathcal{P}}] \\ G_k & =_{\mathsf{st}<n} & (H_{k+1}[\mathsf{E}_k^{\mathcal{P}}])^{e-1}(H_{k+1}[N]G_k), \end{array}$$

since $H_{k+1}[X]$ is always of order 0. Therefore, by Lemma 8.17 we have

$$H_{k+1}[\mathsf{E}_{2k}^{\mathcal{P}}] =_{\mathsf{st}<n} H_{k+1}[N].$$

Applying the induction hypothesis twice, we obtain $M \sim_{\mathcal{P}} \mathsf{E}_k^{\mathcal{P}} \sim_{\mathcal{P}} \mathsf{E}_{2k}^{\mathcal{P}} \sim_{\mathcal{P}} N$. □

Step 6. Harvesting. Let $\mathcal{P} = \{\mathcal{A}_0, \mathcal{A}_1\}$ be an Ershov-partition.

PROPOSITION 8.19. *For $M, N \in \Lambda^o$ one has for H constructed in Definition 8.11*

$$HM =_\beta HN \iff M \sim_{\mathcal{P}} N.$$

PROOF. ($\Leftarrow$) This is Proposition 8.12.

($\Rightarrow$) Suppose $HM =_\beta HN$. Then

$$\begin{array}{rcl} HM & \twoheadrightarrow_\beta & \lambda a.F_1(aM)G_1 = \lambda a.H_1[aM]; \\ HN & \twoheadrightarrow_\beta & \lambda a.F_1(aN)G_1 = \lambda a.H_1[aN]. \end{array}$$

Applying these to I it follows that $H_1[M] =_\beta H_1[N]$. Then $M \sim_{\mathcal{P}} N$ follows from Proposition 8.18. □

This is just the Cluster Theorem 8.9 instantiated to a particular H.

The clustering theorem is taken from Statman and Barendregt (1999) where it was formulated for an r.e. family $\mathcal{S} = \{\mathcal{A}_e\}_{e \in \mathcal{S}}$ of r.e. sets closed under $=_\beta$. The notion of non-overlapping must be defined as

$$\mathcal{A}_i \cap \mathcal{A}_j \neq \emptyset \quad \Rightarrow \quad \mathcal{A}_i = \mathcal{A}_j,$$

since there can be duplications. From the more general statement it actually follows that PLP fails in $\mathcal{M}(\boldsymbol{\lambda})$ for all $n > 1$.

Chapter 9

Bijectivity and invertibility in $\lambda\eta$

Lambda terms can be considered 'as such', giving rise to syntactic considerations in which one studies e.g. reduction between terms, or more basically one counts the number of occurrences of 'λ' in one term. In this chapter we rather consider the behavior of λ-terms seen as definable functions in term models, like $\mathcal{M}(\boldsymbol{\lambda})$, $\mathcal{M}(\boldsymbol{\lambda}\eta)$, or their closed counterparts. From a set-theoretical perspective, there are the following natural properties of functions: injectivity, surjectivity and bijectivity.

DEFINITION 9.1. Let $\mathcal{M}$ be a λ-algebra and $F \in \mathcal{M}$.

(i) F is *injective* in $\mathcal{M}$ if

$$\forall X, Y \in \mathcal{M} \,.\, \mathcal{M} \models FX = FY \quad \Rightarrow \quad \mathcal{M} \models X = Y$$

(ii) F is *surjective* in $\mathcal{M}$ if

$$\forall Y \in \mathcal{M},\ \exists X \in \mathcal{M} \,.\, \mathcal{M} \models FX = Y$$

(iii) F is *bijective* in $\mathcal{M}$ if it is both injective and surjective.

In other words, F is bijective (resp. injective/surjective) in $\mathcal{M}$ whenever the corresponding function mapping $X \in \mathcal{M} \mapsto FX \in \mathcal{M}$ is such. There is also an algebraic perspective, in which an element $F \in \mathcal{M}$ is seen as left-invertible, right-invertible or even invertible (with definable inverses).

DEFINITION 9.2. Let $\mathcal{M}$ be a λ-algebra and $F \in \mathcal{M}$.

(i) F is *left-invertible* in $\mathcal{M}$ if there exists $L \in \mathcal{M}$ such that $\mathcal{M} \models L \circ F = \mathsf{I}$ holds.

(ii) F is *right-invertible* in $\mathcal{M}$ if there exists $R \in \mathcal{M}$ such that $\mathcal{M} \models \mathsf{I} = F \circ R$ holds.

(iii) F is *invertible* in $\mathcal{M}$ if there exists $G \in \mathcal{M}$ such that both

$$\mathcal{M} \models G \circ F = \mathsf{I} \qquad \text{and} \qquad \mathcal{M} \models \mathsf{I} = F \circ G \qquad \text{hold.}$$

Such elements L, R, G are respectively called left-inverse, right-inverse, and inverse of F.

REMARK 9.3. As in all semi-groups an element F is invertible in $\mathcal{M}$ exactly when it is both left- and right-invertible.

REMARK 9.4. In the category **Set** one has for an arrow $f: X \to Y$

(i) f is surjective $\iff$ f is right-invertible (assuming AC, the axiom of choice).

(ii) f is injective $\iff$ f is left-invertible (assuming the axiom of the excluded middle).

When considering λ-terms as elements of the closed (or open) term model of some λ-theory $\mathcal{T}$, the problem becomes more difficult because their inverses are required to be λ-definable. It is natural to wonder whether the correspondence between (in/sur)bijectivity and (left-/right-)invertibility still holds in this setting. One direction is easy as it holds for set-theoretic reasons.

LEMMA 9.5. *Let $\mathcal{T} \in \lambda\mathcal{T}$ and $F \in \Lambda^{(o)}$. In $\mathcal{M}^{(o)}(\mathcal{T})$, we have:*

(i) *If F is left-invertible then it is injective.*

(ii) *If F is right-invertible then it is surjective.*

(iii) *If F is invertible then it is bijective.*

PROOF. (i) Assume that F is left-invertible and let L be its left-inverse. Then F is injective: for all $X, Y \in \Lambda^{(o)}$ satisfying $FX =_{\mathcal{T}} FY$, we have:

$$X =_{\mathcal{T}} \mathsf{I}X =_{\mathcal{T}} (L \circ F)X =_{\mathcal{T}} L(FX) =_{\mathcal{T}} L(FY) =_{\mathcal{T}} (L \circ F)Y =_{\mathcal{T}} \mathsf{I}Y =_{\mathcal{T}} Y.$$

(ii) Assume that F is right-invertible and let R be its right-inverse. Then F is surjective: for all $X \in \Lambda^{(o)}$, the λ-term $Y \triangleq RX$ satisfies

$$FY = F(RX) =_{\mathcal{T}} (F \circ R)X =_{\mathcal{T}} \mathsf{I}X =_{\mathcal{T}} X.$$

(iii) By (i) and (ii). □

We are mainly interested in $\mathcal{M}^{(o)}(\mathcal{T})$ for $\mathcal{T} \in \{\boldsymbol{\lambda}, \boldsymbol{\lambda}\eta\}$. The left-invertibility correspondence is known to fail in $\mathcal{M}^{(o)}(\boldsymbol{\lambda})$, the unsolvable Ω serving as a counterexample. On the contrary, the right-invertibility correspondence holds, the right-invertible λ-terms being those terms having a hnf of shape $\lambda x.xM_1 \cdots M_k$. These results are due to Margaria and Zacchi (1983). The invertibility correspondence is also preserved because the only invertible λ-term is the identity I, as shown by Böhm and Dezani-Ciancaglini (1974). All these results belong to the λ-calculus tradition of finding syntactic characterizations of λ-terms satisfying suitable properties and are surveyed in Section 9.3.

In B[1984], Exercise 21.4.9, the problem was raised whether bijectivity of a λ-definable map on $\mathcal{M}^o(\boldsymbol{\lambda}\eta)$ implies invertibility. For the open term model the result holds trivially. In Dezani-Ciancaglini (1976) invertibility for λ-terms having a $\beta(\eta)$-normal form was characterized by being a finite hereditary permutation (FHP), that is having a Böhm tree that can be obtained from a finite η-expansion of the identity I via a permutation of its branches. This result was strengthened by Bergstra and Klop (1980) in which it was shown that the hypothesis of normalization could be dropped. This gives a tool to study the problem by taking M a FHP, insert a lesion, so that the resulting λ-term M' is no longer a FHP, and wondering whether M' fails to be a bijection. This approach led to the partial result in Batenburg and Velmans (1983) reported below, Proposition 9.43. Based on this result Folkerts (1995) settled the question in a 214 pages long PhD thesis:

invertibility is indeed equivalent to bijectivity in the closed term model $\mathcal{M}^o(\boldsymbol{\lambda}\eta)$. Based on Folkerts (1998), and talks by and with Folkerts, we present here a proof of the result (Theorem 9.52). In the same article, he also shows that the surjectivity does not imply right-invertibility in this term model. At the end of the chapter the counterexample will be presented (Theorem 9.73), together with a survey of what is known about the correspondence between the set-theoretic, algebraic, and λ-calculus syntactic notions.

Outline of the proof of the main result

The main result of this chapter, namely Theorem 9.52, is presented in Section 9.2. The structure of its proof is outlined in Figure 9.1 and involves a series of intermediate notions. We discuss the main steps and provide some intuitions.

Step 1. If $F \in \Lambda^o$ is bijective, then by definition it is both surjective and injective.

Step 2. Since F is surjective it must be 'regular', i.e., when an argument X is applied (as in FX), then X goes in head position and "takes control" of the computation.

Step 3. Thus F is both regular and injective, and this implies that it is also 'faithful'. Intuitively, this means that

$$FX \twoheadrightarrow_{\beta\eta} \lambda\vec{y}.X(P_1[x:=X])\cdots(P_k[x:=X])$$

where the behavior of the P_i's does not get modified by the substitution because either they are unsolvable or they have a free head variable (different from x).

Step 4. We derive that F is both surjective and faithful, and this entails that F is right-invertible. This step is starred as it relies on the existence of infinitely many unsolvables of order 0, pairwise non-equi-unsolvable. This is proved in Section 9.1.

Step 5. Now, as F is right-invertible and injective we conclude that it is also invertible.

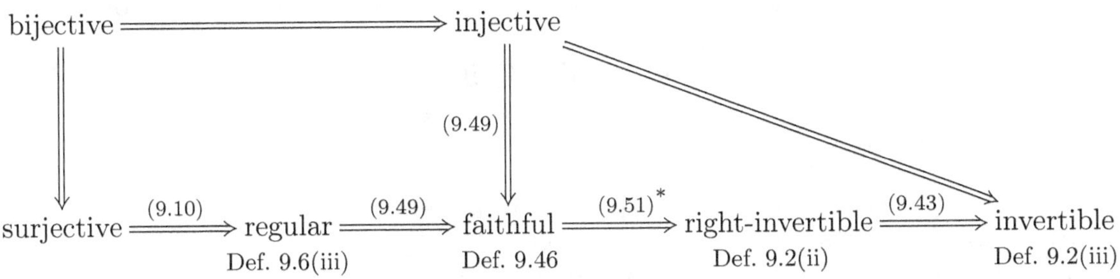

Figure 9.1: The proof that in $\mathcal{M}^o(\boldsymbol{\lambda}\eta)$ bijectivity implies invertibility.

Regular invertibility

We provide some preliminary definitions and results that are needed in the rest of the chapter. We introduce regular terms and λ-terms having empty head functionality, and present some of their properties.

DEFINITION 9.6. Let $M \in \Lambda$.

(i) Define

$$\text{hvar}(M) \triangleq \begin{cases} y, & \text{if } M =_\beta \lambda\vec{x}.yM_1 \cdots M_k \ \& \ y \notin \vec{x}, \\ \uparrow, & \text{otherwise.} \end{cases}$$

In other words, $\text{hvar}(M) = y$ indicates that M is β-convertible to a hnf having y as (free) head variable.

(ii) M has *empty head functionality* if either $M \in \text{UNS}$ or $\text{hvar}(M) = y$.

(iii) M is *regular* whenever $M =_\beta \lambda y\vec{x}.yM_1 \cdots M_k$.

(iv) Write REG for the set of all regular λ-terms.

(v) M is *regularly right-invertible* in $\mathcal{M}^{(o)}(\mathcal{T})$ if M has a regular right-inverse R in $\mathcal{M}^{(o)}(\mathcal{T})$.

Regularly left-invertible and regularly invertible λ-terms can be defined analogously, but we will not need such notions for our purposes.

REMARK 9.7. (i) The notion of empty head functionality is invariant under $=_{\beta\eta}$.

(ii) Similarly, the set REG is closed under $=_{\beta\eta}$.

(iii) The set of λ-terms having empty head functionality and the set REG are disjoint.

(iv) Unsolvables have empty head functionality, so they cannot be regular.

EXAMPLES 9.8. (i) The λ-terms $\Omega, \Omega_3, \mathsf{YI}$ and $\lambda xy.zxy$ have empty head functionality.

(ii) Clearly, $yM_1 \cdots M_k$ has empty head functionality for all $\vec{M} \in \Lambda$.

(iii) The λ-terms $\mathsf{I}, 1, \lambda y.y\mathsf{KI}$ and $\lambda xyz.x\mathsf{SK}\Omega$ are regular.

(iv) If $M =_{\beta\eta} \mathsf{I}$ then M is regular.

(v) For all $n > 0$, $\mathsf{P}_n = \lambda x_1 \ldots x_n z.zx_1 \cdots x_n$ is neither regular nor has empty head functionality.

From Remark 9.7(iv) and Example 9.8(iv), we derive the following inclusions:

$$[\mathsf{I}]_{\beta\eta} \subseteq \text{REG} \subseteq \text{SOL}$$

It is easy to check that the above inclusions are strict.

LEMMA 9.9. (i) *The set* REG *is closed under composition.*

(ii) *If* $M \in \text{REG}$ *then* $M \circ N \in \text{SOL}$ *entails* $N \in \text{SOL}$.

(iii) *Similarly,* $M \in \text{REG}$ *and* $M \circ N \in \text{REG}$ *imply* $N \in \text{REG}$.

PROOF. (i) Easy.

(ii) Since $M \in \mathrm{REG}$, we have $M =_\beta \lambda x_1 \dots x_n.x_1 M_1 \cdots M_k$ for some $k \geq 0$ and $n \geq 1$. Notice that, when head-reducing $M \circ N$, the λ-term N goes in head position:

$$\begin{array}{rll} M \circ N & \twoheadrightarrow_h & \lambda y.M(Ny) \\ & \twoheadrightarrow_h & \lambda y x_2 \dots x_n.NyM_1^\sigma \cdots M_k^\sigma, \quad \text{where } \sigma \triangleq [x_1 := Ny]. \end{array}$$

As a consequence $M \circ N$ can only be solvable if N is.

(iii) Analogous to (ii). □

Recall that a λ-theory $\mathcal{T}$ is semi-sensible if it does not equate a solvable and an unsolvable term. We know that this is the case exactly when $\mathcal{T} \subseteq \mathcal{H}^*$.

LEMMA 9.10. *Let $\mathcal{T} \in \lambda\mathcal{T}$ be semi-sensible and $F \in \Lambda^{(o)}$. If F is surjective in $\mathcal{M}^{(o)}(\mathcal{T})$ then it is regular.*

PROOF. Assume that F is surjective. Then there exists a λ-term X such that $FX =_\mathcal{T} \mathsf{I}$. As $\mathcal{T}$ is semi-sensible, FX and hence F is solvable. Therefore $F =_\beta \lambda x_1 \dots x_n.yM_1 \cdots M_k$ for some $n, k \geq 0$ and $y \in \mathrm{Var}$ (by Theorem 1.23). Also, there exists a $Y \in \Lambda^{(o)}$ such that $FY =_\mathcal{T} \Omega$, which entails $n > 0$ and $y = x_1$. We conclude that F is regular. □

9.1 Equi-unsolvability

In this section we define and study the notion of equi-unsolvability,[1] introduced by Folkerts (1995), and inspired by the work of Statman (1986). For the main result in this chapter we need an infinite set of non-equi-unsolvable terms of order 0. The set is easy to define:

$$\mathcal{X} = \{\mathsf{X}_n \mid n \in \mathbb{N}\},$$

where $\mathsf{X}_n = \mathsf{W}_n \mathsf{W}_n^{\sim n+1}$ for $\mathsf{W}_n = \lambda y_1 \dots y_n x.xxy_1 \cdots y_n$ (see Definition 4.7). To show that these are mutually non-equi-unsolvable, an invariant will be defined separating them.

DEFINITION 9.11. Let $\mathcal{T}$ be a semi-sensible λ-theory.

(i) Define the binary relation $\trianglelefteq$ on UNS by setting, for all $U \in \mathrm{UNS}$ and $P \in \Lambda$:

$$U \trianglelefteq UP$$

Notice that $U \in \mathrm{UNS}$ entails $UP \in \mathrm{UNS}$, so the relation $\trianglelefteq$ is well defined.

(ii) On UNS define the relation $\simeq_\mathcal{T}$ as the least equivalence relation containing $\mathcal{T} \cup \trianglelefteq$.

(iii) Two λ-terms $U, V \in \mathrm{UNS}$ are *equi-unsolvable modulo $\mathcal{T}$*, if $U \simeq_\mathcal{T} V$ holds.

LEMMA 9.12. *For $\mathcal{T} \in \lambda\mathcal{T}$ semi-sensible and $U \in \mathrm{UNS}$, we have:*

(i) *$U \simeq_\mathcal{T} \lambda x.U$, for all $x \in \mathrm{Var}$.*

(ii) *$U \simeq_\mathcal{T} U[x := N]$, for all $x \in \mathrm{Var}$ and $N \in \Lambda$.*

(iii) *$U \simeq_\mathcal{T} GU$, for all $G \in \mathrm{REG}$.*

[1] Originally called *equi-solvability*. We believe equi-unsolvability is more appropriate as it is a property of unsolvable λ-terms.

PROOF. (i) From $\lambda x.U \trianglelefteq (\lambda x.U)x =_\beta U$ we get $\lambda x.U \simeq_\mathcal{T} U$. Conclude by symmetry.

(ii) Applying (i), we get $U \simeq_\mathcal{T} \lambda x.U \trianglelefteq (\lambda x.U)N =_\beta U[x:=N]$.

(iii) Since G is regular, we have $G =_\beta \lambda y x_1 \ldots x_n.y P_1 \cdots P_k$ for some $n, k \geq 0$. Thus, $GU \in \mathrm{UNS}$ follows. We conclude:

$$\begin{array}{rlll} GU & =_\beta & \lambda\vec{x}.UP_1^\sigma \cdots P_k^\sigma, & \text{for } \sigma \triangleq [y:=U], \\ & \simeq_\mathcal{T} & UP_1^\sigma \cdots P_k^\sigma, & \text{by (i)}, \\ & \simeq_\mathcal{T} & U, & \text{by definition of } \trianglelefteq \text{ and symmetry.} \end{array}$$

□

EXAMPLES 9.13. (i) $\Omega\mathsf{K} \simeq_\lambda \Omega\mathsf{F}$.

(ii) $\Omega zz \simeq_\lambda \lambda z.\Omega zz \simeq_\lambda \Omega\mathsf{FK} \simeq_\lambda \Omega$.

(iii) $\Omega \simeq_\lambda \Omega\mathsf{K}^{\sim n}$, for all $n \in \mathbb{N}$.

(iv) If $\mathrm{ord}_\beta(M) \in \mathbb{N}$ and $\mathrm{ord}_\beta(N) = \infty$ then $M \not\simeq_\lambda N$, see Folkerts (1995).

(v) In particular $\Omega \not\simeq_\lambda \mathsf{K}^\star$.

EXERCISE 9.14. (i) Check that $\Omega \not\simeq_\lambda \Omega_3$.

(ii) For $M \triangleq \lambda z.(\lambda xy.xx(zy))(\lambda xy.xx(zy))$ show that $M \simeq_{\lambda\eta} \Omega$ holds, while $M \not\simeq_\lambda \Omega$.

(iii) (Difficult!) For $\mathcal{T} \in \lambda\mathcal{T}$ semi-sensible, define $\mathcal{T}^{\mathrm{eq.uns}}$ as the least λ-theory containing $\mathcal{T} \cup \trianglelefteq$. Since $\mathcal{T}$ is semi-sensible and $\trianglelefteq \subseteq \mathrm{UNS}^2$ we derive $\mathcal{T}^{\mathrm{eq.uns}} \subseteq \mathcal{H}^*$. In particular, $\mathcal{T}^{\mathrm{eq.uns}}$ is consistent. Show that, for $U, V \in \mathrm{UNS}$, the following holds:

$$\mathcal{T}^{\mathrm{eq.uns}} \vdash U = V \iff U \simeq_\mathcal{T} V.$$

From now on, and until the end of the section, we will work on $\mathcal{T} = \lambda\eta$ and $\simeq_{\lambda\eta}$.

The core of λ-terms

We now introduce and study the notion of 'core' of a λ-term. In case M is unsolvable, one can think of its core as the smallest subterm of M responsible for its unsolvability.

DEFINITION 9.15. Let $M \in \Lambda$. The *core of* M, written $\mathrm{core}(M)$, is defined by structural induction on M as follows:

$$\begin{array}{rcl} \mathrm{core}(x) & \triangleq & x; \\ \mathrm{core}(\lambda x.P) & \triangleq & \mathrm{core}(P); \\ \mathrm{core}(PQ) & \triangleq & \begin{cases} \mathrm{core}(P), & \text{if } P \in \mathrm{UNS}, \\ PQ, & \text{else.} \end{cases} \end{array}$$

REMARK 9.16. Let $M \in \Lambda$.

(i) $\mathrm{core}(\mathrm{core}(M)) = \mathrm{core}(M)$.

(ii) If $M = xM_1 \cdots M_k$, then $\mathrm{core}(M) = M$.

(iii) If M is unsolvable, then it has shape $M = \lambda\vec{x}.(\lambda y.N)P_0 \cdots P_n$. In this case

$$\mathrm{core}(M) = \begin{cases} \mathrm{core}(N), & \text{if } N \in \mathrm{UNS}, \\ (\lambda y.N)P_0 \cdots P_{\mu k.[(\lambda y.N)P_0 \cdots P_k \text{ is unsolvable}]}, & \text{otherwise.} \end{cases}$$

DEFINITION 9.17. For $M \in \Lambda$, define a context $H_M[\,]$ by structural induction on M:

$$\begin{array}{lcl} H_x[\,] & \triangleq & [\,]; \\ H_{\lambda x.N}[\,] & \triangleq & \lambda x.H_N[\,]; \\ H_{PQ}[\,] & \triangleq & \begin{cases} H_P[\,]Q, & \text{if } P \in \text{UNS}, \\ [\,], & \text{else.} \end{cases} \end{array}$$

LEMMA 9.18. *Let $M \in \Lambda$. Then M can be uniquely written as $M = H_M[\text{core}(M)]$.*

PROOF. By induction on the structure of M. $\square$

LEMMA 9.19. *Let $U \in$ UNS.*

(i) $\text{core}(U) = \text{core}(\lambda\vec{x}.U) = \text{core}(UP)$, *for all $\vec{x}$ and $P \in \Lambda$.*

(ii) $\text{core}(U) \in$ UNS, *moreover it has shape*

$$\text{core}(U) = U_0 \cdots U_{k+1},$$

with $U_0 = \lambda x.V$, so that U_0 is not an application, $k \geq 0$ and $U_0 \cdots U_k$ solvable.

(iii) $\text{core}(U) = GN$, *for some $G \in$ REG and $N \in \Lambda$.*

PROOF. (i) Straightforward from Definition 9.15.

(ii) By induction on the structure of U, using Remark 9.16(iii).

(iii) By (ii), we have $\text{core}(U) = GN \in \text{UNS}$ where $G \triangleq U_0 \cdots U_k \in \text{SOL}$ and $N \triangleq U_{k+1}$. Therefore $G \twoheadrightarrow_\beta \lambda x_1 \ldots x_n.yG_1 \cdots G_m$. Since $GN \in \text{UNS}$, we get $n \geq 1$ and $y = x_1$. $\square$

DEFINITION 9.20. Let $U \in$ UNS. By Lemma 9.19(ii), we have

$$\text{core}(U) = U_0 \cdots U_{k+1}$$

with $U_0 = (\lambda x.V)$ and $k \geq 0$.

(i) The *quantitative core of* U, in symbols $\text{core}_\#(U)$, is the number $k+1$ of external applications in $\text{core}(U)$.

(ii) The definition of $\text{core}_\#(-)$ may be extended to Λ by $\text{core}_\#(M) \triangleq 0$, for $M \in$ SOL.

(iii) U is in *canonical form* if it has shape $U = \lambda\vec{x}.\text{core}(U)\vec{P}$ for some $\vec{x}$ and $\vec{P} \in \Lambda$.

(iv) For $V \in \Lambda$, write $U =_\# V$ whenever $\text{core}_\#(U) = \text{core}_\#(V)$.

For the sake of generality, the definitions above are stated for arbitrary λ-terms but we will be only interested in $\text{core}(U)$ and $\text{core}_\#(U)$ for U unsolvable.

EXAMPLES 9.21.

U	$\mathrm{core}(U)$	$\mathrm{core}_\#(U)$	canonical form?
Ω	U	1	✓
$\lambda x.\Omega$	Ω	1	✓
$\mathsf{I}^{\sim 99}\Omega$	U	100	✓
$\mathsf{I}\Delta\Delta$	U	2	✓
$(\lambda y.y\Delta\Delta y)\mathsf{I}$	U	1	✓
$(\lambda y.(\lambda x.xxy)(\lambda x.xxy))\mathsf{I}$	$(\lambda x.xxy)(\lambda x.xxy)$	1	✗
$(\lambda y.(\lambda x.yxx)(\lambda x.yxx))\mathsf{I}$	U	1	✓
$(\lambda xy.\Omega)\mathsf{K}^{\sim 3}$	Ω	1	✗
X_n	U	$n+1$	✓

Recall that, for $n \in \mathbb{N}$, we have $\mathsf{X}_n = \mathsf{W}_n \mathsf{W}_n^{\sim n+1}$, where $\mathsf{W}_n = \lambda y_1 \ldots y_n x.xxy_1 \cdots y_n$.

REMARK 9.22. Let $U \in \mathrm{UNS}$. Then we have the following.
  (i) $\mathrm{core}_\#(U) \geq 1$, therefore $U =_\# V$ implies that $V \in \mathrm{UNS}$.
  (ii) By Lemma 9.19(i), we have $U =_\# \lambda \vec{x}.U =_\# UP$, for all $\vec{x}$ and $P \in \Lambda$.
  (iii) By Lemma 9.19(ii), there exists a V such that $U \twoheadrightarrow_\beta V$ and $\mathrm{core}_\#(V) = 1$.

Neither the notion of core nor its quantitative version are stable under β-reduction. For example $\mathsf{II}\Omega \twoheadrightarrow_\beta \Omega$, but $\mathrm{core}(\mathsf{II}\Omega) = \mathsf{II}\Omega$ and $\mathrm{core}_\#(\mathsf{II}\Omega) = 2$, while $\mathrm{core}(\Omega) = \Omega$ and $\mathrm{core}_\#(\Omega) = 1$. We now show that quantitative core is stable under substitution.

LEMMA 9.23. *Let* $U \in \mathrm{UNS}$ *and* σ *be a substitution.*
  (i) *If* $U^\sigma \to_h V'$ *then there exists* $V \in \Lambda$ *such that* $U \to_h V$ *and* $V^\sigma = V'$.

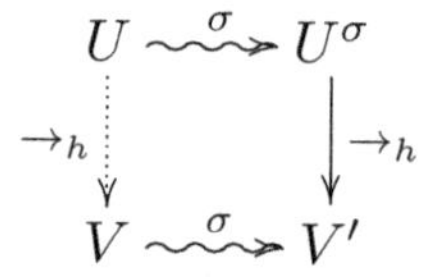

$$\begin{array}{ccc} U & \stackrel{\sigma}{\rightsquigarrow} & U^\sigma \\ {\scriptstyle \to_h}\downarrow & & \downarrow{\scriptstyle \to_h} \\ V & \stackrel{\sigma}{\rightsquigarrow} & V' \end{array}$$

  (ii) $\mathrm{core}(U^\sigma) = \mathrm{core}(U)^\sigma$.
  (iii) $U^\sigma =_\# U$.

PROOF. (i) As $U \in \mathrm{UNS}$ it has shape $U = \lambda \vec{x}.(\lambda y.U_0)U_1 \cdots U_{k+1}$. By the variable convention $\vec{x}, y \notin \mathrm{dom}(\sigma)$, implying

$$U^\sigma = \lambda \vec{x}.(\lambda y.U_0^\sigma)U_1^\sigma \cdots U_{k+1}^\sigma \to_h \lambda \vec{x}.U_0^\sigma[y := U_1^\sigma]U_2^\sigma \cdots U_{k+1}^\sigma = V'$$

Now take $V \triangleq \lambda \vec{x}.U_0[y := U_1]U_2 \cdots U_{k+1}$.

(ii) By Lemma 9.19(ii)-(iii), we have $\mathrm{core}(U) = GN \in \mathrm{UNS}$ for some $G \in \mathrm{REG}$, $N \in \Lambda$. As G is regular $G^\sigma \twoheadrightarrow_{\beta\eta} \lambda y\vec{x}.yM_1^\sigma \cdots M_k^\sigma$, implying $G^\sigma \in \mathrm{SOL}$. As $GN \in \mathrm{UNS}$ and unsolvability is preserved under substitution we get $(GN)^\sigma \in \mathrm{UNS}$. The conclusion follows.

(iii) By (ii). □

The following proposition is clarifying, but not actually used.

PROPOSITION 9.24. *Let $U \in \mathrm{UNS}$, $x \in \mathrm{Var}$ and $P \in \Lambda$. Then*

$$(\lambda x.U)P =_{\#} U =_{\#} U[x:=P].$$

PROOF. As $\mathrm{core}((\lambda x.U)P) = \mathrm{core}(\lambda x.U) = \mathrm{core}(U)$, one obtains the first equality by counting applications. The second equality follows from Lemma 9.23(iii). □

Now we will show that every $U \in \mathrm{UNS}$ reaches a canonical form by head reduction.

NOTATION. (i) *Recall that $\to^{\iota}_{\beta\eta}$ represents inner (namely, non-head) $\beta\eta$-reduction.*

(ii) *Given $M, N \in \Lambda$, a reduction $\rho\colon M \twoheadrightarrow N$ and a context $C[\,]$, we write $C[\rho]$ for the reduction $C[M] \twoheadrightarrow C[N]$ obtained by performing the steps in ρ under the hole of $C[\,]$. E.g., given $\rho : \mathsf{K}x\Omega \to_{\beta} (\lambda y.x)\Omega \to_{\beta} x$, we have $\lambda x.\rho : \lambda x.\mathsf{K}x\Omega \to_{\beta} \lambda x.(\lambda y.x)\Omega \to_{\beta} \mathsf{I}$.*

LEMMA 9.25. *Let $U, V \in \mathrm{UNS}$.*

(i) *There exists a substitution σ and $\vec{x}, \vec{P}$ such that*

$$U \twoheadrightarrow_h U' \triangleq \lambda\vec{x}.\mathrm{core}(U)^{\sigma}\vec{P}$$

and U' is in canonical form.

(ii) *Let U be in canonical form. Then $U \to^{\iota}_{\beta\eta} V$ implies $U =_{\#} V$.*

(iii) *Assume that $U \to^{\iota}_{\beta\eta} V$. Then U is in canonical form if and only if V is.*

PROOF. (i) By structural induction on U. By Lemma 9.18, we have $U = H_U[\mathrm{core}(U)]$.

Case $U = x$. Vacuous as U is unsolvable.

Case $U = \lambda y.V$ for $V \in \mathrm{UNS}$, then $\mathrm{core}(U) = \mathrm{core}(V)$. By definition $H_U[\,] = \lambda y.H_V[\,]$ for $V = H_V[\mathrm{core}(V)]$. By induction hypothesis $\rho : V \twoheadrightarrow_h \lambda\vec{x}.\mathrm{core}(V)^{\sigma}\vec{P}$. Therefore

$$\begin{array}{rcll} U & \twoheadrightarrow_h & \lambda y\vec{x}.\mathrm{core}(V)^{\sigma}\vec{P}, & \text{as } \rho \text{ is head reduction,} \\ & = & \lambda y\vec{x}.\mathrm{core}(U)^{\sigma}\vec{P}, & \text{as } \mathrm{core}(V) = \mathrm{core}(U), \\ & \triangleq & U'. & \end{array}$$

We now check that U' is in canonical form.

$$\begin{array}{rcll} \mathrm{core}(U') & = & \mathrm{core}(\mathrm{core}(U)^{\sigma}), & \text{by Lemma 9.19(i),} \\ & = & \mathrm{core}(\mathrm{core}(U))^{\sigma}, & \text{by Lemma 9.23(ii),} \\ & = & \mathrm{core}(U)^{\sigma}, & \text{by Remark 9.16(i).} \end{array}$$

Case $U = MN$. There are two subcases.

Subcase $M \in \mathrm{UNS}$, so $\mathrm{core}(U) = \mathrm{core}(M)$. By definition $H_U[\,] = H_M[\,]N$ and $M = H_M[\mathrm{core}(M)]$. By induction hypothesis we get $\rho : M \twoheadrightarrow_h \lambda x_1 \ldots x_n.\mathrm{core}(M)^{\sigma_0}\vec{P}$.

If $n = 0$ then $U \twoheadrightarrow_h \mathrm{core}(M)^{\sigma_0}\vec{P}N = \mathrm{core}(U)^{\sigma_0}\vec{P}N \triangleq U'$ which is in canonical form.

If $n > 0$ then the reduction ρ factorizes as $\rho = \rho_0; (\lambda x_1.\rho_1)$ where $\rho_0 : M \twoheadrightarrow_h \lambda x_1.X$ and $\rho_1 : X \twoheadrightarrow_h \lambda x_2 \ldots x_n.\mathrm{core}(M)^{\sigma_0}\vec{P}$. By confluence, setting $\sigma_1 \triangleq [x_1 := N]$, we obtain

$$\begin{array}{ccc}
U = H_M[\mathrm{core}(M)]N & \xrightarrow[h]{\rho_0 N} (\lambda x_1.X)N \xrightarrow[h]{} & X^{\sigma_1} \\
\downarrow \rho N & & \downarrow (\rho_1)^{\sigma_1} \\
(\lambda x_1 \ldots x_n.\mathrm{core}(M)^{\sigma_0}\vec{P})N & \xrightarrow[h]{} & \lambda x_2 \ldots x_n.\mathrm{core}(M)^{\sigma_0\sigma_1}\vec{P}^{\sigma_1}
\end{array}$$

As ρ_1 is head, and head-reductions are closed under substitution, also $(\rho_1)^{\sigma_1}$ is head. Therefore $U \twoheadrightarrow_h \lambda x_2 \ldots x_n.\mathrm{core}(M)^{\sigma_0\sigma_1}\vec{P}^{\sigma_1} = \lambda x_2 \ldots x_n.\mathrm{core}(U)^{\sigma_0\sigma_1}\vec{P}^{\sigma_1} \triangleq U'$.

As above, one can easily check that U' is canonical.

Subcase $M \in \mathrm{SOL}$, so $H_U[] = []$. In zero steps $U \twoheadrightarrow_h U = \mathrm{core}(U)$, so $U' \triangleq U$.

(ii) Since U is in canonical form it has shape $U = \lambda\vec{x}.\mathrm{core}(U)\vec{P}$. As the core of U is unsolvable it must have shape $\mathrm{core}(U) = (\lambda y.U_0)U_1 \cdots U_m$ for some $m \geq 1$, hence the head β-redex of U is $(\lambda y.U_0)U_1$. Assume now that $U \to^{\iota}_{\beta\eta} V$, there are two cases:

(1) the contracted redex occurs in one of the P_i's, in which case $\mathrm{core}(U) = \mathrm{core}(V)$;

(2) the contracted redex occurs within $\mathrm{core}(U)$, but not in head position. That is, $U_i \to_{\beta\eta} V_i$ for some $i \in \{0, \ldots, m\}$ and $\mathrm{core}(V) = (\lambda y.U_0)U_1 \cdots U_{i-1}V_iU_{i+1} \cdots U_m$.

In both cases we get $\mathrm{core}_\#(U) = \mathrm{core}_\#(V) = m$.

(iii) By a similar case analysis on U (respectively V). □

PROPOSITION 9.26. *Let $U, V \in \mathrm{UNS}$ and σ be a substitution. If $U^\sigma \twoheadrightarrow_{\beta\eta} V$ then there exists U' such that*

$$U \twoheadrightarrow_h U' \;\&\; U' =_\# V.$$

PROOF. Let $U^\sigma \twoheadrightarrow_{\beta\eta} V$. As V is unsolvable we have $\mathrm{core}(V) = (\lambda y.V_0)V_1 \cdots V_m$ for some $m > 0$, thus $\mathrm{core}_\#(V) = m$. By Lemma 9.25(i), there are a substitution σ' and λ-terms $\vec{P}$ such that $V \twoheadrightarrow_h \lambda\vec{x}.(\lambda y.V_0^{\sigma'})V_1^{\sigma'} \cdots V_m^{\sigma'}\vec{P}$. By standardization for $\beta\eta$ (Theorem 1.67), from $U^\sigma \twoheadrightarrow_{\beta\eta} \lambda\vec{x}.\mathrm{core}(V^{\sigma'})\vec{P}$ we obtain a head reduction sequence $U^\sigma \twoheadrightarrow_h V'$ for some $V' = \lambda\vec{x}z_1 \ldots z_\ell.(\lambda y.V_0')V_1' \cdots V_m'\vec{P}'Q_1' \cdots Q_\ell'$ such that:

- $V_i' \twoheadrightarrow_{\beta\eta} V_i^{\sigma'}$, for all $i\,(0 \leq i \leq m)$,
- for each $P_i' \in \vec{P}'$ we have $P_i' \twoheadrightarrow_{\beta\eta} P_i$,
- $z_j \notin \mathrm{FV}(\vec{V}'\vec{P}')$ and $Q_j \twoheadrightarrow_{\beta\eta} z_j$, for all $j\,(1 \leq j \leq \ell)$.

Thus $V' \twoheadrightarrow_{\beta\eta} \lambda\vec{x}.(\lambda y.V_0')V_1' \cdots V_m'\vec{P}' \twoheadrightarrow^{\iota}_{\beta\eta} \lambda\vec{x}.\mathrm{core}(V^{\sigma'})\vec{P}$, which is in canonical form. By Lemma 9.25(iii) and (ii), we obtain $\mathrm{core}_\#(\lambda\vec{x}.(\lambda y.V_0')V_1' \cdots V_m'\vec{P}') = m$. Therefore, by Remark 9.22(ii), we derive $\mathrm{core}_\#(V') = m$. Since $U^\sigma \twoheadrightarrow_h V'$, by Lemma 9.23(i) there exists U' such that $U \twoheadrightarrow_h U'$ and $(U')^\sigma = V'$, so we conclude by Lemma 9.23(iii).

In diagrammatic form:

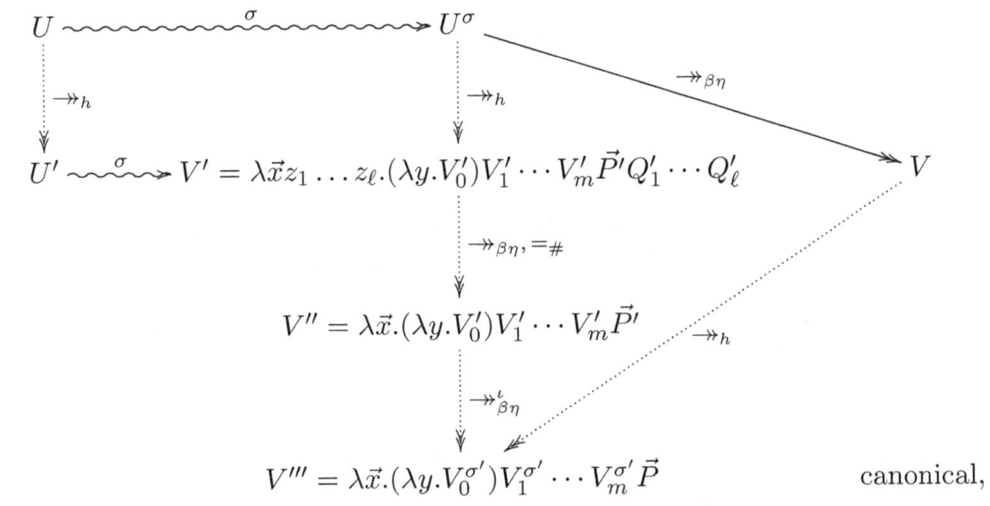

where $\text{core}(V) = (\lambda y.V_0)V_1 \cdots V_m$, hence $\text{core}_\#(V) = m$ by definition. Now

$$
\begin{array}{rcll}
\text{core}_\#(U') & = & \text{core}_\#(V'), & \text{by Lemma 9.23(iii)}, \\
& = & \text{core}_\#(V''), & \text{by Remark 9.22(ii)}, \\
& = & \text{core}_\#(V'''), & \text{by Lemma 9.25(iii), (ii), since } V''' \text{ is canonical}, \\
& = & m, & \text{by Remark 9.22(ii)}, \\
& = & \text{core}_\#(V), & \text{as we saw before.}
\end{array}
$$
□

Extensional order and degree

The notion of order (Definition 2.37), just like that of quantitative core, is not stable under $\beta\eta$-reduction. For instance, we have $\lambda z.\Omega z \to_\eta \Omega$, $\text{ord}_\beta(\lambda z.\Omega z) = 1$ and $\text{ord}_\beta(\Omega) = 0$. We are going to stabilize both notions by performing a sort of 'limit superior'—the former gives rise to the notion of extensional order, the latter of extensional degree.

DEFINITION 9.27. The *extensional order* of an $M \in \Lambda$, written $\text{ord}_{\beta\eta}(M)$, is defined[2] by:

$$\text{ord}_{\beta\eta}(M) \triangleq \mu k \,.\, \exists N \in \mathcal{G}_{\beta\eta}(M) \,.\, \text{ord}_\beta(N) = k$$

REMARK 9.28. The number of external abstractions of a λ-term may only increase along β-reduction. It can decrease along an η-reduction, e.g., $\lambda xy.\Omega(\mathsf{I}y) \to_\beta \lambda xy.\Omega y \to_\eta \lambda x.\Omega$.

EXAMPLES 9.29. (i) $\text{ord}_{\beta\eta}(1) = \text{ord}_{\beta\eta}(\mathsf{I}) = 1$, while $\text{ord}_\beta(1) = 2$, $\text{ord}_\beta(\mathsf{I}) = 1$.
(ii) $\text{ord}_{\beta\eta}(\lambda x.\Omega x) = \text{ord}_{\beta\eta}(\Omega) = 0$.
(iii) For all $n \in \mathbb{N}$, we have $\text{ord}_{\beta\eta}(\mathsf{X}_n) = 0$.
(iv) $\text{ord}_{\beta\eta}(\mathsf{K}^\star) = \text{ord}_\beta(\mathsf{K}^\star) = \infty$.

[2]The relationship between our definitions and the notion of limit superior is discussed in Exercise 9.42.

EXERCISE 9.30. (i) Define $M \in \Lambda^o$ such that $M \twoheadrightarrow_\beta \lambda x.Mx$. Note that, for all $\vec{x}$, one has $M \twoheadrightarrow_\beta \lambda\vec{x}.M\vec{x} \twoheadrightarrow_\eta M$. Show that, nevertheless, $\mathrm{ord}_{\beta\eta}(M) = \infty$.

(ii) Convince yourself through examples that

$$\forall U \in \mathrm{UNS}\,.\,[\,\mathrm{ord}_\beta(U) = \infty \iff \mathrm{ord}_{\beta\eta}(U) = \infty\,].$$

A detailed proof is in Folkerts' PhD thesis.

DEFINITION 9.31. A term $M \in \Lambda$ is a *potential η-redex* if it is of the form $M = \lambda x.N$ and $N \twoheadrightarrow_{\beta\eta} N'x$, with $x \notin \mathrm{FV}(N')$. In that case $M = \lambda x.N \twoheadrightarrow_{\beta\eta} \lambda x.N'x \to_\eta N'$.

LEMMA 9.32. *Let $U \in \mathrm{UNS}$. Assume $\mathrm{ord}_{\beta\eta}(U) \in \mathbb{N}$. Then the following are equivalent.*

1. $\mathrm{ord}_{\beta\eta}(U) = k$;

2. $U \twoheadrightarrow_{\beta\eta} \lambda y_1 \ldots y_k.V$ *for some unsolvable* $V = V_0V_1$ *such that* $\mathrm{ord}_\beta(V) = 0$. *Moreover, if $k > 0$, then $\lambda y_k.V$ is not a potential η-redex.*

In particular, $\mathrm{ord}_{\beta\eta}(U) = 0 \iff \mathrm{ord}_\beta(U) = 0$.

PROOF. ($1 \Rightarrow 2$) By Remark 9.28, $\mathrm{ord}_{\beta\eta}(U) = k$ holds whenever there is a reduction $U \twoheadrightarrow_{\beta\eta} \lambda y_1 \ldots y_k.V$ such that $\mathrm{ord}_\beta(V) = 0$ and k is minimal for this property. If $k = 0$ then we are done, so consider $k > 0$. Then $\lambda y_k.V$ cannot be a potential η-redex as that would contradict the minimality of k.

($2 \Rightarrow 1$) It is easy to verify that (2) entails $\mathrm{ord}_{\beta\eta}(U) \leq k$, so if $k = 0$ we are done. In case $k > 0$, by the supplementary condition, $\lambda y_1 \ldots y_k.V \twoheadrightarrow_{\beta\eta} V'$ implies that V' has k external λ-abstractions. Now, suppose $U' \in \mathcal{G}_{\beta\eta}(U)$. Then, by confluence, U' has at least k external λ-abstractions. We conclude that $\mathrm{ord}_{\beta\eta}(U) = k$. □

These notations are useful for defining and studying the notion of extensional degree.

NOTATION 9.33. Given $W \in \mathrm{UNS}$ and $d \in \mathbb{N}$, write:

(i) $\mathcal{G}_{\beta\eta}(W) \leq_\# d \iff \forall V \in \mathcal{G}_{\beta\eta}(W)\,.\,\mathrm{core}_\#(V) \leq d$.

(ii) $W \rightsquigarrow_\# d \iff \forall V \in \mathcal{G}_{\beta\eta}(W),\ \exists V' \in \mathcal{G}_{\beta\eta}(V)\,.\,\mathrm{core}_\#(V') = d$.

DEFINITION 9.34. The *extensional degree* of $U \in \mathrm{UNS}$, written $\deg_{\beta\eta}(U)$, is defined by

$$\deg_{\beta\eta}(U) \triangleq \mu d\,.\,\exists W \in \mathcal{G}_{\beta\eta}(U)\,[\mathcal{G}_{\beta\eta}(W) \leq_\# d],$$

interpreting the μd in the right way ($= \infty$ if d does not exist).

In case of a finite extensional degree, the minimality condition on d can be expressed using the cofinal notion $W \rightsquigarrow_\# d$ in order to prevent that a $d' < d$ exists. This is shown in the next lemma. As a warm up for the proof, notice that for $d, k \in \mathbb{N}$ one has:

$$k \leq d \quad \Rightarrow \quad [(\forall d' < d\,.\,k > d') \iff k = d].$$

LEMMA 9.35. *Let $U \in \mathrm{UNS}$ and $d \in \mathbb{N}$. Then*

$$\deg_{\beta\eta}(U) = d \iff \exists W \in \mathcal{G}_{\beta\eta}(U) \,.\, [\mathcal{G}_{\beta\eta}(W) \leq_{\#} d \;\&\; W \rightsquigarrow_{\#} d]$$

In diagrammatic form:

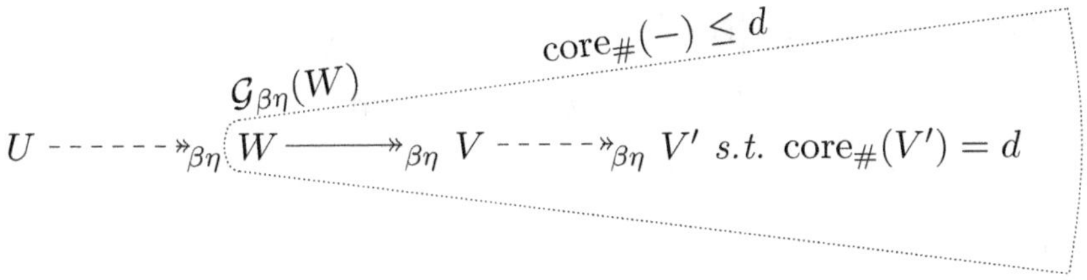

PROOF. ($\Rightarrow$) Assume $\deg_{\beta\eta}(U) = d \in \mathbb{N}$. Then

$$\exists W \in \mathcal{G}_{\beta\eta}(U) \,.\, [\mathcal{G}_{\beta\eta}(W) \leq_{\#} d] \tag{9.1}$$

and

$$\forall d' {<} d,\ \nexists W \in \mathcal{G}_{\beta\eta}(U),\ \forall V \in \mathcal{G}_{\beta\eta}(W) \,.\, \mathrm{core}_{\#}(V) \leq d'.$$

By pulling the negation through the quantifiers one obtains

$$\forall d' {<} d,\ \forall W \in \mathcal{G}_{\beta\eta}(U),\ \exists V \in \mathcal{G}_{\beta\eta}(W) \,.\, \mathrm{core}_{\#}(V) > d'. \tag{9.2}$$

By (9.1) there is a λ-term $W_1 \in \mathcal{G}_{\beta\eta}(U)$ such that $\mathcal{G}_{\beta\eta}(W_1) \leq_{\#} d$. Take $W_2 \in \mathcal{G}_{\beta\eta}(U)$. Note that both W_1, W_2 are $\beta\eta$-reducts of U, so by confluence they have a common reduct W_3. Applying (9.2) to $d' < d$ and W_3, there exists a $V \in \mathcal{G}_{\beta\eta}(W_3) \subseteq \mathcal{G}_{\beta\eta}(W_1) \leq_{\#} d$ such that $d \geq \mathrm{core}_{\#}(V) > d'$. Then, similarly to the warm-up reasoning above, we derive $\mathrm{core}_{\#}(V) = d$. We conclude that also $W_2 \rightsquigarrow_{\#} d$.

($\Leftarrow$) By hypothesis there exists a $W \in \mathcal{G}_{\beta\eta}(U)$ such that $\mathcal{G}_{\beta\eta}(W) \leq_{\#} d$ and

$$W \rightsquigarrow_{\#} d. \tag{9.3}$$

Suppose towards a contradiction that for some $d' < d$ and $W' \in \mathcal{G}_{\beta\eta}(U)$ one has

$$\mathcal{G}_{\beta\eta}(W') \leq_{\#} d'. \tag{9.4}$$

Let V be a common reduct of W, W'. By (9.3) there is $V' \in \mathcal{G}_{\beta\eta}(V)$ such that $\mathrm{core}_{\#}(V') = d$. Then $V' \in \mathcal{G}_{\beta\eta}(W') \leq_{\#} d'$, by (9.4), so that $\mathrm{core}_{\#}(V') \leq d' < d$, a contradiction. $\square$

EXAMPLES 9.36. (i) $\deg_{\beta\eta}(\mathsf{I}\!\mathsf{I}^{\sim 99}\Omega) = \deg_{\beta\eta}(\Omega) = 1$.

(ii) $\mathrm{core}(U) = \Omega$ implies $\deg_{\beta\eta}(U) = 1$.

(iii) The combinators $(\mathsf{X}_n)_{n \in \mathbb{N}}$ satisfy

$$\begin{array}{lll}
\mathsf{X}_n & = & (\lambda y_1 \ldots y_n x.x x y_1 \cdots y_n)\mathsf{W}_n^{\sim n+1} \quad \text{(quantitative core } n+1) \\
& \to_\beta & (\lambda y_2 \ldots y_n x.x x \mathsf{W}_n y_2 \cdots y_n)\mathsf{W}_n^{\sim n} \quad \text{(quantitative core } n) \\
& \to_\beta & (\lambda y_3 \ldots y_n x.x x \mathsf{W}_n \mathsf{W}_n y_3 \cdots y_n)\mathsf{W}_n^{\sim n-1} \quad \text{(quantitative core } n-1) \\
& \cdots & \\
& \twoheadrightarrow_\beta & (\lambda y_n x.x x \mathsf{W}_n \cdots \mathsf{W}_n y_n)\mathsf{W}_n \quad \text{(quantitative core } 1) \\
& \to_\beta & \mathsf{X}_n \quad \text{(quantitative core } n+1)
\end{array}$$

These terms have no η-redex, therefore $\deg_{\beta\eta}(\mathsf{X}_n) = n+1$, for all $n \in \mathbb{N}$. As no λ comes to the front one has $\mathrm{ord}_\beta(\mathsf{X}_n) = 0$, and therefore also $\mathrm{ord}_{\beta\eta}(\mathsf{X}_n) = 0$, by Remark 9.28.

EXERCISE 9.37. (i) For $n \in \mathbb{N}$, compute $\deg_{\beta\eta}(\mathsf{I}^{\sim n}\Delta(\lambda x.xx^{\sim n+1}))$.
(ii) Define an unsolvable U such that $\deg_{\beta\eta}(U) = \infty$.

COROLLARY 9.38. *For $U, V \in \mathrm{UNS}$ having finite extensional degree, we have:*

$$U =_{\beta\eta} V \quad \Rightarrow \quad \deg_{\beta\eta}(U) = \deg_{\beta\eta}(V)$$

PROOF. By Lemma 9.35, using confluence of $\beta\eta$-reduction. □

PROPOSITION 9.39. *Let $U \in \mathrm{UNS}$ and $P \in \Lambda$. Assume $\mathrm{ord}_{\beta\eta}(U), \deg_{\beta\eta}(U) \in \mathbb{N}$. Then*

$$\deg_{\beta\eta}(U) = \deg_{\beta\eta}(UP).$$

PROOF. Let $k = \mathrm{ord}_{\beta\eta}(U)$ and $d = \deg_{\beta\eta}(U)$, with $k \geq 0$ and $d > 0$, by Remark 9.22(i). If $k = 0$ then $\mathrm{ord}_\beta(U) = 0$, by Lemma 9.32. By applying Lemma 2.45(iii) we obtain

$$\mathcal{G}_{\beta\eta}(UP) = \{U'P' \mid U' \in \mathcal{G}_{\beta\eta}(U), P' \in \mathcal{G}_{\beta\eta}(P)\}$$

so the result follows using Remark 9.22(ii). The rest of the proof is devoted to the case $k > 0$. By Lemma 9.35, there exists $W \in \mathcal{G}_{\beta\eta}(U)$ such that $\mathcal{G}_{\beta\eta}(W) \leq_\# d$ and $W \leadsto_\# d$.

Claim 9.39.1. Wlog W has shape $W = \lambda x_1 \ldots x_k.\mathrm{core}(W)\vec{Q}$, with $\mathrm{ord}_\beta(\mathrm{core}(W)\vec{Q}) = 0$. Moreover, if $k > 0$ then $\lambda x_k.\mathrm{core}(W)\vec{Q}$ is not a potential η-redex.

Subproof. By Lemma 9.32, $U \twoheadrightarrow_{\beta\eta} \lambda x_1 \ldots x_k.U_0$ for some $U_0 \in \mathrm{UNS}$ of order 0. Also, if $k > 0$ then $\lambda x_k.U_0$ is not a potential η-redex. Since $U \twoheadrightarrow_{\beta\eta} W$, by $\beta\eta$-confluence there is a common reduct $W \twoheadrightarrow_{\beta\eta} Z \;{}_{\beta\eta}\!\twoheadleftarrow \lambda x_1 \ldots x_k.U_0$, so $Z = \lambda x_1 \ldots x_k.U_1$ for some $U_1 \;{}_{\beta\eta}\!\twoheadleftarrow U_0$. By applying Lemma 2.45(ii) also $\mathrm{ord}_\beta(U_1) = 0$. By Lemma 9.25(i), $U_1 \twoheadrightarrow_{\beta\eta} \mathrm{core}(U_1^\sigma)\vec{Q}$ for some substitution σ and λ-terms $\vec{Q}$. Hence $U \twoheadrightarrow_{\beta\eta} \lambda x_1 \ldots x_k.\mathrm{core}(U_1^\sigma)\vec{Q} \triangleq W_0$, with $\mathrm{core}(W_0)\vec{Q} = \mathrm{core}(U_1^\sigma)\vec{Q} \;{}_{\beta\eta}\!\twoheadleftarrow U_1$ having order 0. Therefore $W_0 \in \mathcal{G}_{\beta\eta}(W) \subseteq \mathcal{G}_{\beta\eta}(U)$ satisfies the conditions for W, with the right form and order. ■

In particular, we can assume that W is in canonical form. We now have:

$$\begin{array}{rcll} UP & \twoheadrightarrow_{\beta\eta} & (\lambda x_1 \ldots x_k.\mathrm{core}(W)\vec{Q})P, & \text{by Claim 9.39.1,} \\ & \to_\beta & \lambda x_2 \ldots x_k.\mathrm{core}(W)^\sigma \vec{Q}^\sigma, & \text{as } k > 0, \text{ setting } \sigma \triangleq [x_1 := P], \\ & = & \lambda x_2 \ldots x_k.\mathrm{core}(W^\sigma)\vec{Q}^\sigma, & \text{by Lemma 9.23(ii),} \\ & \triangleq & W', & \text{say.} \end{array}$$

Claim 9.39.2. $\forall V \in \mathcal{G}_{\beta\eta}(W), \exists V' \in \mathcal{G}_{\beta\eta}(W') \,.\, V' =_\# V$.

Subproof. By Claim 9.39.1, we must have $V = \lambda x_1 \ldots x_k.V_0$ for some unsolvable V_0 such that $\mathrm{core}(W)\vec{Q} \twoheadrightarrow_{\beta\eta} V_0$. Therefore $W' = \lambda x_2 \ldots x_k.\mathrm{core}(W^\sigma)\vec{Q}^\sigma \twoheadrightarrow_{\beta\eta} \lambda x_2 \ldots x_k.V_0^\sigma$. By Lemma 9.19(i), we have $\mathrm{core}(V) = \mathrm{core}(\lambda\vec{x}.V_0) = \mathrm{core}(V_0)$. Taking $V' \triangleq \lambda x_2 \ldots x_k.V_0^\sigma$, we get $\mathrm{core}(V') = \mathrm{core}(V_0^\sigma)$. By Lemma 9.23(iii), we obtain $V_0^\sigma =_\# V_0$. Therefore, we derive $V' =_\# V$. ■

Claim 9.39.3. $\forall V' \in \mathcal{G}_{\beta\eta}(W'), \exists V \in \mathcal{G}_{\beta\eta}(W) \,.\, V =_{\#} V'$.

Subproof. By Proposition 9.26, since $W' = (\lambda x_2 \ldots x_k.\mathrm{core}(W)\vec{Q})^{\sigma}$. ■

In order to show that also $\deg_{\beta\eta}(UP) = d$, by Lemma 9.35 it remains to show that $\mathcal{G}_{\beta\eta}(W') \leq_{\#} d$ and $W' \rightsquigarrow_{\#} d$. This means that for $W' \twoheadrightarrow_{\beta\eta} V'$ we must show that $\mathrm{core}_{\#}(V') \leq d$ and that this reduction can be extended by $V' \twoheadrightarrow_{\beta\eta} X'$ such that $\mathrm{core}_{\#}(X') = d$. This follows from Claims 9.39.3 and 9.39.2 and an easy diagram chase:

$$\begin{array}{ccccc}
W' & \xrightarrow{\beta\eta}\!\!\!\!\rightarrow & V' & \overset{\beta\eta}{\cdots\twoheadrightarrow} & X' \\
 & & =_{\#}\Big| & & =_{\#}\Big| \\
W & \overset{\beta\eta}{\cdots\twoheadrightarrow} & V & \overset{\beta\eta}{\cdots\twoheadrightarrow} & X
\end{array}$$

(Remember that $\mathcal{G}_{\beta\eta}(W) \leq_{\#} d$, thus $\mathrm{core}_{\#}(V) \leq d$, and $W \rightsquigarrow_{\#} d$, thus $W \twoheadrightarrow_{\beta\eta} V$ can be extended to X with $\mathrm{core}_{\#}(X) = d$.) □

THEOREM 9.40. *For all $U, V \in \mathrm{UNS}$ having finite extensional degree, we have:*

$$U \simeq_{\lambda\eta} V \quad \Rightarrow \quad \deg_{\beta\eta}(U) = \deg_{\beta\eta}(V).$$

PROOF. The relation $\simeq_{\lambda\eta}$ is the smallest equivalence relation containing $=_{\beta\eta}$ and $\trianglelefteq$. Corollary 9.38 and Proposition 9.39, respectively, show that the operation $\deg_{\beta\eta}(-)$ is an invariant for these relations. Therefore it is an invariant for $\simeq_{\lambda\eta}$. □

Theorem 9.40 can be generalized to unsolvables having arbitrary extensional degree. The converse however does not hold in general. In his PhD thesis, Folkerts (1995) has shown via an *ad hoc* argument that $(\Delta(\lambda x.xx^{\sim n+1}))_{n \in \mathbb{N}}$ are pairwise non-equi-unsolvable, despite sharing the same (infinite) extensional degree.

COROLLARY 9.41. *There exists an infinite family of combinators $(U_n)_{n \in \mathbb{N}}$ satisfying the following conditions (for all $n, m \geq 0$):*

1. $\mathrm{ord}_{\beta\eta}(U_n) = 0$,
2. *The U_n are mutually not equi-unsolvable, i.e. $U_n \simeq_{\lambda\eta} U_m$ entails $n = m$.*

PROOF. As shown in Example 9.36(iii), the family $(\mathsf{X}_n)_{n \in \mathbb{N}}$ consists of unsolvables of order 0 such that $\deg_{\beta\eta}(\mathsf{X}_n) = n + 1$. By Theorem 9.40, $\mathsf{X}_n \simeq_{\lambda\eta} \mathsf{X}_m$ entails $n = m$. □

The following are equivalent definitions of $\mathrm{ord}_{\beta\eta}(-)$, $\deg_{\beta\eta}(-)$, from Folkerts (1995).

EXERCISE 9.42. Given a function $f : \Lambda \to \mathbb{N}^{\infty}$, define:

$$\begin{array}{rcl}
(\sup_{\beta\eta} f)(M) & \triangleq & \sup\{\, f(N) \mid N \in \mathcal{G}_{\beta\eta}(M) \,\}; \\
(\inf_{\beta\eta} f)(M) & \triangleq & \min\{\, f(N) \mid N \in \mathcal{G}_{\beta\eta}(M) \,\}; \\
\overline{\lim}_{\beta\eta}(f) & \triangleq & (\inf_{\beta\eta} \circ \sup_{\beta\eta})(f).
\end{array}$$

Show that, for all $U \in \mathrm{UNS}$:

(i) $\mathrm{ord}_{\beta\eta}(U) = (\overline{\lim}_{\beta\eta}\, \mathrm{lam}_{\#})(U)$, where $\mathrm{lam}_{\#}(M) = k \iff M$ has k external λ's;
(ii) $\deg_{\beta\eta}(U) = (\overline{\lim}_{\beta\eta}\, \mathrm{core}_{\#})(U)$.

9.2 Invertibility in $\mathcal{M}^o(\lambda\eta)$

The following result was first proved using the partial validity of the ω-rule in $\boldsymbol{\lambda\eta}$ (namely, Theorem 17.3.24 in B[1984]).

PROPOSITION 9.43 (BATENBURG AND VELMANS 1983).
Let $F \in \Lambda^o$. In $\mathcal{M}^o(\boldsymbol{\lambda\eta})$ the following are equivalent:

1. *F is invertible.*

2. *F is right-invertible and injective.*

PROOF. $(1 \Rightarrow 2)$ Trivial.

$(2 \Rightarrow 1)$ (Avoiding the ω-rule.) Assume that F is right-invertible and injective in $\mathcal{M}^o(\boldsymbol{\lambda\eta})$. Then there exists an $R \in \Lambda^o$ such that $F \circ R =_{\beta\eta} \mathsf{I}$. So we have:

$$\begin{array}{rcll} F((R \circ F)\,\Omega) & =_{\beta\eta} & (F \circ (R \circ F))\,\Omega, & \text{by definition of } \circ, \\ & =_{\beta\eta} & ((F \circ R) \circ F)\,\Omega, & \text{by associativity of } \circ, \\ & =_{\beta\eta} & (\mathsf{I} \circ F)\,\Omega, & \text{as } R \text{ is the right-inverse of } F, \\ & =_{\beta\eta} & F\,\Omega, & \text{since } \mathsf{I} \circ F =_{\beta\eta} F. \end{array}$$

This shows that $F((R \circ F)\,\Omega) =_{\beta\eta} F\,\Omega$ holds, so we obtain $(R \circ F)\,\Omega =_{\beta\eta} \Omega$ from the injectivity of F in $\mathcal{M}^o(\boldsymbol{\lambda\eta})$. By Lemma 9.5(ii), F is surjective and therefore regular by Lemma 9.10. Since $F \in \mathrm{REG}$ we have $F =_{\beta\eta} (F \circ R) \circ F =_{\beta\eta} F \circ (R \circ F) \in \mathrm{REG}$ whence $R \circ F \in \mathrm{REG}$ by Lemma 9.9(iii) and this entails $R \circ F =_{\beta} \lambda y x_1 \ldots x_n . y N_1 \cdots N_k$. So we get $(R \circ F)\Omega =_{\beta\eta} \lambda \vec{x}.\Omega(N_1[y := \Omega]) \cdots (N_k[y := \Omega]) =_{\beta\eta} \Omega$ which is only possible if $n = k$ and $N_i[y{:=}\Omega] =_{\beta\eta} x_i$. By the Genericity Lemma for $\boldsymbol{\lambda\eta}$ (Corollary 1.82) we obtain $N_i =_{\beta\eta} x_i$. We conclude that $R \circ F =_{\beta\eta} \mathsf{I}$, so R is also a left-inverse of F.

$(2 \Rightarrow 1)$ [Original proof in Batenburg and Velmans (1983)]. Assuming that F is injective and has a right-inverse R, one has $F \circ R =_{\beta\eta} \mathsf{I}$. Therefore

$$\begin{array}{rcll} F(RZ) & =_{\beta\eta} & Z, & \text{for all } Z \in \Lambda^o, \\ F(R(FZ)) & =_{\beta\eta} & FZ, & \text{for all } Z \in \Lambda^o, \\ R(FZ) & =_{\beta\eta} & Z, & \text{for all } Z \in \Lambda^o \text{ since } F \text{ is injective on } \mathcal{M}^o(\boldsymbol{\lambda\eta}), \\ (R \circ F)\,Z & =_{\beta\eta} & \mathsf{I}Z, & \text{for all } Z \in \Lambda^o, \\ R \circ F & =_{\beta\eta} & \mathsf{I}, & \text{since the } \omega\text{-rule holds in } \boldsymbol{\lambda\eta}, \text{ Theorem 17.3.24} \\ & & & \text{in B[1984], as } \mathsf{I} \text{ is not a universal generator.} \end{array}$$

Therefore R is both right and left inverse of F, hence the latter is invertible. □

LEMMA 9.44. *Let $U, V_0 \in \mathrm{UNS}$, $\mathrm{ord}_\beta(V_0) = 0$ and σ be a substitution. Suppose that*

$$\boldsymbol{\lambda\eta} \vdash U^\sigma = V_0 V_1 \cdots V_n, \text{ with } n \geq 0.$$

Then there exist $U_0 \in \mathrm{UNS}$ and $U_1 \cdots U_n \in \Lambda$ such that

$$\begin{array}{rcll} \boldsymbol{\lambda\eta} \vdash U & = & U_0 U_1 \cdots U_n, & \\ \boldsymbol{\lambda\eta} \vdash U_i^\sigma & = & V_i, & \text{for all } i\ (0 \leq i \leq n). \end{array}$$

PROOF. Wlog, consider the case $n = 2$. As $U^\sigma =_{\beta\eta} V_0V_1V_2$, by $\beta\eta$-confluence these two terms have a common $\beta\eta$-reduct, say W. Since $\mathrm{ord}_\beta(V_0) = 0$, by Lemma 2.45(iii), W must have shape $W = W_0W_1W_2$ with $V_0 \twoheadrightarrow_{\beta\eta} W_0$, $V_1 \twoheadrightarrow_{\beta\eta} W_1$, $V_2 \twoheadrightarrow_{\beta\eta} W_2$, and

$$U^\sigma \twoheadrightarrow_{\beta\eta} W_0W_1W_2 \;{}_{\beta\eta}\!\twoheadleftarrow V_0V_1V_2$$

By the Corollary 1.68 of the Standardization Theorem, the reduction $U^\sigma \twoheadrightarrow_{\beta\eta} W$ factorizes as follows

$$\begin{array}{rll} U^\sigma & \twoheadrightarrow_h & \lambda z_1 \dots z_k.W_0'W_1'W_2'Z_1 \cdots Z_k \;\triangleq N, \text{ say,} \\ & \twoheadrightarrow^{\iota}_{\beta\eta} & \lambda\vec{z}.W_0W_1W_2\vec{z} \\ & \twoheadrightarrow_\eta & W_0W_1W_2 = W. \end{array}$$

with

- $W_i' \twoheadrightarrow_{\beta\eta} W_i$, for all $i\,(0 \le i \le 2)$,
- $Z_j \twoheadrightarrow_{\beta\eta} z_j$ and $z_j \notin \mathrm{FV}(W_0W_1W_2)$, for all $j\,(1 \le j \le k)$.

By Lemma 9.23(i), there exists an unsolvable λ-term L such that $U \twoheadrightarrow_h L$ and $L^\sigma = N$. Therefore

$$L = \lambda z_1 \dots z_k.L_0L_1L_2Z_1' \cdots Z_k'$$

satisfying, for all $i\,(0 \le i \le 2)$ and $j\,(1 \le j \le k)$, the following:

- $L_i^\sigma = W_i'$,
- $L_i \twoheadrightarrow_{\beta\eta} L_i'$, for some L_i' such that $\vec{z} \notin \mathrm{FV}(L_i')$,
- $(Z_j')^\sigma \twoheadrightarrow_{\beta\eta} z_j$ and $z_j \notin \mathrm{FV}(L_0'L_1'L_2')$.

So we can take:

- $U_0 \triangleq \lambda w_1w_2z_1 \dots z_k.L_0w_1w_2Z_1' \cdots Z_k'$, for fresh variables w_1, w_2,
- $U_i \triangleq L_i$, for all $i\,(1 \le i \le 2)$.

Indeed, we have:

$$\begin{array}{rll} U_0U_1U_2 & =_\beta & \lambda\vec{z}.L_0U_1U_2\vec{Z}' \\ & =_{\beta\eta} & \lambda\vec{z}.L_0'L_1'L_2'\vec{z} \\ & =_\eta & L_0'L_1'L_2' \\ & =_{\beta\eta} & L \\ & =_{\beta\eta} & U. \end{array}$$

Moreover $U_0 = \lambda w_1w_2\vec{z}.L_0w_1w_2\vec{Z}' =_{\beta\eta} \lambda w_1w_2\vec{z}.L_0w_1w_2\vec{z} =_\eta L_0$, so that

$$U_0^\sigma =_{\beta\eta} L_0^\sigma = W_0' =_{\beta\eta} W_0 =_{\beta\eta} V_0.$$

Finally $U_1^\sigma = L_1^\sigma =_{\beta\eta} W_1' =_{\beta\eta} W_1 =_{\beta\eta} V_1$ and similarly $U_2^\sigma =_{\beta\eta} V_2$. □

PROPOSITION 9.45 (TRANSLATION LEMMA).
Let $F \in \Lambda^o$. Then the following are equivalent.

1. *F is regularly right-invertible, i.e. for some regular $R \in \Lambda^o$ one has $\boldsymbol{\lambda\eta} \vdash F \circ R = \mathsf{I}$.*
2. *$F \in \mathrm{REG}$ and $\boldsymbol{\lambda\eta} \vdash FU = \Omega$, for some $U \in \mathrm{UNS}$.*
3. *$F \in \mathrm{REG}$ and $\boldsymbol{\lambda\eta} \vdash FU = V$, for some $U, V \in \mathrm{UNS}$ with $\mathrm{ord}_\beta(V) = 0$.*

PROOF. ($1 \Rightarrow 2$) By assumption $F(R\,\mathsf{I}) =_{\beta\eta} \mathsf{I}$, thus F is solvable. Since F is moreover closed, we obtain $F =_{\beta\eta} \lambda x_0 \ldots x_n.x_i P_1 \cdots P_k$ for some $n, k \geq 0$. Assuming $i > 0$ we get

$$\Omega =_{\beta\eta} F(R\,\Omega) =_{\beta\eta} \lambda x_1 \ldots x_n.x_i P_1^\sigma \cdots P_k^\sigma$$

for $\sigma \triangleq [x_0 := R\,\Omega]$, thus contradicting the fact that Ω is unsolvable. Hence, we must have $i = 0$, which entails $F =_{\beta\eta} \lambda x_0 \ldots x_n.x_0 P_1 \cdots P_k \in \mathrm{REG}$. Finally, taking $U \triangleq R\,\Omega$, we obtain $U \in \mathrm{UNS}$ since $R \in \mathrm{REG}$ and $FU = F(R\,\Omega) =_{\beta\eta} \Omega$.

($2 \Rightarrow 3$) Trivial.

($3 \Rightarrow 1$) As $F \in \mathrm{REG}$ we have $F =_\beta \lambda x y_1 \ldots y_n.x P_1 \cdots P_k$. Since $FU =_{\beta\eta} V$ one has

$$\begin{array}{rcll} V y_1 \cdots y_n & =_{\beta\eta} & F U y_1 \cdots y_n & \\ & =_{\beta\eta} & U P_1^{\sigma_U} \cdots P_k^{\sigma_U}, & \text{with } \sigma_U \triangleq [x := U], \\ & = & (U z_1 \cdots z_k)^{\tau_U}, & \text{with } \tau_U \triangleq [z_1 := P_1^{\sigma_U}, \ldots, z_k := P_k^{\sigma_U}]. \end{array}$$

Therefore, we have $(U z_1 \cdots z_k)^{\tau_U} =_{\beta\eta} V y_1 \cdots y_n$ for $U, V \in \mathrm{UNS}$ such that $\mathrm{ord}_\beta(V) = 0$. Setting $W \triangleq U z_1 \cdots z_k$, $V_0 \triangleq V$ and $V_1 \triangleq y_1, \ldots, V_n \triangleq y_n$, we obtain

$$W^{\tau_U} =_{\beta\eta} V_0 V_1 \cdots V_n.$$

By Lemma 9.44, there are $W_0 \in \mathrm{UNS}, W_1, \ldots, W_n \in \Lambda$ such that $W =_{\beta\eta} W_0 W_1 \cdots W_n$ and $W_0^{\tau_U} =_{\beta\eta} V_0, W_1^{\tau_U} =_{\beta\eta} y_1, \ldots, W_n^{\tau_U} =_{\beta\eta} y_n$. By the Genericity Lemma (Corollary 1.82) applied to $\lambda x.W_i[z_1 := P_1, \ldots, z_k := P_k]$, we obtain:[3]

$$W_i^{\tau_M} =_{\beta\eta} y_i, \text{ for all } M \in \Lambda \text{ and } i \, (1 \leq i \leq n). \tag{9.5}$$

We claim that $R \triangleq \lambda w z_1 \ldots z_k.w W_1 \cdots W_n \in \mathrm{REG}$ is the right-inverse of F. Indeed,

$$\begin{array}{rcll} F \circ R & =_{\beta\eta} & \lambda w.F(Rw), & \\ & =_{\beta\eta} & \lambda w.(\lambda x y_1 \ldots y_n.x P_1 \cdots P_k)(Rw), & \text{as } F =_\beta \lambda x y_1 \ldots y_n.x P_1 \cdots P_k, \\ & =_{\beta\eta} & \lambda w y_1 \ldots y_n.Rw P_1^{\sigma_{Rw}} \cdots P_k^{\sigma_{Rw}}, & \text{with } \sigma_{Rw} \triangleq [x := Rw], \\ & =_{\beta\eta} & \lambda w y_1 \ldots y_n.w W_1^{\tau_{Rw}} \cdots W_n^{\tau_{Rw}}, & \text{as } Rw =_\beta \lambda z_1 \ldots z_k.w W_1 \cdots W_n, \\ & & & \text{for } \tau_{Rw} \triangleq [z_1 := P_1^{\sigma_{Rw}}, \ldots, z_k := P_k^{\sigma_{Rw}}], \\ & =_{\beta\eta} & \lambda w y_1 \ldots y_n.w y_1 \cdots y_n, & \text{by (9.5) instantiated to } M \triangleq Rw, \\ & =_{\beta\eta} & \lambda w.w. & \end{array}$$

□

[3] The notation τ_M in (9.5) stands for the substitution τ_U defined above with M replacing U.

DEFINITION 9.46. A λ-term F is *faithful* if it is regular, say $F =_\beta \lambda xy_1 \ldots y_n.xP_1 \cdots P_k$, and $\lambda x.P_j$ has empty head functionality, for all $j\,(1 \le j \le k)$.

EXAMPLES 9.47. (i) The regular λ-terms I, $\mathbf{1}$, $\lambda x.x\Omega$, $\lambda x.xy_1 \cdots y_n$ are faithful.

(ii) The regular λ-terms $\lambda x.xx$, $\lambda x.xx\Omega$, $\lambda x.xy\mathsf{I}$, $\lambda x.x(\lambda z.zx)$ are not faithful.

REMARK 9.48. (i) Consider a regular λ-term $F =_\beta \lambda xy_1 \ldots y_n.xP_1 \cdots P_k \in \Lambda^o$. Then

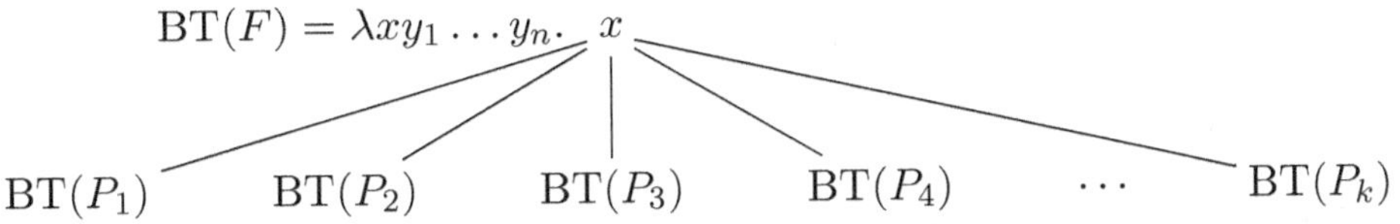

Four different forms are possible for the top level of $\mathrm{BT}(P_i)$: $\lambda\vec{z}.x$, $\lambda\vec{z}.y_i$, $\lambda\vec{z}.z_j$, and $\bot$. For F to be faithful the first and third ($\lambda\vec{z}.x$ and $\lambda\vec{z}.z_j$) are not allowed, the second and fourth ($\lambda\vec{z}.y_i$ and $\bot$) are allowed. So we have

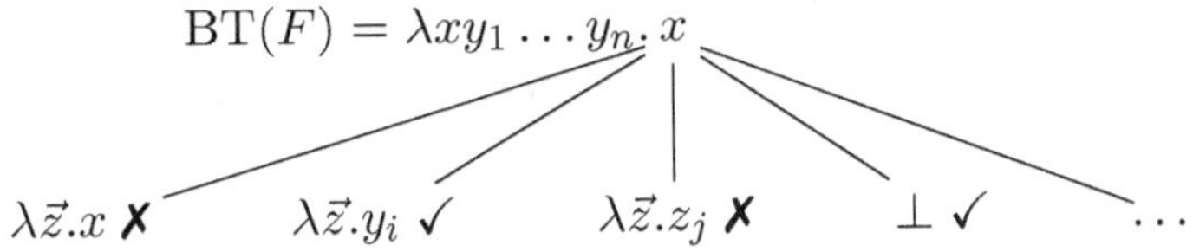

(ii) It is easy to check that the notion of faithfulness is invariant under $\beta\eta$-conversion. This follows from the invariance of solvability and Remark 9.7(i).

LEMMA 9.49. *Assume that $F \in \Lambda^o$ is regular and injective in $\mathcal{M}^o(\boldsymbol{\lambda}\eta)$. Then F is faithful.*

PROOF. By regularity $F =_\beta \lambda xy_1 \ldots y_n.xP_1 \cdots P_k$. We have to show that every $\mathrm{BT}(P_i)$ is of the allowed shape. Suppose towards a contradiction that, for some i, one has either

1. $P_i =_{\beta\eta} \lambda z_1 \ldots z_m.xQ_1 \cdots Q_q$, or
2. $P_i =_{\beta\eta} \lambda z_1 \ldots z_m.z_jQ_1 \cdots Q_q$, with $1 \le j \le m$.

Wlog, we may assume $q \ge k$, by taking η-expansions of P_i. We treat the two cases.

Case 1. Define $N \triangleq \lambda w_1 \ldots w_q.[w_i\mathsf{I}^{\sim m}\mathsf{F}, \mathsf{I}]$. We have:

$$
\begin{array}{lll}
 & FNy_1 \cdots y_nw_{k+1} \cdots w_q & \\
=_{\beta\eta} & NP_1^\sigma \cdots P_k^\sigma w_{k+1} \cdots w_q, & \text{for } \sigma \triangleq [x{:=}N],\\
=_{\beta\eta} & [P_i^\sigma\mathsf{I}^{\sim m}\mathsf{F}, \mathsf{I}], & \\
=_{\beta\eta} & [(\lambda z_1 \ldots z_m.NQ_1^\sigma \cdots Q_q^\sigma)\mathsf{I}^{\sim m}\mathsf{F}, \mathsf{I}], & \\
=_{\beta\eta} & [(NQ_1^{\sigma\sigma_1\cdots\sigma_m} \cdots Q_q^{\sigma\sigma_1\cdots\sigma_m})\mathsf{F}, \mathsf{I}], & \text{where } \sigma_i \triangleq [z_i{:=}\mathsf{I}] \text{ for all } i\,(1 \le i \le m),\\
=_{\beta\eta} & [[Q_i^{\sigma\sigma_1\cdots\sigma_m}\mathsf{I}^{\sim m}\mathsf{F}, \mathsf{I}]\mathsf{F}, \mathsf{I}], & \text{by definition of } N,\\
=_{\beta\eta} & [\mathsf{I}, \mathsf{I}] &
\end{array}
$$

More easily, for $N' \triangleq \lambda w_1 \ldots w_q.[\mathsf{I}, \mathsf{I}]$ we get

$$FN'y_1 \cdots y_nw_{k+1} \cdots w_q =_{\beta\eta} N'(P_1[x{:=}N']) \cdots (P_k[x{:=}N'])w_{k+1} \cdots w_q =_{\beta\eta} [\mathsf{I}, \mathsf{I}].$$

We conclude $FN =_{\beta\eta} \lambda y_1 \ldots y_nw_{k+1} \ldots w_q.[\mathsf{I}, \mathsf{I}] =_{\beta\eta} FN'$ while $N \neq_{\beta\eta} N'$, contradicting the fact that F is injective.

Case 2. Now define $N \triangleq \lambda x_1 \ldots x_i.x_i(\mathsf{K}^\star)^{\sim j}$ and $N' \triangleq \mathsf{K}^\star$, with $\mathsf{K}^\star = \mathsf{YK}$. Again, we have $FN =_{\beta\eta} \mathsf{K}^\star =_{\beta\eta} FN'$, while $N \neq_{\beta\eta} N'$, thus contradicting the injectivity of F.

As we got in both cases a contradiction it follows that F is faithful. □

LEMMA 9.50. *Let $F \triangleq \lambda xy_1 \ldots y_n.xP_1 \cdots P_k \in \Lambda^o$ be faithful. For all $U, V \in \Lambda^o$ satisfying $U \in \mathrm{UNS}$, $V \in \mathrm{SOL}$ and $\boldsymbol{\lambda\eta} \vdash FV = U$ we have, for all j $(1 \leq j \leq k)$:*

$$\boldsymbol{\lambda\eta} \vdash V = \lambda z_1 \ldots z_m.z_j Q_1 \cdots Q_q \quad \Rightarrow \quad U \simeq_{\boldsymbol{\lambda\eta}} P_j.$$

PROOF. Wlog, assume $m \leq k$, by making some η-expansions of F. One has

$$\begin{aligned} Uy_1 \cdots y_n &=_{\beta\eta} FVy_1 \cdots y_n, && \text{hence by the shape of } F, \\ &=_{\beta\eta} VP_1^* \cdots P_k^*, && \text{with } * \triangleq [x := V], \\ &=_{\beta\eta} (\lambda z_1 \ldots z_m.z_j Q_1 \cdots Q_q)P_1^* \cdots P_k^*, && \text{since } V \text{ is solvable}, \\ &=_{\beta\eta} P_j^* Q_1^\sigma \cdots Q_q^\sigma P_{m+1}^* \cdots P_k^*, && \text{with } \sigma \triangleq [z_i := P_i^*]_{1 \leq i \leq m}. \end{aligned}$$

It follows that $\lambda x.P_j$ and therefore P_j^* is unsolvable, as otherwise by faithfulness of F we have $\mathrm{hvar}(\lambda x.P_i) \in \{y_1, \ldots, y_n\}$, contradicting that U, hence $Uy_1 \cdots y_n$, is unsolvable. We conclude as follows:

$$\begin{aligned} U &\simeq_{\boldsymbol{\lambda\eta}} Uy_1 \cdots y_n, && \text{by definition of } \simeq_{\boldsymbol{\lambda\eta}}, \\ &=_{\beta\eta} P_j^* Q_1^\sigma \cdots Q_q^\sigma P_{m+1}^* \cdots P_k^*, && \text{as above}, \\ &\simeq_{\boldsymbol{\lambda\eta}} P_j^*, && \text{using } \trianglelefteq \text{ and symmetry}, \\ &\simeq_{\boldsymbol{\lambda\eta}} P_j, && \text{by Lemma 9.12(ii) as } P_j \in \mathrm{UNS}. \end{aligned}$$

□

PROPOSITION 9.51. *Let $F \in \Lambda^o$ be surjective in $\mathcal{M}^o(\boldsymbol{\lambda\eta})$ and faithful. Then F is right-invertible.*

PROOF. Let $F =_{\beta\eta} \lambda xy_1 \ldots y_n.xP_1 \cdots P_k$. By Corollary 9.41 there exist infinitely many unsolvable combinators of order 0 that are pairwise non equi-unsolvable modulo $\boldsymbol{\lambda\eta}$. Take now an unsolvable $U \in \Lambda^o$ with $\mathrm{ord}_\beta(U) = 0$, such that $U \not\simeq_{\boldsymbol{\lambda\eta}} P_j$, for all j $(1 \leq j \leq k)$. Since F is surjective in $\mathcal{M}^o(\boldsymbol{\lambda\eta})$, there exists $V \in \Lambda^o$ satisfying $FV =_{\beta\eta} U$. By Lemma 9.50, V must be unsolvable. We conclude that F is (regularly) right-invertible by applying the Translation Lemma 9.45. □

THEOREM 9.52. *Let $F \in \Lambda^o$. In $\mathcal{M}^o(\boldsymbol{\lambda\eta})$ the following are equivalent:*

1. *F is invertible.*
2. *F is bijective.*

PROOF. $(1 \Rightarrow 2)$ This implication corresponds to Lemma 9.5(iii).

$(2 \Rightarrow 1)$ Assume that F is bijective in $\mathcal{M}^o(\boldsymbol{\lambda\eta})$. Then F is surjective and thus regular by Lemma 9.10. By bijectivity F is also injective. Hence, F is faithful by Lemma 9.49. Now it follows by Proposition 9.51 that F is right-invertible. Therefore F is invertible by Proposition 9.43. □

9.3 Partial characterizations of {left|right}-invertibility

In this section we present partial characterizations of left-invertibility, right-invertibility and invertibility in the principal term models $\mathcal{M}^{(o)}(\boldsymbol{\lambda})$ and $\mathcal{M}^{(o)}(\boldsymbol{\lambda}\eta)$. As mentioned in the introduction of this chapter, this approach fits in the λ-calculus tradition of finding characterizations in terms of syntactic properties of the λ-term in question.

Characterizations in $\mathcal{M}^{(o)}(\boldsymbol{\lambda})$

The following notions are introduced in Margaria and Zacchi (1983) under a different name.

DEFINITION 9.53. Let $T \in \mathscr{B}$.

(i) A variable x is called *reachable (fruit)* in T if there is an occurrence of x in T that is final (a leaf) and all head variables occurring on the nodes of the branch (from the root) to that occurrence of x (not included) are bound.

(ii) A variable x is called *easily reachable (fruit)* in T if there is a final occurrence of x in T and a branch α to x, such that:

(1) all head variables in the nodes on α are bound variables;

(2) none of the head variables occurring on the branch α is repeated.

(iii) An $F \in \Lambda$ is called *head-reachable* if a fresh variable x is reachable in $\mathrm{BT}(Fx)$.

EXAMPLES 9.54. Consider

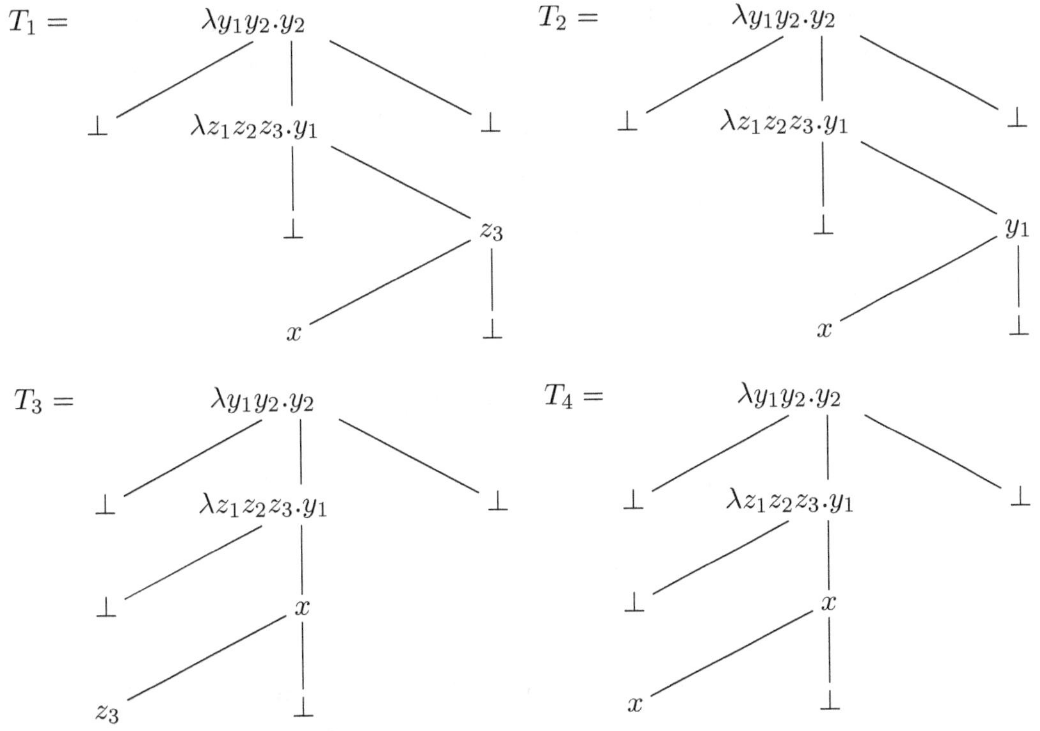

Then x is easily reachable in T_1, reachable in T_2, and not at all reachable in T_3, T_4.

LEMMA 9.55. *Let $F \in \Lambda$ and let x be fresh variable.*

(i) *If there is an easily reachable occurrence of x in* $\mathrm{BT}(Fx)$, *then for some $\vec{P} \in \Lambda^o$*

$$Fx\vec{P} =_\beta x.$$

(ii) *If F is head-reachable, then for some $\vec{Q} \in \Lambda^o$ we have:*

$$Fx\vec{Q} =_\beta x.$$

PROOF. (i) The conclusion will be proved by an exemplary example. Take

$$F \triangleq \lambda xy_1y_2.y_2\Omega(\lambda z_1z_2z_3.y_1\Omega(z_3x\Omega))\Omega,$$

with $\mathrm{BT}(Fx) = T_1$. One can 'guide this term to x' (in $\mathrm{BT}(Fx)$, i.e. x can be reached indeed) by applying suitable combinators:

$$\begin{array}{rll} Fx\mathsf{U}^2_2\mathsf{U}^3_2 & =_\beta & (\lambda xy_1y_2.y_2\Omega(\lambda z_1z_2z_3.y_1\Omega z_3x\Omega)\Omega)x\mathsf{U}^2_2\mathsf{U}^3_2, \\ & =_\beta & \mathsf{U}^3_2\Omega(\lambda z_1z_2z_3.\mathsf{U}^2_2\Omega(z_3x\Omega))\Omega \\ & =_\beta & \lambda z_1z_2z_3.\mathsf{U}^2_2\Omega(z_3x\Omega), \\ & =_\beta & \lambda z_1z_2z_3.z_3x\Omega; \qquad \text{hence} \\ Fx\mathsf{U}^2_2\mathsf{U}^3_2\Omega\Omega\mathsf{U}^2_1 & =_\beta & x. \end{array}$$

(ii) Now consider $F \triangleq \lambda xy_1y_2.y_2\Omega(\lambda z_1z_2z_3.y_1\Omega(y_1x\Omega))\Omega$ with $\mathrm{BT}(Fx) = T_2$ having an occurrence of a less easily reachable x. Using the Böhm-out technique, used to prove the separability of distinct $\beta\eta$-normal forms, take $\mathsf{P}_n = \lambda x_1 \ldots x_n z.zx_1 \cdots x_n$ and compute

$$\begin{array}{rll} Fx\mathsf{P}_3\mathsf{P}_3 & =_\beta & \mathsf{P}_3\Omega(\lambda z_1z_2z_3.\mathsf{P}_3\Omega(\mathsf{P}_3x\Omega))\Omega \\ & =_\beta & \langle \Omega, \lambda z_1z_2z_3.\mathsf{P}_3\Omega(\mathsf{P}_3x\Omega), \Omega\rangle. \end{array}$$

Define $F' \triangleq \lambda x.Fx\mathsf{P}_3\mathsf{P}_3$. Note that there is an easily reachable occurrence of x in

$\mathrm{BT}(F'x) = \mathrm{BT}(Fx\mathsf{P}_3\mathsf{P}_3) =$

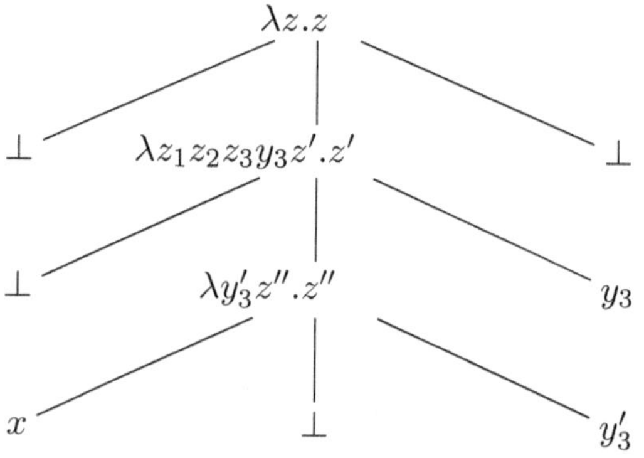

By (i) there exist $\vec{P}$ such that $F'x\vec{P} =_\beta x$. Therefore we can take $[\,]\vec{Q} \triangleq [\,]\mathsf{P}_3\mathsf{P}_3\vec{P}$:

$$Fx\mathsf{P}_3\mathsf{P}_3\vec{P} =_\beta F'x\vec{P} =_\beta x. \qquad \square$$

DEFINITION 9.56. An $F \in \Lambda$ is called *strongly regular* if $F =_\beta \lambda x.x\vec{M}$ for some $\vec{M} \in \Lambda$.

PROPOSITION 9.57 (MARGARIA AND ZACCHI (1983)).
Let $F \in \Lambda^{(o)}$. In $\mathcal{M}^{(o)}(\boldsymbol{\lambda})$, we have:
(i) *F is right-invertible $\iff$ F is strongly regular.*
(ii) *F is left-invertible $\iff$ F is head-reachable.*

PROOF. (i) ($\Rightarrow$) Let R be a right-inverse of F. Then $F \circ R =_\beta \mathsf{I}$ so that $F(Rx) =_\beta x$. In particular $F(R\mathsf{I}) =_\beta \mathsf{I}$, hence F is solvable, implying $F =_\beta \lambda x_1 \ldots x_n.x_i M_1 \cdots M_k$. As $F(R\,\Omega) =_\beta \Omega$, one has $x_i = x_1$, as Ω is unsolvable. Moreover $n = 1$, since Ω has order 0.

($\Leftarrow$) For $F =_\beta \lambda x.x M_1 \cdots M_k$, one has $R \triangleq \lambda x y_1 \ldots y_k.x$ as right-inverse.

(ii) ($\Leftarrow$) Let x be reachable in $\mathrm{BT}(Fx)$. By Lemma 9.55(ii), there exist $\vec{Q} \in \Lambda^o$ such that $Fx\vec{Q} =_\beta x$. Then F has a left-inverse $L \triangleq \lambda z.z\vec{Q}$.

($\Rightarrow$) Suppose that $T = \mathrm{BT}(Fx)$ has no reachable occurrence of x. We show by example that such an F is not left-invertible. There are several reasons why x may not be reachable in a tree; these are displayed in the following examples

$$T_5 = \mathrm{BT}(\lambda z.z), \quad T_6 = \mathrm{BT}(\lambda z.ux), \quad T_7 = \mathrm{BT}(\lambda z.xzz).$$

Using underlining one can show that there is no $L \in \Lambda^o$ such that one of the following equations holds:[4] 1. $L(\lambda z.z) =_\beta x$, 2. $L(\lambda z.ux) =_\beta x$, or 3. $L(\lambda z.xzz) =_\beta x$. Therefore there is no L that is a left-inverse for $\lambda xz.z, \lambda xz.ux$, or $\lambda xz.xzz$. These examples are sufficiently representative to make the conclusion in general. $\square$

The following result characterizes the invertibility in both $\mathcal{M}(\boldsymbol{\lambda})$ and $\mathcal{M}^o(\boldsymbol{\lambda})$.

PROPOSITION 9.58 (BÖHM AND DEZANI-CIANCAGLINI (1974)). *Let $F \in \Lambda^{(o)}$. Then*

$$F \text{ is invertible in } \mathcal{M}^{(o)}(\boldsymbol{\lambda}) \iff F =_\beta \mathsf{I}.$$

PROOF. ($\Rightarrow$) Suppose $F \circ G =_\beta G \circ F =_\beta \mathsf{I}$. Then both F and G are right-invertible in $\mathcal{M}(\boldsymbol{\lambda})$. By Proposition 9.57(i), one has $F =_\beta \lambda x.x\vec{M}$ and $G =_\beta \lambda y.y\vec{N}$. Now, setting $\sigma \triangleq [y := Fx]$, we get

$$x =_\beta G(Fx) =_\beta Fx\vec{N}^\sigma =_\beta x\vec{M}\vec{N}^\sigma,$$

implying that $\vec{M} = \vec{N}^\sigma = \emptyset$. We conclude that $F =_\beta G =_\beta \mathsf{I}$.

($\Leftarrow$) Trivial. $\square$

[4]The non-existence of a λ-term L satisfying one of the three equations can be proved also without underlining as follows. 1. Suppose towards a contradiction that $L\mathsf{I} =_\beta x$ for some L. Substitution $[x := P]$ yields $L\mathsf{I} =_\beta P$ for all P, thus contradicting the consistency of $\boldsymbol{\lambda}$. 2. Now substitution $[u := \mathsf{KI}, x := P]$ in $L(\lambda z.ux) =_\beta x$ similarly yields a contradiction. 3. Using substitution $[x := \lambda y.xy]$ one obtains

$$x =_\beta L(\lambda z.xzz) =_\beta L(\lambda z.(\lambda y.xy)zz) =_\beta \lambda y.xy,$$

therefore $x =_\beta \lambda y.xy$, which is not the case by the Church-Rosser Theorem for β-conversion (even if it is a consistent equation). Other substitutions even leads to a λ-calculus inconsistency:

$$\mathsf{I} =_\beta L(\lambda z.\mathsf{I}zz) =_\beta L(\lambda z.zz) =_\beta L(\lambda z.(\lambda xy.yy)zz) =_\beta \lambda xy.yy,$$

therefore $\mathsf{I} =_\beta \lambda xy.yy$, from which any equation is derivable.

Characterizations in $\mathcal{M}^{(o)}(\boldsymbol{\lambda}\eta)$

LEMMA 9.59 (FOLKERTS (1995)). *For $L, R \in \text{REG}$, we have:*

$$\begin{array}{lll} L \circ R =_{\beta\eta} \mathsf{I} & \iff & \exists w, z_1, \ldots, z_n, \exists L_1, \ldots, L_m \in \Lambda, \\ & & \exists x, y_1, \ldots, y_m, \exists R_1, \ldots, R_n \in \Lambda . \\ & & [\, L =_{\beta\eta} \lambda w z_1 \ldots z_n.w L_1 \cdots L_m \;\& \\ & & \; R =_{\beta\eta} \lambda x y_1 \ldots y_m.x R_1 \cdots R_n \;\; \& \\ & & \; (\lambda\vec{y}.R_1)\vec{L} =_{\beta\eta} z_1 \;\&\; \cdots \;\&\; (\lambda\vec{y}.R_n)\vec{L} =_{\beta\eta} z_n \,]. \end{array}$$

PROOF. ($\Rightarrow$) Since $L, R \in \Lambda^o$ are regular one has

$$\begin{array}{lll} L & =_{\beta\eta} & \lambda w z_1 \ldots z_n.w L_1 \cdots L_m, \\ R & =_{\beta\eta} & \lambda x y_1 \ldots y_{m'}.x R_1 \cdots R_{n'}, \end{array}$$

for some $\vec{L}, \vec{R} \in \Lambda$. By making some η-expansions in L or R we may assume that $m' = m$. By the Variable Convention, we also assume $x, \vec{y} \notin \text{FV}(\vec{L})$ and $w, \vec{z} \notin \text{FV}(\vec{R})$. Now

$$\begin{array}{lll} x & =_{\beta\eta} & (L \circ R)x \\ & =_{\beta\eta} & L(Rx) \\ & =_{\beta\eta} & \lambda z_1 \ldots z_n.Rx L_1^* \cdots L_m^*, \qquad\qquad \text{with } * \triangleq [w := Rx], \\ & =_{\beta\eta} & \lambda z_1 \ldots z_n.x(R_1[\vec{y} := \vec{L}^*]) \cdots (R_{n'}[\vec{y} := \vec{L}^*]), \qquad \text{since } m = m', \\ & =_{\beta\eta} & \lambda z_1 \ldots z_n.x((\lambda\vec{y}.R_1)\vec{L}^*) \cdots ((\lambda\vec{y}.R_{n'})\vec{L}^*) \\ & \Rightarrow & n' = n \;\&\; (\lambda\vec{y}.R_1)\vec{L}^* =_{\beta\eta} z_1 \;\&\; \cdots \;\&\; (\lambda\vec{y}.R_{n'})\vec{L}^* =_{\beta\eta} z_n, \end{array}$$

by the Church-Rosser Theorem for $\twoheadrightarrow_{\beta\eta}$. By replacing Ω for x we obtain, for every j,

$(\lambda\vec{y}.R_j[x := \Omega])(L_1[w := R\,\Omega]) \cdots (L_m[w := R\,\Omega]) =_{\beta\eta} (\lambda w.(\lambda\vec{y}.R_j[x := \Omega])\vec{L})(R\,\Omega) =_{\beta\eta} z_j$.

From $R \in \text{REG}$ we obtain $R\,\Omega \in \text{UNS}$. Therefore the Genericity Lemma for $\boldsymbol{\lambda}\eta$ entails $(\lambda w.(\lambda\vec{y}.R_j[x := \Omega])\vec{L})M =_{\beta\eta} z_j$ for all $M \in \Lambda$, including $M = w$, from which it follows

$$(\lambda\vec{y}.R_j[x := \Omega])\vec{L} =_{\beta\eta} z_j.$$

By applying again the Genericity Lemma to $(\lambda x.(\lambda\vec{y}.R_j)\vec{L})\Omega =_{\beta\eta} (\lambda\vec{y}.R_j[x := \Omega])\vec{L} =_{\beta\eta} z_j$, we conclude $(\lambda\vec{y}.R_j)\vec{L} =_{\beta\eta} z_j$.

($\Leftarrow$) Similar, but simpler, as the form of L and R is already given. □

The following is a partial characterization of left- and right- invertibility in $\mathcal{M}^{(o)}(\boldsymbol{\lambda}\eta)$.

PROPOSITION 9.60. *Let $F \in \Lambda^{(o)}$. In $\mathcal{M}^{(o)}(\boldsymbol{\lambda}\eta)$, the following holds.*

(i) *Suppose F is regular.*

$$\begin{array}{lll} F \text{ is left-invertible} & \iff & \exists w, z_1, \ldots, z_n, \exists L_1, \ldots, L_m \in \Lambda \,(\in \Lambda^o(w, \vec{z}) \text{ in case } F \in \Lambda^o), \\ & & \exists x, y_1, \ldots, y_m, \exists P_1, \ldots, P_n \in \Lambda . \\ & & [\, F =_{\beta\eta} \lambda x y_1 \ldots y_m.x P_1 \cdots P_n \;\; \& \\ & & \; (\lambda\vec{y}.P_1)\vec{L} =_{\beta\eta} z_1 \;\&\; \cdots \;\&\; (\lambda\vec{y}.P_n)\vec{L} =_{\beta\eta} z_n \,]. \\ & & \textit{In this case a left-inverse is } L \triangleq \lambda w z_1 \ldots z_n.w L_1 \cdots L_m. \end{array}$$

(ii) *F is regularly right-invertible* $\iff$

$$\begin{aligned}
&\exists w, z_1, \ldots, z_n, \exists P_1, \ldots, P_m \in \Lambda,\\
&\exists x, y_1, \ldots, y_m, \exists R_1, \ldots, R_n \in \Lambda\,(\in \Lambda^o(x, \vec{y}) \text{ in case } F \in \Lambda^o).\\
&[F =_{\beta\eta} \lambda w z_1 \ldots z_n.wP_1 \cdots P_m \;\&\\
&(\lambda\vec{y}.R_1)\vec{P} =_{\beta\eta} z_1 \;\& \;\cdots\; \& \;(\lambda\vec{y}.R_n)\vec{P} =_{\beta\eta} z_n].
\end{aligned}$$

In this case a right-inverse is $R \triangleq \lambda x y_1 \ldots y_m.xR_1 \cdots R_n$

PROOF. (i) Originally proved in Böhm and Tronci (1991).

($\Rightarrow$) Let $L \circ F =_{\beta\eta} \mathsf{I}$. Then L is right-invertible, hence by assumption, Lemma 9.5(ii) plus Lemma 9.10 both F and L are regular. The conclusion follows from Lemma 9.59.

($\Leftarrow$) Given the $\vec{L}, \vec{P}$ as in the RHS, by Lemma 9.59 for $L = \lambda w z_1 \ldots z_n.wL_1 \cdots L_m$ one has $L \circ F =_{\beta\eta} \mathsf{I}$, that is, F is left-invertible.

(ii) This result is due to Folkerts (2020). The proof is similar, but easier. [Exercise. Find all differences between (i) and (ii).] □

The requirement of regularity in both (i) and (ii) of Proposition 9.60 cannot be dropped. The non-regular $F = \lambda xy.yx$ has a left-inverse $L = \lambda w.w\mathsf{I}$, but the RHS of (i) in Proposition 9.60 does not hold.

DEFINITION 9.61. The set FHP of *finite hereditary permutations* consists of those λ-terms whose Böhm tree results from a (finite) η-expansion of I by permuting around, as in the following example. Consider

$$\mathsf{I}' \triangleq \lambda x y_1 y_2.x y_1(\lambda z_1 z_2 z_3.y_2 z_1 z_2 z_3).$$

Then $\mathsf{I}' =_{\beta\eta} \mathsf{I}$ and

$\mathrm{BT}(\mathsf{I}') =$

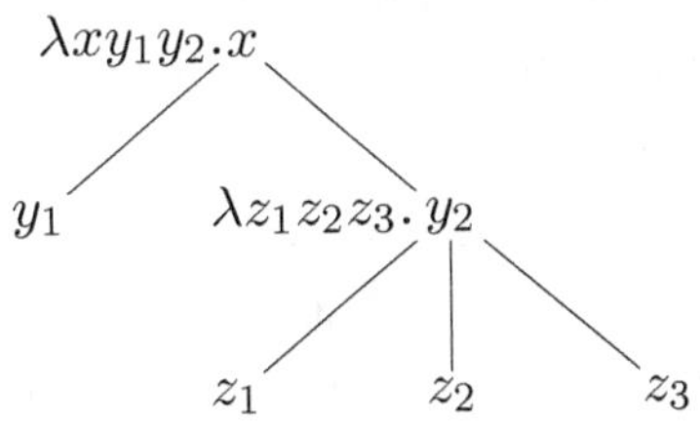

Then $F = \lambda x y_1 y_2.x(\lambda z_1 z_2 z_3.y_2 z_1 z_3 z_2)y_1$ with

$\mathrm{BT}(F) =$

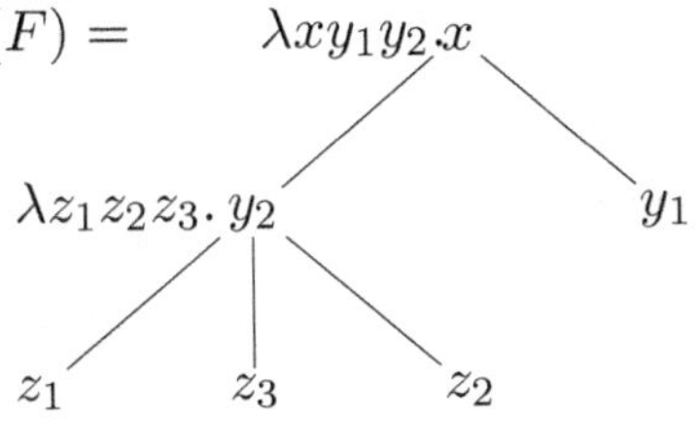

is a finite hereditary permutation.

EXAMPLES 9.62. Examples of $L, R \in \Lambda^o$ such that $L \circ R =_{\beta(\eta)} \mathsf{I}$.

L	R	$L(Rx) =_\beta$	
$L_0 \triangleq \lambda x.x\mathsf{I}$	$R_0 \triangleq \lambda xy.yx$	$\mathsf{I}x$	$=_\beta x$
$L_1 \triangleq \lambda x.xM_1M_2$	$R_1 \triangleq \lambda xy_1y_2.x$	$(\lambda y_1y_2.x)M_1M_2$	$=_\beta x$
$L_2 \triangleq \lambda x.x\mathsf{K}\Omega$	$R_2 \triangleq \lambda xy_1y_2.y_1x\Omega$	$(\lambda y_1y_2.y_1x\Omega)\mathsf{K}\Omega$	$=_\beta x$
$L_3 \triangleq \lambda x.x\mathsf{KF}\Omega$	$R_3 \triangleq \lambda xy_1y_2y_3.y_2\Omega x$	$(\lambda y_1y_2y_3.y_2\Omega x)\mathsf{KF}\Omega$	$=_\beta x$
$L_4 \triangleq \lambda x.x\mathsf{KF}\Omega$	$R_4 \triangleq \lambda xy_1y_2y_3.y_2y_3x$	$(\lambda y_1y_2y_3.y_2y_3x)\mathsf{KF}\Omega$	$=_\beta x$
$L_5 \triangleq \lambda x.x\mathsf{P}_2\mathsf{U}^2_1\mathsf{U}^2_2$	$R_5 \triangleq \lambda xy.y(y\Omega x)\Omega$	$[[\Omega, x], \Omega]\mathsf{U}^2_1\mathsf{U}^2_2$	$=_\beta x$
$L_6 \triangleq \lambda xyz.x[y, z]$	$R_6 \triangleq \lambda xv.x(v\mathsf{K})(v\mathsf{F})$	$\lambda yz.x([y, z]\mathsf{K})([y, z]\mathsf{F})$	$=_{\beta\eta} x$

Recall that $\mathsf{P}_n = \lambda x_1 \ldots x_n z.zx_1 \cdots x_n$, $[M, N] =_\beta \mathsf{P}_2MN$, and $\mathsf{U}^n_k = \lambda x_1 \ldots x_n.x_k$, so that $\mathsf{P}_nM_1 \cdots M_n\mathsf{U}^n_k =_\beta M_k$, for all $k\,(1 \leq k \leq n)$.

The following result characterizes invertibility in $\mathcal{M}^{(o)}(\boldsymbol{\lambda}\eta)$. The result is due to Dezani-Ciancaglini (1976) for normal F, and Bergstra and Klop (1980) in general.

PROPOSITION 9.63. *Let $F \in \Lambda^{(o)}$. Then, in $\mathcal{M}^{(o)}(\boldsymbol{\lambda}\eta)$, we have:*

$$F \textit{ is invertible} \iff F \in \mathrm{FHP}.$$

PROOF. See B[1984], Theorem 21.2.21, for an exposition of the combined result of Dezani and Bergstra-Klop. □

Propositions 9.57-9.63 are summarized in the following table, characterizing (in some cases only partially) the various notions of invertibility in the closed term model of $\boldsymbol{\lambda}(\eta)$.

Property	equivalence in $\mathcal{M}^o(\boldsymbol{\lambda})$	equivalence in $\mathcal{M}^o(\boldsymbol{\lambda}\eta)$
F is left-invertible	F is head-reachable Margaria and Zacchi (1983)	(only for regular F) $\exists w, z_1, \ldots, z_n, \exists L_1, \ldots, L_m \in \Lambda^o(w, \vec{z})\,.$ $[\,F =_{\beta\eta} \lambda xy_1 \ldots y_m.xP_1 \cdots P_n\ \&$ $\forall j\,(1 \leq j \leq n)\,.\,(\lambda\vec{y}.P_j)\vec{L} =_{\beta\eta} z_j\,]$ Böhm and Tronci (1991)
F is right-invertible	$F =_\beta \lambda x.xM_1 \cdots M_n$ for some $\vec{M} \in \Lambda^o(x)$. Margaria and Zacchi (1983)	(only for F regularly right invertible) $\exists x, y_1, \ldots, y_m, \exists R_1, \ldots, R_n \in \Lambda^o(x, \vec{y})\,.$ $[F =_{\beta\eta} \lambda wz_1 \ldots z_n.wP_1 \cdots P_m\ \&$ $\forall j\,(1 \leq j \leq n)\,.\,(\lambda\vec{y}.R_j)\vec{P} =_{\beta\eta} z_j\,]$ Folkerts (2020)
F invertible	$F =_\beta \mathsf{I}$ Böhm and Dezani-Ciancaglini (1974)	$F \in \mathrm{FHP}$ Dezani-Ciancaglini (1976) for nf Bergstra and Klop (1980) in general

A surjective term, not right-invertible in $\mathcal{M}^o(\boldsymbol{\lambda}\eta)$

In this section, we follow Folkerts (1995) showing that in $\mathcal{M}^o(\boldsymbol{\lambda}\eta)$

$$F \text{ is surjective} \quad \not\Rightarrow \quad F \text{ is right-invertibile.}$$

The task is to construct an F that is surjective, i.e.

$$\forall M \in \Lambda^o, \exists N_M \in \Lambda^o \,.\, FN_M =_{\beta\eta} M,$$

but not uniformly so, i.e.

$$\nexists N \in \Lambda^o, \forall M \in \Lambda^o \,.\, F(NM) =_{\beta\eta} M.$$

A first example that comes to mind is the enumerator E that certainly is surjective on Λ^o, as one has $\forall M \in \Lambda^o \,.\, \mathsf{E}\ulcorner M \urcorner = M$. But, even knowing that there is no λ-term Q such that $QM = \ulcorner M \urcorner$, it is not clear how to prove that this surjectivity cannot be made uniform. If R were a right-inverse of E then, for all $M \in \Lambda^o$, we would have $\mathsf{E}(RM) =_{\beta\eta} M =_{\beta\eta} \mathsf{I}M$. By the partial validity of the ω-rule a consequence is

$$\mathsf{E}(Rx) =_{\beta\eta} \mathsf{I}x.$$

It is not clear how a contradiction can be derived from here. In fact E does posses a right-inverse, as follows from the characterization in Proposition 9.57(i).

PROPOSITION 9.64. *$\mathsf{E} \in \Lambda^o$ has a right-inverse R in $\mathcal{M}^{(o)}(\boldsymbol{\lambda})$. That is,* $\mathsf{E} \circ R =_\beta \mathsf{I}$.

PROOF. Since $\mathsf{E}\ulcorner \mathsf{I} \urcorner =_\beta \mathsf{I}$, the term E is solvable:

$$\mathsf{E} =_\beta \lambda x_1 \ldots x_n . x_i P_1 \cdots P_k.$$

Since $\mathsf{E}\ulcorner \Omega \urcorner =_\beta \Omega$, one must have $i = 1$, otherwise Ω would be solvable. Moreover one must have $n = 1$, otherwise Ω would not be of order 0. It follows that

$$\mathsf{E} =_\beta \lambda x_1 . x_1 P_1 \cdots P_k.$$

Then a right-inverse is $R \triangleq \lambda x y_1 \ldots y_k . x$. Indeed

$$\begin{aligned} \mathsf{E} \circ R &=_\beta \lambda x . \mathsf{E}(Rx) \\ &=_\beta \lambda x . Rx P_1^\sigma \cdots P_k^\sigma, \qquad \text{for } \sigma \triangleq [x_1 := Rx], \\ &=_\beta \lambda x . x. \end{aligned}$$

□

A more complex λ-term, related to E, will be defined that is $\beta\eta$-surjective, but not β-surjective, and does not possess a right-inverse.

NOTATION 9.65. (i) For $P \in \Lambda$, define using Turing's fixed point combinator the λ-term

$$\mathcal{G}^P \triangleq \Theta(\lambda g n . [Pn, g(\mathsf{S}^+ n)])$$

satisfying

$$\mathcal{G}^P \twoheadrightarrow_\beta \lambda n . [Pn, \mathcal{G}^P(\mathsf{S}^+ n)].$$

Using Notation 1.35, we obtain $\mathcal{G}^P \mathsf{c}_0 = [P_n]_{n \in \mathbb{N}} = [P_0, P_1, P_2, \ldots]$.

(ii) The meaning of $[P_n]_{n\in\mathbb{N}}$ depends on P, so if it is necessary to be explicit we write

$$[P_0, P_1, P_2, \ldots]^P = [P_n]^P_{n\in\mathbb{N}} = [P\mathsf{c}_n]^P_{n\in\mathbb{N}}.$$

(iii) Instances are $[f\mathsf{c}_n]^f_{n\in\mathbb{N}} = [f\mathsf{c}_0, f\mathsf{c}_1, f\mathsf{c}_2, \ldots]$ and $[Q_{(n)}]^{\mathsf{K}Q}_{n\in\mathbb{N}} = [Q, Q, Q, \ldots]$.

(iv) The particular case $U = [\mathsf{E}\mathsf{c}_n]^{\mathsf{E}}_{n\in\mathbb{N}}$ is a *universal generator* in the sense that every $M\in\Lambda^o$ appears as subterm of a reduct of U. Such λ-terms play a role in the partial validity of the ω-rule, as studied in Barendregt (1971) and in the counterexample to that rule by Plotkin (1974), both treated in Section 17.3 of B[1984].

Remember that a λ-term has a β-nf if and only if it has a $\beta\eta$-nf (B[1984], Cor. 15.1.5). The following proposition and its corollary are due to Folkerts (2020).

PROPOSITION 9.66. *Let $F\in\Lambda$, x_1, x_2, y_1, y_2, z be arbitrary fresh variables, and $\vec{P}_1, \vec{P}_2$ be arbitrary sequences of λ-terms having a β-normal form.*

(i) *One has*

$$F(y_1\vec{P}_1)(y_2\vec{P}_2)\in\mathrm{NF} \iff Fx_1x_2\in\mathrm{NF}.$$

(ii) *In particular*

$$Fz^{\sim 2}\in\mathrm{NF} \iff Fx_1x_2\in\mathrm{NF}.$$

(iii) *Moreover*

$$Fz^{\sim 2} =_{\beta\eta} z \iff F =_{\beta\eta} \lambda x_1x_2.x_i, \text{ for some } i\,(1\leq i\leq 2).$$

(iv) (i)-(iii) *can be generalized to $n\geq 0$ replacing 2.*

PROOF. In all cases ($\Leftarrow$) follows by substitution. We prove the ($\Rightarrow$) direction.

(i) Suppose that $Fx_1x_2\notin\mathrm{NF}$. Then by Curry's normalization reduction strategy (B[1984], Theorem 13.2.2) there is an infinite leftmost reduction starting form this term. But then by substitution $[x_i := y_i\vec{P}_i]$ we obtain an infinite leftmost reduction sequence starting from $F(y_1\vec{P}_1)(y_2\vec{P}_2)$. Therefore this term cannot have a normal form.

(ii) By taking in (i) $y_1 \triangleq y_2 \triangleq z$ and the $\vec{P}_1$, $\vec{P}_2$ empty.

(iii) Suppose $Fzz =_{\beta\eta} z$. Then Fx_1x_2 has a $\beta\eta$-nf, say N, by (ii). By the assumption $N[x_1 := z][x_2 := z] =_{\beta\eta} z$. Therefore $N = x_1$ or $N = x_2$, implying $F =_{\beta\eta} \mathsf{K}$ or $F =_{\beta\eta} \mathsf{F}$.

(iv) Similarly. □

COROLLARY 9.67. (i) *Assume* $\mathrm{FV}(M_1M_2)\subseteq\{z\}$. *Then*

$$\exists F\in\Lambda^o\,.\,FM_1M_2 =_{\beta\eta} z \iff \exists G\in\Lambda^o\,.\,[\,GM_1 =_{\beta\eta} z \vee GM_2 =_{\beta\eta} z\,].$$

(ii) *This can be generalized to $n\geq 1$ replacing 2.*

PROOF. (i) Computing backwards one has

$$\begin{aligned} z &=_{\beta\eta} FM_1M_2 \\ &=_{\beta\eta} F((\lambda z.M_1)z)((\lambda z.M_2)z) \\ &=_{\beta\eta} (\lambda xy.F((\lambda z.M_1)x)((\lambda z.M_2)y))zz. \end{aligned}$$

By Proposition 9.66(iii) one has, say, $(\lambda xy.F((\lambda z.M_1)x)((\lambda z.M_2)y)) =_{\beta\eta} \mathsf{K}$. Then

$$\begin{aligned} z &=_{\beta\eta} \mathsf{K}z\mathsf{I} \\ &=_{\beta\eta} (\lambda xy.F((\lambda z.M_1)x)((\lambda z.M_2)y))z\mathsf{I} \\ &=_{\beta\eta} F((\lambda z.M_1)z)((\lambda z.M_2)\mathsf{I})) \\ &=_{\beta\eta} FM_1(M_2[z:=\mathsf{I}]) \\ &=_{\beta\eta} GM_1, \end{aligned}$$

with $G \triangleq \lambda a.Fa(M_2[z:=\mathsf{I}])$.

(ii) Similarly. □

In the proof of the following lemma we use 'infinite reasoning', that inspired Dana Scott to speak about λ-*calculus*.

LEMMA 9.68. *Let $M \in \Lambda^o$ and let u be a fresh variable. We have*

$$M[u, u, \ldots]^{\mathsf{K}u} \in \mathrm{NF} \quad \Rightarrow \quad M[u\mathsf{c}_0, u\mathsf{c}_1, \ldots]^u \in \mathrm{NF}.$$

PROOF. Write $u_\infty \triangleq [u, u, u, \ldots]^{\mathsf{K}u}$, $u_0 \triangleq \bot$, and $u_{n+1} \triangleq [u, u_n]$. Note that

$$\mathcal{B} \vdash u_\infty = \bigsqcup_{k \in \mathbb{N}} u_k.$$

Suppose $Mu_\infty \in \mathrm{NF}$. Then

$$\begin{aligned} \bigsqcup_{k \in \mathbb{N}} Mu_k &=_{\mathcal{B}} M\Big(\bigsqcup_{k \in \mathbb{N}} u_k\Big), && \text{by continuity (Corollary 2.35)}, \\ &=_{\mathcal{B}} Mu_\infty && \in \mathrm{NF}. \end{aligned}$$

Since the Böhm tree of a nf is an isolated point in the tree topology, for some $k \in \mathbb{N}$,

$$\mathcal{B} \vdash Mu_k = Mu_\infty \in \mathrm{NF}.$$

That is $M[\underbrace{u, \ldots, u}_{k \text{ times}}, \bot] =_\beta (\lambda\vec{x}.M[x_1, \ldots, x_k, \bot])\vec{u} \in \mathrm{NF}$. By Proposition 9.66(i), we get

$$(\lambda\vec{x}.M[x_1, \ldots, x_k, \bot])(u\mathsf{c}_0) \cdots (u\mathsf{c}_k) =_\beta M[u\mathsf{c}_0, \ldots, u\mathsf{c}_k, \bot] \in \mathrm{NF}.$$

But as

$$\mathrm{BT}(M[u\mathsf{c}_0, \ldots, u\mathsf{c}_k, \bot]) \leq_\bot \mathrm{BT}(M[u\mathsf{c}_n]^u_{n \in \mathbb{N}}),$$

and there are no trees above that of a normal form, also

$$M[u\mathsf{c}_n]^u_{n \in \mathbb{N}} \in \mathrm{NF}.$$

□

DEFINITION 9.69. Let $M \in \Lambda$. Define $M^+ \triangleq \lambda xy.x(My)$ for fresh x, y. Then

$$\mathrm{BT}(M^+) = \begin{array}{c} \lambda xy.x \\ | \\ \mathrm{BT}(My) \end{array}$$

From now on, and until the end of the section, we work in $\mathcal{M}^o(\boldsymbol{\lambda}\eta)$.

LEMMA 9.70. *Let $F \in \Lambda^o$.*
(i) *If F is right-invertible then F is regular.*
(ii) *If F^+ is left-invertible then F is right-invertible.*

PROOF. (i) By Lemmas 9.5(ii) and 9.10.

(ii) Suppose $L \circ F^+ =_{\beta\eta} \mathsf{I}$ for some $L \in \Lambda^o$. Then L is right-invertible and $L =_{\beta\eta} \lambda xy_1 \ldots y_n.xL_1 \cdots L_m$ by (i). By making an η-expansion, we may assume that $m \geq 1$. Define $L'_i \triangleq (\lambda x.L_i)F^+ \in \Lambda^o(y_1, \ldots, y_n)$. Then

$$\begin{aligned} xy_1 \cdots y_n &=_{\beta\eta} L(F^+x)y_1 \cdots y_n \\ &=_{\beta\eta} F^+xL'_1 \cdots L'_m \\ &=_{\beta\eta} x(FL'_1)L'_2 \cdots L'_m. \end{aligned}$$

Then $y_1 =_{\beta\eta} FL'_1$ (and $m = n$). Therefore $R \triangleq \lambda y_1.L'_1$ is a right-inverse of F:

$$F \circ R =_{\beta\eta} \lambda y_1.F(Ry_1) =_{\beta\eta} \lambda y_1.FL'_1 =_{\beta\eta} \lambda y_1.y_1 = \mathsf{I}.$$ □

LEMMA 9.71. *Let $F \triangleq \lambda xy_1 \ldots y_n.x(P_1y_1) \cdots (P_ny_n)$, with all $\vec{P} \in \Lambda^o$ regular. Suppose that $FM =_{\beta\eta} \Omega$ for some $M \in \Lambda^o$. Then*
(i) *$M \in$ UNS,*
(ii) *F is regularly right-invertible,*
(iii) *all $\vec{P}$ are left-invertible.*

PROOF. (i) Suppose towards a contradiction that M is solvable, then $M =_{\beta\eta} \lambda z_1 \ldots z_p.z_i\vec{M}$. By assumption $FMy_1 \cdots y_r =_{\beta\eta} \Omega y_1 \cdots y_r$ for all $r \in \mathbb{N}$; taking $r > p$ one has

$$M(P_1y_1) \cdots (P_py_p)y_{p+1} \cdots y_r =_{\beta\eta} \Omega y_1 \cdots y_r.$$

But $M(P_1y_1) \cdots (P_py_p) \cdots =_{\beta\eta} P_iy_i \cdots =_{\beta\eta} y_i \cdots$ is solvable with head variable y_i and cannot be $\Omega y_1 \cdots y_r$.

(ii) That F is regularly right-invertible follows from (i) and Proposition 9.45.

(iii) By (ii), we have $F \circ R =_{\beta\eta} \mathsf{I}$ for some regular $R =_{\beta\eta} \lambda zv_1 \ldots v_\ell.zR_1 \cdots R_m \in \Lambda^o$. By performing η-expansions it may also be assumed that $\ell \geq n$. It follows that

$$\begin{aligned} zx_1 \cdots x_n &=_{\beta\eta} \mathsf{I}zx_1 \cdots x_n, \\ &=_{\beta\eta} (F \circ R)zx_1 \cdots x_n, \\ &=_{\beta\eta} F(Rz)x_1 \cdots x_n, \\ &=_{\beta\eta} Rz(P_1x_1) \cdots (P_nx_n), \\ &=_{\beta\eta} \lambda v_{n+1} \ldots v_\ell.zR^*_1 \cdots R^*_m, \text{ with } R^*_i \triangleq (\lambda zv_1 \ldots v_n.R_i)z(P_1x_1) \cdots (P_nx_n). \end{aligned}$$

Then

$$zx_1 \cdots x_nv_{n+1} \cdots v_\ell =_{\beta\eta} zR^*_1 \cdots R^*_m.$$

It follows that $\ell = m$ and $x_i =_{\beta\eta} R^*_i$, for every $i\,(1 \leq i \leq n)$. But then one can define as left-inverse for each P_i the λ-term

$$L_i \triangleq \lambda x_i.(\lambda zv_1 \ldots v_n.R_i)z(P_1x_1) \cdots (P_{i-1}x_{i-1})x_i(P_{i+1}x_{i+1}) \cdots (P_nx_n),$$

or better a closed version $L_i^\bullet \in \Lambda^o$, where the free variables are replaced by, say, I. Indeed,

$$\begin{aligned} L_i(P_i x_i) &=_{\beta\eta} (\lambda z v_1 \ldots v_n.R_i) z (P_1 x_1) \cdots (P_{i-1} x_{i-1})(P_i x_i)(P_{i+1} x_{i+1}) \cdots (P_n x_n), \\ &=_{\beta\eta} R_i^*, \\ &=_{\beta\eta} x_i. \end{aligned}$$

Hence $L_i^\bullet$ is a left-inverse of P_i. □

PROPOSITION 9.72. *Let $R \in \Lambda^o$ and $X \triangleq \lambda x y_1 y_2.x(R y_1)(R^+ y_2)$. Assume $XM =_{\beta\eta} \Omega$ for some $M \in \Lambda^o$. Then R is invertible and has a normal form.*

PROOF. Define $F \triangleq \lambda x y_1 y_2.x(\mathsf{I} y_1)(R^+ y_2)$. Assume $XM =_{\beta\eta} \Omega$, then

$$F(M \circ R) =_{\beta\eta} \lambda y_1 y_2.(M \circ R) y_1 (R^+ y_2) =_{\beta\eta} \lambda y_1 y_2.M(R y_1)(R^+ y_2) =_{\beta\eta} XM =_{\beta\eta} \Omega.$$

As I and R^+ are regular, R^+ is left-invertible by Lemma 9.71(iii) applied to F. Then by Lemma 9.70(ii)-(i), R is right-invertible and hence regular. Applying Lemma 9.71(iii) to X it follows that R is left-invertible. Therefore R is invertible. By Proposition 9.63, $R \in \mathrm{FHP}$ and hence has a normal form. □

The following result first appeared in Folkerts (1995).

PROPOSITION 9.73. *There exists an $F \in \Lambda^o$ which is surjective, but not right-invertible in $\mathcal{M}^o(\boldsymbol{\lambda}\eta)$.*

PROOF. Define F as follows:

$$\begin{aligned} F &\triangleq G \circ H, \text{ where:} \\ H &\triangleq \lambda m f.m[f\mathsf{c}_n]_{n \in \mathbb{N}}, \\ G &\triangleq \lambda h x y.h\mathsf{E}\big((Lh)x\big)\big((Lh)^+ y\big), \\ L &\triangleq \lambda v.v \circ \mathsf{K}. \end{aligned}$$

Recall that $[f\mathsf{c}_n]_{n \in \mathbb{N}}$ represents a sequence with projections π_i satisfying

$$\pi_i [f\mathsf{c}_n]_{n \in \mathbb{N}} =_\beta f\mathsf{c}_i.$$

Note that

$$FM =_\beta G(HM) =_\beta \lambda x y.HM\mathsf{E}\big(HM(\mathsf{K}x)\big)\big(y \circ HM \circ \mathsf{K}\big).$$

Claim 9.73.1. F is surjective: $\forall N \in \Lambda^o, \exists M_N \in \Lambda^o \,.\, FM_N =_{\beta\eta} N$.

Subproof. Indeed, take $M_N \triangleq \pi_{\#N}$. Then, we have

$$\begin{aligned} FM_N &=_{\beta\eta} \lambda x y.HM_N\mathsf{E}\big(HM_N(\mathsf{K}x)\big)\big(y \circ (HM_N) \circ \mathsf{K}\big) \\ &=_{\beta\eta} \lambda x y.\pi_{\#N}[\mathsf{E}\mathsf{c}_n]_{n \in \mathbb{N}}\big(\pi_{\#N}[\mathsf{K}x\mathsf{c}_n]_{n \in \mathbb{N}}\big)\big(\lambda u.y(HM_N(\mathsf{K}u))\big) \\ &=_{\beta\eta} \lambda x y.\mathsf{E}\mathsf{c}_{\#N}\big(\pi_{\#N}[\mathsf{K}x\mathsf{c}_n]_{n \in \mathbb{N}}\big)\big(\lambda u.y(\pi_{\#N}[\mathsf{K}u\mathsf{c}_n]_{n \in \mathbb{N}})\big) \\ &=_{\beta\eta} \lambda x y.\mathsf{E}\ulcorner N \urcorner(\mathsf{K}x\ulcorner N \urcorner)\big(\lambda u.y(\mathsf{K}u\ulcorner N \urcorner)\big) \\ &=_{\beta\eta} \lambda x y.Nx(\lambda u.yu) \\ &=_{\beta\eta} N, \text{ by three times an } \eta\text{-reduction.} \end{aligned}$$

■

Claim 9.73.2. Suppose $FM =_{\beta\eta} \Omega$. Then HM has a normal form.

Subproof. One has $FM =_{\beta\eta} \lambda y_1 y_2.HM\mathsf{E}(Ry_1)(R^+ y_2)$, with $R =_{\beta\eta} L(HM)$. Define $X \triangleq \lambda xy_1y_2.x(Ry_1)(R^+y_2)$. Then $X(HM\mathsf{E}) =_{\beta\eta} FM =_{\beta\eta} \Omega$. So, by Proposition 9.72 $R \in \text{NF}$. But then also $Ru =_{\beta\eta} L(HM)u =_{\beta\eta} (HM \circ \mathsf{K})u =_{\beta\eta} HM(\mathsf{K}u) =_{\beta\eta} M[u, u, \ldots]^{\mathsf{K}u} \in \text{NF}$. By Lemma 9.68 $HMu =_{\beta\eta} M[u\mathsf{c}_0, u\mathsf{c}_1, \ldots]^u$ has a normal form, and so does HM. ■

Assume towards a contradiction that F has a right-inverse $R \in \Lambda^o$. Then

$$\forall P \in \Lambda^o\,.\,G(H(RP)) =_{\beta\eta} F(RP) =_{\beta\eta} P.$$

Then $F(R\,\Omega) =_{\beta\eta} \Omega$, hence by Claim 9.73.2 $H(R\,\Omega) \in \text{NF}$. By the Genericity Lemma it follows that $H(RP) =_{\beta\eta} H(R\,\Omega)$, for every $P \in \Lambda^o$. We obtain a contradiction

$$\Omega =_{\beta\eta} G(H(R\,\Omega)) =_{\beta\eta} G(H(R\mathsf{I})) =_{\beta\eta} \mathsf{I}.$$ □

This answers a question raised by Böhm in the open problem section of Böhm (1975).

EXERCISE 9.74 (FOLKERTS). Let $P, Q \in \Lambda$. Then in $\mathcal{M}^o(\boldsymbol{\lambda}\eta)$

$$P \circ Q = \mathsf{I} \iff Q^+ \circ P^+ = \mathsf{I}.$$

EXERCISE 9.75 (FOLKERTS). (i) If one tries to simplify the proof of Proposition 9.73 by taking H as before and $G \triangleq \lambda hy.h\mathsf{E}(y \circ h \circ \mathsf{K})$, then the resulting $F =_{\beta\eta} G \circ H$ does have a right-inverse. [Hint. Try $R_p \triangleq \lambda uv.\pi_p vu$, with $\mathsf{Ec}_p =_{\beta\eta} \lambda ab.a(b \circ \mathsf{K})$.]

(ii) Show that there are λ-terms A, B such that $A(\lambda u.x(Bu)) =_{\beta\eta} \Omega x$, but $B \neq_{\beta\eta} \mathsf{I}$. [Hint. Take $A \triangleq \lambda b.\Omega(b \circ \mathsf{K})(=_{\beta\eta}\mathsf{Ec}_p\Omega$, see (i)) and $B \triangleq \lambda c.c\mathsf{I}$.]

EXERCISE 9.76 (FOLKERTS). (i) Define

$$\begin{aligned}
Q &\triangleq \lambda pqr.q(pr)\\
R &\triangleq \lambda ab.ba\mathsf{I}\\
L &\triangleq \lambda abc.ca\\
U &\triangleq \lambda ab.Q(QL)a =_\beta \lambda abc.a(\lambda d.c(\lambda ef.fd))\\
F &\triangleq \lambda xy.xU(Qxy)\\
&=_\beta \lambda xy.x\big(\lambda abc.a(\lambda d.c(\lambda ef.fd))\big)\big(\lambda f.y(xf)\big)
\end{aligned}$$

Then $F \circ R \twoheadrightarrow_{\beta\eta} \mathsf{I}$, so that F is right-invertible in $\mathcal{M}^o(\boldsymbol{\lambda}\eta)$, but R is not regular. Show that F is not regularly right-invertible, i.e. it does not have a regular right-inverse. [Hints. 1. Suppose F, that we write as $F = \lambda wz.wL_1L_2$, has a regular right-inverse R. Then $R =_{\beta\eta} \lambda xy_1y_2.xM$. (This does not follow immediately from the statement of Proposition 9.59, but using a similar proof!) 2. Then by Proposition 9.59 $(\lambda y_1y_2.M)L_1L_2 = z$. 3. Conclude from Corollary 9.67(i) that $\lambda z.L_1$ or $\lambda z.L_2$ is right-invertible. 4. Show that this is impossible for the given L_1, L_2 part of F.]

(ii) Show that also $\lambda hy.h\mathsf{E}(y \circ h \circ \mathsf{K})$ has a right-inverse, but is not regularly right-invertible in $\mathcal{M}^o(\boldsymbol{\lambda}\eta)$.

The state of the art concerning {left|right}-invertibility

To conclude this chapter, we summarize the results we have seen in the chapter. The next proposition presents what is known about the converse implications in Lemma 9.5 in the (extensional) term models $\mathcal{M}^{(o)}(\boldsymbol{\lambda}(\eta))$. See also Figure 9.2 for a graphical representation.

PROPOSITION 9.77. *Consider the term models $\mathcal{M}^{(o)}(\boldsymbol{\lambda})$ and $\mathcal{M}^{(o)}(\boldsymbol{\lambda}\eta)$.*

(i) *In none of these models one has*

$$F \textit{ injective} \quad \Rightarrow \quad F \textit{ left-invertible.}$$

(ii) *In all of these models except $\mathcal{M}^o(\boldsymbol{\lambda}\eta)$ one has*

$$F \textit{ surjective} \quad \Rightarrow \quad F \textit{ right-invertible.}$$

(iii) *In all of these models one has*

$$F \textit{ bijective} \iff F \textit{ invertible.}$$

PROOF. (i) Consider $F \triangleq \Omega$. It is injective by the confluence of $\twoheadrightarrow_{\beta(\eta)}$ and the fact that $\Omega M \twoheadrightarrow_{\beta(\eta)} Z$ implies $Z =_{\beta(\eta)} \Omega M'$, with $M \twoheadrightarrow_{\beta(\eta)} M'$. Suppose towards a contradiction that L is a left-inverse of Ω. Then $L(\Omega \mathsf{I}) =_{\beta(\eta)} \mathsf{I}$. By the Genericity Lemma, we obtain $L(\Omega \mathsf{K}) =_{\beta\eta} \mathsf{I}$, a contradiction.

(ii) For $\mathcal{M}^o(\boldsymbol{\lambda})$. Let F be surjective. Then, by Lemma 9.10, F is regular:

$$F =_\beta \lambda x y_1 \ldots y_n . x P_1 \cdots P_k .$$

One must have $n = 0$, otherwise Ω would not be of order 0. Hence $F =_\beta \lambda y . y P_1 \cdots P_k$. Then a right-inverse is $R \triangleq \lambda x y_1 \ldots y_k . x$. Indeed

$$\begin{aligned} F \circ R &=_\beta \lambda x . F(Rx) \\ &= \lambda x . R x P_1^\sigma \cdots P_k^\sigma, \qquad \text{for } \sigma \triangleq [y := Rx], \\ &= \lambda x . x . \end{aligned}$$

For $\mathcal{M}(\boldsymbol{\lambda})$, $\mathcal{M}(\boldsymbol{\lambda}\eta)$. Let F be surjective. Then $F M_x =_{\beta(\eta)} x$ for some $M_x \in \Lambda$ and

$$\begin{aligned} F \circ (\lambda x . M_x) &=_{\beta(\eta)} \lambda x . F((\lambda x . M_x) x), \\ &=_{\beta(\eta)} \lambda x . F M_x , \\ &=_{\beta(\eta)} \lambda x . x . \end{aligned}$$

For $\mathcal{M}^o(\boldsymbol{\lambda}\eta)$. By Proposition 9.73 there is a surjective term that is not right-invertible.

(iii) For $\mathcal{M}(\boldsymbol{\lambda})$ and $\mathcal{M}(\boldsymbol{\lambda}\eta)$. Suppose that F is bijective. Then by (ii) F has a right-inverse R. It follows that

$$\begin{aligned} F \circ R &=_{\beta(\eta)} \mathsf{I}, && \text{hence} \\ F \circ R \circ F &=_{\beta(\eta)} F, && \text{and} \\ (F \circ R \circ F) x &=_{\beta(\eta)} F x, && \text{so} \\ F(R(Fx)) &=_{\beta(\eta)} F x, && \text{and therefore} \\ (R \circ F) x &=_{\beta(\eta)} x, && \text{by injectivity of } F. \end{aligned}$$

Thus R is both a right and left-inverse of F, i.e. F is invertible.
For $\mathcal{M}^o(\boldsymbol{\lambda}\eta)$. That bijectivity and invertibility are equivalent is the main result of this chapter, namely Theorem 9.52.
For $\mathcal{M}^o(\boldsymbol{\lambda})$. Let F be bijective in this model. Then F is bijective in $\mathcal{M}^o(\boldsymbol{\lambda}\eta)$. Hence, by Theorem 9.52, it is invertible in $\mathcal{M}^o(\boldsymbol{\lambda}\eta)$. Therefore by Proposition 9.63, F is a hereditary finite permutation. Then $F =_\beta \lambda xy_1 \ldots y_n.xP_1 \cdots P_n$. But then, again using M_Ω such that $FM_\Omega =_\beta \Omega$, it follows that $n = 0$, as Ω is of order 0. Hence $F =_\beta \lambda x.x$, which is the unique invertible term in this closed term model by Proposition 9.58. □

Correspondence	Models			
	$\mathcal{M}(\boldsymbol{\lambda})$	$\mathcal{M}^o(\boldsymbol{\lambda})$	$\mathcal{M}(\boldsymbol{\lambda}\eta)$	$\mathcal{M}^o(\boldsymbol{\lambda}\eta)$
left-invertibility $\Longleftrightarrow$ injectivity	✗	✗	✗	✗
right-invertibility $\Longleftrightarrow$ surjectivity	✓	✓	✓	✗ 9.73
invertibility $\Longleftrightarrow$ bijectivity	✓	✓	✓	✓ 9.52

Figure 9.2: Known correspondences for $\mathcal{M}^{(o)}(\boldsymbol{\lambda})$ and $\mathcal{M}^{(o)}(\boldsymbol{\lambda}\eta)$.

In Figures 9.3 and 9.4, we present the partial syntactic characterizations of (left/right) invertibility in $\mathcal{M}^o(\boldsymbol{\lambda})$ and $\mathcal{M}^o(\boldsymbol{\lambda}\eta)$, together with some discriminating examples.

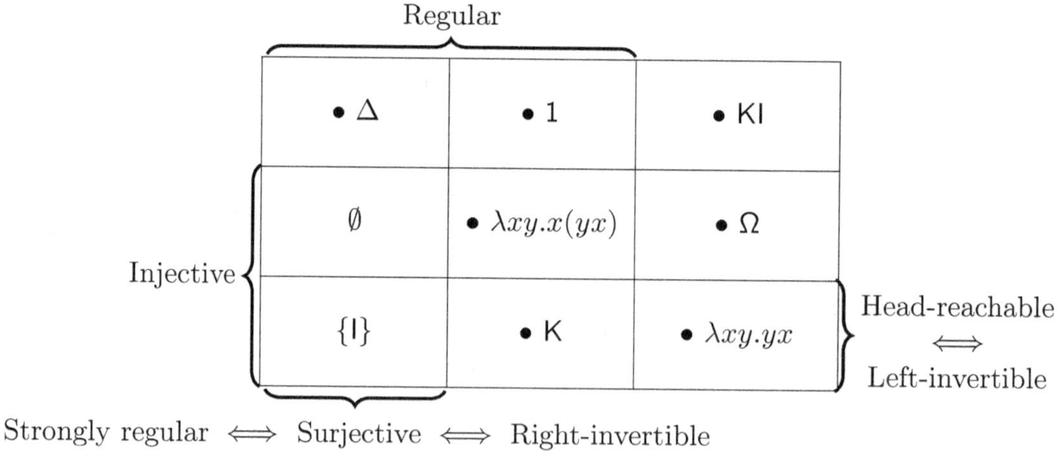

Figure 9.3: Syntactic characterizations in $\mathcal{M}^o(\boldsymbol{\lambda})$.

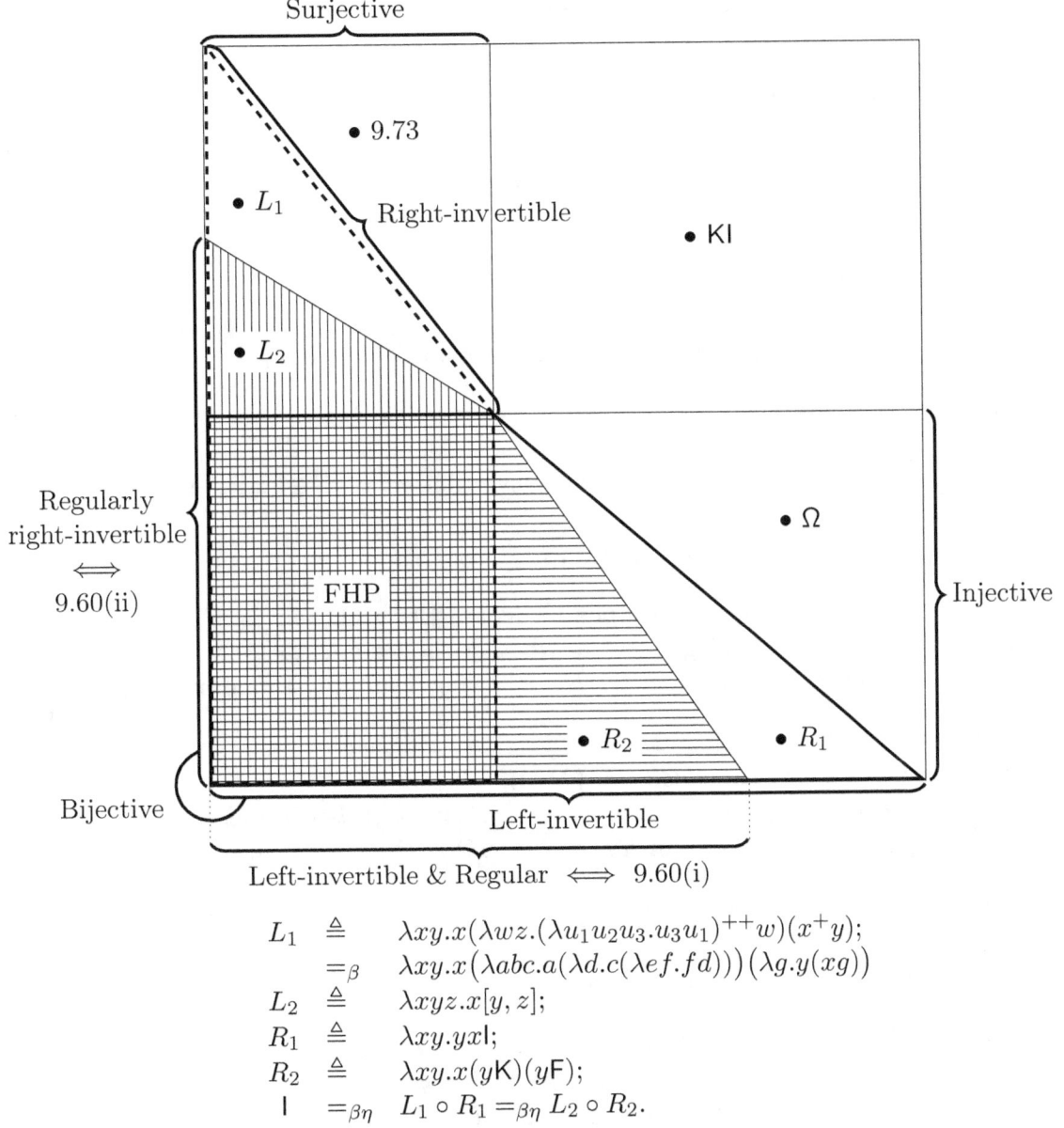

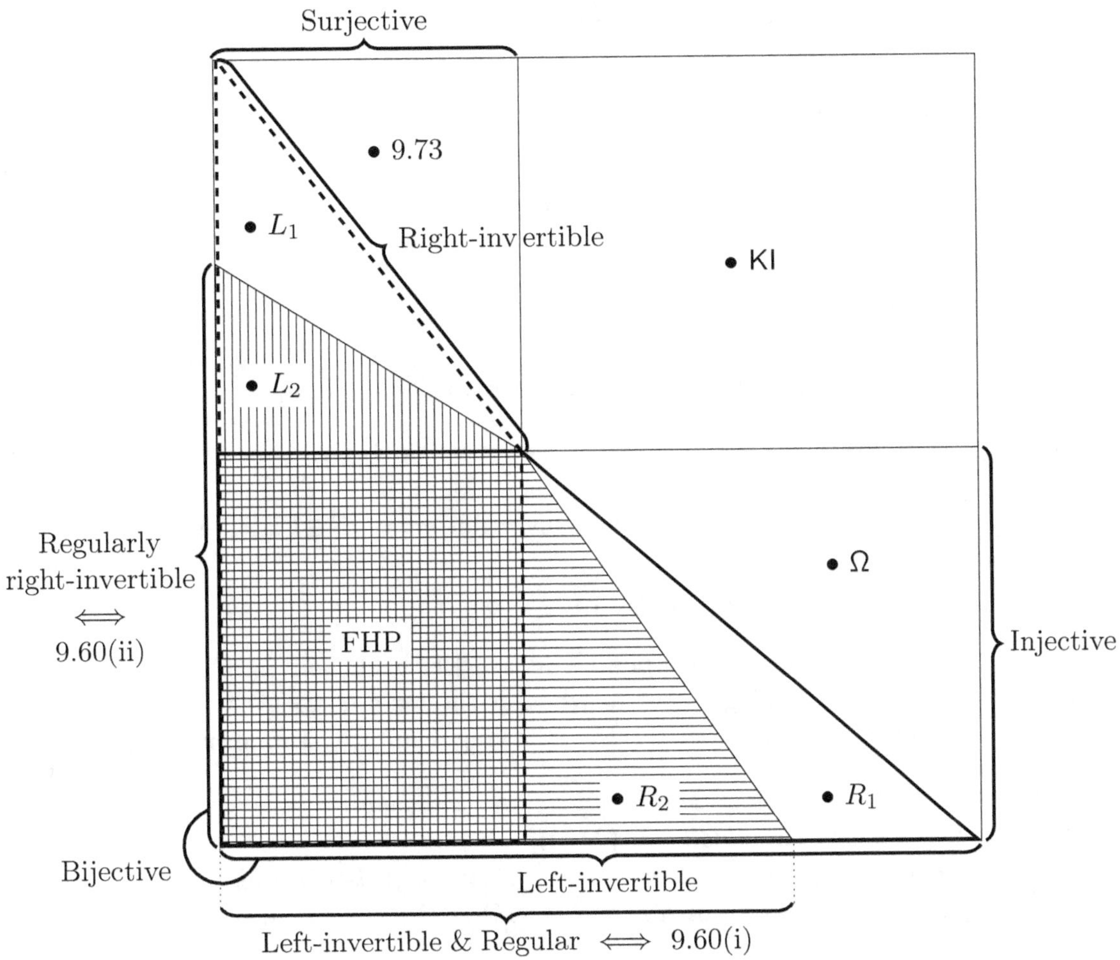

$$\begin{array}{lll}
L_1 & \triangleq & \lambda xy.x(\lambda wz.(\lambda u_1u_2u_3.u_3u_1)^{++}w)(x^+y); \\
 & =_\beta & \lambda xy.x\big(\lambda abc.a(\lambda d.c(\lambda ef.fd))\big)\big(\lambda g.y(xg)\big) \\
L_2 & \triangleq & \lambda xyz.x[y,z]; \\
R_1 & \triangleq & \lambda xy.yx\mathsf{I}; \\
R_2 & \triangleq & \lambda xy.x(y\mathsf{K})(y\mathsf{F}); \\
\mathsf{I} & =_{\beta\eta} & L_1 \circ R_1 =_{\beta\eta} L_2 \circ R_2.
\end{array}$$

Figure 9.4: Syntactic characterizations in $\mathcal{M}^o(\boldsymbol{\lambda}\eta)$.

The state of the art concerning {in|sur|bi}-jectivity

There is only a general syntactic characterization for bijectivity. See Chapter 16 for a statement of open problems.

PROPOSITION 9.78. *Let $F \in \Lambda$. Then*

(i) *F is bijective in $\mathcal{M}^{(o)}(\boldsymbol{\lambda})$* $\iff$ $F =_\beta \mathsf{I}$.

(ii) *F is bijective in $\mathcal{M}^{(o)}(\boldsymbol{\lambda}\eta)$* $\iff$ $F \in \mathrm{FHP}$.

PROOF. (i) By Propositions 9.77(iii) and 9.58 one has in $\mathcal{M}^{(o)}(\boldsymbol{\lambda})$

$$F \text{ is bijective} \iff F \text{ is invertible} \iff F =_\beta \mathsf{I}.$$

(ii) Similarly by Propositions 9.77(iii) and 9.63 in $\mathcal{M}^{(o)}(\boldsymbol{\lambda}\eta)$

$$F \text{ is bijective} \iff F \text{ is invertible} \iff F \in \mathrm{FHP}. \qquad \square$$

Invertibility notions for open terms

It is also interesting to know what open terms are {left|right}-invertible. The characterizations given above are independent of whether the term is open or closed. But since these characterizations are only partial, the following results give some extra information.

PROPOSITION 9.79. *Let $F \in \Lambda$. The following are equivalent.*

(i) *F is {left/right}-invertible in $\mathcal{M}(\boldsymbol{\lambda}(\eta))$.*

(ii) *Every closed substitution instance of F is {left/right}-invertible in $\mathcal{M}^o(\boldsymbol{\lambda}(\eta))$.*

(iii) *Every closed substitution instance of F by Ω is {left/right}-invertible in $\mathcal{M}^o(\boldsymbol{\lambda}(\eta))$.*

PROOF. (i) $\Rightarrow$ (ii). By substitutivity.

(ii) $\Rightarrow$ (iii). Trivial.

(iii) $\Rightarrow$ (i). By the Genericity Lemma. $\square$

PROPOSITION 9.80. *Let $F \in \Lambda^o$. Then*

(i) $\forall M \in \Lambda^o . FM =_{\beta\eta} M \iff \forall M \in \Lambda . FM =_{\beta\eta} M$.

(ii) $\forall M \in \Lambda^o . FM =_{\beta} M \iff \forall M \in \Lambda . FM =_{\beta} M$.

PROOF. (i) Suppose that for all $M \in \Lambda^o$ one has $GM =_{\beta\eta} M$. This can be written as $GM =_{\beta\eta} \mathsf{I}M$. The ω-rule is valid for G and I, since I is not a universal generator, B[1984], Theorem 17.3.24, it follows that $G =_{\beta\eta} \mathsf{I}$. Therefore $\forall M \in \Lambda . GM =_{\beta\eta} M$. The converse implication is trivial.

(ii) Use the partial validity of the term-rule in $\mathcal{M}^o(\boldsymbol{\lambda})$, B[1984], Exercise 17.5.11. $\square$

COROLLARY 9.81. *Let $F \in \Lambda^o$. Then*

(i) *F is {left/right}-invertible in $\mathcal{M}^o(\boldsymbol{\lambda})$ if and only if it is in $\mathcal{M}(\boldsymbol{\lambda})$.*

(ii) *F is {left/right}-invertible in $\mathcal{M}^o(\boldsymbol{\lambda}\eta)$ if and only if it is in $\mathcal{M}(\boldsymbol{\lambda}\eta)$.*

PROOF. (i) Consider the notion $\mathcal{P}(F) \triangleq F$ is R-invertible. Then

$$\begin{array}{rrcll}
\mathcal{M}^o(\boldsymbol{\lambda}) \models \mathcal{P}(F) & \Rightarrow & F \circ R & =_\beta & \mathsf{I}, & \text{for some } R \in \Lambda^o, \\
& \Rightarrow & (F \circ R)M & =_\beta & M, & \text{for all } M \in \Lambda^o, \\
& \Rightarrow & (F \circ R)M & =_\beta & M, & \text{for all } M \in \Lambda \text{ by Proposition 9.80(ii)}, \\
& \Rightarrow & (F \circ R)x & =_\beta & x, & \text{for a fresh variable } x, \\
& \Rightarrow & F \circ R & =_\beta & \mathsf{I}, & \text{in } \mathcal{M}(\boldsymbol{\lambda}), \\
& \Rightarrow & \mathcal{M}(\boldsymbol{\lambda}) & \models & \mathcal{P}(F) & .
\end{array}$$

The converse follows from Proposition 9.79.

The same reasoning holds for left- and right-invertibility.

(ii) Similarly, now using Proposition 9.80(i). □

Part IV

Theories

Henk Barendregt

Giulio Manzonetto

Alexis Saurin

Benedetto Intrigila

Andrew Polonsky

Jakob Grue Simonsen

(Courtesy of the Department of Computer Science, University of Copenhagen)

Chapter 10

Sensible theories

In this chapter we discuss two celebrated problems in λ-calculus, namely the range property and the fixed point property. The range property states that a combinator F, seen as a total function

$$F : \mathcal{M}^o(\boldsymbol{\lambda}) \to \mathcal{M}^o(\boldsymbol{\lambda})$$

has either an infinite range or a singleton range (that is, it is a constant function). This property has been conjectured by Böhm (1968) and proved by Barendregt in Theorem 17.1.16 of B[1984]. The proof constitutes a striking example of the power of the hyperconnectedness property enjoyed by the Visser topology (see Visser (1980)). Some constructive proofs of the range property have been presented by Barendregt (1993).

The fixed point property states that every combinator has either one or infinitely many pairwise β-distinct closed fixed points. The question whether this property holds was first raised by Intrigila and Biasone (2000) and appears as Problem 25 in the TLCA list of open problems (see Intrigila (2000)). Modest advances have been published by Intrigila and Statman (2015) and Manzonetto et al. (2019), but the problem remains open.

Both the range and the fixed point properties generalize to arbitrary λ-theories.

DEFINITION 10.1. Let $\mathcal{T}$ be a λ-theory and $F \in \Lambda^o$.

(i) $\mathsf{Range}_{\mathcal{T}}(F) \triangleq \{[FN]_{\mathcal{T}} \mid N \in \Lambda^o\}$ is the *range of F modulo $\mathcal{T}$*.

(ii) $\mathsf{Fix}_{\mathcal{T}}(F) \triangleq \{[N]_{\mathcal{T}} \mid N \in \Lambda^o \ \&\ FN =_{\mathcal{T}} N\}$ is the set of *fixed points of F modulo $\mathcal{T}$*.

The following examples and exercises are designed to help the reader getting familiar with these notions.

EXAMPLES 10.2. (i) $\mathsf{Range}_{\boldsymbol{\lambda}}(\mathsf{I}) = \mathsf{Fix}_{\boldsymbol{\lambda}}(\mathsf{I}) = \Lambda^o/\beta$.

(ii) $\mathsf{Range}_{\boldsymbol{\lambda}}(\mathsf{F}) = \mathsf{Fix}_{\boldsymbol{\lambda}}(\mathsf{F}) = \{[\mathsf{I}]_\beta\}$.

(iii) For $1_n \triangleq \lambda x_0 \dots x_n.x_0 \cdots x_n$, $\mathsf{Range}_{\boldsymbol{\lambda}}(1_n) = \{[\lambda x_1 \dots x_n.Mx_1 \cdots x_n]_\beta \mid M \in \Lambda^o\}$ and $\mathsf{Fix}_{\boldsymbol{\lambda}}(1_n) = \{[M]_\beta \mid \exists N \in \Lambda, \mathrm{FV}(N) \subseteq \{x_1, \dots, x_n\} \ \&\ M =_\beta \lambda x_1 \dots x_n.N\}$.

EXERCISE 10.3. (i) Find $F \in \Lambda^o$ such that $|\mathsf{Range}_{\boldsymbol{\lambda}}(F)| = \aleph_0$ while $|\mathsf{Range}_{\mathcal{H}}(F)| = 1$.

(ii) Find $F \in \Lambda^o$ such that $|\mathsf{Range}_{\boldsymbol{\lambda}}(F)| = \aleph_0$ while $|\mathsf{Fix}_{\boldsymbol{\lambda}}(F)| = 1$. [Hint: $F \in \mathrm{NF}$.]

(iii) Exhibit $N, F \in \Lambda^o$ such that $[N]_{\mathcal{H}} \notin \mathsf{Fix}_{\mathcal{H}}(F)$ while $[N]_{\mathcal{B}} \in \mathsf{Fix}_{\mathcal{B}}(F)$.

REMARK 10.4. Given $F \in \Lambda^o$ and λ-theories $\mathcal{T}, \mathcal{T}'$ satisfying $\mathcal{T} \subseteq \mathcal{T}'$, we have:
(i) $|\mathsf{Range}_{\mathcal{T}}(F)| \geq |\mathsf{Range}_{\mathcal{T}'}(F)|$,
(ii) $|\mathsf{Fix}_{\mathcal{T}}(F)| \leq |\mathsf{Fix}_{\mathcal{T}'}(F)|$, since $FN =_{\mathcal{T}} N$ entails $FN =_{\mathcal{T}'} N$, while the converse might fail. In fact, the goal of Exercise 10.3(iii) is to find a situation like this.

That is, the function $|\mathsf{Fix}_{\_}(F)|$ is monotonic, while $|\mathsf{Range}_{\_}(F)|$ is anti-monotonic.

DEFINITION 10.5. Let $\mathcal{T}$ be a λ-theory.
(i) $\mathcal{T}$ *satisfies the range property* whenever, for all $F \in \Lambda^o$, either $|\mathsf{Range}_{\mathcal{T}}(F)| = 1$ or $|\mathsf{Range}_{\mathcal{T}}(F)| = \aleph_0$.
(ii) $\mathcal{T}$ *satisfies the fixed point property* whenever, for all $F \in \Lambda^o$, either $|\mathsf{Fix}_{\mathcal{T}}(F)| = 1$ or $|\mathsf{Fix}_{\mathcal{T}}(F)| = \aleph_0$.

Theorem 17.1.16 of B[1984] actually states that the range property holds for all r.e. λ-theories, hence in particular for $\boldsymbol{\lambda}, \boldsymbol{\lambda}\eta$. The fact that the λ-theory $\mathcal{B}$ generated by equating all λ-terms having the same Böhm tree satisfies the range property is easy to check. Indeed, given a closed λ-term of shape $\lambda x.M$, there are only two cases to consider. The first case is that $x \notin \mathrm{FV}(\mathrm{BT}(M))$, and this can happen either because $x \notin \mathrm{FV}(M)$ or because x is pushed into infinity, then the range of $\lambda x.M$ is a singleton. The second case is that x does occur in $\mathrm{BT}(M)$ at some position $\alpha \in \mathbb{N}^*$, namely $M_\alpha =_\beta \lambda \vec{y}.xP_1 \cdots P_k$ for some $k \geq 0$ and $P_1, \dots, P_k \in \Lambda$. Then the range of $\lambda x.M$ can be proved infinite by considering its behavior on the countable set $\{\lambda x_1 \dots x_k.\mathsf{c}_n \mid n \in \mathbb{N}\}$.

The same reasoning applies to every λ-theory $\mathcal{T}$ satisfying $\mathcal{B} \subseteq \mathcal{T} \subseteq \mathcal{H}^*$, thus every λ-theory in this interval satisfies the range property.

As illustrated by these examples, the range property is possessed by most of the λ-theories arising in the study of λ-terms and their operational properties, so one could be tempted to consider this property as a reasonable test to determine if a λ-theory is "free of pathologies". In this context, the question of whether the λ-theory $\mathcal{H}$ satisfies the range property arises naturally. This question remained open for a long time and a positive answer was conjectured in Conjecture 20.2.8 of B[1984] and extensively motivated by Barendregt (2008) and Intrigila and Statman (2007).

In 2011, Barendregt's conjecture has been refuted by Polonsky, who published a counterexample in his PhD thesis (see Polonsky (2011b)); subsequently a more refined version of the proof appeared in the journal paper Polonsky (2012). This unexpected result generated new interesting research (e.g. David and Nour (2014)) and led Intrigila and Statman to conjecture that in the λ-theory $\mathcal{H}$ "a very complicated example could exist with, say, exactly two fixed points" (Intrigila and Statman (2015)). In the same article, the authors manage to construct with an ingenious, but complex technique, a λ-theory satisfying the range property while violating the fixed point property.

Some years later, Manzonetto et al. (2019) confirmed Intrigila and Statman's conjecture by demonstrating that no sensible λ-theory satisfies the fixed point property. As an interesting consequence, we obtain a much more natural example of a λ-theory satisfying the range property while not satisfying the fixed point property, namely the λ-theory $\mathcal{B}$.

CONVENTION 10.6. *In this chapter we mainly work in the λ-theory $\mathcal{H}$, hence we use $=$ to denote $=_{\mathcal{H}}$ thus temporarily overriding the convention on Page 29.*

Some properties of $\mathcal{H}$

We recall the main properties of the λ-theory $\mathcal{H}$, axiomatized by equating all unsolvables. As shown in B[1984], Section 16.1, $\mathcal{H}$ can be characterized in terms of reduction as follows.

DEFINITION 10.7. The notion of *Ω-reduction* is given by the axiom:

$$M \to \Omega, \qquad \text{if } M \in \text{UNS} \ \&\ M \neq \Omega. \tag{Ω}$$

and generates reduction relations $\to_\Omega, \twoheadrightarrow_\Omega, \to_{\beta\Omega}, \twoheadrightarrow_{\beta\Omega}$ and conversions $=_\Omega, =_{\beta\Omega}$.

REMARK 10.8. The reduction relations $\to_\Omega, \to_{\beta\Omega}$ are both confluent, see Lemma 15.2.5 and Theorem 15.2.15 in B[1984]. Hence $M^{(\beta)\Omega\text{-}\mathbf{nf}}$, when it exists, is unique. Moreover, the reduction relation $\to_{\beta\Omega}$ enjoys standardization and $\to_\Omega$ strongly normalization.

PROPOSITION 10.9. *For $M, N \in \Lambda$, the following are equivalent:*

1. $\mathcal{H} \vdash M = N$,
2. $M =_{\beta\Omega} N$,
3. $\exists Z \in \Lambda \,.\, M \twoheadrightarrow_{\beta\Omega} Z \ \&\ N \twoheadrightarrow_{\beta\Omega} Z$.

10.1 The range property fails in $\mathcal{H}$

In this section we present Polonsky's result stating that the λ-theory $\mathcal{H}$ does not satisfy the range property. Our presentation mainly follows the article Polonsky (2012).

The outline of the proof

To prove that the λ-theory $\mathcal{H}$ violates the range property we construct a counterexample, namely a combinator Ψ satisfying $1 < |\mathsf{Range}_{\mathcal{H}}(\Psi)| < \aleph_0$. We start by discussing the abstract properties that must be satisfied by any possible counterexample F. We know from Remark 10.4(i) that $|\mathsf{Range}_{\mathcal{B}}(F)| \leq |\mathsf{Range}_{\mathcal{H}}(F)| \leq |\mathsf{Range}_{\boldsymbol{\lambda}}(F)|$ whence, since $\mathcal{B}$ and $\boldsymbol{\lambda}$ do satisfy the range property, the range of F must be a singleton modulo $\mathcal{B}$ and infinite modulo $\boldsymbol{\lambda}$. This is only possible if $x \notin \text{BT}(Fx)$ while $x \in \text{FV}(M)$ for all $M =_{\beta\Omega} Fx$. In other words, during the growth of the Böhm tree of F the free variable x is pushed into infinity. Moreover, no trace of x towards infinity may occur in a context of shape $xP_1 \cdots P_k$ with k maximal, otherwise the range of F would be infinite by considering $F(\lambda x_1 \ldots x_k.\mathsf{c}_n)$ for all $n \in \mathbb{N}$. The only possibility that is left is that in Fx the free variable x is pushed into infinity and receives an increasing amount of arguments to 'eat'. As an example of this situation, consider a combinator F satisfying the recursive equation $Fx =_\beta \lambda z.z(F(x\Omega)z)$. Easy calculations give:

$$Fx =_\beta \lambda z.z(F(x\Omega)z) =_\beta \lambda z.z(z(F(x\Omega\Omega)z)) =_\beta \cdots =_\beta \lambda z.z^k(F(x\Omega^{\sim k})z) =_\beta \cdots$$

In this particular case, we have $|\mathsf{Range}_{\mathcal{H}}(F)| = 1$ because by Lemma 1.26 every $M \in \Lambda^o$ can be turned into an unsolvable by applying enough Ω's, so sooner or later $M\Omega^{\sim k} =_{\mathcal{H}} \Omega$.

In general, the problem is that x, while being pushed into infinity, may be applied to an infinite sequence $P_1, P_2, P_3, \ldots$ of combinators, called hereafter *tunnel*, and it is unclear what argument M may remain solvable by "eating its way through" a given tunnel. Also, for F to be a counterexample, the infinite tunnel must be non-computable in a relatively strong way:[1] $(P_n)_{n \in \mathbb{N}}$ cannot have any partial computable subsequence $(P_{n_i})_{i \in \mathbb{N}}$.

We now discuss the specific properties of the counterexample Ψ that we build in this section. The combinator Ψ depends on a fixed effective enumeration of closed λ-terms:

$$(\mathsf{L}_n)_{n \in \mathbb{N}} = (\mathsf{E}\,\mathsf{c}_0, \mathsf{E}\,\mathsf{c}_1, \mathsf{E}\,\mathsf{c}_2, \mathsf{E}\,\mathsf{c}_3, \mathsf{E}\,\mathsf{c}_4, \ldots)$$

so L_n represents the closed λ-term having n as code. Intuitively, the λ-term Ψ constructs an infinite tunnel $(P_n)_{n \in \mathbb{N}}$ of combinators whose only 'survivor' can be the identity I. To circumvent the obstacle concerning the non-computability of the sequence, Ψx builds an infinite binary tree where each node is either unsolvable or has the shape of an applicative context $xP_1 \cdots P_k$. This is realized by adding arguments P_i's to the variable x according to the position $\alpha \in 2^*$ of the node in the tree. Let us call $x\vec{P}_\alpha$ the applicative context constructed at position α. When α increases by $\langle 0 \rangle$ (left branch), this context is extended by adding the identity I; when α increases by $\langle 1 \rangle$ (right branch), it is extended by a sequence of arguments depending on α. Specifically, the sequence has the form $\mathsf{U}_{m+1}^{m+1}\Omega^{\sim m}$, where the index m is determined by simulating the head reduction of $\mathsf{L}_{\ell(\alpha)}$, namely the closed λ-term whose code is the length of α, in the context $x\vec{P}_\alpha$ built so far.

There are two possibilities:

- The simulation terminates in a head normal form.

  In this case $\mathsf{L}_{\ell(\alpha)}\vec{P}_\alpha \twoheadrightarrow_h \lambda x_0 \ldots x_n.x_j M_1 \cdots M_k$, so the index m is defined as $n+k$, and the construction of the binary tree continues on the children.

- The simulation gives rise to an unsolvable subterm.

  In this case the node itself becomes Ω (by Ω-reduction) and the construction of the tree only continues in the other branch (at position $\alpha\,;\,0$).

As previously observed, when $\mathsf{L}_{\ell(\alpha)}\vec{P}_\alpha$ diverges, the subtree rooted at $\alpha\,;\,1$ is automatically destroyed, leaving as only option the left branch. Symmetrically, when $\mathsf{L}_{\ell(\alpha)}\vec{P}_\alpha$ converges, we desire the right branch to become the only option. (This is needed to ensure that the infinite tunnel under construction remains unique.) However, even if the left branch $\alpha\,;\,0$ should morally be 'doomed', it does not automatically self-destruct—it extends the context $x\vec{P}_\alpha$ with an I and continues the construction of the subtree. We must therefore take precautions to ensure that all the branches departing from a doomed position eventually become Ω. To this purpose, we introduce in the definition of Ψ a combinator T, called *time-bomb*, which is responsible for harnessing doomed positions.

The time-bomb at position α simulates the head reduction of $\mathsf{L}_{\ell(\beta)}\vec{P}_\beta$ for all previous positions $\beta\,;\,0$ (corresponding to some left branch) by an ever-increasing number of steps. The precise number of steps is actually given by the length of the current position α.

[1] We refer to Polonsky (2012) for more details on these computability issues.

Whenever some subsequence $\beta\,;0 \leq \alpha$ gives rise to a λ-term $\mathsf{L}_{\ell(\beta)}\vec{P}_\beta$ that converges after $\ell(\alpha)$ steps of head reduction, the bomb detonates destroying[2] all children of $\beta\,;0$ at depth $\ell(\alpha)$. Notice that in this case, the position $\beta\,;1$ is certainly 'healthy', because its own simulation has converged. As we will see, it is not difficult to check by induction that there exists only one infinite healthy path in the tree.

The final step of the proof is to verify that the corresponding tunnel $(P_n)_{n\in\mathbb{N}}$ has exactly one survivor, namely I. First, we need to check that the identity always reduces through the tunnel, but this is easy since the latter has been constructed using I and $\mathsf{U}^{m+1}_{m+1}\Omega^{\sim m}$ as building blocks. Second, we need to check that there are no other survivors but this is guaranteed by the peculiar choice of m in the definition of the context associated with $\alpha\,;1$. Suppose indeed that there exists another survivor $M\in\Lambda^o$, and let β be the unique healthy sequence of length $\#M$. Since M is capable of reducing through the whole tunnel, it must in particular remain solvable when applied to the partial tunnel $x\vec{P}_\beta$, so we have $M\vec{P}_\beta \twoheadrightarrow_h \lambda x_0\ldots x_n.x_jM_1\cdots M_k$. This entails that the position $\beta\,;0$ is doomed, whence the next healthy sequence must be $\beta\,;1$. Now, the context $x\vec{P}_\beta$ is extended at position $\beta\,;1$ by adding $\mathsf{U}^{m+1}_{m+1}\Omega^{\sim m}$ with $m=k+n$. So, unless $j=0$ and $n+1\geq k$, the λ-term M is not a survivor after all. In that case, the selector U^{m+1}_{m+1} goes in head position, erases $M_1,\ldots,M_k$ and in order to survive the rest of the tunnel it must coincide with I after eating the next $(n+1-k)$ arguments P_i.

Digging tunnels

From now on, and until the end of the section, we consider fixed the following enumeration of combinators (where $\mathsf{L}_n \triangleq \mathsf{E}\,\mathsf{c}_n$):

$$(\mathsf{L}_n)_{n\in\mathbb{N}} = (\Omega, \mathsf{c}_4, \mathsf{Y}, \mathsf{KS}, \mathsf{U}^{20}_1\mathsf{I}^{\sim 10}, \mathsf{K}^\star, \ldots) \qquad (10.2)$$

This can be done without loss of generality and is useful to produce and discuss examples.

DEFINITION 10.10. (i) For $M\in\Lambda$ and $n\in\mathbb{N}$, write $M{\downarrow^n}$ whenever M reaches a head normal form in at most n steps of head reduction.

(ii) Write $\lfloor M\rfloor \triangleq n+k$ if $M\in\mathrm{SOL}$ and $M^{\mathsf{phnf}} = \lambda x_0\ldots x_n.yM_1\cdots M_k$.

(iii) Given $x\in\mathrm{Var}$ and $\alpha\in 2^*$, define a λ-term M^α_x as follows:

$$\begin{aligned}
\mathsf{M}^{\langle\rangle}_x &\triangleq x;\\
\mathsf{M}^{\alpha\,;0}_x &\triangleq \mathsf{M}^\alpha_x\mathsf{I};\\
\mathsf{M}^{\alpha\,;1}_x &\triangleq \begin{cases}\mathsf{M}^\alpha_x\,\mathsf{U}^{m+1}_{m+1}\,\Omega^{\sim m}, & \text{if } \lfloor \mathsf{M}^\alpha_x[x:=\mathsf{L}_{\ell(\alpha)}]\rfloor = m,\\ \Omega, & \text{if } \mathsf{M}^\alpha_x[x:=\mathsf{L}_{\ell(\alpha)}]\in\mathrm{UNS}.\end{cases}
\end{aligned}$$

(iv) We set $\mathsf{M}^\alpha_N \triangleq \mathsf{M}^\alpha_x[x:=N]$, where $N\in\Lambda^o$, and $\mathsf{M}^\alpha_\natural \triangleq \mathsf{M}^\alpha_x[x:=\mathsf{L}_{\ell(\alpha)}]$.

REMARK 10.11. The λ-term M^α_x satisfies $\mathrm{FV}(\mathsf{M}^\alpha_x)\subseteq\{x\}$. More precisely, if $\mathsf{M}^\alpha_x\in\mathrm{SOL}$ then it has the shape $xP_1\cdots P_k$ where each P_i is either Ω or U^n_n for some $n>0$. We shall think of M^α_x as a closed applicative context having a fresh variable x in the place of $[]$.

[2]In the sense that they become Ω.

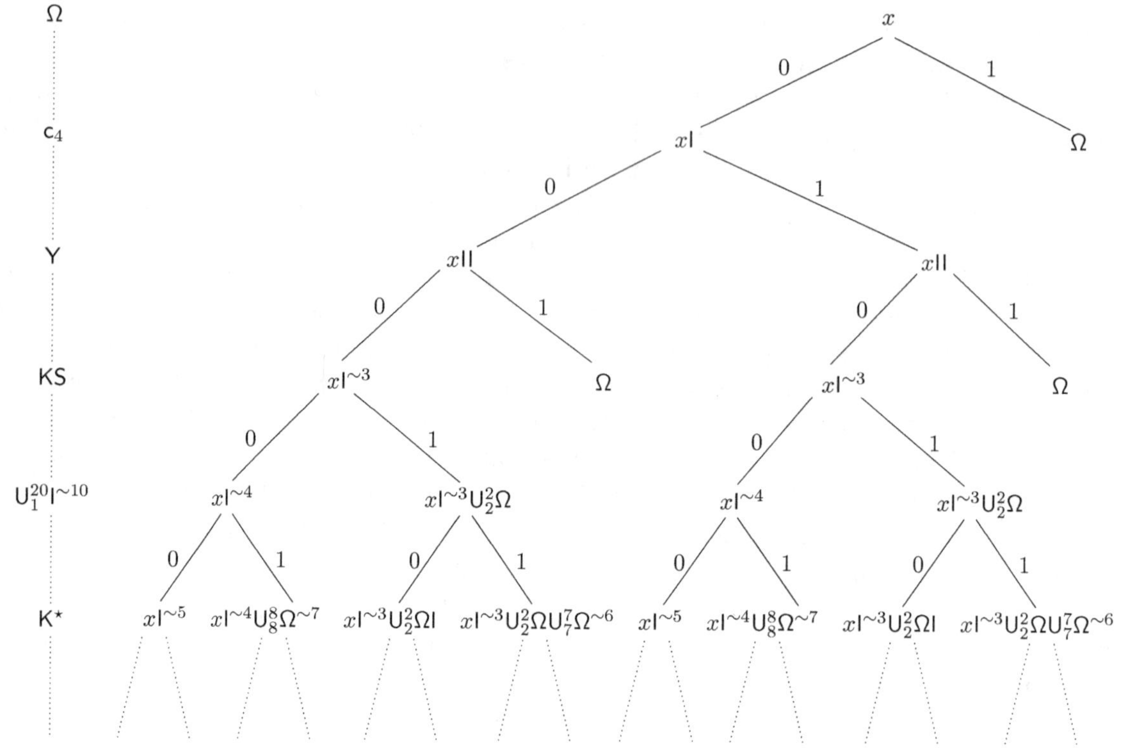

Figure 10.1: The tree generated by M_x^α for $\alpha \in 2^*$.

EXAMPLES 10.12. (i) If M is unsolvable, then $M\!\downarrow^n$ never holds.

(ii) By definition, we have $\mathsf{M}_x^{\langle 0\rangle} = x\mathsf{I}$. Moreover $\mathsf{M}_x^{\langle 1\rangle} = \Omega$ because $\mathsf{M}_{\mathsf{L}_0}^{\langle\rangle} = x[x:=\Omega] = \Omega$ is unsolvable. As a consequence we obtain $\mathsf{M}_x^{\langle 1\rangle\star\alpha} = \Omega$, for every sequence $\alpha \in 2^*$.

(iii) Similarly, $\mathsf{M}_x^{\langle 0,0\rangle} = x\mathsf{II}$. To determine the value of $\mathsf{M}_x^{\langle 0,1\rangle}$ we need to consider $\mathsf{M}_{\mathsf{c}_4}^{\langle 0\rangle}$, which is solvable since $\mathsf{c}_4\mathsf{I} \to_h \lambda z.\mathsf{I}^4(z) \to_h \lambda z.\mathsf{I}^3(z) \to_h \lambda z.\mathsf{I}^2(z) \to_h \lambda z.\mathsf{I}z \to_h \lambda z.z$, so we have $\mathsf{M}_{\mathsf{c}_4}^{\langle 0\rangle}\!\downarrow^n$ for all $n > 4$ and $\lfloor \mathsf{M}_{\mathsf{c}_4}^{\langle 0\rangle} \rfloor = 0$. We conclude that $\mathsf{M}_x^{\langle 0,1\rangle} = x\mathsf{I}\mathsf{U}_1^1\Omega^{\sim 0} = x\mathsf{II}$.

(iv) From $\mathsf{M}_x^{\langle 0,0\rangle} = \mathsf{M}_x^{\langle 0,1\rangle} = x\mathsf{II}$ and the definition, we obtain $\mathsf{M}_x^{\langle 0,0,0\rangle} = \mathsf{M}_x^{\langle 0,1,0\rangle} = x\mathsf{III}$. It is easy to check that YII is unsolvable, whence $\mathsf{M}_x^{\langle 0,0,1\rangle\star\alpha} = \mathsf{M}_x^{\langle 0,1,1\rangle\star\alpha} = \Omega$ for all $\alpha \in 2^*$.

(v) Once again, we have $\mathsf{M}_x^{\langle 0,0,0,0\rangle} = \mathsf{M}_x^{\langle 0,1,0,0\rangle} = x\mathsf{I}^{\sim 4}$. To determine the value of, say, $\mathsf{M}_x^{\langle 0,0,0,1\rangle}$ we need to study the behavior of KS in the context $[\,]\mathsf{III}$. The λ-term KSIII so obtained is solvable since $\mathsf{KSIII} \to_h (\lambda y.\mathsf{S})\mathsf{III} \to_h \mathsf{SII} \to_h (\lambda yx.\mathsf{I}x(yx))\mathsf{I} \to_h \lambda x.\mathsf{I}x(\mathsf{I}x) \to_h \lambda x.x(\mathsf{I}x)$, thus we get $\mathsf{M}_{\mathsf{KS}}^{\langle 0,0,0\rangle}\!\downarrow^n$ for all $n > 4$ and $\lfloor \mathsf{M}_{\mathsf{KS}}^{\langle 0,0,0\rangle} \rfloor = 1$. We conclude that $\mathsf{M}_x^{\langle 0,0,0,1\rangle} = \mathsf{M}_x^{\langle 0,1,0,1\rangle} = x\mathsf{I}^{\sim 3}\mathsf{U}_2^2\Omega$.

EXERCISE 10.13. (i) For $M = \mathsf{U}_1^{20}\mathsf{I}^{\sim 14}$, check that $M\!\downarrow^{14}$ and $\lfloor M \rfloor = 7$.

(ii) For $M = \mathsf{U}_1^{20}\mathsf{I}^{\sim 13}\mathsf{U}_2^2\Omega$, determine the least k such that $M\!\downarrow^k$ and compute $\lfloor M \rfloor$.

(iii) Compute M_x^α for all sequences α of length 5.

Figure 10.1 is obtained by gathering all the information from Example 10.12 and Exercice 10.13, and displays the infinite binary tree of applicative contexts M_x^α for $\alpha \in 2^*$.

As one can see, the left branches of the tree are labelled by 0, while the right ones by 1. The column on the left lists the closed λ-terms from Enumeration (10.2): in other words, at the $\ell(\alpha)$-th level one finds the combinator that should replace x in M_x^α to obtain $\mathsf{M}_\natural^\alpha$. The left child of a node at position α extends the applicative context built at α by adding an I. If $\mathsf{M}_\natural^\alpha$ is unsolvable then the right child of α is Ω (levels 0 and 2), otherwise the context is extended by adding $\mathsf{U}_{m+1}^{m+1}\Omega^{\sim m}$ for a suitable index m (levels 1, 3 and 4).

LEMMA 10.14. *For all $\alpha \in 2^*$, we have:*
(i) $\mathsf{M}_x^{\alpha;1}$ *solvable entails that* $\mathsf{M}_\natural^\alpha$ *is solvable.*
(ii) M_x^α *solvable implies that* $\mathsf{M}_x^{\alpha;0}$ *is solvable.*

PROOF. Both statements follow easily from Definition 10.10(iii). □

We now show how it is possible to make Definition 10.10(iii) effective by going over to the codes. In the proof we exploit the fact that M unsolvable entails $\mathsf{E}M$ unsolvable.

LEMMA 10.15. *There exists a λ-term $F \in \Lambda^o$ such that, for all $\alpha \in 2^*$:*

$$\mathsf{E}(F\ulcorner\alpha\urcorner)x = \mathsf{M}_x^\alpha$$

In particular, $\mathsf{M}_\natural^\alpha = \mathsf{E}(F\ulcorner\alpha\urcorner)\mathsf{L}_{\ell(\alpha)}$.

PROOF. Consider the partial function $f : \mathbb{N} \to \mathbb{N}$ satisfying (for a fresh $z \in \mathrm{Var}$):

$$\begin{array}{lcl} f(\#\langle\rangle) & = & \#\mathsf{I}; \\ f(\#(\alpha\,;0)) & = & \#(\lambda z.\mathsf{L}_{f(\#\alpha)}z\mathsf{I}); \\ f(\#(\alpha\,;1)) & = & \begin{cases} \#(\lambda z.\mathsf{L}_{f(\#\alpha)}z\mathsf{U}_{m+1}^{m+1}\Omega^{\sim m}), & \text{if } \lfloor \mathsf{L}_{f(\#\alpha)}\mathsf{L}_{\ell(\alpha)} \rfloor = m, \\ \uparrow, & \text{if } \mathsf{L}_{f(\#\alpha)}\mathsf{L}_{\ell(\alpha)} \text{ is unsolvable.} \end{cases} \end{array}$$

Moreover $f(\#\alpha)$ is undefined for $\alpha \notin 2^*$. The function f is partial recursive since:
– $\#\alpha$ and $\ell(\alpha)$ can be computed from the code $\#(\alpha\,;i)$ (for $i = 0, 1$);
– the closed λ-term $\mathsf{L}_{f\#(\alpha)}$ can be reconstructed as $\mathsf{E}\,\mathsf{c}_{f(\#\alpha)}$;
– similarly, the closed λ-term $\mathsf{L}_{\ell(\alpha)}$ can be reconstructed as $\mathsf{E}\,\mathsf{c}_{\ell(\alpha)}$;
– the head reduction being an effective strategy, the problem of determining whether $\mathsf{L}_{f(\#\alpha)}\mathsf{L}_{\ell(\alpha)} \twoheadrightarrow_h \lambda x_0 \ldots x_n.yM_1 \cdots M_k$ is semi-decidable, and in that case $m \triangleq n + k$.

The procedure might fail to terminate when $\mathsf{L}_{f(\#\alpha)}\mathsf{L}_{\ell(\alpha)}$ is unsolvable, but in that case $f(\#(\alpha\,;1))$ is also undefined. To conclude, take as F the combinator λ-defining f. □

We introduce some terminology.

DEFINITION 10.16. A sequence $\alpha \in 2^*$ is called:
(i) *doomed* whenever there exists a subsequence $\beta\,;0 \leq \alpha$ such that $\mathsf{M}_\natural^\beta$ is solvable.
(ii) *healthy* if α is not doomed and M_x^α is solvable.

REMARK 10.17. (i) If $\beta \in 2^*$ is healthy and $\alpha \leq \beta$ then the sequence α is also healthy.
(ii) If $\alpha \in 2^*$ is healthy and $\mathsf{M}_\natural^\alpha$ is solvable then $\alpha\,;1$ is healthy as well.

LEMMA 10.18. *For $\alpha \in 2^*$, the following are equivalent:*
1. *α is healthy and $\alpha\,;0$ is doomed;*
2. *$\alpha\,;1$ is healthy.*

PROOF. ($1 \Rightarrow 2$) If $\alpha\,;0$ is doomed then there exists a subsequence $\beta\,;0 \leq \alpha\,;0$ such that $\mathsf{M}_\natural^\beta$ is solvable. As α is healthy we must have $\beta = \alpha$, so we conclude by Remark 10.17(ii).

($2 \Rightarrow 1$) Assume that $\alpha\,;1$ is healthy. By Remark 10.17(i) the sequence α is also healthy. Since $\mathsf{M}_x^{\alpha\,;1}$ is solvable, by Lemma 10.14(i), so is $\mathsf{M}_\natural^\alpha$ whence $\alpha\,;0$ is doomed. □

PROPOSITION 10.19. (i) *For $n \in \mathbb{N}$, there is a unique healthy sequence α_n of length n.*

(ii) *The healthy sequences $(\alpha_n)_{n\in\mathbb{N}}$ form an increasing chain, i.e. $\alpha_i < \alpha_{i+1}$ for $i \in \mathbb{N}$.*

PROOF. We prove (i) by induction on n, and check that (ii) holds at any stage.

Base. The unique sequence of length 0 is $\langle\rangle$, which is healthy because $\mathsf{M}_x^{\langle\rangle} = x$ is solvable.

Induction. Assume that α_n is the unique healthy sequence of length n. We split into cases depending on the solvability of $\mathsf{M}_\natural^{\alpha_n}$.

- If $\mathsf{M}_\natural^{\alpha_n}$ is solvable then $\alpha_n\,;1$ is healthy by Remark 10.17(ii). Moreover, the sequence $\alpha_n\,;0$ is doomed by Lemma 10.18.
- Otherwise $\mathsf{M}_\natural^{\alpha_n}$ is unsolvable. On the one hand this entails that $\mathsf{M}_x^{\alpha_n\,;1}$ is unsolvable so $\alpha_n\,;1$ cannot be healthy. On the other hand, by IH, $\mathsf{M}_\natural^\beta$ is unsolvable for all subsequences $\beta\,;0 \leq \alpha_n$, hence $\alpha_n\,;0$ is not doomed. As α_n is healthy, $\mathsf{M}_x^{\alpha_n}$ is solvable and, by Lemma 10.14(ii), so is $\mathsf{M}_x^{\alpha_n\,;0}$ thus $\alpha_n\,;0$ is healthy.

We have shown that the healthy sequence α_{n+1} of length $n+1$ is uniquely determined. From Remark 10.17(i) and $\alpha_n \neq \alpha_{n+1}$ we conclude that $\alpha_n < \alpha_{n+1}$. □

LEMMA 10.20. *Let $n \in \mathbb{N}$ and α_n be the healthy sequence of length n. Then we have:*

(i) $\mathsf{M}_x^{\alpha_n} \neq \Omega$,

(ii) $\mathsf{M}_\mathsf{I}^{\alpha_n} = \mathsf{I}$.

PROOF. (i) By Definition 10.16(ii). (ii) By a straightforward induction on n. □

The counterexample

Now that we have introduced the tree M_x^α of applicative contexts and studied its key properties, everything is in place to provide the definition of the counterexample Ψ.

DEFINITION 10.21. Using the reducing fixed point combinator Θ, define $\Psi, X, T, S \in \Lambda^o$ satisfying the following recursive equations:

$$\begin{array}{rcl}
\Psi & = & \lambda x.X\ulcorner\langle\rangle\urcorner x; \\
X\ulcorner\alpha\urcorner & = & T\ulcorner\alpha\urcorner[X\ulcorner\alpha\,;0\urcorner(x\mathsf{I}), X\ulcorner\alpha\,;1\urcorner(S\ulcorner\alpha\urcorner x)]; \\
T\ulcorner\alpha\urcorner & = & \begin{cases}\Omega, & \text{if } \exists\beta\in 2^*\,.\,\beta\,;0\leq\alpha \,\&\, \mathsf{M}_\natural^\beta\downarrow^{\ell(\alpha)}, \\ \mathsf{I}, & \text{otherwise;}\end{cases} \\
S\ulcorner\alpha\urcorner & = & \begin{cases}\lambda z.z\mathsf{U}_{m+1}^{m+1}\Omega^m, & \text{if } \lfloor\mathsf{M}_\natural^\alpha\rfloor = m, \\ \Omega, & \text{if } \mathsf{M}_\natural^\alpha \in \mathrm{UNS}.\end{cases}
\end{array}$$

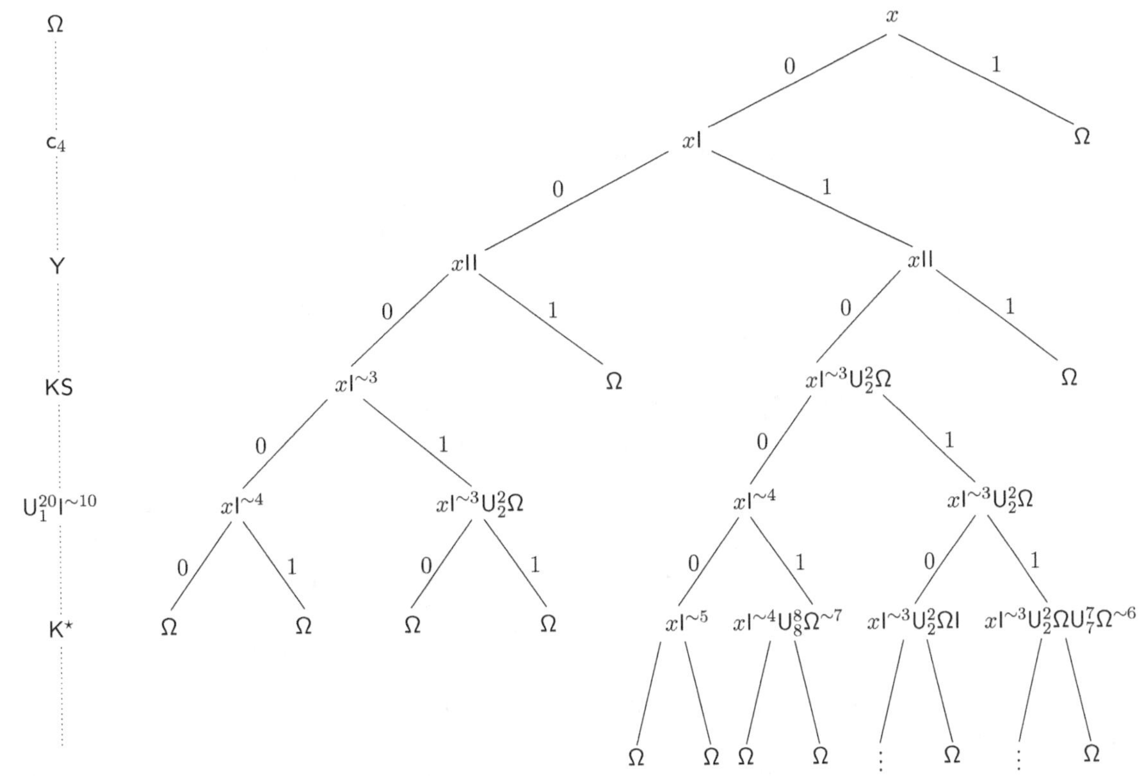

Figure 10.2: The scope tree of x in Ψx.

As Θ is reducing, the λ-terms on the left hand-side actually $\beta\Omega$-reduce to the λ-terms on the right hand-side.

LEMMA 10.22. *The combinators Ψ, X, T, S from Definition 10.21 exist.*

PROOF. References to M_x^α are processed by going over the codes as in Lemma 10.15. Note that in the definition of $T\ulcorner\alpha\urcorner$, it is decidable whether there is a subsequence $\beta\,;0 \leq \alpha$ such that $\mathsf{M}_\natural^\beta \downarrow^{\ell(\alpha)}$ because α is finite and head reduction is an effective and deterministic strategy. Finally, in the definition of $S\ulcorner\alpha\urcorner$, the condition $\mathsf{M}_\natural^\alpha \in \mathrm{SOL}$ is semi-decidable and if it is the case then $\lfloor\mathsf{M}_\natural^\alpha\rfloor$ is computable. □

Before going further, let us compare the tree constructed by Ψx, which is depicted in Figure 10.2, with the tree of applicative contexts M_x^α in Figure 10.1. We observe that a node at position α in the former coincides with the same node M_x^α in the latter, unless it has been destroyed by the time-bomb $T\ulcorner\alpha\urcorner = \Omega$. Moreover, if $\mathsf{M}_\natural^\alpha$ converges after k head reduction steps, then the subtree rooted at $\alpha\,;0$ must be finite—indeed, at depth k, all its nodes get simultaneously destroyed by the time-bomb (levels 1 and 3). Therefore, whenever there is a node different from Ω occurring to the right of α on the same level, the position α itself is doomed and all its descendants are eventually destroyed.

LEMMA 10.23. *For all $\alpha, \beta \in 2^*$, we have:*

$$(X\ulcorner\alpha\urcorner\mathsf{M}_x^\alpha)_\beta = X\ulcorner\alpha\star\beta\urcorner\mathsf{M}_x^{\alpha\star\beta}$$

PROOF. We proceed by induction on β.

Base. By definition we have $(X\ulcorner\alpha\urcorner\mathsf{M}_x^\alpha)_{\langle\rangle} = X\ulcorner\alpha\urcorner\mathsf{M}_x^\alpha = X\ulcorner\alpha\star\langle\rangle\urcorner\mathsf{M}_x^{\alpha\star\langle\rangle}$.

Induction. Assume that β is such that $(X\ulcorner\alpha\urcorner\mathsf{M}_x^\alpha)_\beta = X\ulcorner\alpha\star\beta\urcorner\mathsf{M}_x^{\alpha\star\beta}$ for all $\alpha\in 2^*$, we shall prove that the same holds for $\langle i\rangle\star\beta$ where $i\in\{0,1\}$. Take any $\alpha\in 2^*$. By definition, $X\ulcorner\alpha\urcorner = T\ulcorner\alpha\urcorner[X\ulcorner\alpha\,;\,0\urcorner(x\mathsf{I}), X\ulcorner\alpha\,;\,1\urcorner(S\ulcorner\alpha\urcorner x)]$ and there are two subcases:

- either $T\ulcorner\alpha\urcorner = \mathsf{I}$ and we get

$$\begin{array}{rcl} X\ulcorner\alpha\urcorner\mathsf{M}_x^\alpha & = & [X\ulcorner\alpha\,;\,0\urcorner(\mathsf{M}_x^\alpha\mathsf{I}), X\ulcorner\alpha\,;\,1\urcorner(S\ulcorner\alpha\urcorner\mathsf{M}_x^\alpha)] \\ & = & [X\ulcorner\alpha\,;\,0\urcorner\mathsf{M}_x^{\alpha\,;\,0}, X\ulcorner\alpha\,;\,1\urcorner\mathsf{M}_x^{\alpha\,;\,1}] \end{array}$$

- or $T\ulcorner\alpha\urcorner$ is unsolvable and so is $T\ulcorner\gamma\urcorner$ for all $\gamma\in 2^*$ such that $\alpha\le\gamma$, hence we get

$$X\ulcorner\gamma\urcorner\mathsf{M}_x^\gamma = \Omega = X\ulcorner\alpha\urcorner\mathsf{M}_x^\alpha = (X\ulcorner\alpha\urcorner\mathsf{M}_x^\alpha)_{\langle i\rangle\star\beta}$$

We conclude that $(X\ulcorner\alpha\urcorner\mathsf{M}_x^\alpha)_{\langle i\rangle\star\beta} = (X\ulcorner\alpha\,;\,i\urcorner\mathsf{M}_x^{\alpha\,;\,i})_\beta =_{\mathrm{IH}} X\ulcorner\alpha\star\langle i\rangle\star\beta\urcorner\mathsf{M}_x^{\alpha\star\langle i\rangle\star\beta}$. □

As a short digression, we can wonder how the Böhm tree of Ψx looks like. The next lemma shows that all the information visualized in Figure 10.2 disappears when constructing its Böhm tree because x is pushed into infinity together with its arguments.

LEMMA 10.24. *Let $x\in\mathrm{Var}$ and $\alpha\in\mathrm{BT}(\Psi\, x)$, then for some $T_1, T_2\in\mathscr{B}$:*

$$\mathrm{BT}(X\ulcorner\alpha\urcorner x) \;=\; \begin{array}{c} \lambda y.y \\ \diagup \quad \diagdown \\ T_1 \qquad T_2 \end{array}$$

In particular, $x\notin \mathrm{FV}(\mathrm{BT}(\Psi\, x))$ and $|\mathsf{Range}_{\mathcal{B}}(\Psi)| = 1$.

PROOF. Recall that $\alpha\in\mathrm{BT}(\Psi\, x)$ entails $(\Psi\, x)_\alpha \neq \Omega$, which in its turn implies $T\ulcorner\alpha\urcorner = \mathsf{I}$. So we have:

$$\begin{array}{rcll} (\Psi\, x)_\alpha & = & (X\ulcorner\langle\rangle\urcorner x)_\alpha, & \text{by Definition 10.21,} \\ & = & T\ulcorner\alpha\urcorner[X\ulcorner\alpha\,;\,0\urcorner\mathsf{M}_x^{\alpha\,;\,0},\ X\ulcorner\alpha\,;\,1\urcorner\mathsf{M}_x^{\alpha\,;\,1}], & \text{by Lemma 10.23,} \\ & = & \lambda y.y(X\ulcorner\alpha\,;\,0\urcorner\mathsf{M}_x^{\alpha\,;\,0})(X\ulcorner\alpha\,;\,1\urcorner\mathsf{M}_x^{\alpha\,;\,1}), & \text{as } T\ulcorner\alpha\urcorner = \mathsf{I}. \end{array}$$

We conclude by taking $T_1 \triangleq \mathrm{BT}(X\ulcorner\alpha\,;\,0\urcorner\mathsf{M}_x^{\alpha\,;\,0})$ and $T_2 \triangleq \mathrm{BT}(X\ulcorner\alpha\,;\,1\urcorner\mathsf{M}_x^{\alpha\,;\,1})$. □

Together with Lemma 10.20(i), the next lemma shows that $\mathrm{BT}(\Psi\, x)$ has an infinite path because $\alpha\in\mathrm{BT}(\Psi\, x)$ for every healthy sequence α.

LEMMA 10.25. *Let $n\in\mathbb{N}$ and α_n be the healthy sequence of length n. Then $T\ulcorner\alpha_n\urcorner\neq\Omega$.*

PROOF. By induction on n. □

LEMMA 10.26. *If* $\alpha \in 2^*$ *is not healthy then* $X\ulcorner\alpha\urcorner\mathsf{M}^\alpha_x = X\ulcorner\alpha\urcorner\Omega$.

PROOF. The case where α is not healthy because $\mathsf{M}^\alpha_x = \Omega$ is trivial. Suppose then that M^α_x is solvable and α is doomed. In this case there exist a subsequence $\gamma; 0 \leq \alpha$ and a natural number n such that $\mathsf{M}^\gamma_x\downarrow^n$. Hence, for every γ of length $\ell(\gamma) \geq n - \ell(\alpha)$ we have $T\ulcorner\alpha\star\gamma\urcorner = \Omega$. By applying Lemma 10.23, we obtain

$$(X\ulcorner\alpha\urcorner\mathsf{M}^\alpha_x)_\gamma = X\ulcorner\alpha\star\gamma\urcorner\mathsf{M}^{\alpha\star\gamma}_x = \Omega$$

This entails that $\mathrm{BT}(X\ulcorner\alpha\urcorner\mathsf{M}^\alpha_x)$ is finite since its height is at most $n - \ell(\alpha)$. In other words, $X\ulcorner\alpha\urcorner\mathsf{M}^\alpha_x$ has a $\beta\Omega$-normal form without free occurrences of x (by Lemma 10.24).

We conclude that

$$\begin{array}{rcll} X\ulcorner\alpha\urcorner\mathsf{M}^\alpha_x & = & (X\ulcorner\alpha\urcorner\mathsf{M}^\alpha_x)^{\beta\Omega\text{-}\mathbf{nf}}, & \\ & = & (X\ulcorner\alpha\urcorner\mathsf{M}^\alpha_x)^{\beta\Omega\text{-}\mathbf{nf}}[x{:=}\Omega], & \text{as } x \notin \mathrm{FV}((X\ulcorner\alpha\urcorner\mathsf{M}^\alpha_x)^{\beta\Omega\text{-}\mathbf{nf}}), \\ & = & (X\ulcorner\alpha\urcorner\mathsf{M}^\alpha_\Omega)^{\beta\Omega\text{-}\mathbf{nf}}, & \text{as } X\ulcorner\alpha\urcorner \in \Lambda^o, \\ & = & X\ulcorner\alpha\urcorner\mathsf{M}^\alpha_\Omega. & \end{array}$$

□

LEMMA 10.27. *Let* α_n *be the healthy sequence of length* n. *For all* $N \in \Lambda^o$, *we have:*

$$\Psi N = \Psi\Omega \iff \exists n \in \mathbb{N}\ .\ \mathsf{M}^{\alpha_n}_N = \Omega$$

PROOF. ($\Rightarrow$) Assume $\Psi N = \Psi\Omega$. By Proposition 10.9 (Church-Rosser for $\mathcal{H}$) they have a common $\beta\Omega$-reduct Z, namely $\Psi N \twoheadrightarrow_{\beta\Omega} Z$ and $\Psi\Omega \twoheadrightarrow_{\beta\Omega} Z$. By Lemma 10.25 for every $i \geq 0$ we have $\alpha_i \in \mathrm{BT}(\Psi\, x)$, hence $X\ulcorner\alpha_i\urcorner\mathsf{M}^{\alpha_i}_x$ is always solvable. Take n as the greatest number such that no redex occurs in the syntax tree of Z along the path α_n. Then there exists a subterm W of Z at position α_n having the shape $W = X\ulcorner\alpha_n\urcorner\mathsf{M}^{\alpha_n}_N$ by Lemma 10.23. As the combinators from Definition 10.21 have been defined using a reducing fpc, head reducing W produces $W \twoheadrightarrow_h (\lambda x.W_1)W_2$ where $X\ulcorner\alpha_n\urcorner \twoheadrightarrow_h \lambda x.W_1$, $\mathsf{M}^{\alpha_n}_N \twoheadrightarrow_{\beta\Omega} W_2$ and $\mathsf{M}^{\alpha_n}_\Omega \twoheadrightarrow_{\beta\Omega} W_2$. We conclude because these last two conditions together with $\mathsf{M}^{\alpha_n}_\Omega = \Omega$ entail $\mathsf{M}^{\alpha_n}_N = \Omega$.

($\Leftarrow$) Let $n \in \mathbb{N}$ be such that $\mathsf{M}^{\alpha_n}_N = \Omega$. Then by applying Lemma 10.23, we get

$$(\Psi N)_{\alpha_n} = X\ulcorner\alpha_n\urcorner\mathsf{M}^{\alpha_n}_N = X\ulcorner\alpha_n\urcorner\Omega = (\Psi\Omega)_{\alpha_n} \qquad (10.3)$$

Thanks to Lemma 10.26 we already know that for all non healthy $\alpha \in 2^*$ with $\ell(\alpha) = n$:

$$(\Psi N)_\alpha = X\ulcorner\alpha\urcorner\Omega = (\Psi\Omega)_\alpha \qquad (10.4)$$

Putting together (10.3) and (10.4) we get that the subterms of ΨN and $\Psi\Omega$ at all sequences β of length n relative to their Böhm trees are interconvertible. This entails $\Psi N = \Psi\Omega$ by first unfolding both λ-terms up to depth n and then performing the conversions. □

PROPOSITION 10.28. *For all* $N \in \Lambda^o$, *we have:*

$$\Psi N \neq \Psi\Omega \iff \Psi N = \Psi\mathsf{I}$$

Therefore, $\mathsf{Range}_{\mathcal{H}}(\Psi) = \{[\Omega]_{\mathcal{H}}, [\mathsf{I}]_{\mathcal{H}}\}$.

PROOF. ($\Rightarrow$) Assume that $\Psi N \neq \Psi\Omega$. Let n be the code of N, i.e. $\mathsf{L}_n = N$, and let α_n be the unique healthy sequence of length n. Now, $\mathsf{M}_{\natural}^{\alpha_n} = \mathsf{M}_N^{\alpha_n}$ is solvable by Lemma 10.27, thus $\alpha_{n+1} \triangleq \alpha_n\,;1$ is healthy by Remark 10.17(ii) and $\alpha_n\,;0$ is doomed by Lemma 10.18.

Let $(\mathsf{M}_{\natural}^{\alpha_n})^{\mathsf{phnf}} = \lambda x_0 \dots x_j.x_i P_1 \cdots P_m$ so that $\lfloor \mathsf{M}_{\natural}^{\alpha_n} \rfloor = m + j$ and we have:

$$\begin{aligned} \mathsf{M}_N^{\alpha_{n+1}} &= \mathsf{M}_N^{\alpha_n} \mathsf{U}_{m+j+1}^{m+j+1} \Omega^{\sim m+j}, \\ &= (\lambda x_0 \dots x_j.x_i P_1 \cdots P_m) \mathsf{U}_{m+j+1}^{m+j+1} \Omega^{\sim m+j}. \end{aligned}$$

As $\mathsf{M}_N^{\alpha_{n+1}}$ is solvable we must have $i = 0$ otherwise an Ω would be substituted for the x_i in head position. Easy calculations give the following chain of equalities:

$$\begin{aligned} \mathsf{M}_N^{\alpha_{n+1}} &= (\lambda x_0 \dots x_j.x_0 P_1 \cdots P_m) \mathsf{U}_{m+j+1}^{m+j+1} \Omega^{\sim m+j}, \\ &= (\lambda x_1 \dots x_j.\mathsf{U}_{m+j+1}^{m+j+1} P_1^{\sigma} \cdots P_m^{\sigma}) \Omega^{\sim m+j}, \qquad \text{where } \sigma \triangleq [x_0 := \mathsf{U}_{m+j+1}^{m+j+1}], \\ &= (\lambda x_1 \dots x_j.\mathsf{U}_{j+1}^{j+1}) \Omega^{\sim m+j}, \\ &= \mathsf{U}_{j+1}^{j+1} \Omega^{\sim m}. \end{aligned}$$

Again, $\mathsf{M}_N^{\alpha_{n+1}}$ to be solvable must have $m \leq j$, so we get $\mathsf{M}_N^{\alpha_{n+1}} = \mathsf{U}_{j-m+1}^{j-m+1}$. By Definition 10.10(iii), for any $k \geq 0$ the λ-term $\mathsf{M}_x^{\alpha_{n+1+k}}$ can be written as follows (for $q_\ell > 0$):

$$\mathsf{M}_x^{\alpha_{n+1+k}} = \mathsf{M}_x^{\alpha_{n+1}} \underbrace{Q_1 \cdots Q_{q_1}}_{\text{stage } \alpha_{n+2}} \underbrace{Q_{q_1+1} \cdots Q_{q_1+q_2}}_{\alpha_{n+3}} \cdots \underbrace{Q_{q_1+\dots+q_{k-1}+1} \cdots Q_{q_1+\dots+q_k}}_{\alpha_{n+1+k}}$$

where, for every $\ell \leq k$, $Q_{q_1+\dots+q_{\ell-1}+1} \cdots Q_{q_1+\dots+q_\ell}$ represent the combinators added at position $\alpha_{n+1+\ell}$. In particular, the first element $Q_{q_1+\dots+q_{\ell-1}+1}$ of each sequence is a selector while the other elements are unsolvable. The fact that $\mathsf{M}_N^{\alpha_{n+1}} = \mathsf{U}_{j-m+1}^{j-m+1}$ implies that Q_{j-m+1} must be a selector because otherwise

$$\begin{aligned} \mathsf{M}_N^{\alpha_{n+1+j-m}} &= \mathsf{M}_N^{\alpha_{n+1}} Q_1 \cdots Q_{q_1+\dots+q_{j-m}}, \\ &= \mathsf{U}_{j-m+1}^{j-m+1} Q_1 \cdots Q_{q_1+\dots+q_{j-m}}, \\ &= Q_{j-m+1} \cdots Q_{q_1+\dots+q_{j-m}}. \end{aligned}$$

would be unsolvable. As a consequence, $j - m = q_1 + \dots + q_\ell$ for some $\ell \geq 0$ so that

$$\begin{aligned} \mathsf{M}_N^{\alpha_{n+1+j-m}} &= \mathsf{M}_N^{\alpha_{n+1}} Q_1 \cdots Q_{q_1+\dots+q_\ell}, \\ &= \mathsf{U}_{q_1+\dots+q_\ell+1}^{q_1+\dots+q_\ell+1} Q_1 \cdots Q_{q_1+\dots+q_\ell}, \\ &= \mathsf{I}. \end{aligned}$$

By Lemma 10.20(ii) we obtain $\mathsf{M}_N^{\alpha_{n+1+j-m}} = \mathsf{M}_{\mathsf{I}}^{\alpha_{n+1+j-m}}$. Now, Lemma 10.23 entails that ΨN and $\Psi \mathsf{I}$ have the same subterm at position $\alpha_{n+1+j-m}$ relative to their Böhm trees. By Lemma 10.26, they also agree on all other positions of length $n + 1 + j - m$ whence they are equal in $\mathcal{H}$.

($\Leftarrow$) Indeed, Lemma 10.20(ii) together with Lemma 10.27 entails $\Psi N = \Psi \mathsf{I} \neq \Psi\Omega$. □

As an immediate corollary, we obtain the main result of the section.

THEOREM 10.29. $|\mathsf{Range}_{\mathcal{H}}(\Psi)| = 2$.

COROLLARY 10.30. *$\mathcal{H}\eta$ and $\mathcal{H}\omega$ do not satisfy the range property.*

PROOF. Let $\mathcal{T} \in \{\mathcal{H}\eta, \mathcal{H}\omega\}$. By Remark 10.4(i) we have $|\mathsf{Range}_{\mathcal{T}}(\Psi)| \leq |\mathsf{Range}_{\mathcal{H}}(\Psi)|$ so it is sufficient to check that $\Psi\mathsf{I} \neq_{\mathcal{T}} \Psi\Omega$. Note that no sensible theory can equate a solvable and an unsolvable λ-term, whence $\mathsf{M}_{\natural}^{\alpha_n}$ is solvable in $\mathcal{T}$ exactly when it is solvable in $\mathcal{H}$.

We now discuss the two cases separately:

- If $\mathcal{T} = \mathcal{H}\eta$ then the same proofs establish that the range of Ψ modulo $\mathcal{T}$ is a doubleton. Notice that in the proof of Lemma 10.27, one needs to invoke the Church Rosser property for $\mathcal{H}\eta$, namely Theorem 15.2.15(ii) in B[1984].
- The case $\mathcal{T} = \mathcal{H}\omega$ follows since $\mathsf{M}_{\natural}^{\alpha_n}$ is solvable in $\mathcal{H}$ if and only if it is solvable in $\mathcal{H}\omega$. From this and Lemma 3.24 one can prove by induction on a derivation in $\mathcal{H}\omega$ that also $\Psi\mathsf{I} \neq_{\mathcal{H}\omega} \Psi\Omega$. □

EXERCISE 10.31. Let α_n be the unique healthy sequence of length n (Proposition 10.19). Write $\mathsf{M}_x^{\alpha_{n+1}}$ as $x\vec{Q}$, where $Q_{q_m} \triangleq \mathsf{U}_{k_m+1}^{k_m+1}$, for every $m\,(0 \leq m \leq n)$. For every $n \in \mathbb{N}$, define

$$\begin{array}{rcll} \Psi_n & \triangleq & [F_0, \ldots, F_n], & \text{where} \\ F_m x & = & \Psi(\mathsf{K}^{q_m}(x\Omega^{\sim q_m})), & \text{for } 0 \leq m \leq n. \end{array}$$

Check that $|\mathsf{Range}_{\mathcal{T}}(\Psi_n)| = n + 2$, for every $\mathcal{T} \in \{\mathcal{H}, \mathcal{H}\eta, \mathcal{H}\omega\}$.

[Hint 1: Consider the case $\mathcal{T} \triangleq \mathcal{H}$, then proceed as in the proof of Corollary 10.30]

[Hint 2: First show that $F_m N \in \{\Psi\mathsf{I}, \Psi\Omega\}$, for all $N \in \Lambda^o$. Second, check that $F_m N = F_m\Omega$ and $m < m'$ imply $F_{m'}N = F_{m'}\Omega$. Third, for every $m < n$ provide a λ-term N_m such that $F_m N_m = \Psi\mathsf{I}$ and $F_{m+1}N_m = \Psi\Omega$. Using these facts, verify that the range of Ψ_n consists of $[\Psi\mathsf{I}, \ldots, \Psi\mathsf{I}, \Psi\mathsf{I}], [\Psi\mathsf{I}, \ldots, \Psi\mathsf{I}, \Psi\Omega], \ldots, [\Psi\Omega, \ldots, \Psi\Omega, \Psi\Omega]$.]

10.2 The fixed point property fails in all sensible λ-theories

This section is devoted to prove that no sensible λ-theory $\mathcal{T}$ satisfies the fixed point property (Definition 10.5(ii)), a result originally appeared in Manzonetto et al. (2019). Interestingly, while the counterexample of the range property for $\mathcal{H}$ has been constructed using the full power of λ-calculus self-encoding, the counterexample of the fixed point property is a λ-term A having a much simpler structure:

$$\mathsf{A} \triangleq \lambda xy.x(x(\mathsf{K}y))\Omega \tag{10.5}$$

We first demonstrate that the λ-term A has only two possible fixed points modulo $\mathcal{H}$, subsequently we discuss how the proof generalizes to an arbitrary sensible λ-theory $\mathcal{T}$.

LEMMA 10.32. *$[\Omega]_{\mathcal{H}} \in \mathsf{Fix}_{\mathcal{H}}(\mathsf{A})$, hence $[\Omega]_{\mathcal{T}} \in \mathsf{Fix}_{\mathcal{T}}(\mathsf{A})$ for every sensible λ-theory $\mathcal{T}$.*

PROOF. Indeed, we have $\mathsf{A}\Omega =_{\mathcal{H}} \lambda y.\Omega(\Omega(\mathsf{K}y))\Omega =_{\mathcal{H}} \Omega$. □

We need some auxiliary results. The following is an easy variant of Lemma 1.26.

LEMMA 10.33. *Let $M \in \Lambda^o$ and $y \in \mathrm{Var}$. Assume that $My\Omega^{\sim n}$ is solvable for all $n \in \mathbb{N}$. Then $M =_{\mathcal{H}} \lambda x_0 \dots x_k.x_0 M_1 \cdots M_m$ for some $k, m \geq 0$.*

PROOF. For $n = 0$ we know that My is solvable, therefore M must be solvable as well. As M is moreover closed, we get $M =_{\mathcal{H}} \lambda x_0 \dots x_k.x_i M_1 \cdots M_m$ for some indices $k, m \geq 0$ and $i \leq k$. Assume now $i \neq 0$ towards a contradiction. In this case we get:

$$\begin{aligned} My\Omega^{\sim k} &=_{\mathcal{H}} (x_i M_1 \cdots M_m)^\sigma, \quad \text{for } \sigma \triangleq [x_0 := y][x_1, \dots, x_k := \Omega], \\ &=_{\mathcal{H}} \Omega M_1^\sigma \cdots M_m^\sigma \\ &=_{\mathcal{H}} \Omega. \end{aligned}$$

As a consequence, we must have $x_i = x_0$. □

In the following we consider an arbitrary sensible λ-theory $\mathcal{T}$. We show that the only solvable fixed point of A modulo $\mathcal{T}$ is the identity. We are afraid there are no high level intuitions behind the following demonstration. The proof is just a syntactic, but quite rewarding, *tour de force.*

LEMMA 10.34. *Let $[M]_{\mathcal{T}} \in \mathsf{Fix}_{\mathcal{T}}(\mathsf{A})$. Suppose M is solvable. Then, for some $k, m \geq 0$, we have $M =_{\mathcal{T}} \lambda x_0 \dots x_k.x_0 M_1 \cdots M_m$.*

PROOF. Since M is a fixed point of A in $\mathcal{T}$, we get for fresh variables x, y:

$$My =_{\mathcal{T}} \mathsf{A}My =_{\mathcal{T}} M(M(\mathsf{K}y))\Omega =_{\mathcal{T}} (Mx\Omega)[x := M(\mathsf{K}y)] \tag{10.6}$$

Consider now infinitely many fresh variables $(y_i)_{i \in \mathbb{N}}$ and define $\sigma_i \triangleq [y_{i+1} := M(\mathsf{K}y_i)]$. By iterating Equation (10.6) n times, we obtain the following chain of equivalences:

$$My_0 =_{\mathcal{T}} (My_1\Omega)\sigma_0 =_{\mathcal{T}} (My_2\Omega^{\sim 2})\sigma_1\sigma_0 =_{\mathcal{T}} \cdots =_{\mathcal{T}} (My_n\Omega^{\sim n})\sigma_{n-1} \cdots \sigma_0 \tag{10.7}$$

Since M is solvable also My_0 is solvable and, by Equation (10.7), so must be $My_n\Omega^{\sim n}$ for every $n \in \mathbb{N}$. By Lemma 10.33 we conclude that the head variable of M is x_0. □

LEMMA 10.35. *Let $[M]_{\mathcal{T}} \in \mathsf{Fix}_{\mathcal{T}}(\mathsf{A})$. If $M =_{\mathcal{T}} \lambda x_0.x_0 M_1 \cdots M_m$ then m must be 0.*

PROOF. By exploiting the definition of A we obtain:

$$\begin{aligned} \mathsf{A}M &=_{\mathcal{T}} \lambda y.M(M(\mathsf{K}y))\Omega, && \text{by (10.5)}, \\ &=_{\mathcal{T}} \lambda y.M(\mathsf{K}yM_1^{\sigma_1} \cdots M_m^{\sigma_1})\Omega && \text{for } \sigma_1 \triangleq [x_0 := \mathsf{K}y], \\ &=_{\mathcal{T}} \lambda y.\mathsf{K}yM_1^{\sigma_1} \cdots M_m^{\sigma_1} M_1^{\sigma_2} \cdots M_m^{\sigma_2}\Omega, && \text{for } \sigma_2 \triangleq [x_0 := \mathsf{K}yM_1^{\sigma_1} \cdots M_m^{\sigma_1}], \\ &=_{\mathcal{T}} \lambda y.y\, \underbrace{M_2^{\sigma_1} \cdots M_m^{\sigma_1} M_1^{\sigma_2} \cdots M_m^{\sigma_2}\Omega}_{2m \text{ arguments}}, && \text{as } m > 0. \end{aligned}$$

Since $M =_{\mathcal{T}} \mathsf{A}M$ we must have $m = 2m$, which is impossible for $m > 0$. □

LEMMA 10.36. *Let* $[M]_{\mathcal{T}} \in \mathsf{Fix}_{\mathcal{T}}(\mathsf{A})$. *If* $M =_{\mathcal{T}} \lambda x_0 \ldots x_k.x_0 M_1 \cdots M_k$ *then* $M =_{\mathcal{T}} \mathsf{I}$.

PROOF. We proceed by induction on $k \in \mathbb{N}$.

The base case $k = 0$ is trivial, as M has already the required form.

In the induction case $k > 0$, we have the following chain of equalities:

$$\begin{array}{lll}
M & =_{\mathcal{T}} \mathsf{A}M & \text{as } M =_{\mathcal{T}} \mathsf{A}M \\
& =_{\mathcal{T}} \lambda y.M(M(\mathsf{K}y))\Omega, & \text{by (10.5)}, \\
& =_{\mathcal{T}} \lambda y.M(\lambda x_1 \ldots x_k.\mathsf{K}y M_1^{\sigma_1} \cdots M_k^{\sigma_1})\Omega, & \text{for } \sigma_1 \triangleq [x_0 := \mathsf{K}y], \\
& =_{\mathcal{T}} \lambda y.M(\lambda x_1 \ldots x_k.y M_2^{\sigma_1} \cdots M_k^{\sigma_1})\Omega, & \text{by } (\beta) \text{ as } k > 0, \\
& =_{\mathcal{T}} \lambda y.(\lambda w_0 \ldots w_k.w_0 N_1 \cdots N_k)(\lambda x_1 \ldots x_k.y M_2^{\sigma_1} \cdots M_k^{\sigma_1})\Omega, & \text{by } \alpha\text{-renaming } M, \\
& =_{\mathcal{T}} \lambda y w_2 \ldots w_k.((\lambda x_1 \ldots x_k.y M_2^{\sigma_1} \cdots M_k^{\sigma_1}) N_1^{\sigma_2} \cdots N_k^{\sigma_2})[w_1 := \Omega], & \text{by } (\beta), \\
& \quad \text{where } \sigma_2 \triangleq [w_0 := \lambda x_1 \ldots x_k.y M_2^{\sigma_1} \cdots M_k^{\sigma_1}], & \\
& =_{\mathcal{T}} \lambda y w_2 \ldots w_k.y M_2^{\sigma_3} \cdots M_k^{\sigma_3}, & \text{by } (\beta), \\
& \quad \text{where } \sigma_3 = \sigma_1[x_1 := N_1^{\sigma_2}] \cdots [x_k := N_k^{\sigma_2}][w_1 := \Omega], & \\
& =_{\mathcal{T}} \lambda z_0 \ldots z_{k-1}.z_0 P_1 \cdots P_{k-1}, & \text{by } \alpha\text{-renaming}, \\
& =_{\mathcal{T}} \mathsf{I}, & \text{by IH.} \quad \square
\end{array}$$

PROPOSITION 10.37. *Let* $[M]_{\mathcal{T}} \in \mathsf{Fix}_{\mathcal{H}}(\mathsf{A})$. *If* M *is solvable then* $\mathcal{T} \vdash M = \mathsf{I}$.

PROOF. As M is solvable, Lemma 10.34 gives for some indices $k, m \geq 0$:

$$M =_{\mathcal{T}} \lambda x_0 \ldots x_k.x_0 M_1 \cdots M_m \tag{10.8}$$

In Lemmas 10.35 and 10.36, we have already shown that $M =_{\mathcal{T}} \mathsf{I}$ whenever $k = 0$ or $k = m$. Assume now $k > 0$ and $k \neq m$ towards a contradiction. Let us define $V \triangleq \lambda x.Mx\Omega$ for some variable $x \notin \mathrm{FV}(M)$.

Claim 10.37.1. For all $n \in \mathbb{N}$, we have $My =_{\mathcal{T}} V^n(M(\mathsf{K}^n y))$.

Subproof. We proceed by induction on n.

The case $n = 0$ is trivial because $My =_{\mathcal{T}} M(\mathsf{K}^0 y) =_{\mathcal{T}} V^0(M(\mathsf{K}^0 y))$.

In the case $n + 1$, we have:

$$\begin{array}{rll}
My =_{\mathcal{T}} & \mathsf{A}My, & \text{as } M =_{\mathcal{T}} \mathsf{A}M, \\
=_{\mathcal{T}} & M(M(\mathsf{K}y))\Omega, & \text{by (10.5)}, \\
=_{\mathcal{T}} & V(M(\mathsf{K}y)), & \text{by def. of } V, \\
=_{\mathcal{T}} & V(V^n(M(\mathsf{K}^n(\mathsf{K}y)))), & \text{by IH}, \\
=_{\mathcal{T}} & V^{n+1}(M(\mathsf{K}^{n+1}y)). & \blacksquare
\end{array}$$

Below, we use the following basic properties of K (where $x \in \mathrm{Var}$ and $i, j \geq 0$):

(K1) $\lambda w_1 \ldots w_i.\mathsf{K}^j x = \mathsf{K}^{i+j} x$,

(K2) if $i > j$ then $(\mathsf{K}^i x)P_1 \cdots P_j = \mathsf{K}^{i-j} x$, for any $P_1, \ldots, P_j \in \Lambda$,

(K3) if $i \leq j$ then $(\mathsf{K}^i x)P_1 \cdots P_j = x P_{i+1} \cdots P_j$, for any $P_1, \ldots, P_j \in \Lambda$.

Claim 10.37.2. For all $n \geq m$, we have $My =_{\mathcal{T}} V(V^n(\mathsf{K}^{n+1-m+k}y))$.

Subproof. We establish the following chain of equalities:

$$\begin{array}{rlll} My & =_{\mathcal{T}} & V^{n+1}(M(\mathsf{K}^{n+1}y)), & \text{by Claim 10.37.1,} \\ & =_{\mathcal{T}} & V^{n+1}(\lambda x_1 \ldots x_k.(\mathsf{K}^{n+1}y)M_1^{\sigma} \cdots M_m^{\sigma}), & \text{for } \sigma \triangleq [x_0 := \mathsf{K}^{n+1}y], \text{ by (10.8),} \\ & =_{\mathcal{T}} & V^{n+1}(\lambda x_1 \ldots x_k.\mathsf{K}^{n+1-m}y), & \text{by (K2), since } n+1 > m, \\ & =_{\mathcal{T}} & V^{n+1}(\mathsf{K}^{n+1-m+k}y), & \text{by (K1).} \end{array}$$

■

Remember that we are working under the assumption that $k > 0$. We split into subcases, depending on whether k is greater than m.

Claim 10.37.3. When $k > m$ we have for all $n \in \mathbb{N}$ and for appropriate $Z_i \in \Lambda$:
(i) $V^n(Vy) =_{\mathcal{T}} \lambda z_1 \ldots z_{k-1+(k-1-m)n}.yZ_1 \cdots Z_m$,
(ii) if $n \geq m$ then $My =_{\mathcal{T}} \mathsf{K}^{(2+n)(k-m)}y$.

Subproof. (i) We proceed by induction on n.
– If $n = 0$ then we have:

$$\begin{array}{rlll} Vy & =_{\mathcal{T}} & (\lambda x.Mx\Omega)y, & \text{by def. of } V, \\ & =_{\mathcal{T}} & My\Omega, & \text{by } (\beta), \\ & =_{\mathcal{T}} & (\lambda x_0 \ldots x_k.x_0M_1 \cdots M_m)y\Omega, & \text{by (10.8),} \\ & =_{\mathcal{T}} & \lambda x_2 \ldots x_k.yM_1^{\sigma} \cdots M_m^{\sigma}, & \text{for } \sigma \triangleq [x_0 := y][x_1 := \Omega]. \end{array}$$

– When $n > 0$ we have the following equivalences:

$$\begin{array}{rll} & V^n(Vy) & \\ =_{\mathcal{T}} & V(V^{n-1}(Vy)), & \text{by definition,} \\ =_{\mathcal{T}} & V(\lambda z_1 \ldots z_{k-1+(k-1-m)(n-1)}.yZ_1 \cdots Z_m), & \text{by IH,} \\ =_{\mathcal{T}} & \lambda x_2 \ldots x_k.(\lambda z_1 \ldots z_{k-1+(k-1-m)(n-1)}.yZ_1 \cdots Z_m)M_1^{\sigma_1} \cdots M_m^{\sigma_1}, & \text{by } (\beta), \\ & \text{where } \sigma_1 \triangleq [x_0 := \lambda z_1 \ldots z_{k-1+(k-1-m)(n-1)}.y\vec{Z}][x_1 := \Omega], & \\ =_{\mathcal{T}} & \lambda x_2 \ldots x_k z_{m+1} \ldots z_{k-1+(k-1-m)(n-1)}.yZ_1^{\sigma_2} \cdots Z_m^{\sigma_2}, & \text{as } k > m, \\ & \text{where } \sigma_2 \triangleq [z_1 := M_1^{\sigma_1}] \cdots [z_m := M_m^{\sigma_1}]. & \end{array}$$

So the number of abstractions is $k-1+k-1+(k-1-m)(n-1)-m = k-1+(k-1-m)n$.
(ii) For $n \geq m$ we have the following:

$$\begin{array}{rlll} My & =_{\mathcal{T}} & V^n(V(\mathsf{K}^{n+1-m+k}y)), & \text{by Claim 10.37.2,} \\ & =_{\mathcal{T}} & \lambda z_1 \ldots z_{k-1+(k-1-m)n}.(\mathsf{K}^{n+1-m+k}y)Z_1 \cdots Z_m, & \text{by (i) above,} \\ & =_{\mathcal{T}} & \lambda z_1 \ldots z_{k-1+(k-1-m)n}.\mathsf{K}^{(n+1-m+k)-m}y, & \text{by (K2) as } n \geq m, k > m, \\ & =_{\mathcal{T}} & \mathsf{K}^{k-1+(k-1-m)n+(n+1-m+k)-m}y, & \text{by (K1).} \end{array}$$

Therefore, the number of K's is equal to $k-1+(k-1-m)n+(n+1-m+k)-m = (k-1-m)n+n+2k-2m = (k-1-m+1)n+2(k-m) = (k-m)(n+2)$. ■

In Claim 10.37.3(ii) we have shown that, for all n large enough, My has a hnf with $(k-m)(n+2)$ external λ-abstractions and 0 applications. By Proposition 3.13(iii), we have $(k-m)(n+2) = k-m$ for all such n, which is only possible if this quantity is independent from n. As we are supposing $k > m$ this actually leads to a contradiction.

Claim 10.37.4. When $0 < k < m$ we have for all $n \in \mathbb{N}$ and for appropriate $Z_i \in \Lambda$:

(i) $V^n(Vy) =_{\mathcal{T}} \lambda z_1 \dots z_{k-1}.yZ_1 \cdots Z_{m+(m-k+1)n}$,

(ii) if $n \geq m$ then $My =_{\mathcal{T}} \lambda z_1 \dots z_{k-1}.yZ_1 \cdots Z_{(m-k)n+2m-k-1}$.

Subproof. (i) We proceed by induction on n.

– The case $n = 0$ is straightforward using the definition of V.

– If $n > 0$ then we have the following chain of equivalences:

$$\begin{array}{lll} & V^n(Vy) & \\ =_{\mathcal{T}} & V(V^{n-1}(Vy)), & \text{by definition,} \\ =_{\mathcal{T}} & V(\lambda z_1 \dots z_{k-1}.yZ_1 \cdots Z_{m+(m-k+1)(n-1)}), & \text{by IH,} \\ =_{\mathcal{T}} & \lambda x_2 \dots x_k.(\lambda z_1 \dots z_{k-1}.yZ_1 \cdots Z_{m+(m-k+1)(n-1)})M_1^{\sigma_1} \cdots M_m^{\sigma_1}, & \text{by } (\beta), \\ & \text{where } \sigma_1 \triangleq [x_0 := \lambda z_1 \dots z_{k-1}.yZ_1 \cdots Z_{m+(m-k+1)(n-1)}][x_1 := \Omega], & \\ =_{\mathcal{T}} & \lambda x_2 \dots x_k.yZ_1^{\sigma_2} \cdots Z_{m+(m-k+1)(n-1)}^{\sigma_2} M_k^{\sigma_1} \cdots M_m^{\sigma_1}, & \text{as } k < m, \\ & \text{where } \sigma_2 = [z_1 := M_1^{\sigma_1}] \cdots [z_{k-1} := M_{k-1}^{\sigma_1}]. & \end{array}$$

So the number of applications is $m+(m-k+1)(n-1)+m-k+1 = m+(m-k+1)n$.

– For $n \geq m$ we have the following equivalences:

$$\begin{array}{lll} My =_{\mathcal{T}} & V^n(V(\mathsf{K}^{n+1-m+k}y)), & \text{by Claim 10.37.2,} \\ =_{\mathcal{T}} & \lambda z_1 \dots z_{k-1}.(\mathsf{K}^{n+1-m+k}y)Z_1 \cdots Z_{m+(m-k+1)n}, & \text{by (i) above,} \\ =_{\mathcal{T}} & \lambda z_1 \dots z_{k-1}.yZ_{(n+1-m+k)+1} \cdots Z_{m+(m-k+1)n}, & \text{by (K3) as } n \geq m, k < m. \end{array}$$

So the number of applications is equal to $m + (m-k+1)n - (n+1-m+k) = (m-k+1-1)n + 2m - k - 1 = (m-k)n + 2m - k - 1$. ■

By Claim 10.37.4(ii), for all n large enough, My has a hnf with $(m-k)n+2m-k-1$ applications and $k-1$ external abstractions, so the difference is $(m-k)n+2m-2k$. By Proposition 3.13(iii), we must have $(m-k)n + 2m - 2k = m - k$ for all such n. Again, this is only possible if this quantity is independent from n. As we are supposing $k < m$ this leads to a contradiction.

Since we ruled out all other possibilities, we conclude $k = m = 0$ and $M =_{\mathcal{T}} \mathsf{I}$. □

COROLLARY 10.38. *For every sensible theory $\mathcal{T}$, the set $\mathsf{Fix}_{\mathcal{T}}(\mathsf{A}) = \{[\Omega]_{\mathcal{T}}, [\mathsf{I}]_{\mathcal{T}}\}$ has cardinality 2.*

As a direct consequence, we obtain the main results of the section.

THEOREM 10.39. *No sensible λ-theory $\mathcal{T}$ satisfies the fixed point property.*

COROLLARY 10.40. *The λ-theory $\mathcal{B}$ satisfies the range property, but does not satisfy the fixed point property.*

EXERCISE 10.41. Let $\mathcal{T}$ be a sensible λ-theory. For every $n \in \mathbb{N}$, construct a λ-term A_n such that $|\mathsf{Fix}_{\mathcal{T}}(\mathsf{A}_n)| = n + 2$.

Chapter 11

The kite

Theorem 17.4.16 in B[1984] displays a 'kite' shaped diagram depicting all strict inclusion relations among the λ-theories involved (in the figure, $\mathcal{T}_1$ is above $\mathcal{T}_2$ whenever $\mathcal{T}_1 \subsetneq \mathcal{T}_2$):

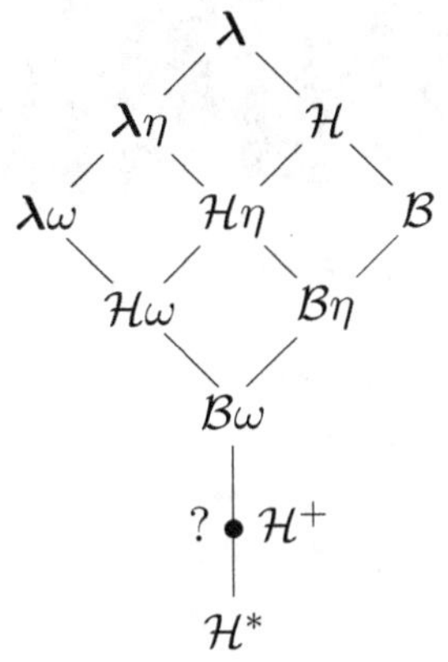

Figure 11.1: Barendregt's kite.

We briefly recall from Section 3.1 how these λ-theories are defined:

- $\boldsymbol{\lambda}$ is the least λ-theory;
- $\mathcal{H}$ is the λ-theory generated by equating all unsolvable λ-terms;
- $\mathcal{B}$ is the λ-theory equating all λ-terms having the same Böhm tree;
- $\boldsymbol{\lambda}\eta$, $\mathcal{H}\eta$, $\mathcal{B}\eta$ are the least extensional λ-theories including $\boldsymbol{\lambda}, \mathcal{H}$ and $\mathcal{B}$, respectively;
- $\boldsymbol{\lambda}\omega$, $\mathcal{H}\omega$, $\mathcal{B}\omega$ denote the closure under the ω-rule of $\boldsymbol{\lambda}, \mathcal{H}$ and $\mathcal{B}$, respectively;
- $\mathcal{H}^+ = \mathcal{T}_{\mathrm{NF}}$ and $\mathcal{H}^* = \mathcal{T}_{\mathrm{SOL}}$ are Morris' λ-theories with observables the β-normalizable and the solvable λ-terms, respectively.

A natural question—raised by Barendregt in the seventies—concerns the position of $\mathcal{H}^+$ in the kite. Indeed, while it is well-known that $\mathcal{B}\eta \subsetneq \mathcal{H}^+ \subsetneq \mathcal{H}^*$, the relationship between the λ-theories $\mathcal{B}\omega$ and $\mathcal{H}^+$ remained unexplored. In the proof of Theorem 17.4.16 in B[1984], it is mentioned that Patrick Sallé conjectured the strict inclusion $\mathcal{B}\omega \subsetneq \mathcal{H}^+$. Sallé's conjecture was probably formulated at the 1979 conference on λ-calculus in Swansea. For almost forty years no progress had been made in that direction.

Figure 11.2: Swansea 1979 λ-calculus conference.

In 2012, Manzonetto noticed that naturally defined relational models induce $\mathcal{H}^+$ as equational theory, and launched a research program in collaboration with Breuvart and Ruoppolo whose aim was to provide a simple semantical description of this λ-theory. Their investigation culminated in the article Breuvart et al. (2018), where a characterization of all relational models inducing that theory is exhibited (for more details, see Section 14.5). A key step to achieve this result is a syntactic weak separation theorem initially suggested by Breuvart. Subsequently, Manzonetto and Polonsky noticed that an easy consequence is the fact that $\mathcal{H}^+$ satisfies the ω-rule, which in its turn entails $\mathcal{B}\omega \subseteq \mathcal{H}^+$. In a larger collaboration including Intrigila, they investigated Sallé's conjecture further and arrived to a refutation first published in Intrigila et al. (2017).

In this chapter we present these results together with an original characterization of $\mathcal{B}\eta$ in terms of an equality between Böhm trees up to countably many η-expansions of bounded size by Intrigila et al. (2019). In Section 11.1 we review the three notions of extensional equality between Böhm trees underlying the λ-theories under consideration, namely $\mathcal{B}\eta, \mathcal{H}^+$ and $\mathcal{H}^*$, we present Sallé's conjecture and discuss its motivations. In Section 11.2 we introduce the Böhm-out technique for proving Breuvart's weak separation theorem and show that $\mathcal{H}^+ \vdash \omega$ holds. Section 11.3 is devoted to the demonstration of $\mathcal{B}\omega = \mathcal{H}^+$, which refutes Sallé's conjecture. Finally, in Section 11.4 we provide the extensional equivalence between Böhm trees which characterizes the equivalence in $\mathcal{B}\eta$. By collecting all these results we obtain some kind of taxonomy for the different degrees of extensionality in the theory of Böhm trees.

11.1 Degrees of extensionality for Böhm trees

The λ-theories $\mathcal{H}^+$ and $\mathcal{H}^*$ have been characterized in the seventies in terms of extensional equalities between Böhm trees. The characterization of $\mathcal{H}^+$ is due to Hyland (1975a) and subsequently improved by Lévy (1978a), while the analogous result for $\mathcal{H}^*$ was found independently by Hyland (1975b) and Wadsworth (1976).

In Section 11.4 we show that a similar characterization can be given for the λ-theory $\mathcal{B}\eta$, a result first appeared in Intrigila et al. (2019).

The η-expansions of the identity

A central role in the study of extensional equivalences on Böhm trees is played by the $\beta\eta$-expansions of the identity. This class of λ-terms satisfies interesting structural properties that have been investigated by Intrigila and Nesi (2003).

DEFINITION 11.1. (i) Let $\mathcal{I}^\eta \triangleq \{Q \in \Lambda \mid Q \twoheadrightarrow_{\beta\eta} \mathsf{I}\}$.
(ii) Define the λ-term 1^n by induction on $n \in \mathbb{N}$: $1^0 \triangleq \mathsf{I}$ and $1^{n+1} \triangleq \lambda xz.x(1^n z)$.

The relation $=_\beta$ partitions the set $\mathcal{I}^\eta$ into equivalence classes. Every λ-term $Q \in \mathcal{I}^\eta$ is β-normalizing and its β-normal form provides a canonical representative of the class. So, in this chapter, we slightly abuse language and simply call the elements of $\mathcal{I}^\eta$ *(finite) η-expansions of the identity*, following Intrigila and Nesi's terminology.

EXAMPLES 11.2. (i) The typical η-expansion of the identity is the combinator $\mathbf{1}$.
(ii) It is easy to check that $1^n \in \mathcal{I}^\eta$ for every $n \in \mathbb{N}$.
(iii) Examples of λ-terms in β-normal form belonging to $\mathcal{I}^\eta$ are $\lambda xy.x(\lambda z.y(\lambda w.zw))$, $\lambda xz_1z_2z_3.xz_1z_2z_3$ and $\lambda xz_1z_2z_3.x(\lambda y.z_1y)z_2(\lambda w_1w_2.z_3w_1w_2)$.

More generally, the η-expansions of the identity admit the following characterizations.

LEMMA 11.3. *For $Q \in \Lambda$, the following are equivalent:*

1. *$Q \in \mathcal{I}^\eta$;*
2. *$Q \twoheadrightarrow_\beta \lambda yz_1 \dots z_m.yQ_1 \cdots Q_m$ for some $m \geq 0$ such that $\lambda z_\ell.Q_\ell \in \mathcal{I}^\eta$ for all $\ell \leq m$;*
3. *$Q \twoheadrightarrow_\beta \lambda y.Q'$ for some $Q' \in \Lambda$ such that $Q' \twoheadrightarrow_{\beta\eta} y$.*

Consider now the combinator $\mathsf{J} = \mathsf{Y}(\lambda jxz.x(jz))$ defined by Wadsworth (1976) that generates the following infinite sequence of β-conversions:

$$\mathsf{J} =_\beta \lambda xz_0.x(\mathsf{J}z_0) =_\beta \lambda xz_0.x(\lambda z_1.z_0(\mathsf{J}z_1)) =_\beta \lambda xz_0.x(\lambda z_1.z_0(\lambda z_2.z_1(\mathsf{J}z_2))) =_\beta \cdots$$

Even if each λ-term of the sequence above 'looks like' an η-expansion of the identity, $\mathsf{J} \notin \mathcal{I}^\eta$ because it is not β-normalizing. Indeed, the λ-term J has an infinite Böhm tree:

$$\begin{array}{rcl} \mathrm{BT}(\mathsf{J}) & = & \lambda xz_0.x \\ & & \quad | \\ & & \quad \lambda z_1.z_0 \\ & & \qquad | \\ & & \qquad \lambda z_2.z_1 \\ & & \qquad\quad | \\ & & \qquad\quad \lambda z_3.z_2 \\ & & \qquad\qquad \vdots \end{array}$$

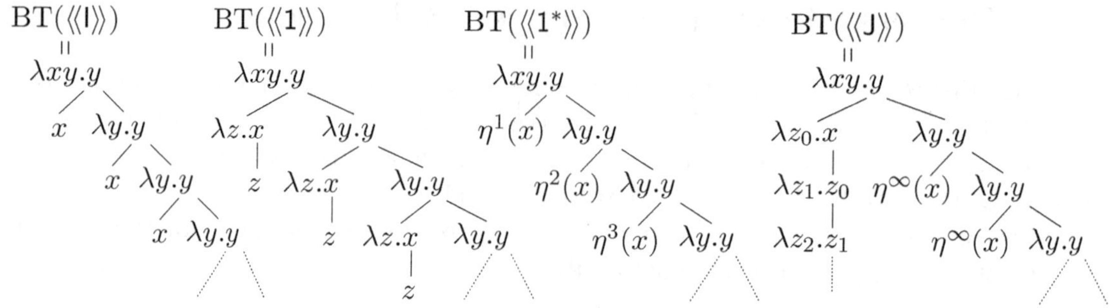

Figure 11.3: The Böhm trees of the streams $\langle\langle \mathsf{I} \rangle\rangle$, $\langle\langle 1 \rangle\rangle$, $\langle\langle 1^* \rangle\rangle$ and $\langle\langle \mathsf{J} \rangle\rangle$.

The λ-term J is actually a so-called *infinite η-expansion of the identity*, a notion that we discuss more thoroughly in Section 11.2.

The following streams will be used in the rest of the section as running examples.

EXAMPLES 11.4. Let $\langle\langle \mathsf{I} \rangle\rangle$, $\langle\langle 1 \rangle\rangle$, $\langle\langle 1^* \rangle\rangle$ and $\langle\langle \mathsf{J} \rangle\rangle$ be the streams

$$\langle\langle \mathsf{I} \rangle\rangle x \triangleq [\mathsf{I}x, \mathsf{I}x, \mathsf{I}x, \dots], \qquad \langle\langle 1 \rangle\rangle x \triangleq [1x, 1x, 1x, \dots],$$

$$\langle\langle 1^* \rangle\rangle x \triangleq [1^1x, 1^2x, 1^3x, \dots], \qquad \langle\langle \mathsf{J} \rangle\rangle x \triangleq [\mathsf{J}x, \mathsf{J}x, \mathsf{J}x, \dots],$$

whose Böhm trees are depicted in Figure 11.3 (where we use the abbreviations $\eta^n(x) \triangleq \mathrm{BT}(1^n x)$ and $\eta^\infty(x) \triangleq \mathrm{BT}(\mathsf{J}x)$). We will see that $\langle\langle \mathsf{I} \rangle\rangle =_{\mathcal{B}\eta} \langle\langle 1 \rangle\rangle =_{\mathcal{H}^+} \langle\langle 1^* \rangle\rangle =_{\mathcal{H}^*} \langle\langle \mathsf{J} \rangle\rangle$.

EXERCISE 11.5. Provide the actual λ-terms corresponding to $\langle\langle \mathsf{I} \rangle\rangle$, $\langle\langle 1 \rangle\rangle$, $\langle\langle 1^* \rangle\rangle$ and $\langle\langle \mathsf{J} \rangle\rangle$. For instance, a possible definition of $\langle\langle \mathsf{I} \rangle\rangle$ is $\mathsf{Y}(\lambda mx.[x, mx])$.

Böhm trees and extensionality

The easiest way of obtaining a theory of extensionality for Böhm trees is to consider the λ-theory $\mathcal{B}\eta$, which is defined as the smallest extensional λ-theory including $\mathcal{B}$. This λ-theory has been little studied in the literature, probably because it does not arise naturally as an observational theory, nor is induced by some non-syntactic denotational model. The results reviewed below are taken from Section 16.4 of B[1984].

To analyze the λ-theory $\mathcal{B}\eta$ it is crucial to understand the interaction between $\to_\eta$ and $\mathrm{BT}(-)$. It turns out that performing a single η-expansion in a λ-term M might generate countably many η-expansions in its Böhm tree.

Consider the streams $\langle\langle \mathsf{I} \rangle\rangle$ and $\langle\langle 1 \rangle\rangle$ defined in Example 11.4. It is easy to check that

$$\langle\langle \mathsf{I} \rangle\rangle =_{\mathcal{B}} \mathsf{Y}(\lambda mx.[x, mx]) \;{}_{\eta}\!\!\leftarrow \mathsf{Y}(\lambda mx.[\lambda z.xz, mx]) =_{\mathcal{B}} \langle\langle 1 \rangle\rangle$$

therefore $\langle\langle \mathsf{I} \rangle\rangle$ and $\langle\langle 1 \rangle\rangle$ are equated in $\mathcal{B}\eta$ despite the fact that their Böhm trees differ by infinitely many η-expansions. More generally, $M \to_\eta N$ entails that $\mathrm{BT}(M)$ can be obtained from $\mathrm{BT}(N)$ by performing *at most* one η-expansion at every position.

LEMMA 11.6 (B[1984], LEMMA 16.4.3). *Let $M, N \in \Lambda$. If $M \to_\eta N$, then* $\mathrm{BT}(M)$ *is obtained from* $\mathrm{BT}(N)$ *by replacing in the latter some*

$$\begin{array}{c}\lambda x_1 \ldots x_n.y \\ \diagup \quad \diagdown \\ T_1 \quad \cdots \quad T_k\end{array} \qquad \text{by} \qquad \begin{array}{c}\lambda x_1 \ldots x_n z.y \\ \diagup \quad \diagdown\diagdown \\ T_1 \quad \cdots \quad T_k \ z\end{array}$$

possibly infinitely often (without changing the new variable z thus created).

As a consequence of this lemma, we obtain that no finite amount of η-expansions in $\langle\!\langle 1 \rangle\!\rangle$ can turn its Böhm tree into the Böhm tree of $\langle\!\langle 1^* \rangle\!\rangle$, because the latter contains infinitely many η-expansions of increasing size.

COROLLARY 11.7. $\mathcal{B}\eta \vdash \langle\!\langle \mathsf{I} \rangle\!\rangle = \langle\!\langle 1 \rangle\!\rangle$, *while* $\mathcal{B}\eta \nvdash \langle\!\langle \mathsf{I} \rangle\!\rangle = \langle\!\langle 1^* \rangle\!\rangle$.

Consider now Morris' λ-theory $\mathcal{H}^+$, defined as the observational theory generated by taking as observables the β-normalizable terms. This λ-theory is sensible and extensional. Moreover, since two λ-terms having the same Böhm tree cannot be distinguished by any context, we have $\mathcal{B}\eta \subseteq \mathcal{H}^+$. The question naturally arising is whether this inclusion is strict and, in this case, what kind of λ-terms different in $\mathcal{B}\eta$ become equal in $\mathcal{H}^+$. It turns out that $\mathcal{H}^+ \vdash M = N$ holds exactly when the Böhm trees of M and N are equal up to countably many η-expansions of finite size. To formalize this notion, we first define coinductively on Böhm-like trees the relation[1] $T \leq^\eta T'$ expressing the fact that T' is obtained from T by performing countably many η-expansions.

DEFINITION 11.8. (i) Let $\leq^\eta$ be the greatest relation on $\mathscr{B}$ such that $T \leq^\eta T'$ entails that either $T = T' = \bot$, or

$$T = \begin{array}{c}\lambda x_1 \ldots x_n.y \\ \diagup \quad \diagdown \\ T_1 \ \cdots \ T_k\end{array} \qquad \text{and} \qquad T' = \begin{array}{c}\lambda x_1 \ldots x_n z_1 \ldots z_m.y \\ \diagup \quad \diagup\diagdown \quad \diagdown \\ T'_1 \ \cdots \ T'_k \ Q_1 \ \cdots \ Q_m\end{array}$$

for some $n, m, k \geq 0$, $T_1, \ldots, T_k, T'_1, \ldots, T'_k, Q_1, \ldots, Q_m \in \mathscr{B}$ such that $T_i \leq^\eta T'_i$ for all $i \leq k$, $z_\ell \notin \mathrm{FV}(y\, T_1 \cdots T_k\, T'_1 \cdots T'_k)$ and $\lambda z_\ell.Q_\ell \in \mathcal{I}^\eta$ for all $\ell \leq m$.

(ii) For $M, N \in \Lambda$, we set $M \sqsubseteq^\eta N$ if and only if $\mathrm{BT}(M) \leq^\eta \mathrm{BT}(N)$.

For example, $\langle\!\langle \mathsf{I} \rangle\!\rangle \sqsubseteq^\eta \langle\!\langle 1^* \rangle\!\rangle$. The next proposition, proved by Böhm-out technique, is a reformulation of Theorem 11.2.20 in Ronchi Della Rocca and Paolini (2004).

PROPOSITION 11.9. *For all $M, N \in \Lambda$, $\mathcal{H}^+ \vdash M = N$ if and only if there exists a Böhm-like tree T such that* $\mathrm{BT}(M) \leq^\eta T \geq^\eta \mathrm{BT}(N)$.

Hence, in general, to prove that M and N are equal in $\mathcal{H}^+$, one might need to η-expand both their Böhm trees in order to find a common 'η-upper bound' T. Such a Böhm-like tree can always be chosen minimally η-expanded and in this case we say that T is the 'η-supremum' of $\mathrm{BT}(M)$ and $\mathrm{BT}(N)$.

COROLLARY 11.10. *As $\mathcal{B}\eta \subseteq \mathcal{H}^+$, we have $\mathcal{H}^+ \vdash \langle\!\langle \mathsf{I} \rangle\!\rangle = \langle\!\langle 1 \rangle\!\rangle$. From Proposition 11.9, we get $\mathcal{H}^+ \vdash \langle\!\langle \mathsf{I} \rangle\!\rangle = \langle\!\langle 1^* \rangle\!\rangle$, while $\mathcal{H}^+ \nvdash \langle\!\langle \mathsf{I} \rangle\!\rangle = \langle\!\langle \mathsf{J} \rangle\!\rangle$.*

---

[1] Contrary to B[1984], where $\leq^\eta$ is more liberal and allows η-expansions of infinite size as well.

A much more famous observational theory is certainly Morris' λ-theory $\mathcal{H}^*$, which arises by considering solvable terms (equivalently, head normalizable λ-terms) as observables. This is, by far, the most well-studied λ-theory since it is induced by Scott's model $\mathcal{D}_\infty$, introduced in the pioneering paper Scott (1972), and it can be also characterized as the (unique) maximal sensible λ-theory as shown by Hyland (Proposition 3.13(i)). It is not difficult to check that $\mathcal{H}^+ \subseteq \mathcal{H}^*$ and that this inclusion is strict: e.g., I and J are equal in $\mathcal{H}^*$ but different in $\mathcal{H}^+$. More precisely, two λ-terms M and N are equal in $\mathcal{H}^*$ when their Böhm trees are equal up to countably many possibly infinite η-expansions.

The next coinductive definition formalizes when a Böhm-like tree T' is obtained from T by performing countably many possibly infinite η-expansions, in symbols $T \leq_\omega^\eta T'$.

DEFINITION 11.11. (i) Let $\leq_\omega^\eta$ be the greatest relation on $\mathscr{B}$ such that $T \leq_\omega^\eta T'$ entails that either $T = T' = \bot$, or

$$T = \begin{array}{c} \lambda x_1 \dots x_n.y \\ \diagup \quad \diagdown \\ T_1 \ \cdots \ T_k \end{array} \qquad \text{and} \qquad T' = \begin{array}{c} \lambda x_1 \dots x_n z_1 \dots z_m.y \\ \diagup \quad \diagdown \\ T_1' \ \cdots \ T_k' \ T_1'' \ \cdots \ T_m'' \end{array}$$

for some $n, m, k \geq 0$, $T_1, \dots, T_k, T_1', \dots, T_k', T_1'', \dots, T_m'' \in \mathscr{B}$ such that $T_i \leq_\omega^\eta T_i'$ for all $i \leq k$, $z_\ell \notin \mathrm{FV}(y T_1 \cdots T_k T_1' \cdots T_k')$ and $z_\ell \leq_\omega^\eta T_\ell''$ for all $\ell \leq m$.
(ii) For $M, N \in \Lambda$, we set $M \sqsubseteq_\omega^\eta N$ if and only if $\mathrm{BT}(M) \leq_\omega^\eta \mathrm{BT}(N)$.

For example we have $\mathsf{I} \sqsubseteq_\omega^\eta \mathsf{J}$ and, by extension, $\langle\!\langle \mathsf{I} \rangle\!\rangle \sqsubseteq_\omega^\eta \langle\!\langle \mathsf{J} \rangle\!\rangle$. The following proposition appears as Theorem 19.2.9 in B[1984].

PROPOSITION 11.12. *For all $M, N \in \Lambda$, $\mathcal{H}^* \vdash M = N$ if and only if there exists a Böhm-like tree T such that* $\mathrm{BT}(M) \leq_\omega^\eta T \geq_\omega^\eta \mathrm{BT}(N)$ *holds.*

Summing up the various results recalled in this section, we obtain this corollary.

COROLLARY 11.13. *The streams $\langle\!\langle \mathsf{I} \rangle\!\rangle, \langle\!\langle 1 \rangle\!\rangle, \langle\!\langle 1^* \rangle\!\rangle, \langle\!\langle \mathsf{J} \rangle\!\rangle$ are all equal in the λ-theory $\mathcal{H}^*$. However, $\mathcal{B}\eta \nvdash \langle\!\langle \mathsf{I} \rangle\!\rangle = \langle\!\langle 1^* \rangle\!\rangle$ and $\mathcal{H}^+ \nvdash \langle\!\langle 1^* \rangle\!\rangle = \langle\!\langle \mathsf{J} \rangle\!\rangle$, so we have $\mathcal{B}\eta \subsetneq \mathcal{H}^+ \subsetneq \mathcal{H}^*$.*

The results surveyed in this section have been described in the literature using several different approaches and we believe that a comparison is mandatory at this point. The presentation of Proposition 11.12 in Hyland (1975b); Wadsworth (1976); Ronchi Della Rocca and Paolini (2004) is equivalent to ours, except for their use of induction rather than coinduction. Similarly for Proposition 11.9 in Hyland (1975a); Lévy (1978a); Ronchi Della Rocca and Paolini (2004). To verify that, say, $\mathrm{BT}(M) \leq_\omega^\eta T \geq_\omega^\eta \mathrm{BT}(N)$ holds for some Böhm-like tree T they consider the finite approximants obtained by cutting $\mathrm{BT}(M)$ and $\mathrm{BT}(N)$ at some level k, check whether they are joinable via η-expansions, and then require that this property is verified at every level $k \in \mathbb{N}$ of the trees.

A different approach consists in defining the 'infinite η-normal form' $\mathrm{nf}_{\eta!}(T)$ of a Böhm-like tree T either coinductively, as in Definition 2.60, or via an infinitary term rewriting system, as in Section 6.4, and then check whether $\mathrm{nf}_{\eta!}(\mathrm{BT}(M)) = \mathrm{nf}_{\eta!}(\mathrm{BT}(N))$. The advantage of our approach is that the η-supremum T in Proposition 11.12 (and, *a fortiori*, the one in Proposition 11.9) can always be chosen λ-definable (see B[1984], Exercise 10.6.7), a key property that we exploit in Section 11.3. On the contrary, there are λ-terms M such that $\mathrm{nf}_{\eta!}(\mathrm{BT}(M))$ is not λ-definable (B[1984], Exercise 10.6.6).

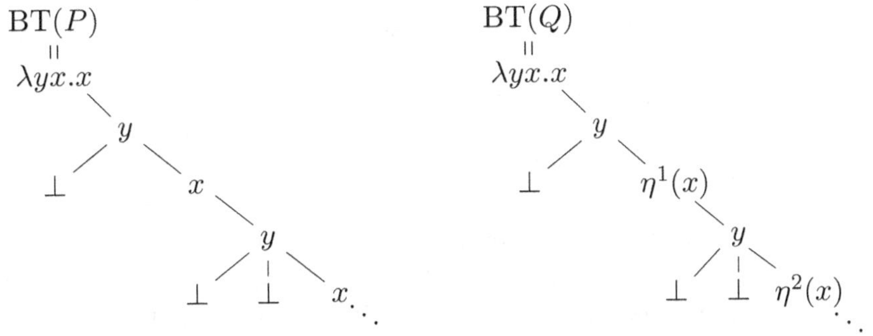

Figure 11.4: The Böhm trees of the λ-terms P, Q from Lemma 16.4.4 in B[1984].

The ω-rule and Sallé's conjecture

In the previous section we reviewed three different notions of extensional equalities on Böhm-like trees. In λ-calculus it is possible to define a strong form of extensionality, known as the *ω-rule*, which has been extensively investigated in connection with several λ-theories (see, e.g. Barendregt (1971, 2020b); Plotkin (1974); Barendregt et al. (1978); Intrigila and Statman (2004, 2009)). It is therefore natural to wonder what λ-terms are equated in the λ-theory $\mathcal{B}\omega$, obtained by closing $\mathcal{B}$ under the ω-rule.

Intuitively the ω-rule mimics the definition of functional equality, namely it states that two λ-terms M and N are equal whenever they coincide on every closed argument P (see Definition 3.21). In general, because of the quantification over all P's, it is difficult to determine what are the λ-terms different in $\mathcal{T}\eta$ that become equal in $\mathcal{T}\omega$. For instance, the counterexample showing that $\boldsymbol{\lambda}\eta \nvdash \omega$ is based on complex universal generators known as *Plotkin terms* (B[1984], Definition 17.3.26). Unfortunately such terms are unsolvable, so they become useless when considering sensible λ-theories. We now discuss the validity of the ω-rule for λ-theories extending $\mathcal{B}$.

Let us call P and Q the λ-terms from Lemma 16.4.4 in B[1984], whose Böhm trees are represented in Figure 11.4. The Böhm trees of P and Q only differ by countably many finite η-expansions of increasing size, a situation analogous to the one already discussed of $\langle\!\langle \mathsf{I} \rangle\!\rangle$ and $\langle\!\langle 1^* \rangle\!\rangle$, therefore they are different in $\mathcal{B}\eta$ but equal in $\mathcal{H}^+$.

FACT 11.14. For the λ-terms P and Q of Figure 11.4, we have $P \sqsubseteq^\eta Q$.

Perhaps surprisingly, P and Q can also be used to prove $\mathcal{B}\eta \subsetneq \mathcal{B}\omega$ since $\mathcal{B}\omega \vdash P = Q$. Indeed, by Lemma 1.26, for every $M \in \Lambda^o$, there exists k such that $M\Omega^{\sim k}$ becomes unsolvable. By inspecting Figure 11.4, we notice that in $\mathrm{BT}(P)$ the variable y is applied to an increasing number of Ω's (represented in the tree by $\perp$). So, when substituting some $M \in \Lambda^o$ for y in $\mathrm{BT}(Py)$, there is a level k of the tree where $M\Omega \cdots \Omega$ becomes $\perp$, thus cutting $\mathrm{BT}(PM)$ at that level. The same holds for $\mathrm{BT}(QM)$. As a consequence $\mathrm{BT}(PM)$ and $\mathrm{BT}(QM)$ are equal up to *finitely many* η-expansions, so, since $\mathcal{B}\eta \subseteq \mathcal{B}\omega$, we obtain that $\mathcal{B}\omega \vdash PM = QM$ holds. Finally, as we considered an arbitrary closed λ-term M, we can apply the ω-rule and conclude that $\mathcal{B}\omega \vdash P = Q$.

In order to find the correct place of the λ-theory $\mathcal{H}^+$ in the kite, one needs to understand how it compares with $\mathcal{B}\omega$. The three possible scenarios are the following:

$$(1)\quad \mathcal{B}\eta \subsetneq \mathcal{B}\omega \subseteq \mathcal{H}^+ \subsetneq \mathcal{H}^*;$$

$$(2)\quad \mathcal{B}\eta \subsetneq \mathcal{H}^+ \subseteq \mathcal{B}\omega \subsetneq \mathcal{H}^*;$$

$$(3)\quad \mathcal{B}\eta \begin{array}{c} \subsetneq \mathcal{B}\omega \subsetneq \\ \subsetneq \mathcal{H}^+ \subsetneq \end{array} \mathcal{H}^*.$$

On the one side, the fact that the observational theory $\mathcal{H}^+$ satisfies a Context Lemma stating that, for closed terms, we can restrict our attention to close applicative contexts suggests that $\mathcal{H}^+$ should satisfy the ω-rule, a result from which scenario (1) would follow. On the other side, the example in Figure 11.4 shows that in some situations $\mathcal{B}\omega$ is able to equate two λ-terms even when their Böhm trees differ by infinitely many η-expansions of increasing size. This looks however hardly generalizable as it is intimately related to the presence of an increasing number of applications to $\bot$ in each level of the tree.

These are probably[2] the considerations that led Sallé to formulate his conjecture.

CONJECTURE 11.15 (SALLÉ'S CONJECTURE). $\mathcal{B}\omega \subsetneq \mathcal{H}^+$.

11.2 $\mathcal{H}^+$ satisfies the ω-rule

This section is devoted to showing that $\mathcal{H}^+$ satisfies the ω-rule. To understand the intuition behind the proof, we start discussing the analogous result for $\mathcal{H}^*$.

Discussion about the proof

The fact that $\mathcal{H}^* \vdash \omega$ is an easy consequence of its maximality. However, there are several direct proofs: see Section 17.2 in B[1984] for a syntactic demonstration and Wadsworth (1976) for a semantic one. A key step in both proofs is to show that, given two closed λ-terms M and N such that the former has a head normal form, while the latter does not, it is possible to find a non-empty closed applicative context $[\,]P_1 \cdots P_n$ that preserves such a property. This follows from Wadsworth's characterization of λ-terms having a head normal form in terms of solvability: it is indeed possible to find $P_1, \dots, P_n \in \Lambda^o$ such that $MP_1 \cdots P_n$ is β-convertible with the identity I, while $NP_1 \cdots P_n$ is unsolvable; moreover, $[\,]P_1 \cdots P_n$ can be chosen of any length by adding copies of I at the end.

To prove that $\mathcal{H}^+$ satisfies the ω-rule, we need to show something similar, namely that if M has a β-normal form and N does not, then we can find a non-empty closed applicative context preserving this property. Interestingly, it is sufficient to consider λ-terms that are equal in the λ-theory $\mathcal{H}^*$; in other words we need to perform a detailed analysis of the equations in $\mathcal{H}^* - \mathcal{H}^+$. We show that when two closed λ-terms M, N are equal in $\mathcal{H}^*$, but different in $\mathcal{H}^+$, their Böhm trees are similar but there exists a (possibly virtual) position α where they differ because of an infinite η-expansion of a variable x, and such an η-expansion follows the structure of some infinite recursive tree. By applying

[2] In a personal communication Sallé told us that, as a longtime has passed, he does not remember his actual motivations. Truth be told, he does not even remember to have formulated this conjecture.

the celebrated Böhm-out technique, we prove that in this case it is possible to extract such a difference by defining a suitable closed applicative context $[\,]P_1 \cdots P_n$ that sends M to the identity and N to some infinite η-expansion of the identity (Theorem 11.28).

This provides a semi-separability result in the spirit of Hyland (1975a); Coppo et al. (1978); Dezani-Ciancaglini and Giovannetti (2001) but the notion of separability that we consider is weaker since it arises from Morris' observational preorder $\sqsubseteq_{\mathrm{NF}}$. We then prove that applying the identity to an infinite η-expansion of the identity, one still ends up with a (possibly different) infinite η-expansion of I. From this closure property we obtain that also in this case the length of the discriminating context $[\,]P_1 \cdots P_n$ can be chosen arbitrarily by adding copies of I at the end. Once this property has been established, the fact that $\mathcal{H}^+$ satisfies the ω-rule follows easily (Theorem 11.31).

Characterizing possibly infinite η-expansions of the identity

We start by extending some of the results in Intrigila and Nesi (2003) to possibly infinite η-expansions of the identity. Recall that $\mathcal{I}^\eta$ has been defined in Definition 11.1(i).

DEFINITION 11.16. (i) We let $\mathcal{I}^\eta_\omega \triangleq \{Q \in \Lambda \mid \mathsf{I} \sqsubseteq^\eta_\omega Q\}$ be the set of all *possibly infinite η-expansions of the identity*.

(ii) We let $\mathcal{I}^\eta_\infty \triangleq \mathcal{I}^\eta_\omega - \mathcal{I}^\eta$ be the set of all *infinite η-expansions of the identity*.

Again, we are abusing language as we are silently considering $\mathcal{B}$-equivalence classes. We associate with every recursive tree $t \in \mathscr{T}_{\mathsf{rec}}$ a combinator $\mathsf{J}_t \in \mathcal{I}^\eta_\omega$ whose Böhm tree follows the structure of t. More precisely, we show that there exists a one-to-one correspondence between $\mathscr{T}_{\mathsf{rec}}$ and $\mathcal{I}^\eta_\omega/\mathcal{B}$, the other direction being given by the injective map sending $[Q]_\mathcal{B}$ to its underlying naked tree $|\mathrm{BT}(Q)|$.

DEFINITION 11.17. Let $t : \mathbb{N}^* \to \mathbb{N}$ be a recursive tree.

(i) Let $f_t : \mathbb{N} \to \mathbb{N}$ be the partial recursive function satisfying $n \in \mathrm{dom}(f_t)$ if and only if $n = \#\alpha$ for some $\alpha \in \mathrm{dom}(t)$, and in this case $f_t(n) \triangleq t(\alpha)$.

(ii) Denote F_t the combinator λ-defining f_t.

(iii) Define a combinator $\mathsf{J}_t \in \Lambda^o$ as follows (for any $X \in \Lambda$):

$$\begin{array}{rcl} \mathsf{J}_t & \triangleq & N\mathsf{c}_{\#\langle\rangle}, \\ N_t & \triangleq & \mathsf{Y}(\lambda gn.M^{g\circ(\mathsf{Cons}\,n)}(F_t\,n)), \\ M^X & \triangleq & \lambda nx.nL^X[\mathsf{c}_0, x]\,\mathsf{F}, \\ L^X & \triangleq & \lambda pz.p(\lambda nwy.z(\mathsf{S}^+n)(w(Xny))), \end{array}$$

where $\mathsf{Cons} \in \Lambda^o$ is such that $\mathsf{Cons}\,\mathsf{c}_{\#\alpha}\,\mathsf{c}_n =_\beta \mathsf{c}_{\#(\alpha\,;\,n)}$, for every $\alpha \in \mathbb{N}^*$ and $n \in \mathbb{N}$.

LEMMA 11.18. *For every $t \in \mathscr{T}_{\mathsf{rec}}$, we have $\mathsf{J}_t \in \mathcal{I}^\eta_\omega$ and $t = |\mathrm{BT}(\mathsf{J}_t)|$.*

PROOF. Consider $t \in \mathscr{T}^\infty_{\mathsf{rec}}$. We prove some properties enjoyed by the auxiliary λ-terms N_t, M^X, L^X occurring in the definition of J_t.

Claim 11.18.1. For every $X \in \Lambda$ and $n \in \mathbb{N}$, we have:

$$(L^X)^n[\mathsf{c}_0, x] =_\beta \lambda zy_1 \ldots y_n.z\mathsf{c}_n(x(X\mathsf{c}_0y_1)\cdots(X\mathsf{c}_{n-1}y_n)).$$

Subproof. We proceed by induction on n, the case $n = 0$ being trivial as $[\mathsf{c}_0, x] = \lambda z.z\mathsf{c}_0 x$. In case $n > 0$, we have:

$$\begin{aligned}
(L^X)^{n+1}[\mathsf{c}_0, x] &= L^X((L^X)^n[\mathsf{c}_0, x]) \\
&=_\beta L^X(\lambda z y_1 \dots y_n.z\mathsf{c}_n(x(X\mathsf{c}_0 y_1)\cdots(X\mathsf{c}_{n-1}y_n))), \qquad \text{by the IH,} \\
&=_\beta \lambda z y_1 \dots y_n.(\lambda nwy.z(\mathsf{S}^+n)(w(Xny)))\mathsf{c}_n(x(X\mathsf{c}_0 y_1)\cdots(X\mathsf{c}_{n-1}y_n)) \\
&=_\beta \lambda z y_1 \dots y_n.(\lambda y.z(\mathsf{S}^+\mathsf{c}_n)((x(X\mathsf{c}_0 y_1)\cdots(X\mathsf{c}_{n-1}y_n))(X\mathsf{c}_n y))) \\
&=_\beta \lambda z y_1 \dots y_{n+1}.z\mathsf{c}_{n+1}(x(X\mathsf{c}_0 y_1)\cdots(X\mathsf{c}_{n-1}y_n)(X\mathsf{c}_n y_{n+1})). \qquad \blacksquare
\end{aligned}$$

Claim 11.18.2. For $n \in \mathbb{N}$, $X \in \Lambda$, we have $M^X\mathsf{c}_n x =_\beta \lambda y_1 \dots y_n.x(X\mathsf{c}_0 y_1)\cdots(X\mathsf{c}_{n-1}y_n)$.

Subproof. Indeed $M^X\mathsf{c}_n x =_\beta \mathsf{c}_n L^X[\mathsf{c}_0, x]\,\mathsf{F} =_\beta (L^X)^n([\mathsf{c}_0, x])\mathsf{F}$. By Claim 11.18.1, we get

$$\begin{aligned}
M^X\mathsf{c}_n x &=_\beta \lambda y_1 \dots y_n.\mathsf{F}\,\mathsf{c}_n(x(X\mathsf{c}_0 y_1)\cdots(X\mathsf{c}_{n-1}y_n)) \\
&=_\beta \lambda y_1 \dots y_n.x(X\mathsf{c}_0 y_1)\cdots(X\mathsf{c}_{n-1}y_n). \qquad \blacksquare
\end{aligned}$$

Claim 11.18.3. For all $\alpha \in \mathbb{N}^*$, if $t(\alpha) = n$ then

$$N_t\mathsf{c}_{\#\alpha}x =_\beta \lambda y_1 \dots y_n.x(N_t\mathsf{c}_{\#(\alpha\,;\,0)}y_1)\cdots(N_t\mathsf{c}_{\#(\alpha\,;\,(n-1))}y_n).$$

Subproof. It is easy to check that N_t satisfies $N_t\mathsf{c}_n =_\beta M^{N_t\circ(\mathsf{Cons}\,\mathsf{c}_n)}(F_t\,\mathsf{c}_n)$, so we get:

$$\begin{aligned}
N_t\mathsf{c}_{\#\alpha}x =_\beta\ & M^{N_t\circ(\mathsf{Cons}\,\mathsf{c}_{\#\alpha})}(F_t\mathsf{c}_{\#\alpha})x \\
=_\beta\ & M^{N_t\circ(\mathsf{Cons}\,\mathsf{c}_{\#\alpha})}(\mathsf{c}_{t(\alpha)})x \\
=_\beta\ & \lambda y_1 \dots y_{t(\alpha)}.x((N_t \circ (\mathsf{Cons}\,\mathsf{c}_{\#\alpha}))\mathsf{c}_0 y_1)\cdots((N_t \circ (\mathsf{Cons}\,\mathsf{c}_{\#\alpha}))\mathsf{c}_{t(\alpha)-1}y_{t(\alpha)}) \\
=_\beta\ & \lambda y_1 \dots y_{t(\alpha)}.x(N_t\mathsf{c}_{\#(\alpha\,;\,0)}y_1)\cdots(N_t\mathsf{c}_{\#(\alpha\,;\,(t(\alpha)-1))}y_{t(\alpha)}),
\end{aligned}$$

where the first equality follows from the definition of N_t, the second from the definition of F_t, the third by Claim 11.18.2 and the fourth by definition of composition. $\blacksquare$

Claim 11.18.4. For all $\alpha \in \mathbb{N}^*$, $\mathcal{B} \vdash \mathsf{J}_{t\restriction_\alpha}x = N_t\mathsf{c}_{\#\alpha}x$.

Subproof. By definition, we have $\mathsf{J}_{t\restriction_\alpha} = N_{t\restriction_\alpha}\mathsf{c}_{\#\langle\rangle}$. Then the claim follows by coinduction, using Claim 11.18.3 and the fact that $(t\restriction_\alpha)\restriction_{\langle i\rangle} = t\restriction_{(\alpha\,;\,i)}$, for all i $(0 \le i < t(\alpha))$. $\blacksquare$

Claim 11.18.5. For all $\alpha \in \mathbb{N}^*$, $x \sqsubseteq^\eta_\omega \mathsf{J}_{t\restriction_\alpha}x$ and $t\restriction_\alpha = |\mathrm{BT}(\mathsf{J}_{t\restriction_\alpha}x)|$.

Subproof. We proceed by coinduction on $\mathrm{BT}(\mathsf{J}_{t\restriction_\alpha}x)$. By Claims 11.18.3 & 11.18.4, we get

$$\mathrm{BT}(\mathsf{J}_{t\restriction_\alpha}x) = \begin{array}{c} \lambda y_1 \dots y_{t(\alpha)}.x \\ \mathrm{BT}(\mathsf{J}_{t\restriction_{(\alpha\,;\,0)}}y_1) \quad \cdots \quad \mathrm{BT}(\mathsf{J}_{t\restriction_{(\alpha\,;\,(t(\alpha)-1))}}y_{t(\alpha)}) \end{array}$$

Conclude as the co-IH gives $y_i \sqsubseteq^\eta_\omega \mathsf{J}_{t\restriction_{(\alpha\,;\,i)}}y_i$ and $t\restriction_{(\alpha\,;\,i)} = |\mathrm{BT}(\mathsf{J}_{t\restriction_{(\alpha\,;\,i)}}y_i)|$, for each i. $\blacksquare$

The main lemma follows from Claim 11.18.5 by taking $\alpha \triangleq \langle\rangle$, since $t\restriction_{\langle\rangle} = t$. Indeed, $|\mathrm{BT}(\mathsf{J}_t x)| = |\mathrm{BT}(\mathsf{J}_t)|$ and $x \sqsubseteq^\eta_\omega \mathsf{J}_t x$ holds if and only if $\mathsf{I} \sqsubseteq^\eta_\omega \mathsf{J}_t$ does. $\square$

The λ-terms J_t's are able to capture all λ-definable η-expansions of the identity.

PROPOSITION 11.19. *For $T \in \mathscr{B}$ satisfying $\mathsf{I} \leq^{\eta}_{\omega} T$ the following are equivalent:*

1. *T is λ-definable,*
2. *there exists a recursive tree t such that $T = \mathrm{BT}(\mathsf{J}_t)$.*

PROOF. ($1 \Rightarrow 2$) Assume that $Q \in \Lambda$ is such that $\mathrm{BT}(Q) = T$ and take $t = |T|$. By Theorem 2.24, the Böhm-like tree T is r.e. and, since $\mathsf{I} \leq^{\eta}_{\omega} T$ entails that T is $\perp$-free, it is even recursive. From this it follows that $t \in \mathscr{T}_{\mathsf{rec}}$, so we conclude by Lemma 11.18.

($2 \Rightarrow 1$) Trivial. □

LEMMA 11.20. *For all $t', t'' \in \mathscr{T}_{\mathsf{rec}}$, there exists $t \in \mathscr{T}_{\mathsf{rec}}$ such that $\mathcal{B} \vdash \mathsf{J}_t = \mathsf{J}_{t'} \circ \mathsf{J}_{t''}$. Moreover if either t' or t'' is infinite then t is infinite as well.*

PROOF. We proceed by coinduction on $\mathrm{BT}(\mathsf{J}_{t'} \circ \mathsf{J}_{t''})$. Let t' have subtrees $t'_1, \dots, t'_n$ and t'' have subtrees $t''_1, \dots, t''_m$. Since t' and t'' are recursive, so are their subtrees. By definition $\mathsf{J}_{t'} =_\beta \lambda x z_1 \dots z_n.x(\mathsf{J}_{t'_1} z_1) \cdots (\mathsf{J}_{t'_n} z_n)$ and $\mathsf{J}_{t''} =_\beta \lambda x y_1 \dots y_m.x(\mathsf{J}_{t''_1} y_1) \cdots (\mathsf{J}_{t''_m} y_m)$.

Consider $n \leq m$, the other case being similar. Easy calculations give:

$$\begin{aligned} \mathsf{J}_{t'} \circ \mathsf{J}_{t''} &=_\beta \lambda x z_1 \dots z_n.(\mathsf{J}_{t''} x)(\mathsf{J}_{t'_1} z_1) \cdots (\mathsf{J}_{t'_n} z_n) \\ &=_\beta \lambda x z_1 \dots z_n y_{n+1} \dots y_m.x(\mathsf{J}_{t''_1}(\mathsf{J}_{t'_1} z_1)) \cdots (\mathsf{J}_{t''_n}(\mathsf{J}_{t'_n} z_n))(\mathsf{J}_{t'_{n+1}} z_{n+1}) \cdots (\mathsf{J}_{t'_m} z_m). \end{aligned}$$

Since $\mathsf{J}_{t''_i}(\mathsf{J}_{t'_i} z_i) =_\beta (\mathsf{J}_{t''_i} \circ \mathsf{J}_{t'_i}) z_i$ for every $i \leq n$, by the co-IH there exists $t_i \in \mathscr{T}_{\mathsf{rec}}$ such that $\mathcal{B} \vdash \mathsf{J}_{t_i} = \mathsf{J}_{t''_i} \circ \mathsf{J}_{t'_i}$ and t_i is infinite whenever either t'_i or t''_i is.

Define t as follows (for all $\alpha \in \mathbb{N}^*$):

$$t(\alpha) \triangleq \begin{cases} m, & \text{if } \alpha = \langle\rangle; \\ t_{k+1}(\beta), & \text{if } \alpha = \langle k \rangle \star \beta \text{ and } 0 \leq k < n; \\ t'_{k+1}(\beta), & \text{if } \alpha = \langle k \rangle \star \beta \text{ and } n \leq k < m; \\ \uparrow & \text{otherwise.} \end{cases}$$

It is routine to check that t is a recursive tree and satisfies the desired properties. We conclude because $\mathsf{J}_{t'} \circ \mathsf{J}_{t''} =_\beta \lambda x z_1 \dots z_m.x(\mathsf{J}_{t_1} z_1) \cdots (\mathsf{J}_{t_m} z_m) =_\beta \mathsf{J}_t$. □

LEMMA 11.21. *Let $t \in \mathscr{T}^\infty_{\mathsf{rec}}$. There exists $t' \in \mathscr{T}^\infty_{\mathsf{rec}}$ such that $\mathcal{B} \vdash \mathsf{J}_t \mathsf{I} = \mathsf{J}_{t'}$.*

PROOF. As t is recursive and non-empty, it must have recursive subtrees $t_0, \dots, t_n$, so that $\mathsf{J}_t \mathsf{I} =_\beta \lambda z_0 \dots z_n.(\mathsf{J}_{t_0} z_0) \cdots (\mathsf{J}_{t_n} z_n)$. If t_0 is empty, then $\mathsf{J}_{t_0} = \mathsf{I}$ and the result follows. Otherwise t_0 has subtrees $s_1, \dots, s_m$ with, say, $m \geq n$. Easy calculations give:

$$\begin{aligned} \mathsf{J}_t \mathsf{I} &=_\beta \lambda z_0 \dots z_n.(\lambda y_1 \dots y_m.z_0(\mathsf{J}_{s_1} y_1) \cdots (\mathsf{J}_{s_m} y_m))(\mathsf{J}_{t_1} z_1) \cdots (\mathsf{J}_{t_n} z_n) \\ &=_\beta \lambda x z_1 \dots z_m.x(\mathsf{J}_{s_1}(\mathsf{J}_{t_1} z_1)) \cdots (\mathsf{J}_{s_n}(\mathsf{J}_{t_n} z_n))(\mathsf{J}_{s_{n+1}} z_{n+1}) \cdots (\mathsf{J}_{s_m} z_m) \end{aligned}$$

Notice that $\mathsf{J}_{s_i}(\mathsf{J}_{t_i} z_i) =_\beta (\mathsf{J}_{s_i} \circ \mathsf{J}_{t_i}) z_i$ for every $i \leq m$, so by Lemma 11.20 there is $t'_i \in \mathscr{T}_{\mathsf{rec}}$ such that $\mathcal{B} \vdash \mathsf{J}_{s_i} \circ \mathsf{J}_{t_i} = \mathsf{J}_{t'_i}$. Define the recursive tree t' by setting (for all $\alpha \in \mathbb{N}^*$):

$$t'(\alpha) \triangleq \begin{cases} m, & \text{if } \alpha = \langle\rangle; \\ t'_{k+1}(\beta), & \text{if } \alpha = \langle k \rangle \star \beta \text{ and } 0 \leq k < n; \\ s_{k+1}(\beta), & \text{if } \alpha = \langle k \rangle \star \beta \text{ and } n \leq k < m; \\ \uparrow, & \text{otherwise.} \end{cases}$$

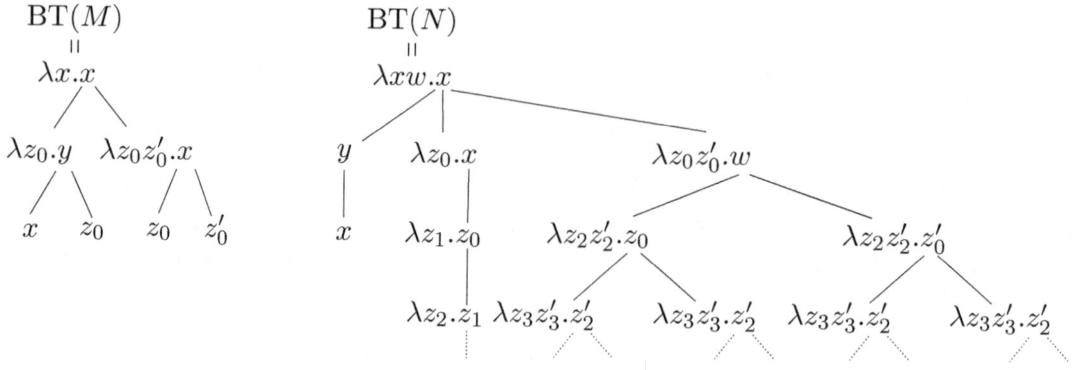

Figure 11.5: Two λ-terms M, N satisfying $M \in \mathrm{NF}$, $\mathcal{H}^* \vdash M = N$, but $\mathcal{H}^+ \nvdash M = N$.

As t is infinite, at least one of its subtrees must be infinite: if $t_k \in \mathscr{T}^\infty_{\mathsf{rec}}$ or $s_k \in \mathscr{T}^\infty_{\mathsf{rec}}$ for some $k \leq n$ then $t'_k \in \mathscr{T}^\infty_{\mathsf{rec}}$, otherwise some s_k with $n \leq k \leq m$ must be infinite. We conclude that $\mathsf{J}_t\mathsf{I} =_\beta \lambda x z_1 \dots z_m.x(\mathsf{J}_{t'_1} z_1) \cdots (\mathsf{J}_{t'_m} z_m) =_\beta \mathsf{J}_{t'}$ with $t' \in \mathscr{T}^\infty_{\mathsf{rec}}$. □

Notice that Lemma 11.21 actually generalizes to arbitrary $t_1, t_2 \in \mathscr{T}^\infty_{\mathsf{rec}}$. We provide here the general statement, its full proof is however beyond the scope of this chapter.

THEOREM 11.22. (i) *The triple* $(\mathcal{I}^\eta_\omega, \circ, \mathsf{I})/\mathcal{B}$ *is an idempotent commutative monoid.*
(ii) *The set* $\mathcal{I}^\eta_\omega$ *is closed under application.*
(iii) *For* $Q_1, Q_2 \in \mathcal{I}^\eta_\omega$, $Q_1 \circ Q_2 \in \mathcal{I}^\eta_\infty$ *and* $Q_1 Q_2 \in \mathcal{I}^\eta_\infty$ *if either* Q_1 *or* Q_2 *belongs to* $\mathcal{I}^\eta_\infty$.

The weak separation theorem

The *Böhm-out technique* is a method for extracting a subterm N, or a substitution instance of it, from a λ-term M by building a suitable context. The selectors U^n_k and the tuplers P_n from Definition 1.16, represent the key ingredients in the construction of such a context. As shown in Section 10.3 of B[1984], this technique can be generalized to extract from $\mathrm{BT}(M)$ an instance of the subtree at position α and constitutes the main tool for proving separability results in the spirit of Proposition 3.16.

Consider for instance the λ-term N whose Böhm tree is represented in Figure 11.5, and imagine that we wish to define an applicative closed context for extracting the subterm of N at $\langle 0 \rangle$ relative to its Böhm tree, namely yx. This is impossible without altering the subterm—the best one can do is to use the context $[\,]\mathsf{U}^3_1 P$, for any $P \in \Lambda^o$, which extracts from N the subterm yx *where* x *is replaced by* U^3_1. This is why we usually speak about a *substitution instance* of the subterm, rather than the subterm itself. When building contexts that work independently from some component P, like $[\,]\mathsf{U}^3_1 P$ above, we will take $P \triangleq \Omega$ as a canonical choice.

In general, to extract the subterm of N at α relative to its Böhm tree, the idea is to replace every variable along the path α with the correct selector. When U^n_k is substituted for x in a λ-term of shape $xM_1 \cdots M_n$, it extracts the instance $M_k[x := \mathsf{U}^n_k]$. An issue arises when the same variable x occurs several times in α and one needs to select a different child in each occurrence. A case in point, to extract $N_{\langle 1,0\rangle}$ from the N of Figure 11.5, the first occurrence of x should be replaced by U_2^3, the second by U_1^1. The problem was originally solved by Böhm by first replacing the occurrences of the same variables along the path by different variables using the tupler, and then replacing each variable by the suitable selector. In the example under consideration, the context $[\,]\mathsf{P}_3\Omega\mathsf{U}_2^3\mathsf{U}_1^1\Omega\Omega\mathsf{U}_1^3$ extracts from N the instance of $N_{\langle 1,0\rangle}$ where z_0 is replaced by U_1^1.

Following an original idea by Breuvart we show that, even when two λ-terms M and N are not semi-separable,[3] if $M \not\sqsubseteq_{\rm NF} N$ then this difference can be Böhmed-out via an appropriate closed head context thus providing a weak-separation result.

DEFINITION 11.23. Let $M, N \in \Lambda$. A sequence $\alpha \in \mathbb{N}^*$ is a *Morris' separator for* M, N, written $\alpha : M \not\sqsubseteq_{\rm NF} N$, if there exists $i > 0$ such that, for some $p \geq i$, we have:

$$M_\alpha =_\beta \lambda x_1 \dots x_n.yM_1 \cdots M_m \qquad \text{and} \qquad N_\alpha =_\beta \lambda x_1 \dots x_{n+p}.yN_1 \cdots N_{m+p}$$

where $\mathcal{B} \vdash N_{m+i} = \mathsf{J}_t x_{n+i}$ for some $t \in \mathscr{T}^\infty_{\mathsf{rec}}$.

REMARK 11.24. It is easy to check that $\langle k\rangle \star \alpha : M \not\sqsubseteq_{\rm NF} N$ entails $\alpha : M_{k+1} \not\sqsubseteq_{\rm NF} N_{k+1}$ where M_{k+1} (resp. N_{k+1}) represents the k-th argument in the phnf of M (resp. N).

Intuitively, a Morris' separator for two $\mathcal{H}^*$-equivalent λ-terms M and N is a common position α of their Böhm trees witnessing the fact that $M \not\sqsubseteq_{\rm NF} N$. For the sake of simplicity, consider the case $M \in \mathrm{NF}$, which entails that $\mathrm{BT}(M)$ is finite and $\bot$-free.[4] By Proposition 11.12 also $\mathrm{BT}(N)$ is $\bot$-free and, at every common position β, M_β and N_β have similar hnf's, which means that the numbers of abstractions and of applications can be matched via suitable η-expansions. Since $M \not\sqsubseteq_{\rm NF} N$, the Böhm tree of N must have an infinite subtree of the form $\mathrm{BT}(\mathsf{J}_t x)$ for some variable x and infinite recursive tree t. Note that $\mathrm{BT}(M)$ might have finite η-expansions that are not present in $\mathrm{BT}(N)$.

EXAMPLES 11.25. The λ-terms M, N whose Böhm trees are represented in Figure 11.5 admit, among others, the following Morris' separators:

(i) The sequence $\langle\rangle$: since $\mathcal{B} \vdash N_{\langle 2\rangle} = \mathsf{J}_{t_2} w$ where t_2 is the complete binary tree.

(ii) The sequence $\langle 1, 0\rangle$: indeed, $\mathcal{B} \vdash N_{\langle 1,0,0\rangle} = \mathsf{J}_{t_1} z_1$ where t_1 is the complete unary tree (i.e., $\mathcal{B} \vdash \mathsf{J}_{t_1} = \mathsf{J}$).

It is well known that finite η-differences can be destroyed along the process of Böhming out. In contrast, we show that infinite η-differences can always be preserved.

PROPOSITION 11.26. *Let* $M, N \in \Lambda$ *be such that* $M \in \mathrm{NF}$, $N \notin \mathrm{NF}$ *and* $\mathcal{H}^* \vdash M = N$. *Then there exists a position* $\alpha \in \mathbb{N}^*$ *such that* $\alpha : M \not\sqsubseteq_{\rm NF} N$.

PROOF. Wlog, we can assume that M is in β-normal form and therefore $\mathrm{BT}(M) = M$. We proceed by induction on the β-normal structure of M.

Base case: $M = \lambda x_1 \dots x_n.y$ for some $n \geq 0$. By Proposition 11.12 there is a Böhm-like tree T such that $M \leq^\eta_\omega T \geq^\eta_\omega \mathrm{BT}(N)$. Notice that in this particular case

[3]Equivalently, when M, N satisfy $\mathcal{H}^* \vdash M = N$.

[4]An example of this situation is given in Figure 11.5.

M is also η-normal and this implies that $M \sqsubseteq_\omega^\eta N$ holds. As a consequence, we get $N =_\beta \lambda x_1 \ldots x_{n+p}.yQ_1 \cdots Q_p$ with $x_{n+j} \sqsubseteq_\omega^\eta Q_j$ for all $j \leq p$. Now, as N does not posses a β-nf, we must have $p > 0$ and some index i ($0 < i \leq p$) such that $\lambda x_{n+i}.Q_i \in \mathcal{I}_\infty^\eta$. By Proposition 11.19, there is $t \in \mathcal{T}_{\text{rec}}^\infty$ such that $\mathcal{B} \vdash \lambda x_{n+i}.Q_i = \mathsf{J}_t$, hence $\mathcal{B} \vdash Q_i = \mathsf{J}_t x_{n+i}$. Therefore the empty sequence $\langle\rangle$ is a Morris' separator for M, N.

Induction case: $M = \lambda x_1 \ldots x_n.yM_1 \cdots M_m$ for $m > 0$, where each M_i is in β-nf. By Proposition 11.12, there exists $T \in \mathscr{B}$ such that $M \leq_\omega^\eta T \geq_\omega^\eta \mathrm{BT}(N)$, say:

$$T = \lambda x_1 \ldots x_{n+p} z_1 \ldots z_k.y \underbrace{T_1 \cdots T_{m+p}}_{} \underbrace{T_1' \cdots T_k'}_{}$$

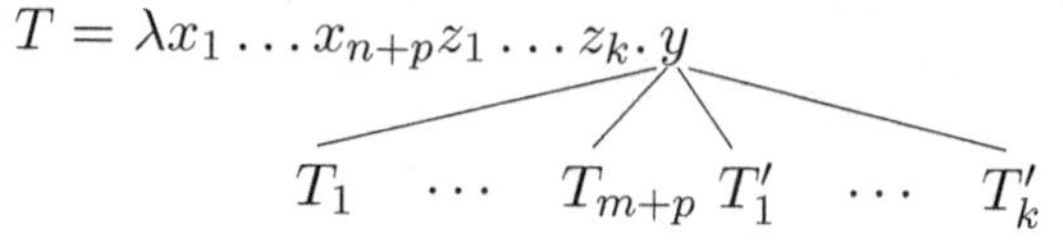

and $N =_\beta \lambda x_1 \ldots x_{n+p}.yN_1 \cdots N_{m+p}$ for some integer p and $k \in \mathbb{N}$. Assume that $p \geq 0$, the case $p < 0$ being symmetrical. We also have $M_j \leq_\omega^\eta T_j \geq_\omega^\eta \mathrm{BT}(N_j)$ for all $j \leq m$, $x_{m+j} \leq_\omega^\eta T_{m+j} \geq_\omega^\eta \mathrm{BT}(N_{m+j})$ for all $j \leq p$ and finally $z_\ell \leq_\omega^\eta T_\ell'$ for all $\ell \leq k$. By Proposition 11.12, this entails that $\mathcal{H}^* \vdash M_j = N_j$ for $j \leq m$ and $\mathcal{H}^* \vdash x_{m+j} = N_{m+j}$ for $j \leq p$. By hypothesis, there exists an index $i \leq m+p$ such that $N_i \notin \mathrm{NF}$.

There are two subcases to consider:

1. $0 < i \leq m$. By the IH there exists $\alpha : M_i \not\sqsubseteq_{\mathrm{NF}} N_i$, so the Morris' separator we are looking for is the position $\langle i-1 \rangle \star \alpha$.

2. $i > m$. By Proposition 11.19, there exists $t \in \mathcal{T}_{\text{rec}}^\infty$ such that $\mathcal{B} \vdash \lambda x_i.N_i = \mathsf{J}_t$, hence $\mathcal{B} \vdash N_i = \mathsf{J}_t x_i$ and the Morris' separator is $\langle\rangle$. $\square$

LEMMA 11.27 (BÖHM-OUT).
Let $M, N \in \Lambda$ be such that $\mathcal{H}^* \vdash M = N$ and assume $\alpha : M \not\sqsubseteq_{\mathrm{NF}} N$. For all $k \geq 0$ large enough, there exist $q \geq 0$, $X_1, \ldots, X_q \in \Lambda^o$ and $t \in \mathcal{T}_{\text{rec}}^\infty$ such that (setting $\vec{y} \triangleq \mathrm{FV}(MN)$):

$$M[\vec{y} := \mathsf{P}_k]X_1 \cdots X_q =_\beta \mathsf{I} \text{ and } N[\vec{y} := \mathsf{P}_k]X_1 \cdots X_q =_\mathcal{B} \mathsf{J}_t.$$

PROOF. We proceed by induction on the Morris' separator α.
Base case $\alpha = \langle\rangle$. Then there exists $i > 0$ such that, for some $p \geq i$, we have:

$$M =_\beta \lambda x_1 \ldots x_n.yM_1 \cdots M_m \text{ and } N =_\beta \lambda x_1 \ldots x_{n+p}.yN_1 \cdots N_{m+p},$$

where $\mathcal{B} \vdash N_{m+i} = \mathsf{J}_t x_{n+i}$ for some $t \in \mathcal{T}_{\text{rec}}^\infty$. For $k \geq n+m+p$, define:

$$\vec{X} \triangleq \mathsf{P}_k^{\sim n} \mathsf{I}^{\sim p} \Omega^{\sim k-m-p} \mathsf{U}_{m+i}^k.$$

We consider the case where the head variable y is free, the other case being analogous. On the one side we have:

$$\begin{aligned}
& (\lambda x_1 \ldots x_n.yM_1 \cdots M_m)[\vec{y} := \mathsf{P}_k]\vec{X} \\
&= (\lambda x_1 \ldots x_n.\mathsf{P}_k M_1^{\sigma_1} \cdots M_m^{\sigma_1})\vec{X}, && \text{where } \sigma_1 \triangleq [\vec{y} := \mathsf{P}_k], \\
&=_\beta (\mathsf{P}_k M_1^{\sigma_2} \cdots M_m^{\sigma_2})\mathsf{I}^{\sim p} \Omega^{\sim k-m-p} \mathsf{U}_{m+i}^k, && \text{where } \sigma_2 \triangleq \sigma_1[\vec{x} := \mathsf{P}_k], \\
&=_\beta (\lambda z.z M_1^{\sigma_2} \cdots M_m^{\sigma_2} \mathsf{I}^{\sim p} \Omega^{\sim k-m-p}) \mathsf{U}_{m+i}^k, && \text{by Remark 1.17(i)}, \\
&=_\beta \mathsf{I}, && \text{by Remark 1.17(ii)}.
\end{aligned}$$

On the other side, we have:

$$\begin{array}{lll} & (\lambda x_1 \dots x_{n+p}.yN_1 \cdots N_{m+p})[\vec{y} := \mathsf{P}_k]\vec{X} & \\ = & (\lambda x_1 \dots x_{n+p}.\mathsf{P}_k N_1^{\sigma_1} \cdots N_{m+p}^{\sigma_1})\vec{X}, & \text{for } \sigma_1 \triangleq [\vec{y} := \mathsf{P}_k], \\ =_\beta & (\lambda x_{n+1} \dots x_{n+p}.\mathsf{P}_k N_1^{\sigma_2} \cdots N_{m+p}^{\sigma_2})\mathsf{I}^{\sim p}\Omega^{\sim k-m-p}\,\mathsf{U}_{m+i}^k, & \text{for } \sigma_2 \triangleq \sigma_1[x_1,\dots,x_n := \mathsf{P}_k], \\ =_\beta & (\mathsf{P}_k N_1^{\sigma_3} \cdots N_{m+p}^{\sigma_3})\Omega^{\sim k-m-p}\,\mathsf{U}_{m+i}^k, & \text{for } \sigma_3 \triangleq \sigma_2[x_{n+1},\dots,x_{n+p} := \mathsf{I}], \\ =_\beta & (\lambda z.zN_1^{\sigma_3} \cdots N_{m+p}^{\sigma_3}\Omega^{\sim k-m-p})\,\mathsf{U}_{m+i}^k, & \text{by Remark 1.17(i)}, \\ =_\beta & N_{m+i}^{\sigma_3} =_{\mathcal{B}} (\mathsf{J}_t\, x_{n+i})[x_{n+i} := \mathsf{I}] = \mathsf{J}_t\,\mathsf{I}, & \text{by Remark 1.17(ii)}, \\ =_{\mathcal{B}} & \mathsf{J}_{t'},\ \text{for some } t' \in \mathscr{T}_{\mathsf{rec}}^\infty, & \text{by Lemma 11.21.} \end{array}$$

Induction case $\alpha = \langle i \rangle \star \beta$. By Proposition 3.13(iii), $\mathcal{H}^* \vdash M = N$ entails $M \sim N$. I.e., for some $n, n', m, m' \geq 0$ such that $n - m = n' - m'$ and $i + 1 \leq \min\{m, m'\}$ we have:

$$M =_\beta \lambda x_1 \dots x_n.yM_1 \cdots M_m, \qquad N =_\beta \lambda x_1 \dots x_{n'}.yN_1 \cdots N_{m'}$$

and either y is free in both λ-terms or $y = x_j$ for some $j \leq \min\{n, n'\}$. Moreover, $\mathcal{H}^* \vdash M_j = N_j$ holds for all $j \leq \min\{m, m'\}$. Suppose that, say, $n \leq n'$. Then there exists $p \geq 0$ such that $n' = n + p$ and $m' = m + p$. Since $\mathcal{H}^* \vdash M_{i+1} = N_{i+1}$ and $\beta : M_{i+1} \not\sqsubseteq_{\mathrm{NF}} N_{i+1}$ we apply the IH and get, for any k' large enough, $\vec{Y} \in \Lambda^o$ such that

$$M_{i+1}[\vec{y}, \vec{x} := \mathsf{P}_{k'}]\vec{Y} =_\beta \mathsf{I} \text{ and } N_{i+1}[\vec{y}, \vec{x} := \mathsf{P}_{k'}]\vec{Y} =_{\mathcal{B}} \mathsf{J}_t \text{ for some } t \in \mathscr{T}_{\mathsf{rec}}^\infty.$$

For $k \geq \max\{k', n + m + p\}$, define $\vec{X} \triangleq \mathsf{P}_k^{\sim n+p}\Omega^{\sim k-m-p}\,\mathsf{U}_{i+1}^k\vec{Y}$. Suppose that y is free, the other case being analogous. On the one side we have:

$$\begin{array}{lll} & (\lambda x_1 \dots x_n.yM_1 \cdots M_m)[\vec{y} := \mathsf{P}_k]\,\vec{X} & \\ = & (\lambda x_1 \dots x_n.\mathsf{P}_k M_1^{\sigma_1} \cdots M_m^{\sigma_1})\mathsf{P}_k^{\sim n+p}\Omega^{\sim k-m-p}\,\mathsf{U}_{i+1}^k\vec{Y}, & \text{where } \sigma_1 \triangleq [\vec{y} := \mathsf{P}_k], \\ =_\beta & (\mathsf{P}_k M_1^{\sigma_2} \cdots M_m^{\sigma_2})\mathsf{P}_k^{\sim p}\Omega^{\sim k-m-p}\,\mathsf{U}_{i+1}^k\vec{Y}, & \text{where } \sigma_2 \triangleq \sigma_1[\vec{x} := \mathsf{P}_k], \\ =_\beta & (\lambda z.zM_1^{\sigma_2} \cdots M_m^{\sigma_2}\mathsf{P}_k^{\sim p}\Omega^{\sim k-m-p})\,\mathsf{U}_{i+1}^k\vec{Y}, & \text{by Remark 1.17(i)}, \\ =_\beta & M_{i+1}^{\sigma_2}\vec{Y} = M_{i+1}[\vec{y}, \vec{x} := \mathsf{P}_k]\vec{Y}, & \text{by Remark 1.17(ii)}, \\ =_\beta & \mathsf{I}, & \text{by the IH.} \end{array}$$

On the other side, we have:

$$\begin{array}{lll} & (\lambda x_1 \dots x_{n+p}.yN_1 \cdots N_{m+p})[\vec{y} := \mathsf{P}_k]\vec{X} & \\ = & (\lambda x_1 \dots x_{n+p}.\mathsf{P}_k N_1^{\sigma_1} \cdots N_{m+p}^{\sigma_2})\vec{X}, & \text{where } \sigma_1 \triangleq [\vec{y} := \mathsf{P}_k], \\ =_\beta & (\mathsf{P}_k N_1^{\sigma_2} \cdots N_{m+p}^{\sigma_2})\Omega^{\sim k-m-p}\,\mathsf{U}_{i+1}^k\vec{Y}, & \text{where } \sigma_2 \triangleq \sigma_1[\vec{x} := \mathsf{P}_k], \\ =_\beta & (\lambda z.zN_1^{\sigma_2} \cdots N_{m+p}^{\sigma_2}\Omega^{\sim k-m-p})\,\mathsf{U}_{i+1}^k\vec{Y}, & \text{by Remark 1.17(i)}, \\ =_\beta & N_{i+1}^{\sigma_2}\vec{Y} = N_{i+1}[\vec{y}, \vec{x} := \mathsf{P}_k]\vec{Y}, & \text{by Remark 1.17(ii)}, \\ =_{\mathcal{B}} & \mathsf{J}_t, & \text{by the IH.} \end{array}$$

□

THEOREM 11.28 (WEAK SEPARATION THEOREM).
If $M, N \in \Lambda$ are such that $\mathcal{H}^* \vdash M = N$ while $M \not\sqsubseteq_{\mathrm{NF}} N$, then there exists a closed head context $H[\,]$ satisfying

$$H[M] =_\beta \mathsf{I} \quad \text{and} \quad H[N] =_{\mathcal{B}} \mathsf{J}_t, \text{ for some } t \in \mathscr{T}_{\mathsf{rec}}^\infty.$$

Moreover, when M and N are closed, the context $H[\,]$ can be chosen applicative.

PROOF. By Lemma 3.15, the condition $M \not\sqsubseteq_{\mathrm{NF}} N$ is equivalent to the existence of a closed head context $H_2[\,]$ satisfying $H_2[M] \in \mathrm{NF}$ and $H_2[N] \notin \mathrm{NF}$. Moreover, if $M, N \in \Lambda^o$ then $H_2[\,]$ can be chosen applicative. Since $\mathcal{H}^*$ is a λ-theory, therefore contextual, we have that $\mathcal{H}^* \vdash M = N$ implies $\mathcal{H}^* \vdash H_2[M] = H_2[N]$. By applying Lemma 11.27 to the closed λ-terms $H_2[M], H_2[N]$, we obtain a closed applicative context $H_1[\,]$ satisfying $H_1[H_2[M]] =_\beta \mathsf{I}$ and $H_1[H_2[N]] =_{\mathcal{B}} \mathsf{J}_t$ for some $t \in \mathscr{T}^{\infty}_{\mathsf{rec}}$. We conclude by setting $H[\,] \triangleq H_1[H_2[\,]]$ since closed head (resp. applicative) contexts are closed under composition. □

The λ-theory $\mathcal{B}\omega$ is included in $\mathcal{H}^+$

The fact that $\mathcal{H}^+$ satisfies the ω-rule follows easily from the Weak Separation Theorem. Recall that we can focus on (ω^0), namely the restriction of ω to closed λ-terms, without loosing generality thanks to Lemma 3.23(ii).

LEMMA 11.29. *Let $M, N \in \Lambda^o$ be such that $M \in \mathrm{NF}$ while $N \notin \mathrm{NF}$. There exist $k \geq 1$ and combinators $Z_1, \ldots, Z_k \in \Lambda^o$ such that $MZ_1 \cdots Z_k \in \mathrm{NF}$ while $NZ_1 \cdots Z_k \notin \mathrm{NF}$.*

PROOF. By hypothesis $M \not\sqsubseteq_{\mathrm{NF}} N$, so there are two possible cases.

Case $\mathcal{H}^* \vdash M = N$. The Weak Separation Theorem entails that there exist $k' \in \mathbb{N}$, $Z_1, \ldots, Z_{k'} \in \Lambda^o$ and $t \in \mathscr{T}^{\infty}_{\mathsf{rec}}$ such that $MZ_1 \cdots Z_{k'} =_\beta \mathsf{I}$ and $NZ_1 \cdots Z_{k'} =_{\mathcal{B}} \mathsf{J}_t$. In case $k' = 0$ just take $Z_1 \triangleq \mathsf{I}$ and conclude since $\mathsf{J}_t\mathsf{I} =_{\mathcal{B}} \mathsf{J}_{t'}$ for some $t' \in \mathscr{T}^{\infty}_{\mathsf{rec}}$ by Lemma 11.21.

Case $\mathcal{H}^* \nvdash M = N$. In particular, since $M \in \mathrm{NF}$ we must have $M \not\sqsubseteq_{\mathrm{SOL}} N$. By Proposition 3.16, M and N are semi-separable, so there are $Z_1, \ldots, Z_{k'} \in \Lambda^o$ such that $M\vec{Z} =_\beta \mathsf{I}$ and $N\vec{Z} =_{\mathcal{H}} \Omega$. If $k' = 0$ we can take again $Z_1 \triangleq \mathsf{I}$ as $\mathsf{II} =_\beta \mathsf{I}$ and $\Omega\mathsf{I} =_{\mathcal{H}} \Omega$. □

LEMMA 11.30. *Let $M, N \in \Lambda^o$. Assume that $\forall Z \in \Lambda^o \,.\, \mathcal{H}^+ \vdash MZ = NZ$ holds, then*

$$\forall \vec{P} \in \Lambda^o \,.\, [M\vec{P} \in \mathrm{NF} \iff N\vec{P} \in \mathrm{NF}].$$

PROOF. By the Context Lemma for $\mathcal{H}^+$ (Lemma 3.15), the hypothesis is equivalent to assume that, for all $Z, \vec{Q} \in \Lambda^o$, $MZ\vec{Q} \in \mathrm{NF}$ if and only if $NZ\vec{Q} \in \mathrm{NF}$.

By case distinction on the length k of $\vec{P} \in \Lambda^o$, we show that

$$MP_1 \cdots P_k \in \mathrm{NF} \iff NP_1 \cdots P_k \in \mathrm{NF}.$$

Case $k = 0$. Since the contrapositive holds by Lemma 11.29.

Case $k > 0$. It follows from $\mathcal{H}^+ \vdash MP_1 = NP_1$, by taking $C[\,] \triangleq [\,]P_2 \cdots P_k$. □

As a consequence, we get the main result of this section.

THEOREM 11.31 (BREUVART ET AL. (2016)).
(i) $\mathcal{H}^+ \vdash \omega$,
(ii) $\mathcal{B}\omega \subseteq \mathcal{H}^+$.

PROOF. (i) Lemma 11.30 actually shows that $\mathcal{H}^+ \vdash \omega^0$, which is equivalent to $\mathcal{H}^+ \vdash \omega$ by Lemma 3.23(ii).

(ii) We know that $\mathcal{B}\eta \subseteq \mathcal{H}^+$, by B[1984], Proposition 16.4.7(i). From this inclusion and Lemma 3.23(i)-(iv), it follows that $\mathcal{B}\omega \subseteq \mathcal{H}^+\omega$. We have proved in (i) that $\mathcal{H}^+ \vdash \omega$, so we conclude $\mathcal{H}^+\omega = \mathcal{H}^+$. □

11.3 Characterizing $\mathcal{H}^+$

In Section 11.2 we have shown that $\mathcal{H}^+ = \mathcal{T}_{\mathrm{NF}}$ satisfies the ω-rule, and we have seen that this entails $\mathcal{B}\omega \subseteq \mathcal{H}^+$. The present section is devoted to the demonstration of the converse inclusion, namely that $\mathcal{B}\omega = \mathcal{H}^+$. This refutes Sallé's Conjecture.

Outline of the proof

To prove $\mathcal{H}^+ \subseteq \mathcal{B}\omega$ we need to show that, whenever two λ-terms M and N are equal in $\mathcal{H}^+$, they are also equal in $\mathcal{B}\omega$. From Proposition 11.9, we know that in this case the Böhm trees of M and N are equal up to countably many η-expansions, so they are compatible w.r.t. $\sqsubseteq^\eta$ and have a common η-supremum $T \in \mathscr{B}$. Our proof has four steps:

1. We show that the η-supremum T satisfying $\mathrm{BT}(M) \leq^\eta T \geq^\eta \mathrm{BT}(N)$ is λ-definable (Proposition 11.41).
2. We apply the ω-rule to equate the Böhm tree of the stream $[\eta_n]_{n \in \mathbb{N}}$ enumerating all finite η-expansions of the identity, and the Böhm tree of the stream $[\mathsf{I}]_{n \in \mathbb{N}}$ containing infinitely many copies of the identity (Corollary 11.46).
3. We define a λ-term Ξ (Definition 11.37) taking as arguments the *codes* $\ulcorner \cdot \urcorner$ of two λ-terms M_1, M_2 and a stream S, and such that $M_1 \sqsubseteq^\eta M_2$ entails:

   (i) $\mathcal{B} \vdash \Xi \ulcorner M_1 \urcorner \ulcorner M_2 \urcorner [\eta_n]_{n \in \mathbb{N}} = M_2$ (Lemma 11.40(i)),
   (ii) $\mathcal{B} \vdash \Xi \ulcorner M_1 \urcorner \ulcorner M_2 \urcorner [\mathsf{I}]_{n \in \mathbb{N}} = M_1$ (Lemma 11.40(ii)).

4. Summing up, if M, N are equal in $\mathcal{H}^+$, then by (1) there is a λ-term P such that $M \sqsubseteq^\eta P \sqsupseteq^\eta N$. Since $\mathcal{B}\omega$ equates all λ-terms having the same Böhm tree, we obtain the following sequence of equalities:

$$\begin{array}{lllllll} M & =_{3(ii)} & \Xi \ulcorner M \urcorner \ulcorner P \urcorner [\mathsf{I}]_{n \in \mathbb{N}} & =_2 & \Xi \ulcorner M \urcorner \ulcorner P \urcorner [\eta_n]_{n \in \mathbb{N}} & =_{3(i)} & P \\ N & =_{3(ii)} & \Xi \ulcorner N \urcorner \ulcorner P \urcorner [\mathsf{I}]_{n \in \mathbb{N}} & =_2 & \Xi \ulcorner N \urcorner \ulcorner P \urcorner [\eta_n]_{n \in \mathbb{N}} & =_{3(i)} & P \end{array}$$

   so M and N are equal in $\mathcal{B}\omega$ (Theorem 11.47).

The intuition behind the λ-term Ξ is that, when it receives as arguments $\ulcorner M \urcorner$, $\ulcorner N \urcorner$ and S, it computes the head normal forms of M and N (if they exist), it compares them to verify that they have a similar structure[5] and, before recurring on their subterms, it applies to the less η-expanded one an element extracted from the stream S in the attempt of matching the structure of the more η-expanded one. There are two possibilities:

- If the stream S contains all possible η-expansions of I then each attempt succeeds, so $\Xi \ulcorner M \urcorner \ulcorner N \urcorner [\eta_n]_{n \in \mathbb{N}}$ computes the η-supremum of $\mathrm{BT}(M)$ and $\mathrm{BT}(N)$.
- If S only contains infinitely many copies of the identity, each non-trivial attempt fails, and $\Xi \ulcorner M \urcorner \ulcorner N \urcorner [\mathsf{I}]_{n \in \mathbb{N}}$ computes their η-infimum.

By showing that the equality $[\eta_n]_{n \in \mathbb{N}} = [\mathsf{I}]_{n \in \mathbb{N}}$ holds in $\mathcal{B}\omega$, we actually obtain that the η-infimum and the η-supremum collapse.

[5] In the sense that one 'looks like' an η-expansion of the other one.

Building the η-supremum by codes and streams

Recall from Example 1.36(i) that $[\mathsf{I}]_{n\in\mathbb{N}} = [\mathsf{I},\mathsf{I},\mathsf{I},\mathsf{I},\dots]$ denotes the stream containing infinitely many copies of the identity.

DEFINITION 11.32. Given an effective enumeration $\vec{\eta} = (\eta_0, \eta_1, \dots, \eta_i, \dots)$ of the set $\mathcal{I}^\eta$, define the corresponding stream $[\eta_n]_{n\in\mathbb{N}} \triangleq [\eta_0, \eta_1, \eta_2, \eta_3, \dots]$.

From now on, and until the end of the chapter, we assume that the enumeration $\vec{\eta}$ and the stream $[\eta_n]_{n\in\mathbb{N}}$ are fixed.

EXERCISE 11.33. Write a λ-term implementing the stream $[\eta_n]_{n\in\mathbb{N}}$ generated by an actual enumeration $\vec{\eta}$ of $\mathcal{I}^\eta$. [Hint: Remember from Intrigila and Nesi (2003) that $\mathcal{I}^\eta/\beta$ is in one-to-one correspondence with the set of all finite trees.]

We now define a λ-term Ξ that builds from the codes of M, N and the stream $[\eta_n]_{n\in\mathbb{N}}$ the η-supremum of $\mathrm{BT}(M)$ and $\mathrm{BT}(N)$, namely

$$M \sqsubseteq^\eta \Xi \ulcorner M \urcorner \ulcorner N \urcorner [\eta_n]_{n\in\mathbb{N}} \sqsupseteq^\eta N$$

whenever such an η-supremum exists, that is when M and N are compatible w.r.t. $\sqsubseteq^\eta$. Intuitively, at every common position α, the λ-term Ξ needs to compare the structure of the subterms of M, N at α and apply the correct expansion η_i taken from $[\eta_n]_{n\in\mathbb{N}}$. Indeed, since every closed η-expansion $Q \in \mathcal{I}^\eta$ is β-normalizable and the enumeration $\vec{\eta}$ is effective, it is possible to decide starting from the code $\#Q$ the index i of Q in $\vec{\eta}$. Moreover, it is possible to choose such an index i minimal as shown in the next lemma.

LEMMA 11.34. *There exists a partial recursive map* $\iota : \mathbb{N} \to \mathbb{N}$ *such that, for all* $M \in \Lambda^o$, *if* $M =_\beta \eta_i$ *and* $M \neq_\beta \eta_k$ *for all* $k < i$ *then* $\iota(\#M) = i$.

PROOF. Let $\delta(m, n)$ be the partial recursive function satisfying for all $M, N \in \Lambda^o \cap \mathrm{NF}$:

$$\delta(\#M, \#N) = \begin{cases} 0, & \text{if } M \text{ and } N \text{ have the same } \beta\text{-normal form,} \\ 1, & \text{otherwise.} \end{cases}$$

Then ι can be defined as $\iota(n) \triangleq \mu k.\delta\big(\#(\pi_k \circ \vec{\eta}), n\big) = 0$. □

Until the end of the chapter, we also consider fixed the function ι which depends on the enumeration $\vec{\eta}$ generating the stream $[\eta_n]_{n\in\mathbb{N}}$. The next definition formalizes the expression *a λ-term N 'looks like' an η-expansion of M*, that we already used informally.

DEFINITION 11.35. Let $M, N \in \Lambda$.

(i) We say that N *looks like an η-expansion of* M, in symbols $M \leq_h N$, if and only if, for some $n, m, k \geq 0$ and $y \in \mathrm{Var}$, we have:

$$M \twoheadrightarrow_h \lambda x_1 \dots x_n.yM_1 \cdots M_k \qquad N \twoheadrightarrow_h \lambda x_1 \dots x_n z_1 \dots z_m.yN_1 \cdots N_k Q_1 \cdots Q_m$$

and $\lambda z_i.Q_i \in \mathcal{I}^\eta$ for all $i \leq m$.

(ii) We write $M \sim_h N$ if and only if both $M \leq_h N$ and $N \leq_h M$ hold.
(iii) We write $M <_h N$ if and only if $M \leq_h N$ holds, but $M \not\sim_h N$.

The fact that N looks like an η-expansion of M does not necessarily mean that it actually is, as witnessed by the following examples.

EXAMPLES 11.36. (i) $\mathsf{I} \leq_h \mathsf{1}$.
(ii) $\lambda z.x\mathsf{F}z \leq_h \lambda z.x\mathsf{K}z$ since we do not require that $\mathsf{F} \leq_h \mathsf{K}$ holds.
(iii) $z \leq_h \lambda z.zz$ since in Definition 11.35(i) we do not check that $z_i \notin \mathrm{FV}(\mathrm{BT}(y\vec{M}\vec{N}))$.

Compared with the relation $\sqsubseteq^\eta$ from Definition 11.8, the relation $\leq_h$ is weaker since it lacks the coinductive calls and the free occurrence check on the z_i's. This is necessary to ensure that $\leq_h$ remains semi-decidabile, a property needed in the proof of Lemma 11.38.

Compared with the relation $\sim$ from Definition 1.22(iv), the relation $\sim_h$ is stronger. On solvable λ-terms, the former requires that their amount of external abstractions and applications coincide, while the latter simply requires that these numbers can be matched via suitable η-expansions.

DEFINITION 11.37. Let $\iota : \mathbb{N} \to \mathbb{N}$ be the partial recursive function from Lemma 11.34. Define $\Xi_\iota \in \Lambda^o$ satisfying the following recursive equations, for all $M, N \in \Lambda^o$ and $s \in \mathrm{Var}$:

1. if $M \leq_h N$ then

$$\Xi_\iota \ulcorner N \urcorner \ulcorner M \urcorner s =_{\mathcal{H}} \lambda \underline{x}_1 \dots \underline{x}_n . \pi_q s (\underline{x}_j (\Upsilon_1 \underline{x}_1 \cdots \underline{x}_n) \cdots (\Upsilon_k \underline{x}_1 \cdots \underline{x}_n))$$

where:

- $s, \underline{x}_1, \dots, \underline{x}_n$ are fresh variables,
- $M \twoheadrightarrow_h \lambda x_1 \dots x_n . x_j M_1 \cdots M_k$,
- $N \twoheadrightarrow_h \lambda x_1 \dots x_n z_1 \dots z_m . x_j N_1 \cdots N_k Q_1 \cdots Q_m$,
- π_q denotes the q-th projection on streams,
- $q \triangleq \iota(\#(\lambda y z_1 \dots z_m . \underline{y} Q_1 \cdots Q_m)^{\mathrm{nf}})$, for $\underline{y}$ fresh, and
- $\Upsilon_i \triangleq \Xi_\iota \ulcorner \lambda x_1 \dots x_n . M_i \urcorner \ulcorner \lambda x_1 \dots x_n . N_i \urcorner s$, for all i $(1 \leq i \leq k)$;

2. if $N <_h M$ then $\Xi_\iota \ulcorner N \urcorner \ulcorner M \urcorner s =_{\mathcal{H}} \Xi_\iota \ulcorner M \urcorner \ulcorner N \urcorner s$;

3. $\Xi_\iota \ulcorner N \urcorner \ulcorner M \urcorner s =_{\mathcal{H}} \Omega$, otherwise.

There are some subtleties in the definition above that deserve a discussion. The closures $\lambda\vec{x}.M_i$ and $\lambda\vec{x}.N_i$ on the recursive calls Υ_i are needed to obtain closed terms (since, e.g., $M \in \Lambda^o$ entails $\mathrm{FV}(M_i) \subseteq \vec{x}$) and that is why the free variables $x_1, \dots, x_n$ are reapplied externally. Although not explicitly written, each Q_i is such that $Q_i \twoheadrightarrow_{\beta\eta} z_i$ since $M \leq_h N$ holds. *A priori* $\lambda z_i.Q_i \in \mathcal{I}^\eta$ might contain some free variables[6], but its β-normal form is always closed. This is the reason why we compute $(\lambda y\vec{z}.yQ_1 \cdots Q_m)^{\mathrm{nf}}$ before applying ι to its code. In particular, ι is defined on all the codes $\#((\lambda y\vec{z}.yQ_1 \cdots Q_m)^{\mathrm{nf}})$.

[6]Consider for instance $\lambda z_i.\mathsf{K}z_i y \to_\beta \mathsf{I}$.

LEMMA 11.38. *The combinator Ξ_ι from Definition 11.37 exists.*

PROOF. Working on the codes of $M, N \in \Lambda^o$ it is possible to semi-decide whether $M \leq_h N$ holds using the following effective procedure:

1. First, head-reduce in parallel M and N until they reach a head normal form. This step can be performed effectively by Remark 1.30.

2. if both head reductions terminate, compare the two principal hnf's and check whether they have the shape of Definition 11.35(i),

3. finally, semi-decide whether $Q_i \twoheadrightarrow_{\beta\eta} z_i$ for all $i \leq m$.

This procedure might fail to terminate when $M \not\leq_h N$.

Hence, the lemma follows from the effectiveness of the encoding $\#(\cdot)$, the fact that ι is partial recursive (Lemma 11.34) and Proposition 1.28. □

The commutation property below is natural considering that $\Xi_\iota\ulcorner M\urcorner\ulcorner N\urcorner[\eta_n]_{n\in\mathbb{N}}$ is supposed to compute the η-join of $\mathrm{BT}(M)$ and $\mathrm{BT}(N)$, which is a commutative operation.

LEMMA 11.39. *For all $M, N \in \Lambda^o$, we have:*

$$\mathcal{B} \vdash \Xi_\iota\ulcorner M\urcorner\ulcorner N\urcorner = \Xi_\iota\ulcorner N\urcorner\ulcorner M\urcorner.$$

PROOF. We prove by coinduction that $\mathrm{BT}(\Xi_\iota\ulcorner M\urcorner\ulcorner N\urcorner) = \mathrm{BT}(\Xi_\iota\ulcorner N\urcorner\ulcorner M\urcorner)$.

The cases $M <_h N$ and $N <_h M$ follow directly from Definition 11.37(2).

The only non-trivial case is $M \sim_h N$. Then we have $M \twoheadrightarrow_h \lambda x_1 \ldots x_n.x_j M_1 \cdots M_k$ and $N \twoheadrightarrow_h \lambda x_1 \ldots x_n.x_j N_1 \cdots N_k$ for $n, k \geq 0$, $1 \leq j \leq n$. Since $\pi_q s \twoheadrightarrow_\beta s\,\mathsf{F}^{\sim q}\,\mathsf{K}$ we have

$$\begin{aligned}\Xi_\iota\ulcorner M\urcorner\ulcorner N\urcorner &=_\beta \lambda\vec{x}s.s\,\mathsf{F}^{\sim q}\,\mathsf{K}(x_j(\Upsilon_1\vec{x})\cdots(\Upsilon_k\vec{x}))\\ \Xi_\iota\ulcorner N\urcorner\ulcorner M\urcorner &=_\beta \lambda\vec{x}s.s\,\mathsf{F}^{\sim q}\,\mathsf{K}(x_j(\Upsilon'_1\vec{x})\cdots(\Upsilon'_k\vec{x})),\end{aligned}$$

where $\Upsilon_i \triangleq \Xi_\iota\ulcorner\lambda\vec{x}.M_i\urcorner\ulcorner\lambda\vec{x}.N_i\urcorner s$ and $\Upsilon'_i \triangleq \Xi_\iota\ulcorner\lambda\vec{x}.N_i\urcorner\ulcorner\lambda\vec{x}.M_i\urcorner s$ for all $i \leq k$. We conclude since, by the co-IH,[7] we get $\mathcal{B} \vdash \Upsilon_i = \Upsilon'_i$ for all $i \leq k$. □

Whenever $M \sqsubseteq^\eta N$ holds, we expect the λ-term $\Xi_\iota\ulcorner M\urcorner\ulcorner N\urcorner[\eta_n]_{n\in\mathbb{N}}$ to compute the Böhm tree of N. Under the same assumption, we can also use $\Xi_\iota\ulcorner M\urcorner\ulcorner N\urcorner$ to retrieve the Böhm tree of M by applying the stream $[\mathsf{I}]_{n\in\mathbb{N}}$.

LEMMA 11.40. *Let $M, N \in \Lambda^o$. Assume $M \sqsubseteq^\eta N$. Then*
(i) $\mathcal{B} \vdash \Xi_\iota\ulcorner M\urcorner\ulcorner N\urcorner[\eta_n]_{n\in\mathbb{N}} = N$.
(ii) $\mathcal{B} \vdash \Xi_\iota\ulcorner M\urcorner\ulcorner N\urcorner[\mathsf{I}]_{n\in\mathbb{N}} = M$.

[7] Here we are silently exploiting the fact that $\mathcal{B}$ is a λ-theory, so $\mathcal{B} \vdash \Upsilon = \Upsilon'$ entails $\mathcal{B} \vdash \Upsilon\vec{x} = \Upsilon'\vec{x}$. This is, in our opinion, a small price to pay in order to have cleaner statements and more readable proofs. The skeptical reader can rephrase the statement as follows: For all $M, N \in \Lambda$ such that $\{x_1, \ldots, x_n\} = \mathrm{FV}(MN)$ and fresh variables $s, y_1, \ldots, y_n$, we have $\mathcal{B} \vdash \Xi_\iota\ulcorner\lambda\vec{x}.M\urcorner\ulcorner\lambda\vec{x}.N\urcorner s\vec{y} = \Xi_\iota\ulcorner\lambda\vec{x}.N\urcorner\ulcorner\lambda\vec{x}.M\urcorner s\vec{y}$.

PROOF. (i) We prove by coinduction that $\mathrm{BT}(\Xi_\iota\ulcorner M\urcorner\ulcorner N\urcorner[\eta_n]_{n\in\mathbb{N}}) = \mathrm{BT}(N)$.

The only non-trivial case is when both M and N are solvable, whence we have

$$M \twoheadrightarrow_h \lambda\vec{x}.x_j M_1 \cdots M_k, \qquad N \twoheadrightarrow_h \lambda\vec{x}z_1\ldots z_m.x_j N_1 \cdots N_k Q_1 \cdots Q_m,$$

where each $z_\ell \notin \mathrm{FV}(\mathrm{BT}(x_j\vec{M}\vec{N}))$, $\lambda z_\ell.Q_\ell \in \mathcal{I}^\eta$ for all $\ell \leq m$ and $M_i \sqsubseteq^\eta N_i$ for all $i \leq k$. In particular $M \leq_h N$ holds, so the first condition of Definition 11.37 applies.

From $\lambda z_\ell.Q_\ell \in \mathcal{I}^\eta$ it follows that $\lambda y\vec{z}.y\vec{Q} \in \mathcal{I}^\eta$, therefore $\iota(\#(\lambda y\vec{z}.y\vec{Q})^{\mathrm{nf}}) = q$ for some index q. Setting $\Upsilon_i \triangleq \Xi_\iota\ulcorner\lambda\vec{x}.M_i\urcorner\ulcorner\lambda\vec{x}.N_i\urcorner[\eta_n]_{n\in\mathbb{N}}$, easy calculations give:

$$\begin{aligned}
\Xi_\iota\ulcorner M\urcorner\ulcorner N\urcorner[\eta_n]_{n\in\mathbb{N}} &=_\beta \lambda\vec{x}.\pi_q[\eta_n]_{n\in\mathbb{N}}(x_j(\Upsilon_1\vec{x})\cdots(\Upsilon_k\vec{x})) \\
&=_\beta \lambda\vec{x}.(\lambda y\vec{z}.y\vec{Q})(x_j(\Upsilon_1\vec{x})\cdots(\Upsilon_k\vec{x})) \\
&=_\beta \lambda\vec{x}\vec{z}.x_j(\Upsilon_1\vec{x})\cdots(\Upsilon_k\vec{x})Q_1\cdots Q_m.
\end{aligned}$$

We conclude as, by the co-IH, we have $\mathcal{B} \vdash \Upsilon_i = \lambda\vec{x}.N_i$ for all $i \leq k$.

(ii) Analogous to the proof of (i) once observed that ι only depends on the enumeration $\vec{\eta}$. Indeed, given $Q \in \mathcal{I}^\eta \cap \Lambda^o$, $\iota(\#Q)$ still provides an index q such that

$$\pi_q[\eta_n]_{n\in\mathbb{N}} =_\beta Q,$$

but when applied to $[\mathsf{I}]_{n\in\mathbb{N}}$ it necessarily gives $\pi_q[\mathsf{I}]_{n\in\mathbb{N}} =_\beta \mathsf{I}$. □

The λ-theories $\mathcal{B}\omega$ and $\mathcal{H}^+$ are equal

We are a few steps away from proving that the λ-theories $\mathcal{B}\omega$ and $\mathcal{H}^+$ coincide. The first[8] step is to verify that the combinator Ξ_ι, when applied to $\mathcal{H}^+$-equivalent λ-terms, actually computes an η-upper bound of their Böhm trees.

PROPOSITION 11.41. *For all* $M, N \in \Lambda^o$, *the following are equivalent:*

1. $M \sqsubseteq^\eta \Xi_\iota\ulcorner M\urcorner\ulcorner N\urcorner[\eta_n]_{n\in\mathbb{N}} \sqsupseteq^\eta N$.

2. $\mathcal{H}^+ \vdash M = N$.

PROOF. $(1 \Rightarrow 2)$ It follows directly from Proposition 11.9.

$(2 \Rightarrow 1)$ By Proposition 11.9, we known that there is a Böhm-like tree T such that $\mathrm{BT}(M) \leq^\eta T \geq^\eta \mathrm{BT}(N)$. As usual, we proceed by coinduction on the Böhm(-like) trees. The only non-trivial case is when M, N are both solvable. In this case we have, say:

$$M \twoheadrightarrow_h \lambda\vec{x}.x_j M_1 \cdots M_k, \qquad N \twoheadrightarrow_h \lambda\vec{x}z_1\ldots z_m.x_j N_1 \cdots N_k Q_1 \cdots Q_m,$$

$$T = \lambda\vec{x}z_1\ldots z_m\ldots z_{m'}.x_j$$

$$T_1 \;\cdots\; T_k \; Q'_1 \;\cdots\; Q'_m \;\cdots\; Q'_{m'}$$

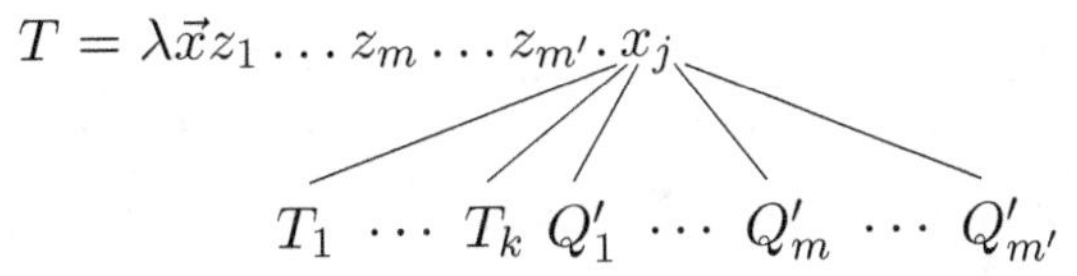

such that

---

[8] We number the steps as in the outline of the proof given on Page 327.

- $z_1, \ldots, z_{m'} \notin \mathrm{FV}(x_j \mathrm{BT}(M_1) \cdots \mathrm{BT}(M_k) T_1 \cdots T_k)$,
- $z_{m+1}, \ldots, z_{m'} \notin \mathrm{FV}(x_j \mathrm{BT}(N_1) \cdots \mathrm{BT}(N_k) Q_1 \cdots Q_m T_1 \cdots T_k Q'_1 \cdots Q'_m)$,
- $\mathrm{BT}(M_i) \leq^\eta T_i$ and $\mathrm{BT}(N_i) \leq^\eta T_i$, for all $i \leq k$,
- $Q_\ell \leq^\eta Q'_\ell$, for all $\ell \leq m$,
- and $\lambda z_{\ell'}.Q'_{\ell'} \in \mathcal{I}^\eta$, for all $\ell' > m$.

By Lemma 11.3, we have $Q'_\ell \twoheadrightarrow_{\beta\eta} z_\ell$ so $Q_\ell \leq^\eta Q'_\ell$ entails $\lambda z_\ell.Q_\ell \in \mathcal{I}^\eta$, hence $M \leq_h N$. Setting $q \triangleq \iota(\#((\lambda y z_1 \ldots z_m.yQ_1 \cdots Q_m)^{\mathtt{nf}}))$ and $\Upsilon_i \triangleq \Xi_\iota \ulcorner \lambda\vec{x}.M_i \urcorner \ulcorner \lambda\vec{x}.N_i \urcorner [\eta_n]_{n\in\mathbb{N}}$, we get:

$$\begin{aligned}\Xi_\iota \ulcorner M \urcorner \ulcorner N \urcorner [\eta_n]_{n\in\mathbb{N}} &=_\beta \lambda\vec{x}.\pi_q[\eta_n]_{n\in\mathbb{N}}(x_j(\Upsilon_1\vec{x})\cdots(\Upsilon_k\vec{x}))\\ &=_\beta \lambda\vec{x}.(\lambda y\vec{z}.yQ_1\cdots Q_m)(x_j(\Upsilon_1\vec{x})\cdots(\Upsilon_k\vec{x}))\\ &=_\beta \lambda\vec{x}\vec{z}.x_j(\Upsilon_1\vec{x})\cdots(\Upsilon_k\vec{x})Q_1\cdots Q_m.\end{aligned}$$

This case follows from the co-IH since $\lambda\vec{x}.M_i \sqsubseteq^\eta \Upsilon_i \sqsupseteq^\eta \lambda\vec{x}.N_i$ for all $i \leq k$. The symmetric case $N <_h M$ is analogous, using Lemma 11.39 to apply the co-IH. □

REMARK 11.42. It is easy to check that in the proof above

$$T \triangleq \mathrm{BT}(\Xi_\iota \ulcorner M \urcorner \ulcorner N \urcorner [\eta_n]_{n\in\mathbb{N}})$$

is minimally η-expanded, so T is actually the η-supremum of $\mathrm{BT}(M)$ and $\mathrm{BT}(N)$.

The second step is to show that the streams $[\mathsf{I}]_{n\in\mathbb{N}}$ and $[\eta_n]_{n\in\mathbb{N}}$ are equated in $\mathcal{B}\omega$. To prove this result, we are going to use the following auxiliary streams.

DEFINITION 11.43. Define the streams:

- $\langle\!\langle \mathsf{I} \rangle\!\rangle^\Omega\, yx \triangleq [yx, y\Omega x, y\Omega^{\sim 2}x, y\Omega^{\sim 3}x, y\Omega^{\sim 4}x, \ldots]$,
- $\langle\!\langle \eta \rangle\!\rangle^\Omega\, yx \triangleq [y(\eta_0 x), y\Omega(\eta_1 x), y\Omega^{\sim 2}(\eta_2 x), y\Omega^{\sim 3}(\eta_3 x), y\Omega^{\sim 4}(\eta_4 x), \ldots]$,

whose Böhm trees are given in Figure 11.6.

The streams $\langle\!\langle \mathsf{I} \rangle\!\rangle^\Omega$ and $\langle\!\langle \eta \rangle\!\rangle^\Omega$ are equal in $\mathcal{B}\omega$, for the same reason the λ-terms P, Q of Figure 11.4 are. The formal argument is the following.

LEMMA 11.44. $\mathcal{B}\omega \vdash \langle\!\langle \mathsf{I} \rangle\!\rangle^\Omega = \langle\!\langle \eta \rangle\!\rangle^\Omega$.

PROOF. Let $M \in \Lambda^o$, by Lemma 1.26 there is $k \geq 0$ such that $M\Omega^{\sim k} =_{\mathcal{B}} \Omega$. So we have:

$$\begin{aligned}\langle\!\langle \mathsf{I} \rangle\!\rangle^\Omega M &=_{\mathcal{B}} \lambda x.[Mx, M\Omega x, \ldots, M\Omega^{\sim k-1}x, \Omega, \Omega, \Omega, \ldots]\\ &=_{\mathcal{B}} \lambda x.[M(\mathsf{I}x), M\Omega(\mathsf{I}x), \ldots, M\Omega^{\sim k-1}(\mathsf{I}x), \Omega, \Omega, \Omega, \ldots]\\ &=_{\lambda\eta} \lambda x.[M(\eta_0 x), M\Omega(\eta_1 x), \ldots, M\Omega^{\sim k-1}(\eta_{k-1}x), \Omega, \Omega, \Omega, \ldots]\\ &=_{\mathcal{B}} \langle\!\langle \eta \rangle\!\rangle^\Omega M,\end{aligned}$$

where the third equality follows from $\mathsf{I} =_{\beta\eta} \eta_i$, for all $i \in \mathbb{N}$. Since M is an arbitrary closed λ-term, we can apply the ω-rule and conclude that $\mathcal{B}\omega \vdash \langle\!\langle \mathsf{I} \rangle\!\rangle^\Omega = \langle\!\langle \eta \rangle\!\rangle^\Omega$. □

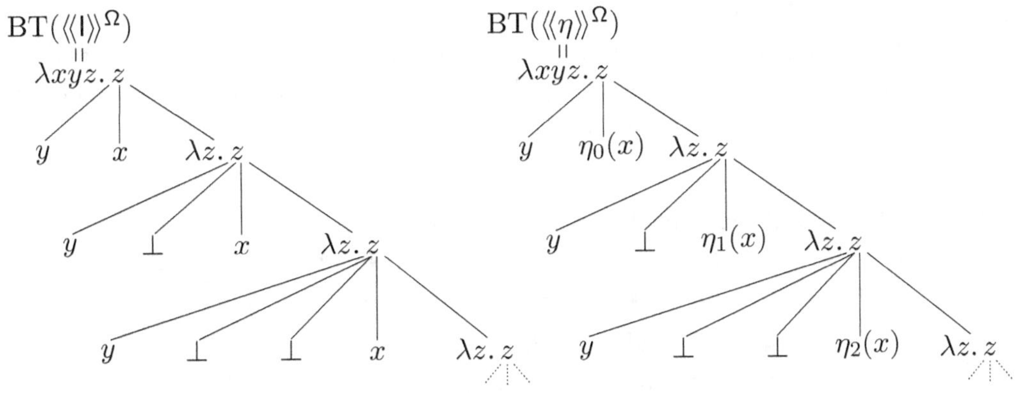

Figure 11.6: The Böhm trees of $\langle\!\langle \mathsf{I} \rangle\!\rangle^\Omega$ and $\langle\!\langle \eta \rangle\!\rangle^\Omega$ where we set $\eta_n(x) \triangleq \mathrm{BT}(\eta_n x)$.

As the variable y occurs in head-position in the λ-terms occurring in the stream $\langle\!\langle \eta \rangle\!\rangle^\Omega yx$ (resp. $\langle\!\langle \mathsf{I} \rangle\!\rangle^\Omega yx$), we can substitute for it a suitably modified projection that erases the Ω's and returns the n-th occurrence of x in $\langle\!\langle \mathsf{I} \rangle\!\rangle^\Omega$ (resp. $(\eta_n x)$ in $\langle\!\langle \eta \rangle\!\rangle^\Omega$).

LEMMA 11.45. *There exists a closed λ-term* Pr *such that, for all* $n \in \mathbb{N}$*:*

$$\mathsf{Pr}\,\mathsf{c}_n \langle\!\langle \mathsf{I} \rangle\!\rangle^\Omega =_\beta \mathsf{I}, \qquad \mathsf{Pr}\,\mathsf{c}_n \langle\!\langle \eta \rangle\!\rangle^\Omega =_\beta \eta_n.$$

PROOF. Let Pr be a combinator satisfying the recursive equation

$$\mathsf{Pr}\, n\, s =_\beta \mathsf{ifz}(n, \lambda z.s\mathsf{I}z\mathsf{K}, \mathsf{Pr}\,(\mathsf{S}^- n)\,(\lambda zw.s(\mathsf{K}z)w\mathsf{F})).$$

By induction on n, we show $\mathsf{Pr}\,\mathsf{c}_n(\lambda yx.[y\Omega^{\sim i}(\eta_{i+k}x)]_{i\in\mathbb{N}}) =_{\mathcal{B}} \eta_{n+k}$ for all $n, k \in \mathbb{N}$. Note that $\eta_i \in \mathcal{I}^\eta$ entails $\lambda z.\eta_i z =_\beta \eta_i$.
If $n = 0$ then $\mathsf{Pr}\,\mathsf{c}_0(\lambda yx.[y\Omega^{\sim i}(\eta_{i+k}x)]_{i\in\mathbb{N}})$ β-reduces to

$$\begin{array}{llll} & \lambda z.(\lambda yx.[y\Omega^{\sim i}(\eta_{i+k}x)]_{i\in\mathbb{N}})\mathsf{I}z\mathsf{K} & =_\beta & \lambda z.[\Omega^{\sim i}(\eta_{i+k}z)]_{i\in\mathbb{N}}\mathsf{K} \\ =_\beta & \lambda z.\mathsf{K}(\eta_k z)[\Omega^{\sim i+1}(\eta_{i+k+1}z)]_{i\in\mathbb{N}} & =_\beta & \lambda z.\eta_k z \qquad =_\beta \eta_k. \end{array}$$

If $n > 0$ then we have $\mathsf{Pr}\,\mathsf{c}_n(\lambda yx.[y\Omega^{\sim i}(\eta_{i+k}x)]_{i\in\mathbb{N}}) =_\beta \mathsf{Pr}\,\mathsf{c}_{n-1}(\lambda yx.[y\Omega^{\sim i}(\eta_{i+k+1}x)]_{i\in\mathbb{N}})$ $=_{\text{(by the IH)}} \eta_{n-1+k+1} = \eta_{n+k}$. Indeed, easy calculations give:

$$\begin{array}{llll} & \lambda zw.(\lambda yx.[y\Omega^{\sim i}(\eta_{i+k}x)]_{i\in\mathbb{N}})(\mathsf{K}z)w\mathsf{F} & =_\beta & \lambda zw.[\mathsf{K}z(\eta_k w), [\mathsf{K}z\Omega^{\sim i+1}(\eta_{i+k+1}w)]_{i\in\mathbb{N}}]\mathsf{F} \\ =_{\mathcal{B}} & \lambda zw.[z, [z\Omega^{\sim i}(\eta_{i+k+1}w)]_{i\in\mathbb{N}}]\mathsf{F} & =_\beta & \lambda zw.\mathsf{F}z[z\Omega^{\sim i}(\eta_{i+k+1}w)]_{i\in\mathbb{N}} \\ =_\beta & \lambda zw.[z\Omega^{\sim i}(\eta_{i+k+1}w)]_{i\in\mathbb{N}} & =_\alpha & \lambda yx.[y\Omega^{\sim i}(\eta_{i+k+1}x)]_{i\in\mathbb{N}}. \end{array}$$

Analogous calculations show $\mathsf{Pr}\,\mathsf{c}_n\langle\!\langle \mathsf{I} \rangle\!\rangle^\Omega =_{\mathcal{B}} \mathsf{I}$. □

COROLLARY 11.46. $\mathcal{B}\omega \vdash [\mathsf{I}]_{n\in\mathbb{N}} = [\eta_n]_{n\in\mathbb{N}}$.

PROOF. From Lemmas 11.45, 2.17 and 11.44 we get:

$$[\mathsf{I}]_{n\in\mathbb{N}} =_{\mathcal{B}} [\mathsf{Pr}\,\mathsf{c}_n\langle\!\langle \mathsf{I} \rangle\!\rangle^\Omega]_{n\in\mathbb{N}} =_{\mathcal{B}\omega} [\mathsf{Pr}\,\mathsf{c}_n\langle\!\langle \eta \rangle\!\rangle^\Omega]_{n\in\mathbb{N}} =_{\mathcal{B}} [\eta_n]_{n\in\mathbb{N}}.$$ □

We have seen that, when $M \sqsubseteq^\eta N$ holds, the λ-term $\Xi_\iota \ulcorner M \urcorner \ulcorner N \urcorner$ computes the Böhm tree of N from $[\eta_n]_{n \in \mathbb{N}}$ (Lemma 11.40(i)) and the Böhm tree of M from $[\mathsf{I}]_{n \in \mathbb{N}}$ (Lemma 11.40(ii)), but now we have proved that $[\eta_n]_{n \in \mathbb{N}} =_{\mathcal{B}\omega} [\mathsf{I}]_{n \in \mathbb{N}}$. As a consequence, we get that M and N are equal in $\mathcal{B}\omega$.

THEOREM 11.47. $\mathcal{B}\omega = \mathcal{H}^+$.

PROOF. ($\subseteq$) This inclusion was shown in Theorem 11.31.

($\supseteq$) By Remark 3.2, it is enough to consider $M, N \in \Lambda^o$. If $\mathcal{H}^+ \vdash M = N$, then by Proposition 11.41 we have $M \sqsubseteq^\eta P \sqsupseteq^\eta N$ for $P \triangleq \Xi_\iota \ulcorner M \urcorner \ulcorner N \urcorner [\eta_n]_{n \in \mathbb{N}}$. Then we have:

$$\begin{array}{llll} M & =_{\mathcal{B}} & \Xi_\iota \ulcorner M \urcorner \ulcorner P \urcorner [\mathsf{I}]_{n \in \mathbb{N}}, & \text{by Lemma 11.40(ii)}, \\ & =_{\mathcal{B}\omega} & \Xi_\iota \ulcorner M \urcorner \ulcorner P \urcorner [\eta_n]_{n \in \mathbb{N}}, & \text{by Corollary 11.46}, \\ & =_{\mathcal{B}} & P, & \text{by Lemma 11.40(i)}, \\ & =_{\mathcal{B}} & \Xi_\iota \ulcorner N \urcorner \ulcorner P \urcorner [\eta_n]_{n \in \mathbb{N}}, & \text{by Lemma 11.40(i)}, \\ & =_{\mathcal{B}\omega} & \Xi_\iota \ulcorner N \urcorner \ulcorner P \urcorner [\mathsf{I}]_{n \in \mathbb{N}}, & \text{by Corollary 11.46}, \\ & =_{\mathcal{B}} & N, & \text{by Lemma 11.40(ii)}. \end{array}$$

We conclude that $\mathcal{B}\omega \vdash M = N$. □

This theorem disproves Sallé's conjecture, so the next theorem should substitute Theorem 17.4.16 in B[1984], where $\mathcal{H}^+$ is denoted by $\mathcal{T}_{\mathrm{NF}}$.

THEOREM 11.48. *The following diagram indicates all possible inclusion relations of the λ-theories involved (if $\mathcal{T}_1$ is above $\mathcal{T}_2$, then $\mathcal{T}_1 \subsetneq \mathcal{T}_2$):*

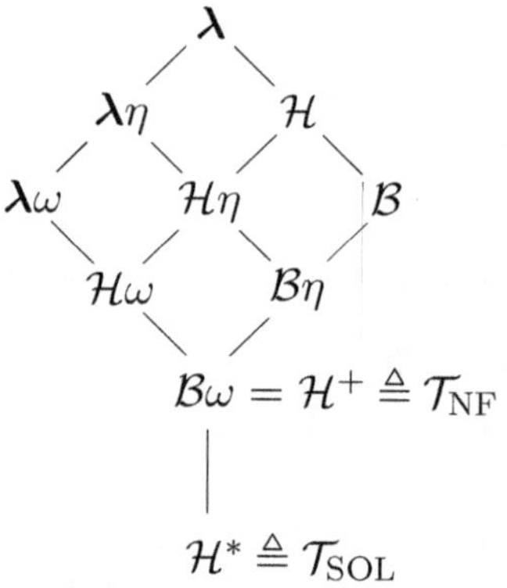

Figure 11.7: Kite revisited.

We conclude this section by discussing the legitimate question of whether it is possible to exploit the λ-term Ξ_ι, or a variation of it, in order to construct an η-supremum that requires possibly infinite η-expansions. The situation we have in mind is given by two λ-terms M and N that are equal in $\mathcal{H}^*$, but different in $\mathcal{H}^+$. Indeed, we know that such an η-supremum is λ-definable (B[1984], Exercise 10.6.7) and—as $\mathscr{T}_{\mathsf{rec}}$ is an r.e. set—so is the stream $[\mathsf{J}_t]_{t \in \mathscr{T}_{\mathsf{rec}}}$ of all possibly infinite η-expansion of the identity. Unfortunately, the function mapping $Q \in \mathcal{I}^\eta_\infty$ to its index in $[\mathsf{J}_t]_{t \in \mathscr{T}_{\mathsf{rec}}}$ is not computable whence Lemma 11.34, and the whole approach, do not generalize to the infinitary case.

11.4 A characterization of $\mathcal{B}\eta$

We have seen in Section 11.1, that $\mathcal{B}\eta$ is strictly included in $\mathcal{B}\omega$ because the λ-terms P, Q from Figure 11.4 are equal in $\mathcal{B}\omega$ but different in $\mathcal{B}\eta$. The proof-technique described in the previous sections opens the way to provide a characterization of $\mathcal{B}\eta$ in terms of equality between Böhm trees up to countably many η-expansions of *bounded* size.

Bounded η-expansions of Böhm trees

We define the *tree size* and the *weight-for-height* of a finite η-expansion Q of the identity. Both quantities are defined looking at the Böhm tree of Q: the tree size is defined as the usual size of a tree, while the weight-for-height as the maximum between the height and the maximal number of branchings in the tree.

DEFINITION 11.49. (i) The *tree size* of $Q \in \mathcal{I}^\eta$, written $\mathrm{tsize}(Q)$, is given by:

$$\mathrm{tsize}(Q) \triangleq \begin{cases} 1, & \text{if } Q^{\mathrm{nf}} = \mathsf{I}, \\ 1 + \sum_{i=1}^{m} \mathrm{tsize}(\lambda z_i.Q_i), & \text{if } Q^{\mathrm{nf}} = \lambda y z_1 \dots z_m.yQ_1 \cdots Q_m. \end{cases}$$

(ii) The *weight-for-height* of $Q \in \mathcal{I}^\eta$, written $\mathrm{wh}(Q)$, is defined as follows:

$$\mathrm{wh}(Q) \triangleq \begin{cases} 0, & \text{if } Q^{\mathrm{nf}} = \mathsf{I}, \\ \max\{m, \max_{i \leq m}\{\mathrm{wh}(\lambda z_i.Q_i)\} + 1\}, & \text{if } Q^{\mathrm{nf}} = \lambda y z_1 \dots z_m.yQ_1 \cdots Q_m. \end{cases}$$

(iii) For $p \in \mathbb{N}$, let $\mathcal{I}^\eta_p \triangleq \{Q \in \mathcal{I}^\eta \mid \mathrm{wh}(Q) < p\}$ be the set of *η-expansions of the identity whose weight-for-height is bounded by p.*

For example, we have $\mathrm{tsize}(1) = 2$ and $\mathrm{wh}(1) = 1$. By definition, two β-convertible Q and Q' have the same tree size and weight-for-height. We show some properties of the weight-for-height, as the tree size is more canonical and better understood.

EXAMPLES 11.50. (i) For each $n \in \mathbb{N}$ the weight-for-height of 1^n is $\mathrm{wh}(1^n) = n$.
(ii) $\mathcal{I}^\eta_0 = \emptyset$, $\mathcal{I}^\eta_1 = \{\mathsf{I}\}^\beta$, $\mathcal{I}^\eta_2 = \{\mathsf{I}, 1\}^\beta$. Moreover, setting $2 \triangleq \lambda xyz.xyz$, we have:

$$\mathcal{I}^\eta_3 = \left\{ \begin{array}{l} \mathsf{I},\, 1,\, \lambda xyz.x(1y)z,\, \lambda xyz.xy(1z),\, \lambda xyz.x(1y)(1z), \\ 1^2, 2,\, \lambda xyz.x(2y)z,\, \lambda xyz.xy(2z),\, \lambda xyz.x(2y)(2z) \end{array} \right\}^\beta$$

and so on. More generally, $\mathcal{I}^\eta_p \subsetneq \mathcal{I}^\eta_{p+1}$ and the sequence $\left(\mathcal{I}^\eta_p\right)_{p \in \mathbb{N}}$ is increasing.

The next lemma is useful for studying the behavior of Ξ_ι, when applied to the codes of λ-terms whose Böhm trees differ because of η-expansions of bounded weight-for-height.

LEMMA 11.51. *Let $m, p \in \mathbb{N}$ and $Q_1, \dots, Q_m \in \Lambda$. If $m \leq p$ and $\lambda z_i.Q_i \in \mathcal{I}^\eta_p$ for all $i \leq m$, then $\lambda y z_1 \dots z_m.yQ_1 \cdots Q_m \in \mathcal{I}^\eta_{p+1}$.*

PROOF. Since $m \leq p$ and $\mathrm{wh}(\lambda z_i.Q_i) < p$ for each $i \leq m$, it follows that

$$\begin{array}{rcl} \mathrm{wh}(\lambda y z_1 \dots z_m.yQ_1 \cdots Q_m) & = & \max\{m, \max\{\mathrm{wh}(\lambda z_1.Q_1), \dots, \mathrm{wh}(\lambda z_m.Q_m)\} + 1\} \\ & < & p + 1. \end{array}$$

□

We now specify when two Böhm-like trees T, T' are such that T' is an η-expansion of T having some quantity bounded by p. This quantity is obtained by applying a generic function f, but it will be subsequently instantiated either to the tree size or to the weight-for-height.

DEFINITION 11.52. (i) Let $p \in \mathbb{N}$ and $f : \mathcal{I}^\eta \to \mathbb{N}$. Define the greatest relation $\leq^\eta_{(f,p)}$ between Böhm-like trees such that $T \leq^\eta_{(f,p)} T'$ entails that:

- either $T = T' = \bot$,
- or there exists $m \in \mathbb{N}$ such that $m \leq p$ and (for some $n, k \geq 0$) we have:

$$T = \begin{array}{c} \lambda x_1 \dots x_n.y \\ \diagup \quad \diagdown \\ T_1 \ \cdots \ T_k \end{array} \quad \text{and} \quad T' = \begin{array}{c} \lambda x_1 \dots x_n z_1 \dots z_m.y \\ \diagup \quad \diagdown \\ T'_1 \ \cdots \ T'_k \ Q_1 \ \cdots \ Q_m \end{array}$$

where $z_\ell \notin \mathrm{FV}(y T_1 \cdots T_k T'_1 \cdots T'_k)$ for all $\ell \leq m$, $\lambda z_1.Q_i \in \mathcal{I}^\eta$ and $f(\lambda z_i.Q_i) < p$ for all $i \leq m$ and $T_j \leq^\eta_{(f,p)} T'_j$ for all $j \leq k$.

(ii) For $M, N \in \Lambda$, we set $M \sqsubseteq^\eta_{(f,p)} N$ if and only if $\mathrm{BT}(M) \leq^\eta_{(f,p)} \mathrm{BT}(N)$.

REMARK 11.53. In the definition of $\leq^\eta_{(f,p)}$ above we verify not only that $f(\lambda z_i.Q_i)$ is bounded by p, but also that the number m of Q_i's is bounded by p. Notice the asymmetry between the strict bound $f(\lambda z_i.Q_i) < p$ and the bound $m \leq p$.

EXAMPLES 11.54. Consider the streams $\langle\!\langle \mathsf{I} \rangle\!\rangle$, $\langle\!\langle 1 \rangle\!\rangle$ and $\langle\!\langle 1^* \rangle\!\rangle$ from Figure 11.3.

(i) $\langle\!\langle \mathsf{I} \rangle\!\rangle \sqsubseteq^\eta_{\mathrm{wh},1} \langle\!\langle 1 \rangle\!\rangle$ and $\langle\!\langle \mathsf{I} \rangle\!\rangle \sqsubseteq^\eta_{\mathrm{tsize},2} \langle\!\langle 1 \rangle\!\rangle$.

(ii) $\langle\!\langle \mathsf{I} \rangle\!\rangle \not\sqsubseteq^\eta_{\mathrm{tsize},1} \langle\!\langle 1 \rangle\!\rangle$.

(iii) $\langle\!\langle \mathsf{I} \rangle\!\rangle \not\sqsubseteq^\eta_{\mathrm{wh},p} \langle\!\langle 1^* \rangle\!\rangle$ and $\langle\!\langle \mathsf{I} \rangle\!\rangle \not\sqsubseteq^\eta_{\mathrm{tsize},p} \langle\!\langle 1^* \rangle\!\rangle$, for all $p \geq 0$.

For technical reasons, in the rest of the section we focus on the weight-for-height. However, by applying the following lemma, at the end of the chapter we draw conclusions for the more standard notion of size as well.

LEMMA 11.55. *For $T, T' \in \mathscr{B}$ the following are equivalent:*

1. *$T \leq^\eta_{(\mathrm{tsize},p)} T'$, for some $p \in \mathbb{N}$;*

2. *$T \leq^\eta_{(\mathrm{wh},p)} T'$, for some $p \in \mathbb{N}$.*

PROOF. A straightforward coinduction proves both ($1 \Rightarrow 2$) and ($2 \Rightarrow 1$).

For ($1 \Rightarrow 2$) observe that, for $Q \in \mathcal{I}^\eta$, $\mathrm{tsize}(Q) < p$ entails that both its height and its number of branchings are smaller than p, which in its turn implies $\mathrm{wh}(Q) < p$.

For ($2 \Rightarrow 1$) notice that the syntax tree of any β-normal $Q \in \mathcal{I}^\eta_{p+1}$ is a subtree of the complete p-ary tree of height p, therefore $\mathrm{tsize}(Q) < 2^{p+1}$. □

NOTATION. *Given two Böhm-like trees T, T', we simply write $T \leq^\eta_p T'$ for $T \leq^\eta_{\mathrm{wh},p} T'$. Similarly, given two λ-terms M, N, we write $M \sqsubseteq^\eta_p N$ for $M \sqsubseteq^\eta_{\mathrm{wh},p} N$.*

Some properties of bounded η-expansions

The properties below follow straightforwardly from the definitions.

LEMMA 11.56. *Let* $M, N \in \Lambda$ *and* $p \in \mathbb{N}$. *Suppose* $M \sqsubseteq^{\eta}_{p} N$. *Then:*

(i) $M \sqsubseteq^{\eta} N$,

(ii) $\mathcal{B} \vdash M = M'$ *and* $\mathcal{B} \vdash N = N'$ *entail* $M' \sqsubseteq^{\eta}_{p} N'$,

(iii) *for all* $p' \geq p$, *we have* $M \sqsubseteq^{\eta}_{p'} N$.

The next lemma describes the interaction between $\sqsubseteq^{\eta}_{p}$ and the weight-for-height wh$(-)$, the intuition being that $\sqsubseteq^{\eta}_{p}$ can increase the weight-for-height of $Q \in \mathcal{I}^{\eta}$ by at most p.

LEMMA 11.57. *Given* $Q, Q' \in \Lambda$ *and* $p, p' \in \mathbb{N}$, *we have*

$$\lambda y.Q \in \mathcal{I}^{\eta}_{p} \ \&\ Q \sqsubseteq^{\eta}_{p'} Q' \quad \Rightarrow \quad \lambda y.Q' \in \mathcal{I}^{\eta}_{p+p'}.$$

PROOF. Wlog, assume that Q is in β-normal form.

We proceed by structural induction on $Q = \lambda z_1 \dots z_m.yQ_1 \cdots Q_m$ (by Lemma 11.3).

From $\lambda y.Q \in \mathcal{I}^{\eta}_{p}$ we get that $m < p$ and the Q_i's are such that $\mathrm{wh}(\lambda z_i.Q_i) < p$. Since $Q \sqsubseteq^{\eta}_{p'} Q'$ holds we have $Q' = \lambda z_1 \dots z_m w_1 \dots w_{m'}.yQ'_1 \cdots Q'_{m+m'}$ where $m' \leq p'$, $Q_i \sqsubseteq^{\eta}_{p'} Q'_i$ for all $i \leq m$ and $\lambda w_j.Q'_j \in \mathcal{I}^{\eta}_{p'}$ for all $j > m$. Therefore $m + m' < p + p'$, $\mathrm{wh}(\lambda w_j.Q'_j) < p'$ for all $j > m$ and, by the IH, $\mathrm{wh}(\lambda z_i.Q'_i) < p + p'$ for all $i \leq m$. We conclude that $\mathrm{wh}(\lambda y.Q') < p + p'$, which entails $\lambda y.Q' \in \mathcal{I}^{\eta}_{p+p'}$. □

The relation $\sqsubseteq^{\eta}_{p}$ enjoys the following 'weighted' transitivity property.

LEMMA 11.58. *For all* $M, N \in \Lambda$, $p_1, p_2 \in \mathbb{N}$ *we have*

$$M \sqsubseteq^{\eta}_{p_1} P \ \&\ P \sqsubseteq^{\eta}_{p_2} N \quad \Rightarrow \quad M \sqsubseteq^{\eta}_{p_1+p_2} N.$$

PROOF. We proceed by coinduction on the Böhm trees of M, N, P. The only interesting case is when they are all solvable λ-terms. In this case, $M \sqsubseteq^{\eta}_{p_1} P$ entails

$$M \twoheadrightarrow_h \lambda\vec{x}.yM_1 \cdots M_k \text{ and } P \twoheadrightarrow_h \lambda\vec{x}z_1 \dots z_{m_1}.yP_1 \cdots P_k Q_1 \cdots Q_{m_1},$$

where $\vec{z} \notin \mathrm{FV}(\mathrm{BT}(y\vec{M}\vec{P}))$, $m_1 \leq p_1$, $\lambda z_i.Q_i \in \mathcal{I}^{\eta}_{p_1}$ for $i \leq m_1$ and $M_j \sqsubseteq^{\eta}_{p_1} P_j$ for $j \leq k$.

From $P \sqsubseteq^{\eta}_{p_2} N$ we obtain, for some $\vec{w} \notin \mathrm{FV}(\mathrm{BT}(y\vec{P}\vec{Q}\vec{N}\vec{Q}'))$:

$$N \twoheadrightarrow_h \lambda\vec{x}z_1 \dots z_{m_1} w_1 \dots w_{m_2}.yN_1 \cdots N_k Q'_1 \cdots Q'_{m_1} Q''_1 \cdots Q''_{m_2}$$

with $m_2 \leq p_2$, $P_j \sqsubseteq^{\eta}_{p_2} N_j$ for $j \leq k$, $Q_i \sqsubseteq^{\eta}_{p_2} Q'_i$ for $i \leq m_1$ and $\lambda w_\ell.Q''_\ell \in \mathcal{I}^{\eta}_{p_2}$ for $\ell \leq m_2$. Note that $m_1 \leq p_1$ and $m_2 \leq p_2$ imply $m_1 + m_2 \leq p_1 + p_2$, and that $\mathcal{I}^{\eta}_{p_1} \cup \mathcal{I}^{\eta}_{p_2} \subseteq \mathcal{I}^{\eta}_{p_1+p_2}$. From $\lambda z_i.Q_i \in \mathcal{I}^{\eta}_{p_1}$ and $Q_i \sqsubseteq^{\eta}_{p_2} Q'_i$ we obtain by Lemma 11.57 that $\lambda z_i.Q'_i \in \mathcal{I}^{\eta}_{p_1+p_2}$ for all $i \leq m_1$. Since, for all $j \leq k$, $M_j \sqsubseteq^{\eta}_{p_1} P_j$ and $P_j \sqsubseteq^{\eta}_{p_2} N_j$ we conclude that $M_j \sqsubseteq^{\eta}_{p_1+p_2} N_j$ holds by applying the co-IH. □

The following is an easy corollary of Lemma 11.6.

COROLLARY 11.59. *If $M \twoheadrightarrow_\eta N$ then there exists a bound $p \in \mathbb{N}$ such that $N \sqsubseteq^\eta_p M$.*

PROOF. We perform an induction loading and prove that if the reduction sequence $M \twoheadrightarrow_\eta N$ has length p, then $N \sqsubseteq^\eta_p M$ holds.

If $p = 0$ then $\mathrm{BT}(M) = \mathrm{BT}(N)$ so we have $N \sqsubseteq^\eta_0 M$.

If $p > 0$ then $M \twoheadrightarrow_\eta N' \to_\eta N$ where $M \twoheadrightarrow_\eta N'$ has length $p-1$. From Lemma 11.6, $N \sqsubseteq^\eta_1 N'$ since at every position α of their Böhm trees the lengths of the abstractions '$\lambda x_1 \dots x_n.$' and '$\lambda x_1 \dots x_n z.$' differ at most by 1, and $\lambda z.z \in \mathcal{I}^\eta_1$. By the IH we have $N' \sqsubseteq^\eta_{p-1} M$, so we conclude by applying Lemma 11.58. □

The behavior of Ξ_ι on bounded η-expansions

Recall that in Definition 11.32 we have fixed an effective enumeration $\vec{\eta} = (\eta_0, \eta_1, \dots)$ of the set $\mathcal{I}^\eta$, together with the corresponding stream $[\eta_n]_{n \in \mathbb{N}} = [\eta_0, \eta_1, \eta_2, \eta_3, \dots]$. Moreover, we consider fixed a map ι satisfying the properties of Lemma 11.34.

By Proposition 11.41, given two λ-terms M and N whose Böhm trees differ because of countably many η-expansions, the λ-term $\Xi_\iota \ulcorner M \urcorner \ulcorner N \urcorner$ builds their η-supremum from the stream $[\eta_n]_{n \in \mathbb{N}}$. We now prove that when the weight-for-height of such η-expansions is bounded by p, then the Böhm trees of M, N also differ from $\mathrm{BT}(\Xi_\iota \ulcorner M \urcorner \ulcorner N \urcorner [\eta_n]_{n \in \mathbb{N}})$ because of η-expansions whose weight-for-height is bounded by p.

LEMMA 11.60. *Let $M, N, P \in \Lambda^o$. If $M \sqsupseteq^\eta_p P \sqsubseteq^\eta_p N$ then:*

$$M \sqsubseteq^\eta_p \Xi_\iota \ulcorner M \urcorner \ulcorner N \urcorner [\eta_n]_{n \in \mathbb{N}} \sqsupseteq^\eta_p N.$$

PROOF. We proceed by coinduction on their Böhm trees.

If P is unsolvable, then also M, N and $\Xi_\iota \ulcorner M \urcorner \ulcorner N \urcorner$ are unsolvable and we are done.

Otherwise $P \twoheadrightarrow_h \lambda \vec{x}.x_j P_1 \cdots P_k$ and from $P \sqsubseteq^\eta_p M$ we obtain for some $m \leq p$:

$$M \twoheadrightarrow_h \lambda \vec{x} z_1 \dots z_m.x_j M_1 \cdots M_k Q_1 \cdots Q_m,$$

where $P_i \sqsubseteq^\eta_p M_i$ for $i \leq k$ and $\lambda z_\ell.Q_\ell \in \mathcal{I}^\eta_p$ for $\ell \leq m$. From $P \sqsubseteq^\eta_p N$ we get, say,

$$N \twoheadrightarrow_h \lambda \vec{x} z_1 \dots z_m w_1 \dots w_{m'}.x_j N_1 \cdots N_k Q'_1 \cdots Q'_{m+m'},$$

where $m + m' \leq p$, $P_i \sqsubseteq^\eta_p N_i$ for $i \leq k$, $\lambda z_\ell.Q'_\ell \in \mathcal{I}^\eta_p$ for $\ell \leq m$ and $\lambda w_\ell.Q'_{m+\ell} \in \mathcal{I}^\eta_p$ for $\ell \leq m'$, so we have $M \leq_h N$. (Notice that the case $N <_h M$ is symmetrical, one just needs to apply Lemma 11.39.) Before going further, we prove the following claim.

Claim 11.60.1. Let $Q, Q' \in \Lambda$, $\mathrm{FV}(QQ') \subseteq \{z_1, \dots, z_n\}$. If $\lambda z_\ell.Q, \lambda z_\ell.Q' \in \mathcal{I}^\eta_p$ for some ℓ then

$$Q \sqsubseteq^\eta_p \Xi_\iota \ulcorner \lambda \vec{z}.Q \urcorner \ulcorner \lambda \vec{z}.Q' \urcorner [\eta_n]_{n \in \mathbb{N}} \vec{z} \sqsupseteq^\eta_p Q'.$$

Subproof. Since Q, Q' are normalizing, we assume they are in β-normal form and proceed by structural induction. We have, say:

$$Q = \lambda \vec{y}.z_\ell Q_1 \cdots Q_m, \qquad Q' = \lambda \vec{y} \vec{w}.z_\ell Q'_1 \cdots Q'_{m+m'}$$

where $m + m' \leq p$, $\lambda y_i.Q_i, \lambda y_i.Q'_i \in \mathcal{I}^\eta_p$ for $i \leq m$ and $\lambda w_j.Q'_j \in \mathcal{I}^\eta_p$ for $j > m$.

We split into two subcases.

- If $m = 0$ then $Q = z_\ell$ and $\lambda\vec{z}.z_\ell \sqsubseteq^\eta_p \lambda\vec{z}.Q'$, therefore by Lemma 11.40(ii) we obtain that $\lambda\vec{z}.Q' =_\mathcal{B} \Xi_\iota\ulcorner\lambda\vec{z}.Q\urcorner\ulcorner\lambda\vec{z}.Q'\urcorner[\eta_n]_{n\in\mathbb{N}}$ and the case follows by applying the variables $\vec{z}$ to both sides.
- Otherwise $m > 0$, so for $q \triangleq \iota(\#(\lambda x\vec{w}.xQ'_{m+1}\cdots Q'_{m+m'}))$ we have:

$$\begin{array}{ll} & \Xi_\iota\ulcorner\lambda\vec{z}.Q\urcorner\ulcorner\lambda\vec{z}.Q'\urcorner[\eta_n]_{n\in\mathbb{N}}\vec{z} \\ =_\beta & \lambda\vec{y}.\pi_q[\eta_n]_{n\in\mathbb{N}}(z_\ell(\Upsilon_1\vec{z}\vec{y})\cdots(\Upsilon_m\vec{z}\vec{y})) \\ =_\beta & \lambda\vec{y}.(\lambda x\vec{w}.xQ'_{m+1}\cdots Q'_{m+m'})(z_\ell(\Upsilon_1\vec{z}\vec{y})\cdots(\Upsilon_m\vec{z}\vec{y})) \\ =_\beta & \lambda\vec{y}w.z_\ell(\Upsilon_1\vec{z}\vec{y})\cdots(\Upsilon_m\vec{z}\vec{y})Q'_{m+1}\cdots Q'_{m+m'}, \end{array}$$

where $\Upsilon_i \triangleq \Xi_\iota\ulcorner\lambda\vec{z}\vec{y}.Q_i\urcorner\ulcorner\lambda\vec{z}\vec{y}.Q'_i\urcorner[\eta_n]_{n\in\mathbb{N}}$ and the case follows by IH, thus concluding the proof of the claim. ■

We now continue with the main proof of the lemma.
For $q \triangleq \iota(\#((\lambda y\vec{w}.yQ'_{m+1}\cdots Q'_{m+m'})^{\mathrm{nf}}))$, we have that:

$$\begin{array}{ll} & \Xi_\iota\ulcorner M\urcorner\ulcorner N\urcorner[\eta_n]_{n\in\mathbb{N}} \\ =_\beta & \lambda\vec{x}\vec{z}.\pi_q[\eta_n]_{n\in\mathbb{N}}(x_j(\Upsilon_1\vec{x}\vec{z})\cdots(\Upsilon_k\vec{x}\vec{z})(\Upsilon'_1\vec{x}\vec{z})\cdots(\Upsilon'_m\vec{x}\vec{z})) \\ =_\beta & \lambda\vec{x}\vec{z}\vec{w}.x_j(\Upsilon_1\vec{x}\vec{z})\cdots(\Upsilon_k\vec{x}\vec{z})(\Upsilon'_1\vec{x}\vec{z})\cdots(\Upsilon'_m\vec{x}\vec{z})Q'_{m+1}\cdots Q'_{m+m'}, \end{array}$$

where $\Upsilon_i \triangleq \Xi_\iota\ulcorner\lambda\vec{x}\vec{z}.M_i\urcorner\ulcorner\lambda\vec{x}\vec{z}.N_i\urcorner[\eta_n]_{n\in\mathbb{N}}$ for $i \leq k$, and $\Upsilon'_\ell \triangleq \Xi_\iota\ulcorner\lambda\vec{x}\vec{z}.Q_\ell\urcorner\ulcorner\lambda\vec{x}\vec{z}.Q'_\ell\urcorner[\eta_n]_{n\in\mathbb{N}}$ for $\ell \leq m$. By co-IH, we obtain $M_i \sqsubseteq^\eta_p \Upsilon_i\vec{x}\vec{z} \sqsupseteq^\eta_p N_i$ and, by applying Claim 11.60.1, we get $Q_\ell \sqsubseteq^\eta_p \Upsilon'_\ell\vec{x}\vec{z} \sqsupseteq^\eta_p Q'_\ell$. Since $m \leq m + m' \leq p$ and $Q'_{m+\ell} \in \mathcal{I}^\eta_p$ we conclude that $M \sqsubseteq^\eta_p \Xi_\iota\ulcorner M\urcorner\ulcorner N\urcorner[\eta_n]_{n\in\mathbb{N}} \sqsupseteq^\eta_p N$ holds. □

A characterization of the equality in $\mathcal{B}\eta$

We will exploit the fact that, when $\mathcal{B}\eta \vdash M = N$, the term $\Xi_\iota\ulcorner M\urcorner\ulcorner N\urcorner$ only depends on a finite restriction of its input stream S and that all finite restrictions of $[\eta_n]_{n\in\mathbb{N}}$ and $[\mathsf{I}]_{n\in\mathbb{N}}$ of the same length are $\beta\eta$-convertible with each other (since they are finite).

DEFINITION 11.61. Let $S \triangleq [S_i]_{i\in\mathbb{N}}$ be a stream of λ-terms. Given $n \in \mathbb{N}$, define the *n-truncation of S* as the following (finitary) stream:

$$S\restriction_n \triangleq [S_0, S_1, \ldots, S_n, \Omega, \Omega, \Omega, \ldots].$$

As shown in Lemma 11.34, $\iota(\#Q)$ corresponds to the *smallest* index i such that Q occurs in $[\eta_n]_{n\in\mathbb{N}}$. The next property is a consequence of such a minimality condition.

LEMMA 11.62. *Let $p \in \mathbb{N}$. There exists an index n such that for every closed λ-term $Q \in \mathcal{I}^\eta_p$ we have $\iota(\#Q) \leq n$.*

PROOF. Since the weight-for-height of each $Q \in \mathcal{I}_p^\eta$ is bounded by p, it follows that the set $[\mathcal{I}_p^\eta]^{\mathrm{nf}} \triangleq \{Q^{\mathrm{nf}} \mid Q \in \mathcal{I}_p^\eta\}$ is finite. By Lemma 11.34, $\iota(\#Q)$ gives the smallest index i such that $Q =_\beta \eta_i$, therefore the set $\iota[\mathcal{I}_p^\eta] \triangleq \{\iota(\#Q) \mid Q \in \mathcal{I}_p^\eta \cap \Lambda^o\}$ is finite.

We conclude by taking $n \triangleq \max(\iota[\mathcal{I}_p^\eta])$. □

As a corollary we get that if $M \sqsubseteq_p^\eta N$ holds, then $\Xi_\iota\ulcorner M\urcorner\ulcorner N\urcorner$ only uses a finite portion of its input stream.

COROLLARY 11.63. *For all $M, N \in \Lambda^o$, if $M \sqsubseteq_p^\eta N$ then there exists an index $n \in \mathbb{N}$ such that for every stream $S = [S_i]_{i \in \mathbb{N}}$ we have:*

$$\mathcal{B} \vdash \Xi_\iota\ulcorner M\urcorner\ulcorner N\urcorner S = \Xi_\iota\ulcorner M\urcorner\ulcorner N\urcorner(S\lceil_n).$$

PROOF. By Lemma 11.62 there exists an $n \in \mathbb{N}$ such that, for every closed $Q \in \mathcal{I}_{p+1}^\eta$, we have $\iota(\#Q) \leq n$. We prove the statement by coinduction for that particular n.

If M or N are unsolvable, than so is $\Xi_\iota\ulcorner M\urcorner\ulcorner N\urcorner$ and we are done.

Otherwise, we have $M \twoheadrightarrow_h \lambda\vec{x}.x_j M_1 \cdots M_k$ and $N \twoheadrightarrow_h \lambda\vec{x}\vec{z}.x_j N_1 \cdots N_k Q_1 \cdots Q_m$ for $m \leq p$ and $\lambda z_\ell.Q_\ell \in \mathcal{I}_p^\eta$ for all $\ell \leq m$. By Lemma 11.51, the λ-term defined as $Q \triangleq (\lambda y\vec{z}.yQ_1 \cdots Q_m)^{\mathrm{nf}}$ belongs to the set $\mathcal{I}_{p+1}^\eta$.

Hence, for some index $q \triangleq \iota(\#Q) \leq n$, we have on the one side:

$$\begin{aligned}\Xi_\iota\ulcorner M\urcorner\ulcorner N\urcorner S &=_\beta \lambda\vec{x}.\pi_q S(x_j(\Upsilon_1\vec{x})\cdots(\Upsilon_k\vec{x}))\\ &=_\beta \lambda\vec{x}.S_q(x_j(\Upsilon_1\vec{x})\cdots(\Upsilon_k\vec{x}))\end{aligned}$$

where $\Upsilon_i \triangleq \Xi_\iota\ulcorner M_i\urcorner\ulcorner N_i\urcorner S$. On the other side, we have:

$$\begin{aligned}\Xi_\iota\ulcorner M\urcorner\ulcorner N\urcorner(S\lceil_n) &=_\beta \lambda\vec{x}.\pi_q(S\lceil_n)(x_j(\Upsilon'_1\vec{x})\cdots(\Upsilon'_k\vec{x}))\\ &=_\beta \lambda\vec{x}.S_q(x_j(\Upsilon'_1\vec{x})\cdots(\Upsilon'_k\vec{x}))\end{aligned}$$

where $\Upsilon'_i \triangleq \Xi_\iota\ulcorner M_i\urcorner\ulcorner N_i\urcorner(S\lceil_n)$. We can conclude since, by co-IH, $\mathrm{BT}(\Upsilon_i) = \mathrm{BT}(\Upsilon'_i)$ holds for all $i \leq k$. □

COROLLARY 11.64. *For all $M, N \in \Lambda^o$, if $M \sqsubseteq^\eta N$ then $\mathcal{B}\eta \vdash M = N$.*

PROOF. Assume $M \sqsubseteq^\eta N$. Then, for some $k \in \mathbb{N}$:

$$\begin{aligned}M &=_{\mathcal{B}} \Xi_\iota\ulcorner M\urcorner\ulcorner N\urcorner[\mathsf{I}]_{n \in \mathbb{N}}, && \text{by Lemma 11.40(ii)},\\ &=_{\mathcal{B}} \Xi_\iota\ulcorner M\urcorner\ulcorner N\urcorner(([\mathsf{I}]_{n \in \mathbb{N}})\lceil_k), && \text{by Corollary 11.63},\\ &=_{\mathcal{B}\eta} \Xi_\iota\ulcorner M\urcorner\ulcorner N\urcorner(([\eta_n]_{n \in \mathbb{N}})\lceil_k), && \text{by } \beta\eta\text{-conversion},\\ &=_{\mathcal{B}} \Xi_\iota\ulcorner M\urcorner\ulcorner N\urcorner[\eta_n]_{n \in \mathbb{N}}, && \text{by Corollary 11.63},\\ &=_{\mathcal{B}} N, && \text{by Lemma 11.40(i)}.\end{aligned}$$

We conclude that $\mathcal{B}\eta \vdash M = N$. □

Since $\mathcal{B}\eta$ is the join of $\mathcal{B}$ and $\boldsymbol{\lambda}\eta$, the equation $M =_{\mathcal{B}\eta} N$ holds exactly when there is an alternating sequence of shape $M =_{\mathcal{B}} M_0 =_\eta M_1 =_{\mathcal{B}} \cdots =_\eta M_k =_{\mathcal{B}} N$. Therefore, alternating sequences as the one in the next lemma naturally arise.

LEMMA 11.65. *Let $M_1, \dots, M_{k+1} \in \Lambda^o$ and let $p \in \mathbb{N}$. If, for some $N_1, \dots, N_k \in \Lambda^o$, there exists a zig-zag sequence such that:*

$$\begin{array}{ccccccccccccccc} M_1 & & & & M_2 & \cdots & M_{k-1} & & & & M_k & & & & M_{k+1} \\ & \sqsupseteq^\eta_p & & \sqsubseteq^\eta_p & & & & \sqsupseteq^\eta_p & & \sqsubseteq^\eta_p & & \sqsupseteq^\eta_p & & \sqsubseteq^\eta_p & \\ & & N_1 & & & N_2 & \cdots & & N_{k-1} & & & & N_k & & \end{array}$$

then there is $P \in \Lambda^o$ such that $M_1 \sqsubseteq^\eta_{kp} P \sqsupseteq^\eta_{kp} M_{k+1}$.

PROOF. We proceed by induction on k.

Case $k = 0$. Trivial, just take $P \triangleq M_1$.

Case $k > 0$. By applying Lemma 11.60 to each pair M_i and M_{i+1} we get, setting $N'_i \triangleq \Xi_\iota \ulcorner M_i \urcorner \ulcorner M_{i+1} \urcorner [\eta_n]_{n \in \mathbb{N}}$, the sequence:

$$M_1 \sqsubseteq^\eta_p N'_1 \sqsupseteq^\eta_p M_2 \cdots \sqsubseteq^\eta_p N'_k \sqsupseteq^\eta_p M_{k+1}.$$

As the subsequence $N'_1 \sqsupseteq^\eta_p M_2 \cdots \sqsubseteq^\eta_p N'_k$ is shorter and satisfies the hypotheses of the lemma, we get from the IH a λ-term $P \in \Lambda^o$ such that $N'_1 \sqsubseteq^\eta_{(k-1)p} P \sqsupseteq^\eta_{(k-1)p} N'_k$. We conclude by Lemma 11.58 since $(k-1)p + p = kp$. □

We are finally able to provide the following characterization of $\mathcal{B}\eta$, which confirms the informal intuition discussed by the first author in Section 16.4 of B[1984].

THEOREM 11.66 (INTRIGILA ET AL. (2019)).
For all $M, N \in \Lambda^o$, the following are equivalent:

1. *$\mathcal{B}\eta \vdash M = N$;*
2. *there exist $P \in \Lambda^o$ and $p \in \mathbb{N}$ such that $M \sqsubseteq^\eta_p P \sqsupseteq^\eta_p N$;*
3. *there exist $P \in \Lambda^o$ and $p \in \mathbb{N}$ such that $M \sqsubseteq^\eta_{(\mathsf{tsize},p)} P \sqsupseteq^\eta_{(\mathsf{tsize},p)} N$.*

PROOF. ($1 \Rightarrow 2$) Since $\mathcal{B}\eta$ is the join of two congruences, namely $=_\mathcal{B}$ and $=_\eta$, we have that $\mathcal{B}\eta \vdash M = N$ holds if and only if there are $M_0, \dots, M_k \in \Lambda^o$ such that

$$M =_\mathcal{B} M_0 =_\eta M_1 =_\mathcal{B} \cdots =_\eta M_k =_\mathcal{B} N$$

Since η-reduction is Church-Rosser and the M_j's are closed, for even indices i (as odd indices correspond to $=_\mathcal{B}$ steps), we have $M_i =_\eta M_{i+1}$ if and only if $M_i \twoheadrightarrow_\eta N_i \;{}_\eta\!\twoheadleftarrow M_{i+1}$ for some $N_i \in \Lambda^o$. By Corollary 11.59 there exist p_i, q_i such that $M_i \sqsupseteq^\eta_{p_i} N_i \sqsubseteq^\eta_{q_i} M_{i+1}$. By Lemma 11.56(iii), setting $p' \triangleq \max_i\{p_i, q_i\}$ we get $M_i \sqsupseteq^\eta_{p'} N_i \sqsubseteq^\eta_{p'} M_{i+1}$. By applying Lemma 11.56(ii) we can get rid of the equality $=_\mathcal{B}$ and obtain the sequence:

$$M \sqsupseteq^\eta_{p'} N_1 \sqsubseteq^\eta_{p'} M_2 \sqsupseteq^\eta_{p'} N_2 \cdots N_{k-1} \sqsubseteq^\eta_{p'} N.$$

Therefore this implication follows from Lemma 11.65.

($2 \Rightarrow 1$) We assume that $M \sqsubseteq^\eta_p P \sqsupseteq^\eta_p N$ holds. From Lemma 11.56(i), we obtain $M \sqsubseteq^\eta P \sqsupseteq^\eta N$ as well. The conclusion follows by applying Corollary 11.64 twice.

($2 \iff 3$) By Lemma 11.55. □

Part V

Models

Thomas Ehrhard

Antonio Bucciarelli

Antonino Salibra

Delia Kesner

Mario Coppo

Simona Ronchi Della Rocca

Maddalena Zacchi

Flavien Breuvart

Domenico Ruoppolo

Chapter 12

Partially ordered models and inequational theories

In Section 3.2 we have seen that every model $\mathcal{M}$ of λ-calculus induces a λ-theory through the kernel congruence relation of the interpretation function. In practice, most of the models introduced in the literature come equipped with some partial ordering $(\mathcal{M}, \leq)$ that allows to compare the denotations of λ-terms. See, e.g. Scott (1972), Hyland (1975b), Wadsworth (1976), Berry (1978), Engeler (1981), Ronchi Della Rocca (1982), Barendregt et al. (1983), Coppo et al. (1987), Dezani-Ciancaglini and Margaria (1987),Plotkin (1993), Di Gianantonio et al. (1999), Berline (2000), Salibra (2003), Bucciarelli and Salibra (2004), Severi and de Vries (2005b), Hyland et al. (2006), Bucciarelli et al. (2007), Berline et al. (2009), Manzonetto (2009), Carraro and Salibra (2012), Breuvart (2014), Breuvart (2016), Paolini et al. (2017), Breuvart et al. (2018), ...

In Chapters 13 and 14 we are going to study two classes of models—respectively the class of *filter models* and the class of *relational graph models*—where λ-terms are interpreted as *sets*. As a consequence, the denotations of λ-terms in these models are naturally ordered by set-theoretical inclusion. From a more abstract, categorical perspective, this is due to the fact that such models can be presented as reflexive objects living in partially ordered (or even cpo) enriched categories.

In this chapter, we present the notion of inequational theory introduced by Breuvart et al. (2018), but silently used in previous literature. Inequational theories are designed to capture contextual preorders on λ-terms arising both from denotational models and from the observation of operational properties, like head or β-normalization. As an example, we show that the different kinds of extensional orderings between Böhm-like trees studied in Section 10.2 of B[1984] induce inequational theories on λ-terms. We go beyond that material and present the extensional theory of program approximation developed by J.-J. Lévy in the 70's. We conclude by recalling the definition of Cartesian closed categories enriched over partial orders and complete partial orders (cpo). For more details on enriched categories, we refer to Amadio and Curien (1998) or Kelly (2005).

12.1 Inequational theories and ordered models

We provide an abstract definition of inequational theories, that can be seen as the inequational counterpart of the definition of a λ-theory (Definition 3.1). As they are meant to describe preorders on λ-terms arising from λ-models, it is natural to require that inequational theories contain the whole β-conversion (not only β-reduction).

DEFINITION 12.1. (i) An *inequational theory* is a compatible preorder on Λ containing β-conversion. (Compatibile relations have been introduced in Definition 1.5(iii).)

(ii) An inequational theory $\sqsubseteq$ is called:

(1) *consistent* whenever it is different from ∇, i.e., $\sqsubseteq \neq \Lambda \times \Lambda$;

(2) *inconsistent* if it is not consistent;

(3) *order sensible* if it is consistent and $M \sqsubseteq N$ holds for all $M \in \mathrm{UNS}$ and $N \in \Lambda$;

(4) *order semi-sensible* if $M \sqsubseteq N$ and $N \in \mathrm{UNS}$ entails $M \in \mathrm{UNS}$.

(iii) Given an inequational theory $\sqsubseteq$ we define the relation $\equiv$ by setting:

$$M \equiv N \iff M \sqsubseteq N \;\&\; N \sqsubseteq M$$

LEMMA 12.2. (i) *Every λ-theory $\mathcal{T}$ is an inequational theory.*

(ii) *If $\sqsubseteq$ is an inequational theory then $\equiv$ is a λ-theory.*

(iii) *If $\sqsubseteq$ is order (semi-)sensible then $\equiv$ is a (semi-)sensible λ-theory.*

PROOF. (i) Trivial.

(ii) Since $\sqsubseteq$ contains the β-conversion, $M =_\beta N$ entails both $M \sqsubseteq N$ and $N \sqsubseteq M$. The compatibility of $\equiv$ follows from the fact that $\sqsubseteq$ enjoys the same property.

(iii) Easy. □

A (semi-)sensible λ-theory $\mathcal{T}$ might not be order (semi-)sensible. E.g., take $\mathcal{T} = \mathcal{H}$. However, any consistent λ-theory is also consistent as an inequational theory.

EXAMPLES 12.3. The following examples are order sensible inequational theories.

(i) The preorder $\sqsubseteq_\perp$ induced on Λ by Böhm tree inclusion, namely

$$M \sqsubseteq_\perp N \iff \mathrm{BT}(M) \leq_\perp \mathrm{BT}(N)$$

(ii) The following preorders[1] (where $\sqsubseteq_{m1}, \sqsubseteq_{m2}$ were defined by Morris (1968)):

$$\begin{array}{llll} M \sqsubseteq_{m1} N & \iff & \forall C[\,].[\, C[M] \text{ has a } \beta\text{-nf} & \Rightarrow \; C[N] \text{ has the same } \beta\text{-nf}\,] \\ M \sqsubseteq_{[\mathsf{I}]_\beta} N & \iff & \forall C[\,].[\, C[M] =_\beta \mathsf{I} & \Rightarrow \; C[N] =_\beta \mathsf{I}\,] \\ M \sqsubseteq_{m2} N & \iff & \forall C[\,].[\, C[M] \text{ has a } \beta\eta\text{-nf} & \Rightarrow \; C[N] \text{ has the same } \beta\eta\text{-nf}\,] \\ M \sqsubseteq_{[\mathsf{I}]_{\beta\eta}} N & \iff & \forall C[\,].[\, C[M] =_{\beta\eta} \mathsf{I} & \Rightarrow \; C[N] =_{\beta\eta} \mathsf{I}\,] \end{array}$$

(iii) The observational preorders $\sqsubseteq_{\mathrm{NF}}$ and $\sqsubseteq_{\mathrm{SOL}}$ introduced in Definition 3.10(i).

EXERCISE 12.4. (i) Check that the relations in Examples 12.3 are inequational theories.

(ii) Prove $\sqsubseteq_\perp \subsetneq \sqsubseteq_{m1} \subsetneq \sqsubseteq_{[\mathsf{I}]_\beta} \subsetneq \sqsubseteq_{m2} \subseteq \sqsubseteq_{[\mathsf{I}]_{\beta\eta}}$.

(iii) Verify that $\sqsubseteq_{m2}$ and $\sqsubseteq_{\mathrm{NF}}$ coincide.

(iv) Verify the strict inclusion $\sqsubseteq_{\mathrm{NF}} \subsetneq \sqsubseteq_{\mathrm{SOL}}$.

[1]The notation $\sqsubseteq_{mi}$ is taken from B[1984], Exercise 16.5.5, and the numbering is consistent.

The next lemma generalizes Proposition 3.13(i) to the inequational case. The argument in the proof is an easy adaptation of the one used for Theorem 16.2.6 in B[1984].

LEMMA 12.5. *The preorder* $\sqsubseteq_{\text{SOL}}$ *is the maximum order sensible inequational theory.*

PROOF. Let $\sqsubseteq$ be an order sensible inequational theory, we need to show that $\sqsubseteq \subseteq \sqsubseteq_{\text{SOL}}$. Assume by contradiction that there exist M, N such that $M \sqsubseteq N$ holds, but $M \not\sqsubseteq_{\text{SOL}} N$. This means that there is a context $C[]$ such that $C[M]$ has a hnf while $C[N]$ does not, in particular $C[M] \not\sqsubseteq_{\text{SOL}} C[N]$. By Proposition 3.16 there is a (closed head) context $H[]$ such that $H[C[M]] =_\beta \mathsf{I}$ and $H[C[N]]$ is unsolvable. Since $\sqsubseteq$ is a compatible relation, $M \sqsubseteq N$ entails $H[C[M]] \sqsubseteq H[C[N]]$, so we have:

$$\mathsf{I} =_\beta H[C[M]] \sqsubseteq H[C[N]] \equiv \Omega_3$$

The rightmost equivalence $\equiv$ holds by Lemma 12.2(iii) because Ω_3 is unsolvable and the inequational theory $\sqsubseteq$ is order sensible. For the same reason, we must have $\Omega_3 \sqsubseteq \mathsf{I}$. By definition of $\equiv$ we obtain $\Omega_3 \equiv \mathsf{I}$, which leads to the following chain of equivalences:

$$\mathsf{I} \equiv \Omega_3 =_\beta \Omega_3(\lambda x.xxx) \equiv \mathsf{I}(\lambda x.xxx) =_\beta \lambda x.xxx$$

Since I and $\lambda x.xxx$ are $\beta\eta$-distinct normal forms this contradicts Proposition 3.13(iii). □

Ordered models of λ-calculus

The main definitions of denotational models of λ-calculus have been recalled in Chapter 3. Here, we focus on models endowed with a partial order agreeing with their operations.

DEFINITION 12.6. (i) A *partially ordered (p.o.) λ-model* is a pair $\mathcal{M} = (\mathcal{C}, \leq)$ such that

- $\mathcal{C}$ is a λ-model, and
- $\leq$ is a partial order on C, which makes the application operator of $\mathcal{C}$ monotonic.

(ii) Given a p.o. λ-model $\mathcal{M} = (\mathcal{C}, \leq)$, we write $\mathcal{M} \models M \sqsubseteq N$ whenever

$$\forall \rho \in \mathrm{Val}_C \,.\, [\![M]\!]^{\mathcal{C}}_\rho \leq [\![N]\!]^{\mathcal{C}}_\rho$$

(iii) The *inequational theory* induced by a p.o. λ-model $\mathcal{M}$ is defined by:

$$\mathrm{Th}_{\sqsubseteq}(\mathcal{M}) \triangleq \{(M, N) \mid \mathcal{M} \models M \sqsubseteq N\}$$

LEMMA 12.7 (MONOTONICITY OF $[\![-]\!]^{\mathcal{C}}_\rho$). *Let* $\mathcal{M} = (\mathcal{C}, \leq)$ *be a p.o. λ-model and* $\rho \in \mathrm{Val}_C$. *For all* $M, N \in \Lambda$ *and contexts* $C[]$, $[\![M]\!]^{\mathcal{C}}_\rho \leq [\![N]\!]^{\mathcal{C}}_\rho$ *entails* $[\![C[M]]\!]^{\mathcal{C}}_\rho \leq [\![C[N]]\!]^{\mathcal{C}}_\rho$.

PROOF. By structural induction on $C[]$. □

LEMMA 12.8. *If* $\mathcal{M} = (\mathcal{C}, \leq)$ *is a p.o. λ-model, then* $\mathrm{Th}_{\sqsubseteq}(\mathcal{M})$ *is an inequational theory.*

PROOF. Since $\mathcal{M}$ is a λ-algebra, we have $\boldsymbol{\lambda} \subseteq \mathrm{Th}_{\sqsubseteq}(\mathcal{M})$ by Proposition 3.36(i). The fact that $\mathrm{Th}_{\sqsubseteq}(\mathcal{M})$ is compatible with application and abstraction follows from the monotonicity of application and the Meyer-Scott axiom. □

In order to discuss partially ordered categorical models of λ-calculus, we introduce some basic concepts of order-enriched category theory. For more details on the subject, the reader may consult Kelly (2005) and Fiore (1995).

DEFINITION 12.9. A *poset-enriched CCC* is given by a locally small CCC $\mathbf{C}$ where:

- every hom-set $\mathbf{C}(A, B)$ is equipped with a partial order $\sqsubseteq_{(A,B)}$;
- composition, pairing and currying are monotonic.

REMARK 12.10. In the definition above, the poset-enrichment is specified as extra structure. That is, being cpo-enriched is not a *property* of a CCC as it depends on the choice of the partial order. Because of the conditions on pairing and currying, a poset-enriched CCC is not just a poset-enriched category that happens to be Cartesian closed.

EXAMPLES 12.11. (i) Any CCC can be poset-enriched using the trivial ordering $=$.

(ii) The category **Poset** of partially ordered sets and monotonic functions is a poset-enriched CCC.

(iii) The category **Rel** of sets and relations, ordered by inclusion, is not a poset-enriched CCC because it is not Cartesian closed.

(iv) Given a category $\mathbf{C}$ and a comonad ! we denote by $\mathbf{C}_!$ its coKleisli. The following are poset-enriched CCCs (taking the appropriate comonads):

(1) $\mathbf{Fin}_!$, where **Fin** is the category of finiteness spaces defined by Ehrhard (2005).

(2) $\mathbf{Cpms}_!$, where **Cpms** is the category from Pagani et al. (2014), having natural numbers as objects and completely positive maps as morphisms. The partial order on morphisms is obtained by considering Löwner ordering.

For more details on the comonads above, see the respective articles. It is worth noticing that $\mathbf{Fin}_!$ contains no non-trivial reflexive objects, because of its finiteness conditions. On the contrary, there are categorical models of λ-calculus living in $\mathbf{Cpms}_!$.

EXERCISE 12.12. (i) Verify that the coKleisli category of Köthe spaces introduced by Ehrhard (2002) is a CCC, but not a poset-enriched one (w.r.t. the pointwise ordering). [Hint: the issue is due to the presence of negative reals.]

(ii) Define a 'positive' version of Köthe spaces, in such a way that the resulting co-Kleisli is a poset-enriched CCC.

DEFINITION 12.13. Let $\mathcal{U}$ be a categorical model living in a poset-enriched CCC $\mathbf{C}$.

(i) We write $\mathcal{U} \models M \sqsubseteq N$ whenever $M, N \in \Lambda^o(x_1, \ldots, x_n)$ and $|M|_{\vec{x}} \sqsubseteq_{(U^n,U)} |N|_{\vec{x}}$.

(ii) The *inequational theory* induced by $\mathcal{U}$ is defined by:

$$\mathrm{Th}_{\sqsubseteq}(\mathcal{U}) \triangleq \{(M, N) \mid \mathcal{U} \models M \sqsubseteq N\}$$

LEMMA 12.14. *If $\mathcal{U}$ is a categorical model living in a poset-enriched CCC, then* $\mathrm{Th}_{\sqsubseteq}(\mathcal{U})$ *is an inequational theory.*

PROOF. From Proposition 3.48(iii), we obtain $\boldsymbol{\lambda} \subseteq \mathrm{Th}_{\sqsubseteq}(\mathcal{U})$. The compatibility with application and abstraction follows easily from the monotonicity of composition, pairing and currying. □

Although the previous result shows that categorical models living in poset-enriched Cartesian closed categories are sufficient to induce inequational theories, the categories considered in practice frequently enjoy further properties. For instance, it is often the case that least upper bounds of directed sets of morphisms exist, and this leads to the notion of cpo-enriched Cartesian closed category.

DEFINITION 12.15. A poset-enriched CCC $\mathbf{C}$ is *cpo-enriched* if:

- every poset $(\mathbf{C}(A,B), \sqsubseteq_{(A,B)})$ is in fact a cpo (whose bottom is denoted by $\bot_{(A,B)}$);
- composition is continuous;
- the following strictness condition holds:

$$\text{(l-strict)} \quad \bot \circ f \;=\; \bot$$

REMARK 12.16. (i) Being a cpo-enriched CCC can be seen as a property of a poset-enriched category, in the sense that we do not require extra structure (cf. Remark 12.10).
(ii) From the axiom (l-strict), the following equations are derivable:

$$\begin{array}{lrcl} \text{(Ev-strict)} & \mathrm{Eval} \circ \langle \bot, f \rangle & = & \bot \\ \text{(Cur-strict)} & \mathrm{Cur}(\bot) & = & \bot \end{array}$$

EXAMPLES 12.17. (i) The following are cpo-enriched CCCs:
(1) The category **Cpo** of complete partial orders and continuous functions.
(2) The category **Alg** of ω-algebraic complete lattices and continuous functions.
(3) The category **MRel**, namely the coKleisli of the category **Rel** of sets and relations with respect to the finite multisets comonad $\mathscr{M}_{\mathrm{f}}(-)$.
(ii) All cpo-enriched CCCs are in particular poset-enriched CCCs.
(iii) The categories $\mathbf{Fin}_!$ and $\mathbf{Cpms}_!$ from Example 12.11(iv) are not cpo-enriched CCCs because the least upper bounds of some directed subsets are missing.

LEMMA 12.18. *In a cpo-enriched CCC, pairing and currying are continuous.*

PROOF. The fact that pairing and currying are continuous is easy to verify. We show the case of currying. Let $\mathcal{F} \subseteq \mathbf{C}(A,B)$ be a directed set. By monotonicity of $\mathrm{Cur}(-)$, the set $\{\mathrm{Cur}(f) \mid f \in \mathcal{F}\}$ is directed. By continuity of composition and pairing, we have

$$\mathrm{Eval} \circ \Big(\bigsqcup\{\mathrm{Cur}(f) \mid f \in \mathcal{F}\} \times \mathrm{Id}\Big) = \bigsqcup\{\mathrm{Eval} \circ (\mathrm{Cur}(f) \times \mathrm{Id}) \mid f \in \mathcal{F}\} = \bigsqcup \mathcal{F}.$$

It follows that $\mathrm{Cur}(\bigsqcup \mathcal{F}) = \bigsqcup\{\mathrm{Cur}(f) \mid f \in \mathcal{F}\}$. □

For the sake of completeness we mention that—when working in concrete[2] CCCs—categorical models admit a simpler interpretation of λ-terms. We use the category **Cpo** as an example, but the method works for arbitrary concrete CCCs. Recall that in the category **Cpo** the exponential object $\mathcal{D} \Rightarrow \mathcal{D}$ is the space of continuous functions, considered as a cpo by pointwise ordering.

[2] A concrete category comes equipped with a faithful functor to the category **Set** of sets and functions. One can loosely think of concrete categories as categories whose objects are sets with additional structure and morphisms are structure-preserving functions.

DEFINITION 12.19. Let $\mathcal{D} = (D, \leq, \bot)$ be a reflexive cpo via the pair (App, Lam).

(i) For $a, b \in D$, define a binary operator $\cdot^{\mathcal{D}}$ by setting:

$$a \cdot^{\mathcal{D}} b \triangleq \mathrm{App}(a)(b)$$

(ii) Define an interpretation function $[\![-]\!]^{\mathcal{D}}_{(\cdot)} : \Lambda \times \mathrm{Val}_D \to D$ by induction as follows:

$$\begin{aligned} [\![x]\!]^{\mathcal{D}}_\rho &\triangleq \rho(x); \\ [\![MN]\!]^{\mathcal{D}}_\rho &\triangleq \mathrm{App}([\![M]\!]^{\mathcal{D}})([\![N]\!]^{\mathcal{D}}_\rho); \\ [\![\lambda x.M]\!]^{\mathcal{D}}_\rho &\triangleq \mathrm{Lam}(\lambda d \in D\,.\,[\![M]\!]^{\mathcal{D}}_{\rho[x:=d]}), \text{ noticing that the map} \\ & \quad\ \lambda d \in D\,.\,[\![M]\!]^{\mathcal{D}}_{\rho[x:=d]} \text{ is continuous (by B[1984], Lemma 5.4.3).} \end{aligned}$$

PROPOSITION 12.20. *Let $\mathcal{D} = (D, \leq)$ be a reflexive cpo. Then the structure*

$$\mathcal{M} = ((D, \cdot^{\mathcal{D}}, [\![\mathsf{K}]\!]^{\mathcal{D}}, [\![\mathsf{S}]\!]^{\mathcal{D}}), \leq),$$

where $\cdot^{\mathcal{D}}$ and $[\![-]\!]^{\mathcal{D}}_{(\cdot)}$ are defined as in Definition 12.19, is a p.o. λ-model.

PROOF. It is easy to check that $[\![-]\!]^{\mathcal{D}}_{(\cdot)}$ satisfies the conditions (a)-(e) of Definition 3.37. Therefore, the pair $((D, \cdot^{\mathcal{D}}), [\![-]\!]^{\mathcal{D}}_{(\cdot)})$ is a syntactic λ-model. By Theorem 3.50(ii), the structure $(D, \cdot^{\mathcal{D}}, [\![\mathsf{K}]\!]^{\mathcal{D}}, [\![\mathsf{S}]\!]^{\mathcal{D}})$ is a λ-model. By continuity of the morphism App, the operator $\cdot^{\mathcal{D}}$ is monotonic. We conclude that $\mathcal{M}$ is a p.o. λ-model. □

12.2 Extensional orders on Böhm trees

In Section 2.3 we have seen that coinductive methods are elegant but unnecessary in the study of λ-definable Böhm trees. Indeed, the Syntactic Approximation Theorem 2.32 states that two λ-terms have the same Böhm tree exactly when their finite approximants coincide. In this section we show that a similar approach can be extended to extensional Böhm trees and Nakajima trees—this is useful for characterizing observational preorders like $\sqsubseteq_{\mathrm{NF}}$ and $\sqsubseteq_{\mathrm{SOL}}$. For the presentation of this material, we follow an unpublished draft by Lévy (1993) as well as his lecture notes (Lévy (2005)).

Recall that $\mathscr{A}$ denotes the set of $\lambda\bot$-terms in $\beta\bot$-nf (Definition 2.26(iii)), Lemma 2.27 provides a syntactic characterization of its elements, and $\leq_\bot$ is introduced in Def. 2.28(i).

The following commutation properties are easy to check.

LEMMA 12.21. *On $\Lambda_\bot$ and $\mathscr{A}$, we have:*

(i) $(\to_\eta ; \leq_\bot) \subseteq (\leq_\bot ; \to_\eta)$,

(ii) $(\leq_\bot ; {}_\eta\!\leftarrow) \subseteq ({}_\eta\!\leftarrow ; \leq_\bot)$,

(iii) $(\to_\eta ; {}_\eta\!\leftarrow) \subseteq ({}_\eta\!\leftarrow ; \to_\eta)$,

where $\mathrm{R}_1 ; \mathrm{R}_2$ stands for relational composition in diagrammatic order.

COROLLARY 12.22. *The η-expansion is Church-Rosser on $\Lambda_\bot$ and $\mathscr{A}$.*

PROOF. From Lemma 12.21(iii), we obtain the diamond property for ${}_\eta\!\leftarrow$. By Proposition 1.10, we conclude that it is confluent. □

EXERCISE 12.23. (i) Show that $(\leq_\perp\,;\, \to_\eta) \not\subseteq (\to_\eta\,;\, \leq_\perp)$.
(ii) Show that $({}_\eta\!\leftarrow\,;\, \leq_\perp) \not\subseteq (\leq_\perp\,;\, {}_\eta\!\leftarrow)$.

LEMMA 12.24. *For $L, P, Q, L', L'_1, L'_2, P', Q' \in \mathscr{A}$, we have:*
(i) $P \leq_\perp L' \twoheadrightarrow_\eta L \leq_\perp Q' \twoheadrightarrow_\eta Q \quad \Rightarrow \quad P \leq_\perp Q'' \twoheadrightarrow_\eta Q$, *for some* $Q'' \in \mathscr{A}$.
(ii) $P \;{}_\eta\!\twoheadleftarrow P' \leq_\perp L' \;{}_\eta\!\twoheadleftarrow L \leq_\perp Q \quad \Rightarrow \quad P \;{}_\eta\!\twoheadleftarrow P'' \leq_\perp Q$, *for some* $P'' \in \mathscr{A}$.
(iii) $P \;{}_\eta\!\twoheadleftarrow P' \leq_\perp L'_1 \twoheadrightarrow_\eta L \;{}_\eta\!\twoheadleftarrow L'_2 \leq_\perp Q' \twoheadrightarrow_\eta Q \quad \Rightarrow \quad P \;{}_\eta\!\twoheadleftarrow P'' \leq_\perp Q'' \twoheadrightarrow_\eta Q$, *for some* $P'', Q'' \in \mathscr{A}$.

PROOF. (i) By Lemma 12.21(i), there exists $Q'' \in \mathscr{A}$ such that $L' \leq_\perp Q'' \twoheadrightarrow_\eta Q'$.
(ii) By Lemma 12.21(ii), there exists $P'' \in \mathscr{A}$ such that $P' \;{}_\eta\!\twoheadleftarrow P'' \leq_\perp L$.
(iii) By applying CR for η-expansion on $\mathscr{A}$ (Corollary 12.22) to $L'_1 \twoheadrightarrow_\eta L \;{}_\eta\!\twoheadleftarrow L'_2$, we obtain an $L'' \in \mathscr{A}$ such that $L'_1 \;{}_\eta\!\twoheadleftarrow L'' \twoheadrightarrow_\eta L'_2$. From $P' \leq_\perp L'_1 \;{}_\eta\!\twoheadleftarrow L''$ we obtain P'' by Lemma 12.21(ii). Similarly, Lemma 12.21(i) applied to $L'' \twoheadrightarrow_\eta L'_2 \leq_\perp Q'$ gives Q''. □

DEFINITION 12.25. (i) Let $\mathscr{A}^{\eta\text{-nf}} \triangleq \{P \in \mathscr{A} \mid P \text{ is in } \eta\text{-nf}\}$.
(ii) Define the following binary relations on $\mathscr{A}$:

$$\begin{array}{llll} P \preceq_{\mathsf{r}} Q & \iff & \exists Q' \in \mathscr{A} \,.\, P \leq_\perp Q' \twoheadrightarrow_\eta Q & (\eta\text{-reduction ordering}) \\ P \preceq_{\mathsf{e}} Q & \iff & \exists P' \in \mathscr{A} \,.\, P \;{}_\eta\!\twoheadleftarrow P' \leq_\perp Q & (\eta\text{-expansion ordering}) \\ P \preceq_\eta Q & \iff & \exists P', Q' \in \mathscr{A} \,.\, P \;{}_\eta\!\twoheadleftarrow P' \leq_\perp Q' \twoheadrightarrow_\eta Q & (\text{extensional ordering}) \end{array}$$

Although clearly symmetrical, the relations $\preceq_{\mathsf{r}}$ and $\preceq_{\mathsf{e}}$ have very different behaviors on approximants in $\mathscr{A}$. This is shown in the following examples.

EXAMPLES 12.26. (i) $\lambda x.y\perp \preceq_{\mathsf{r}} y$, indeed $\lambda x.y\perp \leq_\perp \lambda x.yx \to_\eta y$.
(ii) $x\perp \preceq_{\mathsf{e}} \lambda y.xyy$, since $x\perp \;{}_\eta\!\leftarrow \lambda y.x\perp y \leq_\perp \lambda y.xyy$.
(iii) $\lambda x.y(z\perp)\perp \preceq_\eta y(\lambda w.zww)$ holds. Indeed, we have:

$$\lambda x.y(z\perp)\perp \;{}_\eta\!\leftarrow \lambda x.y(\lambda w.z\perp w)x \leq_\perp \lambda x.y(\lambda w.zww)x \to_\eta y(\lambda w.zww)$$

LEMMA 12.27. *The structures $(\mathscr{A}, \preceq_{\mathsf{r}})$, $(\mathscr{A}, \preceq_{\mathsf{e}})$ and $(\mathscr{A}^{\eta\text{-nf}}, \preceq_\eta)$ are posets.*

PROOF. Reflexivity is trivial. Antisymmetry can be shown by structural induction, using Lemma 2.27. Transitivity is given by Lemma 12.24. □

REMARK 12.28. (i) Clearly, the relations $\preceq_{\mathsf{r}}$ and $\preceq_{\mathsf{e}}$ are refinements of $\preceq_\eta$.
(ii) The relation $\preceq_\eta$ on the whole set $\mathscr{A}$ is reflexive and transitive, but not antisymmetric. For instance, we have $\mathsf{I} \preceq_\eta \mathbf{1}$ and $\mathbf{1} \preceq_\eta \mathsf{I}$, but $\mathsf{I} \neq \mathbf{1}$.
(iii) Therefore, we will consider the relation $\preceq_\eta$ to be restricted to η-nf's, i.e.

$$\preceq_\eta \subseteq \mathscr{A}^{\eta\text{-nf}} \times \mathscr{A}^{\eta\text{-nf}}.$$

DEFINITION 12.29. (i) Let $M \in \Lambda_\perp$. The relations in Definition 12.25(ii), generate three notions of *extensional (finite) approximants of M*:

$$\begin{array}{lll} \mathscr{A}_{\mathsf{e}}(M) & \triangleq & \{P \in \mathscr{A} \mid \exists Q \in \mathscr{A}(M) \,.\, P \preceq_{\mathsf{e}} Q\} \\ \mathscr{A}_{\mathsf{r}}(M) & \triangleq & \{P \in \mathscr{A} \mid \exists Q \in \mathscr{A}(M) \,.\, P \preceq_{\mathsf{r}} Q\} \\ \mathscr{A}_\eta(M) & \triangleq & \{P \in \mathscr{A}^{\eta\text{-nf}} \mid \exists Q \in \mathscr{A}(M) \,.\, P \preceq_\eta Q^{\eta\text{-nf}}\} \end{array}$$

(ii) For all $M, N \in \Lambda_\bot$, define:

$$\begin{array}{lcl} M \sqsubseteq_{\mathsf{r}} N & \iff & \mathscr{A}_{\mathsf{r}}(M) \subseteq \mathscr{A}_{\mathsf{r}}(N) \\ M \sqsubseteq_{\mathsf{e}} N & \iff & \mathscr{A}_{\mathsf{e}}(M) \subseteq \mathscr{A}_{\mathsf{e}}(N) \\ M \sqsubseteq_{\eta} N & \iff & \mathscr{A}_{\eta}(M) \subseteq \mathscr{A}_{\eta}(N) \end{array}$$

EXAMPLES 12.30. (i) Consider the identity I and Wadsworth's combinator J (Figure 2.6, page 77). Notice that:

- $\mathscr{A}(\mathsf{I}) = \{\bot, \mathsf{I}\}$;
- $\mathscr{A}(\mathsf{J}) = \{\bot, \lambda x z_0.x\bot, \lambda x z_0.x(\lambda z_1.z_0\bot), \lambda x z_0.x(\lambda z_1.z_0(\lambda z_2.z_1\bot)), \dots\}$.

(ii) The set $\mathscr{A}_\eta(\mathsf{I})$ includes $\mathscr{A}(\mathsf{I}) \cup \mathscr{A}(\mathsf{J})$ but contains also elements like $\lambda x\vec{z}y.x\vec{z}\bot$ or $\lambda x z_0 \dots z_n.x\bot^{\sim n+1}$. It is easy to check that $\mathscr{A}_\eta(\mathsf{I}) = \mathscr{A}_\eta(\mathsf{J}) \cup \{\mathsf{I}\}$.

(iii) From the examples above, it follows $\mathscr{A}(\mathsf{J}) \neq \mathscr{A}(\mathsf{I})$, and:

$$\mathsf{J} \sqsubseteq_{\mathsf{r}} \mathsf{I}, \qquad \mathsf{J} \not\sqsubseteq_{\mathsf{e}} \mathsf{I}, \qquad \mathsf{J} \sqsubseteq_{\eta} \mathsf{I}.$$

(iv) $\langle \mathsf{I}, \mathsf{J}, \Omega\rangle \sqsubseteq_\eta \langle 1, \mathsf{I}, \Omega\rangle$ holds, but $\langle \mathsf{I}, \mathsf{J}, \Omega\rangle \not\sqsubseteq_{\mathsf{e}} \langle 1, \mathsf{I}, \Omega\rangle$ and $\langle \mathsf{I}, \mathsf{J}, \Omega\rangle \not\sqsubseteq_{\mathsf{r}} \langle 1, \mathsf{I}, \Omega\rangle$.

(v) Given the effective enumeration $(M_n)_{n \in \mathbb{N}}$ defined by

$$M_n = \begin{cases} \mathsf{J}, & \text{if } n \text{ is even,} \\ \Omega, & \text{if } n \text{ is odd,} \end{cases}$$

it is easy to check that $[M_n]_{n\in\mathbb{N}} \sqsubseteq_{\mathsf{r}} [\mathsf{I}]_{n\in\mathbb{N}}$. Similarly for $\sqsubseteq_\eta$.

REMARK 12.31. (i) For all $M \in \Lambda_\bot$, we have $\mathscr{A}(M) \subseteq \mathscr{A}_{\mathsf{r}}(M)$ and $\mathscr{A}(M) \subseteq \mathscr{A}_{\mathsf{e}}(M)$.

(ii) Given $P, Q \in \mathscr{A}$, if P is $\bot$-free then $P \preceq_\eta Q$ entails that Q is $\bot$-free.

(iii) Let $M, N \in \Lambda$. If $M \in \mathrm{SOL}$ and $M \sqsubseteq_{\mathsf{r}} N$ then M looks like an η-expansion of N. Formally, $N \leq_h M$ holds (see Definition 11.35(i)).

(iv) Similarly, if $M \sqsubseteq_{\mathsf{e}} N$ then $M \leq_h N$.

LEMMA 12.32. *The relations* $\sqsubseteq_{\mathsf{r}}, \sqsubseteq_{\mathsf{e}}$, *and* $\sqsubseteq_\eta$ *are inequational theories.*

PROOF. Transitivity and reflexivity are shown as in the proof of Lemma 12.27. Compatibility follows from their respective definitions and the contextuality of Böhm trees. □

At the end of the chapter we will see that these relations arise naturally as inequational theories of denotational models of λ-calculus.

LEMMA 12.33. *Let* $\mathsf{x} \in \{\mathsf{r}, \mathsf{e}, \eta\}$. *For all* $M \in \Lambda$, *the set* $\mathscr{A}_{\mathsf{x}}(M)$ *is an ideal w.r.t.* $\preceq_{\mathsf{x}}$.

PROOF. By definition $\mathscr{A}_{\mathsf{x}}(M)$ is non-empty and downward closed w.r.t. $\preceq_{\mathsf{x}}$. For $\mathsf{x} \in \{\mathsf{r}, \mathsf{e}\}$, the statement follows easily from the fact that $\mathscr{A}(M)$ is an ideal w.r.t. $\leq_\bot \subseteq \preceq_{\mathsf{x}}$ using the transitivity of $\preceq_{\mathsf{x}}$. Consider now $P_1, P_2 \in \mathscr{A}_\eta(M)$. By definition, $P_1 \preceq_\eta {Q_1}^{\eta\text{-nf}}$ for some $Q_1 \in \mathscr{A}(M)$. Similarly, $P_2 \preceq_\eta {Q_2}^{\eta\text{-nf}}$ for some $Q_2 \in \mathscr{A}(M)$. By Lemma 2.31(iii), there is an upper bound $Q \in \mathscr{A}(M)$ of Q_1, Q_2. By definition $Q_1 \leq_\bot Q$ entails ${Q_1}^{\eta\text{-nf}} \preceq_\eta Q^{\eta\text{-nf}}$, so by transitivity we get $P_1 \preceq_\eta Q^{\eta\text{-nf}}$. An analogous reasoning gives $P_2 \preceq_\eta Q^{\eta\text{-nf}}$. □

Let us discuss some syntactic properties of $\sqsubseteq_{\mathsf{r}}$ and $\sqsubseteq_{\mathsf{e}}$.

LEMMA 12.34. *Let* $\mathsf{x} \in \{\mathsf{r}, \mathsf{e}\}$. *For all* $M, N \in \Lambda_\bot$, *the following are equivalent:*

1. $M \sqsubseteq_{\mathsf{x}} N$;
2. $\forall P \in \mathscr{A}(M), \exists Q \in \mathscr{A}(N) \,.\, P \preceq_{\mathsf{x}} Q$.

PROOF. ($1 \Rightarrow 2$) Consider $P \in \mathscr{A}(M)$. By assumption and Remark 12.31(i) we have $P \in \mathscr{A}(M) \subseteq \mathscr{A}_{\mathsf{x}}(M) \subseteq \mathscr{A}_{\mathsf{x}}(N)$, thus $P \in \mathscr{A}_{\mathsf{x}}(N)$. Setting $Q \triangleq P$ we obtain $P \preceq_{\mathsf{x}} Q$.

($2 \Rightarrow 1$) We have to show $\mathscr{A}_{\mathsf{x}}(M) \subseteq \mathscr{A}_{\mathsf{x}}(N)$. Take $L \in \mathscr{A}_{\mathsf{x}}(M)$. By definition, there exists $P \in \mathscr{A}(M)$ such that $L \preceq_{\mathsf{x}} P$. By hypothesis, $P \preceq_{\mathsf{x}} Q$ holds for some $Q \in \mathscr{A}(N)$. By transitivity (Lemma 12.27), we obtain $L \preceq_{\mathsf{x}} Q$. We conclude that $L \in \mathscr{A}_{\mathsf{x}}(N)$. □

We show that $\sqsubseteq_{\mathsf{r}}$ and $\sqsubseteq_{\mathsf{e}}$ respectively coincide with the relations $\lesssim_\eta$ and ${}^\eta\!\sqsubseteq$ given in B[1984], Definition 10.2.32.

PROPOSITION 12.35. *For* $M, N \in \Lambda$, *we have:*
(i) $M \sqsubseteq_{\mathsf{r}} N \iff \exists T \in \mathscr{B} \,.\, \mathrm{BT}(M) \leq_\bot T \geq_\omega^\eta \mathrm{BT}(N)$.
(ii) $M \sqsubseteq_{\mathsf{e}} N \iff \exists T \in \mathscr{B} \,.\, \mathrm{BT}(M) \leq_\omega^\eta T \leq_\bot \mathrm{BT}(N)$.

PROOF. (i) By Lemma 12.34, $M \sqsubseteq_{\mathsf{r}} N$ is equivalent to

$$\forall P \in \mathscr{A}(M), \exists Q \in \mathscr{A}(N), L \in \mathscr{A} \,.\, P \leq_\bot L \twoheadrightarrow_\eta Q. \tag{12.1}$$

($\Rightarrow$) Assume (12.1). Since, for all $P \in \Lambda_\bot$, $P \in \mathscr{A}(M)$ is equivalent to $P \leq_\bot \mathrm{BT}(M)$, we can proceed by coinduction on $\mathrm{BT}(M)$.

If $M \in \mathrm{UNS}$, we conclude because $\mathrm{BT}(M) = \bot \leq_\bot \mathrm{BT}(N) \geq_\omega^\eta \mathrm{BT}(N)$.

Otherwise $M \in \mathrm{SOL}$, so we have $M =_\beta \lambda\vec{x}z_1 \dots z_m.yM_1 \cdots M_k M'_1 \cdots M'_m$. Therefore, every $P \in \mathscr{A}(M)$ different from $\bot$ has shape

$$P = \lambda\vec{x}z_1 \dots z_m.yP_1 \cdots P_k P'_1 \cdots P'_m$$

for some $P_i \in \mathscr{A}(M_i)$ and $P'_j \in \mathscr{A}(M'_j)$ for all i, j ($1 \leq i \leq k$, $1 \leq j \leq m$). By hypothesis, for such i and j, there exist $L_i, L'_j \in \mathscr{A}$ such that

$$P \leq_\bot \lambda\vec{x}z_1 \dots z_m.yL_1 \cdots L_k L'_1 \cdots L'_m \twoheadrightarrow_\eta Q \in \mathscr{A}(N).$$

This entails $Q = \lambda\vec{x}.yQ_1 \cdots Q_k$ and $N =_\beta \lambda\vec{x}.yN_1 \cdots N_k$ with

$Q_i \in \mathscr{A}(N_i)$ and $P_i \leq_\bot L_i \twoheadrightarrow_\eta Q_i$, for all i ($1 \leq i \leq k$),

$P'_j \leq_\bot L'_j \twoheadrightarrow_\eta z_j$ and $z_j \notin \mathrm{FV}(y\vec{P}\vec{L}\vec{Q})$, for all j ($1 \leq j \leq m$).

Since P is an arbitrary approximant of M the condition above hold for all $P_i \in \mathscr{A}(M_i)$ and $P'_j \in \mathscr{A}(M'_j)$. In particular, this entails $\vec{z} \notin \mathrm{FV}(\vec{M})$. Moreover, by co-IH, for each i there exists $T_i \in \mathscr{B}$ such that $\mathrm{BT}(M_i) \leq_\bot T_i \geq_\omega^\eta \mathrm{BT}(N_i)$ and $\vec{z} \notin \mathrm{FV}(T_i)$. Similarly, for each j, there exists a $T'_j \in \mathscr{B}$ satisfying $\mathrm{BT}(M'_j) \leq_\bot T'_j \geq_\omega^\eta z_j$. Therefore we conclude

$$\mathrm{BT}(M) \quad \leq_\perp \quad \lambda x_1 \ldots x_n z_1 \ldots z_m.y \quad \geq^\eta_\omega \quad \mathrm{BT}(N)$$
$$T_1 \;\cdots\; T_k \; T'_1 \;\cdots\; T'_m$$

($\Leftarrow$) By structural induction on P, we show (12.1).

If $P = \perp$, then we can take $Q \triangleq L \triangleq \mathsf{da}(N)$.

Otherwise, Remark 12.31(iii) entails $M =_\beta \lambda x_1 \ldots x_n z_1 \ldots z_m.yM_1 \cdots M_k M'_1 \cdots M'_m$ and $N =_\beta \lambda x_1 \ldots x_n.yN_1 \cdots N_k$. By hypothesis, there exists a Böhm-like tree

$$T = \lambda x_1 \ldots x_n z_1 \ldots z_m.y$$
$$T_1 \;\cdots\; T_k \; T'_1 \;\cdots\; T'_m$$

such that:

$$\mathrm{BT}(M_i) \leq_\perp T_i \geq^\eta_\omega \mathrm{BT}(N_i), \text{ for all } i\,(1 \leq i \leq k),$$

$$\mathrm{BT}(M'_j) \leq_\perp T'_j \geq^\eta_\omega z_j, \text{ for all } j\,(1 \leq j \leq m) \text{ and}$$

$$z_j \notin \mathrm{FV}(y\mathrm{BT}(N_1) \cdots \mathrm{BT}(N_k)), \text{ for all } j\,(1 \leq j \leq m).$$

Thus, we have $P = \lambda \vec{x} z_1 \ldots z_m.yP_1 \cdots P_k P'_1 \cdots P'_m$ where $P_i \in \mathscr{A}(M_i)$ and $P'_j \in \mathscr{A}(M'_j)$. Since $M_i \sqsubseteq_\mathsf{r} N_i$ for all $i\,(1 \leq i \leq k)$ and $M'_j \preceq_\mathsf{r} z_j$ for all $j\,(1 \leq j \leq m)$ we can apply the IH and get $P_i \leq_\perp L_i \twoheadrightarrow_\eta Q_i$, for some $Q_i \in \mathscr{A}(N_i)$ and $M'_j \leq_\perp L'_j \twoheadrightarrow_\eta z_j$. As a consequence, we get $P \leq_\perp \lambda \vec{x} z_1 \ldots z_m.yL_1 \cdots L_k L'_1 \cdots L'_m \twoheadrightarrow_\eta \lambda \vec{x}.yQ_1 \cdots Q_k \in \mathscr{A}(N)$. This shows (12.1), from which $M \sqsubseteq_\mathsf{r} N$ follows.

(ii) Analogous. □

We show that the equational theories induced by $\sqsubseteq_\mathsf{r}$ and $\sqsubseteq_\mathsf{e}$ coincide, and correspond to the λ-theory $\mathcal{B}$.

Corollary 12.36. *For $M, N \in \Lambda$, the following are equivalent:*

1. $\mathcal{B} \vdash M = N$;
2. $M \sqsubseteq_\mathsf{r} N \sqsubseteq_\mathsf{r} M$;
3. $M \sqsubseteq_\mathsf{e} N \sqsubseteq_\mathsf{e} M$.

Proof. Assume $\mathcal{B} \vdash M = N$, i.e., $\mathrm{BT}(M) = \mathrm{BT}(N)$. By Theorem 2.32, this is equivalent to $\mathscr{A}(M) = \mathscr{A}(N)$ from which $(1 \Rightarrow 2)$ and $(1 \Rightarrow 3)$ follow.

$(2 \Rightarrow 1)$ By Proposition 12.35(i), $M \sqsubseteq_\mathsf{r} N \sqsubseteq_\mathsf{r} M$ is equivalent to

$$\mathrm{BT}(M) \leq_\perp T \geq^\eta_\omega \mathrm{BT}(N) \leq_\perp T' \geq^\eta_\omega \mathrm{BT}(M)$$

for some Böhm-like trees T, T'. We proceed by coinduction on the Böhm trees of M, N. We distinguish two cases, depending on the solvability of M.

Case $M \in \mathrm{UNS}$. Then, also N must be unsolvable. Therefore $\mathrm{BT}(M) = \mathrm{BT}(N) = \perp$.

Case $M \in \mathrm{SOL}$. By Remark 12.31(iii), also N is solvable and both $M \leq_h N$ and $N \leq_h M$ hold. Therefore, $M =_\beta \lambda\vec{x}.yM_1 \cdots M_k$ and $N =_\beta \lambda\vec{x}.yN_1 \cdots N_k$ for the same $\vec{x}, y \in \mathrm{Var}$ and $k \in \mathbb{N}$. We derive $T = \lambda\vec{x}.yT_1 \cdots T_k$ with $\mathrm{BT}(M_i) \leq_\perp T_i \geq^\eta_\omega \mathrm{BT}(N_i)$ for every $i\,(1 \leq i \leq k)$ and appropriate T_i. Similarly, $T' = \lambda\vec{x}.yT'_1 \cdots T'_k$ with $\mathrm{BT}(N_i) \leq_\perp T'_i \geq^\eta_\omega \mathrm{BT}(M_i)$. By co-IH, we obtain $\mathrm{BT}(M_i) = \mathrm{BT}(N_i)$ for all such i, from which $\mathrm{BT}(M) = \mathrm{BT}(N)$ follows.

$(3 \Rightarrow 1)$ Analogous. □

Clearly the corollary above does not hold for the inequational theory $\sqsubseteq_\eta$. For instance, one has $\mathsf{I} \sqsubseteq_\eta \mathsf{1} \sqsubseteq_\eta \mathsf{I}$, whilst $\mathrm{BT}(\mathsf{I}) \neq \mathrm{BT}(\mathsf{1})$.

Approximations for extensional Böhm trees

We show that $\sqsubseteq_\eta$ provides an alternative characterization of Morris' observational preorder $\sqsubseteq_{\mathrm{NF}}$. In Proposition 12.40 (below), we give a simpler description of the set $\mathscr{A}_\eta(M)$. First, we need a couple of preliminary lemmas.

LEMMA 12.37. *For $M, N, N' \in \Lambda$, we have:*

(i) $M \to_\beta N \;\&\; N' \to_\eta N \quad \Rightarrow \quad \exists M' \in \Lambda\,.\, M' \twoheadrightarrow_{\beta\eta} N' \;\&\; M' \to_\eta M.$

$$\begin{array}{ccc} M' & \overset{\eta}{\dashrightarrow} & M \\ {\scriptstyle \beta\eta}\downarrow\!\!\!\downarrow & & \downarrow{\scriptstyle \beta} \\ N' & \xrightarrow{\eta} & N \end{array}$$

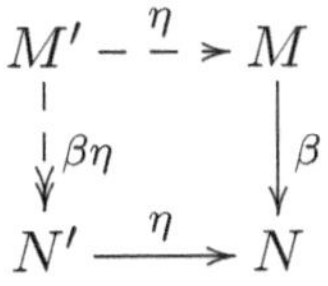

(ii) $M \twoheadrightarrow_\beta N \;\&\; N' \to_\eta N \quad \Rightarrow \quad \exists M' \in \Lambda\,.\, M' \twoheadrightarrow_{\beta\eta} N' \;\&\; M' \to_\eta M.$

$$\begin{array}{ccc} M' & \overset{\eta}{\dashrightarrow} & M \\ {\scriptstyle \beta\eta}\downarrow\!\!\!\downarrow & & \downarrow\!\!\!\downarrow{\scriptstyle \beta} \\ N' & \xrightarrow{\eta} & N \end{array}$$

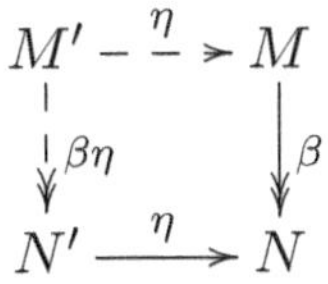

(iii) $M \twoheadrightarrow_\beta N \;\&\; N' \twoheadrightarrow_\eta N \quad \Rightarrow \quad \exists M' \in \Lambda\,.\, M' \twoheadrightarrow_{\beta\eta} N' \;\&\; M' \twoheadrightarrow_\eta M.$

$$\begin{array}{ccc} M' & \overset{\eta}{\dashrightarrow\!\!\!\!\rightarrow} & M \\ {\scriptstyle \beta\eta}\downarrow\!\!\!\downarrow & & \downarrow\!\!\!\downarrow{\scriptstyle \beta} \\ N' & \overset{\eta}{\twoheadrightarrow} & N \end{array}$$

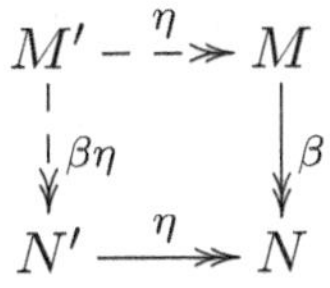

PROOF. (i) By induction on a derivation of $M \to_\beta N$.

Base case $M = (\lambda x.M_0)M_1$ and $N = M_0[x := M_1]$. Write $N = C[M_1, \ldots, M_1]$ where $C[x, \ldots, x] = M_0$ for some n-ary context $C[]$ such that $x \notin \mathrm{FV}(C[])$. Since $N' \to_\eta N$, one of the following cases holds:

- $N' = C[M_1, \ldots, M_1', \ldots, M_1]$ with $M_1' \to_\eta M_1$. Taking $M' = (\lambda x.M_0)M_1'$, we obtain

$$\begin{array}{ccccc}
M' = (\lambda x.M_0)M_1' & \xrightarrow{\beta} & C[M_1', \ldots, M_1'] \xrightarrow{\eta}\!\!\!\!\!\!\to C[M_1, \ldots, M_1', \ldots, M_1] & = N' \\
\downarrow{\scriptstyle \eta} & & \downarrow{\scriptstyle \eta} & \\
M = (\lambda x.M_0)M_1 & \xrightarrow{\beta} & C[M_1, \ldots, M_1] & = N
\end{array}$$

- $N' = C'[M_1, \ldots, M_1]$ with $C'[x, \ldots, x] \to_\eta C[x, \ldots, x] = M_0$. Note that the η-redex contracted in $C'[x, \ldots, x]$ has shape $\lambda y.Py$ with $y \notin P$, so the redex in N' has the form $\lambda y.P[x{:=}M_1]y$, where $y \notin \mathrm{FV}(P[x{:=}M_1])$ by the variable convention. Therefore, it is enough to take $M' \triangleq (\lambda x.C'[x, \ldots, x])M_1$, indeed:

$$\begin{array}{ccccc}
M' = (\lambda x.C'[x, \ldots, x])M_1 & \xrightarrow{\beta} & C'[M_1, \ldots, M_1] & = N' \\
\downarrow{\scriptstyle \eta} & & \downarrow{\scriptstyle \eta} & \\
M = (\lambda x.\underbrace{C[x, \ldots, x]}_{M_0})M_1 & \xrightarrow{\beta} & C[M_1, \ldots, M_1] & = N
\end{array}$$

The inductive cases follow straightforwardly from the IH.

(ii) By (i) and a diagram chase.

$$\begin{array}{ccc}
M' & \overset{\eta}{\dashrightarrow} & M \\
{\scriptstyle \beta\eta}\downarrow & & \downarrow{\scriptstyle \beta} \\
M_1' & \overset{\eta}{\dashrightarrow} & M_1 \\
{\scriptstyle \beta\eta}\downarrow & & \downarrow{\scriptstyle \beta} \\
N' & \xrightarrow{\eta} & N
\end{array}$$

(iii) By (ii) and a diagram chase.

$$\begin{array}{ccccc}
M' & \overset{\eta}{\dashrightarrow} & M_1' & \overset{\eta}{\dashrightarrow} & M \\
{\scriptstyle \beta\eta}\downarrow & & {\scriptstyle \beta\eta}\downarrow & & \downarrow{\scriptstyle \beta} \\
N' & \xrightarrow{\eta} & N_1' & \xrightarrow{\eta} & N
\end{array}$$

□

Recall that the notion of actual η-reduction $\to_{\eta_a}$ has been introduced in Notation 1.65. Folkerts (1998) gives the following description of multi-step actual η-reduction $\twoheadrightarrow_{\eta_a}$.

DEFINITION 12.38. For $m \in \mathbb{N}$, we define $M \overset{m}{\twoheadrightarrow}_{\eta_a} N$ by structural induction as follows. We assume that $\vec{z}$ are fresh variables and write $M \twoheadrightarrow_{\eta_a} N$ when m is irrelevant.

$$\frac{Z_i \twoheadrightarrow_{\eta_a} z_i \quad 1 \le i \le m}{\lambda z_1 \ldots z_m.xZ_1 \cdots Z_m \overset{m}{\twoheadrightarrow}_{\eta_a} x}\ (\mathrm{var}_m) \qquad \frac{M \twoheadrightarrow_{\eta_a} M'}{\lambda x.M \overset{0}{\twoheadrightarrow}_{\eta_a} \lambda x.M'}\ (\mathrm{lam}_0)$$

$$\frac{L \overset{0}{\twoheadrightarrow}_{\eta_a} L' \quad M \twoheadrightarrow_{\eta_a} M' \quad Z_i \twoheadrightarrow_{\eta_a} z_i \quad 1 \le i \le m}{\lambda z_1 \ldots z_m.LMZ_1 \cdots Z_m \overset{m}{\twoheadrightarrow}_{\eta_a} L'M'}\ (\mathrm{app}_m)$$

LEMMA 12.39. *For $N, N' \in \Lambda$, we have:*

$$N \twoheadrightarrow_{\eta_{\mathsf{a}}} N' \quad \Rightarrow \quad \mathsf{da}(N) \twoheadrightarrow_{\eta} \mathsf{da}(N').$$

PROOF. By induction on a derivation of $N \twoheadrightarrow_{\eta_{\mathsf{a}}} N'$. Below, we assume $\vec{z}$ fresh variables.

Case (var$_0$). Trivial, since $N = N' = x$.

Case (var$_m$) with $m > 0$. Then $N = \lambda\vec{z}.x\vec{Z}$, with each $Z_i \twoheadrightarrow_{\eta_{\mathsf{a}}} z_i$, and $N' = x$. Therefore, we have $\mathsf{da}(N) = \lambda\vec{z}.x\,\mathsf{da}(Z_1)\cdots\mathsf{da}(Z_m) \twoheadrightarrow_{\eta} x = \mathsf{da}(N')$ by the IH.

Case (lam$_0$). Then $N = \lambda x.M$ and $N' = \lambda x.M'$ with $M \twoheadrightarrow_{\eta_{\mathsf{a}}} M'$. By IH, we obtain $\mathsf{da}(M) \twoheadrightarrow_{\eta} \mathsf{da}(M')$. Now, if $\mathsf{da}(M) = \bot$ then $\mathsf{da}(N) = \bot \twoheadrightarrow_{\eta} \bot = \mathsf{da}(N')$. Otherwise, we obtain $\mathsf{da}(N) = \lambda x.\mathsf{da}(M) \twoheadrightarrow_{\eta} \lambda x.\mathsf{da}(M') = \mathsf{da}(N')$.

Case (app$_m$). Then $N = \lambda z_1 \dots z_m.LMZ_1 \cdots Z_m$ and $N' = L'M'$, with $L \twoheadrightarrow_{\eta_{\mathsf{a}}} L'$, $M \twoheadrightarrow_{\eta_{\mathsf{a}}} M'$ and $Z_i \twoheadrightarrow_{\eta_{\mathsf{a}}} z_i$ for each i. By IH on $L \twoheadrightarrow_{\eta_{\mathsf{a}}} L'$, we get $\mathsf{da}(L) \twoheadrightarrow_{\eta} \mathsf{da}(L')$ so $\mathsf{da}(L) = \bot$ entails $\mathsf{da}(L') = \bot$ and in this case $\mathsf{da}(N) = \mathsf{da}(N')$. Otherwise $\mathsf{da}(L) \neq \bot$, whence by applying the IH we conclude

$$\begin{aligned}\mathsf{da}(N) = {} & \lambda z_1 \dots z_m.\mathsf{da}(L)\mathsf{da}(M)\mathsf{da}(Z_1)\cdots\mathsf{da}(Z_m) \\ \twoheadrightarrow_{\eta} {} & \lambda\vec{z}.\mathsf{da}(L')\mathsf{da}(M')\vec{z} \twoheadrightarrow_{\eta} \mathsf{da}(L')\mathsf{da}(M') = \mathsf{da}(N').\end{aligned}$$ □

PROPOSITION 12.40. *For $M \in \Lambda$, we have:*

$$\mathscr{A}_{\eta}(M) = \{P^{\eta\text{-}\mathtt{nf}} \mid \exists M' \in \Lambda \;.\; M' \twoheadrightarrow_{\eta} M \;\&\; P \in \mathscr{A}(M')\}$$

PROOF. ($\subseteq$) Take $P \in \mathscr{A}_{\eta}(M)$. By definition, $P \in \mathscr{A}^{\eta\text{-}\mathtt{nf}}$ and there is $Q \in \mathscr{A}(M)$ such that $P \preceq_{\eta} Q^{\eta\text{-}\mathtt{nf}}$. Since $Q \in \mathscr{A}(M)$, by Lemma 2.30 there exists $N_1 \in \Lambda$ satisfying $M \twoheadrightarrow_{\beta} N_1$ and $Q \leq_{\bot} N_1$. By definition, $P \preceq_{\eta} Q^{\eta\text{-}\mathtt{nf}}$ holds whenever $P \;{}_{\eta}\!\twoheadleftarrow P' \leq_{\bot} Q' \twoheadrightarrow_{\eta} Q^{\eta\text{-}\mathtt{nf}}$, for some $P', Q' \in \mathscr{A}$. Since $Q \twoheadrightarrow_{\eta} Q^{\eta\text{-}\mathtt{nf}} \;{}_{\eta}\!\twoheadleftarrow Q'$, there is $Q_2 \in \mathscr{A}$ such that $Q \;{}_{\eta}\!\twoheadleftarrow Q_2 \twoheadrightarrow_{\eta} Q'$ by Corollary 12.22. By Lemma 12.21(i), $Q_2 \leq_{\bot} N_2$ for some $N_2 \twoheadrightarrow_{\eta} N_1$. Therefore, by Lemma 12.37(iii), there is an $M' \in \Lambda$ such that $M' \twoheadrightarrow_{\eta} M$ and $M' \twoheadrightarrow_{\beta\eta} N_2$. By η- and η_{a}-postponement, i.e. Lemma 1.66(iii), we obtain an N_3 satisfying $M' \twoheadrightarrow_{\beta} N_3 \twoheadrightarrow_{\eta_{\mathsf{a}}} N_2$. By Lemma 12.39, $N_3 \twoheadrightarrow_{\eta_{\mathsf{a}}} N_2$ entails $\mathsf{da}(N_3) \twoheadrightarrow_{\eta} \mathsf{da}(N_2) \geq_{\bot} Q_2$ where the inequality follows from $Q_2 \leq_{\bot} N_2$ by Remark 2.29(ii). By Lemma 12.21(ii), $\mathsf{da}(N_3) \geq_{\bot} Q_3$ for some $Q_3 \twoheadrightarrow_{\eta} Q_2 \twoheadrightarrow_{\eta} Q' \geq_{\bot} P'$. By Lemma 12.21(i), there is P_3 such that $Q_3 \geq_{\bot} P_3 \twoheadrightarrow_{\eta} P'$. Summing up, we have:

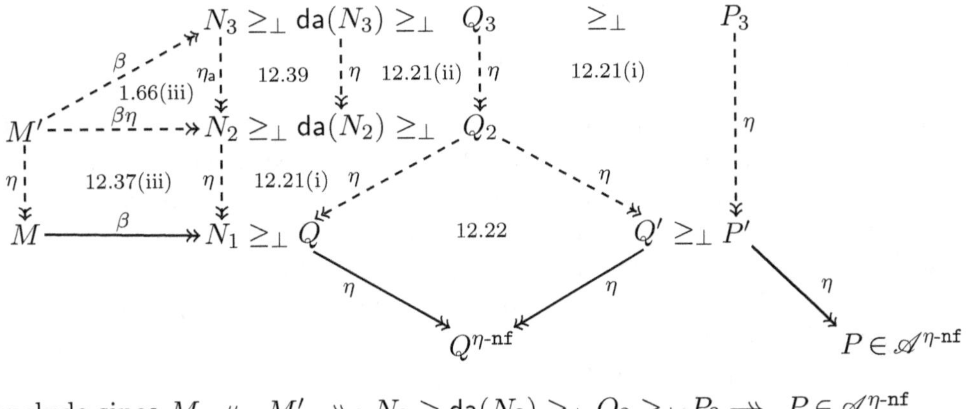

We conclude since $M \;{}_{\eta}\!\twoheadleftarrow M' \twoheadrightarrow_{\beta} N_3 \geq \mathsf{da}(N_3) \geq_{\bot} Q_3 \geq_{\bot} P_3 \twoheadrightarrow_{\eta} P \in \mathscr{A}^{\eta\text{-}\mathtt{nf}}$.

($\supseteq$) Assume that there exists $M' \in \Lambda$ such that $M' \twoheadrightarrow_\eta M$ and $P \in \mathscr{A}(M')$. We need to show that $P^{\eta\text{-nf}} \in \mathscr{A}_\eta(M)$. By Lemma 2.30, $P \in \mathscr{A}(M')$ whenever $P \leq_\bot N$, for some $N \in \Lambda_\bot$ satisfying $M' \twoheadrightarrow_\beta N$. By commutation of β and η (B[1984], Lemma 3.3.8), there exists $Z \in \Lambda_\bot$ such that $M \twoheadrightarrow_\beta Z$ and $N \twoheadrightarrow_\eta Z$. By postponement of actual η, namely Lemma 1.66(iii), there is $N' \in \Lambda_\bot$ such that $N \twoheadrightarrow_\beta N' \twoheadrightarrow_{\eta_\mathsf{a}} Z$. In diagrammatic form:

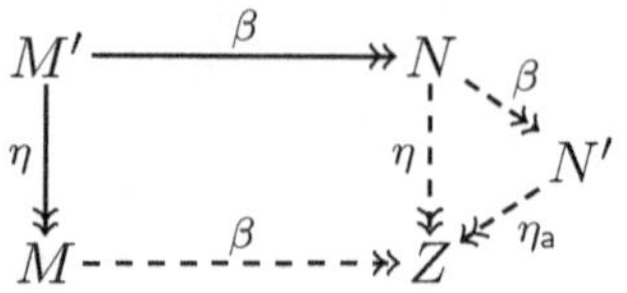

By Remark 2.29(i)-(ii), we have $P \leq_\bot \mathsf{da}(N) \leq_\bot \mathsf{da}(N')$ and $\mathsf{da}(N') \twoheadrightarrow_\eta \mathsf{da}(Z)^{\eta\text{-nf}}$, by Lemma 12.39. For $Q \triangleq \mathsf{da}(Z) \in \mathscr{A}(M)$ we get $P^{\eta\text{-nf}} \preceq_\eta Q^{\eta\text{-nf}}$, i.e., $P^{\eta\text{-nf}} \in \mathscr{A}_\eta(M)$. □

EXERCISE 12.41. (i) Show that $\mathscr{A}_\eta(M) = \{\mathsf{da}(N)^{\eta\text{-nf}} \mid N =_{\beta\eta} M\}$.

(ii) Show that $\lambda x.x\bot\mathsf{J}\mathsf{I} \sqsubseteq_\eta \lambda xz.xz\mathsf{I}\mathbf{1}z$.

(iii) Prove that $M \twoheadrightarrow_{\beta\eta} x$ and $\mathscr{A}_\eta(M) \subseteq \mathscr{A}_\eta(N)$ imply $N \twoheadrightarrow_{\beta\eta} x$.

(iv) Prove that $\mathrm{BT}(M) \geq^\eta_\omega x$ and $\mathscr{A}_\eta(M) \subseteq \mathscr{A}_\eta(N)$ imply $\mathrm{BT}(N) \geq^\eta_\omega x$.

LEMMA 12.42. *For all $M, N \in \Lambda_\bot$, we have:*

$$M^{\beta\text{-nf}} \in \Lambda \;\&\; M \sqsubseteq_\eta N \quad \Rightarrow \quad N^{\beta\text{-nf}} \in \Lambda$$

PROOF. Suppose that M is β-normalizable. Note that $M^{\beta\text{-nf}} \in \mathscr{A}(M) \subseteq \mathscr{A}_\eta(M)$, by Remark 12.31(i). By assumption we have $\mathscr{A}_\eta(M) \subseteq \mathscr{A}_\eta(N)$, whence $M^{\beta\text{-nf}} \in \mathscr{A}_\eta(N)$. Since $M^{\beta\text{-nf}}$ is $\bot$-free, Remark 12.31(ii) entails that $N \in \mathrm{NF}$. □

The following can be seen as an inequational version of Proposition 11.9. Although the definitions are slightly different, this result is basically shown in Hyland (1975a).

PROPOSITION 12.43. *For $M, N \in \Lambda$, the following are equivalent:*

1. $M \sqsubseteq_{\mathrm{NF}} N$;

2. $M \sqsubseteq_\eta N$;

3. $\exists T, T' \in \mathscr{B}.\, \mathrm{BT}(M) \leq^\eta T \leq_\bot T' \geq^\eta_\omega \mathrm{BT}(N)$.

PROOF. ($1 \Rightarrow 2$) We provide a proof sketch. Assume $M \not\sqsubseteq_\eta N$, towards a contradiction. Then, there exists an approximant $P \in \mathscr{A}_\eta(M) - \mathscr{A}_\eta(N)$ witnessing a finite difference between M and N. By applying the Böhm-out technique, one constructs a context $C[]$ such that $C[M] \in \mathrm{NF}$ and $C[N] \notin \mathrm{NF}$. The technical details are worked out in the book of Ronchi Della Rocca and Paolini (2004), Theorem 11.2.40.

($2 \Rightarrow 1$) Assume $M \sqsubseteq_\eta N$. Let $C[]$ be a context such that $C[M] =_\beta M'$ for some β-normal λ-term M'. By Lemma 12.32, we have that $M \sqsubseteq_\eta N$ entails $C[M] \sqsubseteq_\eta C[N]$. By applying Lemma 12.42, we conclude that $C[N] \in \mathrm{NF}$.

($2 \Rightarrow 3$) Assuming $\mathscr{A}_\eta(M) \subseteq \mathscr{A}_\eta(N)$, we construct coinductively T, T' satisfying (3).

The case $M \in \mathrm{UNS}$ is trivial, take $T \triangleq \bot$ and $T' \triangleq \mathrm{BT}(N)$.

If $M \in \mathrm{SOL}$ then it has a hnf. Wlog, assume $\mathrm{nf}_\eta(\mathrm{BT}(M)) = \lambda x_1 \ldots x_n.yT_1 \cdots T_k$. Then, there is a $P \in \mathscr{A}_\eta(M) \subseteq \mathscr{A}_\eta(N)$ of shape $\lambda \vec{x}.y\bot^{\sim k}$. By Proposition 12.40, there exist $M' \twoheadrightarrow_\eta M$ and $N' \twoheadrightarrow_\eta N$ such that $Q \in \mathscr{A}(M') \cap \mathscr{A}(N')$ and $Q^{\eta\text{-nf}} = P$. We have

$$\begin{array}{llll} M' \twoheadrightarrow_h & \lambda \vec{x} z_1 \ldots z_m.yM'_1 \cdots M'_k Z_1 \cdots Z_m, & \forall j \leq m \,.\, [\, z_j \notin \mathrm{BT}(y\vec{M}') \;\&\; Z_j \twoheadrightarrow_{\beta\eta} z_j \,], \\ N' \twoheadrightarrow_h & \lambda \vec{x} z_1 \ldots z_m.yN'_1 \cdots N'_k Z'_1 \cdots Z'_m, & \forall i \leq k \,.\, \mathscr{A}_\eta(M'_i) \subseteq \mathscr{A}_\eta(N'_i) \text{ and} \\ & & \forall j \leq m \,.\, \mathscr{A}_\eta(Z_j) \subseteq \mathscr{A}_\eta(Z'_j), \end{array}$$

where the above inclusions hold by $\mathscr{A}_\eta(M') \subseteq \mathscr{A}_\eta(N')$ and compatibility (Lemma 12.32). As seen in Exercise 12.41(iii), $Z_j^{\beta\eta\text{-nf}} = z_j$ and $\mathscr{A}_\eta(Z_j) \subseteq \mathscr{A}_\eta(Z'_j)$ entail $Z'_j \twoheadrightarrow_{\beta\eta} z_j$, for all $j\,(1 \leq j \leq m)$, so it is easy to construct $U_j \in \mathscr{B}$ such that $\mathrm{BT}(Z_j) \leq^\eta U_j \geq^\eta \mathrm{BT}(Z'_j)$. By co-IH, we construct $V_i, V'_i \in \mathscr{B}$ satisfying $\mathrm{BT}(M'_i) \leq^\eta V_i \leq_\bot V'_i \geq^\eta_\omega \mathrm{BT}(N'_i)$ for every $i\,(1 \leq i \leq k)$. Define the Böhm-like trees

$$T \triangleq \begin{array}{c} \lambda x_1 \ldots x_n z_1 \ldots z_m.y \\ \diagup \;\; \diagdown \\ V_1 \;\cdots\; V_k \; U_1 \;\cdots\; U_m \end{array} \qquad\qquad T' \triangleq \begin{array}{c} \lambda x_1 \ldots x_n z_1 \ldots z_m.y \\ \diagup \;\; \diagdown \\ V'_1 \;\cdots\; V'_k \; U_1 \;\cdots\; U_m \end{array}$$

Gathering all the information, it is easy to check that $\mathrm{BT}(M) \leq^\eta T \leq_\bot T'$. There still might be the possibility that $\mathrm{nf}_{\eta!}(\mathrm{BT}(M)) = \lambda x_1 \ldots x_{n-\ell}.yT_1 \cdots T_{k-\ell}$ for some $\ell > 0$. This happens, for example, when $n, k > 0$, $M'_k = \mathsf{J}x_n$ and $x_n \notin \mathrm{FV}(\mathrm{BT}(yM'_1 \cdots M'_{k-1}))$. By Exercise 12.41(iv), $\mathrm{BT}(M'_k) \geq^\eta_\infty x_n$ and $\mathscr{A}_\eta(M'_k) \subseteq \mathscr{A}_\eta(N'_k)$ imply that $\mathrm{BT}(N'_k) \geq^\eta x_n$. This is the reason why, in general, we only get $T' \geq^\eta_\omega \mathrm{BT}(N)$.

($3 \Rightarrow 2$) Assume (3). If $P \in \mathscr{A}_\eta(M)$ then there is $P_1 \in \mathscr{A}(M)$ such that $P \preceq_\eta P_1{}^{\eta\text{-nf}}$. Since $\mathrm{BT}(M) \leq^\eta T$, there exists $P_2 \twoheadrightarrow_\eta P_1$ such that $P_2 \leq_\bot T \leq_\bot T' \geq^\eta_\omega \mathrm{BT}(N)$. By Proposition 12.35(i) we obtain $P_2 \sqsubseteq_{\mathsf{r}} N$ so, by Lemma 12.34, it follows that $P_2 \preceq_{\mathsf{r}} Q$ holds for some $Q \in \mathscr{A}(N)$. This entails $P_2{}^{\eta\text{-nf}} \preceq_\eta Q^{\eta\text{-nf}}$ and since $P_2{}^{\eta\text{-nf}} = P_1{}^{\eta\text{-nf}}$ we obtain $P \preceq_\eta P_1{}^{\eta\text{-nf}} \preceq_\eta Q^{\eta\text{-nf}}$. We conclude that $P \in \mathscr{A}_\eta(N)$, whence $\mathscr{A}_\eta(M) \subseteq \mathscr{A}_\eta(N)$. □

Approximations for Nakajima trees

One can check that I and J are distinct if compared using the η-reduction ordering or the extensional ordering. This is due to the fact that new least upper bounds are added when considering infinite Böhm trees, even if the terms were already bounded by a finite approximant. Following Courcelle and Raoult (1980), we can add the missing suprema by performing the (Scott-)closed completion of a poset. For a more recent treatment, we refer to Keimel and Lawson (2009) and Zhao and Fan (2010).

We recall here the general definition.

Definition 12.44. Let $\mathcal{P} = (D, \leq)$ be a poset.

(i) A subset $I \subseteq D$ is a *closed ideal* over $\mathcal{P}$ if
  (1) I is an ideal over $\mathcal{P}$;
  (2) for every directed subset $A \subseteq I$, if $\bigvee A$ exists then $\bigvee A \in I$.

(ii) Given an ideal $I \subseteq D$, its *closure* I^o is the least closed ideal including I.

(iii) The set $\mathcal{I}^o(\mathcal{P})$ of all closed ideals over $\mathcal{P}$, ordered by inclusion, is called the *closed ideal completion* of $\mathcal{P}$.

Hereafter, we work in the closed ideal completion $\mathcal{I}^o(\mathscr{A}^{\eta\text{-nf}}, \preceq_\eta)$ and consider $\mathscr{A}_\eta(M)^o$.

DEFINITION 12.45. For $M, N \in \Lambda$, define

$$M \sqsubseteq^o_\eta N \iff \mathscr{A}_\eta(M)^o \subseteq \mathscr{A}_\eta(N)^o$$

EXAMPLES 12.46. (i) For all $P \in \mathscr{A}_\eta(\mathsf{I})$, we have $P \preceq_\eta \mathsf{I}$. In fact $\bigvee_\eta \mathscr{A}_\eta(\mathsf{I}) = \mathsf{I}$.

(ii) It follows that $\mathscr{A}_\eta(\mathsf{I})$ is already a closed ideal, since it contains I, so $\mathscr{A}_\eta(\mathsf{I})^o = \mathscr{A}_\eta(\mathsf{I})$.

(iii) By Examples 12.30(ii), we obtain $\mathscr{A}_\eta(\mathsf{J})^o = \mathscr{A}_\eta(\mathsf{J}) \cup \{\mathsf{I}\} = \mathscr{A}_\eta(\mathsf{I})^o$.

(iv) Note that $\bigvee_\eta \mathscr{A}_\eta(\langle\!\langle \mathsf{J} \rangle\!\rangle)$ would be an infinite Böhm-like tree, namely $\mathrm{BT}(\langle\!\langle \mathsf{I} \rangle\!\rangle)$. Since this supremum does not exist in $\mathscr{A}^{\eta\text{-nf}}$, it does not belong to $\mathscr{A}_\eta(\langle\!\langle \mathsf{J} \rangle\!\rangle)^o$ either.

Recall that, for a Böhm-like tree T and a variable z, the predicate $T \Downarrow_{\eta!} z$ has been introduced in Definition 2.59. This is also equivalent to $T \geq^\eta_\omega z$ and to $\mathrm{nf}_{\eta!}(T) = z$.

LEMMA 12.47. *Let $Z \in \Lambda$ and $z \in \mathrm{Var}$. If $\mathrm{BT}(Z) \Downarrow_{\eta!} z$ then*

(i) $\forall P \in \mathscr{A}(Z) \,.\, P \preceq_{\mathsf{r}} z$;

(ii) $\mathscr{A}_\eta(Z)^o = \mathscr{A}_\eta(z)$.

PROOF. If $\mathrm{BT}(Z) \Downarrow_{\eta!} z$ then $Z \twoheadrightarrow_\beta \lambda y_1 \dots y_m.zY_1 \cdots Y_m$ where $\forall j \,.\, \mathrm{BT}(Y_j) \Downarrow_{\eta!} y_j$.

(i) By structural induction on P. The case $P = \bot$ is trivial. If $P \neq \bot$ then we have $P = \lambda y_1 \dots y_m.zP_1 \cdots P_m$ where $P_j \in \mathscr{A}(Y_j)$ for all such j. By IH we get $P_j \preceq_{\mathsf{r}} y_j$ and by compatibility of $\preceq_{\mathsf{r}}$ we obtain $P \preceq_{\mathsf{r}} \lambda y_1 \dots y_m.zy_1 \cdots y_m \twoheadrightarrow_\eta z$. We conclude $P \preceq_{\mathsf{r}} z$.

(ii) Take $P \in \mathscr{A}_\eta(Z)$. Then there exists $Q \in \mathscr{A}(Z)$ such that $P \preceq_\eta Q^{\eta\text{-nf}}$. Notice that $Q = \lambda y_1 \dots y_m.zQ_1 \cdots Q_m$ for some $Q_j \in \mathscr{A}(Y_j)$. By (i), for all j, we get $Q_j \preceq_{\mathsf{r}} y_j$ whence $Q \preceq_{\mathsf{r}} z$. Since $Q^{\eta\text{-nf}} \;{}_\eta\!\twoheadleftarrow Q \preceq_{\mathsf{r}} z$ we derive $Q^{\eta\text{-nf}} \preceq_\eta z$ and thus $P \preceq_\eta z$, by transitivity. This shows that z is an upper bound of all $P \in \mathscr{A}_\eta(Z)$, whence $\mathscr{A}_\eta(Z)^o \subseteq \mathscr{A}_\eta(z)$. Now, it is easy to check that $P \preceq_\eta z$ and $P \neq z$ entails $P \in \mathscr{A}_\eta(Z)$, whence z is the least upper bound of $P \in \mathscr{A}_\eta(Z)$. We conclude that $\mathscr{A}_\eta(Z)^o = \mathscr{A}_\eta(z)$. □

PROPOSITION 12.48. *For all $M \in \Lambda$, we have:*

$$\mathscr{A}_\eta(M)^o = \mathrm{nf}_{\eta!}(\mathrm{BT}(M))^*{\downarrow_\eta}$$

where $T^* \triangleq \{P \in \mathscr{A} \mid P \leq_\bot T\}$ and $\downarrow_\eta$ denotes the downward closure w.r.t. $\preceq_\eta$.

PROOF. We divide the proof into claims.

Claim 12.48.1. $\forall P \in \mathscr{A}(M), \exists Q \in \mathscr{A} \,.\, P \preceq_{\mathsf{r}} Q^{\eta\text{-nf}} \in \mathrm{nf}_{\eta!}(\mathrm{BT}(M))^*$.

Subproof. By structural induction on P. The only interesting case is $P \neq \bot$. Then

$$\begin{array}{rcl} P & = & \lambda \vec{x} z_1 \dots z_m.yP_1 \cdots P_k P'_1 \cdots P'_m, \\ M & \twoheadrightarrow_\beta & \lambda \vec{x} z_1 \dots z_m.yM_1 \cdots M_k Z_1 \cdots Z_m, \\ \mathrm{nf}_{\eta!}(\mathrm{BT}(M)) & = & \lambda \vec{x}.yT_1 \cdots T_k, \end{array}$$

where

– for all $i \leq k$, $P_i \in \mathscr{A}(M_i)$ and $T_i = \mathrm{nf}_{\eta!}(\mathrm{BT}(M_i))$;
– for all $j \leq m$, $P'_j \in \mathscr{A}(Z_j)$ and $\mathrm{BT}(Z_j) \Downarrow_{\eta!} z_j \notin \mathrm{FV}(\mathrm{BT}(y\vec{M}))$.

By IH, for every $i\,(1 \leq i \leq k)$, there exists $Q_i \in \mathscr{A}$ satisfying $P_i \precsim_{\mathsf{r}} Q_i^{\eta\text{-}\mathbf{nf}} \leq_\perp T_i$. Now, by Lemma 12.47(i), we obtain $P'_j \precsim_{\mathsf{r}} z_j$ for all $j\,(1 \leq j \leq m)$. We conclude by taking $Q \triangleq \lambda\vec{x}z_1 \ldots z_m.yQ_1 \cdots Q_k z_1 \cdots z_m$. Indeed, $P \precsim_{\mathsf{r}} Q$ holds by compatibility of $\precsim_{\mathsf{r}}$ and $Q^{\eta\text{-}\mathbf{nf}} \in \mathrm{nf}_{\eta!}(\mathrm{BT}(M))^*$ by construction. ■

Claim 12.48.2. $\mathscr{A}_\eta(M) \subseteq \mathrm{nf}_{\eta!}(\mathrm{BT}(M))^*{\downarrow_\eta}$.

Subproof. Take $P_1 \in \mathscr{A}_\eta(M)$. By definition, there is a $P_2 \in \mathscr{A}(M)$ such that $P_1 \precsim_\eta P_2^{\eta\text{-}\mathbf{nf}}$. By Claim 12.48.1, we obtain a Q such that $P_2 \precsim_{\mathsf{r}} Q^{\eta\text{-}\mathbf{nf}} \in \mathrm{nf}_{\eta!}(\mathrm{BT}(M))^*$. Notice that $P_2^{\eta\text{-}\mathbf{nf}} {}_\eta\!\twoheadleftarrow P_2 \precsim_{\mathsf{r}} Q^{\eta\text{-}\mathbf{nf}}$ entails $P_2^{\eta\text{-}\mathbf{nf}} \precsim_\eta Q^{\eta\text{-}\mathbf{nf}}$. By transitivity of $\precsim_\eta$, we get $P_1 \precsim_\eta Q^{\eta\text{-}\mathbf{nf}}$. From $Q^{\eta\text{-}\mathbf{nf}} \in \mathrm{nf}_{\eta!}(\mathrm{BT}(M))^*$, we conclude $P_1 \in \mathrm{nf}_{\eta!}(\mathrm{BT}(M))^*{\downarrow}$. ■

Claim 12.48.3. $\mathrm{nf}_{\eta!}(\mathrm{BT}(M))^*{\downarrow_\eta}$ is a closed ideal.

Subproof. Given a directed set $A \subseteq \mathrm{nf}_{\eta!}(\mathrm{BT}(M))^*{\downarrow_\eta}$ such that $\bigvee_\eta A \in \mathscr{A}^{\eta\text{-}\mathbf{nf}}$, it is easy to find $P \in \mathrm{nf}_{\eta!}(\mathrm{BT}(M))^*$ satisfying $\bigvee_\eta A \precsim_\eta P$. Conclude $\bigvee_\eta A \in \mathrm{nf}_{\eta!}(\mathrm{BT}(M))^*{\downarrow_\eta}$. ■

We proved that $\mathrm{nf}_{\eta!}(\mathrm{BT}(M))^*{\downarrow_\eta}$ is a closed ideal containing $\mathscr{A}_\eta(M)$. To show that it is the least, proceed as in the proof of Lemma 12.47(ii). □

The next result has been morally proved by Hyland (1975b) and can be seen as an inequational version of Proposition 11.12. The idea of applying the closed ideal completion to retrieve Nakajima trees is due to Lévy (2005).

PROPOSITION 12.49. *For $M, N \in \Lambda$, the following are equivalent:*

1. $M \sqsubseteq_{\mathrm{SOL}} N$;
2. $\exists T, T' \in \mathscr{B}.\, \mathrm{BT}(M) \leq^\eta_\omega T \leq_\perp T' \geq^\eta_\omega \mathrm{BT}(N)$;
3. $\mathrm{nf}_{\eta!}(M) \leq_\perp \mathrm{nf}_{\eta!}(N)$;
4. $M \sqsubseteq^o_\eta N$.

PROOF. $(1 \iff 2)$ B[1984], Proposition 19.2.9(i).

$(2 \iff 3)$ B[1984], Theorem 10.2.31.

$(3 \iff 4)$ It follows from Proposition 12.48, by exploiting the fact that $\mathscr{A}_\eta(M)^o$ is uniquely determined by $\mathrm{nf}_{\eta!}(\mathrm{BT}(M))$. □

COROLLARY 12.50. *For $M, N \in \Lambda$,*

$$M \in \mathrm{NF} \;\&\; M \sqsubseteq_{\mathrm{SOL}} N \quad \Rightarrow \quad \mathcal{H}^* \vdash M = N.$$

PROOF. Note that $M \in \mathrm{NF}$ entails $\mathrm{BT}(M) = M^{\beta\text{-}\mathbf{nf}}$. From $M \sqsubseteq_{\mathrm{SOL}} N$ and the above proposition, we obtain $M^{\beta\text{-}\mathbf{nf}} \leq^\eta_\omega T \leq_\perp T' \geq^\eta_\omega \mathrm{BT}(N)$ for appropriate $T, T' \in \mathscr{B}$. Since $M^{\beta\text{-}\mathbf{nf}}$ is $\perp$-free and $M \leq^\eta_\omega T$ then also T must be $\perp$-free. Therefore, $T \leq_\perp T'$ entails $T = T'$. Conclude by Proposition 11.12. □

Several inequational theories discussed in this chapter admit a semantical characterization, in the sense that they arise as inequational theories of partially ordered models.

THEOREM 12.51 (SEMANTIC CHARACTERIZATIONS).
Consider the following p.o. λ-models:

- *The graph model $\mathcal{D}_A$,* *(Engeler (1981))*
- *The filter model $\mathcal{F}_{\mathrm{BCD}}$,* *(Barendregt et al. (1983))*
- *The graph model $\mathcal{P}_\omega$,* *(Plotkin (1971) and Scott (1974))*
- *The filter model $\mathcal{F}_{\mathrm{CDZ}}$,* *(Coppo et al. (1987))*
- *The continuous model $\mathcal{D}_\infty$.* *(Scott (1972))*

We have:

(i) $\mathrm{Th}_{\sqsubseteq}(\mathcal{D}_A) = \sqsubseteq_{\perp}$,

(ii) $\mathrm{Th}_{\sqsubseteq}(\mathcal{F}_{\mathrm{BCD}}) = \sqsubseteq_{\mathsf{r}}$,

(iii) $\mathrm{Th}_{\sqsubseteq}(\mathcal{P}_\omega) = \sqsubseteq_{\mathsf{e}}$,

(iv) $\mathrm{Th}_{\sqsubseteq}(\mathcal{F}_{\mathrm{CDZ}}) = \sqsubseteq_{\mathrm{NF}}$,

(v) $\mathrm{Th}_{\sqsubseteq}(\mathcal{D}_\infty) = \sqsubseteq_{\mathrm{SOL}}$.

PROOF. Originally proved in:
(i) Longo (1983).
(ii) Ronchi Della Rocca (1982).
(iii) Hyland (1975b).
(iv) Coppo et al. (1987).
(v) Hyland (1975b) and Wadsworth (1976), independently. □

COROLLARY 12.52. (i) $\mathrm{Th}(\mathcal{D}_A) = \mathrm{Th}(\mathcal{F}_{\mathrm{BCD}}) = \mathrm{Th}(\mathcal{P}_\omega) = \mathcal{B}$.
(ii) $\mathrm{Th}(\mathcal{F}_{\mathrm{CDZ}}) = \mathcal{H}^+$.
(iii) $\mathrm{Th}(\mathcal{D}_\infty) = \mathcal{H}^*$.

PROOF. (i) By Corollary 12.36.
(ii)–(iii) Immediate. □

REMARK 12.53. (i) In the proof of Theorem 12.51, one can find the original articles where these inequational theories have been characterized. For a more recent treatment of some of these results, we invite the reader to consult:
– B[1984], Sections 19.1 for $\mathcal{P}_\omega$ and 19.2 for $\mathcal{D}_\infty$;
– Ronchi Della Rocca and Paolini (2004), Sections 11.1 for $\mathcal{D}_\infty$ and 11.2 for $\mathcal{F}_{\mathrm{CDZ}}$.
(ii) Another famous model is the universal domain $\mathbb{T}^\omega$ introduced by Plotkin (1978). Its inequational theory has been characterized in Barendregt and Longo (1980), where it is shown that $\mathrm{Th}_{\sqsubseteq}(\mathbb{T}^\omega) = \sqsubseteq_{\perp}$, whence $\mathrm{Th}(\mathbb{T}^\omega) = \mathcal{B}$.

	Characterizations		
Preorder	Observational	Tree-like / Coinductive	Semantic
$M \sqsubseteq_{\perp} N$	✗	$\mathrm{BT}(M) \leq_{\perp} \mathrm{BT}(N)$	$\mathcal{D}_A \models M \sqsubseteq N$
$M \sqsubseteq_{\mathsf{r}} N$	✗	$\mathrm{BT}(M) \leq_{\perp} T \geq^{\eta}_{\omega} \mathrm{BT}(N)$	$\mathcal{F}_{\mathrm{BCD}} \models M \sqsubseteq N$
$M \sqsubseteq_{\mathsf{e}} N$	✗	$\mathrm{BT}(M) \leq^{\eta}_{\omega} T \leq_{\perp} \mathrm{BT}(N)$	$\mathcal{P}_{\omega} \models M \sqsubseteq N$
$M \sqsubseteq_{\eta} N$	$M \sqsubseteq_{\mathrm{NF}} N$	$\mathrm{BT}(M) \leq^{\eta} T \leq_{\perp} T' \geq^{\eta}_{\omega} \mathrm{BT}(N)$	$\mathcal{F}_{\mathrm{CDZ}} \models M \sqsubseteq N$
$M \sqsubseteq^{o}_{\eta} N$	$M \sqsubseteq_{\mathrm{SOL}} N$	$\mathrm{BT}(M) \leq^{\eta}_{\omega} T \leq_{\perp} T' \geq^{\eta}_{\omega} \mathrm{BT}(N)$	$\mathcal{D}_{\infty} \models M \sqsubseteq N$

Table 12.1: Characterizations of the main inequational theories.

(iii) The models $\mathcal{F}_{\mathrm{BCD}}$, $\mathcal{F}_{\mathrm{CDZ}}$ and $\mathcal{D}_{\infty}$ (presented as a filter model) will be further analyzed in Chapter 13. In particular, we present an easy proof of the fact that the inequational theory induced by $\mathcal{F}_{\mathrm{CDZ}}$ is exactly $\sqsubseteq_{\mathrm{NF}}$ (see Theorem 13.82).

(iv) In Chapter 14 we will construct simpler models, living in the relational semantics, inducing $\sqsubseteq_{\mathsf{r}}$, $\sqsubseteq_{\mathrm{NF}}$ and $\sqsubseteq_{\mathrm{SOL}}$ as inequational theories.

Let $\mathcal{O} \subseteq \Lambda$ be a set of observables, and $\equiv_{\mathcal{O}}$ (resp. $\sqsubseteq_{\mathcal{O}}$) be the associated observational theory (resp. preorder). As a matter of terminology, we shall say that a model $\mathcal{D}$ is

- *inequationally fully abstract for* $\sqsubseteq_{\mathcal{O}}$ if its inequational theory coincide with $\sqsubseteq_{\mathcal{O}}$;
- *fully abstract for* $\equiv_{\mathcal{O}}$ if it induces the observational theory $\equiv_{\mathcal{O}}$.

For instance, $\mathcal{F}_{\mathrm{CDZ}}$ is inequationally fully abstract for $\sqsubseteq_{\mathrm{NF}}$ and fully abstract for $\mathcal{H}^{+}$. Similarly, $\mathcal{D}_{\infty}$ is inequationally fully abstract for $\sqsubseteq_{\mathrm{SOL}}$ and fully abstract for $\mathcal{H}^{*}$.

We conclude by presenting in Table 12.1 a summary of the main results discussed in this chapter.

Chapter 13

Filter models

Filter models were born from the fruitful interaction of complementary ideas. In the sixties, probably under the influence of Hans Freudenthal, filters were taught at the Department of Mathematics of Utrecht University. Barendregt was a student there, where the following kind of results belonged to the curriculum.

DEFINITION 13.1. A *filter* $\mathcal{F}$ on a set X is a subset $\mathcal{F} \subseteq \mathscr{P}(X)$ such that

1. $X \in \mathcal{F}$ and $\emptyset \notin \mathcal{F}$;
2. $A, B \in \mathcal{F} \quad \Rightarrow \quad A \cap B \in \mathcal{F}$;
3. $B \supseteq A \in \mathcal{F} \quad \Rightarrow \quad B \in \mathcal{F}$.

An *ultrafilter* is a maximal filter.

PROPOSITION 13.2. *Let $\mathcal{U}$ be an ultrafilter on X.*
(i) *For every $A \subseteq X$ one has either $A \in \mathcal{U}$ or $\overline{A} \in \mathcal{U}$.*
(ii) *If $A_1 \cup \cdots \cup A_n \in \mathcal{U}$ then $A_i \in \mathcal{U}$, for some $i\,(1 \leq i \leq n)$.*

PROPOSITION 13.3. *Every filter $\mathcal{F}$ on X can be extended to an ultrafilter $\mathcal{U} \supseteq \mathcal{F}$ on X.*

This fundamental property is often proved by applying Zorn's Lemma (ZL) or the equivalent axiom of choice (AC). But the statement that filters can be extended to ultrafilters is equivalent to the dual Boolean Prime Ideal Theorem (BPI), stating that every Boolean algebra has a prime ideal.

DEFINITION 13.4. Let X be a topological space and $\mathcal{F}$ be a filter on X.
(i) Then $\mathcal{F}$ *converges to* $x \in X$ if every open set O containing x belongs to $\mathcal{F}$.
(ii) $\mathcal{F}$ *converges* if $\mathcal{F}$ converges to some $x \in X$.

PROPOSITION 13.5. *Let X be a topological space. Then X is compact if and only if every ultrafilter $\mathcal{U}$ on X converges.*

THEOREM 13.6 (TYCHONOFF'S THEOREM). *(In* ZFC = ZF + AC*) A general product of topological spaces is compact if and only if each of the component spaces is compact.*

PROPOSITION 13.7. *(In ZF) Tychonoff's theorem is equivalent to* AC.

In the late 1960's Barendregt was under the (false) impression that Tychonoff's theorem could be proved from just the BPI, using Proposition 13.5 in the following way. Suppose that all $\{X_i\}_{i\in I}$ are compact, towards $\prod_{i\in I} X_i$ is compact. Take an ultrafilter $\mathcal{F}$ on the product space in order to show that it converges. Let $\mathcal{F}_i$ be the projection on the coordinate space X_i. Using BPI extend these to ultrafilters $\mathcal{U}_i$. By compactness of X_i these $\mathcal{U}_i$ converge to, say, x_i. Then $\mathcal{F}$ converges to $\lambda\!\!\lambda i.x_i$.

Therefore one seems to have established in ZF set theory that BPI implies AC.

$$\begin{aligned} \text{BPI} &\Rightarrow \text{filters can be extended to ultrafilters} \\ &\Rightarrow \text{Tychonoff's theorem} \\ &\Rightarrow \text{AC}. \end{aligned}$$

On the other hand, in set theory it was known that BPI $\not\Rightarrow$ AC. Halpern (1964) showed this, using a Fraenkel-Mostowski model. (Later in Halpern and Lévy (1971) this was shown by using forcing.) This apparent 'koan' (contradiction) haunted Barendregt, as a student. In 1968, at a summer-school on logic in Varenna, he posed the question to Roberto Magari, who found the solution: *"Using the BPI, Tychonoff's theorem can only be proved for Hausdorff spaces, in which limit points are unique. The existence of a limit point in the product space was concluded from limit points in all coordinate spaces; however, if one deals with non-Hausdorff spaces these limit points are not unique and one needs an application of AC".* This settled the koan. In fact one can prove in ZF set theory

$$\begin{aligned} \text{BPI} &\iff \text{the product of compact Hausdorff spaces is compact.} \\ \text{AC} &\iff \text{the product of compact topological spaces is compact.} \end{aligned}$$

After this 'incident' Barendregt would never forget filters. This became useful in 1980, when he worked with Mario Coppo and Mariangiola Dezani on the intersection type systems they had introduced in Coppo and Dezani-Ciancaglini (1980). Filters provided a dual viewpoint—in the sense of Stone duality—on these. In that context, filters are not subsets of a powerset $\mathscr{P}(X)$, but of the more general meet semilattices arising from intersection type systems. As a result of this interaction of ideas, the first filter model $\mathcal{F}^{\text{BCD}}$ was born, and subsequently described in Barendregt et al. (1983). It is worth mentioning that, although the arguments for proving the properties of (ultra)filters described above are highly non-constructive, the reasoning needed for filter models is constructive. This is due to the fact that the filter models individually introduced in the literature are generated by finite compact elements, namely the intersection types.

The transition from topological spaces to the more flexible *locales* (pointless topology, see Johnstone (1983)) is partly motivated by similar considerations, as it makes results like Tychonoff's theorem much more constructive. Going in another direction, in the theory of λ-calculus models the relation between λ-models and the more general λ-algebras is also stemming from a transition between categories with objects having enough points and categories where not all objects possess this property, see Koymans (1982).

Now that we have set the mood with this historical and slightly anecdotical digression, we can move forward and present the ideas behind intersection types.

Intersection types are obtained from simple types by adding a binary operator $\wedge$, called *intersection* because it shares many properties with set-theoretical intersection. For instance $\wedge$ is assumed to be commutative, associative and idempotent—we will see in the next chapter that removing this last assumption leads to different type systems. Intuitively, a λ-term M has type $\sigma \wedge \tau$ whenever M can be assigned both the type σ and the type τ. Therefore, intersection types allow functions to take arguments having different types simultaneously. To retrieve classical properties of intersection like $\sigma\wedge\tau \leq \sigma$ and $\sigma \wedge \tau \leq \tau$ as well as the usual covariance/contravariance of the arrow, one needs to introduce a subtyping relation $\leq$. Typability and inhabitation become undecidable, the former being equivalent to strong normalization (in absence of a universal type ω) and the latter to λ-definability for models of simple types, as shown by Salvati et al. (2012).

Under certain natural conditions intersection type systems enjoy nice properties like subject reduction and expansion; moreover the set of types assigned to a λ-term can be proved to be a filter. Thus—crossing the order-reversing bridge provided by Stone duality—one obtains a filter model, i.e. a λ-model constituted by filters of types. Under this correspondence each intersection type uniquely identifies a principal filter. Formally, filter models live in the Cartesian closed category $\omega\mathbf{Alg}$ of ω-algebraic complete lattices and continuous functions. We show that classical lattice models defined as an inverse-limit, that are ubiquitous in theoretical computer science, can also be described as intersection type theories and ultimately as filter models. These models are called *$\mathcal{D}_\infty$-models* because their definition mimics the construction of Scott's model $\mathcal{D}_\infty$.

Concerning specific examples of models, we consider:

- $\mathcal{F}_{\mathrm{BCD}}$, corresponding to the original filter model by Barendregt et al. (1983);
- $\mathcal{F}_{\mathrm{Scott}}$, namely the original Scott's model $\mathcal{D}_\infty$ (1972) presented as a filter model;
- $\mathcal{F}_{\mathrm{CDZ}}$, a particular $\mathcal{D}_\infty$-model defined by Coppo et al. (1987).

It is well known that the models above satisfy a Semantic Approximation Theorem stating that the interpretation of a λ-term is equal to the union of the denotations of its finite approximants—this is readily proved using Tait's computability technique. As a direct consequence one obtains that the theories induced by these models include the λ-theory $\mathcal{B}$. As announced in Theorem 12.51, their theories turn out to be different: it is a classical result by Hyland (1975b) and Wadsworth (1976) that $\mathrm{Th}(\mathcal{D}_\infty) = \mathcal{H}^*$, Ronchi Della Rocca (1982) has shown that $\mathrm{Th}(\mathcal{F}_{\mathrm{BCD}}) = \mathcal{B}$ and Coppo et al. (1987) have proved that $\mathrm{Th}(\mathcal{F}_{\mathrm{CDZ}}) = \mathcal{H}^+$. This last result refutes a conjecture by Böhm (1975) (open problem #11.3) stating that all $\mathcal{D}_\infty$-models induce the same λ-theory ($\mathcal{H}^*$). We mainly focus on the model $\mathcal{F}_{\mathrm{CDZ}}$ since it has been largely overlooked in the literature and characterize its theory using Lévy's extensional approximants from Section 12.2.

For a more detailed discussion on intersection type systems and filter models, the reader is invited to consult Barendregt et al. (2013a), Part III. The presentation of the material in this chapter mainly follows that book, although we often employ less general notions in order to simplify the discussion. The price to pay is that some intersection type systems and filter models considered in the literature are not instances of our definitions.

13.1 Intersection type assignment systems

In this section we introduce the intersection type assignment systems, we study their main properties and provide some examples.

Intersection type theories

We start by defining the notions of intersection types, subtyping and type equivalence.

DEFINITION 13.8. (i) Given a set $\mathbb{A}$ of type constants and a distinguished constant $\omega \notin \mathbb{A}$, called the *universal type*, the set of *type atoms* is defined by setting $\mathbb{A}_\omega \triangleq \mathbb{A} \cup \{\omega\}$.

(ii) The set $\mathbb{T}_\wedge$ of *intersection types* over $\mathbb{A}_\omega$ is given by the grammar (for $\xi \in \mathbb{A}$):

$$\sigma ::= \omega \mid \xi \mid \sigma \to \sigma \mid \sigma \wedge \sigma \qquad (\mathbb{T}_\wedge)$$

(iii) We denote intersection types by $\sigma, \tau, \gamma, \delta$, possibly with indices. We assume that the arrow $\to$ associates to the right and has lower precedence than the intersection $\wedge$. When writing $\sigma = \tau$, we mean that the intersection types σ and τ are syntactically equal.

(iv) A *subtyping relation* $\leq$ is any binary relation on $\mathbb{T}_\wedge$ closed under the rules:

$$\frac{}{\sigma \leq \sigma}\ (\text{refl}) \qquad \frac{}{\sigma \wedge \tau \leq \sigma}\ (\text{inc}_\text{L}) \qquad \frac{}{\sigma \wedge \tau \leq \tau}\ (\text{inc}_\text{R}) \qquad \frac{}{\sigma \leq \omega}\ (\text{top})$$

$$\frac{}{(\sigma \to \tau) \wedge (\sigma \to \gamma) \leq \sigma \to (\tau \wedge \gamma)}\ (\text{arr}_\wedge) \qquad \frac{}{\omega \leq \sigma \to \omega}\ (\text{arr}_\omega)$$

$$\frac{\sigma_2 \leq \sigma_1 \quad \tau_1 \leq \tau_2}{\sigma_1 \to \tau_1 \leq \sigma_2 \to \tau_2}\ (\text{arr}_\leq) \qquad \frac{\sigma \leq \tau \quad \sigma \leq \gamma}{\sigma \leq \tau \wedge \gamma}\ (\text{glb}) \qquad \frac{\sigma \leq \tau \quad \tau \leq \gamma}{\sigma \leq \gamma}\ (\text{trans})$$

Every subtyping relation $\leq$ is a preorder since it satisfies reflexivity (refl) and transitivity (trans). The universal type ω is the top element w.r.t. $\leq$ by (top) and can be seen as an arrow using (arr_ω). We write $\sigma < \tau$ whenever $\sigma \leq \tau$ holds, but $\sigma \neq \tau$.

(v) An *intersection type theory* (*itt*) T is determined by a set $\mathbb{A}^T$ of type constants and a subtyping relation $\leq^T$ on the set $\mathbb{T}^T_\wedge$ of intersection types over $\mathbb{A}^T_\omega$, i.e., $T = (\mathbb{A}^T, \leq^T)$.

(vi) Every subtyping relation $\leq$ on $\mathbb{T}_\wedge$ induces a *type equivalence* $\simeq$ by setting:

$$\sigma \simeq \tau \iff \sigma \leq \tau \ \&\ \tau \leq \sigma$$

REMARK 13.9. It is easy to check that the operator $\wedge$ is commutative $\sigma \wedge \tau \simeq \tau \wedge \sigma$, associative $(\sigma \wedge \tau) \wedge \gamma \simeq \sigma \wedge (\tau \wedge \gamma)$, idempotent $\sigma \simeq \sigma \wedge \sigma$ (by (refl) and (glb)), and has the universal type ω as neutral element, i.e., $\sigma \wedge \omega \simeq \sigma$ (using (refl), (top) and (glb)).

The following intersection type theories T will be used as running examples. We describe $\mathbb{A}^T_\omega$ rather than the plain set $\mathbb{A}^T$ to emphasize the presence of ω in these systems.

EXAMPLES 13.10. (i) The itt BCD, introduced in Barendregt et al. (1983), is defined as follows:

- $\mathbb{A}^{\text{BCD}}_\omega \triangleq \{\xi_i \mid i \in \mathbb{N}\} \cup \{\omega\}$ and
- $\leq^{\text{BCD}}$ is the least subtyping relation.

In other words, the relation $\leq^{\text{BCD}}$ is the smallest preorder on $\mathbb{T}_{\wedge}^{\text{BCD}}$ closed under the rules in Definition 13.8(iv).

(ii) The itt Scott, representing the model $\mathcal{D}_\infty$ by Scott (1972), is given by
- $\mathbb{A}_\omega^{\text{Scott}} \triangleq \{\xi, \omega\}$ together with
- the least subtyping relation $\leq^{\text{Scott}}$ closed under the rule:

$$\frac{}{\omega \to \xi \simeq^{\text{Scott}} \xi}\ (\xi)$$

which means that both $\omega \to \xi \leq^{\text{Scott}} \xi$ and $\xi \leq^{\text{Scott}} \omega \to \xi$ hold.

(iii) The itt CDZ, first defined by Coppo et al. (1987), is given by
- $\mathbb{A}_\omega^{\text{CDZ}} \triangleq \{\varepsilon, \star, \omega\}$ and
- the least subtyping relation $\leq^{\text{CDZ}}$ closed under the rules:

$$\frac{}{\varepsilon \leq^{\text{CDZ}} \star}\ (\varepsilon \leq \star) \qquad \frac{}{\star \to \varepsilon \simeq^{\text{CDZ}} \varepsilon}\ (\varepsilon) \qquad \frac{}{\varepsilon \to \star \simeq^{\text{CDZ}} \star}\ (\star)$$

Again, the rule (ε) stands for both $\star \to \varepsilon \leq^{\text{CDZ}} \varepsilon$ and $\varepsilon \leq^{\text{CDZ}} \star \to \varepsilon$. Similarly for $(\star)$.

In order to get familiar with these type theories, we invite the reader to verify the following equalities, inequalities and equivalences that hold in general.

EXERCISE 13.11. (i) $(\sigma \to \tau \to \gamma) \wedge \sigma \to \tau \to \gamma = ((\sigma \to (\tau \to \gamma)) \wedge \sigma) \to (\tau \to \gamma)$;
(ii) $\sigma \to \tau \to \gamma \wedge \sigma \to \tau \to \gamma = (\sigma \to (\tau \to ((\gamma \wedge \sigma) \to (\tau \to \gamma))))$;
(iii) $\omega \leq \sigma \to \omega \leq \omega$, whence $\omega \simeq \sigma \to \omega$ (in particular, $\omega \simeq \omega \to \omega$);
(iv) $(\sigma \to \tau) \wedge (\gamma \to \delta) \leq (\sigma \wedge \tau) \to (\tau \wedge \delta)$;
(v) $(\sigma \to \tau) \wedge (\sigma \to \gamma) \simeq \sigma \to (\tau \wedge \gamma)$.

Points (iv) and (v) above, generalize to any number of intersection types as follow.

LEMMA 13.12. *For $n \geq 1$ and $\sigma_1, \dots, \sigma_n, \tau_1, \dots, \tau_n \in \mathbb{T}_\wedge$, we have:*
(i) $(\sigma_1 \to \tau_1) \wedge \cdots \wedge (\sigma_n \to \tau_n) \leq (\sigma_1 \wedge \cdots \wedge \sigma_n) \to (\tau_1 \wedge \cdots \wedge \tau_n)$.
(ii) $(\sigma \to \tau_1) \wedge \cdots \wedge (\sigma \to \tau_n) \simeq \sigma \to (\tau_1 \wedge \cdots \wedge \tau_n)$.

PROOF. (i) By induction on n, the case $n = 1$ being trivial.
Case $n = 2$. By $(\text{arr}_\leq)$ and $(\text{arr}_\wedge)$, we have:

$$(\sigma_1 \to \tau_1) \wedge (\sigma_2 \to \tau_2) \leq ((\sigma_1 \wedge \sigma_2) \to \tau_1) \wedge ((\sigma_1 \wedge \sigma_2) \to \tau_2) \leq (\sigma_1 \wedge \sigma_2) \to (\tau_1 \wedge \tau_2).$$

Case $n > 2$. Proceed as in the previous case, then apply the IH.
(ii) We have $(\sigma \to \tau_1) \wedge \cdots \wedge (\sigma \to \tau_n) \leq \sigma \to (\tau_1 \wedge \cdots \wedge \tau_n)$ by (i) and the idempotence of $\wedge$. For the other direction, use the rule $(\text{arr}_\leq)$ to show $\sigma \to (\tau_1 \wedge \cdots \wedge \tau_n) \leq (\sigma \to \tau_i)$ for all $i\,(1 \leq i \leq n)$ and then apply $(n-1)$-times the rule (glb). □

Thanks to the rule (glb) one can show that $\sigma \wedge \tau$ actually represents the greatest lower bound of σ and τ.

LEMMA 13.13. *For all $\sigma_1, \sigma_2, \tau_1, \tau_2 \in \mathbb{T}_\wedge$, we have:*
(i) $\sigma_1 \leq \tau_1 \;\&\; \sigma_1 \leq \tau_2 \iff \sigma_1 \leq \tau_1 \wedge \tau_2$.
(ii) $\sigma_1 \leq \tau_1 \;\&\; \sigma_2 \leq \tau_2 \Rightarrow \sigma_1 \wedge \sigma_2 \leq \tau_1 \wedge \tau_2$.

PROOF. Both points follow from (inc_R), (inc_L), (trans) and (glb). □

PROPOSITION 13.14. *Every type equivalence $\simeq$ is a congruence, namely it is compatible with the operators $\wedge$ and $\to$.*

PROOF. Consider $\sigma_1 \simeq \sigma_2$ and $\tau_1 \simeq \tau_2$. By applying (trans), (inc$_R$), (inc$_L$) and (glb) we derive $\sigma_1 \wedge \tau_1 \simeq \sigma_2 \wedge \tau_2$. The fact that $\sigma_1 \to \tau_1 \simeq \sigma_2 \to \tau_2$ holds follows from (arr$_\leq$). □

In order to associate a type to an open λ-term we need to introduce the notion of type environment, a finitary function associating each variable with a type.

DEFINITION 13.15. (i) The *support* of a function $\Gamma : \mathrm{Var} \to \mathbb{T}_\wedge$ is defined as follows

$$\mathsf{supp}(\Gamma) \triangleq \{x \in \mathrm{Var} \mid \Gamma(x) \not\simeq \omega\}$$

(ii) A *type environment* Γ is a function from Var to $\mathbb{T}_\wedge$ whose support is finite.

(iii) Write $x_1 : \sigma_1, \ldots, x_n : \sigma_n$ for the type environment Γ defined by (for all $y \in \mathrm{Var}$):

$$\Gamma(y) \triangleq \begin{cases} \sigma_i, & \text{if } y = x_i \text{ for some } i \in \{1, \ldots, n\}, \\ \omega, & \text{otherwise.} \end{cases}$$

(iv) The type equivalence $\simeq$ *extends to type environments* Γ_1, Γ_2 by setting:

$$\Gamma_1 \simeq \Gamma_2 \iff \forall x \in \mathrm{Var} \,.\, \Gamma_1(x) \simeq \Gamma_2(x)$$

(v) Given two type environments Γ_1, Γ_2, define their *intersection* $\Gamma_1 \wedge \Gamma_2$ by setting:

$$(\Gamma_1 \wedge \Gamma_2)(x) \triangleq \Gamma_1(x) \wedge \Gamma_2(x), \text{ for all } x \in \mathrm{Var}.$$

Notice that $\Gamma_1 \wedge \Gamma_2$ does not represent the intersection of Γ_1 and Γ_2 seen as functions, but rather the pointwise extension of the intersection operator $\wedge$ to type environments. In particular, $\Gamma_1 \wedge \Gamma_2$ is actually more defined than both Γ_1 and Γ_2 in the sense that $\mathsf{supp}(\Gamma_1 \wedge \Gamma_2) = \mathsf{supp}(\Gamma_1) \cup \mathsf{supp}(\Gamma_2)$. This might seem counterintuitive at first sight.

Intersection type assignment systems

Given an intersection type theory we need to specify the rules for assigning types to λ-terms in appropriate type environments.

DEFINITION 13.16. (i) An *intersection type assignment system* $\boldsymbol{\lambda}_\wedge$ is given by a set $\mathbb{A}_\omega$ of type atoms (including ω), a subtyping relation $\leq$ and the rules displayed in Figure 13.1. In the rules (ax) and ($\to_I$) we assume without loss of generality that $x \notin \mathsf{supp}(\Gamma)$.

(ii) Given an itt $T = (\mathbb{A}^T, \leq^T)$, we write $\boldsymbol{\lambda}_\wedge^T$ for the associated type assignment system.

(iii) *Type judgements* of an intersection type assignment system are triples (Γ, M, σ) that will be written as

$$\Gamma \vdash_\wedge M : \sigma$$

where Γ is a type environment, M is a λ-term and σ is an intersection type. In accordance with the notation introduced in Definition 13.15(iii), the 'empty' type environment Γ satisfying $\mathsf{supp}(\Gamma) = \emptyset$ will be omitted, so we simply write $\vdash_\wedge M : \sigma$.

$$\dfrac{}{\Gamma, x : \sigma \vdash_\wedge x : \sigma}\,(\mathrm{ax}) \qquad\qquad \dfrac{}{\Gamma \vdash_\wedge M : \omega}\,(\mathrm{U})$$

$$\dfrac{\Gamma \vdash_\wedge M : \sigma \to \tau \quad \Gamma \vdash_\wedge N : \sigma}{\Gamma \vdash_\wedge MN : \tau}\,(\to_{\mathrm{E}}) \qquad\qquad \dfrac{\Gamma \vdash_\wedge M : \sigma \quad \Gamma \vdash_\wedge M : \tau}{\Gamma \vdash_\wedge M : \sigma \wedge \tau}\,(\wedge_{\mathrm{I}})$$

$$\dfrac{\Gamma, x : \sigma \vdash_\wedge M : \tau}{\Gamma \vdash_\wedge \lambda x.M : \sigma \to \tau}\,(\to_{\mathrm{I}}) \qquad\qquad \dfrac{\Gamma \vdash_\wedge M : \tau \quad \tau \le \sigma}{\Gamma \vdash_\wedge M : \sigma}\,(\le)$$

Figure 13.1: Intersection type assignment system.

(iv) A *derivation* is a finite tree whose nodes are instances of the rules in Figure 13.1 such that the premises of a rule are consequences of its children. If the root of a derivation Π has $\Gamma \vdash_\wedge M : \sigma$ as a consequence we say that Π *is a derivation of* $\Gamma \vdash_\wedge M : \sigma$. When writing $\Gamma \vdash_\wedge M : \sigma$, we assume that there exists a derivation of this type judgment.

EXAMPLES 13.17. The itt's from Examples 13.10 induce type assignment systems $\boldsymbol{\lambda}_\wedge^{\mathrm{BCD}}$, $\boldsymbol{\lambda}_\wedge^{\mathrm{Scott}}$ and $\boldsymbol{\lambda}_\wedge^{\mathrm{CDZ}}$, respectively.

Let us construct some examples of derivations in the type assignment systems discussed above.

EXAMPLES 13.18. (i) For all λ-terms M we have $\vdash_\wedge M : \omega$ by applying the rule (U).

(ii) For all $\sigma, \tau \in \mathbb{T}_\wedge$, setting $\Gamma \triangleq x : (\sigma \to \tau) \wedge \sigma$, we have:

$$\dfrac{\dfrac{\dfrac{\dfrac{}{\Gamma \vdash_\wedge x : \Gamma(x)}\,(\mathrm{ax}) \quad \dfrac{}{(\sigma \to \tau) \wedge \sigma \le \sigma \to \tau}\,(\mathrm{inc_L})}{\Gamma \vdash_\wedge x : \sigma \to \tau}\,(\le) \quad \dfrac{\dfrac{}{\Gamma \vdash_\wedge x : \Gamma(x)}\,(\mathrm{ax}) \quad \dfrac{}{(\sigma \to \tau) \wedge \sigma \le \sigma}\,(\mathrm{inc_R})}{\Gamma \vdash_\wedge x : \sigma}\,(\le)}{\Gamma \vdash_\wedge xx : \tau}\,(\to_{\mathrm{E}})}{\vdash_\wedge \lambda x.xx : ((\sigma \to \tau) \wedge \sigma) \to \tau}\,(\to_{\mathrm{I}})$$

(iii) The derivations above are valid in any intersection type assignment system. In the system $\boldsymbol{\lambda}_\wedge^{\mathrm{Scott}}$ we can also assign $\lambda x.xx$ the type $\xi \to \xi$ by exploiting the axiom (ξ):

$$\dfrac{\dfrac{\dfrac{\dfrac{}{x : \xi \vdash_\wedge^{\mathrm{Scott}} x : \xi}\,(\mathrm{ax}) \quad \xi \le^{\mathrm{Scott}} \omega \to \xi}{x : \xi \vdash_\wedge^{\mathrm{Scott}} x : \omega \to \xi}\,(\le) \quad \dfrac{}{x : \xi \vdash_\wedge^{\mathrm{Scott}} x : \omega}\,(\mathrm{U})}{x : \xi \vdash_\wedge^{\mathrm{Scott}} xx : \xi}\,(\to_{\mathrm{E}})}{\vdash_\wedge^{\mathrm{Scott}} \lambda x.xx : \xi \to \xi}\,(\to_{\mathrm{I}})$$

(iv) In the itt CDZ we have the following derivation Π of $\varepsilon \le^{\mathrm{CDZ}} \varepsilon \to \varepsilon$:

$$\dfrac{\dfrac{}{\varepsilon \le^{\mathrm{CDZ}} \star \to \varepsilon}\,(\varepsilon) \quad \dfrac{\dfrac{}{\varepsilon \le^{\mathrm{CDZ}} \star}\,(\varepsilon \le \star) \quad \dfrac{}{\varepsilon \le^{\mathrm{CDZ}} \varepsilon}\,(\mathrm{refl})}{\star \to \varepsilon \le^{\mathrm{CDZ}} \varepsilon \to \varepsilon}\,(\mathrm{arr}_\le)}{\varepsilon \le^{\mathrm{CDZ}} \varepsilon \to \varepsilon}\,(\mathrm{trans})$$

In the system $\boldsymbol{\lambda}_\wedge^{\mathrm{CDZ}}$, using Π and setting $\Gamma \triangleq x:\varepsilon, y:\star$, we can construct the derivation:

$$
\dfrac{
\dfrac{
\dfrac{\dfrac{(\mathrm{ax})}{\Gamma \vdash_\wedge^{\mathrm{CDZ}} x:\varepsilon} \quad \dfrac{\Pi}{\varepsilon \leq^{\mathrm{CDZ}} \varepsilon \to \varepsilon}}{\Gamma \vdash_\wedge^{\mathrm{CDZ}} x:\varepsilon \to \varepsilon}\,(\leq)
\quad
\dfrac{\dfrac{\dfrac{(\mathrm{ax})}{\Gamma \vdash_\wedge^{\mathrm{CDZ}} x:\varepsilon} \quad \dfrac{(\varepsilon)}{\varepsilon \leq \star \to \varepsilon}}{\Gamma \vdash_\wedge^{\mathrm{CDZ}} x:\star \to \varepsilon}\,(\leq) \quad \dfrac{(\mathrm{ax})}{\Gamma \vdash_\wedge^{\mathrm{CDZ}} y:\star}}{\Gamma \vdash_\wedge^{\mathrm{CDZ}} xy:\varepsilon}\,(\to_{\mathrm{E}})
}{
\dfrac{\Gamma \vdash_\wedge^{\mathrm{CDZ}} x(xy):\varepsilon}{x:\varepsilon \vdash_\wedge^{\mathrm{CDZ}} \lambda y.x(xy):\star \to \varepsilon}\,(\to_{\mathrm{I}})
}\,(\to_{\mathrm{E}})
\quad \star \to \varepsilon \leq^{\mathrm{CDZ}} \varepsilon
}{x:\varepsilon \vdash_\wedge^{\mathrm{CDZ}} \lambda y.x(xy):\varepsilon}\,(\leq)
$$

(v) Still in the system $\boldsymbol{\lambda}_\wedge^{\mathrm{CDZ}}$, setting $\Gamma \triangleq x:\varepsilon, y:\varepsilon$ and writing $\leq$ for $\leq^{\mathrm{CDZ}}$, we have:[1]

$$
\dfrac{
\dfrac{
\dfrac{
\dfrac{
\dfrac{\dfrac{(\mathrm{ax})}{\Gamma \vdash_\wedge^{\mathrm{CDZ}} x:\varepsilon} \quad \varepsilon \leq \star \to \varepsilon}{\Gamma \vdash_\wedge^{\mathrm{CDZ}} x:\star \to \varepsilon}\,(\leq)
\quad
\dfrac{\dfrac{(\mathrm{ax})}{\Gamma \vdash_\wedge^{\mathrm{CDZ}} y:\varepsilon} \quad \varepsilon \leq \star}{\Gamma \vdash_\wedge^{\mathrm{CDZ}} y:\star}\,(\leq)
}{\Gamma \vdash_\wedge^{\mathrm{CDZ}} xy:\varepsilon}\,(\to_{\mathrm{E}})
\quad \varepsilon \leq \star
}{\Gamma \vdash_\wedge^{\mathrm{CDZ}} xy:\star}\,(\leq)
}{\vdash_\wedge^{\mathrm{CDZ}} \lambda xy.xy:\varepsilon \to \varepsilon \to \star}\,(\to_{\mathrm{I}})
\quad \varepsilon \to \varepsilon \to \star \leq \star
}{\vdash_\wedge^{\mathrm{CDZ}} \lambda xy.xy:\star}\,(\leq)
$$

Definition 13.19. An arbitrary rule R is called:
(i) *derivable* if there exists a derivation tree having the same premises and conclusion;
(ii) *admissible* if adding R to the system does not affect the set of derivable judgments.

Clearly all derivable rules are admissible, while the converse does not hold in general.

Examples 13.20. The following rules are derivable by applying the rule $(\leq)$:

$$
\dfrac{\Gamma \vdash_\wedge M:\sigma \wedge \tau}{\Gamma \vdash_\wedge M:\sigma}\,(\wedge_{\mathrm{E}_\ell}) \qquad \dfrac{\Gamma \vdash_\wedge M:\sigma \wedge \tau}{\Gamma \vdash_\wedge M:\tau}\,(\wedge_{\mathrm{E}_r})
$$

Proposition 13.21. *The following rules are admissible:*

$$
\dfrac{\Gamma, x:\tau \vdash_\wedge M:\sigma \quad \Gamma \vdash_\wedge N:\tau}{\Gamma \vdash_\wedge M[x{:=}N]:\sigma}\,(\mathrm{cut}) \qquad \dfrac{\Gamma, x:\tau \vdash_\wedge M:\sigma \quad x \notin \mathrm{FV}(M)}{\Gamma \vdash_\wedge M:\sigma}\,(\text{strenghtening})
$$

$$
\dfrac{\Gamma, x:\sigma \vdash_\wedge M:\tau}{\Gamma, x:\sigma \wedge \gamma \vdash_\wedge M:\tau}\,(\wedge_{\mathrm{L}}) \qquad \dfrac{\Gamma \vdash_\wedge M:\sigma \quad x \notin \mathsf{supp}(\Gamma)}{\Gamma, x:\tau \vdash_\wedge M:\sigma}\,(\text{weakening})
$$

$$
\dfrac{\Gamma, x:\tau \vdash_\wedge M:\sigma \quad \gamma \leq \tau}{\Gamma, x:\gamma \vdash_\wedge M:\sigma}\,(\leq_{\mathrm{L}}) \qquad \dfrac{\Gamma, y:\tau \vdash_\wedge M:\sigma \quad \Gamma \vdash_\wedge N:\gamma}{\Gamma, x:\gamma \to \tau \vdash_\wedge M[y{:=}xN]:\sigma}\,(\to_{\mathrm{L}})
$$

Proof. By induction on a derivation, one shows that these rules can be postponed and eventually eliminated. □

[1] In this chapter, the double line in a derivation stands for multiple applications of the corresponding rule. In this particular case, we contracted two applications of the rule $(\to_{\mathrm{I}})$.

We present a technical lemma that describes under what conditions a type judgement of the form $\Gamma \vdash_\wedge M : \sigma$ holds, depending on the structure of M. Statements of this kind are often called Inversion Lemmas and they constitute convenient tools for performing proofs by induction on the structure of M, rather than by induction on the derivation. Another example of an Inversion Lemma is given in Theorem 13.25 (below).

THEOREM 13.22 (INVERSION LEMMA).

$$
\begin{array}{llcl}
\text{(i)} & \Gamma \vdash_\wedge x : \sigma & \iff & \Gamma(x) \le \sigma. \\
\text{(ii)} & \Gamma \vdash_\wedge MN : \sigma & \iff & \exists k \ge 1, \exists \tau_1, \dots, \tau_k, \gamma_1, \dots, \gamma_k \in \mathbb{T}_\wedge \\
& & & \gamma_1 \wedge \cdots \wedge \gamma_k \le \sigma \ \& \\
& & & \forall i\,(1 \le i \le k)\,.\,[\Gamma \vdash_\wedge M : \tau_i \to \gamma_i \ \&\ \Gamma \vdash_\wedge N : \tau_i]. \\
\text{(iii)} & \Gamma \vdash_\wedge \lambda x.M : \sigma & \iff & \exists k \ge 1, \exists \tau_1, \dots, \tau_k, \gamma_1, \dots, \gamma_k \in \mathbb{T}_\wedge \\
& & & (\tau_1 \to \gamma_1) \wedge \cdots \wedge (\tau_k \to \gamma_k) \le \sigma \ \& \\
& & & \forall i\,(1 \le i \le k)\ .\ \Gamma, x : \tau_i \vdash_\wedge M : \gamma_i.
\end{array}
$$

PROOF. The proof of ($\Leftarrow$) is easy, so we only verify that the implication ($\Rightarrow$) holds. In each point we proceed by structural induction on the derivation.

The only possible cases are the following.

(i) Case (ax). In this case we have $\Gamma(x) = \sigma$.

Case (U). Trivial, since ω is the top element of $\mathbb{T}_\wedge$.

Cases ($\le$) and ($\wedge_{\mathrm{I}}$). Straightforward from the IH.

(ii) Case ($\to_{\mathrm{E}}$). In this case $\Gamma \vdash_\wedge M : \tau \to \sigma$ and $\Gamma \vdash N : \tau$ are derivable, so we take $k \triangleq 1$, $\tau_1 \triangleq \tau$ and $\gamma_1 \triangleq \sigma$.

Case (U). Since $\omega \simeq \omega \to \omega$, we may take $k \triangleq 1$ and $\tau_1 \triangleq \gamma_1 \triangleq \omega$.

Case ($\le$). In this case $\Gamma \vdash_\wedge MN : \tau$ for some $\tau \le \sigma$, so we apply the IH and (trans).

Case ($\wedge_{\mathrm{I}}$). In this case $\sigma = \sigma_1 \wedge \sigma_2$ and $\Gamma \vdash_\wedge M : \sigma_j$ holds for $j \in \{1,2\}$. By IH, there exist $k_j \ge 1$ and $\tau_{j1}, \dots, \tau_{jk_j}, \gamma_{j1}, \dots, \gamma_{jk_j} \in \mathbb{T}_\wedge$ such that $\gamma_{j1} \wedge \cdots \wedge \gamma_{jk_j} \le \sigma_j$, $\Gamma \vdash_\wedge M : \tau_{ji} \to \gamma_{ji}$ and $\Gamma \vdash_\wedge N : \tau_{ji}$ for all $i\,(1 \le i \le k_j)$. By Lemma 13.13(ii), we derive $\gamma_{11} \wedge \cdots \wedge \gamma_{1k_1} \wedge \gamma_{21} \wedge \cdots \wedge \gamma_{2k_2} \le \sigma_1 \wedge \sigma_2$, and we are done.

(iii) Case ($\to_{\mathrm{I}}$). In this case $\sigma = \tau \to \gamma$ and $\Gamma, x : \tau \vdash_\wedge M : \gamma$, so we take $k \triangleq 1$, $\tau_1 \triangleq \tau$ and $\gamma_1 \triangleq \gamma$.

Case (U). Trivial since $\omega \simeq \omega \to \omega$ and $x \notin \mathsf{supp}(\Gamma)$ whence $\Gamma = \Gamma, x : \omega$.

Case ($\le$). By IH and (trans).

Case ($\wedge_{\mathrm{I}}$). In this case we have $\sigma = \sigma_1 \wedge \sigma_2$. For every $j \in \{1,2\}$, the IH gives $k_j \ge 1$ and $\tau_{j1}, \dots, \tau_{jk_j}, \gamma_{j1}, \dots, \gamma_{jk_j} \in \mathbb{T}_\wedge$ such that $(\tau_{j1} \to \gamma_{j1}) \wedge \cdots \wedge (\tau_{jk_j} \to \gamma_{jk_j}) \le \sigma_j$ and $\Gamma, x : \tau_i \vdash_\wedge M : \gamma_i$ for all $i\,(1 \le i \le k_j)$. We conclude by Lemma 13.13(ii). □

DEFINITION 13.23. An itt T (equivalently, $\boldsymbol{\lambda}^T_\wedge$) is called:

(i) *β-sound* if it satisfies the following property, for all intersection types σ, σ_i, τ_i $(1 \le i \le n \ge 1)$ and $\tau \not\simeq \omega$:

$$
\begin{gathered}
(\sigma_1 \to \tau_1) \wedge \cdots \wedge (\sigma_n \to \tau_n) \le \sigma \to \tau \\
\Downarrow \\
\exists k > 0, \exists \{i_1, \dots, i_k\} \subseteq \{1, \dots, n\}\,.\,[(\sigma_{i_1} \wedge \cdots \wedge \sigma_{i_k}) \ge \sigma \ \&\ (\tau_{i_1} \wedge \cdots \wedge \tau_{i_k}) \le \tau]
\end{gathered}
$$

(ii) *η-sound* whenever, for every $\sigma \in \mathbb{T}_\wedge$, there are $n \geq 1$ and $\tau_1, \ldots, \tau_n, \gamma_1, \ldots, \gamma_n \in \mathbb{T}_\wedge$ satisfying

$$\sigma \simeq (\tau_1 \to \gamma_1) \wedge \cdots \wedge (\tau_n \to \gamma_n)$$

(iii) *βη-sound* if it is both β-sound and η-sound.

REMARK 13.24. β-soundness can be expressed more compactly through the following condition (for all $\sigma, \sigma_i, \tau, \tau_i \in \mathbb{T}_\wedge$):

$$\bigwedge_{i=1}^{n}(\sigma_i \to \tau_i) \leq \sigma \to \tau \quad \Rightarrow \quad \bigwedge_{\{i \mid \sigma \leq \sigma_i\}} \tau_i \leq \tau$$

It is easy to check that the two formulations are equivalent.

The β(η)-soundness ensures that two β(η)-convertible λ-terms share the same types in the same type environments. It is customary not to include such a condition directly in the definition of an intersection type assignment system because there are systems that are interesting despite not being β-sound. Examples are the model defined by Alessi (1993) (see also Alessi et al. (2003)), and the model ABD in Section 16.2 of Barendregt et al. (2013a). Concerning η-soundness, it is only satisfied by extensional models.

When considering intersection type assignment systems that do satisfy β-soundness, we can refine the Inversion Lemma into the following statement.

THEOREM 13.25 (INVERSION LEMMA II). *Assuming that* $\boldsymbol{\lambda}_\wedge$ *is β-sound, we have:*

$$\begin{array}{lrcl} \text{(i)} & \Gamma \vdash_\wedge MN : \sigma & \iff & \exists \tau \in \mathbb{T}_\wedge . [\Gamma \vdash_\wedge M : \tau \to \sigma \ \&\ \Gamma \vdash_\wedge N : \tau]. \\ \text{(ii)} & \Gamma \vdash_\wedge \lambda x.M : \sigma \to \tau & \iff & \Gamma, x : \sigma \vdash_\wedge M : \tau \end{array}$$

PROOF. Again, the proof of (⇐) is easy so we only treat the implication (⇒).

(i) Assume $\Gamma \vdash_\wedge MN : \sigma$. By Theorem 13.22(ii) there are $\tau_1, \ldots, \tau_k, \gamma_1, \ldots, \gamma_k \in \mathbb{T}_\wedge$, for some $k \geq 1$, such that $\gamma_1 \wedge \cdots \wedge \gamma_k \leq \sigma$, $\Gamma \vdash_\wedge M : \tau_i \to \gamma_i$ and $\Gamma \vdash_\wedge N : \tau_i$ for all $i\,(1 \leq i \leq k)$. Therefore $\Gamma \vdash_\wedge N : \tau_1 \wedge \cdots \wedge \tau_k$ and $\Gamma \vdash_\wedge M : (\tau_1 \to \gamma_1) \wedge \cdots \wedge (\tau_k \to \gamma_k)$. By applying Lemma 13.12(i) and $(\mathrm{arr}_\leq)$, respectively, we have:

$$\begin{array}{rcl} (\tau_1 \to \gamma_1) \wedge \cdots \wedge (\tau_k \to \gamma_k) & \leq & (\tau_1 \wedge \cdots \wedge \tau_k) \to (\gamma_1 \wedge \cdots \wedge \gamma_k) \\ & \leq & (\tau_1 \wedge \cdots \wedge \tau_k) \to \sigma \end{array}$$

So we conclude by taking $\tau \triangleq (\tau_1 \wedge \cdots \wedge \tau_k)$.

(ii) Suppose $\Gamma \vdash_\wedge \lambda x.M : \tau \to \gamma$. By Theorem 13.22(iii), we have $k \geq 1$ and $\tau_1, \ldots, \tau_k$, $\gamma_1, \ldots, \gamma_k$ such that $(\tau_1 \to \gamma_1) \wedge \cdots \wedge (\tau_k \to \gamma_k) \leq \tau \to \gamma$, and $\Gamma, x : \tau_i \vdash_\wedge M : \gamma_i$ for all $i\,(1 \leq i \leq k)$. Assume that $\gamma \not\simeq \omega$, the other case being trivial. By β-soundness there exist indices $1 \leq i_1, \ldots, i_p \leq k$, $p \geq 1$ such that $\tau \leq \tau_{i_1} \wedge \cdots \wedge \tau_{i_p}$ and $\gamma_{i_1} \wedge \cdots \wedge \gamma_{i_p} \leq \gamma$. By applying the admissible rule $(\leq_{\mathrm{L}})$ we derive $\Gamma, x : \tau \vdash_\wedge M : \gamma_{i_j}$, for all $1 \leq j \leq p$, from which it follows $\Gamma, x : \tau \vdash_\wedge M : \gamma_{i_1} \wedge \cdots \wedge \gamma_{i_p} \leq \gamma$. □

Notice that, in general, $\Gamma \vdash_\wedge \lambda x.M : \sigma$ does not imply that $\sigma \simeq \tau \to \gamma$ with $\Gamma, x : \tau \vdash_\wedge M : \gamma$. A counterexample is $\vdash_\wedge^{\mathrm{BCD}} \mathsf{I} : (\xi_1 \to \xi_1) \wedge (\xi_2 \to \xi_2)$ for atoms $\xi_1 \neq \xi_2$.

LEMMA 13.26 (SUBSTITUTION LEMMA).
For all $M, N \in \Lambda$ and $x \in \mathrm{Var}$, we have:

$$\Gamma \vdash_\wedge M[x:=N] : \sigma \iff \exists \tau \in \mathbb{T}_\wedge . [\Gamma, x : \tau \vdash_\wedge M : \sigma \;\&\; \Gamma \vdash_\wedge N : \tau]$$

PROOF. ($\Rightarrow$) By induction on a derivation of $\Gamma \vdash_\wedge M[x:=N] : \sigma$. In case $M = x$ we have $M[x:=N] = N$ so we choose $\tau \triangleq \sigma$. Indeed, both $\Gamma, x : \sigma \vdash_\wedge x : \sigma$ and $\Gamma \vdash N : \sigma$ hold. Otherwise, we look at the last rule of the derivation.

Case (ax). In this case $M = y \neq x$ and $\Gamma(y) = \sigma$, so we take $\tau \triangleq \omega$.

Case (U). Here $\sigma = \omega$, so we choose $\tau \triangleq \omega$ again.

Case ($\to_{\mathrm{E}}$). In this case $M = M_1 M_2$ and we have $\Gamma \vdash_\wedge M_1 : \gamma \to \sigma$ and $\Gamma \vdash_\wedge M_2 : \gamma$. By IH, there exist $\tau_1, \tau_2 \in \mathbb{T}_\wedge$ such that $\Gamma, x : \tau_1 \vdash_\wedge M_1 : \gamma \to \sigma$ and $\Gamma, x : \tau_2 \vdash_\wedge M_2 : \gamma$ as well as $\Gamma \vdash_\wedge N : \tau_i$ for every $i \in \{1, 2\}$. Using the admissible rule ($\leq_{\mathrm{L}}$), we get $\Gamma, x : \tau_1 \wedge \tau_2 \vdash_\wedge M_1 : \gamma \to \sigma$ and $\Gamma, x : \tau_1 \wedge \tau_2 \vdash_\wedge M_2 : \gamma$ whence $\Gamma, x : \tau_1 \wedge \tau_2 \vdash_\wedge M_1 M_2 : \gamma$. Moreover, by applying ($\to_{\mathrm{E}}$) we have $\Gamma \vdash_\wedge N : \tau_1 \wedge \tau_2$, so we conclude taking $\tau \triangleq \tau_1 \wedge \tau_2$.

Cases ($\to_{\mathrm{I}}$) and ($\leq$). They follow easily from the IH.

Case ($\wedge_{\mathrm{I}}$). In this case $\sigma = \sigma_1 \wedge \sigma_2$ and $\Gamma \vdash_\wedge M[x:=N] : \sigma_i$ for $i \in \{1, 2\}$. By IH, there exist τ_1, τ_2 such that $\Gamma, x : \tau_i \vdash_\wedge M : \sigma_i$ and $\Gamma \vdash_\wedge N : \tau_i$ are derivable. Using the admissible rule ($\leq_{\mathrm{L}}$) we infer $\Gamma, x : \tau_1 \wedge \tau_2 \vdash_\wedge M : \sigma_i$ and using ($\wedge_{\mathrm{I}}$) we obtain $\Gamma \vdash_\wedge N : \tau_1 \wedge \tau_2$, so we may choose $\tau \triangleq \tau_1 \wedge \tau_2$.

($\Leftarrow$) By the admissible rule (cut). □

PROPOSITION 13.27. *Assume that $\boldsymbol{\lambda}_\wedge$ is β-sound and consider $M, N \in \Lambda$.*

(i) *(Subject reduction) If $M \to_\beta N$ and $\Gamma \vdash_\wedge M : \sigma$ then $\Gamma \vdash_\wedge N : \sigma$.*

(ii) *(Subject expansion) If $M \to_\beta N$ and $\Gamma \vdash_\wedge N : \sigma$ then $\Gamma \vdash_\wedge M : \sigma$.*

PROOF. By the Inversion Lemma II and the Substitution Lemma above, we have for all λ-terms P, Q that $\Gamma \vdash_\wedge (\lambda x.P)Q : \sigma$ holds if and only if $\Gamma \vdash_\wedge P[x:=Q] : \sigma$ does. Hence, both (i) and (ii) follow by an easy induction on the single-hole context $C[]$ such that $M = C[(\lambda x.P)Q]$ and $N = C[P[x:=Q]]$ using the Inversion Lemmas. □

PROPOSITION 13.28. *Assume that $\boldsymbol{\lambda}_\wedge$ is $\beta\eta$-sound and consider $M, N \in \Lambda$.*

(i) *(Extensional subject reduction) If $M \to_\eta N$ and $\Gamma \vdash_\wedge M : \sigma$ then $\Gamma \vdash_\wedge N : \sigma$.*

(ii) *(Extensional subject expansion) If $M \to_\eta N$ and $\Gamma \vdash_\wedge N : \sigma$ then $\Gamma \vdash_\wedge M : \sigma$.*

PROOF. By η-soundness, there exists an $n > 0$ such that $\sigma \simeq (\sigma_1 \to \tau_1) \wedge \cdots \wedge (\sigma_n \to \tau_n)$ for appropriate $\sigma_i, \tau_i \in \mathbb{T}_\wedge$. Since $M \to_\eta N$, there exists a single-hole context $C[]$ such that M and N have shape $M = C[\lambda x.Px]$ and $N = C[P]$, respectively, with $x \notin \mathrm{FV}(P)$. We proceed by structural induction on $C[]$. The only interesting case is $C[] = []$.

(i) Assume $\Gamma \vdash \lambda x.Px : \sigma$. By applying ($\leq$) and the rules in Example 13.20, we obtain $\Gamma \vdash \lambda x.Px : \sigma_i \to \tau_i$ for all i ($1 \leq i \leq n$). From the Inversion Lemma II we derive $\Gamma, x : \sigma_i \vdash P : \sigma_i \to \tau_i$, thus $\Gamma \vdash P : \sigma_i \to \tau_i$ by (strenghtening). By applying ($\wedge_{\mathrm{I}}$) we get $\Gamma \vdash P : (\sigma_1 \to \tau_1) \wedge \cdots \wedge (\sigma_n \to \tau_n)$, from which we conclude $\Gamma \vdash P : \sigma$ by ($\leq$).

(ii) Assume $\Gamma \vdash P : \sigma$ by ($\leq$). Proceeding as above, we obtain $\Gamma \vdash P : \sigma_i \to \tau_i$ for all $i\,(1 \leq i \leq n)$. By (weakening) we have $\Gamma, x : \sigma_i \vdash P : \sigma_i \to \tau_i$, for all $x \notin \mathsf{supp}(\Gamma)$, and since $\Gamma, x : \sigma_i \vdash P : \sigma_i$ holds, we obtain $\Gamma, x : \sigma_i \vdash Px : \tau_i$. We can now apply rule ($\to_{\mathrm{I}}$) and derive $\Gamma \vdash \lambda x.Px : \sigma_i \to \tau_i$. Conclude $\Gamma \vdash \lambda x.Px : \sigma$ by ($\leq$).

All the other cases follow easily from the IH by applying the Inversion Lemmas. □

DEFINITION 13.29. For $T \in \{\mathrm{BCD}, \mathrm{Scott}, \mathrm{CDZ}\}$ and $\sigma \in \mathbb{T}_\wedge^T$, define the *expanded form* $\mathrm{ef}_T(\sigma)$ *of* σ satisfying $\sigma \simeq^T \mathrm{ef}_T(\sigma)$ by induction on σ:

$$
\begin{array}{lcll}
\mathrm{ef}_{\mathrm{BCD}}(\xi_i) & \triangleq & \xi_i & \text{(for all } \xi_i \in \mathbb{A}^{\mathrm{BCD}}) \\
\mathrm{ef}_{\mathrm{Scott}}(\xi) & \triangleq & \omega \to \xi & \\
\mathrm{ef}_{\mathrm{CDZ}}(\star) & \triangleq & \varepsilon \to \star & \\
\mathrm{ef}_{\mathrm{CDZ}}(\varepsilon) & \triangleq & \star \to \varepsilon & \\
\mathrm{ef}_T(\omega) & \triangleq & \omega \to \omega & \\
\mathrm{ef}_T(\tau \wedge \gamma) & \triangleq & \mathrm{ef}_T(\tau) \wedge \mathrm{ef}_T(\gamma) & \\
\mathrm{ef}_T(\tau \to \gamma) & \triangleq & \tau \to \gamma &
\end{array}
$$

THEOREM 13.30. (i) *The type systems* $\boldsymbol{\lambda}_\wedge^{\mathrm{BCD}}$, $\boldsymbol{\lambda}_\wedge^{\mathrm{Scott}}$ *and* $\boldsymbol{\lambda}_\wedge^{\mathrm{CDZ}}$ *are* β*-sound.*
(ii) $\boldsymbol{\lambda}_\wedge^{\mathrm{Scott}}$ *and* $\boldsymbol{\lambda}_\wedge^{\mathrm{CDZ}}$ *are* η*-sound, while* $\boldsymbol{\lambda}_\wedge^{\mathrm{BCD}}$ *is not.*

PROOF. (i) Let $T \in \{\mathrm{BCD}, \mathrm{Scott}, \mathrm{CDZ}\}$. We perform an induction loading and prove the following statement. Let $\sigma, \sigma' \in \mathbb{T}_\wedge^T$ satisfying $\sigma \leq^T \sigma'$ and having expanded forms[2]

$$
\begin{array}{lcl}
\mathrm{ef}_T(\sigma) & = & \xi_1 \wedge \cdots \wedge \xi_n \wedge (\tau_1 \to \gamma_1) \wedge \cdots \wedge (\tau_k \to \gamma_k) \\
\mathrm{ef}_T(\sigma') & = & \xi'_1 \wedge \cdots \wedge \xi'_{n'} \wedge (\tau'_1 \to \gamma'_1) \wedge \cdots \wedge (\tau'_{k'} \to \gamma'_{k'})
\end{array}
$$

where $n, n', k, k' \geq 0$, $n + k, n' + k' \geq 1$ and $\xi_i, \xi'_j \in \mathbb{A}^T$ for all indices $i\,(1 \leq i \leq k)$ and $j\,(1 \leq j \leq k')$. Then, for all $j \in \{1, \ldots, k'\}$, we have that $\gamma'_j \not\simeq^T \omega$ entails

$$\exists p \geq 1, \exists i_1, \ldots, i_p \in \{1, \ldots, k\} . [\, \tau'_j \leq^T \tau_{i_1} \wedge \cdots \wedge \tau_{i_p} \;\&\; \gamma_{i_1} \wedge \cdots \wedge \gamma_{i_p} \leq^T \gamma'_j \,].$$

Note that we may only have $k = 0$ or $k' = 0$ when $T = \mathrm{BCD}$ because in the other type assignment systems the expanded form of an atom is always an arrow type.

We proceed by induction on a derivation of $\sigma \leq \sigma'$.

The specific cases (ξ) for $T = \mathrm{Scott}$ and ($\varepsilon \leq \star$), (ε), ($\star$) for $T = \mathrm{CDZ}$ are easy.

Case (refl). In this case $\sigma = \sigma'$ and $\mathrm{ef}_T(\sigma) = \mathrm{ef}_T(\sigma')$, so we take $p \triangleq 1$ and $i_p \triangleq j$.

Case ($\mathrm{inc}_{\mathrm{L}}$). This case is immediate since $\sigma = \sigma' \wedge \delta$ for some $\delta \in \mathbb{T}_\wedge$.

Case ($\mathrm{inc}_{\mathrm{R}}$). Analogous to the previous case.

Cases (top) and (arr_ω). Since $\mathrm{ef}_T(\omega) = \omega \to \omega$, there is nothing to prove.

Case ($\mathrm{arr}_\wedge$). Immediate.

Cases ($\mathrm{arr}_\leq$) and (glb). These cases follow from the IH.

Case (trans). In this case there exists $\delta \in \mathbb{T}_\wedge$ such that $\sigma \leq^T \delta$ and $\delta \leq^T \sigma'$, then

$$\mathrm{ef}_T(\delta) = \xi''_1 \wedge \cdots \wedge \xi''_{n''} \wedge (\tau''_1 \to \gamma''_1) \wedge \cdots \wedge (\tau''_{k''} \to \gamma''_{k''})$$

[2] The equality here is considered up to commutativity as well.

for some $n'', k'' \geq 0$. For $T = \mathrm{BCD}$, $k'' \geq 1$ holds because the atoms in $\mathbb{A}^{\mathrm{BCD}}$ are incomparable with arrow types $\tau \to \gamma$, unless $\gamma \simeq^{\mathrm{BCD}} \omega$. For the other type systems, the fact that $k'' \geq 1$ follows directly from the definition of expanded form. By IH, for every $h \in \{1, \ldots, k'\}$, if $\gamma'_h \not\simeq^T \omega$ then there is a non-empty $\mathcal{I}_h \subseteq \{1, \ldots, k''\}$ such that

$$\tau'_h \leq^T \bigwedge_{i \in \mathcal{I}_h} \tau''_i \quad \& \quad \bigwedge_{i \in \mathcal{I}_h} \gamma''_i \leq^T \gamma'_h$$

By IH, for every $i \in \mathcal{I}_h$, if $\gamma''_i \not\simeq^T \omega$ then there is a non-empty $\mathcal{J}_i \subseteq \{1, \ldots, k\}$ such that

$$\tau''_h \leq^T \bigwedge_{j \in \mathcal{J}_i} \tau_j \quad \& \quad \bigwedge_{j \in \mathcal{J}_i} \gamma_j \leq^T \gamma''_i$$

By Lemma 13.13(ii) and the fact that ω is the neutral element of $\wedge$, we obtain setting

$$\mathcal{L}_h \triangleq \bigcup_{i \in \mathcal{I}_h} \mathcal{J}_i \qquad (\text{for } h \in \{1, \ldots, k'\})$$

that $\tau'_h \leq^T \bigwedge_{\ell \in \mathcal{L}} \tau_\ell$ and $\bigwedge_{\ell \in \mathcal{L}} \gamma_\ell \leq^T \gamma'_h$. This concludes the proof of (i).

(ii) For $T \in \{\mathrm{Scott}, \mathrm{CDZ}\}$, verify by induction on σ that $\mathrm{ef}_T(\sigma)$ is an intersection of arrow types. Conclude using the fact that $\sigma \simeq^T \mathrm{ef}_T(\sigma)$.

The fact that $\boldsymbol{\lambda}_\wedge^{\mathrm{BCD}}$ is not η-sound is given as an exercise below. □

EXERCISE 13.31. Consider $\boldsymbol{\lambda}_\wedge^{\mathrm{BCD}}$.

(i) Show that the extensional subject reduction holds.

(ii) Show that the extensional subject expansion does not hold.

13.2 Filter models in logical form

Intersection type theories are tightly related to filter models, namely models of λ-calculus having as underlying set a collection of filters of types. Indeed, another way of understanding intersection types is in terms of the domain $(\mathcal{F}, \sqsubseteq)$ which they serve to define. More precisely, one can define an interpretation function $(\cdot)^* : \mathbb{T}_\wedge \to \mathcal{F}$ mapping

- atomic types $\xi \in \mathbb{A}_\omega$ to compact elements $\xi^* \in \mathcal{K}(\mathcal{F})$, the correspondence being bijective but anti-monotonic ($\xi_1 \leq \xi_2 \iff \xi_2^* \sqsubseteq \xi_1^*$). In particular, $\omega^* = \bot_{\mathcal{F}}$;
- arrow types $\sigma \to \tau$ to step functions $(\sigma^* \mapsto \tau^*)$;
- intersections $\sigma \wedge \tau$ to joins $\sigma^* \sqcup \tau^*$.

The precise correspondence is formally given by the so-called Stone duality, as already noticed in Barendregt et al. (1983). This duality has been subsequently explored by Abramsky (1991), Alessi et al. (2004), and Amadio and Curien (1998), Chapter 10.

In this section we show how to present intersection type theories and $\mathcal{D}_\infty$-models as filter models. We mainly work in the category $\omega\mathbf{Alg}$ of ω-algebraic complete lattices and continuous functions. This category is defined in the technical Appendix A.3.

Intersection type theories as filter models

Intersection type theories (itt's) $T = (\mathbb{T}^T_\wedge, \leq^T)$ have been introduced in Definition 13.8(v). For the sake of readability, we often omit the superscript T and simply write $\mathbb{T}_\wedge$ and $\leq$.

DEFINITION 13.32. Let T be an itt and $F \subseteq \mathbb{T}^T_\wedge$.

(i) We say that F is a *filter of types* if:

(1) $\omega \in F$;

(2) $\sigma, \tau \in F$ entails $\sigma \wedge \tau \in F$;

(3) $\sigma \in F$ and $\sigma \leq \tau$ implies $\tau \in F$.

(ii) Define $\mathcal{F}_T \triangleq \{F \subseteq \mathbb{T}^T_\wedge \mid F \text{ is a filter of types }\}$.

(iii) For $X \subseteq \mathbb{T}^T_\wedge$, write $X\!\uparrow$ for the smallest filter of types including X.

(iv) For $\sigma \in \mathbb{T}^T_\wedge$, write $\sigma\!\uparrow \triangleq \{\tau \in \mathbb{T}^T_\wedge \mid \sigma \leq \tau\}$ for the principal filter generated by σ.

REMARK 13.33. (i) For every $\sigma \in \mathbb{T}_\wedge$, $\sigma\!\uparrow$ is a filter.

(ii) Given $X \subseteq \mathbb{T}_\wedge$, we have

$$X\!\uparrow = \bigcup\{(\sigma_1 \wedge \cdots \wedge \sigma_n)\!\uparrow \mid n \in \mathbb{N} \;\&\; \sigma_1, \ldots, \sigma_n \in X\}$$

In particular, for a finite $X = \{\sigma_1, \ldots, \sigma_n\}$, we have $X\!\uparrow = (\sigma_1 \wedge \cdots \wedge \sigma_n)\!\uparrow$. If X is the empty set, then $X\!\uparrow = \emptyset\!\uparrow = [\omega]_{\simeq}$.

PROPOSITION 13.34. *Let T be an itt.*

(i) *$(\mathcal{F}_T, \subseteq)$ is an ω-algebraic complete lattice having $\mathbb{T}^T_\wedge$ as top and $\omega\!\uparrow$ as bottom.*

(ii) *For $F, G \in \mathcal{F}_T$, the join $\sqcup$ and meet $\sqcap$ are respectively given by*

$$\begin{aligned} F \sqcup G &= (F \cup G)\!\uparrow \\ F \sqcap G &= F \cap G \end{aligned}$$

(iii) *Similarly, for $\mathcal{X} \subseteq \mathcal{F}_T$, its supremum is $\bigsqcup \mathcal{X} = (\bigcup \mathcal{X})\!\uparrow$.*

(iv) *The set $\mathcal{K}(\mathcal{F}_T)$ of compact elements of $\mathcal{F}_T$ is given by*

$$\mathcal{K}(\mathcal{F}_T) = \{\sigma\!\uparrow \mid \sigma \in \mathbb{T}_\wedge\}$$

(v) *For $\sigma, \tau \in \mathbb{T}_\wedge$, we have $\sigma\!\uparrow \sqcup \tau\!\uparrow = (\sigma \wedge \tau)\!\uparrow$.*

PROOF. Easy. □

REMARK 13.35. (i) In the category ω**Alg**, the exponential object $\mathcal{F}_T \Rightarrow \mathcal{F}_T$ is given by the set of continuous functions from $\mathcal{F}_T$ to $\mathcal{F}_T$, endowed with pointwise ordering $\sqsubseteq$.

(ii) A continuous function $f : \mathcal{D} \to \mathcal{E}$ between ω-algebraic complete lattices is uniquely determined by its behavior on the compact elements in $\mathcal{K}(\mathcal{D})$.

(iii) The compact elements populating $\mathcal{F}_T \Rightarrow \mathcal{F}_T$ are finite suprema of step functions $(\sigma\!\uparrow \mapsto \tau\!\uparrow)$ for $\sigma, \tau \in \mathbb{T}_\wedge$. In other words, every $f \in \mathcal{K}(\mathcal{F}_T \Rightarrow \mathcal{F}_T)$ has the shape

$$f = (\sigma_1\!\uparrow \mapsto \tau_1\!\uparrow) \sqcup \cdots \sqcup (\sigma_n\!\uparrow \mapsto \tau_n\!\uparrow)$$

for appropriate $\sigma_i, \tau_i \in \mathbb{T}_\wedge$ $(1 \leq i \leq n)$.

DEFINITION 13.36. Given an itt T, define two maps

$$\begin{array}{rl} \mathrm{App}^T: & \mathcal{F}_T \to (\mathcal{F}_T \Rightarrow \mathcal{F}_T) \\ \mathrm{Lam}^T: & (\mathcal{F}_T \Rightarrow \mathcal{F}_T) \to \mathcal{F}_T \end{array}$$

by setting (for all $F, G \in \mathcal{F}_T$ and $f : \mathcal{F}_T \to \mathcal{F}_T$ continuous):

$$\begin{array}{rcl} \mathrm{App}^T(F)(G) & \triangleq & \{\tau \in \mathbb{T}^T_\wedge \mid \exists \sigma \in G . (\sigma \to \tau) \in F\} \\ \mathrm{Lam}^T(f) & \triangleq & \{\sigma \to \tau \mid \tau \in f(\sigma\uparrow)\}\uparrow \end{array}$$

The triple $(\mathcal{F}_T, \mathrm{App}^T, \mathrm{Lam}^T)$ is called the *filter structure* induced by T.

We check that App^T and Lam^T are actually morphisms in the category $\omega\mathbf{Alg}$. As a matter of notations, we write $F \cdot^T G$ for $\mathrm{App}^T(F)(G)$.

LEMMA 13.37. *Let T be an itt.*
(i) *For every $F, G \in \mathcal{F}_T$, $F \cdot^T G$ is a filter.*
(ii) App^T *and* Lam^T *are continuous.*

PROOF. (i) We verify the three conditions of Definition 13.32(i).
(1) $\omega \in (F \cdot^T G)$ since $\omega \leq \omega \to \omega$ entails $\omega \to \omega \in F$.
(2) Given $\tau_1, \tau_2 \in (F \cdot^T G)$, there are $\sigma_1, \sigma_2 \in G$ such that $\sigma_i \to \tau_i \in F$, for $i = 1, 2$. Since F is a filter, we get $(\sigma_1 \to \tau_1) \wedge (\sigma_2 \to \tau_2) \in F$ (condition (2) of Definition 13.32(i)). By Lemma 13.12(i) and condition (3), we obtain $(\sigma_1 \wedge \sigma_2) \to (\tau_1 \wedge \tau_2) \in F$, from which it follows $\tau_1 \wedge \tau_2 \in (F \cdot^T G)$ by definition of application $- \cdot^T -$.
(3) From the covariance of the arrow given in $(\mathrm{arr}_\leq)$.
(ii) Easy. □

LEMMA 13.38. *Let T be an itt, $\sigma, \tau \in \mathbb{T}^T_\wedge$, $F, G \in \mathcal{F}_T$ and $f : \mathcal{F}_T \to \mathcal{F}_T$ continuous.*
(i) $\tau \in F \cdot^T (\sigma\uparrow) \iff (\sigma \to \tau) \in F$.
(ii) $\tau \in f(\sigma\uparrow) \Rightarrow (\sigma \to \tau) \in \mathrm{Lam}^T(f)$.
(iii) $\mathrm{Lam}^T(\sigma\uparrow \mapsto \tau\uparrow) = (\sigma \to \tau)\uparrow$.

PROOF. (i) By definition, $\tau \in F \cdot^T (\sigma\uparrow)$ if and only if there exists a type $\sigma' \in \sigma\uparrow$ such that $(\sigma' \to \tau) \in F$. By covariance $(\mathrm{arr}_\leq)$, this is equivalent to $(\sigma \to \tau) \in F$.
(ii) Easy.
(iii) We have $\mathrm{Lam}^T(\sigma\uparrow \mapsto \tau\uparrow) = \{(\gamma \to \delta) \mid \delta \in (\sigma\uparrow \mapsto \tau\uparrow)(\gamma\uparrow)\}\uparrow = (\sigma \to \tau)\uparrow$. We discuss the last equality.

($\supseteq$) This inclusion holds since $(\sigma \to \tau)$ is one of the $\gamma \to \delta$;

($\subseteq$) Take a $\gamma \to \delta$ such that $\delta \in (\sigma\uparrow \mapsto \tau\uparrow)(\gamma\uparrow)$. This means that both $\gamma \leq \sigma$ and $\tau \leq \delta$ hold, from which $\sigma \to \tau \leq \gamma \to \delta$ follows by $(\mathrm{arr}_\leq)$. □

PROPOSITION 13.39. *Let T be an itt. Then $(\mathrm{App}^T, \mathrm{Lam}^T)$ constitutes a Galois connection, which means:*
(i) $\mathrm{App}^T \circ \mathrm{Lam}^T \sqsupseteq \mathrm{Id}_{\mathcal{F}_T \Rightarrow \mathcal{F}_T}$;
(ii) $\mathrm{Lam}^T \circ \mathrm{App}^T \sqsubseteq \mathrm{Id}_{\mathcal{F}_T}$.

PROOF. (i) We need to show $f(F) \subseteq \mathrm{Lam}^T(f) \cdot^T F$, for all $f \in \mathcal{F}_T \Rightarrow \mathcal{F}_T$ and $F \in \mathcal{F}_T$. By continuity, it is sufficient to consider $F \in \mathcal{K}(\mathcal{F}_T)$, namely $F = \sigma\!\uparrow$ for some $\sigma \in \mathbb{T}^T_\wedge$.

$$\begin{array}{rcll} f(\sigma\!\uparrow) & = & \{\tau \mid \tau \in f(\sigma\!\uparrow)\} & \\ & \subseteq & \{\tau \mid \sigma \to \tau \in \mathrm{Lam}^T(f)\}, & \text{Lemma 13.38(ii)}, \\ & = & \mathrm{Lam}^T(f) \cdot^T \sigma\!\uparrow, & \text{Lemma 13.38(i)}. \end{array}$$

(ii) For $F \in \mathcal{F}_T$, we have:

$$\begin{array}{rcll} \mathrm{Lam}^T(\mathrm{App}^T(F)) & = & \{\sigma \to \tau \mid \tau \in F \cdot \sigma\!\uparrow\}\!\uparrow & \\ & = & \{\sigma \to \tau \mid \sigma \to \tau \in F\}\!\uparrow, & \text{by Lemma 13.38(i)}, \\ & \subseteq & F. & \end{array}$$

□

In general the object $\mathcal{F}_T$ is not reflexive because $(\mathrm{App}^T, \mathrm{Lam}^T)$ only provides a Galois connection. We show that $\mathcal{F}_T$ is a categorical model whenever the itt T is β-sound.

PROPOSITION 13.40. *Let T be β-sound. Then*

(i) *$\mathcal{F}_T$ is a reflexive object in ω**Alg** via $(\mathrm{App}^T, \mathrm{Lam}^T)$.*

(ii) *If T is η-sound, then the categorical model $\mathcal{F}_T$ is extensional.*

(iii) *The interpretation $[\![-]\!]^{\mathcal{F}_T}_{(\cdot)} : \Lambda \times \mathrm{Val}_{\mathcal{F}_T} \to \mathcal{F}_T$ given in Definition 12.19*(ii) *can be characterized as follows:*

$$\begin{array}{rcl} [\![x]\!]^{\mathcal{F}_T}_\rho & = & \rho(x); \\ [\![MN]\!]^{\mathcal{F}_T}_\rho & = & \{\tau \in \mathbb{T}^T_\wedge \mid \exists \sigma \in [\![N]\!]^{\mathcal{F}_T}_\rho . \sigma \to \tau \in [\![M]\!]^{\mathcal{F}_T}_\rho\}; \\ [\![\lambda x.M]\!]^{\mathcal{F}_T}_\rho & = & \{\sigma \to \tau \in \mathbb{T}^T_\wedge \mid \tau \in [\![M]\!]^{\mathcal{F}_T}_{\rho[x:=\sigma\uparrow]}\}\!\uparrow. \end{array}$$

PROOF. (i) By Remark 13.35(ii), it is enough to show $\mathrm{Lam}^T(f) \cdot^T \gamma\!\uparrow \subseteq f(\gamma\!\uparrow)$ for all $f \in \mathcal{K}(\mathcal{F}_T \Rightarrow \mathcal{F}_T)$ and $\gamma \in \mathbb{T}_\wedge$. Thus, we may assume $f = (\sigma_1\!\uparrow \mapsto \tau_1\!\uparrow) \sqcup \cdots \sqcup (\sigma_n\!\uparrow \mapsto \tau_n\!\uparrow)$ for appropriate $\sigma_j, \tau_j \in \mathbb{T}_\wedge$ $(1 \le j \le n)$. We proceed as follows:

$$\begin{array}{rll} & \mathrm{Lam}^T(f) \cdot^T \gamma\!\uparrow & \\ = & (\bigsqcup_{j=1}^n \mathrm{Lam}^T(\sigma_j\!\uparrow \mapsto \tau_j\!\uparrow)) \cdot^T (\gamma\!\uparrow), & \text{by Lemma 13.37(ii)}, \\ = & (\bigsqcup_{j=1}^n (\sigma_j \to \tau_j)\!\uparrow) \cdot^T (\gamma\!\uparrow), & \text{by Lemma 13.38(iii)}, \\ = & (\bigwedge_{j=1}^n (\sigma_j \to \tau_j))\!\uparrow \cdot^T (\gamma\!\uparrow), & \text{by Proposition 13.34(v)}, \\ = & \{\delta \mid \exists \gamma' \in \gamma\!\uparrow . (\gamma' \to \delta) \in (\bigwedge_{j=1}^n (\sigma_j \to \tau_j))\!\uparrow\}, & \text{by definition of } \mathrm{App}^T, \\ = & \{\delta \mid \exists \gamma' \ge \gamma . \bigwedge_{j=1}^n (\sigma_j \to \tau_j) \le (\gamma' \to \delta)\} & \\ = & \{\delta \mid (\sigma_1 \to \tau_1) \wedge \cdots \wedge (\sigma_n \to \tau_n) \le (\gamma \to \delta)\}, & \text{by } (\mathrm{arr}_\le), \\ = & \{\delta \mid \exists k > 0, i_1, \ldots, i_k \in \{1, \ldots, n\} . & \\ & \quad \tau_{i_1} \wedge \cdots \wedge \tau_{i_k} \le \delta \;\&\; \gamma \le \sigma_{i_1} \wedge \cdots \wedge \sigma_{i_k} \le \delta\} \cup \omega\!\uparrow, & \text{by } \beta\text{-soundness}, \\ = & (\bigwedge_{\{i \mid \gamma \le \sigma_i \,\&\, 1 \le i \le n\}} \tau_i)\!\uparrow, & \text{by Remark 13.24}, \\ = & \bigsqcup_{\{i \mid \gamma \le \sigma_i \,\&\, 1 \le i \le n\}} \tau_i\!\uparrow, & \text{by Proposition 13.34(v)}, \\ = & \bigsqcup \{\tau_j\!\uparrow \mid \sigma_j\!\uparrow \subseteq \gamma\!\uparrow \;\&\; 1 \le j \le n\} & \\ = & f(\gamma\!\uparrow) & \end{array}$$

(ii) By Proposition 13.39(ii), it is enough to show $F \subseteq \mathrm{Lam}^T(\mathrm{App}^T(F))$ for every $F \in \mathcal{F}_T$. Take $\sigma \in F$. By η-soundness, there exist $\sigma_j, \tau_j \in \mathbb{T}_\wedge$ $(1 \leq j \leq k > 0)$ such that $\sigma \simeq \bigwedge_{j=1}^k(\sigma_j \to \tau_j) \in F$. By upward closure, we get $\sigma_j \to \tau_j \in F$ for all such j. Thus:

$$\begin{array}{rll} \sigma \in & (\bigwedge_{j=1}^k(\sigma_j \to \tau_j))\uparrow, & \text{by } \bigwedge_{j=1}^k(\sigma_j \to \tau_j) \leq \sigma, \\ \subseteq & \{(\bigwedge_{i=1}^n(\sigma_i \to \tau_i))\uparrow \mid \forall i \in \{1,\dots,n\} \,.\, \sigma_i \to \tau_i \in F\}, & \text{since } \sigma_j \to \tau_j \in F, \\ = & \{\sigma \to \tau \mid \sigma \to \tau \in F\}\uparrow, & \text{by Remark 13.33(ii)}, \\ = & \{\sigma \to \tau \mid \tau \in F \cdot \sigma\uparrow\}\uparrow, & \text{by Lemma 13.38(i)}, \\ = & \mathrm{Lam}^T(\mathrm{App}^T(F)). & \end{array}$$

(iii) The first case is immediate. Application and abstraction follow easily from Definition 12.19(ii) and the definition of App^T and Lam^T, respectively. □

DEFINITION 13.41. Let T be an itt, Γ be a type environment and $\rho \in \mathrm{Val}_{\mathcal{F}_T}$.

(i) We say that Γ *agrees with* ρ, written $\Gamma \models \rho$, if for all $x \in \mathrm{dom}(\Gamma)$ we have

$$\Gamma(x) = \sigma \quad \Rightarrow \quad \sigma \in \rho(x)$$

(ii) We write $\Gamma \backslash x$ for the restriction of Γ to $\mathrm{dom}(\Gamma) - \{x\}$.

LEMMA 13.42. (i) $\Gamma \models \rho$ *and* $\Gamma' \models \rho$ *entail* $\Gamma \wedge \Gamma' \models \rho$.

(ii) *If* $\Gamma \models \rho[x := \sigma\uparrow]$ *then* $\Gamma\backslash x \models \rho$.

PROOF. Immediate. □

THEOREM 13.43. *Let T be a β-sound itt. For all $M \in \Lambda$ and $\rho \in \mathcal{F}_T$, we have:*

$$[\![M]\!]_\rho^{\mathcal{F}_T} = \{\sigma \in \mathbb{T}_\wedge^T \mid \Gamma \vdash_\wedge M : \sigma \text{ for some } \Gamma \models \rho\}$$

PROOF. By structural induction on M.

Case $M = x$. We have

$$\begin{array}{rcl} [\![x]\!]_\rho^{\mathcal{F}_T} & = & \rho(x), \\ & = & \{\sigma \mid \sigma \in \rho(x)\} \\ & = & \{\sigma \mid x : \sigma \vdash x : \sigma \;\&\; \sigma \in \rho(x)\}, \quad \text{by (ax)}. \end{array}$$

Case $M = PQ$. We have

$$\begin{array}{rcl} [\![PQ]\!]_\rho^{\mathcal{F}_T} & = & [\![P]\!]_\rho^{\mathcal{F}_T} \cdot^T [\![Q]\!]_\rho^{\mathcal{F}_T} \\ & = & \{\sigma \mid \exists \tau \in [\![Q]\!]_\rho^{\mathcal{F}_T} \,.\, (\tau \to \sigma) \in [\![P]\!]_\rho^{\mathcal{F}_T}\} \\ & = & \{\sigma \mid \exists k > 0, \exists \tau_1, \dots, \tau_k, \gamma_1, \dots, \gamma_k \,.\, \forall i\,(1 \leq i \leq k) \\ & & [(\tau_i \to \gamma_i) \in [\![P]\!]_\rho^{\mathcal{F}_T} \;\&\; \tau_i \in [\![Q]\!]_\rho^{\mathcal{F}_T} \;\&\; (\wedge_{i=1}^k \gamma_i) \leq \sigma]\} \cup \{\omega\}\uparrow, \\ & & \text{by Lemma 13.37(i) and Remark 13.33(ii)}, \\ & = & \{\sigma \mid \exists k > 0, \exists \tau_1, \dots, \tau_k, \gamma_1, \dots, \gamma_k, \Gamma_1, \dots, \Gamma_k, \Gamma'_1, \dots, \Gamma'_k \,. \\ & & \forall i\,(1 \leq i \leq n)\,[\Gamma_i, \Gamma'_i \models \rho \;\&\; \Gamma_i \vdash_\wedge P : \tau_i \to \gamma_i \\ & & \;\&\; \Gamma'_i \vdash Q : \tau_i \;\&\; (\wedge_{i=1}^k \gamma_i) \leq \sigma]\} \cup \{\omega\}\uparrow, \qquad \text{by IH}, \\ & = & \{\sigma \mid \bigwedge_{i=1}^k(\Gamma_i \wedge \Gamma'_i) \vdash_\wedge PQ : \sigma \;\&\; \bigwedge_{i=1}^k(\Gamma_i \wedge \Gamma'_i) \models \rho\}, \end{array}$$

where $\Gamma_i, \Gamma'_i \models \rho$ is short for $\Gamma_i \models \rho$ and $\Gamma'_i \models \rho$, and the last equality holds by Theorem 13.22(ii) and Lemma 13.42(i).

Case $M = \lambda x.N$. We have:

$$\begin{array}{rcll} [\![\lambda x.N]\!]^{\mathcal{F}_T}_\rho & = & \{\tau \to \gamma \mid \gamma \in [\![N]\!]^{\mathcal{F}_T}_{\rho[x:=\tau\uparrow]}\}\uparrow, & \text{by Prop. 13.40(iii)}, \\ & = & \{\sigma \mid \exists k > 0, \exists \tau_1, \ldots, \tau_k, \gamma_1, \ldots, \gamma_k, \Gamma_1, \ldots, \Gamma_k \,. & \\ & & [\Gamma_i \models \rho[x := \tau_i\uparrow] \;\&\; \Gamma_i, x : \tau_i \vdash_\wedge N : \gamma_i & \\ & & \&\; \bigwedge_{i=1}^k (\tau_i \to \gamma_i) \leq \sigma]\}, & \text{by IH}, \\ & = & \{\sigma \mid \Gamma \backslash x \vdash \lambda x.N : \sigma \;\&\; \Gamma \models \rho\}, & \text{for } \Gamma \triangleq \bigwedge_{i=1}^k \Gamma_i, \end{array}$$

where the last equality holds by Theorem 13.22(iii) and Lemma 13.42(ii). □

COROLLARY 13.44. *Let T be a β-sound itt. For all $M \in \Lambda^o$, we have*

$$[\![M]\!]^{\mathcal{F}_T}_\rho = \{\sigma \in \mathbb{T}^T_\wedge \mid \;\vdash_\wedge M : \sigma\}$$

$\mathcal{D}_\infty$-models as filter models

We describe a class of categorical models living in ω**Alg** that are built by mimicking the order-theoretic limit construction originally employed by Scott (1972) to solve the fundamental domain equation $\mathcal{D} \cong (\mathcal{D} \Rightarrow \mathcal{D})$, where $\cong$ stands for an order-isomorphism. We call the models obtained in this way *$\mathcal{D}_\infty$-models*. Our aim is to provide a reasonably self-contained presentation in order to discuss some examples in the next section, but without the ambition of developing all the proofs of standard properties. For a more detailed presentation of this material, the reader is spoilt for choice: B[1984], Chapter 18.2; Amadio and Curien (1998), Chapter 3; Plotkin (1982); Schmidt (1986) and Barendregt et al. (2013a), Chapter 16. Our presentation mainly follows this last reference.

For the general properties of the category ω**Alg**, consult Appendix A.3.

DEFINITION 13.45. Let $\mathcal{D}, \mathcal{E}$ be ω-algebraic complete lattices.

(i) An *embedding-projection pair* from $\mathcal{D}$ to $\mathcal{E}$, written $(\mathsf{i}, \mathsf{j}) : \mathcal{D} \to \mathcal{E}$, is given by a pair of morphisms

$$\begin{array}{l} \mathsf{i} : \mathcal{D} \to \mathcal{E} \\ \mathsf{j} : \mathcal{E} \to \mathcal{D} \end{array}$$

satisfying

$$\begin{array}{rcl} \mathsf{i} \circ \mathsf{j} & \sqsubseteq & \mathrm{Id}_\mathcal{E} \\ \mathsf{j} \circ \mathsf{i} & = & \mathrm{Id}_\mathcal{D}. \end{array}$$

We call i the *embedding* and j the *projection* of the pair.

(ii) Embedding-projection pairs $(\mathsf{i}_0, \mathsf{j}_0) : \mathcal{D} \to \mathcal{D}'$ and $(\mathsf{i}_1, \mathsf{j}_1) : \mathcal{D}' \to \mathcal{E}$ compose as follows

$$(\mathsf{i}_0, \mathsf{j}_0) \circ (\mathsf{i}_1, \mathsf{j}_1) \triangleq (\mathsf{i}_1 \circ \mathsf{i}_0, \mathsf{j}_0 \circ \mathsf{j}_1) : \mathcal{D} \to \mathcal{E}$$

(iii) Embedding-projection pairs of the form $(\mathsf{i}, \mathsf{j}) : \mathcal{D} \to (\mathcal{D} \Rightarrow \mathcal{D})$ are called *functional* and are denoted by the triple $(\mathcal{D}, \mathsf{i}, \mathsf{j})$.

REMARK 13.46. In an embedding-projection pair $(\mathsf{i},\mathsf{j}) : \mathcal{D} \to \mathcal{E}$:
(i) the embedding i is uniquely determined by the projection j, and *vice versa*;
(ii) both i and j are *strict*, i.e., $\mathsf{i}(\bot_{\mathcal{D}}) = \bot_{\mathcal{E}}$ and $\mathsf{j}(\bot_{\mathcal{E}}) = \bot_{\mathcal{D}}$.

The following construction allows to embed any ω-algebraic complete lattice $\mathcal{D}_0$ into an ω-algebraic complete lattice $\mathcal{D}_\infty$ satisfying $\mathcal{D}_\infty \cong (\mathcal{D}_\infty \Rightarrow \mathcal{D}_\infty)$.

DEFINITION 13.47. Let $(\mathcal{D}_0, \mathsf{i}_0, \mathsf{j}_0)$ be a functional embedding-projection pair.
(i) The *tower* generated by $(\mathsf{i}_0, \mathsf{j}_0)$ is the sequence of embedding-projection pairs

$$(\mathsf{i}_n, \mathsf{j}_n) : \mathcal{D}_n \to \mathcal{D}_{n+1}$$

where, for every $n \in \mathbb{N}$, we set:

$$\begin{array}{lcll} \mathcal{D}_{n+1} & \triangleq & \mathcal{D}_n \Rightarrow \mathcal{D}_n, & \\ \mathsf{i}_{n+1}(f) & \triangleq & \mathsf{i}_n \circ f \circ \mathsf{j}_n, & \text{for all } f \in \mathcal{D}_n, \\ \mathsf{j}_{n+1}(g) & \triangleq & \mathsf{j}_n \circ g \circ \mathsf{i}_n, & \text{for all } g \in \mathcal{D}_{n+1}. \end{array}$$

(ii) The *inverse limit* $\mathcal{D}_\infty$ of a tower $(\mathsf{i}_n, \mathsf{j}_n)_{n \in \mathbb{N}}$ as above is given by

$$\mathcal{D}_\infty \triangleq \{\langle d_n \rangle_{n \in \mathbb{N}} \mid \forall n \in \mathbb{N} . [\, d_n \in \mathcal{D}_n \;\&\; \mathsf{j}_n(d_{n+1}) = d_n \,]\}$$

partially ordered by the relation

$$\langle d_n \rangle_{n \in \mathbb{N}} \sqsubseteq_\infty \langle e_n \rangle_{n \in \mathbb{N}} \iff \forall n \in \mathbb{N} . d_n \sqsubseteq e_n$$

In this context $(\mathcal{D}_0, \mathsf{i}_0, \mathsf{j}_0)$ is called the *initial pair* of $\mathcal{D}_\infty$.
(iii) For $m \in \mathbb{N}$, the *standard embedding-projection pair* $(\mathsf{i}_{m\infty}, \mathsf{j}_{\infty m}) : \mathcal{D}_m \to \mathcal{D}_\infty$ is defined as follows:

$$\begin{array}{rcll} \mathsf{i}_{m\infty}(d) & \triangleq & \langle \Phi_{mn}(d) \rangle_{n \in \mathbb{N}}, & \text{for } d \in \mathcal{D}_m; \\ \mathsf{j}_{\infty m}(\langle d_n \rangle_{n \in \mathbb{N}}) & \triangleq & d_m, & \end{array}$$

where Φ_{mn} is given by (for all $m, n \in \mathbb{N}$):

$$\Phi_{mn} \triangleq \begin{cases} \mathsf{i}_{n-1} \circ \cdots \circ \mathsf{i}_m, & \text{if } m < n, \\ \mathrm{Id}_{\mathcal{D}_n}, & \text{if } m = n, \\ \mathsf{j}_n \circ \cdots \circ \mathsf{j}_{m-1}, & \text{if } m > n. \end{cases}$$

Notice that the inverse limit $\mathcal{D}_\infty$ is uniquely determined by its initial pair $(\mathcal{D}_0, \mathsf{i}_0, \mathsf{j}_0)$.

EXERCISE 13.48. Show the following basic properties.
(i) For all $n \in \mathbb{N}, c, d \in \mathcal{D}_n, \mathsf{i}_{n+1}(c \mapsto d) = (\mathsf{i}_n(c) \mapsto \mathsf{i}_n(d))$.
(ii) For all $m, n \in \mathbb{N}$, $\Phi_{mn} = \mathsf{j}_{\infty n} \circ \mathsf{i}_{m\infty}$.
(iii) $\bigsqcup_{n \in \mathbb{N}} \mathsf{i}_{n\infty} \circ \mathsf{j}_{\infty n} = \mathrm{Id}_{\mathcal{D}_\infty}$.

LEMMA 13.49.

$$\mathcal{K}(\mathcal{D}_\infty) = \bigsqcup_{n\in\mathbb{N}} \mathsf{i}_{n\infty}(\mathcal{K}(\mathcal{D}_n))$$

PROOF. For $d = \langle d_n\rangle_{n\in\mathbb{N}}$, we show

$$d\in\mathcal{K}(\mathcal{D}_\infty) \iff \exists k\in\mathbb{N}, e\in\mathcal{K}(\mathcal{D}_k)\,.\,\mathsf{i}_{k\infty}(e) = d$$

($\Rightarrow$) By Exercise 13.48(iii), we have $d = \bigsqcup_{n\in\mathbb{N}} \mathsf{i}_{n\infty}(d_n)$. Since $d\in\mathcal{K}(\mathcal{D})$, there exists an index $k\in\mathbb{N}$ such that $d = \mathsf{i}_{k\infty}(d_k)$. Now, for every directed $X\subseteq\mathcal{D}_k$, we have

$$\begin{array}{llll} d_k\sqsubseteq\bigsqcup X & \Rightarrow & d\sqsubseteq_\infty \mathsf{i}_{k\infty}(\bigsqcup X) & \\ & \Rightarrow & d\sqsubseteq_\infty \bigsqcup \mathsf{i}_{k\infty}(X), & \text{by continuity of } \mathsf{i}_{k\infty}, \\ & \Rightarrow & d\sqsubseteq_\infty \mathsf{i}_{k\infty}(x), & \text{for some } x\in X \text{ since } d \text{ is compact,} \\ & \Rightarrow & \mathsf{j}_{\infty k}(d)\sqsubseteq \mathsf{j}_{\infty k}(\mathsf{i}_{k\infty}(x)) & \end{array}$$

This shows that $d_k\in\mathcal{K}(\mathcal{D}_k)$.

($\Leftarrow$) By induction on $k\in\mathbb{N}$, one shows that $e\in\mathcal{K}(\mathcal{D}_k)$ entails $\mathsf{i}_k(e)\in\mathcal{K}(\mathcal{D}_{k+1})$ and $\Phi_{km}(e)\in\mathcal{K}(\mathcal{D}_m)$ for all $m\geq k$. It follows that $\mathsf{i}_{k\infty}(e) = d\in\mathcal{K}(\mathcal{D}_\infty)$. □

THEOREM 13.50 (SCOTT (1972)). *Let $(\mathcal{D}_0, \mathsf{i}_0, \mathsf{j}_0)$ be a functional embedding-projection pair. Then, the corresponding $\mathcal{D}_\infty$ is an extensional categorical model via*

$$(\mathrm{App}_\infty, \mathrm{Lam}_\infty) : \mathcal{D}_\infty \to (\mathcal{D}_\infty \Rightarrow \mathcal{D}_\infty)$$

so defined:

$$\begin{array}{lll} \mathrm{App}_\infty(\langle d_n\rangle_{n\in\mathbb{N}}) & \triangleq & \bigsqcup_{n\in\mathbb{N}}(\mathsf{i}_{n\infty}\circ d_{n+1}\circ \mathsf{j}_{n\infty}) \\ \mathrm{Lam}_\infty(f) & \triangleq & \mathsf{i}_{(n+1)\infty}(\mathsf{j}_{\infty n}\circ f\circ \mathsf{i}_{n\infty}) \end{array}$$

PROOF. The original proof by Scott considers the specific embedding-projection pair given by $\mathsf{i}_0(d)\triangleq\lambda\!\!\lambda e.d$ and $\mathsf{j}_0(f)\triangleq f(\bot)$, see B[1984], Theorem 18.2.16. The general case is developed in Plotkin (1982). □

We will show that every $\mathcal{D}_\infty$-model can be presented as a filter model (Theorem 13.54), via an order-reversing isomorphism. As an intermediate step, we define an itt having as type atoms the compact elements of $\mathcal{D}_0$ and a subtyping relation inherited from $\sqsubseteq_{\mathcal{D}_0}$.

DEFINITION 13.51. Let $(\mathcal{D}_0, \mathsf{i}_0, \mathsf{j}_0)$ be a functional embedding-projection pair.

(i) Consider a set $\mathbb{A}^\infty$ of type constants representing the compact elements of $\mathcal{D}_0$ different from $\bot$. In other words, assuming $\xi_\bot = \omega$, we have

$$\mathbb{A}^\infty_\omega \triangleq \{\xi_d \mid d\in\mathcal{K}(\mathcal{D}_0)\}$$

(ii) Write $\mathbb{T}^\infty_\wedge$ for the set of intersection types over $\mathbb{A}^\infty_\omega$ (as in Definition 13.8(ii)).

(iii) Define $\leq^\infty\subseteq\mathbb{T}^\infty_\wedge\times\mathbb{T}^\infty_\wedge$ as the least subtyping relation closed under the rules:

$$\frac{e\sqsubseteq d}{\xi_d\leq^\infty\xi_e}\ (\sqsubseteq) \qquad \frac{c = d\sqcup e}{\xi_c\simeq^\infty\xi_d\wedge\xi_e}\ (\sqcup) \qquad \frac{\mathsf{i}_0(d) = (c_1\mapsto d_1)\sqcup\cdots\sqcup(c_n\mapsto d_n)}{\bigwedge_{i=1}^n(\xi_{c_i}\to\xi_{d_i})\simeq^\infty\xi_d}\ (\mathsf{i}_0)$$

where $c, d, e, c_i, d_i\in\mathcal{K}(\mathcal{D}_0)$, and $\xi_d\simeq^\infty\xi_e$ stands for $\xi_d\leq^\infty\xi_e\leq^\infty\xi_d$, as usual.

(iv) The itt associated with $\mathcal{D}_0$ is given by $\mathrm{T}^{\mathcal{D}_0} \triangleq (\mathbb{A}^\infty, \leq^\infty)$.

DEFINITION 13.52. (i) Define a map $(\cdot)^* : \mathbb{T}^\infty_\wedge \to \mathcal{K}(\mathcal{D}_\infty)$ by induction as follows:

$$\begin{array}{lcll} (\xi_d)^* & \triangleq & \mathsf{i}_{0\infty}(d), & \text{for } d \in \mathcal{K}(\mathcal{D}_0), \\ (\sigma \to \tau)^* & \triangleq & (\sigma^* \mapsto \tau^*) & \\ (\sigma \wedge \tau)^* & \triangleq & \sigma^* \sqcup \tau^* & \end{array}$$

(ii) The *rank* of a type $\sigma \in \mathbb{T}^\infty_\wedge$, in symbols $\mathrm{rank}(\sigma)$, is defined by induction on σ:

$$\begin{array}{lcll} \mathrm{rank}(\xi_d) & \triangleq & 0, & \text{for } d \in \mathcal{K}(\mathcal{D}_0), \\ \mathrm{rank}(\sigma \to \tau) & \triangleq & \max\{\mathrm{rank}(\sigma), \mathrm{rank}(\tau)\} + 1, & \\ \mathrm{rank}(\sigma \wedge \tau) & \triangleq & \max\{\mathrm{rank}(\sigma), \mathrm{rank}(\tau)\}. & \end{array}$$

Intuitively, if the type σ has rank n then σ^* represents an element of $\mathcal{K}(\mathcal{D}_n)$.

LEMMA 13.53. (i) *The map* $(\cdot)^* : \mathbb{T}^\infty_\wedge \to \mathcal{K}(\mathcal{D}_\infty)$ *is surjective.*
(ii) *For all* $\sigma, \tau \in \mathbb{T}^\infty_\wedge$, *we have:* $\sigma \leq^\infty \tau \iff \tau^* \sqsubseteq_\infty \sigma^*$.

PROOF. (i) We show that, for all $n \in \mathbb{N}, e \in \mathcal{K}(\mathcal{D}_n)$, one has $\mathsf{i}_{n\infty}(e) = \sigma^*$ for some $\sigma \in \mathbb{T}^\infty_\wedge$. We proceed by induction on n.

Case $n = 0$. By definition.

Case $n > 0$. We split into subcases depending on e.

Subcase $e = \bot$. Then $\mathsf{i}_{n\infty}(\bot) = \omega^*$ by construction, using Remark 13.46(ii).

Subcase $e = (c \mapsto d)$ for some $c, d \in \mathcal{K}(\mathcal{D}_{n-1})$. By IH, there exist σ, τ such that $\mathsf{i}_{(n-1)\infty}(c) = \sigma^*$ and $\mathsf{i}_{(n-1)\infty}(d) = \tau^*$. So, we have:

$$\begin{array}{lcll} \mathsf{i}_{n\infty}(c \mapsto d) & = & (\mathsf{i}_{(n-1)\infty}(c) \mapsto \mathsf{i}_{(n-1)\infty}(d)), & \text{(cf. Exercise 13.48(i))}, \\ & = & (\sigma^* \mapsto \tau^*) & \\ & = & (\sigma \to \tau)^*. & \end{array}$$

Subcase $e = (c_1 \mapsto d_1) \sqcup \cdots \sqcup (c_k \mapsto d_k)$ where $c_i, d_i \in \mathcal{K}(\mathcal{D}_{n-1})$ for all $i \, (1 \leq i \leq k)$. Proceeding as above, the IH gives $\mathsf{i}_{n\infty}(c_i \mapsto d_i) = (\sigma_i \to \tau_i)^*$ for every i and appropriate $\sigma_i, \tau_i \in \mathbb{T}^\infty_\wedge$. We conclude $\mathsf{i}_{n\infty}(e) = ((\sigma_1 \to \tau_1) \wedge \cdots \wedge (\sigma_k \to \tau_k))^*$.

The surjectivity follows because, by Lemma 13.49, for every $d \in \mathcal{K}(\mathcal{D}_\infty)$ there exist $k \in \mathbb{N}, e \in \mathcal{K}(\mathcal{D}_k)$ satisfying $\mathsf{i}_{k\infty}(e) = d$.

(ii) ($\Rightarrow$) By induction on a derivation of $\sigma \leq^\infty \tau$, by cases on the last applied rule. The only interesting case from Definition 13.8(iv) is the following.

Case ($\mathrm{arr}_\leq$). Then $\sigma = \gamma_1 \to \delta_1$, $\tau = \gamma_2 \to \delta_2$, $\gamma_2 \leq^\infty \gamma_1$ and $\delta_1 \leq^\infty \delta_2$. We have

$$\begin{array}{lcll} \gamma_2 \leq^\infty \gamma_1 \;\&\; \delta_1 \leq^\infty \delta_2 & \Rightarrow & \gamma_1^* \sqsubseteq \gamma_2^* \;\&\; \delta_2^* \sqsubseteq \delta_1^*, & \text{by IH}, \\ & \Rightarrow & (\gamma_2^* \mapsto \delta_2^*) \sqsubseteq (\gamma_1^* \mapsto \delta_1^*) & \\ & \Rightarrow & \tau^* \sqsubseteq \sigma^*. & \end{array}$$

We continue with the cases from Definition 13.51(iii).

Case ($\sqsubseteq$). In this case $\sigma = \xi_d$ and $\tau = \xi_e$ for $d, e \in \mathcal{K}(\mathcal{D}_0)$ such that $e \sqsubseteq d$. Then $(\xi_e)^* = \mathsf{i}_{0\infty}(e) \sqsubseteq_\infty \mathsf{i}_{0\infty}(d) = (\xi_d)^*$, by monotonicity of $\mathsf{i}_{0\infty}$.

Case ($\sqcup$). Similar.

Case (i_0). Then $\sigma = \bigwedge_{i=1}^n (\xi_{c_i} \to \xi_{d_i})$, $\tau = \xi_d$ and $\mathsf{i}_0(d) = (c_1 \mapsto d_1) \sqcup \cdots \sqcup (c_n \mapsto d_n)$. Clearly σ^* and τ^* represent the same compact element.

($\Leftarrow$) Proceed by induction on the rank $r = \mathrm{rank}(\sigma \wedge \tau)$.

Case $r = 0$. Then $\sigma = \bigwedge_{i \in I} \xi_{d_i}$ and $\tau = \bigwedge_{j \in J} \xi_{e_j}$ for some finite $I, J \subseteq \mathbb{N}$. We have

$$\begin{array}{rcl}
\tau^* \sqsubseteq_\infty \sigma^* & \Rightarrow & \bigsqcup_{j \in J} \mathsf{i}_{0\infty}(e_j) \sqsubseteq_\infty \bigsqcup_{i \in I} \mathsf{i}_{0\infty}(d_i) \\
& \Rightarrow & \mathsf{j}_{\infty 0}(\bigsqcup_{j \in J} \mathsf{i}_{0\infty}(e_j)) \sqsubseteq \mathsf{j}_{\infty 0}(\bigsqcup_{i \in I} \mathsf{i}_{0\infty}(d_i)) \\
& \Rightarrow & \bigsqcup_{j \in J} (\mathsf{j}_{\infty 0} \circ \mathsf{i}_{0\infty})(e_j) \sqsubseteq \bigsqcup_{i \in I} (\mathsf{j}_{\infty 0} \circ \mathsf{i}_{0\infty})(d_i) \\
& \Rightarrow & \bigsqcup_{j \in J} e_j \sqsubseteq \bigsqcup_{i \in I} d_i \\
& \Rightarrow & \sigma \leq^\infty \tau.
\end{array}$$

Case $r > 0$. It is sufficient to prove $\sigma' \leq^\infty \tau'$ for some $\sigma' \simeq^\infty \sigma$ and $\tau' \simeq^\infty \tau$. By construction of $\mathcal{D}_\infty$ and the rule (i_0), for every $c \in \mathcal{K}(\mathcal{D}_0)$, we have $\xi_c \simeq^\infty \bigwedge_{i=1}^n (\xi_{d_i} \to \xi_{e_i})$ for appropriate $d_i, e_i \in \mathcal{K}(\mathcal{D}_0)$. Moreover, $\mathrm{rank}(\bigwedge_{i=1}^n (\xi_{d_i} \to \xi_{e_i})) = 1$. Therefore, we can construct σ' by expanding all the atoms in σ as above, in such a way that $\mathrm{rank}(\sigma') \leq r$. The type τ' is constructed from τ using the same recipe. Summing up, we have:

$$\sigma' = \bigwedge_{i \in I} \gamma_i \to \gamma'_i, \qquad \tau' = \bigwedge_{j \in J} \delta_j \to \delta'_j$$

for appropriate $\gamma_i, \gamma'_i, \delta_j, \delta'_j$ having rank strictly smaller than r. Assume now

$$(\tau')^* = \bigsqcup_{j \in J} ((\delta_j)^* \mapsto (\delta'_j)^*) \sqsubseteq_\infty \bigsqcup_{i \in I} ((\gamma_i)^* \mapsto (\gamma'_i)^*) = (\sigma')^*$$

This entails that, for every $j \in J$, we have $((\delta_j)^* \mapsto (\delta'_j)^*) \sqsubseteq_\infty \bigsqcup_{i \in I} ((\gamma_i)^* \mapsto (\gamma'_i)^*) = (\sigma')^*$. By Lemma A.61, assuming $(\delta'_j)^* \neq \bot_{\mathcal{D}_\infty}$, we obtain a non-empty $I' \subseteq I$ such that:

$$\Big(\bigwedge_{i \in I'} \gamma_i\Big)^* = \bigsqcup_{i \in I'} (\gamma_i)^* \sqsubseteq_\infty (\delta_j)^* \qquad \& \qquad (\delta'_j)^* \sqsubseteq_\infty \bigsqcup_{i \in I'} (\gamma'_i)^* = \Big(\bigwedge_{i \in I'} \gamma'_i\Big)^*.$$

By applying the IH, we get $\delta_j \leq^\infty \bigwedge_{i \in I'} \gamma_i$ and $\bigwedge_{i \in I'} \gamma'_i \leq^\infty \delta'_j$, respectively. From ($\mathrm{arr}_\leq$) and Lemma 13.12, it follows that $\sigma' \leq^\infty \delta_j \to \delta'_j$. In case $(\delta'_j)^* = \bot_{\mathcal{D}_\infty}$, we easily obtain $\sigma' \leq^\infty \delta_j \to \delta'_j$ by applying Remark 13.46(ii). We conclude $\sigma \simeq \sigma' \leq^\infty \tau \simeq \tau$. □

THEOREM 13.54. *Let* $(\mathcal{D}_0, \mathsf{i}_0, \mathsf{j}_0)$ *be a functional embedding-projection pair. We have*

$$\mathcal{D}_\infty \simeq \mathcal{F}_{\mathrm{T}^{\mathcal{D}_0}}$$

PROOF. Define two functions $\varphi : \mathcal{D}_\infty \to \mathcal{F}_{\mathrm{T}^{\mathcal{D}_0}}$ and $\psi : \mathcal{F}_{\mathrm{T}^{\mathcal{D}_0}} \to \mathcal{D}_\infty$ as follows:

$$\begin{array}{rcl}
\varphi(d) & = & \{\sigma^* \mid \sigma \in \mathbb{T}^\infty_\wedge \ \& \ \sigma^* \sqsubseteq_\infty d\}\!\uparrow \\
\psi(F) & = & \bigsqcup\{\sigma^* \mid \sigma \in F\}
\end{array}$$

By Lemma 13.53, φ and ψ realize the desired isomorphism. □

Theory	Atoms $\neq \omega$	Subtyping	Axioms
BCD	$(\xi_i)_{i \in \mathbb{N}}$	✗	✗
Scott	ξ	$\xi \simeq \omega \to \xi$	(ξ)
CDZ	$\varepsilon, \star$	$\varepsilon \leq \star$ $\star \to \varepsilon \simeq \varepsilon$ $\varepsilon \to \star \simeq \star$	$(\varepsilon \leq \star)$ (ε) $(\star)$

Table 13.1: Examples of intersection type theories.

13.3 Filter models: some case-studies

We study the filter models $\mathcal{F}_{\mathrm{BCD}}, \mathcal{F}_{\mathrm{Scott}}, \mathcal{F}_{\mathrm{CDZ}}$ generated by the itt's from Example 13.17, whose definition is recalled in Figure 13.1 for convenience. We start by showing that $\mathcal{F}_{\mathrm{Scott}}$ and $\mathcal{F}_{\mathrm{CDZ}}$ can be constructed as $\mathcal{D}_\infty$-models, up to order-reversing isomorphism.

DEFINITION 13.55. Define the following functional embedding-projection pairs.

(i) $\mathcal{D}_0^{\mathrm{Scott}} = \{\omega, \xi\}$ with $\omega \sqsubseteq^{\mathrm{Scott}} \xi$. The injection $\mathsf{i}_0^{\mathrm{Scott}}$ is given by:

$$\mathsf{i}_0^{\mathrm{Scott}}(d) \triangleq \begin{cases} (\omega \mapsto \omega), & \text{if } d = \omega, \\ (\omega \mapsto \xi), & \text{if } d = \xi. \end{cases}$$

(ii) $\mathcal{D}_0^{\mathrm{CDZ}} = \{\omega, \varepsilon, \star\}$ with $\omega \sqsubseteq^{\mathrm{CDZ}} \star \sqsubseteq^{\mathrm{CDZ}} \varepsilon$. The injection $\mathsf{i}_0^{\mathrm{CDZ}}$ is given by:

$$\mathsf{i}_0^{\mathrm{CDZ}}(d) \triangleq \begin{cases} (\omega \mapsto \omega), & \text{if } d = \omega, \\ (\varepsilon \mapsto \star), & \text{if } d = \star, \\ (\star \mapsto \varepsilon), & \text{if } d = \varepsilon. \end{cases}$$

In both cases, i.e. for $T \in \{\mathrm{Scott}, \mathrm{CDZ}\}$, the projection j_0^T is obtained from i_0^T by setting

$$\mathsf{j}_0^T(f) \triangleq \bigsqcup\{d \in \mathcal{D}_0^T \mid \mathsf{i}_0^T(d) \sqsubseteq f\} \text{ for } f \in (\mathcal{D}_0^T \Rightarrow \mathcal{D}_0^T)$$

THEOREM 13.56. *For $T \in \{\mathrm{Scott}, \mathrm{CDZ}\}$, call $\mathcal{D}_\infty^T$ the $\mathcal{D}_\infty$-model generated by the initial pair $(\mathcal{D}_0^T, \mathsf{i}_0^T, \mathsf{j}_0^T)$. Then, we have*

$$\mathcal{D}_\infty^T \simeq \mathcal{F}_T$$

PROOF. Consider the case $T = \mathrm{Scott}$. Note that the itt $\mathrm{T}^{\mathcal{D}_0^{\mathrm{Scott}}}$ associated with $\mathcal{D}_0^{\mathrm{Scott}}$ (Definition 13.51) is equivalent to the itt Scott itself. Indeed, since they have the same type atoms, also the generated sets of intersection types coincide. Then, by induction on $\mathrm{rank}(\sigma \wedge \tau)$, one checks that $\sigma \leq^\infty \tau$ holds if and only if $\sigma \leq^{\mathrm{Scott}} \tau$ does. Therefore, we conclude by Theorem 13.54. The case $T = \mathrm{CDZ}$ is analogous. □

REMARK 13.57. Clearly, $\mathcal{F}_{\mathrm{BCD}}$ cannot be constructed as a $\mathcal{D}_\infty$-model because it is not extensional. However, Coppo et al. (1984) did prove that the filter model $\mathcal{F}_{\mathrm{BCD}}$ is isomorphic to the minimal solution of the domain equation

$$\mathcal{D} \cong (\mathcal{D} \Rightarrow \mathcal{D}) \times \mathscr{P}(\mathbb{A}^{\mathrm{BCD}})$$

Intersection type assignment systems and the interpretation function in the corresponding filter models can be easily extended to $\lambda\bot$-terms. The filter models under consideration in this section satisfy an Approximation Theorem stating that the interpretation of a λ-term coincides with the union of the interpretations of its approximants.

DEFINITION 13.58. (i) An intersection type assignment system $\boldsymbol{\lambda}_\wedge^T$ can be extended to $\lambda\bot$-terms by considering $M, N \in \Lambda_\bot$ in the rules of Figure 13.1 and by adding the axiom:

$$\frac{}{\Gamma \vdash_\wedge^T \bot : \omega}\ (\mathrm{U}_\bot)$$

(ii) Similarly, the interpretation function $[\![-]\!]_{(\cdot)}$ from Definition 12.19(ii) is extended to $\lambda\bot$-terms by setting $[\![\bot]\!]_\rho^{\mathcal{D}} = \bot$. In case of a filter model $\mathcal{F}_T$, this gives $[\![\bot]\!]_\rho^{\mathcal{F}_T} = [\omega]_{\simeq^T}$.

REMARK 13.59. Let $M \in \Lambda$ and $P \in \Lambda_\bot$. By structural induction on P, one easily verifies that $\Gamma \vdash_\wedge^T P : \sigma$ and $P \leq_\bot M$ entail $\Gamma \vdash_\wedge^T M : \sigma$.

The Approximation Theorem below is traditionally proved through Tait's computability predicates or—equivalently—using Girard's reducibility candidates or Krivine's saturated sets. The reader who is interested in these techniques may consult Tait (1967), Girard (1989) and Krivine (1993), respectively. Here we provide the statement without its proof, but an analogous theorem is shown for $\mathcal{F}_{\mathrm{CDZ}}$ at the end of the chapter.

THEOREM 13.60 (SEMANTIC APPROXIMATION THEOREM FOR Scott AND BCD). *Let $T \in \{\mathrm{Scott}, \mathrm{BCD}\}$. For all λ-terms M and environments ρ, we have:*

$$[\![M]\!]_\rho^{\mathcal{F}_T} = \bigcup_{P \in \mathscr{A}(M)} [\![P]\!]_\rho^{\mathcal{F}_T}$$

As an immediate consequence, we obtain a logical characterization of solvability.

COROLLARY 13.61 (CHARACTERIZATION OF SOL).
For all λ-terms M, we have:

$$M \in \mathrm{SOL} \iff \exists\Gamma, \exists\sigma \not\leq^{\mathrm{BCD}} \omega \,.\, \Gamma \vdash_\wedge^{\mathrm{BCD}} M : \sigma$$

PROOF. ($\Rightarrow$) Let $M \in \mathrm{SOL}$. By Theorem 1.23, $M \twoheadrightarrow_h \lambda x_1 \ldots x_n . y M_1 \cdots M_k$ for some $n, k \geq 0$. We consider the case where y is free, the other case being analogous. By (U), we have $\Gamma \vdash_\wedge^{\mathrm{BCD}} M_i : \omega$ for all Γ, whence $y : \omega^k \to \xi_i \vdash y M_1 \cdots M_k : \xi_i$ by ($\to_{\mathrm{E}}$). By (weakening) and ($\to_{\mathrm{I}}$), we obtain $y : \omega^k \to \xi_i \vdash \lambda\vec{x}.y\vec{M} : \xi_i$ for $\omega \not\leq^{\mathrm{BCD}} \xi_i \in \mathbb{A}^{\mathrm{BCD}}$. We conclude by subject expansion, that holds by Theorem 13.30(i) and Proposition 13.27(ii).

($\Leftarrow$) Assume $\Gamma \vdash_\wedge^{\mathrm{BCD}} M : \sigma$. By the Semantic Approximation Theorem for BCD, $\Gamma \vdash_\wedge^{\mathrm{BCD}} P : \sigma$ holds for some $P \in \mathscr{A}(M)$. By Lemma 2.30, $M \twoheadrightarrow_\beta N$ such that $P \leq_\bot N$. By Remark 13.59, we obtain $\Gamma \vdash_\wedge^{\mathrm{BCD}} P : \sigma$. Now, if $\sigma \not\leq^{\mathrm{BCD}} \omega$ then we must have $P \neq \bot$. We conclude that $\mathrm{BT}(M) \neq \bot$, equivalently, $M \in \mathrm{SOL}$. □

By Theorem 13.56, $\mathcal{F}_{\mathrm{Scott}}$ and Scott's original model $\mathcal{D}_\infty$ share the same inequational theory, namely Morris' observational preorder $\sqsubseteq_{\mathrm{SOL}}$. As already mentioned in Theorem 12.51, the inequational theory of $\mathcal{F}_{\mathrm{BCD}}$ is $\sqsubseteq_{\mathsf{r}}$.

THEOREM 13.62. (i) $\mathrm{Th}_\sqsubseteq(\mathcal{F}_{\mathrm{Scott}}) = \sqsubseteq_{\mathrm{SOL}}$ *and* $\mathrm{Th}(\mathcal{F}_{\mathrm{Scott}}) = \mathcal{H}^*$;

(ii) $\mathrm{Th}_\sqsubseteq(\mathcal{F}_{\mathrm{BCD}}) = \sqsubseteq_{\mathsf{r}}$ *and* $\mathrm{Th}(\mathcal{F}_{\mathrm{BCD}}) = \mathcal{B}$.

CDZ: A filter model fully abstract for $\mathcal{H}^+$

We study the model $\mathcal{F}_{\mathrm{CDZ}}$ presented in logical form, i.e. through the intersection type assignment system $\boldsymbol{\lambda}_\wedge^{\mathrm{CDZ}}$. We show that the (closed) λ-terms inhabiting the type $\star$ in $\boldsymbol{\lambda}_\wedge^{\mathrm{CDZ}}$ are exactly the β-normalizing ones (Proposition 13.81(ii)). We exploit this fact to prove that the inequational theory of $\mathcal{F}_{\mathrm{CDZ}}$ is $\sqsubseteq_{\mathrm{NF}}$, whence its equational theory is $\mathcal{H}^+$.

CONVENTION 13.63. *From now on, and until the end of the section, we consider fixed the system* $\boldsymbol{\lambda}_\wedge^{\mathrm{CDZ}}$ *therefore we omit the superscript* CDZ *and simply write* $\mathbb{A}_\omega$, $\mathbb{T}_\wedge$, $\leq$, $\simeq$. *We will still write* $\vdash_\wedge^{\mathrm{CDZ}}$ *as a reminder, so that no confusion may possibly arise.*

As a preliminary step, we verify that the three atoms $\varepsilon, \star, \omega \in \mathbb{A}_\omega$ are distinguished modulo $\simeq$. We start by characterizing those intersection types that are equivalent to ω.

LEMMA 13.64. *Let* $\gamma \in \mathbb{T}_\wedge$. *Then* $\gamma \simeq \omega$ *if and only if* γ *belongs to the set* $\mathbb{U} \subseteq \mathbb{T}_\wedge$ *whose elements are generated by the following grammar (for* $\sigma \in \mathbb{T}_\wedge$*):*

$$\gamma ::= \ \omega \mid \sigma \to \gamma \mid \gamma \wedge \gamma \tag{$\mathbb{U}$}$$

PROOF. Remark that, because of the rule (top), $\sigma \simeq \omega$ holds exactly when $\omega \leq \sigma$ does.

($\Rightarrow$) Let us show that $\mathbb{U}$ is upward closed w.r.t. $\leq$, in other words:

$$\gamma \leq \sigma \ \&\ \gamma \in \mathbb{U} \quad \Rightarrow \quad \sigma \in \mathbb{U} \tag{13.1}$$

We proceed by induction on a derivation of $\gamma \leq \sigma$.

Cases ($\varepsilon \leq \star$), (ε) and ($\star$). Impossible.

Cases (refl), (top) and (arr_ω). Trivial.

Cases ($\mathrm{inc_L}$) and ($\mathrm{inc_R}$). Easy, since $\gamma = \gamma_1 \wedge \gamma_2 \in \mathbb{U}$ entails $\gamma_1, \gamma_2 \in \mathbb{U}$.

Case ($\mathrm{arr}_\wedge$). In this case $\gamma = (\tau \to \gamma_1) \wedge (\tau \to \gamma_2)$ and $\sigma = \tau \to (\gamma_1 \wedge \gamma_2)$. Notice that $\gamma \in \mathbb{U}$ implies $\gamma_1, \gamma_2 \in \mathbb{U}$ while τ is an arbitrary intersection type. We may conclude because $\gamma_1 \wedge \gamma_2 \in \mathbb{U}$ and $\tau \to (\gamma_1 \wedge \gamma_2) \in \mathbb{U}$ for all $\tau \in \mathbb{T}_\wedge$.

Case ($\mathrm{arr}_\leq$). In this case $\gamma = \sigma_1 \to \tau_1$ and $\sigma = \sigma_2 \to \tau_2$ with $\sigma_2 \leq \sigma_1$ and $\tau_1 \leq \tau_2$. By definition $\sigma_1 \to \tau_1 \in \mathbb{U}$ implies $\tau_1 \in \mathbb{U}$, by IH we have $\tau_2 \in \mathbb{U}$ so we infer $\sigma_2 \to \tau_2 \in \mathbb{U}$.

Cases (glb) and (trans). Easy, by IH. This concludes the proof of (13.1).

As a consequence we have: $\sigma \simeq \omega \iff \omega \leq \sigma \Rightarrow \sigma \in \mathbb{U}$.

($\Leftarrow$) Assume $\gamma \in \mathbb{U}$. We show $\gamma \simeq \omega$ by structural induction on γ.

Case $\gamma = \omega$. Trivial.

Case $\gamma = \gamma_1 \to \gamma_2$ with $\gamma_2 \in \mathbb{U}$. By (arr_ω) and IH we derive $\omega \leq \gamma_1 \to \omega \simeq \gamma_1 \to \gamma_2$.

Case $\gamma = \gamma_1 \wedge \gamma_2$. Both $\gamma_1, \gamma_2 \in \mathbb{U}$, so we conclude by IH and idempotence of $\wedge$. □

REMARK 13.65. We consider the intersection type assignment system $\boldsymbol{\lambda}_\wedge^{\mathrm{CDZ}}$ extended to $\lambda\bot$-terms as in Definition 13.58. By the preceding lemma, we have that

$$\Gamma \vdash_\wedge^{\mathrm{CDZ}} \bot : \sigma \iff \sigma \simeq \omega$$

Intuitively, the role of $\bot$ in the extended system is already played by Ω in the system restricted to Λ. Hence, all previous results like the Inversion Lemmas, Subject Reduction and Expansion hold in the extended system and will be used accordingly.

LEMMA 13.66. *Let $\mathbb{E}, \mathbb{S}$ be the sets of types $\gamma \in \mathbb{E}$ and $\delta \in \mathbb{S}$ respectively generated by (for $\sigma \in \mathbb{T}_\wedge$):*

$$(\mathbb{E}) \qquad \gamma \ ::= \ \varepsilon \mid \delta \to \gamma \mid \gamma \wedge \sigma \mid \sigma \wedge \gamma$$

$$(\mathbb{S}) \qquad \delta \ ::= \ \star \mid \omega \mid \sigma \to \delta \mid \gamma \to \sigma \mid \delta \wedge \delta$$

Then we have (i) $\Rightarrow$ (ii) $\Rightarrow$ (iii) *for:*

(i) $\sigma \leq \tau$,

(ii) $\sigma \in \mathbb{S} \Rightarrow \tau \in \mathbb{S}$,

(iii) $\tau \in \mathbb{E} \Rightarrow \sigma \in \mathbb{E}$.

PROOF. By induction on a derivation of $\sigma \leq \tau$. □

COROLLARY 13.67. *For $\sigma \in \mathbb{T}_\wedge$ we have:*

$$\begin{array}{ccc} \sigma \leq \varepsilon & \Rightarrow & \sigma \in \mathbb{E} \\ \star \leq \sigma & \Rightarrow & \sigma \in \mathbb{S} \end{array}$$

Therefore $\varepsilon < \star < \omega$.

PROOF. Both implications follow directly from the lemma above. The conclusion follows using the rules ($\varepsilon \leq \star$) and (top), the fact that $\varepsilon \notin \mathbb{S}, \star \notin \mathbb{E}$, and Lemma 13.64. □

Intuitively, the type ε corresponds to λ-terms having a β-nf which is 'persistent' in the sense specified below. For the precise property, we refer to Proposition 13.81(iii).

DEFINITION 13.68. (i) A λ-term M has a *persistent normal form* if and only if, for all $k \geq 0$ and $N_1, \ldots, N_k \in \mathrm{NF}$, we have that $M\vec{N} \in \mathrm{NF}$.

(ii) We write PNF for the set of λ-terms having a persistent normal form.

Taking $k = 0$ in (i), we have that $M \in \mathrm{PNF}$ entails $M \in \mathrm{NF}$, whence $\mathrm{PNF} \subseteq \mathrm{NF}$.

EXAMPLES 13.69. (i) Consider a λ-term $M \in \mathrm{NF}$ such that $M^{\mathrm{nf}} = \lambda x_1 \ldots x_n.yM_1 \cdots M_k$. If $x_1, \ldots, x_n \notin \mathrm{FV}(yM_1 \cdots M_k)$ then $M \in \mathrm{PNF}$.

(ii) As a particular case, we have $x\vec{M} \in \mathrm{PNF}$ for all $x \in \mathrm{Var}$ and $\vec{M} \in \Lambda$.

(iii) The λ-term $M \triangleq \lambda z.y(z\Delta\Delta)$ is in β-normal form but does not belong to PNF because $M\mathsf{I} \twoheadrightarrow_\beta y\Omega \notin \mathrm{NF}$.

LEMMA 13.70. *For all $M \in \Lambda$, we have:*

(i) $M \in \mathrm{NF} \iff \forall k \geq 0, \forall N_1, \ldots, N_k \in \mathrm{PNF} \,.\, M\vec{N} \in \mathrm{NF}$.

(ii) $M \in \mathrm{PNF} \iff \forall k \geq 0, \forall N_1, \ldots, N_k \in \mathrm{NF} \,.\, M\vec{N} \in \mathrm{PNF}$.

PROOF. In both (i) and (ii), the direction ($\Leftarrow$) is straightforward: just take $k = 0$.

(i) ($\Rightarrow$) Consider $M \in \mathrm{NF}$ and proceed by induction on $M^{\mathrm{nf}} = \lambda x_1 \ldots x_n.yM_1 \cdots M_m$. Given arbitrary $N_1, \ldots, N_k \in \Lambda$, define $\ell \triangleq \min\{k, n\}$ and $* \triangleq [x_1 := N_1, \ldots, x_\ell := N_\ell]$. Notice that for all $i\,(1 \leq i \leq m)$ the size of $\lambda x_1 \ldots x_\ell.M_i$ is strictly smaller than the size of M^{nf}, so by IH $(\lambda x_1 \ldots x_\ell.M_i)\vec{N}$ is β-normalizing, from which it follows $M_i^* \in \mathrm{NF}$ because the set NF is closed under $=_\beta$. We consider several subcases.

Case $k \leq n$. In this case $M\vec{N} =_\beta \lambda x_{k+1} \ldots x_n.y^* M_1^* \cdots M_m^*$. Now, if $y \notin \{x_1, \ldots, x_n\}$ then the thesis follows directly from the IH, otherwise $y = x_j$ for some $j\,(1 \leq j \leq n)$ so $y^* = N_j$ and the thesis follows from the IH and the fact that $N_j \in \mathrm{PNF}$.

Case $k > n$. In this case $M\vec{N} =_\beta y^* M_1^* \cdots M_m^* N_{n+1} \cdots N_k$. Similarly, if $y \in \mathrm{FV}(M^{\mathrm{nf}})$ then the lemma follows by IH, otherwise by IH and the definition of PNF.

(ii) ($\Rightarrow$) Let $M \in \mathrm{PNF}$. By contradiction, assume that there are $N_1, \ldots, N_k \in \mathrm{NF}$ such that $M\vec{N} \in \mathrm{NF} - \mathrm{PNF}$. Then there exist $P_1, \ldots, P_n \in \mathrm{NF}$ such that $M\vec{N}\vec{P} \notin \mathrm{NF}$, which is impossible because we assumed $M \in \mathrm{PNF}$. □

REMARK 13.71. (i) Note that $\mathrm{PNF} \subseteq \mathrm{NF}$ is coherent with $\varepsilon \leq \star$. Moreover:

(ii) $\star \simeq \varepsilon \to \cdots \to \varepsilon \to \star$ (cf. Lemma 13.70(i));

(iii) $\varepsilon \simeq \star \to \cdots \to \star \to \varepsilon$ (cf. Lemma 13.70(ii)).

LEMMA 13.72. *Let $M \in \Lambda$, $\sigma, \tau \in \mathbb{T}_\wedge$ and Γ be a type environment.*

(i) *If $M \twoheadrightarrow_\beta N$ and $\Gamma \vdash_\wedge^{\mathrm{CDZ}} \mathsf{da}(M) : \sigma$ then $\Gamma \vdash_\wedge^{\mathrm{CDZ}} \mathsf{da}(N) : \sigma$.*

(ii) *If $P_1, P_2 \in \mathscr{A}(M)$ satisfy $\Gamma \vdash_\wedge^{\mathrm{CDZ}} P_1 : \sigma$ and $\Gamma \vdash_\wedge^{\mathrm{CDZ}} P_2 : \tau$ then there exists $Q \in \mathscr{A}(M)$ such that $\Gamma \vdash_\wedge^{\mathrm{CDZ}} Q : \sigma \wedge \tau$.*

PROOF. (i) By induction on M. We distinguish two cases.

Case $M = \lambda x_1 \ldots x_n.(\lambda y.M')M_0 \cdots M_k$. By definition we have $\mathsf{da}(M) = \bot$ hence, by Remark 13.65, we have $\sigma \simeq \omega$. It is therefore enough to apply (U).

Case $M = \lambda x_1 \ldots x_n.yM_1 \cdots M_k$. The reduction $M \twoheadrightarrow_\beta N$ must take place within the arguments, i.e. $N = \lambda x_1 \ldots x_n.yN_1 \cdots N_k$ where $M_i \twoheadrightarrow_\beta N_i$ for all $i\,(1 \leq i \leq k)$. So, this case follows from the IH using the Inversion Lemma and the definition of $\mathsf{da}(-)$.

(ii) By definition there exist $M_1, M_2 \in \Lambda$ such that $M =_\beta M_1 =_\beta M_2$ with $P_1 = \mathsf{da}(M_1)$ and $P_2 = \mathsf{da}(M_2)$. By the Church-Rosser property M_1 and M_2 have a common reduct, namely $M_1 \twoheadrightarrow_\beta N$ and $M_2 \twoheadrightarrow_\beta N$ for some $N \in \Lambda$. By (i), we can take $Q = \mathsf{da}(N)$. □

DEFINITION 13.73. Let $\sigma \in \mathbb{T}_\wedge$ and Γ be a type environment.

(i) Let $\mathscr{A}_\Gamma^\sigma \triangleq \{M \in \Lambda \mid \exists P \in \mathscr{A}(M) \,.\, \Gamma \vdash_\wedge^{\mathrm{CDZ}} P : \sigma\}$.

(ii) Define the subset $[\![\sigma]\!]_\Gamma \subseteq \Lambda$ by induction on σ as follows:

$$\begin{aligned}
[\![\xi]\!]_\Gamma &\triangleq \mathscr{A}_\Gamma^\xi, \text{ for } \xi \in \{\star, \omega\}, \\
[\![\varepsilon]\!]_\Gamma &\triangleq \{M \mid \forall \Gamma', \forall n \in \mathbb{N}, \forall N_1, \ldots, N_n \in \mathscr{A}_{\Gamma'}^\star \,.\, M\vec{N} \in \mathscr{A}_{\Gamma \wedge \Gamma'}^\varepsilon\}, \\
[\![\tau_1 \to \tau_2]\!]_\Gamma &\triangleq \{M \mid \forall \Gamma', \forall N \in [\![\tau_1]\!]_{\Gamma'} \,.\, MN \in [\![\tau_2]\!]_{\Gamma \wedge \Gamma'}\}, \\
[\![\tau_1 \wedge \tau_2]\!]_\Gamma &\triangleq [\![\tau_1]\!]_\Gamma \cap [\![\tau_2]\!]_\Gamma.
\end{aligned}$$

Notice that for all type environments Γ we have $[\![\omega]\!]_\Gamma = \mathscr{A}_\Gamma^\omega = \Lambda$.

LEMMA 13.74. *For $M \in \Lambda$, $y \notin \mathrm{FV}(M)$, $\sigma, \tau \in \mathbb{T}_\wedge$ and a type environment Γ, we have:*

$$My \in \mathscr{A}_{\Gamma, y:\tau}^\sigma \quad \Rightarrow \quad M \in \mathscr{A}_\Gamma^{\tau \to \sigma}$$

PROOF. By definition, there exists N such that $My =_\beta N$ and $\Gamma, y : \tau \vdash_\wedge^{\mathrm{CDZ}} \mathsf{da}(N) : \sigma$. By the Church-Rosser property, My and N must have a common reduct whose direct approximant satisfies the same type judgement (by Lemma 13.72(i)). For simplicity, we just assume that $My \twoheadrightarrow_\beta N$ so there are two possibilities to consider:

1. Case $My \twoheadrightarrow_\beta M'y = N$ with $M \twoheadrightarrow_\beta M'$. We split into two subcases.

   If $M' = \lambda x.M''$ then $\mathsf{da}(N) = \bot$ and $\sigma \simeq \omega$ by Remark 13.65. By the rule (arr_ω) we have $\omega \leq \tau \to \omega$, so we can derive $M \in \mathscr{A}_\Gamma^{\tau\to\omega}$ by applying $(\leq)$ and (U).

   Otherwise $M' = xM_1 \cdots M_k$ is such that $\mathsf{da}(M') = x\mathsf{da}(M_1)\cdots\mathsf{da}(M_k) \in \mathscr{A}(M)$. The case $\sigma \simeq \omega$ is treated as above, so we assume $\sigma \not\simeq \omega$. Hence by Theorem 13.25(i) we must have $\Gamma, y : \tau \vdash^{\mathrm{CDZ}} \mathsf{da}(M') : \gamma \to \sigma$ and $\Gamma, y : \tau \vdash^{\mathrm{CDZ}} y : \gamma$, for some $\gamma \in \mathbb{T}_\wedge$. By Theorem 13.22(i) we obtain $\tau \leq \gamma$, therefore we can construct a derivation of $\Gamma \vdash_\wedge^{\mathrm{CDZ}} \mathsf{da}(M') : \tau \to \sigma$ using the rules $(\leq)$ and (strenghtening).

2. Case $My \twoheadrightarrow_\beta (\lambda x.M')y \to_\beta M'[x{:=}y] \twoheadrightarrow_\beta N$ with $M \twoheadrightarrow_\beta \lambda x.M'$. Also here there are two subcases to consider.

   If $N = \lambda\vec{x}.(\lambda z.N')N_0 \cdots N_k$ then $\mathsf{da}(N) = \bot$ and $\sigma \simeq \omega$, so we proceed as above.

   Otherwise $N = \lambda\vec{x}.zN_1 \cdots N_k$ with $\mathsf{da}(N) = \lambda\vec{x}.z\mathsf{da}(N_1)\cdots\mathsf{da}(N_k) \in \mathscr{A}(M[x{:=}y])$, therefore $\lambda y.\mathsf{da}(N) = \mathsf{da}(M)$ satisfies $\Gamma \vdash_\wedge^{\mathrm{CDZ}} \lambda y.\mathsf{da}(N) : \tau \to \sigma$ by $(\to_\mathrm{I})$. □

LEMMA 13.75. *Let $\sigma \in \mathbb{T}_\wedge$ and Γ be a type environment.*
(i) *If $M \in [\![\sigma]\!]_\Gamma$ and $N \twoheadrightarrow_\beta M$ then $N \in [\![\sigma]\!]_\Gamma$.*
(ii) *For every type environment Γ' we have :*
  (1) $[\![\sigma]\!]_\Gamma \subseteq [\![\sigma]\!]_{\Gamma\wedge\Gamma'}$;
  (2) $\Gamma \simeq \Gamma'$ entails $[\![\sigma]\!]_\Gamma = [\![\sigma]\!]_{\Gamma'}$.

PROOF. (i) By structural induction on σ.

Cases $\sigma = \varepsilon, \star, \omega$. These cases follow easily from the definition of $\mathscr{A}_\Gamma^\sigma$.

Case $\sigma = \tau \to \gamma$. In this case we know that $MN' \in [\![\tau]\!]_{\Gamma\wedge\Gamma'}$, for all Γ' and $N' \in [\![\tau]\!]_{\Gamma'}$. Since $N \twoheadrightarrow_\beta M$ entails $NN' \twoheadrightarrow_\beta MN'$, we derive $NN' \in [\![\tau]\!]_{\Gamma\wedge\Gamma'}$ from the IH, so we conclude $N \in [\![\tau \to \gamma]\!]_\Gamma$.

Case $\sigma = \tau \wedge \gamma$. Straightforward from the IH.

(ii) Both (1) and (2) follow by an easy induction on σ. □

LEMMA 13.76. *Let $\sigma \in \mathbb{T}_\wedge$ and Γ be a type environment.*
(i) *For all $x \in \mathrm{Var}$, $k \in \mathbb{N}$ and $M_1, \ldots, M_k \in \Lambda$, if $x\vec{M} \in \mathscr{A}_\Gamma^\sigma$ then $x\vec{M} \in [\![\sigma]\!]_\Gamma$.*
(ii) *$[\![\sigma]\!]_\Gamma \subseteq \mathscr{A}_\Gamma^\sigma$.*

PROOF. We prove (i) and (ii) simultaneously by induction on σ.

(i) The cases $\sigma = \star$ and $\sigma = \omega$ are trivial.

Case $\sigma = \varepsilon$. By hypothesis there exists $x\vec{P} \in \mathscr{A}(x\vec{M})$ such that $\Gamma \vdash_\wedge^{\mathrm{CDZ}} x\vec{P} : \varepsilon$. Clearly, for all $N_1, \ldots, N_n$ and $Q_i \in \mathscr{A}(N_i)$ we have $x\vec{P}\vec{Q} \in \mathscr{A}(x\vec{M}\vec{N})$. Moreover, if there exists Γ' such that $\Gamma' \vdash_\wedge^{\mathrm{CDZ}} Q_i : \star$ holds for all $i\,(1 \leq i \leq n)$, then it is easy to construct a derivation of $\Gamma \wedge \Gamma' \vdash_\wedge^{\mathrm{CDZ}} x\vec{P}\vec{Q} : \varepsilon$ by exploiting the fact that $\varepsilon \simeq \star^n \to \varepsilon$.

Case $\sigma = \tau \to \gamma$. Assume that $x\vec{P} \in \mathscr{A}(x\vec{M})$ satisfies $\Gamma \vdash_\wedge^{\mathrm{CDZ}} x\vec{P} : \tau \to \gamma$ and take $N \in [\![\tau]\!]_{\Gamma'}$. By IH(ii) we have $[\![\tau]\!]_{\Gamma'} \subseteq \mathscr{A}_{\Gamma'}^\tau$, whence there exists a $Q \in \mathscr{A}(N)$ such that $\Gamma' \vdash_\wedge^{\mathrm{CDZ}} Q : \tau$ holds. This entails $\Gamma \wedge \Gamma' \vdash_\wedge^{\mathrm{CDZ}} x\vec{P}Q : \sigma$, so we conclude $x\vec{M}N \in [\![\gamma]\!]_{\Gamma\wedge\Gamma'}$.

Case $\sigma = \tau \wedge \gamma$. It follows from the IH(i) and the admissible rules $(\wedge_{\mathrm{E}_\ell})$ and $(\wedge_{\mathrm{E}_r})$.

(ii) Case $\sigma \in \{\star, \varepsilon, \omega\}$. Immediate.

Case $\sigma = \tau \to \gamma$. Take $M \in [\![\tau \to \gamma]\!]_\Gamma$ and let y be a fresh variable. Since $y \in \mathscr{A}^{\tau}_{y:\tau}$ we obtain $y \in [\![\tau]\!]_{y:\tau}$ by IH(i), therefore we have $My \in [\![\gamma]\!]_{\Gamma,y:\tau}$ by Definition 13.73(ii). Now, by IH(ii) we get $My \in \mathscr{A}^{\gamma}_{\Gamma,y:\tau}$, so we conclude by Lemma 13.74.

Case $\sigma = \tau \wedge \gamma$. It follows from the IH(ii), using Lemma 13.72(ii). □

LEMMA 13.77. *Let $\sigma, \tau \in \mathbb{T}_\wedge$ and Γ be a type environment.*

(i) *If $M \in \mathscr{A}^{\sigma}_{\Gamma,y:\varepsilon}$ and $N \in [\![\varepsilon]\!]_\Gamma$ then $M[y{:=}N] \in \mathscr{A}^{\sigma}_{\Gamma}$.*

(ii) *If $\sigma \leq \tau$ then $[\![\sigma]\!]_\Gamma \subseteq [\![\tau]\!]_\Gamma$.*

PROOF. (i) Recall that $\mathscr{A}^{\omega}_{\Gamma} = \Lambda$, so we may focus our attention on the case $\sigma \not\simeq \omega$. By Definition 13.73(i), if $M \in \mathscr{A}^{\sigma}_{\Gamma,y:\varepsilon}$ then there is a $P \in \mathscr{A}(M)$ satisfying $\Gamma, y : \varepsilon \vdash^{\mathrm{CDZ}}_{\wedge} P : \sigma$. We proceed by structural induction on P.

Case $P = \bot$. This case is vacuous by Remark 13.65.

Case $P = \lambda x.P'$. In this case $M =_\beta \lambda x.M'$ for some $M' \in \Lambda$ such that $P' \in \mathscr{A}(M')$. By Theorem 13.22(iii) there are $\tau_1, \ldots, \tau_k, \gamma_1, \ldots, \gamma_k \in \mathbb{T}_\wedge$ satisfying $\bigwedge_{i=1}^{k} \tau_i \to \gamma_i \leq \sigma$ and such that $\Gamma, y : \varepsilon, x : \tau_i \vdash^{\mathrm{CDZ}}_{\wedge} P' : \gamma_i$ is derivable for every $i\,(1 \leq i \leq k)$. By IH, for all such i, there exists an approximant $P_i \in \mathscr{A}(M'[x{:=}N])$ satisfying $\Gamma, x : \tau_i \vdash^{\mathrm{CDZ}}_{\wedge} P_i : \gamma_i$. This means that there exist $M_1, \ldots, M_k \in \Lambda$ such that $M[y := N] =_\beta M_i$ and $P_i = \mathsf{da}(M_i)$; let us call M'' their common reduct. For $P'' \triangleq \mathsf{da}(M'')$, we have $P'' \in \mathscr{A}(M'[y := N])$ and, by Lemma 13.72(i), $\Gamma, x : \tau_i \vdash^{\mathrm{CDZ}}_{\wedge} P'' : \gamma_i$ is derivable for all $i\,(1 \leq i \leq k)$. This entails that $\lambda x.P'' \in \mathscr{A}(M[y := N])$ and we can easily construct a derivation of $\Gamma \vdash^{\mathrm{CDZ}}_{\wedge} \lambda x.P'' : \bigwedge_{i=1}^{k} \tau_i \to \gamma_i$ from which $\Gamma \vdash^{\mathrm{CDZ}}_{\wedge} \lambda x.P'' : \bigwedge_{i=1}^{k} \sigma$ follows by the rule $(\leq)$.

Case $P = xP_1 \cdots P_k$. Then $M =_\beta xM_1 \cdots M_k$ with $P_i \in \mathscr{A}(M_i)$ for all $i\,(1 \leq i \leq k)$. By an iterated application of Theorem 13.25(i), we obtain types $\tau_1, \ldots, \tau_k \in \mathbb{T}_\wedge$ such that $\Gamma, y : \varepsilon \vdash^{\mathrm{CDZ}}_{\wedge} x : \tau_1 \to \cdots \to \tau_k \to \sigma$ and $\Gamma, y : \varepsilon \vdash^{\mathrm{CDZ}}_{\wedge} P_i : \tau_i$ are derivable for all i, whence from the IH we get $P'_i \in \mathscr{A}(M_i[y := N])$ such that $\Gamma \vdash^{\mathrm{CDZ}}_{\wedge} P'_i : \tau_i$. Now, if $x \neq y$, we are done because $x\vec{P}' \in \mathscr{A}(M[y := N])$ and we can derive $\Gamma \vdash^{\mathrm{CDZ}}_{\wedge} x\vec{P}' : \sigma$ using $(\to_{\mathrm{E}})$. Otherwise we have $x = y$ and $\Gamma, y : \varepsilon \vdash^{\mathrm{CDZ}}_{\wedge} y : \tau_1 \to \cdots \to \tau_k \to \sigma$ implies $\varepsilon \leq \vec{\tau} \to \sigma$ by Theorem 13.22(i). Since $\varepsilon \simeq \vec{\star} \to \varepsilon$ by Remark 13.71(iii), from β-soundness (Theorem 13.30(i)) we obtain $\tau_i \leq \star$ and $\varepsilon \leq \sigma$ by Remark 13.24. As a consequence, we have $\Gamma \vdash^{\mathrm{CDZ}}_{\wedge} P'_i : \star$ which entails $M_i[y{:=}N] \in \mathscr{A}^{\star}_{\Gamma}$. From $N \in \mathscr{A}^{\varepsilon}_{\Gamma}$ and $M_i[y{:=}N] \in \mathscr{A}^{\star}_{\Gamma}$ we obtain $M[y{:=}N] \in \mathscr{A}^{\varepsilon}_{\Gamma}$. Since $\varepsilon \leq \sigma$ we conclude $M[y{:=}N] \in \mathscr{A}^{\sigma}_{\Gamma}$.

(ii) By induction on a derivation of $\sigma \leq \tau$.

Case $(\varepsilon \leq \star)$. By definition, $M \in [\![\varepsilon]\!]_\Gamma$ entails that there is an approximant $P \in \mathscr{A}(M)$ such that $\Gamma \vdash^{\mathrm{CDZ}}_{\wedge} P : \varepsilon$, so we get a derivation of $\Gamma \vdash^{\mathrm{CDZ}}_{\wedge} P : \star$ by applying the rule $(\leq)$.

Case $(\star)$. The inclusion $[\![\star]\!]_\Gamma \subseteq [\![\varepsilon \to \star]\!]_\Gamma$ is immediate. For the converse inclusion:

$$\begin{array}{rcll} [\![\varepsilon \to \star]\!]_\Gamma & \subseteq & \mathscr{A}^{\varepsilon\to\star}_{\Gamma}, & \text{by Lemma 13.76(ii)}, \\ & = & \mathscr{A}^{\star}_{\Gamma}, & \text{by } \varepsilon \to \star \simeq \star \text{ and rule } (\leq), \\ & = & [\![\star]\!]_\Gamma, & \text{by Definition 13.73(ii)}. \end{array}$$

Case (ε). The inclusion $[\![\varepsilon]\!]_\Gamma \subseteq [\![\star \to \varepsilon]\!]_\Gamma$ is immediate. To check the converse inclusion, take $M \in [\![\star \to \varepsilon]\!]_\Gamma$. We need to show that for all $n \in \mathbb{N}$, type environments Γ' and $N_1, \ldots, N_n \in \mathscr{A}^{\star}_{\Gamma\wedge\Gamma'}$, we have $M\vec{N} \in \mathscr{A}^{\varepsilon}_{\Gamma\wedge\Gamma'}$. We split into cases depending on n.

- For $n = 0$ we have to show $M \in \mathscr{A}^{\varepsilon}_{\Gamma\wedge\Gamma'}$. This follows from:

$$\begin{array}{rcll} [\![\star \to \varepsilon]\!]_\Gamma & \subseteq & \mathscr{A}^{\star\to\varepsilon}_{\Gamma}, & \text{by Lemma 13.76(ii)}, \\ & = & \mathscr{A}^{\varepsilon}_{\Gamma}, & \text{by } \star \to \varepsilon \simeq \varepsilon \text{ and rule } (\leq), \\ & \subseteq & \mathscr{A}^{\varepsilon}_{\Gamma\wedge\Gamma'}, & \text{by Lemma 13.75(ii)(1)}. \end{array}$$

- If $n > 0$ then by definition of $[\![\star \to \varepsilon]\!]_\Gamma$ we have that $MN_1 \in [\![\varepsilon]\!]_{\Gamma\wedge\Gamma'}$, for all type environments Γ' and $N_1 \in [\![\star]\!]_{\Gamma'} \subseteq \mathscr{A}^{\star}_{\Gamma'}$, where the inclusion holds by Lemma 13.75(ii)(1). By definition of $[\![\varepsilon]\!]_{\Gamma\wedge\Gamma'}$, we conclude $MN_1 \cdots N_n \in \mathscr{A}^{\varepsilon}_{\Gamma\wedge\Gamma'}$ for all $N_2, \ldots, N_n \in \mathscr{A}^{\star}_{\Gamma'}$.

Cases (refl), (inc_L), (inc_R), (top) and (arr_ω). Trivial.

Cases (glb) and (trans). Immediate from the IH.

Case ($\text{arr}_\wedge$). In this case we have $\sigma = (\sigma_1 \to \gamma_1)\wedge(\sigma_1 \to \gamma_2)$ and $\tau = \sigma_1 \to (\gamma_1\wedge\gamma_2)$. If $M \in [\![(\sigma_1 \to \gamma_1)\wedge(\sigma_1 \to \gamma_2)]\!]_\Gamma = [\![\sigma_1 \to \gamma_1]\!]_\Gamma \cap [\![\sigma_1 \to \gamma_2]\!]_\Gamma$ then for all Γ' and $N \in [\![\sigma_1]\!]_{\Gamma'}$ we have both $MN \in [\![\gamma_1]\!]_{\Gamma\wedge\Gamma'}$ and $MN \in [\![\gamma_2]\!]_{\Gamma\wedge\Gamma'}$. From this it follows $MN \in [\![\gamma_1 \wedge \gamma_2]\!]_{\Gamma\wedge\Gamma'}$, so we conclude $M \in [\![\sigma_1 \to (\gamma_1 \wedge \gamma_2)]\!]_\Gamma$.

Case ($\text{arr}_\leq$). Assume $\sigma = \sigma_1 \to \tau_1$, $\tau = \sigma_2 \to \tau_2$ with $\sigma_2 \leq \sigma_1$ and $\tau_1 \leq \tau_2$. Consider any type environment Γ' and take $N \in [\![\sigma_2]\!]_{\Gamma'} \subseteq [\![\sigma_1]\!]_{\Gamma'}$ where the inclusion holds by IH. If $M \in [\![\sigma_1 \to \tau_1]\!]_\Gamma$ then $MN \in [\![\tau_1]\!]_{\Gamma\wedge\Gamma'} \subseteq [\![\tau_2]\!]_{\Gamma\wedge\Gamma'}$, by IH, thus $M \in [\![\sigma_2 \to \tau_2]\!]_\Gamma$. □

PROPOSITION 13.78. *Let $M \in \Lambda$ and $x_1, \ldots, x_n$ be such that* $\text{FV}(M) \subseteq \{x_1, \ldots, x_n\}$. *For all intersection types $\sigma_1, \ldots, \sigma_n \in \mathbb{T}_\wedge$, type environments $\Gamma_1, \ldots, \Gamma_n$ and λ-terms $N_i \in [\![\sigma_i]\!]_{\Gamma_i}$, for all i $(1 \leq i \leq n)$:*

$$x_1 : \sigma_1, \ldots, x_n : \sigma_n \vdash^{\text{CDZ}}_{\wedge} M : \tau \quad \Rightarrow \quad M[\vec{x} := \vec{N}] \in [\![\tau]\!]_{\Gamma_1\wedge\cdots\wedge\Gamma_n}$$

PROOF. Let us call $\Gamma \triangleq x_1 : \sigma_1, \ldots, x_n : \sigma_n$. We proceed by induction on a derivation of $\Gamma \vdash^{\text{CDZ}}_{\wedge} M : \tau$. We split into cases, depending on the last rule applied.

Case (ax). In this case $M = x_i$ and $\tau = \sigma_i$, so by hypothesis we have $x_i[\vec{x} := \vec{N}] = N_i \in [\![\sigma_i]\!]_{\Gamma_i} \subseteq [\![\sigma_i]\!]_{\bigwedge_i \Gamma_i}$, where the last inclusion holds by Lemma 13.75(ii)(1).

Case (U). Immediate since $\tau = \omega$ and $[\![\omega]\!]_{\bigwedge_i \Gamma_i} = \Lambda$.

Case ($\to_\text{E}$). In this case $M = M_1 M_2$ and there is $\gamma \in \mathbb{T}_\wedge$ such that $\Gamma \vdash_\wedge M_1 : \gamma \to \sigma$ and $\Gamma \vdash M_2 : \gamma$ are derivable. By IH, we know that $M_1[\vec{x} := \vec{N}] \in [\![\gamma \to \tau]\!]_{(\bigwedge_i \Gamma_i)}$, therefore $M_1[\vec{x} := \vec{N}]N' \in [\![\tau]\!]_{(\bigwedge_i \Gamma_i)\wedge\Gamma'}$ for all $N' \in [\![\gamma]\!]_{\Gamma'}$. Since $M_2[\vec{x} := \vec{N}] \in [\![\gamma]\!]_{(\bigwedge_i \Gamma_i)}$ by IH, we get

$$(M_1 M_2)[\vec{x} := \vec{N}] = M_1[\vec{x} := \vec{N}](M_2[\vec{x} := \vec{N}]) \in [\![\tau]\!]_{(\bigwedge_i \Gamma_i)\wedge(\bigwedge_i \Gamma_i)}$$

We may conclude because $(\bigwedge_i \Gamma_i) \wedge (\bigwedge_i \Gamma_i) \simeq \bigwedge_i \Gamma_i$ (by idempotence of $\wedge$), therefore $[\![\tau]\!]_{(\bigwedge_i \Gamma_i)\wedge(\bigwedge_i \Gamma_i)} = [\![\tau]\!]_{(\bigwedge_i \Gamma_i)}$ by Lemma 13.75(ii)(2).

Case ($\wedge_\text{I}$). Straightforward from the IH.

Case ($\to_\text{I}$). In this case $M = \lambda y.M'$, $\tau = \tau_1 \to \tau_2$ and $\Gamma, y : \tau_1 \vdash_\wedge M' : \tau_2$. From the IH, we obtain that $M'[\vec{x} := \vec{N}, y := N'] \in [\![\tau_2]\!]_{(\bigwedge_i \Gamma_i)\wedge\Gamma'}$ for all $N' \in [\![\tau_1]\!]_{\Gamma'}$. Since $(\lambda y.M'[\vec{x} := \vec{N}])N' \to_\beta M'[\vec{x} := \vec{N}, y := N']$ we derive $(\lambda y.M'[\vec{x} := \vec{N}])N' \in [\![\tau_2]\!]_{(\bigwedge_i \Gamma_i)\wedge\Gamma'}$, by Lemma 13.75(i), for all such Γ', N'. Thus $(\lambda y.M')[\vec{x} := \vec{N}] \in [\![\tau_1 \to \tau_2]\!]_{(\bigwedge_i \Gamma_i)}$.

Case ($\leq$). By IH and Lemma 13.77(ii). □

PROPOSITION 13.79. *For all λ-terms M, we have:*

$$\Gamma \vdash_{\wedge}^{\mathrm{CDZ}} M : \tau \iff \exists P \in \mathscr{A}(M) \,.\, \Gamma \vdash_{\wedge}^{\mathrm{CDZ}} P : \tau$$

PROOF. ($\Rightarrow$) Assume $x_1 : \sigma_1, \dots, x_n : \sigma_n \vdash_{\wedge}^{\mathrm{CDZ}} M : \tau$. For all $i\,(1 \le i \le n)$ we have $x_i \in \mathscr{A}_{\Gamma}^{\sigma_i}$, whence $x_i \in [\![\sigma_i]\!]_{x_i:\sigma_i}$ by Lemma 13.76(i). By Proposition 13.78, we have

$$M[x_1 := x_1] \cdots [x_n := x_n] = M \in [\![\tau]\!]_{(x_1:\sigma_1)\wedge\cdots\wedge(x_n:\sigma_n)} = [\![\tau]\!]_{\Gamma},$$

so we conclude because the inclusion $[\![\tau]\!]_{\Gamma} \subseteq \mathscr{A}_{\Gamma}^{\tau}$ holds by Lemma 13.76(ii).

($\Leftarrow$) Let $P \in \mathscr{A}(M)$. By definition, there exists $M' =_{\beta} M$ such that $P = \mathsf{da}(M')$. By induction on a derivation, it is possible to show that $\Gamma \vdash_{\wedge}^{\mathrm{CDZ}} P : \tau$ entails $\Gamma \vdash_{\wedge}^{\mathrm{CDZ}} M' : \tau$. The only interesting case is when the rule $(\mathrm{U}_{\perp})$ is applied to give $P = \perp$ the type ω and, in that case, we derive $\Gamma \vdash M' : \omega$ by applying (U). Since M and M' are β-convertible, we conclude that $\Gamma \vdash_{\wedge}^{\mathrm{CDZ}} M : \tau$ is derivable by Proposition 13.27. □

THEOREM 13.80 (SEMANTIC APPROXIMATION THEOREM FOR CDZ).
For all λ-terms M and environments ρ, we have:

$$[\![M]\!]_{\rho}^{\mathcal{F}_{\mathrm{CDZ}}} = \bigcup_{P \in \mathscr{A}(M)} [\![P]\!]_{\rho}^{\mathcal{F}_{\mathrm{CDZ}}}$$

PROPOSITION 13.81. *For all $M \in \Lambda$ and $\{x_1, \dots, x_n\} \supseteq \mathrm{FV}(M)$, we have:*
(i) *$M \in \mathrm{SOL}$ if and only if there exist Γ and $\sigma \not\simeq \omega$ such that $\Gamma \vdash_{\wedge}^{\mathrm{CDZ}} M : \sigma$.*
(ii) *$M \in \mathrm{NF}$ if and only if $x_1 : \varepsilon, \dots, x_n : \varepsilon \vdash_{\wedge}^{\mathrm{CDZ}} M : \star$.*
(iii) *$x_1 : \varepsilon, \dots, x_n : \varepsilon \vdash_{\wedge}^{\mathrm{CDZ}} M : \varepsilon$ implies $M \in \mathrm{PNF}$.*

PROOF. (i) ($\Rightarrow$) Since M is solvable, it has a principle head normal form M^{phnf}. By Proposition 13.27, one can proceed by induction on a derivation of $\Gamma \vdash_{\wedge}^{\mathrm{CDZ}} M^{\mathsf{phnf}} : \sigma$.

($\Leftarrow$) If $M \notin \mathrm{SOL}$ then $\mathscr{A}(M) = \{\perp\}$, so the result follows from Proposition 13.79.

(ii) ($\Rightarrow$) Assume that $M \in \mathrm{NF}$ and proceed by induction on the β-normal structure of $M^{\mathtt{nf}} = \lambda y_1 \dots y_m . z M_1 \cdots M_k$ where $z \in \{\vec{x}, \vec{y}\}$. Let $\Gamma \triangleq \vec{x} : \varepsilon, \vec{y} : \varepsilon$, by IH we have a derivation Π_j of $\Gamma \vdash_{\wedge}^{\mathrm{CDZ}} M_j : \varepsilon$, for all $j\,(1 \le j \le k)$. We can construct the derivation

$$\dfrac{\dfrac{\dfrac{\dfrac{\Gamma \vdash_{\wedge}^{\mathrm{CDZ}} z : \varepsilon \quad \varepsilon \simeq \star^k \to \varepsilon}{\Gamma \vdash_{\wedge}^{\mathrm{CDZ}} z : \star^k \to \varepsilon}\,(\le) \quad \dfrac{\dfrac{\Pi_j}{\Gamma \vdash_{\wedge}^{\mathrm{CDZ}} M_j : \varepsilon} \quad \varepsilon \le \star}{\Gamma \vdash_{\wedge}^{\mathrm{CDZ}} M_j : \star}\,(\le) \quad 1 \le j \le k}{\Gamma \vdash_{\wedge}^{\mathrm{CDZ}} z M_1 \cdots M_k : \star}\,(\to_{\mathrm{E}})}{x_1 : \varepsilon, \dots, x_n : \varepsilon \vdash_{\wedge}^{\mathrm{CDZ}} M^{\mathtt{nf}} : \varepsilon^m \to \star}\,(\to_{\mathrm{I}}) \quad \varepsilon^m \to \star \simeq \star}{x_1 : \varepsilon, \dots, x_n : \varepsilon \vdash_{\wedge}^{\mathrm{CDZ}} M^{\mathtt{nf}} : \star}\,(\le)$$

($\Leftarrow$) Set $\Gamma \triangleq \vec{x} : \varepsilon$ and assume that $\Gamma \vdash_{\wedge}^{\mathrm{CDZ}} M : \star$ is derivable. By Proposition 13.79, there exists $P \in \mathscr{A}(M)$ such that $\Gamma \vdash_{\wedge}^{\mathrm{CDZ}} P : \star$ is derivable as well. It is enough to show that $\perp$ does not occur in P, a result that we prove by structural induction on P.

Case $P = \perp$. Impossible since $\perp$ is only typable with $\sigma \simeq \omega$ (see Remark 13.65).

Case $P = \lambda y_1 \dots y_m.zP_1 \cdots P_k$ for some $m, k \geq 0$ and $z \in \{\vec{x}, \vec{y}\}$. Now, $\Gamma \vdash_\wedge^{\mathrm{CDZ}} P : \star$ and $\star \simeq \varepsilon^m \to \star$ imply, by rule $(\leq)$, that $\Gamma \vdash_\wedge^{\mathrm{CDZ}} P : \varepsilon^m \to \star$ is derivable, and therefore $\Gamma, \vec{y} : \varepsilon \vdash_\wedge^{\mathrm{CDZ}} zP_1 \cdots P_k : \star$ by Theorem 13.25(ii). Now, by Theorem 13.25(i), there exist $\tau_1, \dots, \tau_k$ such that $\Gamma, \vec{y} : \varepsilon \vdash_\wedge^{\mathrm{CDZ}} z : \tau_1 \to \cdots \to \tau_k \to \star$ and $\Gamma, \vec{y} : \varepsilon \vdash_\wedge^{\mathrm{CDZ}} P_i : \tau_i$, for all $i\,(1 \leq i \leq k)$. The former type judgement entails $\varepsilon \leq \tau_1 \to \cdots \to \tau_k \to \star$ by Theorem 13.22(i), therefore $\tau_i \leq \star$ by soundness (Theorem 13.30(i)) and $\varepsilon \simeq \star^k \to \varepsilon$. The latter type judgement, by applying $(\leq)$, gives $\Gamma, \vec{y} : \varepsilon \vdash_\wedge^{\mathrm{CDZ}} P_i : \star$. By IH, there are no occurrences of $\perp$ in any P_i, from which it follows that $P = M^{\mathbf{nf}}$. So, we are done.

(iii) Consider $N_1, \dots, N_k \in \mathrm{NF}$ and variables $y_1, \dots, y_m$ including $\mathrm{FV}(\vec{N})$. Assume that $\vec{x} : \varepsilon \vdash_\wedge^{\mathrm{CDZ}} M : \varepsilon$ is derivable, then there is a derivation Π_0 of $\vec{x} : \varepsilon,\ \vec{y} : \varepsilon \vdash_\wedge^{\mathrm{CDZ}} M : \varepsilon$ by weakening. Now, by (ii), there exists a derivation Π_i of $\vec{x} : \varepsilon,\ \vec{y} : \varepsilon \vdash_\wedge^{\mathrm{CDZ}} N_i : \star$ for all $i\,(1 \leq i \leq k)$, hence setting $\Gamma \triangleq \vec{x} : \varepsilon, \vec{y} : \varepsilon$ we can construct the following derivation

$$\dfrac{\dfrac{\dfrac{\Pi_0}{\Gamma \vdash_\wedge^{\mathrm{CDZ}} M : \varepsilon} \quad \varepsilon \simeq \star^k \to \varepsilon}{\Gamma \vdash_\wedge^{\mathrm{CDZ}} M : \star^k \to \varepsilon}\,(\leq) \qquad \dfrac{\Pi_i}{\Gamma \vdash_\wedge^{\mathrm{CDZ}} N_i : \star} \quad 1 \leq i \leq k}{\dfrac{\Gamma \vdash_\wedge^{\mathrm{CDZ}} M\vec{N} : \varepsilon \qquad\qquad \varepsilon \leq \star}{\Gamma \vdash_\wedge^{\mathrm{CDZ}} M\vec{N} : \star}\,(\leq)}\,(\to_{\mathrm{E}})$$

By (ii) we obtain $M\vec{N} \in \mathrm{NF}$, and since $\vec{N}$ is arbitrary we conclude $M \in \mathrm{PNF}$. □

THEOREM 13.82 (MORRIS' INEQUATIONAL FULL ABSTRACTION).
For all $M, N \in \Lambda$, we have

$$M \sqsubseteq_{\mathrm{NF}} N \iff \mathcal{F}_{\mathrm{CDZ}} \models M \sqsubseteq N$$

In particular, $\mathrm{Th}(\mathcal{F}_{\mathrm{CDZ}}) = \mathcal{H}^+$.

PROOF. $(\Rightarrow)$ For the interpretation of M in $\mathcal{F}_{\mathrm{CDZ}}$ w.r.t. an environment ρ, we have:

$$\begin{aligned} [\![M]\!]_\rho^{\mathcal{F}_{\mathrm{CDZ}}} &= \bigcup_{M' \twoheadrightarrow_\eta M} [\![M']\!]_\rho^{\mathcal{F}_{\mathrm{CDZ}}}, && \text{as } \mathcal{F}_{\mathrm{CDZ}} \text{ is extensional,}\\ &= \bigcup_{M' \twoheadrightarrow_\eta M} \bigcup_{P \in \mathscr{A}(M')} [\![P]\!]_\rho^{\mathcal{F}_{\mathrm{CDZ}}}, && \text{by the Approximation Theorem,}\\ &= \bigcup_{M' \twoheadrightarrow_\eta M} \bigcup_{P \in \mathscr{A}(M')} [\![P^{\eta\text{-nf}}]\!]_\rho^{\mathcal{F}_{\mathrm{CDZ}}}, && \text{by extensionality,}\\ &= \bigcup_{Q \in \mathscr{A}_\eta(M)} [\![Q]\!]_\rho^{\mathcal{F}_{\mathrm{CDZ}}}, && \text{by Proposition 12.40.} \end{aligned}$$

Similarly, we have $[\![N]\!]_\rho^{\mathcal{F}_{\mathrm{CDZ}}} = \bigcup_{Q \in \mathscr{A}_\eta(N)} [\![Q]\!]_\rho^{\mathcal{F}_{\mathrm{CDZ}}}$. Now, from Lévy's characterization of the observational preorder $\sqsubseteq_{\mathrm{NF}}$, namely Proposition 12.43, we obtain the inclusion $\mathscr{A}_\eta(M) \subseteq \mathscr{A}_\eta(N)$ which in its turn entails $[\![M]\!]_\rho^{\mathcal{F}_{\mathrm{CDZ}}} \subseteq [\![N]\!]_\rho^{\mathcal{F}_{\mathrm{CDZ}}}$.

$(\Leftarrow)$ By Lemma 3.15 it is enough to consider contexts $C[]$ such that $C[M], C[N] \in \Lambda^o$. Assume now that $C[M] \in \mathrm{NF}$. By Proposition 13.81 and the monotonicity of the interpretation, we obtain $\star \in [\![C[M]]\!]^{\mathcal{F}_{\mathrm{CDZ}}} \subseteq [\![C[N]]\!]^{\mathcal{F}_{\mathrm{CDZ}}}$ which is only possible if $C[N] \in \mathrm{NF}$. □

REMARK 13.83. Notice that in the proof of $\sqsubseteq_{\mathrm{NF}} \subseteq \mathrm{Th}_{\sqsubseteq}(\mathcal{F}_{\mathrm{CDZ}})$ we only use the fact that the model is extensional and satisfies the semantic approximation theorem.

Chapter 14

Relational models

In this chapter we present the relational semantics of λ-calculus whose origin is intertwined with the discovery of *linear logic* by Girard (1987). The key idea behind linear logic is the decomposition of intuitionistic implication $A \to B$ into a *linear implication* $\multimap$ and an *exponential modality* $!(\cdot)$, also called *promotion*, as in

$$A \to B \cong\, !A \multimap B$$

Intuitively $F : A \multimap B$ represents a linear map from A to B, where the mathematical notion of linearity has a clear counterpart in computer science: F needs to consume exactly one input of type A in order to produce an output of type B. In the setting of λ-calculus, this means that F cannot erase nor duplicate its argument along the reduction. The exponential modality $!(\cdot)$ can be thought of as a replication-operator, whence erasure and duplication are still possible provided that the argument has a promoted type $!A$.

It was first noticed by Girard (1988) that the category **Rel** of sets and relations constitutes a particularly simple quantitative model of linear logic, where $A \multimap B$ is given by the Cartesian product $A \times B$, and $!A$ represents the set of all finite multisets over A. This model is however 'degenerate' in the sense that it identifies multiplicative conjunction ($\otimes$) and disjunction ($\parr$), i.e. $A \otimes B = A \parr B$, and this induces a series of collapses among which $A^{\perp} = A$. For this reason, the relational semantics was overlooked for decades, even if it remained a typical piece of folklore in the linear logic community. The idea of performing a systematic study of models of λ-calculus living in the relational semantics of linear logic is due to Thomas Ehrhard, although they were also considered (independently) by Hyland et al. (2006) with a different purpose. The first articles on the subject are Hyland et al. (2006), De Carvalho[1] (2018), and Bucciarelli et al. (2007).

We focus on a particular subclass of relational models, called relational graph models because their definition mimics the one of graph models living inside Scott-continuous semantics (see Berline (2000), for a survey). The interest on relational graph models is motivated by the variety of interesting properties that they enjoy. First, not only they

[1]The article De Carvalho (2018) had a troubled reviewing process and its publication date might be confusing. The content was already developed by De Carvalho during his PhD (2004-2007) under the supervision of Ehrhard and a preprint was available online much earlier, see De Carvalho (2009).

allow to characterize qualitative properties of programs, like observing their termination, but also expose intensional, quantitative properties like the amount of resources needed during the evaluation to compute a value. De Carvalho (2018) has indeed shown that, from an element in the interpretation of a λ-term M, it is possible not only to deduce that M is solvable but also to extract an upper bound on the number of head-steps necessary to reach its principal hnf. By generalizing his technique, Breuvart et al. (2018) were able to prove that relational graph models satisfy an Approximation Theorem via a simple induction on $\mathbb{N}^2$, so without the need for computability predicates that are usually employed in filter models, as discussed in Section 13.3. As a consequence, the class of relational graph models is poor in representable theories, called relational graph theories, since they are all sensible and include $\mathcal{B}$. However, it contains simple models of $\mathcal{H}^*$ and $\mathcal{H}^+$, as shown in Manzonetto (2009) and Manzonetto and Ruoppolo (2014), respectively. Breuvart et al. (2018) were able to identify $\sqsubseteq_{\mathsf{r}}$ as the least inequational graph theory, as well as characterize all relational graph models that are inequationally fully abstract for $\sqsubseteq_{\mathrm{SOL}}$ (resp. $\sqsubseteq_{\mathrm{NF}}$), and therefore fully abstract for $\mathcal{H}^*$ (resp. $\mathcal{H}^+$). This represents a breakthrough compared with previous works that only provided sufficient—but not necessary—conditions for a model to have theory $\mathcal{H}^*$ (see, e.g., B[1984], Theorem 19.2.9; Gouy (1995); Di Gianantonio et al. (1999); Franco (2001), but also Manzonetto (2008)).

In some sense, relational graph models can be seen as a resource sensitive version of filter models. In fact, just like filter models correspond to intersection type assignment systems, also relational graph models can be presented as type systems where the types are constructed from type atoms using the linear arrow $\multimap$ and the linear logic tensor $\otimes$. As one might expect, the interpretation of a closed λ-term in a relational graph model corresponds to the set of its types in the corresponding tensor type assignment system. This correspondence was known since De Carvalho's pioneering work, but subsequently clarified by Paolini et al. (2017). Intuitively, the tensor $\otimes$ can be seen as a non-idempotent intersection operator since $\sigma \otimes \sigma \neq \sigma$, so the corresponding type system is *linear*, i.e. it does not allow weakening nor contraction. It is worth mentioning that type systems with these properties were introduced already in the nineties by Gardner (1994), independently from the relational semantics of λ-calculus. In these systems a closed λ-term M is typable with $(\sigma_1 \otimes \cdots \otimes \sigma_k) \multimap \tau$ whenever M needs to call its argument k times to produce an output of type τ, once with type σ_1, once with type σ_2, and so on. Due to the associativity and commutativity of the tensor product, the type $\sigma_1 \otimes \cdots \otimes \sigma_k$ can be seen as a finite multiset $[\sigma_1, \ldots, \sigma_k]$, not as a finite set $\{\sigma_1, \ldots, \sigma_k\}$ because of the lack of idempotence. We will see that tensor type assignment systems satisfy usual properties, like an Inversion Lemma, subject reduction and subject expansion (even quantitative versions of them). Interestingly, Bucciarelli et al. (2018) have shown that the inhabitation problem (IHP), namely the problem of finding a closed λ-term inhabiting a type σ, is decidable in the setting of tensor type assignment systems. This is perhaps unexpected considering that the undecidability of the IHP for intersection types is a well-established result of Urzyczyn (1999), but the lack of idempotence really is a fundamental change.

Concerning the presentation of this material, we follow Breuvart et al. (2018) for the relational graph models and Bucciarelli et al. (2018) for the decidability proof of IHP.

Multisets

Multisets are like sets, except for the fact that they may contain several occurrences of the same element. Given a set A, a subset $B \subseteq A$ is identified by its characteristic function $\chi_B : A \to \{0, 1\}$ associating an element α with 1 if it does belong to B, and with 0 if it does not. The definition of a multiset given below generalizes this notion.

DEFINITION 14.1. (i) A *multiset* over a set A is a function $a : A \to \mathbb{N}$, where $a(\alpha)$ gives the number of occurrences of α in a, which is called its *multiplicity*.

(ii) The *support* of a multiset a is the set $\mathrm{supp}(a) \triangleq \{\alpha \in A \mid a(\alpha) > 0\}$.

(iii) The *union* of two multisets a_1, a_2 over A is defined by $(a_1 + a_2)(\alpha) \triangleq a_1(\alpha) + a_2(\alpha)$.

(iv) A multiset a is called *finite* if its support is finite; *infinite* otherwise.

(v) We let $\mathscr{M}_\mathrm{f}(A)$ denote the set of all finite multisets over A.

NOTATION. *A finite multiset a is represented as an unordered list*

$$a = [\underbrace{\alpha_1, \ldots, \alpha_1}_{n_1}, \ldots, \underbrace{\alpha_k, \ldots, \alpha_k}_{n_k}]$$

where each n_i is the multiplicity of α_i in a. The empty multiset is denoted by $[\,]$.

EXAMPLES 14.2. (i) The function $a(n) \triangleq n$ is an infinite multiset with $\mathsf{supp}(a) = \mathbb{N} - \{0\}$.

(ii) $[1], [0, 1], [0, 0, 1] \in \mathscr{M}_\mathrm{f}(2)$; $[6], [10], [3, 5, 5, 8, 8, 8, 8] \in \mathscr{M}_\mathrm{f}(\mathbb{N})$; $[\,] \in \mathscr{M}_\mathrm{f}(A)$, for any A.

(iii) Given multisets $a_0 \triangleq [5]$, $a_1 \triangleq [3, 3, 6]$, and $a_2 \triangleq [2, 6]$, we have $\mathsf{supp}(a_1) = \{3, 6\}$, $a_1 = [3, 3, 6] = [6, 3, 3] = [3, 6, 3]$. Moreover, $a_0 + a_0 = [5, 5]$ and $a_1 + a_2 = [2, 3, 3, 6, 6]$.

14.1 The class of relational graph models

The category **Rel** is a Seely category (see Definition A.27), whence the coKleisli category **MRel** of the comonad $\mathscr{M}_\mathrm{f}(-)$ of finite multisets is Cartesian closed (Theorem A.29).

The relational semantics

We provide here a direct description of the Cartesian closed category **MRel**.

DEFINITION 14.3. The *category* **MRel** is defined as follows:

- The *objects* of **MRel** are all the sets.
- A *morphism from A to B* is a relation between $\mathscr{M}_\mathrm{f}(A)$ and B, that is

$$\mathbf{MRel}(A, B) \triangleq \mathscr{P}(\mathscr{M}_\mathrm{f}(A) \times B).$$

- The *identity* of A is the relation $\mathrm{Id}_A \triangleq \{([\alpha], \alpha) \mid \alpha \in A\} \in \mathbf{MRel}(A, A)$.
- The *composition* of $\mathrm{R}_1 \in \mathbf{MRel}(B, C)$ and $\mathrm{R}_2 \in \mathbf{MRel}(A, B)$ is defined by:

$$\mathrm{R}_1 \circ \mathrm{R}_2 \triangleq \{(a, \gamma) \mid \exists k \in \mathbb{N},\ \exists (a_1, \beta_1), \ldots, (a_k, \beta_k) \in \mathrm{R}_2 \text{ such that } a = a_1 + \cdots + a_k \text{ and } ([\beta_1, \ldots, \beta_k], \gamma) \in \mathrm{R}_1 \}.$$

For the morphisms of this particular category we privilege the notation $\mathrm{R} \in \mathbf{MRel}(A, B)$, because $\mathrm{R} : A \to B$ is too evocative of a function.

LEMMA 14.4. **MRel** *has all small products.*

PROOF. It is easy to check that the categorical product of two sets A_1 and A_2 is given by their disjoint union:

$$A_1 \uplus A_2 \triangleq (\{1\} \times A_1) \cup (\{2\} \times A_2)$$

whose projection π_i, for $i \in \{1, 2\}$, is defined by setting:

$$\pi_i \triangleq \{([(i, a)], a) \mid a \in A_i\} \in \mathbf{MRel}(A_1 \uplus A_2, A_i).$$

The neutral element of the product is given by the terminal object $\mathbb{1}$, which in this case is the empty set. Indeed, the unique morphism belonging to $\mathbf{MRel}(A, \emptyset)$ is the empty relation $\emptyset$. Given a set B and two relations $\mathrm{R}_1 \in \mathbf{MRel}(B, A_1)$ and $\mathrm{R}_2 \in \mathbf{MRel}(B, A_2)$, the pairing morphism $\langle \mathrm{R}_1, \mathrm{R}_2 \rangle \in \mathbf{MRel}(B, A_1 \uplus A_2)$ is given by:

$$\langle \mathrm{R}_1, \mathrm{R}_2 \rangle \triangleq \bigcup_{i \in \{1,2\}} \{(b, (i, \alpha)) \mid (b, \alpha) \in \mathrm{R}_i\}.$$

These definitions extend easily to arbitrary I-indexed families $(A_i)_{i \in I}$ of sets. □

CONVENTION 14.5. *Note that there is a canonical bijection between the Cartesian product* $\mathscr{M}_{\mathrm{f}}(A) \times \mathscr{M}_{\mathrm{f}}(B)$ *and the set* $\mathscr{M}_{\mathrm{f}}(A \uplus B)$, *mapping the pair*

$$([\alpha_1, \ldots, \alpha_n], [\beta_1, \ldots, \beta_m])$$

to the multiset

$$[(1, \alpha_1), \ldots, (1, \alpha_n), (2, \beta_1), \ldots, (2, \beta_m)].$$

Hereafter, we confuse this bijection with an equality and write (a, b) *for the corresponding element of* $\mathscr{M}_{\mathrm{f}}(A \uplus B)$.

These bijections are sometimes called *Seely isomorphisms*, because they were first introduced in a categorical setting by Seely (1989).

EXERCISE 14.6. Show that the isomorphism described in Convention 14.5 does not extend to infinite products. [Hint: The property fails already for countable products.]

THEOREM 14.7. *The category* **MRel** *is Cartesian closed.*

PROOF. Given two sets A and B, the exponential object $A \Rightarrow B$ is given by $\mathscr{M}_{\mathrm{f}}(A) \times B$ and the evaluation morphism by:

$$\mathrm{Eval}_{AB} \triangleq \{((([(a, \beta)], a), \beta) \mid a \in \mathscr{M}_{\mathrm{f}}(A) \ \&\ \beta \in B\} \in \mathbf{MRel}((A \Rightarrow B) \uplus A, B)\,.$$

Again, it is easy to check that in this way we defined an exponentiation. Indeed, given any set C and any morphism $\mathrm{R} \in \mathbf{MRel}(C \uplus A, B)$, there exists exactly one morphism

$$\mathrm{Cur}(\mathrm{R}) \triangleq \{(c, (a, \beta)) \mid ((c, a), \beta) \in \mathrm{R}\} \in \mathbf{MRel}(C, A \Rightarrow B)$$

satisfying $\mathrm{Eval}_{AB} \circ (\mathrm{Cur}(\mathrm{R}) \times \mathrm{Id}_A) = \mathrm{R}$, where $\mathrm{Cur}(\mathrm{R}) \times \mathrm{Id}_A = \langle \mathrm{Cur}(\mathrm{R}) \circ \pi_1, \pi_2 \rangle$. □

LEMMA 14.8. *The Cartesian closed category* **MRel** *is cpo-enriched.*

PROOF. Clearly, for all sets A and B, the hom-set $(\mathbf{MRel}(A, B), \subseteq, \emptyset)$ constitutes a complete partial order. It is easy to check that composition is continuous, pairing and currying are monotonic, and the strictness conditions hold. □

REMARK 14.9. Since $\mathscr{M}_f(\mathbb{1}) = \mathscr{M}_f(\emptyset) = \{[]\}$ is a singleton, the points of an object A are simply the subsets of A. Indeed, we have $\mathbf{MRel}(\mathbb{1}, A) = \mathscr{P}(\{[]\} \times A) \cong \mathscr{P}(A)$.

PROPOSITION 14.10. *The category* **MRel** *is not well-pointed.*

PROOF. We prove a stronger property, namely that for every $A, B \neq \mathbb{1}$ there exist $\mathrm{R}_1, \mathrm{R}_2 \in \mathbf{MRel}(A, B)$ such that $\mathrm{R}_1 \neq \mathrm{R}_2$, but $\mathrm{R}_1 \circ \mathrm{R} = \mathrm{R}_2 \circ \mathrm{R}$ for all $\mathrm{R} \in \mathbf{MRel}(\mathbb{1}, A)$. Indeed, by definition of composition, we have

$$\mathrm{R}_1 \circ \mathrm{R} = \{([], \beta) \mid \exists k \geq 0, \exists \alpha_1, \ldots, \alpha_k \in A, ([], \alpha_i) \in \mathrm{R}, ([\alpha_1, \ldots, \alpha_k], \beta) \in \mathrm{R}_1\}$$

Similarly for $\mathrm{R}_2 \circ \mathrm{R}$. Hence it is sufficient to choose $\mathrm{R}_1 = \{(a_1, \beta)\}$ and $\mathrm{R}_2 = \{(a_2, \beta)\}$ for two different multisets a_1 and a_2 having the same support. □

As a direct consequence of the above proof, we obtain the following property.

COROLLARY 14.11. *No object* $A \neq \mathbb{1}$ *is well-pointed in* **MRel**.

LEMMA 14.12. *For two objects* A, B *of* **MRel**, *the following are equivalent.*

1. *A and B are isomorphic $A \cong B$ via $\mathrm{R}_1 \in \mathbf{MRel}(A, B)$ and $\mathrm{R}_2 \in \mathbf{MRel}(B, A)$.*

2. *There exists a bijective function $\iota : A \to B$.*

PROOF. $(1 \Rightarrow 2)$ First, note that $(a, \beta) \in \mathrm{R}_1$ entails $a \neq []$ because otherwise such a pair would belong to $\mathrm{R}_1 \circ \mathrm{R}_2 = \mathrm{Id}_B$. Similarly, R_2 does not contain any pair $([], \alpha)$. Thus:

$$\mathrm{R}_1 \circ \mathrm{R}_2 = \{([\beta], \beta) \mid \exists \alpha \in A \,.\, ([\beta], \alpha) \in \mathrm{R}_2 \;\&\; ([\alpha], \beta) \in \mathrm{R}_1\}.$$

Since by hypothesis $\mathrm{R}_2 \circ \mathrm{R}_1 = \mathrm{Id}_A$ we have that for all $\alpha \in A$ there is a $\beta \in B$ such that $([\beta], \alpha) \in \mathrm{R}_2$. By contradiction, assume that there is a $([\alpha_1, \ldots, \alpha_k], \beta) \in \mathrm{R}_1$ for $k > 1$. From the property above there are $\beta_1, \ldots, \beta_k \in B$ such that $([\beta_i], \alpha_i) \in \mathrm{R}_2$ for $1 \leq i \leq k$, thus we would have $([\beta_1, \ldots, \beta_k], \beta) \in \mathrm{R}_1 \circ \mathrm{R}_2 = \mathrm{Id}_B$, which is impossible. Similarly, if $(b, \alpha) \in \mathrm{R}_2$ then b is a singleton. It remains to be checked that $([\alpha_1], \beta), ([\alpha_2], \beta) \in \mathrm{R}_1$ entail $\alpha_1 = \alpha_2$, and $([\alpha], \beta_1), ([\alpha], \beta_2) \in \mathrm{R}_1$ entail $\beta_1 = \beta_2$. Both properties are easy.

$(2 \Rightarrow 1)$ Define $\mathrm{R}_1 \triangleq \{([\alpha], \iota(\alpha)) \mid \alpha \in A\}$ and $\mathrm{R}_2 \triangleq \{([\beta], \iota^{-1}(\beta)) \mid \beta \in B\}$ and verify that they provide an isomorphism in **MRel**. □

DEFINITION 14.13. The class of all reflexive objects living in **MRel** is called the *relational semantics* of λ-calculus.

Relational graph models

The class of graph models constitutes the simplest subclass of the Scott-continuous semantics. Famous models belonging to this subclass are Engeler's graph model $\mathscr{E}$ and Plotkin's graph model $\mathscr{P}_\omega$, respectively defined in Engeler (1981) and Plotkin (1993). This class has been extensively studied in the literature and proved to be useful to obtain several results concerning the operational behavior of λ-terms and the semantical representability of λ-theories. The interested reader is warmly advised to consult Berline (2000) for a quite elaborate survey.

A graph model is given by a set A and a total injection $\iota : \mathscr{P}_{\mathrm{f}}(A)\times A \to A$, and induces $\mathscr{P}(A)$ as a reflexive object in the category of domains and continuous functions. It is well-known that—even when the function ι is bijective—it does not induce an isomorphism between the set $\mathscr{P}(A)$ and the space of continuous functions $\mathscr{P}(A) \Rightarrow \mathscr{P}(A)$. More precisely, no graph model $\mathscr{G}$ can ever be extensional because the inequality $\mathscr{G} \models \mathsf{1} \sqsubseteq \mathsf{I}$ is never satisfied. This is shown, for instance, in Section 5.5 of Berline (2000).

The definition of a relational graph model mimics the one of a graph model while replacing finite sets with finite multisets. We will see that relational graph models only capture a particular subclass of reflexive objects populating **MRel**.

DEFINITION 14.14. (i) A *relational graph model* (*rgm*) $\mathcal{D}$ is given by an infinite set D and a total injection $\iota : \mathscr{M}_{\mathrm{f}}(D) \times D \to D$.

(ii) We say that a relational graph model $\mathcal{D} = (D, \iota)$ is *extensional* when ι is bijective.

REMARK 14.15. (i) Given an rgm $\mathcal{D}$, the elements of its underlying set D can be thought of as types and the elements of $\mathscr{M}_{\mathrm{f}}(D) \times D$ as (linear) arrow types $\multimap$.

(ii) The equality $\iota(a, \alpha) = \beta$ indicates that the arrow type $a \multimap \alpha$ is equivalent to the type β. In particular, when $\mathcal{D}$ is extensional, every type can be seen as an arrow type.

As shown in the next proposition, the reflexive object induced by a relational graph model (D, ι) is D itself—this property allows to define extensional models as well.

PROPOSITION 14.16. *Given an rgm* $\mathcal{D} = (D, \iota)$, *we have that:*

(i) $\mathcal{D}$ *induces a reflexive object* $(D, \mathrm{App}, \mathrm{Lam})$ *where*

$$\begin{aligned}\mathrm{Lam} &\triangleq \{([(a,\alpha)], \iota(a,\alpha)) \mid a \in \mathscr{M}_{\mathrm{f}}(D), \alpha \in D\} \in \mathbf{MRel}(D \Rightarrow D, D),\\ \mathrm{App} &\triangleq \{([\iota(a,\alpha)], (a,\alpha)) \mid a \in \mathscr{M}_{\mathrm{f}}(D), \alpha \in D\} \in \mathbf{MRel}(D, D \Rightarrow D).\end{aligned}$$

(ii) *If* $\mathcal{D}$ *is an extensional rgm, then also the induced reflexive object is.*

PROOF. (i) By definition of composition, we have

$$\begin{aligned}\mathrm{App} \circ \mathrm{Lam} &= \{([(a,\alpha)], (a,\alpha)) \mid ([(a,\alpha)], \iota(a,\alpha)) \in \mathrm{Lam} \ \&\ ([\iota(a,\alpha)], (a,\alpha)) \in \mathrm{App}\},\\ &= \{([(a,\alpha)], (a,\alpha)) \mid a \in \mathscr{M}_{\mathrm{f}}(D) \ \&\ \alpha \in D\} = \mathrm{Id}_{D \Rightarrow D}.\end{aligned}$$

(ii) Since ι is bijective, for every $\beta \in D$ there exist $a \in \mathscr{M}_{\mathrm{f}}(D)$ and $\alpha \in D$ such that $\beta = \iota(a, \alpha)$. So we have:

$$\begin{aligned}\mathrm{Lam} \circ \mathrm{App} &= \{([\iota(a,\alpha)], \iota(a,\alpha)) \mid ([\iota(a,\alpha)], (a,\alpha)) \in \mathrm{App} \ \&\ ([(a,\alpha)], \iota(a,\alpha)) \in \mathrm{Lam}\},\\ &= \{([\iota(a,\alpha)], \iota(a,\alpha)) \mid a \in \mathscr{M}_{\mathrm{f}}(D) \ \&\ \alpha \in D\} = \mathrm{Id}_D.\end{aligned}$$

□

LEMMA 14.17. *A reflexive object $\mathcal{U}$ living in* **MRel** *is extensional if and only if it is induced by some extensional rgm.*

PROOF. ($\Rightarrow$) By Lemma 14.12.
($\Leftarrow$) By Proposition 14.16(ii). □

Recall that the interpretation of a λ-term M in a reflexive object has been defined in Definition 3.45. The following characterization works for relational graph models.

PROPOSITION 14.18. *Let $\mathcal{D}$ be an rgm, $x_1, \dots x_n \in \mathrm{Var}$, for some $n \in \mathbb{N}$, and $M \in \Lambda^o(\vec{x})$. The* categorical interpretation of M in $\mathcal{D}$ (w.r.t. $\vec{x}$) *is equal*[2] *to the relation*

$$|M|^{\mathcal{D}}_{\vec{x}} \subseteq \mathscr{M}_{\mathrm{f}}(D)^n \times D$$

defined as follows:
(i) $|x_i|^{\mathcal{D}}_{\vec{x}} \triangleq \{(([], \dots, [], [\alpha], [], \dots, []), \alpha) \mid \alpha \in D\}$,
where $[\alpha]$ *stands in i-th position;*
(ii) $|\lambda y.N|^{\mathcal{D}}_{\vec{x}} \triangleq \{((a_1, \dots, a_n), \iota(a, \alpha)) \mid ((a_1, \dots, a_n, a), \alpha) \in |N|^{\mathcal{D}}_{\vec{x},y}\}$,
where we assume without loss of generality that $y \notin \vec{x}$*;*
(iii)

$$\begin{array}{lll} |PQ|^{\mathcal{D}}_{\vec{x}} \triangleq \{((a_1, \dots, a_n), \alpha) \mid \exists k \in \mathbb{N}, & & \\ & \exists a_{j1}, \dots, a_{jn} \in \mathscr{M}_{\mathrm{f}}(D), & \textit{for } j = 0, \dots, k, \\ & \exists \alpha_1, \dots, \alpha_k \in D & \textit{such that} \\ & a_i = a_{0\,i} + \dots + a_{ki}, & \textit{for } i = 1, \dots, n, \\ & ((a_{\ell 1}, \dots, a_{\ell n}), \alpha_\ell) \in |Q|^{\mathcal{D}}_{\vec{x}}, & \textit{for } \ell = 1, \dots, k, \\ & ((a_{01}, \dots, a_{0n}), \iota([\alpha_1, \dots, \alpha_k], \alpha)) \in |P|^{\mathcal{D}}_{\vec{x}} & \}. \end{array}$$

PROOF. By structural induction on M. □

The interpretation above extends to $\lambda\bot$-terms by setting $|\bot|^{\mathcal{D}}_{\vec{x}} \triangleq \emptyset$. Hereafter, we will treat the isomorphism described in Remark 14.9 as an equality. Also, for M closed, we consider $|M|^{\mathcal{D}}_{\langle\rangle} \in \mathbf{MRel}(\mathbb{1}, D)$ as a subset $|M|^{\mathcal{D}} \subseteq D$ and write $\alpha \in |M|^{\mathcal{D}}$ for $([], \alpha) \in |M|^{\mathcal{D}}$.

EXERCISE 14.19. Let $\mathcal{D}$ be an rgm. Check the following equalities:
(i) $|\lambda x.x|^{\mathcal{D}} = \{\iota([\alpha], \alpha) \mid \alpha \in D\}$.
(ii) $|\lambda x.xx|^{\mathcal{D}} = \{\iota(a + [\iota(a, \alpha)], \alpha) \mid a \in \mathscr{M}_{\mathrm{f}}(D) \ \&\ \alpha \in D\}$.
(iii) $|\Omega|^{\mathcal{D}} = \emptyset$. [Hint: use (ii) above.]
(iv) $|x\Omega|^{\mathcal{D}}_{x} = \{([\iota([], \alpha)], \alpha) \mid \alpha \in D\}$. [Hint: use (iii) above.]
(v) $|\mathsf{J}|^{\mathcal{D}} = \{\iota([\alpha], \alpha) \mid \alpha \in A\}$, where A is the least subset of D closed under the rules:

$$\frac{\alpha \in D}{\iota([], \alpha) \in A} \qquad\qquad \frac{\alpha \in D \quad a \in \mathscr{M}_{\mathrm{f}}(A)}{\iota(a, \alpha) \in A}$$

Relational graph models—just like the regular ones—can be built by performing the free completion of a partial pair or using the forcing technique presented in the article by Berline and Salibra (2006). Here we are interested in the former technique.

[2]Up to the Seely isomorphisms described in Convention 14.5.

DEFINITION 14.20. (i) A *partial pair* $\mathcal{A}$ is a pair (A, g) where A is a non-empty set of elements, called *atoms*, and $g : \mathscr{M}_f(A) \times A \to A$ is a partial injection.

(ii) A partial pair $\mathcal{A}$ is called *extensional* if g is a bijection between $\mathrm{dom}(g)$ and A.

The following partial pairs will be used as running examples in the rest of the section.

EXAMPLES 14.21. (i) $(\mathbb{N}, \emptyset)$, where $\emptyset$ represents the empty function, is a non-extensional partial pair.

(ii) $(\{\varepsilon\}, \{([], \varepsilon) \mapsto \varepsilon\})$, where $\{([], \varepsilon) \mapsto \varepsilon\}$ represents the partial injection

$$g(a, \alpha) \triangleq \begin{cases} \varepsilon, & \text{if } a = [] \ \&\ \alpha = \varepsilon, \\ \uparrow, & \text{otherwise,} \end{cases}$$

is an extensional partial pair.

(iii) $(\{\star\}, \{([\star], \star) \mapsto \star\})$ is an extensional partial pair.

Concerning partial pairs, we adopt the following convention.

CONVENTION 14.22. *Hereafter, we consider partial pairs $\mathcal{A}$ whose underlying set A does not contain any pair of elements. In other words, we suppose $(\mathscr{M}_f(A) \times A) \cap A = \emptyset$.*

The convention above is actually assumed without loss of generality because partial pairs can be considered up to isomorphism.

DEFINITION 14.23. The *free completion* $\overline{\mathcal{A}}$ of a partial pair $\mathcal{A} = (A, g)$ is the pair $(\overline{A}, \overline{g})$ defined as follows:

(i) let $A_0 \triangleq A$ and $A_{n+1} \triangleq \big((\mathscr{M}_f(A_n) \times A_n) - \mathrm{dom}(g)\big) \cup A$, then

$$\overline{A} \triangleq \bigcup_{n \in \mathbb{N}} A_n$$

(ii) the completion of g is the total function $\overline{g} : \mathscr{M}_f(\overline{A}) \times \overline{A} \to \overline{A}$ defined by setting

$$\overline{g}(a, \alpha) \triangleq \begin{cases} g(a, \alpha), & \text{if } (a, \alpha) \in \mathrm{dom}(g), \\ (a, \alpha), & \text{otherwise.} \end{cases}$$

PROPOSITION 14.24. (i) *If $\mathcal{A}$ is a partial pair then $\overline{\mathcal{A}}$ is an rgm.*

(ii) *If $\mathcal{A}$ is an extensional partial pair then the rgm $\overline{\mathcal{A}}$ is extensional.*

(iii) *Every rgm $\mathcal{D}$ is isomorphic to its own free completion $\overline{\mathcal{D}} \cong \mathcal{D}$.*

PROOF. (i) The proof of the fact that $\overline{\mathcal{A}}$ is a relational graph model is analogous to the one for regular graph models, see Berline (2000).

(ii) It is easy to check that when g is bijective, also its completion $\overline{g}$ is bijective.

(iii) Easy. □

The following relational graph models are built by free completion from the partial pairs in Examples 14.21.

EXAMPLES 14.25. (i) The first example of a relational graph model $\mathcal{E} \triangleq \overline{(\mathbb{N}, \emptyset)}$ has been introduced by Hyland et al. (2006). See also the article by De Carvalho (2018).

(ii) The rgm $\mathcal{D}_\omega \triangleq \overline{(\{\varepsilon\}, \{([\,], \varepsilon) \mapsto \varepsilon\})}$ has been first defined (up to isomorphism) by Bucciarelli et al. (2007).

(iii) The rgm $\mathcal{D}_\star \triangleq \overline{(\{\star\}, \{([\,\star\,], \star) \mapsto \star\})}$ has been introduced by Manzonetto and Ruoppolo (2014).

We now compare the interpretations of some λ-terms in these relational graph models. Such comparisons furnish interesting insights on the induced λ-theories.

EXAMPLES 14.26. (i) For all rgm's $\mathcal{D}$, we have $|\mathsf{J}|^{\mathcal{D}} \subseteq |\mathsf{1}|^{\mathcal{D}} \subseteq |\mathsf{I}|^{\mathcal{D}}$, where I is the identity, $\mathsf{1}$ is a finite η-expansion of I, and J is Wadsworth's combinator.

(ii) For all $n \in \mathbb{N}$, we have $([n], n) \in |\mathsf{I}|^{\mathcal{E}} - |\mathsf{1}|^{\mathcal{E}}$. In fact, $\mathcal{E}$ is not extensional.

(iii) $|\mathsf{I}|^{\mathcal{D}_\omega} = |\mathsf{1}|^{\mathcal{D}_\omega}$, indeed $\mathcal{D}_\omega$ is an extensional rgm. Moreover, $|\mathsf{I}|^{\mathcal{D}_\omega} = |\mathsf{J}|^{\mathcal{D}_\omega}$ holds.

(iv) Also $\mathcal{D}_\star$ is extensional, thus $|\mathsf{I}|^{\mathcal{D}_\star} = |\mathsf{1}|^{\mathcal{D}_\star}$ holds. However, we have $\star \in |\mathsf{I}|^{\mathcal{D}_\star} - |\mathsf{J}|^{\mathcal{D}_\star}$.

14.2 Tensor type assignment systems

Relational graph models can be presented as type assignment systems whose types are constructed using the linear arrow and the symmetric tensor product from Linear Logic. These type systems are called *linear* because weakening and contraction are forbidden. This is related to the fact that symmetric tensor—unlike intersection—is not idempotent. For this reason, some authors like Bucciarelli et al. (2017), Paolini et al. (2017), and De Carvalho (2018) speak about a 'non-idempotent' intersection operator.

DEFINITION 14.27. Given a set $\mathbb{A}$ of *type atoms*, the set $\mathbb{T}_\mathrm{R}$ of *relational types* and the set $\mathbb{T}_\otimes$ of *tensor types* over $\mathbb{A}$ are defined by the grammar (for $\xi \in \mathbb{A}$):

$$\sigma ::= \xi \mid \mu \multimap \sigma \qquad (\mathbb{T}_\mathrm{R})$$

$$\mu ::= \mu \otimes \mu \mid \sigma \mid 1 \qquad (\mathbb{T}_\otimes)$$

We denote relational types by σ, τ, γ and tensor types by μ, ν, possibly with indices. As usual, the linear arrow $\multimap$ associates to the right, e.g. $\xi_1 \multimap \xi_2 \multimap \xi_3 = \xi_1 \multimap (\xi_2 \multimap \xi_3)$.

All relational types are in particular tensor types, while the converse does not hold because symmetric tensors may only appear at the left-hand side of a linear arrow. Intuitively $\xi_1 \otimes (\xi_2 \otimes \xi_3)$ and $((1 \otimes \xi_1) \otimes \xi_2) \otimes \xi_3$ correspond to the multiset $[\xi_1, \xi_2, \xi_3]$, so we need to impose extra equalities among all tensor types representing the same multiset.

CONVENTION 14.28. *We suppose that the symmetric tensor $\otimes$ has higher precedence than the arrow $\multimap$, is commutative, associative and possesses 1 as neutral element. Thus, we assume that the syntactic equality $=$ on relational and tensor types satisfies the rules:*

$$\begin{aligned} \mu \otimes \nu &= \nu \otimes \mu \\ (\mu_1 \otimes \mu_2) \otimes \mu_3 &= \mu_1 \otimes (\mu_2 \otimes \mu_3) \\ \mu \otimes 1 &= \mu \end{aligned}$$

$$\frac{\sigma \simeq \tau}{x:\sigma \vdash_{\otimes} x:\tau}\ (\mathrm{ax}) \qquad \frac{\Gamma, x:\mu \vdash_{\otimes} M:\sigma}{\Gamma \vdash_{\otimes} \lambda x.M : \mu \multimap \sigma}\ (\multimap_{\mathrm{I}}) \qquad \frac{\Gamma \vdash_{\otimes} M:\tau \quad \tau \simeq \sigma}{\Gamma \vdash_{\otimes} M:\sigma}\ (\simeq)$$

$$\frac{\Gamma_0 \vdash_{\otimes} M : (\sigma_1 \otimes \cdots \otimes \sigma_k) \multimap \tau \quad \Gamma_1 \vdash_{\otimes} N : \sigma_1 \quad \cdots \quad \Gamma_k \vdash_{\otimes} N : \sigma_k \quad (k \geq 0)}{\bigotimes_{i=0}^{k} \Gamma_i \vdash_{\otimes} MN : \tau}\ (\multimap_{\mathrm{E}})$$

Figure 14.1: Tensor type assignment system $\boldsymbol{\lambda}_{\otimes}$.

NOTATION. (i) *For* $k \in \mathbb{N}$, *we write* $\otimes_{i=1}^{k}\sigma_i$ *for* $\sigma_1 \otimes \cdots \otimes \sigma_k$, *under the usual proviso that* $\otimes_{i=1}^{0}\sigma_i \triangleq 1$. *In particular,* $(\otimes_{i=1}^{k}\sigma_i) \multimap \tau$ *stands for* $(\sigma_1 \otimes \cdots \otimes \sigma_k) \multimap \tau$. *Since each* $\sigma_i \neq 1$, *the index* k *in* $\otimes_{i=1}^{k}\sigma_i$ *corresponds to the cardinality of the multiset* $[\sigma_1, \ldots, \sigma_k]$.

(ii) *We push this analogy forward and write* $\sigma \in \mu$ *if* $\mu = \sigma \otimes \nu$, *for some* $\nu \in \mathbb{T}_{\otimes}$.

DEFINITION 14.29. (i) A *type environment* Γ is a function from Var to $\mathbb{T}_{\otimes}$ whose *support*, defined as the set

$$\mathsf{supp}(\Gamma) \triangleq \{x \in \mathrm{Var} \mid \Gamma(x) \neq 1\},$$

is finite.

(ii) Let $x_1 : \mu_1, \ldots, x_n : \mu_n$ be the type environment Γ defined by (for all $y \in \mathrm{Var}$):

$$\Gamma(y) \triangleq \begin{cases} \mu_i, & \text{if } y = x_i \text{ for some } i \in \{1, \ldots, n\}, \\ 1, & \text{otherwise.} \end{cases}$$

(iii) Given type environments Γ_1, Γ_2, define their *tensor product* $\Gamma_1 \otimes \Gamma_2$ by setting:

$$(\Gamma_1 \otimes \Gamma_2)(x) \triangleq \Gamma_1(x) \otimes \Gamma_2(x), \text{ for all } x \in \mathrm{Var}.$$

Unlike intersection types, relational types are discretely ordered. However, it is possible to define an equivalence relation between types respecting suitable properties.

DEFINITION 14.30. (i) A *type equivalence* $\simeq$ is any congruence on relational and tensor types including the syntactic equality and satisfying the conditions:

(1) $\mu \simeq \nu \iff \exists k \in \mathbb{N} \,.\, \mu = \sigma_1 \otimes \cdots \otimes \sigma_k \;\&\; \nu = \tau_1 \otimes \cdots \otimes \tau_k \;\&\; \forall i \, (1 \leq i \leq k) \,.\, \sigma_i \simeq \tau_i$;

(2) $\mu \multimap \sigma \simeq \nu \multimap \tau \iff \mu \simeq \nu \;\&\; \sigma \simeq \tau$.

(ii) The equivalence $\simeq$ is extended to type environments Γ_1, Γ_2 in the obvious way:

$$\Gamma_1 \simeq \Gamma_2 \iff \forall x \in \mathrm{Var} \,.\, \Gamma_1(x) \simeq \Gamma_2(x)$$

Condition (1) of Definition 14.30(i) is a consequence of the symmetric tensor non-idempotence. In other words, we want to prevent a type equivalence from modifying the multisets cardinalities. Condition (2) requires that two arrow types are equivalent whenever they are equivalent componentwise. However, equivalent types $\sigma \simeq \tau$ may contain a different number of linear arrows, e.g., by considering a $\simeq$ satisfying $\xi \simeq \xi \multimap \xi$.

DEFINITION 14.31. (i) A *tensor type assignment system* $\boldsymbol{\lambda}_{\otimes}$ is given by a set $\mathbb{A}$ of type atoms, a type equivalence $\simeq$ and the rules displayed in Figure 14.1. Whenever $\sigma = \tau$, we simply write the axiom (ax) with no premises, i.e., $x : \sigma \vdash_{\otimes} x : \sigma$.

System	Atoms	Equivalences
HNPR	$(\xi_i)_{i \in \mathbb{N}}$	✗
BEM	ε	$1 \multimap \varepsilon \simeq \varepsilon$
MR	$\star$	$\star \multimap \star \simeq \star$

Table 14.1: Examples of tensor type assignment systems.

(ii) *Type judgements* of $\boldsymbol{\lambda}_\otimes$ are triples of the form

$$\Gamma \vdash_\otimes M : \sigma$$

where Γ is a type environment, M is a $\lambda(\bot)$-term and σ is a relational type.

(iii) A *derivation* is a finite tree obtained by composing the rules from Figure 14.1 as described in Definition 13.16(iv). Also in this chapter, when writing $\Gamma \vdash_\otimes M : \sigma$, we assume that there exists a derivation of this type judgment.

The tensor type assignment systems below correspond to the relational graph models in Examples 14.25 and will be used as running examples in the rest of the section.

EXAMPLES 14.32. (i) Let $\boldsymbol{\lambda}_\otimes^{\text{HNPR}}$ be the tensor type assignment system having:
- a denumerable set of atoms: $\mathbb{A}^{\text{HNPR}} \triangleq \{\xi_i\}_{i \in \mathbb{N}}$,
- the syntactic equality as type equivalence: $\simeq^{\text{HNPR}} \triangleq =$.

This system is an instance of de Carvalho's System R and corresponds to the relational model introduced by Hyland, Nagayama, Power and Rosolini.

(ii) Let $\boldsymbol{\lambda}_\otimes^{\text{BEM}}$ be the tensor type assignment system having:
- a single atom: $\mathbb{A}^{\text{BEM}} \triangleq \{\varepsilon\}$,
- the type equivalence $\simeq^{\text{BEM}}$ generated by setting $\varepsilon \simeq 1 \multimap \varepsilon$.

This system was introduced by Bucciarelli, Ehrhard and Manzonetto.

(iii) Let $\boldsymbol{\lambda}_\otimes^{\text{MR}}$ be the tensor type assignment system having:
- a single atom: $\mathbb{A}^{\text{MR}} \triangleq \{\star\}$,
- the type equivalence $\simeq^{\text{MR}}$ generated by setting $\star \simeq \star \multimap \star$.

This system was first defined by Manzonetto and Ruoppolo.

The following are examples of derivations in tensor type assignment systems.

EXAMPLES 14.33. (i) The $\lambda\bot$-term $\bot$ is not typable in any $\boldsymbol{\lambda}_\otimes$.

(ii) Similarly, no unsolvable λ-term is typable in any $\boldsymbol{\lambda}_\otimes$ (cf. Proposition 14.54).

(iii) For all $k \geq 0$, we have the following derivation in $\boldsymbol{\lambda}_\otimes^{\text{HNPR}}$:

$$\dfrac{\dfrac{x : (\otimes_{i=1}^k \tau_i) \multimap \sigma \vdash_\otimes^{\text{HNPR}} x : (\otimes_{i=1}^k \tau_i) \multimap \sigma \quad x : \tau_i \vdash_\otimes^{\text{HNPR}} x : \tau_i \quad 1 \leq i \leq k}{x : ((\otimes_{i=1}^k \tau_i) \multimap \sigma) \otimes \tau_1 \otimes \cdots \otimes \tau_k \vdash_\otimes^{\text{HNPR}} xx : \sigma}\,(\multimap_{\text{E}})}{\vdash_\otimes^{\text{HNPR}} \lambda x.xx : (((\otimes_{i=1}^k \tau_i) \multimap \sigma) \otimes \tau_1 \otimes \cdots \otimes \tau_k) \multimap \sigma}\,(\multimap_{\text{I}})$$

(iv) Note that in the previous example, when $k = 0$, the rightmost occurrence of x is not actually typed. We can exploit this fact to obtain the following derivation:

$$\dfrac{\dfrac{x : 1 \multimap \sigma \vdash_\otimes^{\mathrm{HNPR}} x : 1 \multimap \sigma}{x : 1 \multimap \sigma \vdash_\otimes^{\mathrm{HNPR}} x\Omega : \sigma}\,(\multimap_{\mathrm{E}})}{\vdash_\otimes^{\mathrm{HNPR}} \lambda x.x\Omega : (1 \multimap \sigma) \multimap \sigma}\,(\multimap_{\mathrm{I}})$$

The same holds for the corresponding $\lambda\bot$-term $\lambda x.x\bot$.

(v) For the same reason, we can derive the type judgement $x : 1 \multimap \sigma \vdash_\otimes^{\mathrm{HNPR}} xy : \sigma$.

(vi) All derivations above work in the systems $\boldsymbol{\lambda}_\otimes^{\mathrm{MR}}$ and $\boldsymbol{\lambda}_\otimes^{\mathrm{BEM}}$ as well.

(vii) In the tensor type assignment system $\boldsymbol{\lambda}_\otimes^{\mathrm{MR}}$, we have:

$$\dfrac{\dfrac{\dfrac{\dfrac{\star \simeq^{\mathrm{MR}} \star \multimap \star}{x : \star \vdash_\otimes^{\mathrm{MR}} x : \star \multimap \star}\,(\mathrm{ax}) \qquad y : \star \vdash_\otimes^{\mathrm{MR}} y : \star}{x : \star, y : \star \vdash_\otimes^{\mathrm{MR}} xy : \star}\,(\multimap_{\mathrm{E}})}{\vdash_\otimes^{\mathrm{MR}} \lambda xy.xy : \star \multimap \star \multimap \star}\,(\multimap_{\mathrm{I}}) \qquad \star \multimap \star \multimap \star \simeq^{\mathrm{MR}} \star}{\vdash_\otimes^{\mathrm{MR}} \lambda xy.xy : \star}\,(\simeq)$$

(viii) In the tensor type assignment system $\boldsymbol{\lambda}_\otimes^{\mathrm{BEM}}$, we have the derivation Π:

$$\dfrac{\dfrac{\dfrac{\dfrac{\varepsilon \simeq^{\mathrm{BEM}} 1 \multimap \varepsilon}{x : \varepsilon \vdash_\otimes^{\mathrm{BEM}} x : 1 \multimap \varepsilon}\,(\mathrm{ax})}{x : \varepsilon \vdash_\otimes^{\mathrm{BEM}} x(\mathsf{J}y) : \varepsilon}\,(\multimap_{\mathrm{E}})}{x : \varepsilon \vdash_\otimes^{\mathrm{BEM}} \lambda y.x(\mathsf{J}y) : 1 \multimap \varepsilon}\,(\multimap_{\mathrm{I}}) \qquad 1 \multimap \varepsilon \simeq^{\mathrm{BEM}} \varepsilon}{x : \varepsilon \vdash_\otimes^{\mathrm{BEM}} \lambda y.x(\mathsf{J}y) : \varepsilon}\,(\simeq)$$

(ix) From the derivation Π in (viii), we obtain the following:

$$\dfrac{\Pi}{\vdash_\otimes^{\mathrm{BEM}} \lambda xy.x(\mathsf{J}y) : \varepsilon \multimap \varepsilon}\,(\multimap_{\mathrm{I}})$$

Qualitative properties

We now show some standard qualitative properties of tensor type assignment systems. Expected properties like subject reduction and expansion hold, but we are only interested in the former that will be formulated in a quantitative fashion and proved subsequently. Most of these properties can be found in Bucciarelli et al. (2017) but we also show an original one, i.e. the existence of a canonical form for type derivations in these systems.

PROPOSITION 14.34 (INVERSION LEMMA).

(i) $\Gamma \vdash_\otimes x : \sigma$ *if and only if* $\Gamma \simeq x : \sigma$.

(ii) $\Gamma \vdash_\otimes \lambda x.M : \sigma$ *if and only if* $\sigma \simeq \mu \multimap \tau$ *and* $\Gamma, x : \mu \vdash_\otimes M : \tau$.

(iii) $\Gamma \vdash_\otimes MN : \sigma$ *if and only if there exist* $n \in \mathbb{N}$, *a decomposition* $\Gamma = \bigotimes_{j=0}^n \Gamma_j$ *and* $\tau_1, \ldots, \tau_n \in \mathbb{T}_R$ *such that* $\Gamma_0 \vdash_\otimes M : (\otimes_{i=1}^n \tau_i) \multimap \sigma$ *and* $\Gamma_i \vdash_\otimes N : \tau_i$, *for all* i $(1 \le i \le n)$.

PROOF. Easy. □

LEMMA 14.35. *Assume* $\Gamma \vdash_{\otimes} M : \sigma$. *We have:*

(i) $\mathsf{supp}(\Gamma) \subseteq \mathrm{FV}(M)$.

(ii) $\Gamma \simeq \Gamma'$ *and* $\sigma \simeq \tau$ *imply* $\Gamma' \vdash_{\otimes} M : \tau$.

PROOF. By induction on a derivation of $\Gamma \vdash_{\otimes} M : \sigma$. Both items follow by applying the Inversion Lemma above. □

The inclusion in Lemma 14.35(i) can possibly be strict. E.g., in Example 14.33(v), we have $x : 1 \multimap \xi \vdash_{\otimes}^{\mathrm{HNPR}} xy : \xi$ and $\mathsf{supp}(x : 1 \multimap \xi) = \{x\} \subsetneq \{x, y\} = \mathrm{FV}(xy)$.

LEMMA 14.36. *For* $M \in \Lambda_{\perp}$ *and* $x \notin \mathrm{FV}(M)$, $\Gamma \vdash_{\otimes} \lambda x.Mx : \sigma$ *implies* $\Gamma \vdash_{\otimes} M : \sigma$.

PROOF. We apply the Inversion Lemma repeatedly. Indeed, by Proposition 14.34(ii) $\Gamma \vdash_{\otimes} \lambda x.Mx : \sigma$ implies that $\sigma \simeq \mu \multimap \tau$ such that $\Gamma, x : \mu \vdash_{\otimes} Mx : \tau$ is derivable. Let $\mu = \otimes_{i=1}^{n} \tau_i$, then by Proposition 14.34(iii) there exists a decomposition $(\Gamma, x : \mu) = \bigotimes_{j=0}^{n} \Gamma_j$ such that $\Gamma_0 \vdash_{\otimes} M : \mu \multimap \tau$ and $\Gamma_i \vdash_{\otimes} x : \tau_i$ are derivable for all $i\,(1 \leq i \leq n)$. By Proposition 14.34(i), we have $\Gamma_i \simeq x : \tau_i$ which entails $\Gamma_0 = \Gamma$. We conclude by applying Lemma 14.35(ii). □

In general, applications of the rule $(\simeq)$ may appear arbitrarily in a derivation of a type judgement. Interestingly, tensor type assignment systems admit a notion of canonical derivation, having a single application of $(\simeq)$ at the root of the derivation tree.

DEFINITION 14.37. Let Π be a derivation of $\Gamma \vdash_{\otimes} M : \sigma$.

(i) We say that Π is *strongly canonical* if it contains no applications of the rule $(\simeq)$.

(ii) We say that Π is *canonical* if it is of the form

$$\dfrac{\dfrac{\Pi'}{\Gamma \vdash_{\otimes} M : \tau} \quad \sigma \simeq \tau}{\Gamma \vdash_{\otimes} M : \sigma}\,(\simeq)$$

for some strongly canonical derivation Π'.

EXAMPLES 14.38. (i) The derivations in Examples 14.33(iii)-(iv) are strongly canonical.

(ii) Examples 14.33(vii)-(viii) are canonical derivations, but not strongly canonical.

(iii) The derivation in Examples 14.33(ix) is not canonical, so *a fortiori* not strongly canonical. However, it can be turned into a canonical one:

$$\dfrac{\dfrac{\dfrac{\dfrac{\varepsilon \simeq^{\mathrm{BEM}} 1 \multimap \varepsilon}{x : \varepsilon \vdash_{\otimes}^{\mathrm{BEM}} x : 1 \multimap \varepsilon}\,(\mathrm{ax})}{x : \varepsilon \vdash_{\otimes}^{\mathrm{BEM}} x(\mathsf{J}y) : \varepsilon}\,(\multimap_{\mathrm{E}})}{\vdash_{\otimes}^{\mathrm{BEM}} \lambda xy.x(\mathsf{J}y) : \varepsilon \multimap 1 \multimap \varepsilon}\,(\multimap_{\mathrm{I}}) \qquad \varepsilon \multimap \varepsilon \simeq^{\mathrm{BEM}} \varepsilon \multimap 1 \multimap \varepsilon}{\vdash_{\otimes}^{\mathrm{BEM}} \lambda xy.x(\mathsf{J}y) : \varepsilon \multimap \varepsilon}\,(\simeq)$$

In Proposition 14.42 below, we show that $\Gamma \vdash_{\otimes} M : \sigma$ always admits a canonical derivation, which is strongly canonical if the relational type σ is 'expanded enough' in the sense that it contains enough linear arrows. Let us formalize this intuitive notion.

DEFINITION 14.39. Given a type equivalence $\simeq$, the relation $\sigma \preccurlyeq \tau$ is defined by:

$$\frac{\xi \in \mathbb{A} \quad \xi \simeq \tau}{\xi \preccurlyeq \tau} \qquad \frac{\sigma_1 \preccurlyeq \tau_1 \quad \cdots \quad \sigma_{n+1} \preccurlyeq \tau_{n+1}}{(\sigma_1 \otimes \cdots \otimes \sigma_n) \multimap \sigma_{n+1} \preccurlyeq (\tau_1 \otimes \cdots \otimes \tau_n) \multimap \tau_{n+1}}$$

If $\sigma \preccurlyeq \tau$ holds then we say that τ *is more expanded than* σ.

The reader surprised by the fact that $\multimap$ is covariant both in its left and in its right argument should keep in mind that $\preccurlyeq$ is not intended to be a subtyping relation.

EXAMPLES 14.40. Consider the relation $\preccurlyeq$ generated by the type equivalence $\simeq^{\mathrm{MR}}$.
(i) $\star \preccurlyeq \star$ and $\star \preccurlyeq \star \multimap \star \preccurlyeq \star \multimap \star \multimap \star$, but $\star \multimap \star \not\preccurlyeq \star$ and $(\star \multimap \star) \multimap \star \not\preccurlyeq \star \multimap \star$.
(ii) $(\star \multimap \star) \multimap \star \preccurlyeq (\star \multimap \star) \multimap \star \multimap \star \succcurlyeq \star \multimap \star$.

LEMMA 14.41. (i) *The relation* $\preccurlyeq$ *is a preorder.*
(ii) *If* $\sigma \preccurlyeq \tau$ *then* $\sigma \simeq \tau$.
(iii) *If* $\sigma \simeq \tau$ *then there exists* $\gamma \in \mathbb{T}_{\mathrm{R}}$ *such that* $\sigma \preccurlyeq \gamma \succcurlyeq \tau$.

PROOF. (i) It is easy to check that $\preccurlyeq$ is reflexive and transitive.
(ii) By a straightforward induction on a derivation of $\sigma \preccurlyeq \tau$.
(iii) We proceed by structural induction on σ and τ.
Base cases. If $\sigma = \xi$ then $\sigma \preccurlyeq \tau$, so we take $\gamma \triangleq \tau$. The case $\tau = \xi$ is symmetric.
Induction case. Assume $\sigma = (\otimes_{i=1}^{k}\sigma_i) \multimap \sigma_{k+1}$ and $\tau = (\otimes_{i=1}^{n}\tau_i) \multimap \tau_{n+1}$. Since $\sigma \simeq \tau$ we must have $n = k$ and $\sigma_j \simeq \tau_j$, for all $j\,(1 \le j \le k+1)$, by Definition 14.30. By IH, for each j, there exists $\gamma_j \in \mathbb{T}_{\mathrm{R}}$ satisfying $\sigma_j \preccurlyeq \gamma_j \succcurlyeq \tau_j$ and therefore $\sigma_j \simeq \gamma_j \simeq \tau_j$ by (ii). Taking $\gamma \triangleq (\otimes_{i=1}^{k}\gamma_i) \multimap \gamma_{k+1}$, we conclude that $\sigma \preccurlyeq \gamma \succcurlyeq \tau$ holds. □

PROPOSITION 14.42. *Let* Π *be a derivation of* $\Gamma \vdash_{\otimes} M : \sigma$.
(i) *There exists* $\tau \succcurlyeq \sigma$ *such that, for all* $\gamma \succcurlyeq \tau$,

$$\Gamma \vdash_{\otimes} M : \gamma$$

has a strongly canonical derivation Π'.
(ii) *There exists a canonical derivation* Π' *of* $\Gamma \vdash_{\otimes} M : \sigma$.

PROOF. (i) We proceed by induction on the derivation Π. The only interesting case is when the last applied rule is $(\multimap_{\mathrm{E}})$, namely when $M = PQ$ and Π has shape:

$$\frac{\dfrac{\Pi_0}{\Gamma_0 \vdash_{\otimes} P : (\sigma_1 \otimes \cdots \otimes \sigma_k) \multimap \sigma} \quad \dfrac{\Pi_j}{\Gamma_j \vdash_{\otimes} Q : \sigma_j} \quad 1 \le j \le k}{\bigotimes_{i=0}^{k}\Gamma_i \vdash_{\otimes} PQ : \sigma}\,(\multimap_{\mathrm{E}})$$

By IH, for all $j\,(1 \le j \le k)$ there exists $\tau_j \succcurlyeq \sigma_j$ such that for every $\gamma_j \succcurlyeq \tau_j$, the type judgement $\Gamma_j \vdash_{\otimes} Q : \gamma_j$ has a strongly canonical derivation. Moreover, there exists a relational type $\tau_0 \succcurlyeq (\otimes_{j=1}^{k}\sigma_j) \multimap \sigma$ such that, for every $\gamma_0 \succcurlyeq \tau_0$, $\Gamma_0 \vdash_{\otimes} P : \gamma_0$ has a strongly canonical derivation. By Definition 14.39, the relational type τ_0 must have shape $(\otimes_{j=1}^{k}\tau'_j) \multimap \tau$ with $\tau'_j \succcurlyeq \sigma_j$ and $\tau \succcurlyeq \sigma$. Now, Lemma 14.41(ii) gives $\tau_j \simeq \sigma_j \simeq \tau'_j$, so we can apply Lemma 14.41(iii) and obtain $\tau_j \preccurlyeq \gamma'_j \succcurlyeq \tau'_j$, for some $\gamma'_j \in \mathbb{T}_{\mathrm{R}}$. Therefore, for every $\gamma \succcurlyeq \tau \succcurlyeq \sigma$, the derivation

$$\dfrac{\dfrac{\Pi'_0}{\Gamma_0 \vdash_\otimes P : (\gamma'_1 \otimes \cdots \otimes \gamma'_k) \multimap \gamma} \quad \dfrac{\Pi'_j}{\Gamma_j \vdash_\otimes Q : \gamma'_j} \quad 1 \le j \le k}{\bigotimes_{i=0}^{k} \Gamma_i \vdash_\otimes PQ : \gamma} \ (\multimap_{\mathrm{E}})$$

is strongly canonical, for strongly canonical derivations Π'_i $(0 \le i \le k)$ that exist by IH.

(ii) By (i), there exists a strongly canonical derivation Π' of $\Gamma \vdash_\otimes M : \tau$ for some relational type $\tau \simeq \sigma$ (by Lemma 14.41(ii)). Whence, the following derivation is canonical:

$$\dfrac{\dfrac{\Pi'}{\Gamma \vdash_\otimes M : \tau} \quad \tau \simeq \sigma}{\Gamma \vdash_\otimes M : \sigma} \ (\simeq)$$

□

Quantitative properties

We now prove several quantitative properties enjoyed by tensor type assignment systems, that are inherited from the relational semantics—a quantitative model of Linear Logic. We show that a derivation Π of $\Gamma \vdash_\otimes M : \sigma$ carries information about the number of head reduction steps from M to its principal hnf and allows to reconstruct its approximants of type σ. To retrieve this information we simply count the number of applications of the rule $(\multimap_{\mathrm{E}})$ in Π, but one could equivalently consider the size of Π seen as a tree.

DEFINITION 14.43. (i) A *$\lambda\bot$-context* is a context possibly containing occurrences of $\bot$.

(ii) Consider $M \in \Lambda_\bot$. An *occurrence of a β-redex $R = (\lambda x.P)Q$ in M* is identified by a $\lambda\bot$-context $C[\,]$ having a single occurrence of the hole $[\,]$ and such that $M = C[R]$. In this case, we write $M[R{:=}\bot]$ to denote the $\lambda\bot$-term $C[\bot]$.

(iii) For $M \in \Lambda_\bot$, we let $\mathtt{red}_\#(M)$ be the number of occurrences of β-redexes in M.

(iv) We extend Definition 1.40(i) to $M, N \in \Lambda_\bot$ and write $M \xrightarrow{R}_\beta N$ if N is obtained from M by contracting the β-redex occurrence R in M.

(v) Assume that Π is a derivation of $\Gamma \vdash_\otimes M : \sigma$, then $@_\#(\Pi)$ represents the number of applications of the rule $(\multimap_{\mathrm{E}})$ in Π.

REMARK 14.44. By inspecting the proof of Proposition 14.42, one can see that a canonical derivation Π' is obtained from Π by expanding enough the types in the axioms and merging all applications of $(\simeq)$ into a single application at the root. As these transformations preserve the number of applications of $(\multimap_{\mathrm{E}})$, we conclude that $@_\#(\Pi) = @_\#(\Pi')$.

LEMMA 14.45 (WEIGHTED SUBSTITUTION LEMMA).
Let $M, N \in \Lambda_\bot$. Consider a derivation Π_0 of

$$\Gamma_0, x : \tau_1 \otimes \cdots \otimes \tau_n \vdash_\otimes M : \sigma$$

and derivations Π_i of $\Gamma_i \vdash_\otimes N : \tau_i$ for all $i\,(1 \le i \le n)$. There exists a derivation Π of

$$\Gamma_0 \otimes \cdots \otimes \Gamma_n \vdash_\otimes M[x{:=}N] : \sigma$$

such that $@_\#(\Pi) = \sum_{i=0}^{n} @_\#(\Pi_i)$.

PROOF. By induction on Π_0, splitting into cases depending on the last applied rule.

Case (ax). In this case $M = y$ and $@_{\#}(\Pi_0) = 0$. There are two possibilities:

1. If $y \neq x$ then $y[x := N] = y$, $n = 0$, and $\Gamma_0 \simeq y : \sigma$. Simply take $\Pi \triangleq \Pi_0$.
2. If $y = x$ then $x[x := N] = N$, $n = 1$, $\mathsf{supp}(\Gamma_0) = \emptyset$ and $\tau_1 \simeq \sigma$. We construct the derivation Π as follows:

$$\dfrac{\dfrac{\Pi_1}{\Gamma_1 \vdash_\otimes N : \tau_1} \quad \tau_1 \simeq \sigma}{\Gamma_1 \vdash_\otimes N : \sigma}\ (\simeq)$$

Since there are no additional $(\multimap_{\mathrm{E}})$, we get $@_{\#}(\Pi) = @_{\#}(\Pi_0) + @_{\#}(\Pi_1) = @_{\#}(\Pi_1)$.

Case $(\multimap_{\mathrm{E}})$. Then $M = PQ$, and there exist a decomposition $\Gamma_0 = \bigotimes_{i=0}^{k} \Gamma_{0\,i}$, a partition $\{\mathcal{J}_i\}_{i=0}^{k}$ of the set $\{1, \ldots, n\}$ and derivations $\Pi_{0\,i}$, for all $i\,(0 \leq i \leq k)$, such that

$$\dfrac{\dfrac{\Pi_{00}}{\Gamma_{00}, x : \otimes_{j \in \mathcal{J}_0} \tau_j \vdash_\otimes P : (\gamma_1 \otimes \cdots \otimes \gamma_k) \multimap \sigma} \quad \dfrac{\Pi_{0\,\ell}}{\Gamma_{0\,\ell}, x : \otimes_{j \in \mathcal{J}_\ell} \tau_j \vdash_\otimes Q : \gamma_\ell}\ (1 \leq \ell \leq k)}{\Gamma_0, x : \tau_1 \otimes \cdots \otimes \tau_n \vdash_\otimes PQ : \sigma}\ (\multimap_{\mathrm{E}})$$

By IH, there are derivations Π_i' $(0 \leq i \leq k)$ such that

$$\dfrac{\Pi_0'}{\Gamma_{00} \otimes (\bigotimes_{j \in \mathcal{J}_0} \Gamma_j) \vdash_\otimes P[x := N] : (\gamma_1 \otimes \cdots \otimes \gamma_k) \multimap \sigma}$$

$$\dfrac{\Pi_\ell'}{\Gamma_{0\,\ell} \otimes (\bigotimes_{j \in \mathcal{J}_\ell} \Gamma_j) \vdash_\otimes Q[x := N] : \gamma_\ell}\ (1 \leq \ell \leq k)$$

and $@_{\#}(\Pi_i') = @_{\#}(\Pi_{0\,i}) + \sum_{j \in \mathcal{J}_i} @_{\#}(\Pi_j)$ for all $i\,(0 \leq i \leq k)$. We can combine these derivations together and get, setting $\Gamma_i' \triangleq \Gamma_{0\,i} \otimes (\bigotimes_{j \in \mathcal{J}_i} \Gamma_j)$, the following derivation Π:

$$\dfrac{\dfrac{\Pi_0'}{\Gamma_0' \vdash_\otimes P[x := N] : (\gamma_1 \otimes \cdots \otimes \gamma_k) \multimap \sigma} \quad \dfrac{\Pi_\ell'}{\Gamma_\ell' \vdash_\otimes Q[x := N] : \gamma_\ell}\ (1 \leq \ell \leq k)}{\bigotimes_{i=0}^{k} \Gamma_i' \vdash_\otimes (P[x := N])(Q[x := N]) : \sigma}\ (\multimap_{\mathrm{E}})$$

Notice that the following chain of equalities holds:

$$\begin{array}{rcl} \bigotimes_{i=0}^{k} \Gamma_i' & = & \bigotimes_{i=0}^{k} (\Gamma_{0\,i} \otimes (\bigotimes_{j \in \mathcal{J}_i} \Gamma_j)) \\ & = & (\bigotimes_{i=0}^{k} \Gamma_{0\,i}) \otimes (\bigotimes_{i=0}^{k} \bigotimes_{j \in \mathcal{J}_i} \Gamma_j) \\ & = & \Gamma_0 \otimes (\bigotimes_{i=1}^{n} \Gamma_i) \\ & = & \bigotimes_{i=0}^{n} \Gamma_i \end{array}$$

We conclude that

$$\begin{array}{rcl} @_{\#}(\Pi) & = & 1 + \sum_{i=0}^{k} @_{\#}(\Pi_i') \\ & = & 1 + \sum_{i=0}^{k} (@_{\#}(\Pi_{0\,i}) + \sum_{j \in \mathcal{J}_i} @_{\#}(\Pi_j)) \\ & = & (1 + \sum_{i=0}^{k} @_{\#}(\Pi_{0\,i})) + \sum_{i=0}^{k} \sum_{j \in \mathcal{J}_i} @_{\#}(\Pi_j) \\ & = & @_{\#}(\Pi_0) + \sum_{i=1}^{n} @_{\#}(\Pi_i) \\ & = & \sum_{i=0}^{n} @_{\#}(\Pi_i) \end{array}$$

All other cases follow straightforwardly from the IH. □

COROLLARY 14.46 (β-SOUNDNESS). *For $M, N \in \Lambda_\perp$, if Π is a derivation of*

$$\Gamma \vdash_\otimes (\lambda x.M)N : \sigma$$

then there exists a derivation Π' of

$$\Gamma \vdash_\otimes M[x:=N] : \sigma$$

such that $@_\#(\Pi') = @_\#(\Pi) - 1$.

PROOF. By Proposition 14.42 and Remark 14.44 we can assume wlog that σ is sufficiently expanded and Π is strongly canonical. Therefore, the derivation Π has shape:

$$\dfrac{\dfrac{\dfrac{\Pi_0}{\Gamma_0, x : \tau_1 \otimes \cdots \otimes \tau_n \vdash_\otimes M : \sigma}}{\Gamma_0 \vdash_\otimes \lambda x.M : (\tau_1 \otimes \cdots \otimes \tau_n) \multimap \sigma}\,(\multimap_{\mathrm{I}}) \quad \dfrac{\Pi_i}{\Gamma_i \vdash_\otimes N : \tau_i} \quad 1 \leq i \leq n \geq 0}{\Gamma \vdash_\otimes (\lambda x.M)N : \sigma}\,(\multimap_{\mathrm{E}})$$

where $\Gamma = \bigotimes_{j=0}^{n} \Gamma_j$. By Lemma 14.45, there exists a derivation Π' of $\Gamma \vdash_\otimes M[x:=N] : \sigma$ such that $@_\#(\Pi') = \sum_{j=0}^{n} @_\#(\Pi_j) = @_\#(\Pi) - 1$. □

The same holds if the β-redex is placed in a head context, therefore the amount of applications of the rule $(\multimap_{\mathrm{E}})$ decreases by 1 after every head reduction step.

LEMMA 14.47. *Let $M, N \in \Lambda_\perp$. If Π is a derivation of $\Gamma \vdash_\otimes M : \sigma$ and $M \to_h N$ then there exists a derivation Π' of $\Gamma \vdash_\otimes N : \sigma$ satisfying $@_\#(\Pi') = @_\#(\Pi) - 1$.*

PROOF. By induction on the head context $H[]$ such that $M = H[(\lambda x.P)Q]$ and $N = H[P[x:=Q]]$, using Corollary 14.46 for the base case. □

Hence, as observed by De Carvalho (2018), the measure $@_\#(-)$ provides an upper bound on the number of steps necessary to reach the principal hnf of a solvable term.

COROLLARY 14.48. *If Π is a derivation of $\Gamma \vdash_\otimes M : \sigma$ then the head reduction of M has length at most $@_\#(\Pi)$.*

DEFINITION 14.49. We say that two derivation trees Π and Π' *look alike*, in symbols $\Pi \approx \Pi'$, whenever the trees obtained by removing all $\lambda\perp$-terms from Π and Π' coincide.

EXAMPLES 14.50. Let Π and Π' be the following derivation trees (respectively):

$$\dfrac{\dfrac{x : \sigma \vdash_\otimes x : \sigma}{x : \sigma \vdash_\otimes \lambda y.x : 1 \multimap \sigma}}{x : \sigma \vdash_\otimes (\lambda y.x)\mathsf{K} : \sigma} \qquad \dfrac{\dfrac{x : \sigma \vdash_\otimes x : \sigma}{x : \sigma \vdash_\otimes \lambda y.x : 1 \multimap \sigma}}{x : \sigma \vdash_\otimes (\lambda y.x)\mathsf{F} : \sigma}$$

We observe that $\Pi \approx \Pi'$ because, once all λ-terms are erased, they both become

$$\dfrac{\dfrac{x : \sigma \vdash_\otimes \quad : \sigma}{x : \sigma \vdash_\otimes \quad : 1 \multimap \sigma}}{x : \sigma \vdash_\otimes \quad : \sigma}$$

Intuitively, the equivalence $\Pi \approx \Pi'$ captures the fact that Π and Π' roughly represent the same derivation tree, but they may be used to type possibly distinct $\lambda\bot$-terms.

LEMMA 14.51 (WEIGHTED SUBJECT REDUCTION).
Let $M, M' \in \Lambda_\bot$ be such that $M \xrightarrow{R}_\beta M'$. If Π is a derivation of $\Gamma \vdash_\otimes M : \sigma$ then there exists a derivation Π' of $\Gamma \vdash_\otimes M' : \sigma$ such that one of the following cases holds:

1. *either $@_\#(\Pi') < @_\#(\Pi)$,*
2. *or $\Pi' \approx \Pi$ and there is a derivation Π'' of $\Gamma \vdash_\otimes M[R{:=}\bot] : \sigma$ such that $\Pi'' \approx \Pi$.*

PROOF. Assume wlog that the derivation Π is strongly canonical, and proceed by induction on the $\lambda\bot$-context $C[\,]$ corresponding to the redex occurrence R in M, i.e., $M = C[R]$ and $M' = C[R']$ for $R \to_\beta R'$.

Case $C[\,] = [\,]$. By Corollary 14.46, a Π' satisfying $@_\#(\Pi') < @_\#(\Pi)$ exists.

Case $C[\,] = N(C'[\,])$. In this case the derivation Π has the form (for some $n \geq 0$):

$$\dfrac{\dfrac{\Pi_0}{\Gamma_0 \vdash_\otimes N : (\tau_1 \otimes \cdots \otimes \tau_n) \multimap \sigma} \quad \dfrac{\Pi_i}{\Gamma_i \vdash_\otimes C'[R] : \tau_i} \quad 1 \leq i \leq n}{\Gamma \vdash_\otimes N(C'[R]) : \sigma}\ (\multimap\mathrm{E})$$

where $\Gamma = \bigotimes_{j=0}^n \Gamma_j$. In case $n = 0$, both derivations Π' and Π'' have the following shape (for $X = C'[R']$ and $X = C'[\bot]$, respectively):

$$\dfrac{\dfrac{\Pi_0}{\Gamma \vdash_\otimes N : (\tau_1 \otimes \cdots \otimes \tau_n) \multimap \sigma}}{\Gamma \vdash_\otimes NX : \sigma}\ (\multimap\mathrm{E})$$

Otherwise $n > 0$ and we can construct Π' from the derivations Π'_i obtained from the IH:

$$\dfrac{\dfrac{\Pi_0}{\Gamma_0 \vdash_\otimes N : (\tau_1 \otimes \cdots \otimes \tau_n) \multimap \sigma} \quad \dfrac{\Pi'_i}{\Gamma_i \vdash_\otimes C'[R'] : \tau_i} \quad 1 \leq i \leq n}{\bigotimes_{j=0}^n \Gamma_j \vdash_\otimes N(C'[R']) : \sigma}\ (\multimap\mathrm{E})$$

where, for each index $i\,(1 \leq i \leq n > 0)$, Π'_i satisfies either (i) $@_\#(\Pi'_i) < @_\#(\Pi_i)$ or (ii) $\Pi'_i \approx \Pi_i \approx \Pi''_i$, for some derivation Π''_i of $\Gamma_i \vdash_\otimes C'[\bot] : \tau_i$. Now, if there exists a Π'_i satisfying (i), then $@_\#(\Pi') < @_\#(\Pi)$ and we are in case (1). Otherwise $\Pi' \approx \Pi$ and, by substituting $\bot$ for R' and the Π''_is for the Π'_is in Π', we obtain a derivation $\Pi'' \approx \Pi$.

The cases $C[\,] = (C'[\,])N$ and $C[\,] = \lambda x.C'[\,]$ are treated similarly. □

EXERCISE 14.52. (i) Propose a statement for the weighted subject expansion.
(ii) Show that the weighted subject expansion holds. Conclude that every $\boldsymbol{\lambda}_\otimes$ is sound.

Recall that $\mathsf{da}(M)$ stands for the direct approximant of M (Definition 2.28(iii)).

LEMMA 14.53. *Let $M \in \Lambda_\bot$. If M is in β-nf and $\Gamma \vdash_\otimes M : \sigma$ then $\Gamma \vdash_\otimes \mathsf{da}(M) : \sigma$.*

PROOF. By a straightforward induction on the β-normal structure of M. □

Note that the hypothesis that M is in β-normal form is needed to prove Lemma 14.53. Indeed, for $M \triangleq \lambda x.\mathsf{I}x$ we have $\vdash_\otimes M : \sigma \multimap \sigma$, whereas $\mathsf{da}(M) = \bot$ cannot be typed.

PROPOSITION 14.54. *For all $M \in \Lambda_\bot$, we have:*

$$\Gamma \vdash_\otimes M : \sigma \iff \exists P \in \mathscr{A}(M) \ .\ \Gamma \vdash_\otimes P : \sigma$$

PROOF. ($\Rightarrow$) Let Π be derivation of $\Gamma \vdash_\otimes M : \sigma$. We proceed by induction on the pair

$$\langle @_\#(\Pi), \mathtt{red}_\#(M)\rangle \in \mathbb{N} \times \mathbb{N},$$

lexicographically ordered.

Case $\mathtt{red}_\#(M) = 0$. As M is in β-nf, we apply Lemma 14.53 and take $P \triangleq \mathsf{da}(M)$.

Case $\mathtt{red}_\#(M) \geq 1$. As M contains at least an occurrence of a β-redex R, we consider M' such that $M \xrightarrow{R}_\beta M'$. By Lemma 14.51, one can find a derivation Π' of $\Gamma \vdash_\otimes M' : \sigma$ such that either

1. $@_\#(\Pi') < @_\#(\Pi)$, or
2. there is a derivation Π'' of $\Gamma \vdash_\otimes M[R := \bot] : \sigma$ satisfying $\Pi'' \approx \Pi$, from which $@_\#(\Pi'') = @_\#(\Pi)$ and $\mathtt{red}_\#(M[R{:=}\bot]) < \mathtt{red}_\#(M)$ follow.

In both cases the IH gives an approximant $P \in \Lambda_\bot$ such that $\Gamma \vdash_\otimes P : \sigma$. In Case (1) we have $P \in \mathscr{A}(M') = \mathscr{A}(M)$. In Case (2) $P \in \mathscr{A}(M[R{:=}\bot]) \subseteq \mathscr{A}(M)$, by Lemma 2.33(i).

($\Leftarrow$) Let $P \in \mathscr{A}(M)$ and proceed by induction on a derivation Π of $\Gamma \vdash_\otimes P : \sigma$, recalling that the $\beta\bot$-normal structure of P is described in Lemma 2.27.

The case $P = \bot$ is vacuous, because $\bot$ is not typable.

Suppose $P = \lambda x_1 \dots x_n.yP_1 \cdots P_m$, where $n, m \geq 0$ and the P_i's are $\beta\bot$-normal. Assuming without loss of generality that the derivation Π is strongly canonical, it must have the form

$$\dfrac{\dfrac{\dfrac{\Gamma_0 \simeq y : \nu_1 \multimap \cdots \multimap \nu_m \multimap \tau}{\Gamma_0 \vdash_\otimes y : \nu_1 \multimap \cdots \multimap \nu_m \multimap \tau}\,(\mathrm{ax}) \qquad \dfrac{\Pi_{ij}}{\Gamma_{ij} \vdash_\otimes P_i : \tau_{ij}} \quad \begin{array}{l} 1 \leq i \leq m, \\ 1 \leq j \leq k_i \end{array}}{\Gamma, x_1 : \mu_1, \dots, x_n : \mu_n \vdash_\otimes yP_1 \cdots P_m : \tau}\,(\multimap_{\mathrm{E}})}{\Gamma \vdash_\otimes \lambda x_1 \dots x_n.yP_1 \cdots P_m : \mu_1 \multimap \cdots \multimap \mu_n \multimap \tau}\,(\multimap_{\mathrm{I}})$$

where

- $\sigma = \mu_1 \multimap \cdots \multimap \mu_n \multimap \tau$;
- $\Gamma, x_1 : \mu_1, \dots, x_n : \mu_n = \Gamma_0 \otimes \big(\bigotimes_{i=1}^m \bigotimes_{j=1}^{k_i} \Gamma_{ij}\big)$;
- $\nu_i = \tau_{i1} \otimes \cdots \otimes \tau_{ik_i}$, for all $i\,(1 \leq i \leq m)$;
- the derivations Π_{ij} are strongly canonical.

Since $P \in \mathscr{A}(M)$, it must have shape $P = \mathsf{da}(M')$ for some $\lambda\bot$-term $M' =_\beta M$. By Definition 2.28(iv), we have $M' = \lambda x_1 \dots x_n.yM_1 \cdots M_m$ with $P_i = \mathsf{da}(M_i)$ for all $i \leq m$. From the IH we get $\Gamma_{ij} \vdash_\otimes M_i : \tau_{ij}$, for all appropriate i, j. By replacing each P_i by M_i in the derivation tree above, we can construct a derivation of $\Gamma \vdash_\otimes M' : \sigma$. Since $M' =_\beta M$, we conclude that $\Gamma \vdash_\otimes M : \sigma$ is derivable by soundness. □

14.3 $\boldsymbol{\lambda}_{\otimes}^{\mathrm{HNPR}}$ — A case study

We now study the tensor type system $\boldsymbol{\lambda}_{\otimes}^{\mathrm{HNPR}}$, having infinitely many atoms $\mathbb{A}^{\mathrm{HNPR}} = \{\xi_i \mid i \in \mathbb{N}\}$ and no non-trivial equivalences between atoms and arrow types (Table 14.1). Since $\simeq^{\mathrm{HNPR}}$ coincides with the equality, we can safely forget the premise of the rule (ax) and overlook the rule ($\simeq$) entirely.

A logical characterization of normalizability

We have seen that the intersection type system $\boldsymbol{\lambda}_{\wedge}^{\mathrm{BCD}}$ allows to characterize solvable terms (Corollary 13.61). Similarly, in $\boldsymbol{\lambda}_{\wedge}^{\mathrm{CDZ}}$, a closed λ-term is β-normalizable if and only if it is typable with $\star$ (Proposition 13.81(ii)). We show that using $\boldsymbol{\lambda}_{\otimes}^{\mathrm{HNPR}}$ it is possible to derive a logical characterization of β-normalizability, by checking whether a closed λ-term admits a type σ where either the tensor type 1 does not occur, or it only occurs with the correct 'polarity'.

NOTATION. *Given a* polarity $p \in \{+,-\}$, *respectively called* positive/negative polarity, *we denote by* $\neg p$ *the opposite polarity, namely:*

$$\begin{aligned} \neg(-) &\triangleq + \\ \neg(+) &\triangleq - \end{aligned}$$

DEFINITION 14.55. Let $\sigma \in \mathbb{T}_{\mathrm{R}}^{\mathrm{HNPR}}$, $\mu \in \mathbb{T}_{\otimes}^{\mathrm{HNPR}}$ and $p \in \{+,-\}$.

(i) Define the relations $1 \in^p \sigma$ and $1 \in^p \mu$ by mutual induction as follows:

$$\frac{}{1 \in^- 1 \multimap \tau} \qquad \frac{1 \in^p \tau}{1 \in^p \mu \multimap \tau} \qquad \frac{1 \in^{\neg p} \mu}{1 \in^p \mu \multimap \tau} \qquad \frac{1 \in^p \tau}{1 \in^p \tau \otimes \mu}$$

(ii) We say that 1 *occurs positively* in σ (resp. μ) if $1 \in^+ \sigma$ (resp. $1 \in^+ \mu$).

(iii) We say that 1 *occurs negatively* in σ (resp. μ) if $1 \in^- \sigma$ (resp. $1 \in^- \mu$).

(iv) We write $1 \notin^p \sigma$ (resp. $1 \notin^p \mu$) if 1 does not occur in σ (resp. μ) with polarity p.

(v) Given a type environment Γ, we write $1 \in^p \Gamma$ if there is $x \in \mathrm{Var}$ such that $1 \in^p \Gamma(x)$. Specularly, $1 \notin^p \Gamma$ indicates that there exists no $x \in \mathrm{Var}$ such that $1 \in^p \Gamma(x)$.

EXAMPLES 14.56. (i) $1 \notin^p (\xi_0 \otimes \xi_1) \multimap \xi_2$, for all $p \in \{+,-\}$.

(ii) $1 \in^- 1 \multimap \xi$, $1 \in^+ (1 \multimap \xi) \multimap \xi$, whence $1 \notin^+ 1 \multimap \xi$ and $1 \notin^- (1 \multimap \xi) \multimap \xi$.

(iii) For $\sigma \triangleq ((1 \multimap \xi) \otimes ((1 \multimap \xi) \multimap \xi)) \multimap \xi$, we have both $1 \in^+ \sigma$ and $1 \in^- \sigma$.

The following theorem, first shown by Manzonetto and Ruoppolo (2014), is the analogous for tensor type assignment systems of Theorem 3.10 in Krivine (1993).

THEOREM 14.57 (CHARACTERIZATION OF NF).
For $M \in \Lambda$, the following are equivalent:

1. $M \in \mathrm{NF}$;

2. $\Gamma \vdash_{\otimes}^{\mathrm{HNPR}} M : \sigma$ *for some environment* Γ *and type* σ *such that* $1 \notin^+ \sigma$ *and* $1 \notin^- \Gamma$.

PROOF. ($1 \Rightarrow 2$) By structural induction on $M^{\beta\text{-}\mathbf{nf}}$, one constructs a derivation of $\Gamma \vdash_{\otimes}^{\mathrm{HNPR}} M^{\beta\text{-}\mathbf{nf}} : \sigma$ satisfying these criteria. Conclude by subject expansion.

($2 \Rightarrow 1$) Consider a derivation Π of $\Gamma \vdash_{\otimes}^{\mathrm{HNPR}} M : \sigma$. Proceed by induction on $@_{\#}(\Pi)$. By Lemma 14.2, there is a derivation Π' of $\Gamma \vdash_{\otimes}^{\mathrm{HNPR}} M^{\mathsf{phnf}} : \sigma$, with $@_{\#}(\Pi') \leq @_{\#}(\Pi)$. Write $M^{\mathsf{phnf}} = \lambda x_1 \dots x_n . y M_1 \cdots M_k$, for some $n, k \geq 0$. By the Inversion Lemma 14.34, this entails $\sigma = \mu_1 \multimap \cdots \multimap \mu_n \multimap \gamma$. Moreover, the derivation Π' is of the form:

$$\dfrac{\dfrac{\dfrac{}{\Gamma_0 \vdash_{\otimes}^{\mathrm{HNPR}} y : \nu_1 \multimap \cdots \multimap \nu_k \multimap \gamma}\,(\mathrm{ax}) \qquad \dfrac{\Pi_{ij}}{\Gamma_{ij} \vdash_{\otimes}^{\mathrm{HNPR}} M_i : \tau_{ij}} \quad \begin{array}{l} 1 \leq i \leq k, \\ 1 \leq j \leq k_i \end{array}}{\Gamma, x_1 : \mu_1, \dots, x_n : \mu_n \vdash_{\otimes}^{\mathrm{HNPR}} y M_1 \cdots M_k : \gamma}\,(\multimap_{\mathrm{E}})}{\Gamma \vdash_{\otimes}^{\mathrm{HNPR}} \lambda x_1 \dots x_n . y M_1 \cdots M_k : \mu_1 \multimap \cdots \multimap \mu_n \multimap \gamma}\,(\multimap_{\mathrm{I}})$$

where each Π_{ij} satisfies $@_{\#}(\Pi_{ij}) < @_{\#}(\Pi') \leq @_{\#}(\Pi)$, and

- $\Gamma_0 = y : \nu_1 \multimap \cdots \multimap \nu_k \multimap \gamma$;
- $\nu_i = \tau_{i1} \otimes \cdots \otimes \tau_{ik_i}$, for all $i\,(1 \leq i \leq k)$;
- $\Gamma, x_1 : \mu_1, \dots, x_n : \mu_n = \Gamma_0 \otimes (\bigotimes_{i=1}^{k} \bigotimes_{j=1}^{k_i} \Gamma_{ij})$.

If $1 \notin^{+} \sigma$ and $1 \notin^{-} \Gamma$ then, for every $i\,(1 \leq i \leq k)$, we get $\nu_i \neq 1$, equivalently, $k_i > 0$. Moreover, we must have $1 \notin^{+} \tau_{ij}$ and $1 \notin^{-} \Gamma_{ij}$, for all $j \leq k_i$. From the IH, we obtain that all the M_i's (if any) have a β-normal form. We conclude that $M \in \mathrm{NF}$. □

Manzonetto and Ruoppolo (2014) proved that Theorem 14.57 generalizes to all tensor type systems whose type equivalence $\simeq$ preserves the polarities of 1, in the sense that

$$1 \in^{p} \sigma \;\&\; \sigma \simeq \tau \quad \Rightarrow \quad 1 \in^{p} \tau$$

Inhabitation is decidable

The *inhabitation problem* (*IHP*) for a type assignment system requires to determine, for every type environment Γ and type σ, whether there is a λ-term M satisfying $\Gamma \vdash M : \sigma$. Urzyczyn (1999) has shown that IHP is undecidable for the type system introduced by Coppo et al. (1981), a result that generalizes to most intersection type systems. Subsequently, Salvati et al. (2012) showed that IHP for intersection types is Turing-equivalent to the definability problem for the simply typed λ-calculus and strengthened Urzyczyn's theorem by applying results from Joly (2003). Following Bucciarelli et al. (2018) we show that, on the contrary, IHP is decidable for $\boldsymbol{\lambda}_{\otimes}^{\mathrm{HNPR}}$. We describe a non-deterministic algorithm that takes (Γ, σ) as input and returns as output the set of all approximants A satisfying $\Gamma \vdash_{\otimes}^{\mathrm{HNPR}} A : \sigma$ as well as the minimality condition (2) below.

DEFINITION 14.58. (i) An approximant $A \in \mathscr{A}$ is *minimal for* (Γ, σ) whenever

(1) $\Gamma \vdash_{\otimes}^{\mathrm{HNPR}} A : \sigma$, and

(2) $\forall A' \leq_{\perp} A\,.\,[\Gamma \vdash_{\otimes}^{\mathrm{HNPR}} A' : \sigma \quad \Rightarrow \quad A = A']$.

(ii) A is *minimal for* $(\Gamma, \otimes_{i=1}^{n} \sigma_i)$ if there exist a decomposition $\Gamma = \bigotimes_{i=1}^{n} \Gamma_i$ and approximants $A_1, \dots, A_n \in \mathscr{A}$ such that $A = \bigvee_{i=1}^{n} A_i$ and every A_i is minimal for (Γ_i, σ_i).

In order to show that the inhabitation algorithm is complete and terminating, we need to prove that, for fixed Γ and σ, there exists a finite number of approximants that are minimal for (Γ, σ) and that the algorithm non-deterministically constructs them all.

LEMMA 14.59. *Let $A \in \mathscr{A}$ be minimal for (Γ, σ).*

(i) *If $A = \lambda x.A_1$ then $\sigma = \mu \multimap \tau$ and A_1 is minimal for $((\Gamma, x : \mu), \tau)$.*

(ii) *If $A = A_1 A_2$ then there exist $\mu \in \mathbb{T}_{\otimes}^{\text{HNPR}}$ and a decomposition $\Gamma = \Gamma_1 \otimes \Gamma_2$ such that A_1 is minimal for $(\Gamma_1, \mu \multimap \sigma)$ and A_2 is minimal for (Γ_2, μ).*

PROOF. (i) The result follows from Proposition 14.34(ii) and the fact that, for all $A' \in \mathscr{A}$, $A' \leq_{\perp} \lambda x.A_1$ and $\Gamma \vdash_{\otimes}^{\text{HNPR}} A' : \mu \multimap \tau$ imply $A' = \lambda x.A'_1$, for some $A'_1 \leq_{\perp} A_1$.

(ii) Assume that $A_1 A_2$ is minimal for (Γ, σ). By Proposition 14.34(iii), there exist $\mu = \otimes_{i=1}^{n} \tau_i$ and a decomposition $\Gamma = \bigotimes_{j=0}^{n} \Gamma_j$ such that $\Gamma_0 \vdash_{\otimes}^{\text{HNPR}} A_1 : \mu \multimap \sigma$ and $\forall i\, (1 \leq i \leq n)\, .\, \Gamma_i \vdash_{\otimes}^{\text{HNPR}} A_2 : \tau_i$. It is easy to check that A_1 is minimal for $(\Gamma_0, \mu \multimap \sigma)$. Since each (Γ_i, τ_i) is inhabited, it must have a minimal inhabitant $A_{2,i} \leq_{\perp} A_2$. Hence, $A'_2 \triangleq \bigvee_{i=1}^{n} A_{2,i} \leq_{\perp} A_2$ and $\Gamma_i \vdash_{\otimes}^{\text{HNPR}} A'_2 : \tau_i$, for all such i. Therefore $\Gamma \vdash_{\otimes}^{\text{HNPR}} A_1 A'_2 : \sigma$. If $A_2 \not\leq_{\perp} A'_2$ then we also have $A_1 A'_2 \lneq_{\perp} A_1 A_2$, thus contradicting the minimality of A. We conclude that A_2 is minimal for $(\bigotimes_{i=1}^{n} \Gamma_i, \mu)$. □

EXAMPLES 14.60. (i) $x \perp x$ is minimal for $(x : (1 \multimap \sigma \multimap \xi_0) \otimes \sigma,\ \xi_0)$.

(ii) $xx\perp$ is minimal for $(x : (\tau \multimap 1 \multimap \xi_1) \otimes \tau,\ \xi_1)$.

(iii) xxx is minimal for $(x : (1 \multimap \sigma \multimap \xi_0) \otimes \sigma \otimes (\tau \multimap 1 \multimap \xi_1) \otimes \tau,\ \xi_0 \otimes \xi_1)$.

(iv) xx is not minimal for $(x : 1 \multimap \sigma,\ \sigma)$ since $x : 1 \multimap \sigma \vdash_{\otimes}^{\text{HNPR}} x\perp : \sigma$ and $x\perp \lneq_{\perp} xx$.

Finding the minimal approximants inhabiting (Γ, σ) is sufficient for solving the original inhabitation problem since $\Gamma \vdash_{\otimes}^{\text{HNPR}} M : \sigma$ holds exactly when there is an $A \in \mathscr{A}(M)$ minimal for (Γ, σ). We present the inhabitation algorithm 'RT' as a deductive system. When writing $A \in \text{RT}(\Gamma;\ \sigma)$, we intend that RT is called with Γ and σ as inputs, and provides A as output. The premises of an inference rule give the conditions that need to be checked and the recursive calls necessary to construct the output in the conclusion.

DEFINITION 14.61. (i) The *inhabitation algorithm* $\text{RT}(\Gamma;\ \sigma)$ for the type assignment system $\boldsymbol{\lambda}_{\otimes}^{\text{HNPR}}$ is given in Figure 14.2, via an auxiliary predicate $\text{TT}(\Gamma;\ \mu)$, for $\mu \in \mathbb{T}_{\otimes}^{\text{HNPR}}$. In the rule ($\mathsf{sup}_n$), we write $\circ(A_1, \ldots, A_n)$ to indicate that the approximants $A_1, \ldots, A_n$ are *compatible*, i.e., they have a common upper bound $A \in \mathscr{A}$.

(ii) A *run* of the algorithm is a deduction tree built bottom-up by applying the rules in Figure 14.2 in such a way that every node is an instance of a rule (as in Examples 14.63).

(iii) We say that *a run of the algorithm terminates* if the corresponding tree is finite.

(iv) The *inhabitation algorithm terminates* if it needs to execute a finite number of different terminating runs.

REMARK 14.62. (i) The base cases of the algorithm are obtained by taking $n = 0$:

$$\frac{}{x \in \text{RT}(x : \sigma;\ \sigma)}\ (\mathsf{head}_0) \qquad\qquad \frac{}{\perp \in \text{TT}(\emptyset;\ 1)}\ (\mathsf{sup}_0)$$

(ii) The algorithm in Figure 14.2 is clearly non-deterministic. E.g., when $\sigma = \nu \multimap \tau$, both rules ($\mathsf{abs}$) and ($\mathsf{head}_n$) might be applicable. Moreover, in (head_n) and (sup_n), the type environment can be decomposed in several different ways.

(iii) Note that $A \in \text{RT}(\Gamma;\ \sigma)$ (resp. $A \in \text{TT}(\Gamma;\ \mu)$) implies $\text{FV}(A) \subseteq \mathsf{supp}(\Gamma)$.

$$\frac{A \in \mathrm{RT}(\Gamma, x : \mu;\, \sigma)}{\lambda x.A \in \mathrm{RT}(\Gamma;\, \mu \multimap \sigma)}\ (\mathsf{abs})$$

$$\frac{\Gamma = \bigotimes_{i=1}^{n} \Gamma_i \quad A_i \in \mathrm{TT}(\Gamma_i;\, \mu_i) \quad 1 \le i \le n}{x A_1 \cdots A_n \in \mathrm{RT}(\Gamma \otimes (x : \mu_1 \multimap \cdots \multimap \mu_n \multimap \sigma);\, \sigma)}\ (\mathsf{head}_n)$$

$$\frac{\Gamma = \bigotimes_{i=1}^{n} \Gamma_i \quad A_i \in \mathrm{RT}(\Gamma_i;\, \sigma_i) \quad 1 \le i \le n \quad \frown(A_1, \dots, A_n) \quad A = \bigvee_{i=1}^{n} A_i}{A \in \mathrm{TT}(\Gamma;\, \otimes_{i=1}^{n} \sigma_i)}\ (\mathsf{sup}_n)$$

Figure 14.2: The inhabitation algorithm for $\boldsymbol{\lambda}_{\otimes}^{\mathrm{HNPR}}$.

EXAMPLES 14.63. (i) Let $\sigma \triangleq (\xi \multimap \xi) \multimap \xi \multimap \xi$, for some $\xi \in \mathbb{A}^{\mathrm{HNPR}}$. Given the input $(\emptyset, \sigma)$, the algorithm has two successful runs, generating respectively the deduction trees:

$$\dfrac{\dfrac{}{x \in \mathrm{RT}(x : \xi \multimap \xi;\, \xi \multimap \xi)}\ (\mathsf{head}_0)}{\lambda x.x \in \mathrm{RT}(\emptyset;\, \sigma)}\ (\mathsf{abs}) \qquad \dfrac{\dfrac{\dfrac{\dfrac{\dfrac{}{y \in \mathrm{RT}(y : \xi;\, \xi)}\ (\mathsf{head}_0)}{y \in \mathrm{TT}(y : \xi;\, \xi)}\ (\mathsf{sup}_1)}{xy \in \mathrm{RT}(x : \xi \multimap \xi, y : \xi;\, \xi)}\ (\mathsf{head}_1)}{\lambda y.xy \in \mathrm{RT}(x : \xi \multimap \xi;\, \xi \multimap \xi)}\ (\mathsf{abs})}{\lambda xy.xy \in \mathrm{RT}(\emptyset;\, \sigma)}\ (\mathsf{abs})$$

Note that the type σ does not correspond to the simple type $(\alpha \to \alpha) \to \alpha \to \alpha$ representing the data type of Church numerals, thus having an infinite number of inhabitants. In fact, in the type assignment system $\boldsymbol{\lambda}_{\otimes}^{\mathrm{HNPR}}$, there is no common type for all the Church numerals, since each numeral c_n has type $(\otimes_{i=1}^{n}(\xi \multimap \xi)) \multimap \xi \multimap \xi$ (among others).

(ii) Given the input $(\emptyset, (1 \multimap \xi) \multimap \xi)$, a successful run of the inhabitation algorithm is given by:

$$\dfrac{\dfrac{\dfrac{}{\bot \in \mathrm{TT}(\emptyset;\, 1)}\ (\mathsf{sup}_0)}{x\bot \in \mathrm{RT}(x : 1 \multimap \xi;\, \xi)}\ (\mathsf{head}_1)}{\lambda x.x\bot \in \mathrm{RT}(\emptyset;\, (1 \multimap \xi) \multimap \xi)}\ (\mathsf{abs})$$

(iii) Given input $(\emptyset, \xi_0 \multimap \xi_1)$, the only applicable rule is (abs), then the algorithm fails since (head_1) is not applicable because $\xi_0 \neq \xi_1$. It follows that $\mathrm{RT}(\emptyset;\, \xi_0 \multimap \xi_1)$ is empty.

REMARK 14.64. (i) The algorithm in Figure 14.2 was first presented in Bucciarelli et al. (2018) and constitutes a simplified version of the algorithm originally described by the same authors in Bucciarelli et al. (2014b). The price to pay for this simplification is the loss of the linearity property enjoyed by the latter.

(ii) It follows from Example 14.63(i) that this algorithm is not a conservative extension of the classical inhabitation algorithm for simple types (see Ben-Yelles (1979) and Hindley (2008)). In particular, when restricted to simple types, this algorithm finds all the β-nf's inhabiting a given type, while the original one constructs only the 'η-long' nf's.

We are going to prove that the inhabitation algorithm RT is terminating, sound and complete. In order to show that the algorithm terminates we define appropriate measures, on types, on type environments, and on the judgements of the inhabitation algorithm.

DEFINITION 14.65. (i) Define two *measures* on relational and tensor types

$$\begin{array}{rl}(\cdot)^\circ : & \mathbb{T}_{\mathrm{R}}^{\mathrm{HNPR}} \to \mathbb{N} \\ (\cdot)^\bullet : & \mathbb{T}_{\otimes}^{\mathrm{HNPR}} \to \mathbb{N}\end{array}$$

by structural induction as follows (for $\xi \in \mathbb{A}^{\mathrm{HNPR}}$ and $n > 0$):

$$\begin{array}{rclrcl}\xi^\circ & \triangleq & 1, & (\mu \multimap \sigma)^\circ & \triangleq & \mu^\bullet + \sigma^\circ + 1, \\ 1^\bullet & \triangleq & 0, & (\sigma_1 \otimes \cdots \otimes \sigma_n)^\bullet & \triangleq & \sum_{i=1}^n \sigma_i^\circ.\end{array}$$

(ii) The measure $(\cdot)^\bullet$ is extended to type environments Γ by

$$\Gamma^\bullet \triangleq \sum_{x \in \mathsf{supp}(\Gamma)} \Gamma(x)^\bullet$$

(iii) We define a measure $\#(\cdot)$ on judgements $\mathrm{RT}(-;\,-)$ and $\mathrm{TT}(-;\,-)$ by setting

$$\begin{array}{rcl}\#(\mathrm{RT}(\Gamma;\,\sigma)) & \triangleq & \Gamma^\bullet + \sigma^\circ, \\ \#(\mathrm{TT}(\Gamma;\,\mu)) & \triangleq & \Gamma^\bullet + \mu^\bullet + 1.\end{array}$$

(iv) Given $M \in \Lambda_\perp$, the *size of its syntax-tree*, written $\mathrm{tsize}(M)$, is defined by structural induction:

$$\begin{array}{rcl}\mathrm{tsize}(\perp) & \triangleq & 0, \\ \mathrm{tsize}(x) & \triangleq & 0, \\ \mathrm{tsize}(\lambda x.P) & \triangleq & \mathrm{tsize}(P) + 1, \\ \mathrm{tsize}(PQ) & \triangleq & \mathrm{tsize}(P) + \mathrm{tsize}(Q) + 1.\end{array}$$

REMARK 14.66. (i) For all $\sigma \in \mathbb{T}_{\mathrm{R}}^{\mathrm{HNPR}}$, we have $\sigma^\circ > 0$ and $\sigma^\bullet = \sigma^\circ$.

(ii) Given $\mu \in \mathbb{T}_{\otimes}^{\mathrm{HNPR}}$ and a type environment Γ, we have $\mu^\bullet, \Gamma^\bullet \geq 0$. E.g., $1^\bullet = 0 = \emptyset^\bullet$.

(iii) Given type environments $\Gamma, \Gamma_1, \ldots, \Gamma_n$ we have that

$$\Gamma = \bigotimes_{i=1}^n \Gamma_i \quad \Rightarrow \quad \Gamma^\bullet = \sum_{i=1}^n \Gamma_i^\bullet.$$

Note that the notion of size in Definition 14.65(iv) is related to, but different from, the notion of tree size in Definition 11.49(i). We are *de facto* overloading the function tsize(−) but this should not lead to any confusion.

EXAMPLES 14.67. (i) $((\xi_1 \multimap \xi_1) \otimes \xi_1)^\circ = (\xi_1 \multimap \xi_1)^\circ + \xi_1^\circ = \xi_1^\bullet + \xi_1^\circ + 2 = 4$.

(ii) $(1 \multimap (\xi_1 \otimes \xi_1) \multimap \xi_0)^\circ = 1^\bullet + (\xi_1 \otimes \xi_1)^\bullet + 2 + \xi_0^\circ = 0 + 2 \cdot \xi_1^\circ + 3 = 5$.

(iii) For $\Gamma \triangleq x : (\xi_1 \multimap \xi_1) \otimes \xi_1, y : (1 \multimap (\xi_1 \otimes \xi_1) \multimap \xi_0)$, we have $\Gamma^\bullet = 4 + 5 = 9$.

(iv) $\mathrm{tsize}(\lambda xyz.x(yz)(y\perp)\perp) = 3 + \mathrm{tsize}(x(yz)(y\perp)\perp) = 6 + \mathrm{tsize}(yz) + \mathrm{tsize}(y\perp) = 8$.

LEMMA 14.68. *Every run of the inhabitation algorithm terminates.*

PROOF. We need to prove that every run is a finite tree. We show that the measure $\#(\cdot)$ calculated on each premise of a rule, is strictly smaller than the measure associated with its conclusion. The cases (head_0) and (sup_0) are vacuous, so we consider $n > 0$.

Case (abs). In this case, we have:

$$\begin{array}{rcl} \#(\mathrm{RT}(\Gamma, x : \mu;\ \sigma)) & = & \Gamma^\bullet + \mu^\bullet + \sigma^\circ \\ & < & \Gamma^\bullet + \mu^\bullet + \sigma^\circ + 1 \\ & = & \#(\mathrm{RT}(\Gamma;\ \mu \multimap \sigma)) \end{array}$$

Case (head_n). Recall that we assume $n > 0$. By Remark 14.66(i), we have $\sigma^\circ > 0$. Thus, for each i $(1 \le i \le n)$, we obtain:

$$\begin{array}{rcl} \#(\mathrm{TT}(\Gamma_i;\ \mu_i)) & = & \Gamma_i^\bullet + \mu_i^\bullet + 1 \\ & < & \sum_{i=1}^n \Gamma_i^\bullet + \sum_{i=1}^n \mu_i^\bullet + n + 2 \cdot \sigma^\circ \\ & = & \#(\mathrm{RT}((\bigotimes_{i=1}^n \Gamma_i) \otimes x : \mu_1 \multimap \cdots \multimap \mu_n \multimap \sigma;\ \sigma)) \end{array}$$

Case (sup_n). In this case, easy calculations give:

$$\begin{array}{rcl} \#(\mathrm{RT}(\Gamma_i;\ \sigma_i)) & = & \Gamma_i^\bullet + \sigma_i^\circ \\ & < & \sum_{i=1}^n (\Gamma_i^\bullet + \sigma_i^\circ) + 1 \\ & = & \#(\mathrm{TT}(\bigotimes_{i=1}^n \Gamma_i;\ \otimes_{i=1}^n \sigma_i)). \end{array}$$ □

We show that, if a run of $\mathrm{RT}(\Gamma;\ \sigma)$ is successful, then $\#(\mathrm{RT}(\Gamma;\ \sigma))$ provides an upper bound on the size of the syntax-tree of the produced output.

LEMMA 14.69. *Let Γ be a type environment, $\sigma \in \mathbb{T}_{\mathrm{R}}^{\mathrm{HNPR}}$ and $\mu \in \mathbb{T}_{\otimes}^{\mathrm{HNPR}}$. We have:*

(i) $A \in \mathrm{RT}(\Gamma;\ \sigma) \quad \Rightarrow \quad \mathrm{tsize}(A) < \#\mathrm{RT}(\Gamma;\ \sigma)$.

(ii) $A \in \mathrm{TT}(\Gamma;\ \mu) \quad \Rightarrow \quad \mathrm{tsize}(A) < \#\mathrm{TT}(\Gamma;\ \mu)$.

PROOF. We prove (i) and (ii) by induction on a run of $A \in \mathrm{RT}(\Gamma;\ \sigma)$ and $A \in \mathrm{TT}(\Gamma;\ \mu)$, respectively. We call IH(i) and IH(ii) the corresponding induction hypotheses.

Case (abs). In this case $\sigma = \mu \multimap \tau$ and $A = \lambda x.A'$ with $A' \in \mathrm{RT}(\Gamma, x : \mu;\ \tau)$. Then:

$$\begin{array}{rcll} \mathrm{tsize}(\lambda x.A') & = & \mathrm{tsize}(A') + 1, & \text{by Definition 14.65(iv)}, \\ & < & \Gamma^\bullet + \mu^\bullet + \tau^\circ + 1, & \text{since } \mathrm{tsize}(A') < \Gamma^\bullet + \mu^\bullet + \tau^\circ, \text{ by IH(i)}, \\ & = & \Gamma^\bullet + (\mu \multimap \tau)^\circ, & \text{by Definition 14.65(i)}, \\ & = & \#(\mathrm{RT}(\Gamma;\ \sigma)), & \text{by Definition 14.65(iii)}. \end{array}$$

Case (head_n). In this case $\Gamma = (\bigotimes_{j=1}^n \Gamma_j) \otimes x : \mu_1 \multimap \cdots \multimap \mu_n \multimap \sigma$, $A = xA_1 \cdots A_n$ with every $A_i \in \mathrm{TT}(\Gamma_i;\ \mu_i)$. By IH(ii), we have $\mathrm{tsize}(A_i) < \Gamma_i^\bullet + \mu_i^\bullet + 1$. Then

$$\begin{array}{rcll} \mathrm{tsize}(xA_1 \cdots A_n) & = & \sum_{i=1}^n \mathrm{tsize}(A_i) + n + 1, & \text{by Definition 14.65(iv)}, \\ & \le & \sum_{i=1}^n (\Gamma_i^\bullet + \mu_i^\bullet) + n + 1, & \text{since } \mathrm{tsize}(A_i) \le \Gamma_i^\bullet + \mu_i^\bullet, \\ & < & \sum_{i=1}^n (\Gamma_i^\bullet + \mu_i^\bullet) + n + 2 \cdot \sigma^\bullet, & \text{since } \sigma^\bullet > 0, \text{ by Remark 14.66(i)}, \\ & = & \#(\mathrm{TT}(\Gamma;\ \sigma)), & \text{by Definition 14.65(iii)}. \end{array}$$

Case (sup_n). If $n = 0$ then $\text{tsize}(\bot) = 0 < 1 = \emptyset^\bullet + 1^\bullet + 1 = \#(\text{TT}(\emptyset;\ 1))$. If $n > 0$ then $\mu = \otimes_{i=1}^n \sigma_i$, $\Gamma = \sum_{i=1}^n \Gamma_i$ and $A = \bigvee_{i=1}^n A_i$ where every $A_i \in \text{RT}(\Gamma_i;\ \sigma_i)$. Easy calculations give:

$$\begin{array}{rcll} \text{tsize}(A) & \leq & \sum_{i=1}^n \text{tsize}(A_i) & \\ & < & \sum_{i=1}^n (\Gamma_i^\bullet + \sigma_i^\circ), & \text{by IH(i) and } n > 0, \\ & < & \Gamma^\bullet + \mu^\bullet + 1 & \\ & = & \#(\text{TT}(\Gamma;\ \mu)), & \text{by Definition 14.65(iii).} \end{array}$$

□

THEOREM 14.70 (TERMINATION). *The inhabitation algorithm terminates.*

PROOF. Fix an input (Γ, σ). By Lemma 14.68, every run of $A \in \text{RT}(\Gamma;\ \sigma)$ terminates. Now, the set

$$\{A \in \mathscr{A} \mid \text{FV}(A) \subseteq \mathsf{supp}(\Gamma)\ \&\ \text{tsize}(A) \leq \Gamma^\bullet + \sigma^\circ\}$$

is finite, because one cannot add variables or $\bot$ without adding applications as well. By Lemma 14.69, the algorithm needs to execute a finite number of runs. This is reflected in the rules of Figure 14.2 by the fact that there are finitely many decompositions of Γ in the premises of (head_n) and (sup_n). Moreover, in (head_n) there are finitely many possible choices of '$x : \mu_1 \multimap \cdots \multimap \mu_n \multimap \sigma$' from the type environment taken as input. □

Completeness is achieved by exploiting the non-determinism of the algorithm—by collecting all possible runs, we recover all minimal approximants for (Γ, σ).

LEMMA 14.71. *Let Γ be a type environment and $\sigma \in \mathbb{T}_{\text{R}}^{\text{HNPR}}$. For all $A \in \mathscr{A}$, the following are equivalent:*

1. *$A \in \text{RT}(\Gamma;\ \sigma)$;*
2. *A is minimal for (Γ, σ).*

PROOF. As an induction loading, we prove simultaneously the equivalence for $\mu \in \mathbb{T}_{\otimes}^{\text{HNPR}}$:

$$A \in \text{TT}(\Gamma;\ \mu) \iff A \text{ is minimal for } (\Gamma, \mu). \tag{14.1}$$

($1 \Rightarrow 2$) We proceed by induction on a run of $A \in \text{RT}(\Gamma;\ \sigma)$ (resp. $A \in \text{TT}(\Gamma;\ \mu)$).

Case (abs). We must have $A = \lambda x.A'$ and $\sigma = \mu \multimap \tau$, for some $A' \in \text{RT}(\Gamma, x : \mu;\ \tau)$. Therefore, this case follows easily from the IH.

Case (sup_n). In this case we have $\Gamma = \otimes_{i=1}^n \Gamma_i$, $\mu = \otimes_{i=1}^n \sigma_i$ and $A = \bigvee_{i=1}^n A_i$, with each $A_i \in \text{RT}(\Gamma_i;\ \sigma_i)$. By IH, every A_i is minimal for (Γ_i, σ_i), therefore we obtain (14.1).

Case (head_n). Then $A = xA_1 \cdots A_n$, $\Gamma = (\bigotimes_{i=1}^n \Gamma_i) \otimes x : \mu_1 \multimap \cdots \multimap \mu_n \multimap \sigma$ with $A_i \in \text{TT}(\Gamma_i;\ \mu_i)$, for all $i\,(1 \leq i \leq n)$. Now, each μ_i must have shape $\mu_i = \otimes_{j=1}^{m_i} \tau_{ij}$ for some $\tau_{ij} \in \mathbb{T}_{\text{R}}^{\text{HNPR}}$. By IH, there are $A_{i1}, \ldots, A_{im_i} \in \mathscr{A}$ and a decomposition $\Gamma_i = \otimes_{j=1}^{m_i} \Gamma_{ij}$ such that $A_i = \bigvee_{j=1}^{m_i} A_{ij}$ and each A_{ij} is minimal for (Γ_{ij}, τ_{ij}). By Proposition 14.54, $\Gamma_{ij} \vdash_{\otimes}^{\text{HNPR}} A_{ij} : \tau_{ij}$ and $A_{ij} \leq_\bot A_i$ entail $\Gamma_{ij} \vdash_{\otimes}^{\text{HNPR}} A_i : \tau_{ij}$. We obtain the derivation:

$$\frac{x : \mu_1 \multimap \cdots \multimap \mu_n \multimap \sigma \vdash_{\otimes}^{\text{HNPR}} x : \mu_1 \multimap \cdots \multimap \mu_n \multimap \sigma \qquad \forall i, j\,.\,\Gamma_{ij} \vdash_{\otimes}^{\text{HNPR}} A_i : \tau_{ij}}{\Gamma \vdash_{\otimes}^{\text{HNPR}} xA_1 \cdots A_n : \sigma}\ (\multimap_{\text{E}})$$

Any typable $A' \leq_{\perp} A$ has shape $xA'_1 \cdots A'_n$, with $A'_i \leq_{\perp} A_i$ such that $\Gamma_{ij} \vdash_{\otimes}^{\mathrm{HNPR}} A'_i : \tau_{ij}$. Since A_{ij} is minimal for (Γ_{ij}, τ_{ij}) we derive $A_{ij} \leq_{\perp} A'_i$, for all $j\,(1 \leq j \leq m_i)$. As $\bigvee$ constructs the lub we obtain $A_i = A'_i$, for all $i\,(1 \leq i \leq n)$. We conclude $A = A'$.

($2 \Rightarrow 1$) Let A be minimal for (Γ, σ). We proceed by structural induction on A.

Case $A = \lambda x.A'$. By Lemma 14.59(i), we know that $\sigma = \mu \multimap \tau$ and that A' is minimal for $((\Gamma, x : \mu), \tau)$. From the IH, we obtain a successful run $A' \in \mathrm{RT}(\Gamma, x : \mu;\ \tau)$. By applying (abs), we conclude $\lambda x.A' \in \mathrm{RT}(\Gamma;\ \mu \multimap \tau)$.

Case $A = xA_1 \cdots A_n$. From n applications of Lemma 14.59(ii), we obtain $\mu_1, \ldots, \mu_n$ and a decomposition $\Gamma = \otimes_{j=0}^{n} \Gamma_j$ such that x is minimal for $(\Gamma_0, \mu_1 \multimap \cdots \multimap \mu_n \multimap \sigma)$ and each A_i is minimal for (Γ_i, μ_i). So, we have $\Gamma_0 = x : \mu_1 \multimap \cdots \multimap \mu_n \multimap \sigma$. The IH on A_i gives $A_i \in \mathrm{TT}(\Gamma_i;\ \mu_i)$, for all such i. Using (head_n) we conclude $A \in \mathrm{RT}(\Gamma;\ \sigma)$. □

THEOREM 14.72 (SOUNDNESS AND COMPLETENESS).

(i) *If $A \in \mathrm{RT}(\Gamma;\ \sigma)$ then, for all $M \in \Lambda$ satisfying $A \in \mathscr{A}(M)$, we have $\Gamma \vdash_{\otimes}^{\mathrm{HNPR}} M : \sigma$.*

(ii) *If $\Gamma \vdash_{\otimes}^{\mathrm{HNPR}} M : \sigma$ then there exists $A \in \mathrm{RT}(\Gamma;\ \sigma)$ such that $A \in \mathscr{A}(M)$.*

PROOF. (i) By Lemma 14.71, we have $\Gamma \vdash_{\otimes}^{\mathrm{HNPR}} A : \sigma$. Assuming $A \in \mathscr{A}(M)$, we derive $\Gamma \vdash_{\otimes}^{\mathrm{HNPR}} M : \sigma$ by Proposition 14.54.

(ii) By Proposition 14.54, there is an $A' \in \mathscr{A}(M)$ satisfying $\Gamma \vdash_{\otimes}^{\mathrm{HNPR}} A' : \sigma$. Then, there exists an approximant $A \leq_{\perp} A'$ which is minimal for (Γ, σ). By Lemma 14.71, we obtain $A \in \mathrm{RT}(\Gamma;\ \sigma)$ and since $\mathscr{A}(M)$ is downward closed we conclude $A \in \mathscr{A}(M)$. □

COROLLARY 14.73 (BUCCIARELLI ET AL. (2014B)).
The inhabitation problem for $\boldsymbol{\lambda}_{\otimes}^{\mathrm{HNPR}}$ is decidable.

PROOF. Let (Γ, σ) be effectively given.

> Case (Γ, σ) is inhabited. By completeness (Theorem 14.72(ii)) there is a successful run $A \in \mathrm{RT}(\Gamma;\ \sigma)$ of the algorithm, which is found in a finite number of attempts (by Theorem 14.70). By soundness (Theorem 14.72(i)) we obtain $\Gamma \vdash_{\otimes}^{\mathrm{HNPR}} A : \sigma$. By Proposition 14.54, the λ-term $A[\perp := \Omega]$ inhabits (Γ, σ).
>
> Case (Γ, σ) is not inhabited. In this case all attempts of constructing a successful run fail. Since the algorithm is terminating (by Theorem 14.70), its execution returns a negative response. We conclude by completeness (Theorem 14.72(ii)). □

A discrimination property

We conclude this section by showing that $\boldsymbol{\lambda}_{\otimes}^{\mathrm{HNPR}}$ allows to discriminate λ-terms M, N such that $M \not\sqsubseteq_{\mathsf{r}} N$, even when $M \sqsubseteq_{\mathrm{SOL}} N$ does hold. This property will be crucial in Section 14.5 to characterize the minimal inequational relational graph theory.

DEFINITION 14.74. Let $\sigma \in \mathbb{T}_{\mathrm{R}}^{\mathrm{HNPR}}$.

(i) We say that $\xi_i \in \mathbb{A}^{\mathrm{HNPR}}$ is the *range of* σ, in symbols $\mathrm{rg}(\sigma) \triangleq \xi_i$, whenever

$$\sigma = \mu_1 \multimap \cdots \multimap \mu_n \multimap \xi_i$$

(ii) We write $1^k \multimap \sigma$ for $1 \multimap \cdots \multimap 1 \multimap \sigma$ (with k occurrences of 1).

LEMMA 14.75. *Let $M, N \in \Lambda$ be such that $M \sqsubseteq_{\mathrm{SOL}} N$, while $M \not\sqsubseteq_{\mathsf{r}} N$. For some type environment Γ and $\sigma \in \mathbb{T}_{\mathrm{R}}^{\mathrm{HNPR}}$, we have that:*
(i) *there exists $P \in \mathscr{A}(M)$, satisfying*
a) $\Gamma \vdash_{\otimes}^{\mathrm{HNPR}} P : \sigma$, with $\mathrm{rg}(\sigma) = \xi_{\mathrm{tsize}(P)}$,
b) $\forall \tau \in \Gamma . [\,\mathrm{rg}(\tau) = \xi_j \quad \Rightarrow \quad j \leq \mathrm{tsize}(P)\,]$;
(ii) *for all $Q \in \mathscr{A}(N)$, we have $\Gamma \not\vdash_{\otimes}^{\mathrm{HNPR}} Q : \sigma$.*

PROOF. Assume $M \not\sqsubseteq_{\mathsf{r}} N$. This entails $M \in \mathrm{SOL}$. Moreover, suppose that $M \sqsubseteq_{\mathrm{SOL}} N$. By Propositions 12.35(i) and 12.49, only two cases are possible.

Case 1) $M =_\beta \lambda\vec{x}.x_i M_1 \cdots M_k$ and $N =_\beta \lambda\vec{x}z_1 \ldots z_m.x_i N_1 \cdots N_k Z_1 \cdots Z_m$, where $\vec{x} = x_1, \ldots, x_n$ for some $k, n \geq 0$ and $m > 0$. Wlog, suppose that x_i is free, that is, $i > n$. This case follows easily by taking

$$\begin{array}{rcl} P & \triangleq & \lambda\vec{x}.x_i \bot^{\sim k} \in \mathscr{A}(M), \quad \text{with } \mathrm{tsize}(P) = n + k, \\ \Gamma & \triangleq & x_i : 1^k \multimap \xi_{n+k}, \\ \sigma & \triangleq & 1^n \multimap \xi_{n+k}. \end{array}$$

The fact that $\Gamma \not\vdash_{\otimes}^{\mathrm{HNPR}} Q : \sigma$, for all $Q = \lambda\vec{x}\vec{z}.x_i Q_1 \cdots Q_{k+m} \in \mathscr{A}(N)$, follows from $m > 0$ and the fact that ξ_{n+k} is an atom, hence different from any arrow type.

Case 2) $M =_\beta \lambda x_1 \ldots x_n z_1 \ldots z_m.x_i M_1 \cdots M_k Z_1 \cdots Z_m$ and $N =_\beta \lambda\vec{x}.x_i N_1 \cdots N_k$ where $M_j \sqsubseteq_{\mathrm{SOL}} N_j$, for all j $(1 \leq j \leq k)$, and $Z_\ell \sqsubseteq_{\mathrm{SOL}} z_\ell$, for every ℓ $(1 \leq \ell \leq m)$. Since this last condition entails $Z_\ell \sqsubseteq_{\mathsf{r}} z_\ell$, $M \not\sqsubseteq_{\mathsf{r}} N$ is only possible if $M_q \not\sqsubseteq_{\mathsf{r}} N_q$ for some $q \leq k$. We suppose that x_i is free, the other case being analogous. By IH, there exist Γ_q, σ_q and an approximant $P_q \in \mathscr{A}(M_q)$ such that

a) $\Gamma_q \vdash_{\otimes}^{\mathrm{HNPR}} P_q : \sigma_q$ with $\mathrm{rg}(\sigma_q) = \xi_{\mathrm{tsize}(P_q)}$,

b) for all $\tau \in \Gamma_q$, we have $\mathrm{rg}(\tau) = \xi_j$, for some $j \leq \mathrm{tsize}(P_q)$,

and $\Gamma_q \not\vdash_{\otimes}^{\mathrm{HNPR}} Q_q : \sigma_q$, for all $Q_q \in \mathscr{A}(N_q)$. Since the variables $\vec{x}$ might occur free in P_q, we obtain $\Gamma_q = \Gamma_1, x_1 : \mu_1, \ldots, x_n : \mu_n$. Setting $P \triangleq \lambda\vec{x}\vec{z}.x_i \bot^{\sim q-1} P_q \bot^{\sim k+m-q} \in \mathscr{A}(M)$ and $\Gamma_0 \triangleq x_i : 1^{q-1} \multimap \sigma_q \multimap 1^{k+m-q} \multimap \xi_{\mathrm{tsize}(P)}$, we may construct the derivation:

$$\dfrac{\dfrac{\Gamma_0 \vdash_{\otimes}^{\mathrm{HNPR}} x_i : 1^{q-1} \multimap \sigma_q \multimap 1^{k+m-q} \multimap \xi_{\mathrm{tsize}(P)} \qquad \Gamma_q \vdash_{\otimes}^{\mathrm{HNPR}} P_q : \sigma_q}{\Gamma_0 \otimes \Gamma_q \vdash_{\otimes}^{\mathrm{HNPR}} x_i \bot^{\sim q-1} P_q \bot^{\sim k+m-q} : \xi_{\mathrm{tsize}(P)}}\,(\multimap_{\mathrm{E}})}{\Gamma_0 \otimes \Gamma_1 \vdash_{\otimes}^{\mathrm{HNPR}} P : \mu_1 \multimap \cdots \multimap \mu_n \multimap 1^m \multimap \xi_{\mathrm{tsize}(P)}}\,(\multimap_{\mathrm{I}})$$

Now, every $Q \in \mathscr{A}(N) - \{\bot\}$ must have the shape $Q = \lambda\vec{x}.x_i Q_1 \cdots Q_k$ and each derivation of $\Gamma_0 \otimes \Gamma_1 \vdash_{\otimes}^{\mathrm{HNPR}} Q : \xi_{\mathrm{tsize}(P)}$ requires to derive $\Gamma_q \vdash_{\otimes}^{\mathrm{HNPR}} Q_q : \sigma_q$. Indeed, since all $\tau \in \Gamma_q$ satisfy $\mathrm{rg}(\tau) \leq \mathrm{tsize}(P_q) < \mathrm{tsize}(P)$, they cannot be used to produce a $\xi_{\mathrm{tsize}(P)}$ so the decomposition $\Gamma_0 \otimes \Gamma_q$ is in fact unique:

$$\dfrac{\dfrac{\Gamma_0 \vdash_{\otimes}^{\mathrm{HNPR}} x_i : 1^{q-1} \multimap \sigma_q \multimap 1^{k+m-q} \multimap \xi_{\mathrm{tsize}(P)} \qquad \Gamma_q \vdash_{\otimes}^{\mathrm{HNPR}} Q_q : \sigma_q}{\Gamma_0 \otimes \Gamma_q \vdash_{\otimes}^{\mathrm{HNPR}} x_i Q_1 \cdots Q_k : 1^m \multimap \xi_{\mathrm{tsize}}(P)}\,(\multimap_{\mathrm{E}})}{\Gamma_0 \otimes \Gamma_1 \vdash_{\otimes}^{\mathrm{HNPR}} Q : \mu_1 \multimap \cdots \multimap \mu_n \multimap 1^m \multimap \xi_{\mathrm{tsize}(P)}}\,(\multimap_{\mathrm{I}})$$

By IH it is impossible to find a derivation of $\Gamma_q \vdash_{\otimes}^{\mathrm{HNPR}} Q_q : \sigma$, so we conclude

$$\Gamma_0 \otimes \Gamma_1 \not\vdash_{\otimes}^{\mathrm{HNPR}} Q : \mu_1 \multimap \cdots \multimap \mu_n \multimap 1^m \multimap \xi_{\mathrm{tsize}(P)}. \qquad \square$$

14.4 Relational graph models in logical form

In this section we show that any relational graph model $\mathcal{D}$ can be presented as a tensor type system $\vdash_{\otimes}^{\mathcal{D}}$. In this setting, the interpretation of a λ-term M in $\mathcal{D}$ is given by the set of all pairs (Γ,σ) such that $\Gamma \vdash_{\otimes}^{\mathcal{D}} M : \sigma$.

DEFINITION 14.76. Let $\mathcal{D} = (D, \iota)$ be an rgm.

(i) Let $\mathbb{T}_{\mathrm{R}}^{\mathcal{D}}$ (resp. $\mathbb{T}_{\otimes}^{\mathcal{D}}$) be the set of relational types (resp. tensor types) over the set of type atoms

$$\mathbb{A}_{\mathcal{D}} \triangleq D - (\mathscr{M}_{\mathrm{f}}(D) \times D).$$

(ii) Every relational type $\sigma \in \mathbb{T}_{\mathrm{R}}^{\mathcal{D}}$ (resp. tensor type $\mu \in \mathbb{T}_{\otimes}^{\mathcal{D}}$) is associated with an element $\sigma^{\circledast} \in D$ (resp. multiset $\mu^{\circledast} \in \mathscr{M}_{\mathrm{f}}(D)$) by setting:

$$\begin{array}{rcl} \xi^{\circledast} & \triangleq & \xi, \\ (\mu \multimap \sigma)^{\circledast} & \triangleq & \iota(\mu^{\circledast}, \sigma^{\circledast}), \\ 1^{\circledast} & \triangleq & [\,], \\ (\sigma_1 \otimes \cdots \otimes \sigma_k)^{\circledast} & \triangleq & [\sigma_1^{\circledast}, \ldots, \sigma_k^{\circledast}]. \end{array}$$

(iii) The *tensor type assignment system associated with* $\mathcal{D}$ is defined by taking as type atoms the elements of $\mathbb{A}_{\mathcal{D}}$ and as type equivalence the relation induced by $(\cdot)^{\circledast}$:

$$\begin{array}{rcl} \sigma \simeq^{\mathcal{D}} \tau & \iff & \sigma^{\circledast} = \tau^{\circledast}, \\ \mu \simeq^{\mathcal{D}} \nu & \iff & \mu^{\circledast} = \nu^{\circledast}. \end{array}$$

We write $\Gamma \vdash_{\otimes}^{\mathcal{D}} M : \sigma$ for the corresponding type judgements.

(iv) The *logical interpretation* of a λ-term M in $\mathcal{D}$ is defined as follows:

$$[\![M]\!]^{\mathcal{D}} \triangleq \{(\Gamma, \sigma) \mid \Gamma \vdash_{\otimes}^{\mathcal{D}} M : \sigma\}$$

When $\mathcal{D}$ is clear from the context we may omit it and simply write $[\![-]\!]$ for $[\![-]\!]^{\mathcal{D}}$.

(v) The definition of interpretation in (iv) extends to $\lambda\bot$-terms in the obvious way, and to arbitrary subsets $\mathcal{X} \subseteq \Lambda_{\bot}$ by setting

$$[\![\mathcal{X}]\!]^{\mathcal{D}} \triangleq \{\, [\![P]\!]^{\mathcal{D}} \mid P \in \mathcal{X}\,\}$$

Thus, for every $M \in \Lambda_{\bot}$, the interpretations $[\![\mathscr{A}(M)]\!]^{\mathcal{D}}$ and $[\![\mathscr{A}_{\eta}(M)]\!]^{\mathcal{D}}$ are well-defined.

REMARK 14.77. (i) The tensor type assignment system associated with the rgm $\mathcal{E}$ by the above construction is equivalent to $\boldsymbol{\lambda}_{\otimes}^{\mathrm{HNPR}}$.

(ii) Similarly, the rgm's $\mathcal{D}_{\star}$ and $\mathcal{D}_{\omega}$ correspond to $\boldsymbol{\lambda}_{\otimes}^{\mathrm{MR}}$ and $\boldsymbol{\lambda}_{\otimes}^{\mathrm{BEM}}$, respectively.

We now show that the logical interpretation $[\![-]\!]^{\mathcal{D}}$ is equivalent to the categorical one $|-|_{-}^{\mathcal{D}}$ in the sense that they induce the same (in)equalities between λ-terms.

PROPOSITION 14.78 (EQUIVALENCE OF THE INTERPRETATIONS). *Let $\mathcal{D}$ be an rgm and $M \in \Lambda^{o}(x_1, \ldots, x_n)$. Then*

1. $[\![M]\!]^{\mathcal{D}} = \{((x_1 : \mu_1, \ldots, x_n : \mu_n), \sigma) \mid ((\mu_1^{\circledast}, \ldots, \mu_n^{\circledast}), \sigma^{\circledast}) \in |M|_{\vec{x}}^{\mathcal{D}}\}$;
2. $|M|_{\vec{x}}^{\mathcal{D}} = \{((a_1, \ldots, a_n), \alpha) \mid ((x_1 : \mu_1, \ldots, x_n : \mu_n), \sigma) \in [\![M]\!]^{\mathcal{D}}$ *where* $\forall i\,(1 \le i \le n)\,.\,\mu_i^{\circledast} = a_i \;\&\; \sigma^{\circledast} = \alpha\}$.

PROOF. (1) By structural induction on M, we prove that $x_1 : \mu_1, \ldots, x_n : \mu_n \vdash_{\otimes}^{\mathcal{D}} M : \sigma$ holds if and only if $((x_1 : \mu_1^{\circledast}, \ldots, x_n : \mu_n^{\circledast}), \sigma^{\circledast}) \in [\![M]\!]$ does.

The only interesting case is the application.

Case $M = PQ$. By Lemma 14.34(iii) we have that $\Gamma \vdash_{\otimes}^{\mathcal{D}} PQ : \sigma$ is derivable if and only if there exist relational types $\tau_1, \ldots, \tau_k$ and a decomposition $\Gamma = \bigotimes_{j=0}^{k} \Gamma_j$ such that $\Gamma_0 \vdash_{\otimes}^{\mathcal{D}} P : (\otimes_{i=1}^{k} \tau_i) \multimap \sigma$ and $\Gamma_i \vdash_{\otimes}^{\mathcal{D}} Q : \tau_i$ are derivable for all $i\,(1 \leq i \leq k)$. By IH this is equivalent to require that

- $\mathsf{supp}(\Gamma_j) \subseteq \{x_1, \ldots, x_n\}$, for all $j\,(0 \leq j \leq k)$,
- $((\Gamma_0(x_1)^{\circledast}, \ldots, \Gamma_0(x_n)^{\circledast}), \iota([\tau_1^{\circledast}, \ldots, \tau_k^{\circledast}], \sigma^{\circledast})) \in |P|_{\vec{x}}$ and
- $((\Gamma_i(x_1)^{\circledast}, \ldots, \Gamma_i(x_n)^{\circledast}), \tau_i^{\circledast}) \in |Q|_{\vec{x}}$, for all $i\,(1 \leq i \leq k)$.

By definition, this is equivalent to $((\Gamma(x_1)^{\circledast}, \ldots, \Gamma(x_n)^{\circledast}), \sigma^{\circledast}) \in |PQ|_{\vec{x}}$.

(2) Similar to (1). One proceeds by structural induction on M, using the fact that every element $((a_1, \ldots, a_n), \alpha) \in \mathscr{M}_{\mathrm{f}}(D)^n \times D$ has the form $((\mu_1^{\circledast}, \ldots, \mu_n^{\circledast}), \sigma^{\circledast})$ for some $\mu_1, \ldots, \mu_n \in \mathbb{T}_{\otimes}$ and $\sigma \in \mathbb{T}_{\mathrm{R}}$. □

COROLLARY 14.79. *Let $\mathcal{D}$ be an rgm. For all $M, N \in \Lambda$, the following are equivalent:*

1. $[\![M]\!]^{\mathcal{D}} \subseteq [\![N]\!]^{\mathcal{D}}$.
2. $|M|_{\vec{x}}^{\mathcal{D}} \subseteq |N|_{\vec{x}}^{\mathcal{D}}$.

NOTATION. (i) *As usual, we write $\mathcal{D} \models M \sqsubseteq N$ whenever the above inequalities hold.*
(ii) *Similarly, we write $\mathcal{D} \models M = N$ if both $\mathcal{D} \models M \sqsubseteq N$ and $\mathcal{D} \models N \sqsubseteq M$ hold.*

COROLLARY 14.80 (MONOTONICITY OF $[\![-]\!]$).
Let $\mathcal{D}$ be an rgm. For all contexts $C[\,]$ and λ-terms M, N, we have:

$$[\![M]\!]^{\mathcal{D}} \subseteq [\![N]\!]^{\mathcal{D}} \quad \Rightarrow \quad [\![C[M]]\!]^{\mathcal{D}} \subseteq [\![C[N]]\!]^{\mathcal{D}}$$

PROOF. By an easy induction on $C[\,]$. □

LEMMA 14.81. *Let $\mathcal{D}$ be an rgm.*
(i) *If $x \notin \mathrm{FV}(M)$ then $\mathcal{D} \models \lambda x.Mx \sqsubseteq M$,*
(ii) *If $M \to_{\eta} N$ then $\mathcal{D} \models M \sqsubseteq N$.*

PROOF. (i) By Example 14.26(i), we have $\mathcal{D} \models \mathsf{1} \sqsubseteq \mathsf{I}$. By Corollary 14.80 (monotonicity) we obtain $\mathcal{D} \models \mathsf{1}M \sqsubseteq \mathsf{I}M$. We conclude $\mathcal{D} \models \lambda x.Mx \sqsubseteq M$ by β-soundness.
(ii) By (i) and Corollary 14.80. □

THEOREM 14.82 (SEMANTIC APPROXIMATION THEOREM FOR RGMS).
Let $\mathcal{D}$ be an rgm. For all λ-terms M, we have:

$$[\![M]\!]^{\mathcal{D}} = \bigcup [\![\mathscr{A}(M)]\!]^{\mathcal{D}}.$$

PROOF. By Proposition 14.54. □

14.5 Relational graph theories

The Approximation Theorem has important consequences on the possible (in)equational theories that can be induced by relational graph models.

COROLLARY 14.83 (CHARACTERIZATION OF SOL).
Let $\mathcal{D}$ be an rgm. For all λ-terms M, we have

$$M \in \mathrm{SOL} \iff [\![M]\!]^{\mathcal{D}} \neq \emptyset.$$

PROOF. ($\Rightarrow$) Assume $M \in \mathrm{SOL}$. By Theorem 1.23, the λ-term M has an head normal form $M =_\beta \lambda x_1 \dots x_n.yM_1 \cdots M_k$. We consider the case y is free, the other one being analogous. In this case it is straightforward to verify that (for any $\xi \in \mathbb{A}$)

$$(y : 1^k \multimap \xi, 1^n \multimap \xi) \in [\![M]\!],$$

whence $[\![M]\!] \neq \emptyset$.

($\Leftarrow$) By the way of contradiction, assume that $M \in \mathrm{UNS}$. Then $\mathscr{A}(M) = \{\bot\}$ and, by the Approximation Theorem 14.82, we conclude $[\![\bot]\!] = \emptyset$. □

COROLLARY 14.84. *For all rgm's $\mathcal{D}$, we have:*

(i) $\mathrm{Th}_{\sqsubseteq}(\mathcal{D})$ *is order sensible and* $\mathrm{Th}(\mathcal{D})$ *is sensible.*
(ii) $\mathrm{Th}_{\sqsubseteq}(\mathcal{D}) \subseteq \sqsubseteq_{\mathrm{SOL}}$.
(iii) $\mathcal{B} \subseteq \mathrm{Th}(\mathcal{D}) \subseteq \mathcal{H}^*$.

PROOF. (i) By Corollary 14.83.
(ii) By (i) and Lemma 12.5.
(iii) Assume that $\mathcal{B} \vdash M = N$, then $\mathscr{A}(M) = \mathscr{A}(N)$ by Theorem 2.32.

$$\begin{array}{rcll} [\![M]\!] & = & \bigcup [\![\mathscr{A}(M)]\!], & \text{by the Approximation Theorem 14.82,} \\ & = & \bigcup [\![\mathscr{A}(N)]\!], & \text{since } \mathscr{A}(M) = \mathscr{A}(N), \\ & = & [\![N]\!], & \text{by the Approximation Theorem 14.82, again.} \end{array}$$

This shows $\mathcal{B} \subseteq \mathrm{Th}(\mathcal{D})$. The inclusion $\mathrm{Th}(\mathcal{D}) \subseteq \mathcal{H}^*$ simply follows from the maximality of $\mathcal{H}^*$ among sensible λ-theories, namely Proposition 3.13(i). □

COROLLARY 14.85. *Let $\mathcal{D}$ be an extensional rgm. Then we have:*

(i) $\sqsubseteq_{\mathrm{NF}} \subseteq \mathrm{Th}_{\sqsubseteq}(\mathcal{D})$,
(ii) $\mathcal{H}^+ \subseteq \mathrm{Th}(\mathcal{D})$.

PROOF. (i) See the 'if' implication in the proof of Theorem 13.82 (cf. Remark 13.83).
(ii) It is an easy consequence of (i). □

The minimal inequational relational graph theory

We show that the inequational theory $\sqsubseteq_{\mathsf{r}}$ introduced in Definition 12.29(ii) is minimal among the inequational theories representable by rgm's. The fact that its definition involves η-expansions and that $\sqsubseteq_{\mathsf{r}} \subseteq \mathrm{Th}_{\sqsubseteq}(\mathcal{D})$ also holds for non-extensional relational graph models should not be surprising. Indeed, when considering the original graph model $\mathscr{P}_\omega$ defined by Plotkin (1971) and Scott (1976) the situation is analogous, although symmetrical[3], and we know that no graph model is extensional.

PROPOSITION 14.86. *Let $M, N \in \Lambda$. If $M \sqsubseteq_{\mathsf{r}} N$ then $\mathcal{D} \models M \sqsubseteq N$ for every rgm $\mathcal{D}$.*

PROOF. Assume $M \sqsubseteq_{\mathsf{r}} N$. Take any $P \in \mathscr{A}(M)$. By Definition 12.29(ii), there are $Q \in \mathscr{A}(N)$ and $Q' \in \mathscr{A}$ such that $P \leq_{\perp} Q' \twoheadrightarrow_\eta Q$. Since $P \leq_{\perp} Q'$, we derive $P \in \mathscr{A}(Q')$. Thus, by the Approximation Theorem 14.82, we obtain $[\![P]\!] \subseteq [\![Q']\!]$. By Lemma 14.81(ii) we have that $[\![Q']\!] \subseteq [\![Q]\!]$ holds. Since P is an arbitrary element of $\mathscr{A}(M)$, it follows that $[\![\mathscr{A}(M)]\!] \subseteq [\![\mathscr{A}(N)]\!]$. We conclude by applying the Approximation Theorem again. □

We still need to prove that $\sqsubseteq_{\mathsf{r}}$ is actually represented by some relational graph model. We show that this is the case for the rgm $\mathcal{E}$ from Examples 14.25(i), corresponding to the tensor type assignment system $\boldsymbol{\lambda}_{\otimes}^{\mathrm{HNPR}}$ (by Remark 14.77(i)).

THEOREM 14.87 (BREUVART ET AL. (2018)). *For all $M, N \in \Lambda$ we have*

$$M \sqsubseteq_{\mathsf{r}} N \iff \mathcal{E} \models M \sqsubseteq N$$

Therefore $\mathrm{Th}_{\sqsubseteq}(\mathcal{E}) = \sqsubseteq_{\mathsf{r}}$ *and* $\mathrm{Th}(\mathcal{E}) = \mathcal{B}$.

PROOF. ($\Rightarrow$) It follows immediately from Proposition 14.86.

($\Leftarrow$) By the way of contradiction suppose that $M \not\sqsubseteq_{\mathsf{r}} N$, although $\mathcal{E} \models M \sqsubseteq N$. By Corollary 14.84(ii), the latter entails $M \sqsubseteq_{\mathrm{SOL}} N$. By Lemma 14.75, there exist $P \in \mathscr{A}(M)$ and Γ, σ such that $\Gamma \vdash_{\otimes}^{\mathrm{HNPR}} P : \sigma$, while $\Gamma \nvdash_{\otimes}^{\mathrm{HNPR}} Q : \sigma$, for all $Q \in \mathscr{A}(N)$. By the Approximation Theorem 14.82, we have $(\Gamma, \sigma) \in [\![M]\!]^{\mathcal{E}} - [\![N]\!]^{\mathcal{E}}$. Contradiction. □

The above proof-technique could be suitably generalized to prove that all non-extensional relational graph models induce the same inequational theory, namely $\sqsubseteq_{\mathsf{r}}$.

COROLLARY 14.88 (BREUVART ET AL. (2018)).

(i) *The relation $\sqsubseteq_{\mathsf{r}}$ is the minimal inequational relational graph theory.*

(ii) *The λ-theory $\mathcal{B}$ is the minimal relational graph theory.*

Characterizing rgm's that are fully abstract for $\mathcal{H}^+$

We provide necessary and sufficient conditions for an rgm to induce $\sqsubseteq_{\mathrm{NF}}$ as inequational theory, whence $\mathcal{H}^+$ as λ-theory. This characterization was first found by Breuvart et al. (2018), inspired by the semantic characterization of $\mathcal{H}^*$ in Breuvart (2014) which is recalled at the end of the section.

[3]By Theorem 12.51(iii), $\mathrm{Th}_{\sqsubseteq}(\mathscr{P}_\omega) = \sqsubseteq_{\mathsf{e}}$, whose definition is symmetrical w.r.t. $\sqsubseteq_{\mathsf{r}}$.

DEFINITION 14.89. Let $\mathcal{D}$ be an rgm and $t \in \mathscr{T}^{\infty}_{\mathsf{rec}}$.

(i) Let $f : \mathbb{N} \to \mathbb{N}$. We say that a relational type $\sigma \in \mathbb{T}^{\mathcal{D}}_{\mathrm{R}}$ *follows* f *starting from* n whenever there are tensor types $\mu_0, \dots, \mu_{f(n)} \in \mathbb{T}^{\mathcal{D}}_{\otimes}$ and a relational type $\gamma \in \mathbb{T}^{\mathcal{D}}_{\mathrm{R}}$ such that

$$\sigma \simeq^{\mathcal{D}} \mu_0 \multimap \cdots \multimap \mu_{f(n)} \multimap \gamma$$

and there exists $\tau \in \mu_{f(n)}$ following f starting from $n+1$.

(ii) Given $f \in [t]_\infty$ and $n \in \mathbb{N}$, we let $\mathcal{W}_{(\mathcal{D},f,n)}(t)$ be the set of all relational types $\sigma \in \mathbb{T}^{\mathcal{D}}_{\mathrm{R}}$ following the infinite branch f starting from n.

(iii) The set $\mathcal{W}_{\mathcal{D}}(t)$, whose elements are called *witnesses for* t *in* $\mathcal{D}$, is defined by:

$$\mathcal{W}_{\mathcal{D}}(t) \triangleq \bigcup_{f \in [t]_\infty} \mathcal{W}_{(\mathcal{D},f,0)}(t)$$

Note that Definition 14.89(i) has a coinductive nature. We show that the relational types of shape $\sigma \multimap \sigma$, where σ is a witness for t, are not suitable for typing J_t.

LEMMA 14.90. *Let* $\mathcal{D}$ *be an rgm and* $t \in \mathscr{T}^{\infty}_{\mathsf{rec}}$.

(i) *If* $\Gamma \vdash^{\mathcal{D}}_{\otimes} \mathsf{J}_t x : \sigma$ *then* $\Gamma \simeq^{\mathcal{D}} x : \sigma$.

(ii) *For all* $f \in [t]_\infty$ *and* $n \in \mathbb{N}$, *we have:*

$$\sigma \in \mathcal{W}_{(\mathcal{D},f,n)}(t) \quad \Rightarrow \quad (x : \sigma, \sigma) \notin [\![\mathsf{J}_{t\restriction_{\hat{f}(n)}} x]\!]^{\mathcal{D}}$$

PROOF. (i) By Proposition 14.54 there exists $P \in \mathscr{A}(\mathsf{J}_t x)$ such that $\Gamma \vdash^{\mathcal{D}}_{\otimes} P : \sigma$. Since $\mathsf{J}_t x =_\beta \lambda z_0 \dots z_{k-1}.x(\mathsf{J}_{t\restriction_{\langle 0\rangle}} z_0) \cdots (\mathsf{J}_{t\restriction_{\langle k-1\rangle}} z_{k-1})$ where $k \triangleq t(\langle\rangle)$, we have that P is either $\bot$ or has shape $P = \lambda z_0 \dots z_{k-1}.xP_0 \cdots P_{k-1}$, where $P_i \in \mathscr{A}(\mathsf{J}_{t\restriction_{\langle i\rangle}} z_i)$ for all $i\,(0 \le i < k)$. It is straightforward to prove by induction on a canonical derivation Π of $\Gamma \vdash^{\mathcal{D}}_{\otimes} P : \sigma$ that Γ must have shape $x : \tau$ for some $\sigma \simeq^{\mathcal{D}} \tau$, so we conclude by applying Proposition 14.54.

(ii) Fix $t' \triangleq t\restriction_{\hat{f}(n)}$ and let k be the number of children at the root of t', i.e. $t(\hat{f}(n))$. Notice that $k > 0$ since $f \in [t]_\infty$. Now, assume by contradiction that $(x : \sigma, \sigma) \in [\![\mathsf{J}_{t'} x]\!]$. The Approximation Theorem entails that there is $P \in \mathscr{A}(\mathsf{J}_{t'} x)$ such that $x : \sigma \vdash^{\mathcal{D}}_{\otimes} P : \sigma$ has a canonical derivation Π. Wlog, assume that P is minimal among all approximants sharing this property. Since P is typable it cannot be $\bot$, whence it must be of shape

$$P = \lambda z_0 \dots z_{k-1}.xP_0 \cdots P_{k-1}$$

where each $P_i \in \mathscr{A}(\mathsf{J}_{t_i} z_i)$, for $t_i \triangleq t\restriction_{\hat{f}(n);i}$ and $i\,(0 \le i < k)$. Therefore Π has the form

$$\dfrac{\dfrac{\dfrac{\sigma \simeq^{\mathcal{D}} \mu_0 \multimap \cdots \multimap \mu_{k-1} \multimap \gamma}{x : \sigma \vdash^{\mathcal{D}}_{\otimes} x : \mu_0 \multimap \cdots \multimap \mu_{k-1} \multimap \gamma} \quad \dfrac{\Pi_{ij}}{\Gamma_{ij} \vdash^{\mathcal{D}}_{\otimes} P_i : \tau_{ij}}}{\left(\bigotimes_{i=0}^{k-1} \bigotimes_{j=0}^{k_i} \Gamma_{ij}\right), x : \sigma \vdash^{\mathcal{D}}_{\otimes} xP_0 \cdots P_{k-1} : \gamma} \quad \sigma \simeq^{\mathcal{D}} \mu_0 \multimap \cdots \multimap \mu_{k-1} \multimap \gamma}{x : \sigma \vdash^{\mathcal{D}}_{\otimes} \lambda z_0 \dots z_{k-1}.xP_0 \cdots P_{k-1} : \sigma}$$

where $\mu_i \simeq^{\mathcal{D}} \otimes_{j=1}^{k_i} \tau_{ij}$ and, by (i), $\Gamma_{ij} \simeq^{\mathcal{D}} z_i : \tau_{ij}$ for all $i\,(0 \le i < k)$ and $j\,(1 \le j \le k_i)$. Since $\sigma \in \mathcal{W}_{(\mathcal{D},f,n)}(t)$, we must have that $f(n) < k$ (in particular, $k_{f(n)} > 0$) and there exists a relational type $\tau \in \mu_{f(n)}$ following f starting from $n+1$, namely $\tau \in \mathcal{W}_{(\mathcal{D},f,n+1)}(t)$. We conclude that $P_{f(n)}$ satisfies $z_{f(n)} : \tau \vdash P_{f(n)} : \tau$ while having a size strictly smaller than P. This contradicts our assumption concerning the minimality of P. □

PROPOSITION 14.91. *For any extensional rgm $\mathcal{D}$ and any tree $t \in \mathscr{T}^{\infty}_{\mathsf{rec}}$:*

$$\mathcal{W}_{\mathcal{D}}(t) = \{\sigma \in \mathbb{T}_{\mathrm{R}} \mid \sigma \multimap \sigma \notin [\![\mathsf{J}_t]\!]^{\mathcal{D}}\}$$

PROOF. ($\subseteq$) Consider any $\sigma \in \mathcal{W}_{\mathcal{D}}(t)$. By definition there is an infinite branch $f \in [t]_\infty$ such that $\sigma \in \mathcal{W}_{(\mathcal{D},f,0)}(t)$, so we conclude that $\sigma \multimap \sigma \notin [\![\mathsf{J}_t]\!]$ by Lemma 14.90(ii).

($\supseteq$) Let $\sigma \in \mathbb{T}_{\mathrm{R}}$ be such that $(x : \sigma, \sigma) \notin [\![\mathsf{J}_t\, x]\!]$. We coinductively construct an infinite branch f through t such that $\sigma \in \mathcal{W}_{(\mathcal{D},f,0)}(t)$. Since t is infinite, its root must have at least 1 child. Equivalently, we have $t(\langle\rangle) > 0$. Therefore, for $k \triangleq t(\langle\rangle) - 1$, we obtain:

$$\mathsf{J}_t\, x =_{\mathcal{B}} \lambda z_0 \ldots z_k.x(\mathsf{J}_{t\restriction_{\langle 0\rangle}} z_0) \cdots (\mathsf{J}_{t\restriction_{\langle k\rangle}} z_k)$$

and since $\mathcal{D}$ is extensional $\sigma \simeq^{\mathcal{D}} \mu_0 \multimap \cdots \multimap \mu_k \to \gamma$, for some $\mu_0, \ldots, \mu_k \in \mathbb{T}_{\otimes}, \gamma \in \mathbb{T}_{\mathrm{R}}$. From $(x : \sigma, \sigma) \notin [\![\mathsf{J}_t\, x]\!]$ and the soundness of the model we derive:

$$(x : \sigma, \sigma) \notin [\![\lambda z_0 \ldots z_k.x(\mathsf{J}_{t\restriction_{\langle 0\rangle}} z_0) \cdots (\mathsf{J}_{t\restriction_{\langle k\rangle}} z_k)]\!]$$

As a consequence, there exists an index $i \leq k$ such that $\mu_i \neq 1$ and a relational type $\tau \in \mu_i$ such that $(z_i : \tau, \tau) \notin [\![\mathsf{J}_{t\restriction_{\langle i\rangle}} z_i]\!]$. In particular, this entails that the subtree $t\restriction_{\langle i\rangle}$ is infinite because $(z_i : \tau, \tau)$ belongs to the interpretation of any finite η-expansion of z_i. Therefore we set $f(0) \triangleq k$ and, for all $n \in \mathbb{N}$, $f(n+1) \triangleq g(n)$ where g is the branch through $t\restriction_{\langle i\rangle}$ given by the co-IH and satisfying $\tau \in \mathcal{W}_{(\mathcal{D},g,0)}(t\restriction_{\langle i\rangle})$. By construction of f, we conclude that $\sigma \in \mathcal{W}_{(\mathcal{D},f,0)}(t)$ which entails $\sigma \in \mathcal{W}_{\mathcal{D}}(t)$. □

It should now be intuitively clear that an extensional rgm $\mathcal{D}$ is fully abstract for $\mathcal{H}^+$ when every λ-definable infinite η-expansion of the identity has an element in $\mathcal{D}$ witnessing one of its infinite paths, which exists by König's lemma. Symmetrically, if none of these witnesses exists, its theory is $\mathcal{H}^*$. This argument leads to the following definitions.

DEFINITION 14.92. A relational graph model $\mathcal{D}$ is called:

$$\begin{array}{lcl} \lambda\text{-}\textit{König} & \iff & \forall t \in \mathscr{T}^{\infty}_{\mathsf{rec}}\,.\,\mathcal{W}_{\mathcal{D}}(t) \neq \emptyset \\ \textit{hyperimmune} & \iff & \forall t \in \mathscr{T}^{\infty}_{\mathsf{rec}}\,.\,\mathcal{W}_{\mathcal{D}}(t) = \emptyset. \end{array}$$

The notion of hyperimmunity was introduced by Breuvart (2014) for Krivine's so-called K-models and it is related to the classical notion of hyperimmune-free trees in recursion theory (see e.g. Odifreddi (1989), Section V.5). Adopting the standard nomenclature, hyperimmune models should be actually called 'not hyperimmune-free'.

EXAMPLES 14.93. (i) The relational graph model $\mathcal{D}_\star$ is λ-König. In fact, the element $\star$ follows every function $f : \mathbb{N} \to \mathbb{N}$ starting from any $n \in \mathbb{N}$ since:

$$\star \simeq^{\mathcal{D}_\star} \underbrace{\star \multimap \cdots \multimap \star}_{f(n)} \multimap \star$$

It follows that $\star \in \mathcal{W}_{\mathcal{D}_\star}(t) \neq \emptyset$, for all trees $t \in \mathscr{T}^{\infty}_{\mathsf{rec}}$.

rgm $\mathcal{D}$	$\mathrm{Th}_{\sqsubseteq}(\mathcal{D})$	$\mathrm{Th}(\mathcal{D})$	by
$\mathcal{E}$	$\sqsubseteq_{\mathsf{r}}$	$\mathcal{B}$	Theorem 14.87
$\mathcal{D}_\star$	$\sqsubseteq_{\mathrm{NF}}$	$\mathcal{H}^+$	Corollary 14.96
$\mathcal{D}_\omega$	$\sqsubseteq_{\mathrm{SOL}}$	$\mathcal{H}^*$	Corollary 14.98

Table 14.2: The relational graph theories of some rgm's.

(ii) The relational graph model $\mathcal{D}_\omega$ is hyperimmune. Indeed, using the Approximation Theorem 14.82, one verifies that $(x : \sigma, \sigma) \in [\![\mathsf{J}_t x]\!]^{\mathcal{D}_\omega}$, for all $\sigma \in \mathbb{T}_{\mathrm{R}}^{\mathcal{D}_\omega}$ and $t \in \mathscr{T}_{\mathsf{rec}}^{\infty}$. Equivalently, we have $[\![\mathsf{J}_t]\!]^{\mathcal{D}_\omega} = \{\sigma \multimap \sigma \mid \sigma \in \mathbb{T}_{\mathrm{R}}^{\mathcal{D}_\omega}\}$ and therefore, by Proposition 14.91, we conclude $\mathcal{W}_{\mathcal{D}_\omega}(t) = \emptyset$.

THEOREM 14.94 (MORRIS' INEQUATIONAL FULL ABSTRACTION).
Let $\mathcal{D}$ be an extensional λ-König rgm, then:

$$M \sqsubseteq_{\mathrm{NF}} N \iff \mathcal{D} \models M \sqsubseteq N$$

PROOF. ($\Rightarrow$) This implication follows directly from Corollary 14.85(i).

($\Leftarrow$) As $\mathcal{D}$ is extensional and $\mathcal{D} \models M \sqsubseteq N$ we obtain $M \sqsubseteq_{\mathrm{SOL}} N$ by Corollary 14.85(i), whence $C[M] \sqsubseteq_{\mathrm{SOL}} C[N]$, for all contexts $C[\,]$. Assume by contradiction that $M \not\sqsubseteq_{\mathrm{NF}} N$. In this case there exists a context $C[\,]$ such that $C[M] \in \mathrm{NF}$, while $C[N] \notin \mathrm{NF}$. Now, by Corollary 12.50, we have that $C[M] \in \mathrm{NF}$ and $C[M] \sqsubseteq_{\mathrm{SOL}} C[N]$ entail $C[M] =_{\mathcal{H}^*} C[N]$. Therefore, we can apply Proposition 11.26 plus Lemma 11.27 to infer the existence of a head context $H[\,]$ satisfying

$$H[C[M]] =_\beta \mathsf{I} \qquad \text{and} \qquad H[C[N]] =_{\mathcal{B}} \mathsf{J}_t$$

for some $t \in \mathscr{T}_{\mathsf{rec}}^{\infty}$. By monotonicity of the interpretation, we obtain

$$[\![\mathsf{I}]\!] = [\![H[C[M]]]\!] \subseteq [\![H[C[N]]]\!] = [\![\mathsf{J}_t]\!].$$

We derive a contradiction by applying Proposition 14.91. □

THEOREM 14.95 (BREUVART ET AL. (2018)).
For an rgm $\mathcal{D}$, the following are equivalent:

1. *$\mathcal{D}$ is extensional and λ-König;*
2. *$\mathcal{D}$ is inequationally fully abstract for $\sqsubseteq_{\mathrm{NF}}$;*
3. *$\mathcal{D}$ is fully abstract for $\mathcal{H}^+$.*

PROOF. ($1 \Rightarrow 2$) By Theorem 14.94.

($2 \Rightarrow 3$) Trivial, since $\mathcal{H}^+ \vdash M = N$ if and only if $M \sqsubseteq_{\mathrm{NF}} N$ and $N \sqsubseteq_{\mathrm{NF}} M$.

($3 \Rightarrow 1$) Obviously $\mathcal{D}$ must be extensional because $\mathcal{H}^+$ is an extensional λ-theory. Suppose now, by the way of contradiction, that $\mathcal{D}$ is not λ-König. In this case there exists $t \in \mathscr{T}_{\mathsf{rec}}^{\infty}$ such that $\mathcal{W}_{\mathcal{D}}(t) = \emptyset$ thus, by Proposition 14.91, we obtain $\mathcal{D} \models \mathsf{I} = \mathsf{J}_t$. We derive a contradiction since $\mathsf{I} \not\sqsubseteq_{\mathrm{NF}} \mathsf{J}_t$. □

COROLLARY 14.96. $\mathrm{Th}_{\sqsubseteq}(\mathcal{D}_\star) = \sqsubseteq_{\mathrm{NF}}$, *whence* $\mathrm{Th}(\mathcal{D}_\star) = \mathcal{H}^+$.

We show the characterization of full abstraction for $\mathcal{H}^*$ based on hyperimmunity, whose proof is omitted. Table 14.2 summarizes the main results of the section.

THEOREM 14.97 (BREUVART ET AL. (2018)).
For an rgm $\mathcal{D}$, the following are equivalent:

1. *$\mathcal{D}$ is extensional and hyperimmune;*
2. *$\mathcal{D}$ is inequationally fully abstract for $\sqsubseteq_{\mathrm{SOL}}$;*
3. *$\mathcal{D}$ is fully abstract for $\mathcal{H}^*$.*

COROLLARY 14.98. $\mathrm{Th}_{\sqsubseteq}(\mathcal{D}_\omega) = \sqsubseteq_{\mathrm{SOL}}$, *whence* $\mathrm{Th}(\mathcal{D}_\omega) = \mathcal{H}^*$.

Chapter 15

Church algebras for λ-calculus

Although all axioms of the λ-calculus are expressed in the form of equations, it cannot be considered as a genuine equational theory, since it is not first order because of the variable-binding properties of λ-abstraction that prevent its variables from behaving as real algebraic variables. Consequently, the general methods developed in universal algebra for defining the semantics of an arbitrary algebraic theory are not directly applicable. As discussed in Section 3.2 several attempts have been made at reformulating λ-calculus as a purely algebraic theory—the earliest and best known being based on λ-algebras and λ-models that are particular combinatory algebras. It is well known that combinatory algebras are never commutative, associative, finite or recursive (Proposition 3.32) and this led to a common belief that λ-calculus and combinatory logic are somewhat 'pathological' from the algebraic point of view. For example, in Scott (1975) it was stated that

> "During the symposium [(Böhm (1975))] Barendregt exhorted us to study combinatory algebras as algebras just as we do in mathematics with other structures. It is a good question whether anyone outside the 'club' will want to listen to us, but that is a different problem."

Nevertheless an algebraic approach has been subsequently developed and fruitfully applied to the λ-calculus. A starting point, implicitly already present in B[1984], is the consideration that the lattice $\lambda\mathcal{T}$ of λ-theories is isomorphic to the congruence lattice of the term algebra of $\boldsymbol{\lambda}$. For example in the kite depicted in Chapter 17 (also present in this book as Figure 11.1) the natural ordering of λ-theories as subsets is 'drawn upside down' to indicate homomorphisms between the corresponding term algebras.[1] Moreover, in Engeler (1981) it was shown that every algebra $(\mathcal{A}, \cdot)$ can be embedded[2] into some λ-model $\mathcal{M}$, where multiplication $\cdot$ is interpreted as application. The algebraic approach

[1]Similarly, trying to settle the range property for the λ-theory $\mathcal{H}$ one starts reasoning as follows. Let $F \in \Lambda^o$ have e.g. a range of cardinality 2 in $\mathcal{M}^o(\mathcal{H})$. Since $\mathcal{M}(\boldsymbol{\lambda})$ and $\mathcal{M}(\mathcal{B})$ do satisfy the range property, the range of F modulo $=_\beta$ must be a singleton, whereas modulo $=_\mathcal{B}$ it must be infinite.

[2]In B[1984], Exercise 6.8.15, due to Barendregt, Klop, and Dezani, it is shown that for computable (and hence for finite) $(\mathcal{A}, \cdot)$ the embedding can be done even into the closed term model $\mathcal{M}^o(\boldsymbol{\lambda})$. As a consequence, combinators with application can represent arbitrary computable algebras.

was considerably extended by Salibra, who launched in the 90s a research program aiming at exploring λ-calculus using techniques from universal algebra and algebraic logic. In Salibra (2000), he has shown that the variety generated by the term algebra of $\boldsymbol{\lambda}$ is axiomatized by the finite schema of identities characterizing *lambda abstraction algebras*. The equational theory of lambda abstraction algebras, introduced by Pigozzi and Salibra (1995), constitutes a purely algebraic theory of λ-calculus in the same spirit that cylindric and polyadic (Boolean) algebras constitute an algebraic theory of the first-order predicate logic. One feature of lambda abstraction algebras that sets them apart from combinatory algebras is the way variables in the λ-calculus are abstracted—this provides each lambda abstraction algebra with an implicit coordinate system. Another peculiar feature is the algebraic reformulation of β-conversion as the definition of abstract substitution.

Salibra and his coauthors over the years obtained a number of fascinating results.[3] Therefore it can be asserted that this program was successful. Not only techniques from universal algebra proved to be useful for studying the λ-calculus, but programming languages have been a source of inspiration for defining classes of algebras general enough to encompass combinatory algebras, as well as more common algebraic structures. More precisely, Manzonetto and Salibra (2008) introduced the class of Church algebras,[4] i.e., algebras $\mathcal{A}$ over a signature including two nullary terms 0, 1 and a ternary term $\text{ite}(x, y, z)$ modeling the 'if-then-else' instruction in the sense that they validate the equations:

$$\begin{aligned} \mathcal{A} \models \ \text{ite}(1, x, y) &= x, \\ \mathcal{A} \models \ \text{ite}(0, x, y) &= y. \end{aligned}$$

Combinatory algebras, but also Boolean algebras, Heyting algebras and rings with unity are instances of the above definition. The class of Church algebras over a signature Σ being equational, it forms a variety by Birkhoff's theorem, and hence it is closed under subalgebras, homomorphic images, and (direct) products. More importantly, a variety $\mathbb{V}$ of Church algebras whose directly indecomposable members form a universal class,[5] satisfies the following form of the Stone Representation Theorem. Each Church algebra $\mathcal{A} \in \mathbb{V}$ contains a Boolean algebra of central elements, inducing a representation of $\mathcal{A}$ as a weak Boolean product of directly indecomposable algebras.[6] By generalizing the notion of easiness from λ-calculus to universal algebra, it is also possible to prove that if a Church algebra $\mathcal{A} \in \mathbb{V}$ possesses a finite number k of easy elements, then its congruence lattice $\text{Con}(\mathcal{A})$ admits the lattice reduct of the free Boolean algebra with k generators as a lattice interval. These results are presented in this chapter. Other properties of the variety of Church algebras can be found in Manzonetto and Salibra (2008, 2009) and Cvetko-Vah and Salibra (2015).

As shown by Manzonetto and Salibra, the algebraic theory developed for Church algebras has important consequences for the study of λ-calculus models and λ-theories.

---

[3] Some are mentioned in Section 3.1, others are presented in this chapter.

[4] In honor of Alonzo Church, father of λ-calculus.

[5] In particular, this is the case for the variety $\mathbb{CA}$ of combinatory algebras.

[6] I.e., algebras that cannot be decomposed as the Cartesian product of two other non-trivial algebras.

1. AN ALGEBRAIC INCOMPLETENESS THEOREM.

   Several models of λ-calculus have been introduced in various categories of domains, and they have been classified into *semantics* according to the nature of their representable functions. For instance, the *Scott-continuous semantics* indicates the class of λ-models living in the Cartesian closed category having cpo's as objects and continuous functions as morphisms.[7] More generally, we call 'a semantics of λ-calculus' any uniformly presented class of models. A semantic $\mathbb{C}$ is called 'equationally incomplete' if there exists a λ-theory $\mathcal{T}$ which is not of the form $\mathrm{Th}(\mathcal{M})$, for some model $\mathcal{M} \in \mathbb{C}$. Incompleteness proofs[8] were given by Honsell and Ronchi Della Rocca (1992) for the Scott-continuous semantics, and by Bastonero and Gouy (1999) for Berry's stable semantics.

   Consider now the indecomposable semantics of λ-calculus, i.e. the class $\mathbb{IND}$ of λ-models that are directly indecomposable as combinatory algebras. We will see that $\mathbb{IND}$ includes all Scott-continuous, stable and strongly stable models. By showing that the indecomposable semantics is equationally incomplete, one obtains a uniform proof of the equational incompleteness of the main semantics of λ-calculus, thus implying the two results mentioned above.

2. FINITE BOOLEAN SUBLATTICES OF $\lambda\mathcal{T}$.

   A (lattice) interval determined by two λ-theories $\mathcal{T}_1, \mathcal{T}_2$ consists of

$$[\mathcal{T}_1, \mathcal{T}_2] = \{\mathcal{T} \in \lambda\mathcal{T} \mid \mathcal{T}_1 \subseteq \mathcal{T} \subseteq \mathcal{T}_2\}.$$

   In general these intervals are not linearly ordered. In the literature intervals are described with cardinality either 1, 2 or $2^{\aleph_0}$. For example $|[\mathcal{H}^*, \nabla]| = 2$, as $\mathcal{H}^*$ is maximally consistent, by Proposition 3.13(i); if $\mathcal{T}_1, \mathcal{T}_2$ are r.e., then $|[\mathcal{T}_1, \mathcal{T}_2]| = 2^{\aleph_0}$, by Theorem 3.5(vii); in Barendregt et al. (1978) it is also proved that $|[\mathcal{H}\eta, \mathcal{H}^*]| = 2^{\aleph_0}$, even though $\mathcal{H}\eta, \mathcal{H}^*$ are not r.e. λ-theories. No lattice interval of cardinality, say, 4 was known. The aforementioned properties of Church algebras and the existence of infinitely many easy terms entail that $\lambda\mathcal{T}$ contains all finite Boolean lattice intervals.

The chapter is organized as follows. In Section 15.1 we recall some well-known results concerning different kinds of decompositions of algebras; other preliminary notions on universal algebra are available in Appendix A.4. In Section 15.2 we present the definition of Church algebras and prove the corresponding Stone Representation Theorem 15.23. In Section 15.3 we study the congruence lattice of a Church algebra and prove that it contains finite Boolean lattice intervals. Finally, Sections 15.4 and 15.5 are devoted to present the applications of this theory to the study of λ-calculus and λ-theories. For the presentation of the material we mainly follow Manzonetto and Salibra (2008, 2010).

[7] For more details on this classification, see Section 15.5.

[8] Barendregt (1971) gave an incompleteness result for the class of $\mathcal{D}_\infty$-models, but the incompleteness there has a slightly different meaning: it was shown that the *first-order* theories $\mathrm{Th}_1(\mathcal{D}_\infty)$ do not cover all first-order theories of (extensional) λ-models.

15.1 Algebras and Boolean products

The representation of algebras by Boolean products is a very general problem in universal algebra. The algebras we consider in this chapter do not satisfy—in general—this kind of representation, but admit a weaker representation using weak Boolean products.

Given two algebras $\mathcal{A}, \mathcal{B}$, we denote by $\mathcal{A} \times \mathcal{B}$ their *(direct) product* and we let $\mathcal{A} \cong \mathcal{B}$ mean that they are isomorphic. We let $\mathrm{Con}(\mathcal{A})$ denote the lattice of congruences of $\mathcal{A}$.

DEFINITION 15.1. An algebra $\mathcal{A}$ is called:

(i) *trivial* if its underlying set is a singleton;

(ii) *simple* whenever $\mathrm{Con}(\mathcal{A}) = \{\Delta^{\mathcal{A}}, \nabla^{\mathcal{A}}\}$;

(iii) *directly decomposable* if there are two non-trivial algebras $\mathcal{B}, \mathcal{C}$ such that $\mathcal{A} \cong \mathcal{B} \times \mathcal{C}$;

(iv) *directly indecomposable* if it is not directly decomposable;

(v) a *subdirect product* of the algebras $(\mathcal{B}_i)_{i \in I}$, written

$$\mathcal{A} \leq \prod_{i \in I} \mathcal{B}_i,$$

if there exists an embedding f of $\mathcal{A}$ into the direct product $\prod_{i \in I} \mathcal{B}_i$ such that the projection $\pi_i \circ f : \mathcal{A} \to \mathcal{B}_i$ is onto for every $i \in I$.

DEFINITION 15.2. Let $\mathcal{A}$ be an algebra.

(i) Given a subset $X \subseteq A \times A$, we denote by $\vartheta(X)$ the least congruence in $\mathrm{Con}(\mathcal{A})$ including X. When X is a singleton $\{(a, b)\}$, we simply write $\vartheta(a, b)$ for $\vartheta(X)$.

(ii) Given $a \in A$ and $Y \subseteq A$, we define $\vartheta(a, Y) \triangleq \vartheta(\{(a, b) \mid b \in Y\})$. In particular, we assume that $\vartheta(1, \emptyset) = \vartheta(0, \emptyset) = \Delta^{\mathcal{A}}$.

(iii) A congruence $\varphi \in \mathrm{Con}(\mathcal{A})$ is called *consistent* if $\varphi \neq \nabla^{\mathcal{A}}$; *compact* if $\varphi = \vartheta(X)$, for some finite set $X \subseteq A \times A$.

Factor congruences

We present some methods for decomposing an algebra $\mathcal{A}$ into a product of simpler algebras, and prove that they are equivalent. The first method exploits the so-called factor congruences, living in $\mathrm{Con}(\mathcal{A})$, whose definition is given below.

DEFINITION 15.3. A congruence ϑ on an algebra $\mathcal{A}$ is a *factor congruence* if there exists another congruence $\overline{\vartheta}$ such that

$$\vartheta \wedge \overline{\vartheta} = \Delta^{\mathcal{A}}, \qquad \vartheta \circ \overline{\vartheta} = \nabla^{\mathcal{A}}.$$

We call $(\vartheta, \overline{\vartheta})$ as above a *pair of complementary factor congruences* or *cfc-pair*, for short.

The set of factor congruences of $\mathcal{A}$ is not, in general, a sublattice of $\mathrm{Con}(\mathcal{A})$. Moreover, the homomorphism $f : \mathcal{A} \to \mathcal{A}/\vartheta \times \mathcal{A}/\overline{\vartheta}$ defined by setting $f(x) \triangleq (x/\vartheta, x/\overline{\vartheta})$ is:

- injective $\iff \vartheta \wedge \overline{\vartheta} = \Delta^{\mathcal{A}}$;
- surjective $\iff \vartheta \circ \overline{\vartheta} = \nabla^{\mathcal{A}} \iff \overline{\vartheta} \circ \vartheta = \nabla^{\mathcal{A}}$.

As a direct consequence, we obtain the lemma below.

LEMMA 15.4. *Given an algebra $\mathcal{A}$ and $\vartheta, \overline{\vartheta} \in \mathrm{Con}(\mathcal{A})$, the following are equivalent:*

1. *$(\vartheta, \overline{\vartheta})$ is a cfc-pair of $\mathcal{A}$;*
2. *The algebra homomorphism $x \mapsto (x/\vartheta, x/\overline{\vartheta})$ constitutes an isomorphism $\mathcal{A} \cong \mathcal{B} \times \mathcal{C}$, where $\mathcal{B} \cong \mathcal{A}/\vartheta$ and $\mathcal{C} \cong \mathcal{A}/\overline{\vartheta}$.*

Therefore, the existence of non-trivial factor congruences is just another way of saying that an algebra is decomposable as a direct product of simpler algebras. The factor congruences $\Delta^{\mathcal{A}}$ and $\nabla^{\mathcal{A}}$ are called *trivial* since they induce the decomposition $\mathcal{A} \cong \mathcal{A} \times 1$, where $\mathcal{A} \cong \mathcal{A}/\Delta^{\mathcal{A}}$ and $1 \cong \mathcal{A}/\nabla^{\mathcal{A}}$ is the trivial algebra.

LEMMA 15.5. *An algebra $\mathcal{A}$ is directly indecomposable if it admits only the two trivial factor congruences $\Delta^{\mathcal{A}}$ and $\nabla^{\mathcal{A}}$.*

Clearly, every simple algebra is directly indecomposable, while there are algebras that are directly indecomposable but not simple—they have congruences which, however, do not split the algebra up neatly as a Cartesian product.

Decomposition operators

Factor congruences can be characterized in terms of certain algebra homomorphisms called decomposition operators (see also McKenzie et al. (1987), Definition 4.32).

DEFINITION 15.6. A *decomposition operator* for an algebra $\mathcal{A}$ is an algebra homomorphism $f : \mathcal{A} \times \mathcal{A} \to \mathcal{A}$ such that:

1. $f(x, x) = x$ and
2. $f(f(x, y), z) = f(x, z) = f(x, f(y, z))$.

There exists a bijective correspondence between cfc-pairs and decomposition operators, and thus, between decomposition operators and factorizations like $\mathcal{A} \cong \mathcal{B} \times \mathcal{C}$. One can think about $f(x, y)$ as '$\langle \pi_B(x), \pi_C(y) \rangle$', where π_B and π_C project onto the first and second component, respectively, and $\langle \cdot, \cdot \rangle$ forms tuples modulo the isomorphism.

PROPOSITION 15.7. (i) *Given a decomposition operator f for an algebra $\mathcal{A}$, the binary relations ϑ and $\overline{\vartheta}$ defined by setting:*

$$\begin{aligned} x \,\vartheta\, y \quad &\iff \quad f(x, y) = y, \\ x \,\overline{\vartheta}\, y \quad &\iff \quad f(x, y) = x, \end{aligned}$$

form a cfc-pair.

(ii) *Conversely, given a cfc-pair $(\vartheta, \overline{\vartheta})$ of $\mathcal{A}$, the function f defined by:*

$$f(x, y) \triangleq u \iff x \,\vartheta\, u \,\overline{\vartheta}\, y, \tag{15.1}$$

is a decomposition operator.

PROOF. The proof is easy. See McKenzie et al. (1987), Theorem 4.33. Note that the definition of f is sound because, for all $x, y \in A$, there exists exactly one element $u \in A$ satisfying $x \,\vartheta\, u \,\overline{\vartheta}\, y$. □

Boolean and factorable congruences

Consider two algebras $\mathcal{A}$ and $\mathcal{B}$ over the same signature. The *product congruence* of $\varphi_1 \in \mathrm{Con}(\mathcal{A})$ and $\varphi_2 \in \mathrm{Con}(\mathcal{B})$ is the congruence $\varphi_1 \times \varphi_2 \in \mathrm{Con}(\mathcal{A} \times \mathcal{B})$ defined by

$$(a,b)\ (\varphi_1 \times \varphi_2)\ (a',b') \iff a\ \varphi_1\ a'\ \&\ b\ \varphi_2\ b'.$$

We denote by $\mathrm{Con}(\mathcal{A}) \times \mathrm{Con}(\mathcal{B})$ the sublattice of $\mathrm{Con}(\mathcal{A} \times \mathcal{B})$ constituted by all product congruences.

DEFINITION 15.8. (i) An algebra has *Boolean factor congruences* if its factor congruences form a Boolean sublattice of the congruence lattice.

(ii) A variety $\mathbb{V}$ of algebras has *factorable congruences* if, for all algebras $\mathcal{A}, \mathcal{B} \in \mathbb{V}$:

$$\mathrm{Con}(\mathcal{A} \times \mathcal{B}) \cong \mathrm{Con}(\mathcal{A}) \times \mathrm{Con}(\mathcal{B}).$$

Most known examples of varieties in which all algebras have Boolean factor congruences are those with factorable congruences. This is the case, for example, of the congruence distributive varieties, and congruence permutable varieties in which the universal congruences are compact (e.g. the variety of rings with unity).

LEMMA 15.9. *If a variety $\mathbb{V}$ has factorable congruences, then every $\mathcal{A} \in \mathbb{V}$ has Boolean factor congruences.*

PROOF. See Bigelow and Burris (1990), Corollary 1.4. □

The Boolean product construction allows to transfer numerous fascinating properties of Boolean algebras into other varieties of algebras (see Burris and Sankappanavar (2012), Chapter IV). Actually, this construction has been presented for several years as "the algebra of global sections of sheaves of algebras over Boolean spaces" (see Comer (1971); Johnstone (1982)). However, these notions were unnecessarily complex and we prefer to adopt here the following equivalent presentation (see Burris and Werner (1979)).

Recall that a Boolean space is a compact, Hausdorff and totally disconnected topological space.

DEFINITION 15.10. (i) A *weak Boolean product* of a family $(\mathcal{A}_i)_{i \in I}$ of algebras is a subdirect product $\mathcal{A} \leq \prod_{i \in I} \mathcal{A}_i$, where I can be endowed with a Boolean space topology such that, for all $a, b \in A$:

(1) the set $\{i \in I \mid a_i = b_i\}$ is open;

(2) if X is a clopen (i.e., both open and closed) subset of I, then the element c defined by

$$c_i \triangleq \begin{cases} a_i, & \text{if } i \in X, \\ b_i, & \text{if } i \in I - X, \end{cases}$$

belongs to A.

(ii) A *Boolean product* is a weak Boolean product such that the set $\{i \in I \mid a_i = b_i\}$ is clopen for all $a, b \in A$.

15.2 Church algebras

In this section we define the class of Church algebras, first introduced by Manzonetto and Salibra (2008), and study their algebraic properties.

DEFINITION 15.11. (i) An algebra $\mathcal{A}$ over a signature Σ is a *Church algebra* if there are nullary terms $0, 1 \in A$ and a ternary term $\mathrm{ite}(a, x, y)$ satisfying the following identities:

$$\begin{array}{rcl} \mathrm{ite}(1, x, y) & = & x, \\ \mathrm{ite}(0, x, y) & = & y. \end{array} \tag{15.2}$$

(ii) A variety $\mathbb{V}$ over a signature Σ is called a *Church variety* if every algebra in $\mathbb{V}$ is a Church algebra with respect to the same term $\mathrm{ite}(e, x, y)$ and nullary terms $0, 1$.

Intuitively, the above identities formalize the 'if-then-else' conditional statement, which is a basic construct of programming languages. Perhaps surprisingly, many famous algebraic structures are instances of this definition, because they possess term operations satisfying the identities in (15.2). These identities imply strong algebraic properties, that we will apply to the study of λ-models and λ-theories.

EXAMPLES 15.12. The following are examples of Church algebras:

Algebras	$\mathrm{ite}(a, x, y)$	0	1
Combinatory algebras	$(a \cdot x) \cdot y$	$\mathbf{f}$	$\mathbf{t}$
Boolean algebras	$(a \vee y) \wedge (a^- \vee x)$	0	1
Heyting algebras	$(a \vee y) \wedge ((a \to 0) \vee x)$	0	1
Rings with unity	$ax + (1 - a)y$	0	1

where $\mathbf{f} \triangleq \mathbf{sk}$ and $\mathbf{t} \triangleq \mathbf{k}$ in combinatory algebras, a^- represents the complement of a in Boolean algebras.

PROPOSITION 15.13. (i) *The class of all Church algebras over a signature Σ is a variety.*
(ii) *The variety of Church algebras over a signature Σ has factorable congruences.*

PROOF. (i) Church algebras are defined by equations (15.2), whence they constitute an equational class. As shown by Birkhoff (1935), the two notions coincide.

(ii) Let $\mathcal{A}, \mathcal{B}$ be Church algebras. By Definition 15.8(ii), we need to show

$$\mathrm{Con}(\mathcal{A} \times \mathcal{B}) \cong \mathrm{Con}(\mathcal{A}) \times \mathrm{Con}(\mathcal{B})$$

It is easy to check that, up to isomorphism, $\mathrm{Con}(\mathcal{A}) \times \mathrm{Con}(\mathcal{B}) \subseteq \mathrm{Con}(\mathcal{A} \times \mathcal{B})$ holds. Conversely, take $\varphi \in \mathrm{Con}(\mathcal{A} \times \mathcal{B})$. The 'projections' φ_1, φ_2 of φ are the binary relations on $\mathcal{A}$ and $\mathcal{B}$, respectively, defined as follows:

$$\begin{array}{rcl} a_1 \, \varphi_1 \, a_2 & \iff & \exists b_1, b_2 \in B \,.\, (a_1, b_1) \, \varphi \, (a_2, b_2), \\ b_1 \, \varphi_2 \, b_2 & \iff & \exists a_1, a_2 \in A \,.\, (a_1, b_1) \, \varphi \, (a_2, b_2). \end{array}$$

Clearly $\varphi \subseteq \varphi_1 \times \varphi_2$, so we prove the opposite inclusion. Assume $(a_1, b_1)(\varphi_1 \times \varphi_2)(a_2, b_2)$, for some $a_1, a_2 \in A$, $b_1, b_2 \in B$. By definition of $\varphi_1 \times \varphi_2$, we have $a_1 \, \varphi_1 \, a_2$ and $b_1 \, \varphi_2 \, b_2$. Hence, there exist $a_3, a_4 \in A$, $b_3, b_4 \in B$ such that $(a_1, b_3) \, \varphi \, (a_2, b_4)$ and $(a_3, b_1) \, \varphi \, (a_4, b_2)$. From $(1, 0) \, \varphi \, (1, 0)$ and the compatibility of φ, we obtain:

$$(a_1, b_1) = (\mathrm{ite}(1, a_1, a_3), \mathrm{ite}(0, b_3, b_1)) \, \varphi \, (\mathrm{ite}(1, a_2, a_4), \mathrm{ite}(0, b_4, b_2)) = (a_2, b_2).$$

We conclude $\varphi = \varphi_1 \times \varphi_2$. It is easy to check that φ_1, φ_2 are reflexive, symmetric and compatible. We now show that φ_1 is also transitive. Let $a_1 \, \varphi_1 \, a_2 \, \varphi_1 \, a_3$, then there exist b_1, b_2, b_3, b_4 such that $(a_1, b_1) \, \varphi \, (a_2, b_2)$ and $(a_2, b_3) \, \varphi \, (a_3, b_4)$. From the symmetry of φ we have also $(a_3, b_4) \, \varphi \, (a_2, b_3)$. Since $(1, 0) \, \varphi \, (1, 0)$ and φ is a compatible relation, we derive:

$$(a_1, b_4) = (\mathrm{ite}(1, a_1, a_3), \mathrm{ite}(0, b_1, b_4)) \, \varphi \, (\mathrm{ite}(1, a_2, a_2), \mathrm{ite}(0, b_2, b_3)) = (a_2, b_3).$$

Finally, from $(a_1, b_4) \, \varphi \, (a_2, b_3)$ and $(a_2, b_3) \, \varphi \, (a_3, b_4)$ we infer $(a_1, b_4) \, \varphi \, (a_3, b_4)$ and, hence, $a_1 \, \varphi_1 \, a_3$. It follows that $\varphi_1 \in \mathrm{Con}(\mathcal{A})$. An analogous reasoning gives $\varphi_2 \in \mathrm{Con}(\mathcal{B})$. It is now easy to conclude that $\mathrm{Con}(\mathcal{A} \times \mathcal{B}) \cong \mathrm{Con}(\mathcal{A}) \times \mathrm{Con}(\mathcal{B})$. □

COROLLARY 15.14. *All Church algebras have Boolean factor congruences.*

PROOF. It follows from Proposition 15.13(ii), by applying Lemma 15.9. □

The Stone Representation Theorem for Church Algebras

We are going to show that Church algebras satisfy a theorem similar to the Stone representation theorem for Boolean algebras. The observation that Boolean algebras could be regarded as rings is due to Stone and admits a generalization, due to Pierce (1967), to commutative rings with unity (see also Johnstone (1982), Chapter V). To help the reader get familiar with the argument, we outline now Pierce's construction.

Let $\mathcal{A} = (A, +, \cdot, 0, 1)$ be a commutative ring with unity and let

$$\mathrm{IE}(\mathcal{A}) \triangleq \{a \in A \mid a \cdot a = a\}$$

be the set of its *idempotent elements*. One defines a structure of Boolean algebra on $\mathrm{IE}(\mathcal{A})$ by setting, for all $a, b \in \mathrm{IE}(\mathcal{A})$:

$$\begin{array}{rcl} a \wedge b & \triangleq & a \cdot b; \\ a^- & \triangleq & 1 - a. \end{array}$$

Then it is possible to show that every $a \in \mathrm{IE}(A)$ induces a cfc-pair $(\vartheta(1, a), \vartheta(a, 0))$. In other words, the ring $\mathcal{A}$ can be decomposed as

$$\mathcal{A} \cong \mathcal{A}/\vartheta(1, a) \times \mathcal{A}/\vartheta(a, 0).$$

If $a \neq 0, 1$ then the associated cfc-pair is non-trivial, whence the decomposition is also non-trivial. Moreover $\mathcal{A}$ is directly indecomposable whenever $\mathrm{IE}(\mathcal{A}) = \{0, 1\}$. Hence, Pierce's representation theorem for commutative rings with unity can be stated as follows:

> "Every commutative ring with unity is isomorphic to a weak Boolean product of directly indecomposable rings."

If $\mathcal{A}$ is a Boolean ring, we get the Stone representation theorem for Boolean algebras, because the ring of truth values is the unique directly indecomposable Boolean ring.

Vaggione (1996b) generalized idempotent elements to any universal algebra whose top congruence ∇ is compact, and called them *central elements.* Central elements were used, among other things, to investigate the closure of varieties of algebras under Boolean products. Here we give a new characterization based on decomposition operators.

DEFINITION 15.15. Let $\mathcal{A}$ be a Church algebra.

(i) Given $e \in A$, define the function $f_e(x, y) \triangleq \mathrm{ite}(e, x, y)$.

(ii) Given $e \in A$, define the congruences ϑ_e and $\overline{\vartheta}_e$ by setting:

$$\vartheta_e \triangleq \vartheta(1, e) \qquad \text{and} \qquad \overline{\vartheta}_e \triangleq \vartheta(e, 0).$$

(iii) We say that an element e of $\mathcal{A}$ is *central* if $(\vartheta_e, \overline{\vartheta}_e)$ constitutes a cfc-pair.

(iv) We denote by $\mathrm{CE}(\mathcal{A})$ the collection of all central elements of $\mathcal{A}$.

(v) A central element e is called *non-trivial* if $e \neq 0, 1$.

We now show how to internally represent in a Church algebra factor congruences as central elements. We start with a lemma.

LEMMA 15.16. *Let $\mathcal{A}$ be a Church algebra and $e \in A$. Then, for all $x, y \in A$:*

(i) $x \,\vartheta_e\, \mathrm{ite}(e, x, y)\, \overline{\vartheta}_e\, y$.

(ii) $x \,\vartheta_e\, y \iff \mathrm{ite}(e, x, y)\ (\vartheta_e \wedge \overline{\vartheta}_e)\ y$.

(iii) $x \,\overline{\vartheta}_e\, y \iff \mathrm{ite}(e, x, y)\ (\vartheta_e \wedge \overline{\vartheta}_e)\ x$.

(iv) $\vartheta_e \circ \overline{\vartheta}_e = \overline{\vartheta}_e \circ \vartheta_e = \nabla^{\mathcal{A}}$.

PROOF. (i) It follows from $1 \,\vartheta_e\, e \,\overline{\vartheta}_e\, 0$ and the compatibility of ϑ_e and $\overline{\vartheta}_e$.

(ii) By (i), we have $x \,\vartheta_e\, \mathrm{ite}(e, x, y)$. Therefore $x \,\vartheta_e\, y$ if and only if $\mathrm{ite}(e, x, y) \,\vartheta_e\, y$.

(iii) Analogous to (ii).

(iv) By (i). □

PROPOSITION 15.17. *Let $\mathcal{A}$ be a Church algebra. For all $e \in A$, the following conditions are equivalent:*

1. *$e \in \mathrm{CE}(\mathcal{A})$;*
2. *$\vartheta_e \wedge \overline{\vartheta}_e = \Delta^{\mathcal{A}}$;*
3. *For all x and y, $\mathrm{ite}(e, x, y)$ is the unique element such that $x \,\vartheta_e\, \mathrm{ite}(e, x, y)\, \overline{\vartheta}_e\, y$;*
4. *e satisfies the following identities:*

$$\mathrm{ite}(e, x, x) = x \tag{15.3}$$

$$\mathrm{ite}(e, \mathrm{ite}(e, x, y), z) = \mathrm{ite}(e, x, z) = \mathrm{ite}(e, x, \mathrm{ite}(e, y, z)) \tag{15.4}$$

$$\mathrm{ite}(e, f(\vec{x}), f(\vec{y})) = f(\mathrm{ite}(e, x_1, y_1), \ldots, \mathrm{ite}(e, x_n, y_n)), \quad \textit{for all } f \in \Sigma_n, \tag{15.5}$$

$$e = \mathrm{ite}(e, 1, 0) \tag{15.6}$$

5. *The function f_e is a decomposition operator such that $f_e(1, 0) = e$.*

PROOF. ($1 \Leftrightarrow 2$) From Lemma 15.16(iv).

($2 \Rightarrow 3$) By Lemma 15.16(iv), the congruences ϑ_e and $\overline{\vartheta}_e$ form a cfc-pair, so this implication follows from Lemma 15.16(i).

($3 \Rightarrow 2$) First note that $\mathrm{ite}(e, x, x) = x$. If $x\ (\vartheta_e \wedge \overline{\vartheta}_e)\ y$ then $x\, \vartheta_e\, y\, \overline{\vartheta}_e\, x$, from which it follows that $y = \mathrm{ite}(e, x, x) = x$.

($4 \Leftrightarrow 5$) By Proposition 15.7. It is a simple exercise to verify that e satisfies the identities in (4) exactly when f_e is a decomposition operator satisfying $f_e(1, 0) = e$.

($1 \Rightarrow 5$) We already proved that (1) is equivalent to (3). Thus, f_e is a decomposition operator because $(\vartheta_e, \overline{\vartheta}_e)$ is a cfc-pair and $\mathrm{ite}(e, x, y)$ is the unique element satisfying $x\, \vartheta_e\, \mathrm{ite}(e, x, y)\, \overline{\vartheta}_e\, y$. Finally, $f_e(1, 0) = \mathrm{ite}(e, 1, 0) = e$ follows from $1\, \vartheta_e\, e\, \overline{\vartheta}_e\, 0$.

($5 \Rightarrow 1$) Let $(\varphi, \overline{\varphi})$ be the cfc-pair associated with f_e. From Proposition 15.7 and from $f_e(1, 0) = \mathrm{ite}(e, 1, 0) = e$, it follows that $1\, \varphi\, e\, \overline{\varphi}\, 0$, so that $\vartheta_e, \overline{\vartheta}_e \subseteq \varphi$. For the opposite direction, let $x\, \varphi\, y$, which is equivalent to $\mathrm{ite}(e, x, y) = y$ by Proposition 15.7. Then, since $1\, \vartheta_e\, e$, we derive $x = \mathrm{ite}(1, x, y)\, \vartheta_e\, \mathrm{ite}(e, x, y) = y$. Similarly, for $\overline{\varphi}$. □

COROLLARY 15.18. *Let $\mathcal{A}$ be a Church algebra.*

(i) *If h is a decomposition operator on $\mathcal{A}$, then $h(1, 0) \in A$ is a central element such that $f_{h(1,0)} = h$.*

(ii) *The map $e \in \mathrm{CE}(\mathcal{A}) \mapsto f_e$ gives a bijective correspondence between central elements and decomposition operators.*

(iii) *The map $e \in \mathrm{CE}(\mathcal{A}) \mapsto \vartheta_e$ (resp. $e \in \mathrm{CE}(\mathcal{A}) \mapsto \overline{\vartheta}_e$) gives a bijective correspondence between central elements and factor congruences.*

PROOF. (i) Let h be a decomposition operator on $\mathcal{A}$. Then, we have:

$$\begin{aligned}
\mathrm{ite}(h(1,0), x, y) &= \mathrm{ite}(h(1,0), h(x,x), h(y,y)) \\
&= h(\mathrm{ite}(1, x, y), \mathrm{ite}(0, x, y)) \\
&= h(x, y)
\end{aligned}$$

We conclude that $h(1, 0)$ is a central element by Proposition 15.17(5).

(ii) and (iii) are easy consequences of (i). □

COROLLARY 15.19. *Let $\mathcal{A}$ be a Church algebra, $e \in A$ and $\varphi \in \mathrm{Con}(\mathcal{A})$. Then*

(i) *$[e]_\varphi$ is central in $\mathcal{A}/\varphi \iff \vartheta_e \wedge \overline{\vartheta}_e \leq \varphi$;*

(ii) *$[e]_{\vartheta_e \wedge \overline{\vartheta}_e}$ is a non-trivial central element in $\mathcal{A}/(\vartheta_e \wedge \overline{\vartheta}_e) \iff \vartheta_e \neq \nabla^{\mathcal{A}}, \Delta^{\mathcal{A}}$.*

COROLLARY 15.20. *For a Church algebra $\mathcal{A}$, the following are equivalent:*

1. *$\mathcal{A}$ is directly indecomposable;*

2. *$\mathrm{CE}(\mathcal{A}) = \{0, 1\}$;*

3. *$\vartheta_e \wedge \overline{\vartheta}_e \neq \Delta^{\mathcal{A}}$, for all $e \neq 0, 1$.*

EXAMPLES 15.21. (i) All elements of a Boolean algebra are central by characterization (4) of Proposition 15.17 and Example 15.12.

(ii) An element is central in a commutative ring with unity if and only if it is idempotent. This characterization does not hold for non-commutative rings with unity.

(iii) Given $M \in \Lambda^o$, consider the λ-theory $\mathcal{T}_M$ generated by the equation $M = \Omega$. Since Ω is an easy term, the λ-theories $\mathcal{T}_\mathsf{K}$ and $\mathcal{T}_\mathsf{F}$ are consistent, whence so is $\mathcal{T} \triangleq \mathcal{T}_\mathsf{K} \cap \mathcal{T}_\mathsf{F}$. By Corollary 15.19(ii), we conclude that $[\Omega]_\mathcal{T}$ is a non-trivial central element in the term algebra of $\mathcal{T}$ (see Definition 15.30).

We now show that the partial ordering on the central elements defined by

$$e \sqsubseteq d \iff \overline{\vartheta}_e \subseteq \overline{\vartheta}_d \tag{15.7}$$

is a Boolean ordering and that the meet, join and complementation are internally representable. The elements 0 and 1 are respectively the bottom and top of this ordering.

THEOREM 15.22. *Let $\mathcal{A}$ be a Church algebra. The algebra* $(\mathrm{CE}(\mathcal{A}), \sqcap, \sqcup, ^-, 0, 1)$ *of central elements of $\mathcal{A}$, defined by*

$$\begin{aligned} e \sqcap d &\triangleq \mathrm{ite}(e, d, 0), \\ e \sqcup d &\triangleq \mathrm{ite}(e, 1, d), \\ e^- &\triangleq \mathrm{ite}(e, 0, 1), \end{aligned}$$

is a Boolean algebra isomorphic to the Boolean algebra of factor congruences of $\mathcal{A}$.

PROOF. By Corollary 15.18(iii) the map $e \in \mathrm{CE}(\mathcal{A}) \mapsto \overline{\vartheta}_e$ is bijective. By Corollary 15.14 $\mathcal{A}$ has Boolean factor congruences, therefore the partial ordering on central elements defined in (15.7) is a Boolean ordering. We need to show that, for all $e, d \in \mathrm{CE}(\mathcal{A})$, the elements e^-, $e \sqcap d$ and $e \sqcup d$ are central and are respectively associated with the cfc-pairs $(\overline{\vartheta}_e, \vartheta_e)$, $(\vartheta_e \sqcup \vartheta_d, \overline{\vartheta}_e \sqcap \overline{\vartheta}_d)$ and $(\vartheta_e \sqcap \vartheta_d, \overline{\vartheta}_e \sqcup \overline{\vartheta}_d)$.

We start by checking the details for e^-. Since e is central then $(\vartheta_e, \overline{\vartheta}_e)$ is a cfc-pair. The complement of $(\vartheta_e, \overline{\vartheta}_e)$ is the pair $(\overline{\vartheta}_e, \vartheta_e)$. By Lemma 15.16(i), we have that e^- is the unique element such that $0\, \vartheta_e\, e^-\, \overline{\vartheta}_e\, 1$. Then $1\, \overline{\vartheta}_e\, e^-\, \vartheta_e\, 0$ holds for the pair $(\overline{\vartheta}_e, \vartheta_e)$. This means that e^- is the central element associated with $(\overline{\vartheta}_e, \vartheta_e)$.

Consider now $e \sqcup d = \mathrm{ite}(e, 1, d)$. First of all, we show that $\mathrm{ite}(e, 1, d) = \mathrm{ite}(d, 1, e)$. By Lemma 15.16(i) we have that $1\, \vartheta_e\, \mathrm{ite}(e, 1, d)\, \overline{\vartheta}_e\, d$, while $1\, \vartheta_e\, \mathrm{ite}(d, 1, e)\, \overline{\vartheta}_e\, d$ can be obtained as follows:

$$\begin{array}{rcll} 1 & = & \mathrm{ite}(d, 1, 1), & \text{by (15.3) in Proposition 15.17,} \\ \mathrm{ite}(d, 1, 1) & \vartheta_e & \mathrm{ite}(d, 1, e), & \text{by } 1\, \vartheta_e\, e, \\ \mathrm{ite}(d, 1, e) & \overline{\vartheta}_e & \mathrm{ite}(d, 1, 0), & \text{by } e\, \overline{\vartheta}_e\, 0, \\ \mathrm{ite}(d, 1, 0) & = & d, & \text{by (15.6) in Proposition 15.17.} \end{array}$$

By Proposition 15.17(3), there exists a unique element $c \in A$ such that $1\, \vartheta_e\, c\, \overline{\vartheta}_e\, d$ holds. We conclude that $\mathrm{ite}(e, 1, d) = \mathrm{ite}(d, 1, e)$. Now we must show that $\mathrm{ite}(e, 1, d)$ is the central element associated with the factor congruence $\vartheta_e \sqcap \vartheta_d$, i.e.,

$$1\ (\vartheta_e \sqcap \vartheta_d)\ \mathrm{ite}(e, 1, d)\ (\overline{\vartheta}_e \sqcup \overline{\vartheta}_d)\ 0.$$

From $\mathrm{ite}(d, 1, e) = \mathrm{ite}(e, 1, d)$ we easily get that $1\, \vartheta_e\, \mathrm{ite}(e, 1, d)$ and $1\, \vartheta_d\, \mathrm{ite}(e, 1, d)$, that is, $1\ (\vartheta_e \sqcap \vartheta_d)\ \mathrm{ite}(e, 1, d)$. Finally, by Lemma 15.16, we have $\mathrm{ite}(e, 1, d)\, \overline{\vartheta}_e\, d = \mathrm{ite}(d, 1, 0)\, \overline{\vartheta}_d\, 0$, so we conclude $\mathrm{ite}(e, 1, d)\ (\overline{\vartheta}_e \sqcup \overline{\vartheta}_d)\ 0$. Similar reasonings work for $e \sqcap d$. □

NOTATION. *Let $\mathcal{A}$ be a Church algebra.*

(i) *Let I be a maximal ideal of the Boolean algebra* $\mathrm{CE}(\mathcal{A})$. *Then, we define*

$$\varphi_I \triangleq \bigcup_{e\in I} \vartheta_e$$

It is easy to check that $\varphi_I \in \mathrm{Con}(\mathcal{A})$.

(ii) *Let $\mathcal{S}_\mathcal{A}$ be the Boolean space of maximal ideals of the Boolean algebra* $\mathrm{CE}(\mathcal{A})$.

The next theorem first appeared in Salibra et al. (2013) and corrects a partly erroneous statement from Manzonetto and Salibra (2008).

THEOREM 15.23 (STONE REPRESENTATION THEOREM FOR CHURCH ALGEBRAS). *Consider a Church variety $\mathbb{V}$ over a signature Σ.*
(i) *For all $\mathcal{A}\in\mathbb{V}$, the function*

$$f : A \to \prod_{I\in\mathcal{S}_\mathcal{A}} (A/\varphi_I),$$

defined by

$$f(x) \triangleq ([x]_{\varphi_I} \mid I\in\mathcal{S}_\mathcal{A}),$$

gives a weak Boolean product representation of $\mathcal{A}$.

(ii) *If the class $\mathbb{V}_{\mathrm{IND}}$ of directly indecomposable members of $\mathbb{V}$ is a universal class, then the quotient algebras $\mathcal{A}/\varphi_I$ (for $I\in\mathcal{S}_\mathcal{A}$) are directly indecomposable.*

PROOF. (i) By Theorem 15.22 the factor congruences of $\mathcal{A}$ constitute a Boolean sublattice of $\mathrm{Con}(\mathcal{A})$. As shown by Comer (1971), under these hypotheses, f gives a weak Boolean product representation of $\mathcal{A}$.

(ii) It is easy to check that every Church variety $\mathbb{V}$ forms a Pierce variety[9]. Assuming $\mathbb{V}_{\mathrm{IND}}$ is a universal class, we can apply Theorem 8 in Vaggione (1996b) and conclude $\mathcal{A}/\varphi_I \in \mathbb{V}_{\mathrm{IND}}$. For a direct proof, see Salibra et al. (2013). □

The map f of the theorem above does not give—in general—a Boolean product representation. For combinatory algebras, this follows from Proposition 15.13(ii) and two results by Vaggione (1996a) and Plotkin-Simpson (Selinger (1997), Lemma 3.14). Vaggione has shown that, if a variety has factorable congruences and every member of the variety can be represented as a Boolean product of directly indecomposable algebras, then the variety is a discriminator variety. Discriminator varieties satisfy very strong algebraic properties, in particular they are congruence permutable. Plotkin and Simpson have shown that this last property is inconsistent with combinatory logic, hence by Vaggione's theorem not all combinatory algebras have a Boolean product representation.

[9] By a *Pierce variety* we mean here a variety of algebras having two constants $0, 1$ and a 4-ary term $t(x, y, w, z)$ satisfying $t(x, y, 0, 1) = x$ and $t(x, y, 1, 0) = y$. See Vaggione (1996b) for the general definition.

15.3 Easiness in universal algebra

As recalled in Chapter 3, in λ-calculus there are easy terms, i.e., λ-terms that can be consistently equated with any other closed λ-term. One could generalize the notion of easiness to subsets and elements of a Church algebra, but for our purposes a weaker property is sufficient. We show that any Church algebra $\mathcal{A}$ with a weakly easy set of cardinality n admits a congruence φ such that the interval $[\varphi, \nabla^{\mathcal{A}}]$ of all congruences greater than φ is isomorphic to the lattice reduct of the free Boolean algebra with n generators.

DEFINITION 15.24. Let $\mathcal{A}$ be a Church algebra and $X \subseteq A$.

(i) For every $Y \subseteq X$, define

$$\delta_Y \triangleq \vartheta(1, Y) \vee \vartheta(0, X - Y) \tag{15.8}$$

Recall from Definition 15.2(ii), that $\vartheta(1, \emptyset) = \vartheta(0, \emptyset) = \Delta^{\mathcal{A}}$.

(ii) We say that X is a *weakly easy subset of* $\mathcal{A}$ if $\delta_Y \neq \nabla^{\mathcal{A}}$, for every $Y \subseteq X$.

(iii) An element $a \in A$ is called *weakly easy* if $\{a\}$ is a weakly easy subset of $\mathcal{A}$.

Note that a is weakly easy if and only if the congruences ϑ_a and $\overline{\vartheta}_a$ are both different from $\nabla^{\mathcal{A}}$. The relationship with easiness in λ-calculus is discussed in Section 15.4.

EXAMPLES 15.25. (i) A finite subset X of a Boolean algebra $\mathcal{B}$ is a weakly easy subset of $\mathcal{B}$ if the following conditions hold:

(1) $\bigvee X \neq 1$;

(2) $\bigwedge X \neq 0$;

(3) for all $Y \subsetneq X$, we have $\bigvee Y \not\geq \bigwedge(X - Y)$.

For example, $\{\{1,2\},\{2,3\}\}$ is a weakly easy subset of the powerset of $\{1,2,3,4\}$.

(ii) The term model of every r.e. λ-theory has a countable infinite weakly easy set. This will be shown in Section 15.4 (Lemma 15.35).

The following lemmas are used in the proof of Theorem 15.29.

LEMMA 15.26. *The congruences of a Church algebra $\mathcal{A}$ permute with its factor congruences, i.e., $\varphi \circ \psi = \psi \circ \varphi$ for every congruence φ and factor congruence ψ.*

PROOF. Since ψ is a factor congruence we have $\psi = \vartheta_e$, for some element $e \in \mathrm{CE}(\mathcal{A})$. Consider $a, c \in A$ such that $a \,\varphi\, b \,\vartheta_e\, c$, for some $b \in A$. We show $a \,\vartheta_e\, \mathrm{ite}(e,a,c) \,\varphi\, c$. First, we have $a \,\vartheta_e\, \mathrm{ite}(e,a,c)$, by Lemma 15.16(i). Since $a \varphi b$, we obtain $\mathrm{ite}(e,a,c) \,\varphi\, \mathrm{ite}(e,b,c)$. From $b \,\vartheta_e\, c$ and Proposition 15.17(3) we conclude $\mathrm{ite}(e,b,c) = c$. □

For the definition of the Zipper condition, see Definition A.79 in the appendix.

LEMMA 15.27. *Let $\mathcal{A}$ be a Church algebra. Then* $\mathrm{Con}(\mathcal{A})$ *satisfies the Zipper condition.*

PROOF. It was shown by Lampe (1986) that the congruence lattice of every algebra having a binary term with a right-unit and a right-zero satisfies the Zipper condition. □

LEMMA 15.28. *Let $\mathcal{A}$ be a Church algebra and $\varphi \in \mathrm{Con}(\mathcal{A})$. Then $\mathcal{A}/\varphi$ is also a Church algebra and the map $c_\varphi : \mathrm{CE}(\mathcal{A}) \to \mathrm{CE}(\mathcal{A}/\varphi)$, defined by*

$$c_\varphi(x) \triangleq [x]_\varphi$$

is a homomorphism of Boolean algebras.

PROOF. It is not difficult to show that c_φ is a homomorphism with respect to the Boolean operations defined in Theorem 15.22. □

THEOREM 15.29 (MANZONETTO AND SALIBRA (2008)). *Let $\mathcal{A}$ be a Church algebra, $X \subseteq A$ be a weakly easy subset of $\mathcal{A}$ and $\mathcal{B}(X)$ be the free Boolean algebra over the set X of generators. Then there exists a congruence $\varphi_X \in \mathrm{Con}(\mathcal{A})$ satisfying the following conditions:*

1. *The lattice reduct of $\mathcal{B}(X)$ can be embedded into the interval sublattice $[\varphi_X, \nabla^{\mathcal{A}}]$ of $\mathrm{Con}(\mathcal{A})$;*

2. *If X has finite cardinality n then the above embedding is an isomorphism, hence $[\varphi_X, \nabla^{\mathcal{A}}]$ has cardinality 2^{2^n}.*

PROOF. 1) Let $\psi \triangleq \bigwedge_{Y \subseteq X} \delta_Y$, where δ_Y has been defined in (15.8). We define φ_X as any maximal element of the set of all congruences φ which contain ψ and are compatible with each δ_Y, i.e., $\varphi \vee \delta_Y \neq \nabla^{\mathcal{A}}$. Note that φ_X exists by Zorn's Lemma.

Claim 15.29.1. For every $x \in X$, $[x]_{\varphi_X} \in \mathrm{CE}(\mathcal{A}/\varphi_X)$.

Subproof. If we prove that $[x]_\psi$ is central in $\mathcal{A}/\psi$, then by $\psi \subseteq \varphi_X$ and by Lemma 15.28 we get the conclusion of the claim. Since $x \in X$ is equivalent either to 1 or to 0 in each congruence δ_Y, then $[x]_{\delta_Y} \in \mathrm{CE}(\mathcal{A}/\delta_Y)$. Now, $\mathcal{A}/\psi$ is a subdirect product of $\prod_{Y \subseteq X} \mathcal{A}/\delta_Y$, whence $[x]_\psi$ is central in $\mathcal{A}/\psi$ by Theorem 5(e) in Vaggione (1996b). ■

Consider now the unique Boolean homomorphism $f_X : \mathcal{B}(X) \to \mathrm{CE}(\mathcal{A}/\varphi_X)$ satisfying, for all $x \in X$:

$$f_X(x) = [x]_{\varphi_X}.$$

Claim 15.29.2. f_X is an embedding.

Subproof. Let $Y \subseteq X$. Recall that $\varphi_X \vee \delta_Y \neq \nabla^{\mathcal{A}}$. By Lemma 15.28 there exists a Boolean homomorphism, denoted here by h_Y, from $\mathrm{CE}(\mathcal{A}/\varphi_X)$ into $\mathrm{CE}(\mathcal{A}/\varphi_X \vee \delta_Y)$. Since $(x, 1) \in \varphi_X \vee \delta_Y$ for every $x \in Y$, and $(y, 0) \in \varphi_X \vee \delta_Y$ for every $y \in X - Y$, then the kernel of $h_Y \circ f_X$ is an maximal ideal of $\mathcal{B}(X)$. By the arbitrariness of $Y \subseteq X$, every maximal ideal of $\mathcal{B}(X)$ can be the kernel of a suitable $h_Y \circ f_X$. This is possible only if f_X is an embedding. ■

It follows that the lattice reduct of $\mathcal{B}(X)$ can be embedded into the interval $[\varphi, \nabla^{\mathcal{A}}]$.

2) Hereafter, we assume that X has finite cardinality n. Then $\mathcal{B}(X)$ is finite, atomic, has n generators, 2^n atoms, 2^n coatoms and size 2^{2^n}. Recall that the lattice $\mathrm{Con}(\mathcal{A}/\varphi_X)$ is isomorphic to $[\varphi_X, \nabla^{\mathcal{A}}]$. Let us denote by At_X be the set of atoms of $\mathrm{Con}(\mathcal{A}/\varphi_X)$.

Claim 15.29.3. $\bigvee\{\alpha \in \mathrm{At}_X \mid \alpha \text{ is a factor congruence}\} = \nabla^{\mathcal{A}}$.

Subproof. Let $v \in \mathrm{CE}(\mathcal{A}/\varphi_X)$ be such that $v = f_X(u)$, for some atom $u \in \mathcal{B}(X)$. Now, consider $\tau \in \mathrm{Con}(\mathcal{A})$ such that $\tau/\varphi_X = \vartheta(v,0) \in \mathrm{Con}(\mathcal{A}/\varphi_X)$. We claim that $\tau/\varphi_X \in \mathrm{At}_X$. By the way of contradiction, assume that there exists $\sigma \in \mathrm{Con}(\mathcal{A})$ such that $\varphi_X \subsetneq \sigma \subsetneq \tau$. By Lemma 15.28, we have a sequence of Boolean homomorphisms:

$$\mathcal{B}(X) \xrightarrow{f_X} \mathrm{CE}(\mathcal{A}/\varphi_X) \xrightarrow{c_\sigma} \mathrm{CE}(\mathcal{A}/\sigma) \xrightarrow{c_{\tau/\sigma}} \mathrm{CE}(\mathcal{A}/\tau)$$

and a Boolean homomorphism $c_\tau : \mathrm{CE}(\mathcal{A}/\varphi_X) \xrightarrow{c_\tau} \mathrm{CE}(\mathcal{A}/\tau)$ such that $c_\tau = c_{\tau/\sigma} \circ c_\sigma$. Since u is an atom of $\mathcal{B}(X)$, then the set $\{0,u\}$ is the Boolean ideal associated with the kernel of $c_\tau \circ f_X$. If $c_\sigma(v) = 0$, then σ/φ_X contains the pair $(v,0)$, i.e., $\sigma = \tau$. It follows that $c_\sigma(v) \neq 0$ and that the following map is an embedding:

$$c_\sigma \circ f_X : \mathcal{B}(X) \to \mathrm{CE}(\mathcal{A}/\sigma).$$

Since $\mathcal{B}(X)$ is free over X, for every $Y \subseteq X$ there exists an atom $w \in \mathcal{B}(X)$ such that

$$w = \Big(\bigwedge Y\Big) \wedge \Big(\bigwedge\{x^- \mid x \in X - Y\}\Big).$$

Let $w' = c_\sigma(f_X(w)) \in \mathcal{A}/\sigma$ be the corresponding non-trivial central element. By definition of w, the non-triviality of the factor congruence $\vartheta(w',1) \in \mathrm{Con}(\mathcal{A}/\sigma)$ is equivalent to require that $\sigma \vee \delta_Y \neq \nabla^{\mathcal{A}}$. The arbitrariness of Y and the strict inclusion $\varphi_X \subsetneq \sigma$ contradict the maximality of φ_X. In conclusion $\tau/\varphi_X \in \mathrm{At}_X$. Finally, the claim follows because the join of all atoms of $\mathcal{B}(X)$ is the top element. ■

Let Co_X be the set of coatoms of $\mathrm{Con}(\mathcal{A}/\varphi_X)$. We say that the coatoms *form a finite irredundant decomposition* of $\Delta^{\mathcal{A}}$ if

- Co_X is finite,
- $\bigwedge \mathrm{Co}_X = \Delta^{\mathcal{A}}$, and
- $\bigwedge(\mathrm{Co}_X - \{\psi\}) \neq \Delta^{\mathcal{A}}$, for every $\psi \in \mathrm{Co}_X$.

Claim 15.29.4. $\mathrm{Con}(\mathcal{A}/\varphi_X)$ is pseudo-complemented, complemented, atomic, and the coatoms form a finite irredundant decomposition of Δ.

Subproof. By Lemma 15.27, $\mathrm{Con}(\mathcal{A}/\varphi_X)$ satisfies the Zipper condition. By Claim 15.29.3, we have $\bigvee \mathrm{At}_X = \nabla^{\mathcal{A}}$. Then, by Proposition 2 in Diercks et al. (1994), $\mathrm{Con}(\mathcal{A}/\varphi_X)$ is complemented, atomic and every coatom has a complement which is an atom. It is also pseudo-complemented by Proposition 1 (*ibidem*). As the top element $\nabla^{\mathcal{A}}$ is compact, by Proposition 3 (*ibidem*), the coatoms form a finite irredundant decomposition of $\Delta^{\mathcal{A}}$. ■

Claim 15.29.5. Let $\xi \in \mathrm{Con}(\mathcal{A}/\varphi_X)$ be a non-trivial congruence and assume

$$\gamma = \bigvee\{\delta \in \mathrm{At}_X \mid \delta \leq \xi\}.$$

If $\alpha \in \mathrm{At}_X$ is a factor congruence and $\alpha \not\leq \xi$, then $\xi \wedge (\alpha \vee \gamma) = \gamma$.

Subproof. We always have $\gamma \leq \xi \wedge (\alpha \vee \gamma)$. We now show the opposite direction. Consider $(x, y) \in \xi \wedge (\alpha \vee \gamma)$, i.e., $x \,\xi\, y$ and $x \,(\alpha \vee \gamma)\, y$. We have to show that $x \,\gamma\, y$. Since α is a factor congruence, by Lemma 15.26 we have $\alpha \vee \gamma = \alpha \circ \gamma$. Then $x \,\alpha\, z \,\gamma\, y$ for some z. Since $\gamma \leq \xi$ and $z \,\gamma\, y$ then $z \,\xi\, y$, that together with $x \,\xi\, y$ implies $x \,\xi\, z$. Hence $x(\xi \wedge \alpha)z$. Since α is an atom and $\alpha \not\leq \xi$, then $\xi \wedge \alpha = \Delta^{\mathcal{A}}$, so that $x = z$. This last equality and $z \,\gamma\, y$ imply $x \,\gamma\, y$. In other words, $\xi \wedge (\alpha \vee \gamma) = \gamma$. ■

Claim 15.29.6. Every $\xi \in \mathrm{Con}(\mathcal{A}/\varphi_X)$ is a join of atoms.

Subproof. Let At_ξ be the set of atoms included in ξ. We are going to show that $\xi = \bigvee \mathrm{At}_\xi$. Let $\gamma = \bigvee \mathrm{At}_\xi$. By Claim 15.29.5 we have:

$$\begin{aligned}\bigvee\{\nu \mid \xi \wedge \nu = \gamma\} &\geq \bigvee\{\alpha \vee \gamma \mid \alpha \in \mathrm{At}_X, \alpha \not\leq \xi,\ \alpha \text{ is a factor congruence}\}\\ &\geq \bigvee\{\alpha \mid \alpha \in \mathrm{At}_X \text{ is a factor congruence}\}\end{aligned}$$

By Claim 15.29.3 this last element is equal to $\nabla^{\mathcal{A}}$, therefore we get $\bigvee\{\nu \mid \xi \wedge \nu = \gamma\} = \nabla^{\mathcal{A}}$. By applying the Zipper condition (see Definition A.79) this entails $\xi = \gamma$. ■

Claim 15.29.7. $\mathrm{Con}(\mathcal{A}/\varphi_X)$, and hence $[\varphi_X, \nabla^{\mathcal{A}}]$, is isomorphic to the power set of At_X.

Subproof. By Claim 15.29.4 $\mathrm{Con}(\mathcal{A}/\varphi_X)$ is atomic and pseudo-complemented, so that each atom is completely join-prime. By this fact and Claim 15.29.6, every element is univocally represented as a join of atoms. The conclusion follows because every join of atoms exists by completeness. ■

Claim 15.29.8. $\mathrm{Con}(\mathcal{A}/\varphi_X)$, and hence $[\varphi_X, \nabla^{\mathcal{A}}]$, has 2^n coatoms and 2^n atoms.

Subproof. Since $\varphi_X \vee \delta_Y \neq \nabla^{\mathcal{A}}$ for every $Y \subseteq X$, $[\varphi_X, \nabla^{\mathcal{A}}]$ has at least 2^n coatoms. For every $Y \subseteq X$, let c_Y be a coatom including $\varphi_X \vee \delta_Y$. Assume now that there is a coatom ξ distinct from each c_Y, for every $Y \subseteq X$. Consider the intersection $\bigwedge(\mathrm{Co}_X - \{\xi\})$, where Co_X is the set of coatoms of $[\varphi_X, \nabla^{\mathcal{A}}]$. By Claim 15.29.4, we have $\bigwedge(\mathrm{Co}_X - \{\xi\}) \neq \varphi_X$. This contradicts the maximality of φ_X among the congruences which contains $\bigwedge_{Y \leq X} \delta_Y$ and are compatible with δ_Y. Thus, there are 2^n coatoms, and hence 2^n atoms. ■

This concludes the proof of the main theorem. □

15.4 Applications to λ-calculus models and theories

We have seen in Example 15.12 that every combinatory algebra is a Church algebra, by setting $\mathrm{ite}(x, y, z) \triangleq (x \cdot y) \cdot z$, $1 \triangleq \mathbf{t}$ and $0 \triangleq \mathbf{f}$. Therefore, we can apply the algebraic theory developed in the previous sections to the study of models and theories of λ-calculus.

Finite Boolean sublattices of $\lambda\mathcal{T}$

We are going to show that a consequence of Theorem 15.29 is the fact that $\lambda\mathcal{T}$ admits (at the top) Boolean lattice intervals of cardinality 2^k, for every $k \in \mathbb{N}$. This theorem is applicable since $\lambda\mathcal{T}$ is isomorphic to the congruence lattice of the term algebra of $\boldsymbol{\lambda}$.

DEFINITION 15.30. The *term algebra of a λ-theory* $\mathcal{T}$ is the algebraic structure

$$\mathcal{A}(\mathcal{T}) = (\Lambda/\mathcal{T}, \cdot, \lambda_x, \underline{x})_{x \in \mathrm{Var}}$$

where $\cdot$ is a binary operator, $\langle \lambda_x \rangle_{x \in \mathrm{Var}}$ are unary operators and $\langle \underline{x} \rangle_{x \in \mathrm{Var}}$ are nullary operators defined as follows (for all $M, N \in \Lambda$ and $x \in \mathrm{Var}$):

$$\begin{aligned} \underline{x} &\triangleq [x]_{\mathcal{T}}; \\ [M]_{\mathcal{T}} \cdot [N]_{\mathcal{T}} &\triangleq [MN]_{\mathcal{T}}; \\ \lambda_x([M]_{\mathcal{T}}) &\triangleq [\lambda x.M]_{\mathcal{T}}. \end{aligned}$$

Salibra (2000) has shown that the variety generated by the term algebra of $\boldsymbol{\lambda}$ is axiomatized by the finite schema of identities characterizing lambda abstraction algebras. For more information, see Pigozzi and Salibra (1998), Salibra and Goldblatt (1999), Salibra (2001a), Lusin and Salibra (2004) and Manzonetto and Salibra (2010).

The following lattice isomorphisms straightforwardly ensue from Definition 15.30.

LEMMA 15.31. (i) *For every λ-theory* $\mathcal{T}$, *we have* $\mathrm{Con}(\mathcal{A}(\mathcal{T})) \cong [\mathcal{T}, \nabla] \subseteq \lambda\mathcal{T}$.

(ii) *In particular,* $\lambda\mathcal{T} \cong \mathrm{Con}(\mathcal{A}(\boldsymbol{\lambda}))$.

The notions of weakly easy elements and weakly easy subsets of a Church algebra given in Definition 15.24 will play a central role in this section, once rephrased for λ-terms. Recall that, given a set $X \subseteq \Lambda \times \Lambda$ (intended as a set of equations), we let $\mathcal{T}(X)$ denote the least λ-theory containing $\mathcal{T} \cup X$, and we write $\mathcal{T}(M = N)$ for $\mathcal{T}(\{(M, N)\})$.

DEFINITION 15.32. Let $\mathcal{T} \in \lambda\mathcal{T}$.

(i) A λ-term $M \in \Lambda^o$ is *$\mathcal{T}$-easy* if $\mathcal{T}(M = N)$ is consistent, for all $N \in \Lambda^o$.

(ii) We denote by $\Lambda^e_{\mathcal{T}}$ the set of all $\mathcal{T}$-easy terms.

(iii) A λ-term $M \in \Lambda^o$ is *weakly $\mathcal{T}$-easy* if $[M]_{\mathcal{T}}$ is a weakly easy element of $\mathcal{A}(\mathcal{T})$.

(iv) A set $X \subseteq \Lambda^o$ is *weakly $\mathcal{T}$-easy* if $X/\mathcal{T}$ is a weakly easy subset of $\mathcal{A}(\mathcal{T})$.

Note that a λ-term is $\boldsymbol{\lambda}$-easy exactly when it is easy in the sense of Definition 3.7(i). In fact, our definition of $\mathcal{T}$-easiness coincides with the one given in B[1984], p. 434. Clearly every $\mathcal{T}$-easy term is also weakly easy, while the converse is false, as suggested by the terminology. Berarducci and Intrigila (1993) proved that there exists a λ-term M which is not easy, despite the fact that $\boldsymbol{\lambda}(M = \mathsf{T})$ and $\boldsymbol{\lambda}(M = \mathsf{F})$ are both consistent.

REMARK 15.33. (i) If $M \in \Lambda^o$ (resp. $X \subseteq \Lambda^o$) is weakly $\mathcal{T}$-easy then it is also a weakly $\mathcal{T}$-easy element (resp. set) of the combinatory reduct of $\mathcal{A}(\mathcal{T})$, namely $\mathcal{M}(\mathcal{T})$.

(ii) If $[M]_{\mathcal{T}}$ is a central element of $\mathcal{A}(\mathcal{T})$, then it is also a central element of $\mathcal{M}(\mathcal{T})$. This follows from the combinatory completeness of combinatory algebras.

PROPOSITION 15.34. *Let $\mathcal{T}$ be an r.e. λ-theory.*

(i) *Consider $k \in \mathbb{N}$ and $P_i, Q_i \in \Lambda^o$ such that $\mathcal{T} \nvdash P_i = Q_i$, for all $i\,(1 \le i \le k)$. Then*

$$\exists M \in \Lambda^o, \forall N \in \Lambda^o, \forall i \in \{1, \dots, k\} \,.\, \mathcal{T}(M = N) \nvdash P_i = Q_i.$$

(ii) $\Lambda^e_{\mathcal{T}} \neq \emptyset$.

PROOF. (i) In B[1984], Proposition 17.1.9 (due to Visser), the particular case $k = 1$ is shown. The general case follows from a similar argument.

(ii) By (i). □

LEMMA 15.35. *Let $\mathcal{T} \in \lambda\mathcal{T}$ be r.e. Then, there exists an infinite weakly $\mathcal{T}$-easy set.*

PROOF. Since $\mathcal{T}$ is r.e., we can apply Proposition 15.34(ii) and obtain some $P \in \Lambda^e_{\mathcal{T}}$. We prove that $X \triangleq \{P\mathsf{c}_n \mid n \in \mathbb{N}\}$ is a weakly $\mathcal{T}$-easy set. For every $A \subseteq \mathbb{N}$, we need to show

$$\mathcal{T}_A \triangleq \mathcal{T}(\{P\mathsf{c}_n = \mathsf{T} \mid n \in A\} \cup \{P\mathsf{c}_n = \mathsf{F} \mid n \in (\mathbb{N} - A)\}) \neq \nabla$$

We prove that $\mathcal{T}(\{P\mathsf{c}_n = \mathsf{T} \mid n \in B \cap A\} \cup \{P\mathsf{c}_n = \mathsf{F} \mid n \in B \cap (\mathbb{N} - A)\})$ is consistent, for every $B \in \mathscr{P}_{\mathrm{f}}(\mathbb{N})$. Since B is finite, there exists a λ-term $Q \in \Lambda^o$ satisfying

$$Q\mathsf{c}_n =_\beta \begin{cases} \mathsf{T}, & \text{if } n \in B \cap A, \\ \mathsf{F}, & \text{if } n \in B \cap (\mathbb{N} - A). \end{cases}$$

From the $\mathcal{T}$-easiness of P, it follows that $\mathcal{T}' \triangleq \mathcal{T}(P = Q)$ is consistent. Then we have $P\mathsf{c}_n =_{\mathcal{T}'} Q\mathsf{c}_n =_\beta \mathsf{T}$, for all $n \in B \cap A$, and $P\mathsf{c}_n =_{\mathcal{T}'} Q\mathsf{c}_n =_\beta \mathsf{F}$, for all $n \in B \cap (\mathbb{N} - A)$. We conclude that $\mathcal{T}_A$ is consistent by a compactness argument. □

THEOREM 15.36 (MANZONETTO AND SALIBRA (2008)). *Let $\mathcal{T} \in \lambda\mathcal{T}$ and $k \in \mathbb{N}$. If $\mathcal{T}$ is r.e., then there exists a λ-theory $\mathcal{T}_k \supseteq \mathcal{T}$ such that the lattice interval $[\mathcal{T}_k, \nabla]$ is isomorphic to the finite Boolean lattice with 2^k elements.*

PROOF. By Lemma 15.35 and Theorem 15.29, there exists a λ-theory $\mathcal{S}_k$ such that $\mathcal{S}_k \supseteq \mathcal{T}$ and $[\mathcal{S}_k, \nabla]$ is isomorphic to the free Boolean algebra with 2^{2^k} elements. The λ-theory $\mathcal{T}_k$ can be defined by using $\mathcal{S}_k$ and the following facts:

(a) every filter of a finite Boolean algebra is principle, and therefore a Boolean lattice;

(b) the free Boolean algebra with 2^{2^k} elements has filters of arbitrary cardinality 2^i (for $i \le 2^k$). □

Note that the λ-theory $\mathcal{T}_k$ in the statement of Theorem 15.36 cannot be r.e., otherwise the lattice interval $[\mathcal{T}_k, \nabla]$ would have a continuum of elements by Theorem 3.5(vii). The next proposition shows that the result above cannot be generalized to the infinite case.

PROPOSITION 15.37 (MANZONETTO AND SALIBRA (2008)). *There exists no λ-theory $\mathcal{T}$ such that the interval sublattice $[\mathcal{T}, \nabla]$ is isomorphic to an infinite Boolean lattice.*

PROOF. By Proposition 4 in Diercks et al. (1994), a complete coatomic Boolean lattice satisfying the Zipper condition, and whose top element is compact, must be finite. □

The indecomposable semantics

In order to apply the Representation Theorem 15.23 to the variety $\mathbb{CA}$ of combinatory algebras, we need to show that the subclass $\mathbb{CA}_{\mathrm{IND}}$ of directly indecomposable combinatory algebras is a universal class.

Recall from Proposition 15.17 that any factor congruence can be internally represented by a central element. In particular, a combinatory algebra $\mathcal{C}$ is directly indecomposable if and only if $\mathrm{CE}(\mathcal{C}) = \{\mathbf{t}, \mathbf{f}\}$. From this, we now deduce that "being a directly indecomposable combinatory algebra" can be expressed by a universal formula.

PROPOSITION 15.38. *The class $\mathbb{CA}_{\mathrm{IND}}$ is a universal class.*

PROOF. Define the following combinatory terms:

$$\begin{aligned}\mathbf{z} &\triangleq \lambda^* e.[\lambda^* x.exx, \lambda^* xyz.e(exy)z, \lambda^* xyz.exz, \lambda^* xyzu.e(xy)(zu), e\mathbf{tf}];\\ \mathbf{u} &\triangleq \lambda^* e.[\lambda^* x.x, \lambda^* xyz.exz, \lambda^* xyz.ex(eyz), \lambda^* xyzu.exz(eyu), e].\end{aligned}$$

By Proposition 15.17, an element e is central if and only if the equation $\mathbf{z}e = \mathbf{u}e$ holds. Therefore the class $\mathbb{CA}_{\mathrm{IND}}$ can be axiomatized by the following universal formula Ψ:

$$\Psi \triangleq \forall e.\Big(\big(\mathbf{z}e = \mathbf{u}e \Rightarrow (e = \mathbf{t} \vee e = \mathbf{f})\big) \wedge \neg(\mathbf{t} = \mathbf{f})\Big).$$

It follows that $\mathbb{CA}_{\mathrm{IND}}$ is a universal class. □

It is well known that every universal class enjoys the following properties.

COROLLARY 15.39. *The class $\mathbb{CA}_{\mathrm{IND}}$ is closed under subalgebras and ultraproducts.*

The Representation Theorem below, first proved by Manzonetto and Salibra (2006), is an immediate consequence of Theorem 15.23 and Proposition 15.38. Recall that $\mathcal{S}_{\mathcal{C}}$ denotes the Boolean space of maximal ideals of $\mathrm{CE}(\mathcal{C})$ and $\varphi_I = \bigcup_{e \in I} \vartheta_e$, for $I \in \mathcal{S}_{\mathcal{C}}$.

THEOREM 15.40 (REPRESENTATION THEOREM FOR COMBINATORY ALGEBRAS). *Let $\mathcal{C}$ be a combinatory algebra. Then the map*

$$f : C \to \prod_{I \in \mathcal{S}_{\mathcal{C}}} (C/\varphi_I),$$

defined by

$$f(x) \triangleq ([x]_{\varphi_I} \mid I \in \mathcal{S}_{\mathcal{C}}),$$

gives a weak Boolean product representation of $\mathcal{C}$, where $\mathcal{C}/\varphi_I$ are directly indecomposable.

The above theorem can be roughly summarized as follows:

> The directly indecomposable combinatory algebras constitute the 'building blocks' in the variety of combinatory algebras.

Therefore, it is natural to investigate the models of λ-calculus that are directly indecomposable as combinatory algebras. We call this class the *indecomposable semantics*.

DEFINITION 15.41 (INDECOMPOSABLE SEMANTICS). Let $\mathbb{IND} \subseteq \mathbb{CA}_{\mathrm{IND}}$ be the class of λ-models that are directly indecomposable as combinatory algebras.

Let us consider the following questions that arise naturally when introducing a new semantics[10] $\mathbb{C}$ of λ-calculus:

Q1. Is the semantics $\mathbb{C}$ complete? In other words, does every λ-theory arise as the theory of a model living[11] in this semantics?

Q2. If the semantics $\mathbb{C}$ is not complete, what can we say about the size of its incompleteness?

Q3. Does $\mathbb{C}$ contain any previously known model of λ-calculus?

In order to formalize precisely the mathematical content of the questions above, we introduce some notions and terminology.

DEFINITION 15.42. Given a class $\mathbb{C}$ of λ-models and a λ-theory $\mathcal{T}$, we say that:
(i) $\mathbb{C}$ *represents* $\mathcal{T}$ if there is some $\mathcal{M} \in \mathbb{C}$ such that $\mathrm{Th}(\mathcal{M}) = \mathcal{T}$.
(ii) $\mathbb{C}$ *omits* $\mathcal{T}$ if $\mathbb{C}$ does not represent $\mathcal{T}$.
(iii) $\mathbb{C}$ *omits* a subset $\mathcal{X} \subseteq \lambda\mathcal{T}$ whenever $\mathbb{C}$ omits all the λ-theories belonging to $\mathcal{X}$.
We denote by $\lambda\mathbb{C}$ the set of λ-theories that are represented in $\mathbb{C}$.

Recall that non-trivial λ-models may only represent consistent λ-theories.

DEFINITION 15.43. A semantics $\mathbb{C}$ is called:
(i) *equationally complete* if it represents all consistent λ-theories;
(ii) *equationally incomplete* if it is not equationally complete, i.e., if it omits a consistent λ-theory.

We start by showing that the indecomposable semantics is equationally incomplete. The closure of $\mathbb{CA}_{\mathrm{IND}}$ under subalgebras is crucial for proving the next lemma, from which the incompleteness of the indecomposable semantics follows.

PROPOSITION 15.44. *Given a λ-theory $\mathcal{T}$ the following conditions are equivalent:*

1. *$\mathcal{M}_{\mathcal{T}}$ has a non-trivial central element;*
2. *$\mathcal{M}_{\mathcal{T}}$ is directly decomposable;*
3. *all λ-models $\mathcal{M}$ such that $\mathrm{Th}(\mathcal{M}) = \mathcal{T}$ are directly decomposable;*
4. *the indecomposable semantics $\mathbb{IND}$ omits $\mathcal{T}$.*

PROOF. (1 ⇔ 2) By Proposition 15.17.

(2 ⇔ 3) Given a λ-model $\mathcal{M}$, $\mathrm{Th}(\mathcal{M}) = \mathcal{T}$ holds exactly when the term model $\mathcal{M}(\mathcal{T})$ is isomorphic to the minimal subalgebra of $\mathcal{M}$. Thus, the equivalence follows by Corollary 15.39, since the class $\mathbb{CA}_{\mathrm{IND}}$ is closed under subalgebras.

(3 ⇔ 4) By Definitions 15.41 and 15.42(ii). □

[10]By 'semantics' in this chapter we intend a uniformly presented class of (non-trivial) λ-models.
[11]We say that a model $\mathcal{C}$ lives in the semantics $\mathbb{C}$ whenever $\mathcal{C}$ belongs to the class $\mathbb{C}$.

REMARK 15.45. In every λ-model the λ-terms $\mathsf{T} = \lambda xy.x$ and $\mathsf{F} = \lambda xy.y$ are respectively interpreted as **t** and **f**, whence they play the role of trivial central elements.

We now exploit the properties of $\mathcal{T}$-easy terms in order to construct λ-theories whose term model possesses non-trivial central elements. We mainly consider the usual notion of easiness since it has been largely studied in the literature, but in most cases the existence of weakly $\mathcal{T}$-easy terms would be enough for our purposes.

LEMMA 15.46. *Let $M \in \Lambda^o$ and $\mathcal{T}_1, \mathcal{T}_2$ be two consistent λ-theories satisfying*

$$\mathcal{T}_1 \vdash M = \mathsf{T}, \qquad \mathcal{T}_2 \vdash M = \mathsf{F}.$$

Then, for $\mathcal{T} \triangleq \mathcal{T}_1 \cap \mathcal{T}_2$, $[M]_\mathcal{T}$, is a non-trivial central element of $\mathcal{M}(\mathcal{T})$.

PROOF. Since $[\mathsf{T}]_{\mathcal{T}_1}$ and $[\mathsf{F}]_{\mathcal{T}_2}$ are central elements in $\mathcal{M}(\mathcal{T}_1)$ and $\mathcal{M}(\mathcal{T}_2)$, respectively, the λ-theory $\mathcal{T}$ contains all the equations (15.3)–(15.6) of Proposition 15.17, for $e \triangleq [M]_\mathcal{T}$. As a consequence, we obtain $[M]_\mathcal{T} \in \mathrm{CE}(\mathcal{M}(\mathcal{T}))$. Moreover, the central element $[M]_\mathcal{T}$ is non-trivial since $\mathcal{T} \nvdash M = \mathsf{T}$ and $\mathcal{T} \nvdash M = \mathsf{F}$. Conclude by Corollary 15.19(ii). □

The following theorem was first proved by Manzonetto and Salibra (2006).

THEOREM 15.47 (ALGEBRAIC INCOMPLETENESS).
The indecomposable semantics is equationally incomplete.

PROOF. By Proposition 15.44 it is sufficient to construct a λ-theory $\mathcal{T}$ such that $\mathcal{M}(\mathcal{T})$ has a non-trivial central element. Since Ω is an easy term (Example 3.8(i)), the λ-theories $\mathcal{T}_1 \triangleq \boldsymbol{\lambda}(\Omega = \mathsf{T})$ and $\mathcal{T}_2 \triangleq \boldsymbol{\lambda}(\Omega = \mathsf{F})$ are consistent. By Lemma 15.46, $[\Omega]_\mathcal{T}$ is a non-trivial central element of the term model $\mathcal{M}_\mathcal{T}$, for $\mathcal{T} \triangleq \mathcal{T}_1 \cap \mathcal{T}_2$. □

It turns out that, although the indecomposable semantics is equationally incomplete, it is large enough to represent all semi-sensible λ-theories.

LEMMA 15.48. *Let $\mathcal{T}$ be a consistent λ-theory and $M \in \Lambda^o$. If $[M]_\mathcal{T}$ is a non-trivial central element of $\mathcal{M}(\mathcal{T})$, then $[M]_\mathcal{T} \subseteq \mathrm{UNS}$.*

PROOF. Let $e \triangleq [M]_\mathcal{T} \in \mathrm{CE}(\mathcal{M}(\mathcal{T}))$ be non-trivial. Then, the congruences $\vartheta_e = \vartheta(e, [\mathsf{T}]_\mathcal{T})$ and $\overline{\vartheta}_e = \vartheta(e, [\mathsf{F}]_\mathcal{T})$ on $\mathcal{M}_\mathcal{T}$ are non-trivial. It follows that the λ-theories $\mathcal{T}(U = \mathsf{T})$ and $\mathcal{T}(U = \mathsf{F})$ are consistent, for every λ-term $U \in e$. If U is solvable then its hnf should be simultaneously similar to T and to F, i.e., $\mathsf{T} \sim U \sim \mathsf{F}$. Contradiction. □

THEOREM 15.49. *The indecomposable semantics represents all semi-sensible λ-theories.*

PROOF. Let $\mathcal{T}$ be a semi-sensible λ-theory. By the way of contradiction, assume that its term model $\mathcal{M}(\mathcal{T})$ is directly decomposable. By Proposition 15.44, $\mathcal{M}(\mathcal{T})$ admits a non-trivial central element e. By (15.3), we know that $\mathcal{M}_\mathcal{T} \models exx = x$. In particular, for all closed λ-term $U \in e$, we have $\mathcal{T} \vdash U\mathsf{II} = \mathsf{I}$. By Lemma 15.48, U is unsolvable and this entails $U\mathsf{II} \in \mathrm{UNS}$. Since $\mathcal{T} \vdash U\mathsf{II} = \mathsf{I}$, $\mathcal{T}$ cannot be semi-sensible. Contradiction. □

15.5 The main semantics are hugely incomplete

The categorical models of λ-calculus can be classified into 'semantics' according to the nature of the morphisms of their underlying categories. We adopt the terminology introduced by Manzonetto and Salibra (2006) and use 'the main semantics' to indicate collectively the following three classes of models:

1. THE SCOTT-CONTINUOUS SEMANTICS (Scott (1972)).
   This class contains all λ-models arising as reflexive objects living in the category of cpo's and (Scott-)continuous functions. This semantics encompasses several subclasses of *webbed models* that are built from lower level structures called 'web':

   (a) graph models: isolated in the seventies by Plotkin (1971), Scott (1974) and Engeler (1981), they constitute the simplest class of (non-syntactical) models of λ-calculus. The word 'graph' refers to the fact that the continuous functions are encoded in the model via a sufficient fragment of their graphs. Their web is given by a non-empty set D together with an injection $\iota : \mathscr{P}_{\mathrm{f}}(D) \times D \to D$ and generates $(\mathscr{P}(D), \subseteq)$ as a reflexive object.

   (b) K-models: introduced by Krivine (1993), this class contains all graph models as well as extensional models, such as Scott's $\mathcal{D}_\infty$. The domain underlying a K-model is the complete lattice $(\mathscr{S}(D), \subseteq)$, where $\mathscr{S}(D)$ stands for the set of all the initial segments of some preordered set $(D, \preceq)$. One retrieves graph models by considering a trivial preorder on D.

   (c) pcs-models: defined by Berline (2000), is the simplest class including all K-models and allowing to work outside the framework of complete lattices. Their underlying domain is $(\mathscr{S}_{\mathrm{coh}}(D), \subseteq)$, where $\mathscr{S}_{\mathrm{coh}}(D)$ denotes the set of all coherent initial segments of some preordered set with coherences $(D, \preceq, \frown)$. All binary prime algebraic domains can be described in this manner.

   (d) filter models: introduced by Barendregt et al. (1983), they have been extensively treated in Chapter 13. We simply recall that their underlying domain is of the form $(\mathcal{F}_D, \subseteq)$, where $\mathcal{F}_D$ is the set of all filters of a poset $(D, \preceq)$.

2. THE STABLE SEMANTICS (Berry (1979)).
   This class corresponds to the models living in the category whose objects are particular Scott domains called *dI-domains* and morphisms are the stable functions. In a dI-domain $a \sqcap b$ is defined for all pairs (a, b) of compatible elements. A function between dI-domains is *stable* if it is continuous and commutes with inf's of compatible elements. Stability was introduced as an approximation of sequentiality. This class contains Girard's G-models (reflexive coherences) as class of webbed models.

3. THE STRONGLY STABLE SEMANTICS (Bucciarelli and Ehrhard (1991)).
   This semantics includes all the models living in the category whose objects are *dI-domains with coherences* and morphisms are the strongly stable functions, i.e., stable function preserving the coherence. It contains Ehrhard's H-models (reflexive hyper-coherences) as class of webbed models.

For a comprehensive survey presenting these classes of models, see Berline (2000). All these semantics are structurally and equationally rich—in each of them it is possible to construct $2^{\aleph_0}$ models representing pairwise distinct, and even incomparable, λ-theories. Unfortunately, the main semantics are equationally incomplete: this was demonstrated by Honsell and Ronchi Della Rocca (1992) for the Scott-continuous semantics, by Bastonero and Gouy (1999) for the stable semantics and by Bastonero (1996) for H-models. Salibra (2001b) proved that the class of all λ-models involving monotonicity with respect to some partial order and having a bottom element is equationally incomplete, thus giving a first uniform proof of incompleteness encompassing the three semantics.

A uniform proof of incompleteness

We now show that the equational incompleteness of the main semantics follows from the Algebraic Incompleteness Theorem 15.47. Recall that a λ-model $\mathcal{C}$ is *simple* when it has just the two trivial congruences, whence it is directly indecomposable. We show that all λ-models living in the main semantics are simple algebras.

PROPOSITION 15.50 (MANZONETTO AND SALIBRA (2006)).
(i) *All λ-models living in the Scott-continuous semantics are simple combinatory algebras.*
(ii) *All λ-models living in the stable or strongly stable semantics are simple combinatory algebras.*

PROOF. Consider a partially ordered λ-model $(\mathcal{C}, \sqsubseteq)$.

(i) Suppose that $\mathcal{C}$ belongs to the Scott-continuous semantics. It is easy to check that, for all $b, c \in C$, the following function is continuous:

$$g_{b,c}(x) \triangleq \begin{cases} c, & \text{if } x \not\sqsubseteq b, \\ \bot, & \text{otherwise.} \end{cases}$$

Let $\vartheta \in \mathrm{Con}(\mathcal{C})$ and suppose that there exist distinct $a, d \in C$ such that $a \;\vartheta\; d$. Since $a \neq d$, either $a \not\sqsubseteq d$ or $d \not\sqsubseteq a$. Consider the former case, the latter being analogous. Now, take $c \in C$. The continuous function $g_{d,c}$ is representable in $\mathcal{C}$, so we have:

$$\bot = g_{d,c}(a) \;\vartheta\; g_{d,c}(d) = c$$

i.e. $c \,\vartheta\, \bot$. Since c is arbitrary, the congruence ϑ is trivial. We conclude that $\mathcal{C}$ is simple. Note that $g_{d,c}$ is neither stable nor strongly stable, so it cannot be used for proving (ii).

(ii) Suppose that $\mathcal{C}$ is a stable (resp. strongly stable) λ-model. Consider two elements $a, b \in C$ such that $a \neq b$. Wlog, assume $a \not\sqsubseteq b$. Then there exists a compact element $d \in \mathcal{K}(\mathcal{C})$ such that $d \sqsubseteq a$ and $d \not\sqsubseteq b$. The step function $(d \mapsto c)$ defined by:

$$(d \mapsto c)(x) = \begin{cases} c, & \text{if } d \sqsubseteq x, \\ \bot, & \text{otherwise,} \end{cases}$$

is stable (resp. strongly stable) for every element $c \in C$. This function $(d \mapsto c)$ can be used to show that every congruence on $\mathcal{C}$ is trivial as in the proof of (i). □

As a consequence, we obtain a uniform proof of incompleteness for the main semantics.

COROLLARY 15.51 (MANZONETTO AND SALIBRA (2006)). *The Scott-continuous, the stable and the strongly stable semantics are incomplete.*

PROOF. By Proposition 15.50 these semantics are subclasses of $\mathbb{IND}$. We conclude by the Algebraic Incompleteness Theorem 15.47. □

The incompleteness of $\mathbb{IND}$ is as large as possible

To prove some results concerning the size of the equational incompleteness of $\mathbb{IND}$, we need to recall some known facts concerning the size of particular subintervals of $\lambda\mathcal{T}$.

DEFINITION 15.52. Let κ be a cardinal. A subset $\mathcal{X} \subseteq \lambda\mathcal{T}$, is called:
(i) κ-*high* if $\mathcal{X}$ contains a chain of cardinality κ;
(ii) κ-*wide* if $\mathcal{X}$ contains an antichain of cardinality κ;
(iii) κ-*broad* if $\mathcal{X}$ contains a strong antichain of cardinality κ.

REMARK 15.53. (i) If $\mathcal{X} \subseteq \lambda\mathcal{T}$ is κ-broad, then it is also κ-wide.
(ii) Since $\lambda\mathcal{T}$ is a (non-trivial) complete lattice, if $\mathcal{X}$ is dense then it is also $2^{\aleph_0}$-high.

In the next theorem, the results concerning r.e. λ-theories $\mathcal{T}$ are proven by exploiting the existence of $\mathcal{T}$-easy terms (namely, Proposition 15.34).

THEOREM 15.54. (i) *The set of all sensible λ-theories is $2^{\aleph_0}$-high and $2^{\aleph_0}$-wide.*
(ii) *If $\mathcal{S}, \mathcal{T}$ are r.e. λ-theories, then the interval $[\mathcal{S}, \mathcal{T}]$ is $2^{\aleph_0}$-high.*
(iii) *The set of all r.e. λ-theories is dense, whence $\lambda\mathcal{T}$ is $2^{\aleph_0}$-high.*
(iv) *There exists an r.e. λ-theory $\mathcal{T}$ such that $[\mathcal{T}, \nabla]$ is $2^{\aleph_0}$-broad.*
(v) *The set of λ-theories represented by graph models is $2^{\aleph_0}$-broad.*

PROOF. (i)-(ii) See B[1984], Sections 16.3 and 17.1.
(iii) Follows from (ii) and Remark 15.53(ii).
(iv)-(v) By Berline and Salibra (2006). □

The following results first appeared in the PhD thesis of Manzonetto (2008). Let us start with a positive result, concerning the size of the λ-theories representable by indecomposable models. Recall that, given a class $\mathbb{C}$ of λ-models, $\lambda\mathbb{C} = \{\mathcal{T} \mid \exists \mathcal{C} \in \mathbb{C} \,.\, \mathrm{Th}(\mathcal{C}) = \mathcal{T}\}$.

THEOREM 15.55. (i) *$\lambda\mathbb{IND}$ is $2^{\aleph_0}$-high and $2^{\aleph_0}$-wide.*
(ii) *$\lambda\mathbb{IND}$ $2^{\aleph_0}$-broad.*

PROOF. (i) By Theorem 15.49, $\lambda\mathbb{IND}$ contains the interval $[\mathcal{H}, \mathcal{H}^*]$ which is $2^{\aleph_0}$-high and $2^{\aleph_0}$-wide by Theorem 15.54(i).
(ii) By Proposition 15.50(i), $\lambda\mathbb{IND}$ also contains the set of all graph theories, which is $2^{\aleph_0}$-broad by Theorem 15.54(v). □

We show that also the incompleteness of the indecomposable semantics is as large as possible. The next results follow from the Algebraic Incompleteness Theorem 15.47.

THEOREM 15.56. *Let $\mathcal{T}$ be an r.e. λ-theory. Then, the lattice interval $[\mathcal{T}, \nabla]$ contains a subinterval $[\mathcal{T}_1, \mathcal{T}_2]$ satisfying the following conditions:*

1. *$\mathcal{T}_1$ and $\mathcal{T}_2$ are distinct r.e. λ-theories;*
2. *every $\mathcal{S} \in [\mathcal{T}_1, \mathcal{T}_2]$ is omitted by the indecomposable semantics;*
3. *the cardinality of $[\mathcal{T}_1, \mathcal{T}_2]$ is $2^{\aleph_0}$.*

PROOF. Let $\mathcal{T}$ be an r.e. λ-theory. By Proposition 15.34(ii), there is a λ-term $M \in \Lambda^e_{\mathcal{T}}$. In particular, $\mathcal{T} \nvdash M = \mathsf{T}$ and $\mathcal{T} \nvdash M = \mathsf{F}$. Consider the r.e. λ-theory $\mathcal{T}_1$ defined by setting $\mathcal{T}_1 \triangleq \mathcal{T}(M = \mathsf{T}) \cap \mathcal{T}(M = \mathsf{F})$. Since M is $\mathcal{T}$-easy, the λ-theory $\mathcal{T}_1$ is consistent. By Lemma 15.46, we obtain that $[M]_{\mathcal{T}_1}$ is a non-trivial central element of $\mathcal{M}(\mathcal{T}_1)$. By applying Proposition 15.34(i) to the λ-theory $\mathcal{T}_1$ and to the equations $M = \mathsf{T}$ and $M = \mathsf{F}$, we obtain a $\mathcal{T}_1$-easy term $P \in \Lambda^o$ such that

$$\mathcal{T}_1(P = Q) \nvdash M = \mathsf{T} \qquad \& \qquad \mathcal{T}_1(P = Q) \nvdash M = \mathsf{F},$$

for all λ-terms $Q \in \Lambda^o$. Let $\mathcal{T}_2 \triangleq \mathcal{T}_1(P = \mathsf{I})$. Since P is $\mathcal{T}_1$-easy, we have that $\mathcal{T}_1 \subsetneq \mathcal{T}_2$. The term model $\mathcal{M}_{\mathcal{T}_2}$ is a homomorphic image of $\mathcal{M}_{\mathcal{T}_1}$, whence every equation satisfied by $\mathcal{M}_{\mathcal{T}_1}$ is also satisfied by $\mathcal{M}_{\mathcal{T}_2}$. In particular, the equations (15.3)–(15.6) expressing the fact that M is a central element. Finally, $[M]_{\mathcal{T}_2}$ is non-trivial as a central element because $\mathcal{T}_2 \nvdash M = \mathsf{T}$ and $\mathcal{T}_2 \nvdash M = \mathsf{F}$. Thus, for every $\mathcal{S} \in [\mathcal{T}_1, \mathcal{T}_2]$, $[M]_{\mathcal{S}}$ is non-trivial central element of $\mathcal{M}(\mathcal{S})$. We conclude that $[\mathcal{T}_1, \mathcal{T}_2] = 2^{\aleph_0}$ by Theorem 15.54(ii). □

From Proposition 15.44 it follows that all the λ-models $\mathcal{C}$ such that $\mathrm{Th}(\mathcal{C})$ belongs to the interval $[\mathcal{T}_1, \mathcal{T}_2]$ above, are directly decomposable.

THEOREM 15.57. *Let $\mathbb{DEC}$ be the class of all directly decomposable λ-models. Then*
(i) *$\lambda\mathbb{DEC}$ is $2^{\aleph_0}$-broad;*
(ii) *$\lambda\mathbb{DEC}$ is $2^{\aleph_0}$-high, and even contains countably many 'pairwise incompatible' $2^{\aleph_0}$-high intervals.*

PROOF. (i) Let $U_n \triangleq \Omega(\mathsf{U}^{n+1}_{n+1})$. Given a permutation $\sigma : \mathbb{N} \to \mathbb{N}$ and $M \in \{\mathsf{T}, \mathsf{F}\}$, define a set of equations:

$$X^M_\sigma \quad \triangleq \quad \{U_0 = M\} \cup \{U_n = \mathsf{c}_{\sigma(n-1)} \mid n \geq 1\}$$

Obviously, no equation in X^{T}_σ (resp. X^{F}_σ) is consequence of the other ones. By Theorem 22 in Berline and Salibra (2006), the λ-theories $\mathcal{T}_\sigma = \mathcal{T}(X^{\mathsf{T}}_\sigma)$ and $\mathcal{S}_\sigma = \mathcal{T}(X^{\mathsf{F}}_\sigma)$ are consistent. Therefore, by Lemma 15.46, $[U_0]_{\mathcal{T}_\sigma \cap \mathcal{S}_\sigma}$ is a non-trivial central element of $\mathcal{M}(\mathcal{T}_\sigma \cap \mathcal{S}_\sigma)$. Thus, for all permutations σ, we obtain $\mathcal{T}_\sigma \cap \mathcal{S}_\sigma \in \lambda\mathbb{DEC}$ by Proposition 15.44. Now, if σ_1, σ_2 are two distinct permutations, then $\mathcal{T}_{\sigma_1} \cap \mathcal{S}_{\sigma_1}$ and $\mathcal{T}_{\sigma_2} \cap \mathcal{S}_{\sigma_2}$ are incompatible because it is inconsistent to equate $\mathsf{c}_n = \mathsf{c}_m$, for every $n \neq m$. We conclude since there exist $2^{\aleph_0}$ permutations $\sigma : \mathbb{N} \to \mathbb{N}$, giving rise to pairwise incompatible λ-theories $\mathcal{T}_\sigma \cap \mathcal{S}_\sigma \in \lambda\mathbb{DEC}$.

(ii) Let $\sigma : \mathbb{N} \to \mathbb{N}$ be a permutation and $\mathcal{T}_\sigma, \mathcal{S}_\sigma$ be as in the proof of (i). Assume that σ is recursive, then both $\mathcal{T}_\sigma$ and $\mathcal{S}_\sigma$ are r.e. λ-theories, so also $\mathcal{T}_\sigma \cap \mathcal{S}_\sigma \in \lambda\mathbb{DEC}$ is r.e. Thus, by Theorem 15.56, the interval $[\mathcal{T}_\sigma \cap \mathcal{S}_\sigma, \nabla]$ contains an interval of $2^{\aleph_0}$ λ-theories belonging to $\lambda\mathbb{DEC}$. We conclude because there are countably many computable σ's. □

Corollary 15.58. *The indecomposable semantics omits a set of λ-theories which is $2^{\aleph_0}$-broad, $2^{\aleph_0}$-high, and contains countably many pairwise incompatible $2^{\aleph_0}$-high intervals.*

Corollary 15.59. *The Scott-continuous, the stable and the strongly stable semantics omit a set of λ-theories which is $2^{\aleph_0}$-broad, $2^{\aleph_0}$-high, and contains countably many pairwise incompatible $2^{\aleph_0}$-high intervals.*

We conclude by showing that none of the sets of λ-theories represented by the classes of models discussed at the beginning of the section (page 456) constitutes a sublattice of $\lambda\mathcal{T}$. This property follows from the next proposition and the existence of easy terms.

Proposition 15.60. *Let $\mathbb{C} \subseteq \mathbb{IND}$ and let $\mathcal{T} \in \lambda\mathbb{C}$ be consistent. If there exists a weakly $\mathcal{T}$-easy term M, then*
(i) $\lambda\mathbb{C}$ is not closed under intersection,
(ii) $\lambda\mathbb{C}$ is not a sublattice of $\lambda\mathcal{T}$.

Proof. (i) Let $\mathcal{T}_1 \triangleq \mathcal{T}(M = \mathsf{T})$ and $\mathcal{T}_2 = \mathcal{T}(M = \mathsf{F})$. By Definition 15.24(iii), $\mathcal{T}_1 \vee \mathcal{T}_2 \neq \nabla$, whence $\mathcal{T}_1$ and $\mathcal{T}_2$ are both consistent. By Lemma 15.46, the equivalence class $[M]_{\mathcal{T}_1 \cap \mathcal{T}_2}$ is a non-trivial central element of $\mathcal{M}(\mathcal{T}_1 \cap \mathcal{T}_2)$. Conclude that $\mathcal{T}_1 \cap \mathcal{T}_2 \notin \lambda\mathbb{C}$.
(ii) By (i). □

Corollary 15.61. *Let $\mathbb{C}$ be one of the following semantics:*

1. *graph semantics,*
2. *G-semantics,*
3. *H-semantics,*
4. *filter semantics,*
5. *Scott-continuous semantics,*
6. *stable semantics,*
7. *strongly stable semantics.*

Then $\lambda\mathbb{C}$ is not a sublattice of $\lambda\mathcal{T}$.

Proof. Semantic proofs of the easiness of Ω were given for the classes of graph, G- and H-models (see Berline (2000)), and hence still hold for all the larger classes. Then, the conclusion follows from Propositions 15.60(ii) and 15.50. □

Part VI

Open Problems

Chapter 16

Open Problems

We present below a list of open problems on λ-calculus (and combinatory logics) that we consider of interest for the scientific community. For the reader's convenience we classify them into problems concerning reduction and conversion, and problems concerning models and theories, although in some cases the distinction is not so neat. Some are longstanding open problems that already appeared in the literature—in which case we specify their origin—while others are original. We also collect here the problems and conjectures already mentioned along the book.

We hope that this chapter will stimulate the scientific curiosity of some young minds and that in forty years someone will write another book presenting the solutions.

16.1 Reduction and conversion

Possibly non-existing combinators

In Statman (1993a), the author presents several notions of combinators that are not supposed to exist in λ-calculus, but whose existence is difficult to disprove. The list includes hyper-recurrent λ-terms, uniform universal generators and double fixed-point combinators. The first two notions have already been discussed at the end of Chapter 4, therefore we simply present the corresponding questions.

PROBLEM 1 (STATMAN (1993C)). Do hyper-recurrent λ-terms exist?

PROBLEM 2 (CF. PROBLEM 11). Do uniform universal generators exist in λ-calculus?

Recall that all hyper-recurrent λ-terms are uniform universal generators, whence a negative solution of Problem 2 would entail a negative solution of Problem 1 as well. We believe that the notion of 'core' of an unsolvable (Definition 9.15) developed in Chapter 9, might turn out to be useful for investigating the above problems.

*
* *

We now discuss the notion of a double fixed point combinator. (*Nota bene*: this notion is *not* related to the Double Fixed Point Theorem.) Recall that $\boldsymbol{\delta} = \lambda yx.x(yx)$. As remarked by Böhm and van der Mey (see Fact 6.8) Y is a fixed point combinator iff $\boldsymbol{\delta}Y =_\beta Y$. It follows that, if Y is a fixed point combinator, then both $\boldsymbol{\delta}Y$ and $Y\boldsymbol{\delta}$ are fixed point generators. A *double fixed point combinator* is a λ-term Y satisfying

$$\boldsymbol{\delta}Y =_\beta Y =_\beta Y\boldsymbol{\delta}.$$

PROBLEM 3 (STATMAN (1993B)). Does there exist a double fixed point combinator?

This problem appears as Problem 52 in the RTA list of open problems, and it is marked as 'solved' since the appearance of Intrigila (1997). However, in 2011, Endrullis has discovered a gap in a crucial case of the argument and the problem should therefore be considered as open. Klop considers this problem one of the most interesting problems in term rewriting, and Endrullis et al. (2017) have developed a clocked mechanism in the hope of distinguishing every Y from $Y\boldsymbol{\delta}$, but their attempts were unsuccessful.

For more information about this problem and other suggestions for a proof strategy, we refer to Manzonetto et al. (2019).

*
* *

The Fixed Point Property

A λ-theory $\mathcal{T}$ satisfies the fixed point property if every combinator has either one, or infinitely many pairwise $\mathcal{T}$-distinct closed fixed points. The question whether this property holds for $\mathcal{T} = \boldsymbol{\lambda}$ was first raised by Intrigila and Biasone (2000). In Chapter 10 we have seen that no sensible λ-theory does satisfy the fixed point property (Theorem 10.39), however the conjecture is that the following questions have positive answers.

PROBLEM 4 (INTRIGILA (2000)). Does $\boldsymbol{\lambda}$ satisfy the fixed point property?

PROBLEM 5 (MANZONETTO ET AL. (2019)). Does $\boldsymbol{\lambda}\eta$ satisfy the fixed point property?

PROBLEM 6 (MANZONETTO). Is there an r.e. λ-theory satisfying the fixed point property?

As both $\boldsymbol{\lambda}$ and $\boldsymbol{\lambda}\eta$ are r.e., a negative response to Problem 6 would solve the three problems at once. For some advances on related questions, we refer to Intrigila and Statman (2015) and Manzonetto et al. (2019).

*
* *

Easiness

Recall that a combinator M is called *easy* if, for every $N \in \Lambda^o$, the equation $M = N$ can be consistently added to $\boldsymbol{\lambda}$. That is, $M \in \Lambda^o$ is easy whenever $\forall N \in \Lambda^o . \boldsymbol{\lambda}(M = N) \neq \nabla$. It is well known that easy terms must be unsolvable, in fact, they can be considered as programs having a completely uninformative computational content.

The following are longstanding open problems.

PROBLEM 7 (JACOPINI AND VENTURINI ZILLI (1985)). Is $\mathsf{Y}\Omega_3$ easy?

PROBLEM 8 (BERARDUCCI AND INTRIGILA (1993)). Does $X =_\beta \Omega_3 X$ entail that X is easy?

The difference between Problem 7 and 8 is that the former asks whether Curry's fixed point combinator applied to Ω_3 is easy, the latter asks whether *any* fixed point of Ω_3 is easy. The reader can find preliminary results towards a proof of easiness of $\mathsf{Y}\Omega_3$ in Kuper (1999). In Berarducci and Intrigila (1993), the authors proved that $\mathsf{Y}\Omega_3$ can be consistently equate with any closed β-nf, and with any closed λ-term which is not of the form $\lambda x.X_1X_2X_3$, where $X_i \in \mathrm{UNS} \cup \{x\}$ for all $i \in \{1, 2, 3\}$. These results led these authors to conjecture a positive answer for both problems. The most recent and developed attempt at a full proof of easiness for $\mathsf{Y}\Omega_3$ is probably the one in Bertini (2005). In the conclusions of his thesis, Bertini even suggests how to generalize his statements to cover the missing cases—the corresponding proofs are however difficult to adapt.

*
* *

Infinitary reductions

In Chapter 6 we have shown that the Böhm reduction relative to a set $\mathcal{U}$ of meaningless terms enjoys infinitary weak normalization. By taking as $\mathcal{U}$ the set of terms without a head/weak/top normal form, respectively, this allows to retrieve Böhm trees, Lévy-Longo trees and Berarducci trees. We have seen that such an infinitary reduction can be defined either by taking limits of strong convergent sequences $\xrightarrow{\infty}_{\beta\bot_\mathcal{U}}$, or coinductively $\twoheadrightarrow\!\!\!\!\!\twoheadrightarrow_{\beta\bot_\mathcal{U}}$. As discussed at the end of that chapter, taking as $\mathcal{U}$ unsolvable terms and adding extensionality in the form of the rules (η) or $(\eta!)$ one also retrieves η-Böhm trees and Nakajima trees, but the corresponding coinductive definitions of $\twoheadrightarrow\!\!\!\!\!\twoheadrightarrow_{\beta\bot\eta}$ and $\twoheadrightarrow\!\!\!\!\!\twoheadrightarrow_{\beta\bot\eta!}$ are more complicated and the following properties should be established.

PROBLEM 9. Does $\Lambda^\infty_\bot$ endowed with $\twoheadrightarrow\!\!\!\!\!\twoheadrightarrow_{\beta\bot\eta}$ enjoy CR^∞ and WN^∞?

PROBLEM 10. Does $\Lambda^\infty_\bot$ endowed with $\twoheadrightarrow\!\!\!\!\!\twoheadrightarrow_{\beta\bot\eta!}$ enjoy CR^∞ and WN^∞?

More generally, the question is whether Theorems 6.82 and 6.85 can be presented in a uniform way, using coinductive definitions.

*
* *

Combinatory logic

We start by mentioning that Problem 1 and 2 admit an analogous version for Combinatory Logic. However, Statman (1991) already proved that hyper-recurrent combinatory terms do not exist, nor uniform universal generators if we restrict their definition to contexts of shape $C[\,] \triangleq P[\,]$, for $P \in \mathrm{CL}$. The general problem remains open.

PROBLEM 11 (STATMAN (1993D)). In CL, is there a uniform universal generator?

Also in this case, Statman conjectured a negative answer.

*
* *

Recall from Chapter 4 that the Plane Property requires that, if a term can leave a plane at one of its points, then it can be left at any of its points. We have seen in the same chapter that the λ-calculus does not satisfy the Plane Property—it remains open whether CL does. Klop conjectured a positive answer in Klop (1980b), Conjecture 3.6.1.

PROBLEM 12. Does combinatory logic satisfy the Plane Property?

*
* *

Recall from Chapter 7 that, given a set $\mathcal{C} \subseteq \Lambda^o$ of combinators, $\mathcal{C}^\bullet$ denotes the set of all applicative combinations of members of $\mathcal{C}$. Moreover, a combinator $M \in \Lambda^o$ is called proper if, for some $n > 0$ and some applicative combination Q of $x_1, \ldots, x_n$, one has:

$$Mx_1 \cdots x_n \twoheadrightarrow_\beta Q$$

We say that a proper combinator M has order n, if n is minimal for the property above. The *word problem* for $\mathcal{C}$ consists in determining whether, given $M, N \in \mathcal{C}^\bullet$, $M =_{\beta\eta} N$ holds. The problem was originally proposed by Statman (1989), and occurs both in the TLCA list (#7) and in the RTA list (#96) of open problems.

PROBLEM 13 (STATMAN (1989)). Is the word problem for the set of all proper combinators of orders less than 3 decidable?

For more information concerning this problem, and related results, the reader may consult Statman (1986, 1988a,b, 1989) and Statman (2000). In particular, this last work led Statman to conjecture a positive answer.

*
* *

The S-fragment of combinatory logic

The following problems concern the **S**-fragment of CL studied in Chapter 7. We start from the oldest and most important one, which is known in the community as the *word problem for* **S**. Recall that the problems of determining whether an **S**-term is strongly normalizing or head normalizing are both decidable. This was shown in Theorem 7.22 and Corollary 7.47, respectively.

PROBLEM 14 (THE WORD PROBLEM FOR **S**, BARENDREGT (1975)). Is the problem of determining for all $P, Q \in \mathfrak{S}_{\mathrm{CL}}$ whether $P =_w Q$ decidable?

Recall from Section 7.5 that w-conversion does not coincide with the equality induced by Berarducci trees. This was shown by Padovani in this book, Theorem 7.49.

*

* *

We have seen in Proposition 7.19 that there are no spiralling **S**-terms. This means that there is no $P \in \mathfrak{S}_{\mathrm{CL}}$ admitting a non-empty reduction $P \twoheadrightarrow_w C[P]$, for some context $C[\,]$. By substituting w-conversion for w-reduction, one obtains the following open problem.

PROBLEM 15. Is there $P \in \mathfrak{S}_{\mathrm{CL}}$ such that $C[P] =_w P$, for some context $C[\,] \neq [\,]$?

In this perspective, the non-existence of **S**-definable quasi-identities shown in Proposition 7.16 might be seen as a partial negative result towards a solution of this more general problem. We do not feel confident enough to conjecture a negative answer, though.

*

* *

A set $\mathcal{R} \subseteq \mathfrak{S}_{\mathrm{CL}}$ is called *rational* if its elements are generated by a rational (aka. regular) expression. It is easy to check that, for every rational set $\mathcal{R}$ of **S**-terms, there exists a finite set of pairs of rational sets

$$\Delta(\mathcal{R}) \triangleq \{(\mathcal{R}_1^1, \mathcal{R}_2^1), \ldots, (\mathcal{R}_1^n, \mathcal{R}_2^n)\}$$

such that $\mathcal{R} \cap (*\,*) = \bigcup_{i=1}^n (\mathcal{R}_1^i\, \mathcal{R}_2^i)$, where we recall that $*$ is used as an abbreviation for $\mathfrak{S}_{\mathrm{CL}}$ inside an expression. Padovani worked on the following problem:

> Given a rational set $\mathcal{R}_0$, is there a simple way to compute a set $\mathcal{E}(\mathcal{R}_0)$ of rational sets whose union is the set of all w-expansions of elements of $\mathcal{R}_0$?

In other words, $\mathcal{E}(\mathcal{R}_0)$ should satisfy $\bigcup \mathcal{E}(\mathcal{R}_0) = \mathcal{R}_0^{\leftarrow}$. The following naive method to compute $\mathcal{E}(\mathcal{R}_0)$ and the two additional rules presented below that prevents this set from being infinite in the case of $\mathcal{R}_0 = \mathbf{S}**$ were proposed by Padovani around 2015. Starting from $\mathcal{R}_0$, repeatedly apply the following treatment:

> For each rational set $\mathcal{R}$ derived from $\mathcal{R}_0$ so far, for each $(\mathcal{R}_1, \mathcal{R}_2) \in \Delta(\mathcal{R})$, for each pair $(\mathcal{R}'_1, \mathcal{R}'_2)$ such that $\mathcal{R}'_i \in \mathcal{E}(\mathcal{R}_i)$ (where $\mathcal{E}(\mathcal{R}_i)$ is computed with the same method starting from $\mathcal{R}_i$ instead of $\mathcal{R}_0$) and for each $((\mathcal{U}, \mathcal{W}_1), (\mathcal{V}, \mathcal{W}_2)) \in \Delta(\mathcal{R}'_1) \times \Delta(\mathcal{R}'_2)$, define
> $$\mathcal{W} \triangleq \mathcal{W}_1 \cap \mathcal{W}_2.$$
> If $\mathbf{S}\mathcal{U}\mathcal{V}\mathcal{W} \not\subseteq \mathcal{R}$, then add $\mathbf{S}\mathcal{U}\mathcal{V}\mathcal{W}$ to the set of rational sets derived from $\mathcal{R}_0$.

Clearly, the set $\mathcal{E}(\mathcal{R}_0)$ yielded by this method can be finite (for instance, $\mathcal{E}(*) = \{*\}$), but also infinite. Let us see some examples.

For $\mathcal{R}_0 \triangleq \mathbf{S} * *$ the method produces $\mathbf{S}_k * *$, for every positive integer k. More generally, for $\mathcal{R}_0 \triangleq \mathcal{R} * *$, it produces $(\mathbf{S}^k(\mathcal{R})) * *$ for every $k > 0$. However, this infinite computation can be prevented by adding to the above method the following rule:

> (Rule 1) Whenever a rational set of the form $\mathcal{R} * *$ appears in the course of a derivation, immediately replace it with $\mathcal{R}' * *$, where the rational set $\mathcal{R}'$ is the least set satisfying both $\mathcal{R} \subsetneq \mathcal{R}'$ and $\mathbf{S}\mathcal{R}' \subsetneq \mathcal{R}'$, provided that $\mathcal{R}' \not\subseteq \mathcal{R}$.

Accordingly $\mathcal{E}(\mathbf{S} * *) = \mathcal{E}(\mathcal{P} * *)$, where $\mathcal{P}$ is the set introduced in Definition 7.39(ii). In the computation of $\mathcal{E}(\mathcal{P} * *)$, the set $\mathbf{S}\mathcal{P} * * \subsetneq \mathcal{P} * *$ will be rejected because of the last condition of the method, whereas the set $\mathcal{P}' * *$, where $\mathcal{P}'$ is the least set such that $\mathcal{P} \subsetneq \mathcal{P}'$ and $\mathbf{S}\mathcal{P}' \subsetneq \mathcal{P}'$, will be rejected because of the last condition of Rule 1.

The set $\mathcal{E}(\mathcal{P} *)$ is infinite because $(\mathbf{S}_2 *)\mathcal{P} \in \mathcal{E}(\mathcal{P} *)$ and $(\mathbf{S}_2 *)\mathbf{S} \in \mathcal{E}(\mathbf{S}_2 *)$. However, this problem can be easily fixed by defining the rational set $\mathcal{U}$ as the least set such that $(\mathbf{S}_2 *) \subsetneq \mathcal{U}$ and $\mathcal{U} \subsetneq \mathcal{U}\,\mathbf{S}$, and by adding to the method another additional rule:

> (Rule 2) Whenever the set $\mathbf{S}_2 *$ appears in the course of a derivation, immediately replace it with $\mathcal{U}$.

Accordingly, $\mathcal{E}(\mathbf{S}_2 *) = \mathcal{E}(\mathcal{U}) = \{\mathcal{U}\}$. Now, let us restart the computation of $\mathcal{E}(\mathbf{S} * *)$ with these two new rules. The result is depicted on Figure 16.1. The full (and finite) set $\mathcal{E}(\mathbf{S} * *)$ can be retrieved from the expression $\mathcal{P} * *$ derived from $\mathbf{S} * *$ at the top of the figure by freely replacing a sub-expression for which there exists an auxiliary derivation

$$(\mathcal{P} * *,\ \mathcal{P} *,\ \mathbf{S}\mathcal{U} *, \dots)$$

by any of its derived form. This computation was performed and carefully checked by Padovani several times by hand, then performed again with the help of a small OCaml program. Padovani firmly believes that it is correct, i.e, that $\mathcal{E}(\mathbf{S} * *)$ effectively describes the expansion of the set $\mathbf{S} * *$ and accordingly that the set of head-normalizing terms is rational. He also tends to believe that the two *ad-hoc* rules introduced to prevent infinite derivations of $\mathbf{S} * *$ are just special cases of a more general set of rules that could prevent infinite derivations starting from any rational set.

PROBLEM 16 (PADOVANI). Given a rational set $\mathcal{R}$ of terms, is the closure by w-expansion $\mathcal{R}^{\leftarrow}$ also rational?

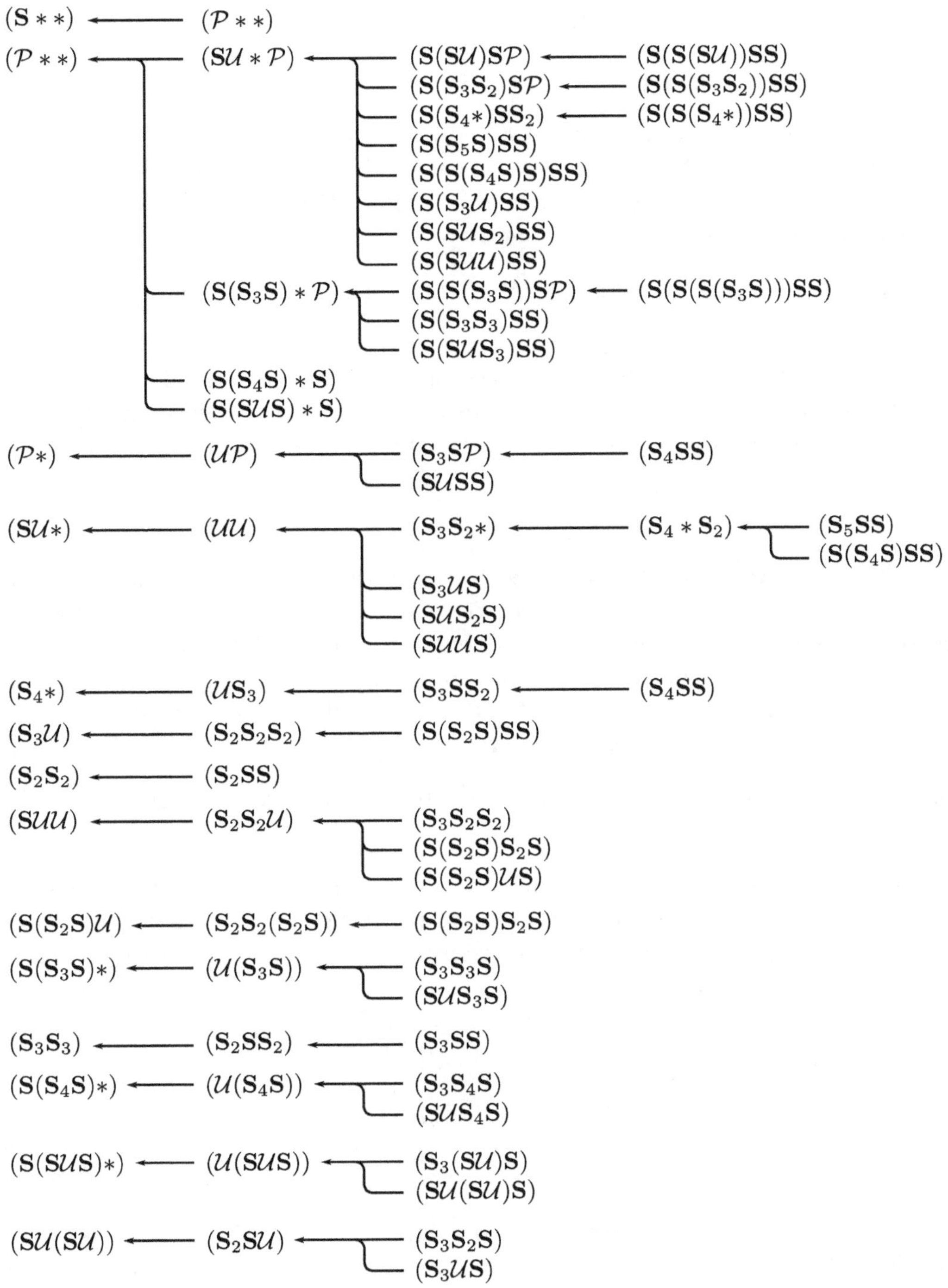

Figure 16.1: The expansion of $\mathcal{P}**$ according to Padovani's method.

16.2 Models and theories

The lattice of λ-theories

The lattice $\lambda\mathcal{T}$ of λ-theories constitutes a very rich mathematical structure. As already mentioned in Section 3.1, $\lambda\mathcal{T}$ does satisfy some non-trivial quasi-identities, like the Et and the Zipper conditions. However, at the end of the nineties, Salibra raised the following question and conjectured that it has a negative answer.

PROBLEM 17 (SALIBRA). Does the lattice $\lambda\mathcal{T}$ of λ-theories satisfy any non-trivial lattice identity?

It was shown by Lusin and Salibra (2004) that, for every non-trivial lattice identity e, there exists $n \in \mathbb{N}$ such that the identity e fails in the lattice of λ-theories over a language of λ-calculus extended with n constants. The general problem remains open.

*
* *

Another interesting problem is related to the lattices that are embeddable into $\lambda\mathcal{T}$. Manzonetto and Salibra (2008) have shown that every finite Boolean lattice can be embedded at the top of $\lambda\mathcal{T}$—a result presented as Theorem 15.36, in Chapter 15 of this book. Moreover, it was shown by Salibra (2001a) that the non-modular pentagon N_5 is a sublattice of $\lambda\mathcal{T}$. Finally, Visser (1980) has proved that every countable partially ordered set embeds into $\lambda\mathcal{T}$ by an order-preserving map.

PROBLEM 18 (MANZONETTO AND SALIBRA (2008)). Is every finite lattice embeddable into $\lambda\mathcal{T}$?

Notice that a positive answer does not follow immediately from Visser's result, since a lattice embedding needs not only to preserve the order but also the lattice structure. Consider, for instance, the 4 elements Boolean lattice $\mathcal{B}_4$, together with the embedding $\iota : \mathcal{B}_4 \to \lambda\mathcal{T}$ defined as follows:

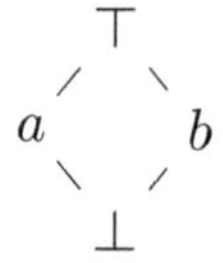

$$\begin{aligned} \iota(\top) &\triangleq \mathcal{H}^*, \\ \iota(a) &\triangleq \boldsymbol{\lambda}\eta, \\ \iota(b) &\triangleq \mathcal{B}, \\ \iota(\bot) &\triangleq \boldsymbol{\lambda}, \end{aligned}$$

We know that $\boldsymbol{\lambda} \subsetneq \mathcal{B} \subsetneq \mathcal{H}^*$ and $\boldsymbol{\lambda} \subsetneq \boldsymbol{\lambda}\eta \subsetneq \mathcal{H}^*$ hold, therefore ι is indeed a poset embedding. However the lattice structure is not preserved since, by Theorem 11.66, we have

$$\boldsymbol{\lambda}\eta \vee \mathcal{B} = \mathcal{B}\eta \neq \mathcal{H}^*$$

Therefore, ι is not a lattice embedding.

*
* *

Meet irreducible elements give important information on the structure of a lattice. It is therefore natural to investigate what λ-theories are meet irreducible.

PROBLEM 19 (MANZONETTO AND SALIBRA (2008)). Is the least λ-theory $\boldsymbol{\lambda}$ meet irreducible?

Salibra conjectured that Problems 18 and 19 have a positive answer.

*
* *

Order incompleteness and absolute unorderability

One of the most important open problems concerning $\lambda\mathcal{T}$ is whether every λ-theory arises as the equational theory of a non-trivially partially ordered model. (We say that a p.o. λ-model $(\mathcal{C}, \leq)$ is trivially partially ordered model if $\leq$ coincides with the equality.)

PROBLEM 20 (SELINGER (1996)). Is there a λ-theory which is omitted by all non-trivially partially ordered λ-models? Equivalently, is the semantics of λ-calculus given in terms of non-trivially partially ordered models complete?

Selinger gave a syntactical characterization of the order-incomplete λ-theories[1] in terms of so-called *generalized Mal'cev operators*. Following this approach, the order incompleteness problem can be roughly stated as follows:

> Are there $n > 0$ and $M_1, \ldots, M_n \in \Lambda^o$ such that the λ-theory $\mathcal{T}_n$ axiomatized by the system of equations:
>
> $$\begin{array}{rcl} x & = & M_1xyy \\ M_1xxy & = & M_2xyy \\ & \vdots & \\ M_{n-1}xxy & = & M_nxyy \\ M_nxxy & = & y \end{array}$$
>
> is consistent?

Salibra (2005) proved that if λ-terms $M_1, \ldots, M_n$ as above actually exist, then they must be unsolvable. Plotkin and Simpson have shown that $\mathcal{T}_1$ is inconsistent, while Plotkin and Selinger obtained the same result for $\mathcal{T}_2$. Both results are presented in Selinger (1997). Salibra and his PhD students have spent a considerable amount of time working on this problem, but only obtained partial results: see Salibra (2001c); Lusin and Salibra (2003); Salibra (2003); Carraro and Salibra (2012, 2013).

*
* *

Selinger (2003) has shown that the order-incompleteness problem is also related to the following question raised by Plotkin.

[1] I.e. the λ-theories that are not induced by any non-trivially ordered model.

PROBLEM 21 (PLOTKIN (1993)). Is there an absolutely unorderable combinatory algebra, i.e. a combinatory algebra which cannot be embedded in any non-trivially partially ordered combinatory algebra?

*
* *

Injectivity and invertibility

In Chapter 9, we have studied the invertibility correspondences in the open and closed term models of $\boldsymbol{\lambda}$ and $\boldsymbol{\lambda}\eta$. We would like to have all syntactic characterizations of those $F \in \Lambda^o$ that are injective or surjective in $\mathcal{M}^{(o)}(\boldsymbol{\lambda}(\eta))$. For some of these term models, such characterizations are already known: see Figure 9.2, for a summary of the situation.

The problems that remain open are the following.

PROBLEM 22. Is there a characterizations of those $F \in \Lambda^o$ that are injective in $\mathcal{M}(\boldsymbol{\lambda})$?

PROBLEM 23. Is there a characterizations of those $F \in \Lambda^o$ that are injective in $\mathcal{M}(\boldsymbol{\lambda}\eta)$?

PROBLEM 24. Is there a characterizations of those $F \in \Lambda^o$ that are injective in $\mathcal{M}^o(\boldsymbol{\lambda})$?

PROBLEM 25. Is there a characterizations of those $F \in \Lambda^o$ that are injective in $\mathcal{M}^o(\boldsymbol{\lambda}\eta)$?

PROBLEM 26. Is there a characterizations of those $F \in \Lambda^o$ that are surjective in $\mathcal{M}^o(\boldsymbol{\lambda}\eta)$?

*
* *

Note that the techniques developed by Folkerts cannot be applied to the term models of $\mathcal{H}\eta$, because all unsolvables are equated.

PROBLEM 27. Does the invertibility correspondence hold in $\mathcal{M}^{(o)}(\mathcal{T})$ for $\mathcal{T} \in [\mathcal{H}\eta, \mathcal{H}^*]$?

[Hint: try to check whether bijective terms have finite Böhm trees and *vice versa*.]

*
* *

The Perpendicular Lines Property

The Perpendicular Lines Property (PLP) states that, if a λ-definable function F of a λ-algebra $\mathcal{M}$ is constant along k perpendicular lines, then F is constant everywhere. (For the precise definition, see Definition 8.1.) We have shown in Chapter 8 that $\mathcal{M}^o(\boldsymbol{\lambda})$, $\mathcal{M}^o(\mathcal{B})$ and $\mathcal{M}(\mathcal{B})$ satisfy this property, while $\mathcal{M}(\boldsymbol{\lambda})$ does not.

PROBLEM 28. Investigate the validity of $\mathcal{M}^{(o)}(\mathcal{T}) \models \mathrm{PLP}$, where $\mathcal{T}$ is one of the theories in the kite, Figure 11.1.

PROBLEM 29. Is there a λ-algebra $\mathcal{M}$ satisfying the Perpendicular Lines Property with a (non λ-definable) element $F \in \mathcal{M}$ that is constant on some perpendicular lines but not globally constant?

*
* *

The hyper-kite

A considerable part of this book is devoted to analyzing the (ω) rule and its interaction with λ-theories like $\boldsymbol{\lambda}, \mathcal{B}$ and $\mathcal{H}$. The *term rule* $(\mathbf{tr})$ represents another form of extensionality for the λ-calculus, given by the following derivation rule:

$$\frac{FZ = GZ \text{ for all } Z \in \Lambda^o}{Fx = Gx \text{ for some fresh variable } x}\ (\mathbf{tr})$$

In Barendregt (1971) this rule was introduced with the observation that

$$(\omega) = (\mathbf{tr}) + (\eta) \tag{16.1}$$

In fact, Plotkin's famous counterexample to (ω), see Plotkin (1974) or Theorem 17.3.30(i) in B[1984], is actually a counterexample to $(\mathbf{tr})$. See also Exercise 17.5.11 in B[1984].

Consider the 'hyper-kite' obtained from the kite in Figure 11.1 by splitting the ω-rule according to equation (16.1). The hyper-kite so obtained is depicted in Figure 16.2 (recall that $\mathcal{H}^+ \triangleq \mathcal{T}_{\text{NF}} = \mathcal{B}\omega$ is the main result of Chapter 11, and that $\mathcal{H}^* \triangleq \mathcal{T}_{\text{SOL}}$).

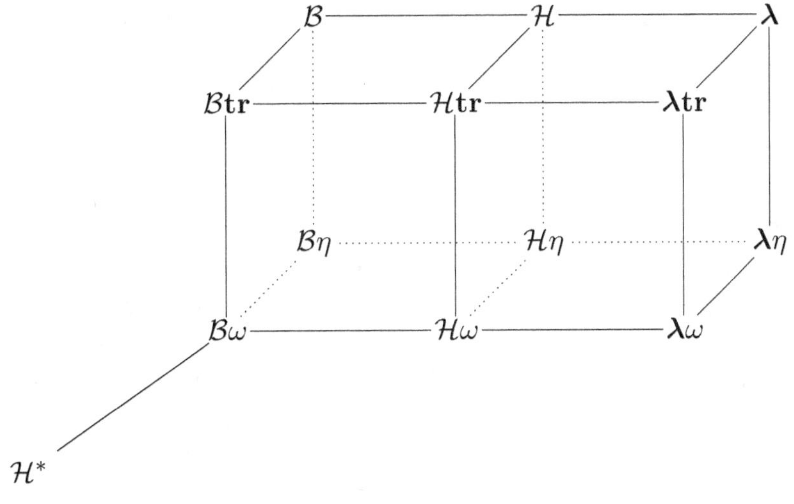

Figure 16.2: The hyper-kite of λ-theories.

PROBLEM 30. Are there distinguishing models for the λ-theories in the hyper-kite?

*
* *

Representability of λ-theories

The question of whether there exists a continuous model of $\boldsymbol{\lambda}$ or $\boldsymbol{\lambda}\eta$ was raised by Honsell and Ronchi Della Rocca (1992), and appears as Problem 22 of the TLCA list. More generally, one can ask for a non-syntactical model of these λ-theories. By 'non-syntactical' we mean a model whose construction does not involve the syntax of λ-terms. For instance, Di Gianantonio et al. (1995) succeeded in building a model of $\boldsymbol{\lambda}\eta$ living in some 'weakly continuous' semantics—however their construction starts from the term model $\mathcal{M}(\boldsymbol{\lambda}\eta)$, whence it cannot be considered as purely non-syntactical. Furthermore, the problem of whether there exists a model of $\boldsymbol{\lambda}$ or $\boldsymbol{\lambda}\eta$ living in one of the main semantics (as defined in Section 15.5) remains completely open.

PROBLEM 31 (CF. HONSELL (2000)). Is there a non-syntactical model $\mathcal{M}$ whose theory is exactly $\mathrm{Th}(\mathcal{M}) = \boldsymbol{\lambda}$ or $\mathrm{Th}(\mathcal{M}) = \boldsymbol{\lambda}\eta$?

Subsequently, Berline et al. (2009) studied the more general question of whether the (in)equational theory of a non-syntactical model of λ-calculus living in one of the main semantics can be recursively enumerable. (See also the PhD thesis of Manzonetto (2008).) They introduced an appropriate notion of 'effective' model, which covers in particular all the models individually introduced in the literature, and proved that the inequational theory of an effective model is never r.e., whence its equational theory cannot be $\boldsymbol{\lambda}$ or $\boldsymbol{\lambda}\eta$. We refer to Carraro and Salibra (2009) for further generalizations of these results.

PROBLEM 32. Is there a non-syntactical model $\mathcal{M}$ such that $\mathrm{Th}(\mathcal{M})$ or $\mathrm{Th}_{\sqsubseteq}(\mathcal{M})$ is r.e.?

Salibra conjectured that both questions have a negative answer. Since both $\boldsymbol{\lambda}$ and $\boldsymbol{\lambda}\eta$ are r.e., a negative solution of Problem 32 would negatively solve Problem 31 as well.

*
* *

Relational graph theories

We have seen that the λ-theory $\mathcal{H}^*$ satisfies the (ω) rule, and the analogous theorem was proved by Breuvart et al. (2016) for $\mathcal{H}^+$. Their result appears in Chapter 11 of the present book as Theorem 11.31. Moreover, by Corollary 14.85, the λ-theory induced by any extensional relational graph model $\mathcal{D}$ belongs to the interval $[\mathcal{H}^+, \mathcal{H}^*]$.

PROBLEM 33 (BREUVART ET AL. (2018)). Does every extensional relational graph model satisfy the ω-rule?

PROBLEM 34 (BREUVART ET AL. (2018)). Are all λ-theories in the interval $[\mathcal{H}^+, \mathcal{H}^*]$ relational graph theories?

In case Problem 34 has a negative answer, it would be interesting to provide a characterization of the representable relational graph theories.

*
* *

Representability of inequational theories

Among the inequational theories defined in Example 12.3(ii), the following have never been seriously studied in the literature:

$$M \sqsubseteq_{m1} N \iff \forall C[\,].\,[\,C[M] \text{ has a } \beta\text{-nf} \Rightarrow C[N] \text{ has the same } \beta\text{-nf}\,]$$
$$M \sqsubseteq_{[\mathsf{I}]_\beta} N \iff \forall C[\,].\,[\,C[M] =_\beta \mathsf{I} \Rightarrow C[N] =_\beta \mathsf{I}\,]$$

As a warm-up, we propose the following exercise.

EXERCISE 16.1. (i) Show that the equational theory induced by $\sqsubseteq_{m1}$ is $\mathcal{B}$, i.e. $\equiv_{m1} = \mathcal{B}$.
(ii) Similarly, for $\sqsubseteq_{[\mathsf{I}]_\beta}$. That is, prove $\equiv_{[\mathsf{I}]_\beta} = \mathcal{B}$.

The next considerations result from several discussions among Barbarossa, Breuvart, Kerinec and Manzonetto. Recall that $M \sqsubseteq_\bot N$ holds if and only if $\mathrm{BT}(M) \leq_\bot \mathrm{BT}(N)$. It is easy to check that $\sqsubseteq_\bot \subseteq \sqsubseteq_{m1} \subseteq \sqsubseteq_{[\mathsf{I}]_\beta} \subseteq \sqsubseteq_{\mathrm{NF}}$, while it is more subtle to verify that these inclusions are strict. Recall that $1x =_\beta \lambda z.xz =_\eta x$.

1. Consider $X \triangleq \lambda zx.xzx$ and $Y \triangleq \lambda zx.xz(1x)$. Every context $C[\,]$ satisfying $C[X\Omega] =_\beta P \in \mathrm{NF}$ needs to erase Ω by applying some functional term $\lambda y.M$, as in $C[\,] \triangleq [\,]\mathsf{F}$. But then $C[Y\Omega] =_\beta P$, since $(1x)[x := \lambda y.M] =_\beta \lambda y.M$ holds for any λ-term M. Thus $X\Omega \equiv_{m1} Y\Omega$ holds, which entails $X\Omega \equiv_{[\mathsf{I}]_\beta} Y\Omega$, while $\mathrm{BT}(X) \neq \mathrm{BT}(Y)$. This shows $\sqsubseteq_\bot \subsetneq \sqsubseteq_{m1}$.

2. Similarly, every closed applicative context $C[\,]$ such that $C[\lambda x.xx] =_\beta \mathsf{I}$ also satisfies $C[\lambda x.x(1x)] =_\beta \mathsf{I}$. Once proved that the same holds for non-applicative contexts, one concludes $\lambda x.xx \equiv_{[\mathsf{I}]_\beta} \lambda x.x(1x)$. On the contrary, it is enough to take $C[\,] \triangleq [\,]$ to get $\lambda x.xx \not\sqsubseteq_{m1} \lambda x.x(1x)$. Therefore, we obtain $\sqsubseteq_{m1} \subsetneq \sqsubseteq_{[\mathsf{I}]_\beta}$.

3. Finally, $1 \not\sqsubseteq_{[\mathsf{I}]_\beta} \mathsf{I}$ and $1 \equiv_{\mathrm{NF}} \mathsf{I}$ entail $\sqsubseteq_{[\mathsf{I}]_\beta} \subsetneq \sqsubseteq_{\mathrm{NF}}$.

Another distinguishing example for (2) are the λ-terms X and Y from (1). Be aware that the generalization to non-applicative contexts in (2) is necessary, since $\sqsubseteq_{[\mathsf{I}]_\beta}$ does not satisfy the analogue of Context Lemma 3.15. E.g., $\lambda xy.y\mathsf{I}(yx)$ and $\lambda xy.y\mathsf{I}(1(yx))$ are distinguished by $C[] = \lambda z.[]z\mathsf{I}$ but there is no applicative context separating them.

PROBLEM 35 (MANZONETTO). Are there inequationally fully abstract denotational models for $\sqsubseteq_{m1}$?

PROBLEM 36 (MANZONETTO). Are there inequationally fully abstract denotational models for $\sqsubseteq_{[\mathsf{I}]_\beta}$?

Preliminary investigations made by Manzonetto in collaboration with Barbarossa, Breuvart and Kerinec suggest that inequationally fully abstract models for these theories might exist in the strongly stable semantics (Bucciarelli and Ehrhard (1991)).

Notice that any model inequationally fully abstract for $\sqsubseteq_{m1}$ or $\sqsubseteq_{[\mathsf{I}]_\beta}$, necessarily induces $=_{\mathcal{B}}$ as its equational theory.

16.3 $\lambda(\mathcal{H})\omega$ in the analytical hierarchy

The complexity of the most relevant λ-theories has been discussed at the end of Section 3.1. Concerning $\mathcal{H}\omega$, Barendregt et al. (1978) proved that for any Π^1_1-predicate P there exist λ-terms $\mathsf{B}^n_0, \mathsf{B}^n_1$ such that

$$\mathrm{P}(n) \quad \Rightarrow \quad \mathcal{H}\omega \vdash \mathsf{B}^n_0 = \mathsf{B}^n_1,$$

in which the fact that $\mathrm{P}(n)$ holds seems to have been properly used (which is not the case if one takes e.g. $\mathsf{B}^n_0 = \mathsf{I} = \mathsf{B}^n_1$). See also B[1984], Theorem 17.4.14. This result motivated the conjecture that the λ-theory $\mathcal{H}\omega$ is Π^1_1-complete. This conjecture first appeared in Barendregt et al. (1978) and was subsequently reproposed in B[1984], Conjecture 17.4.15.

The construction of the B^n_i follows notions involving the principle of Bar Induction, closely related to Π^1_1-predicates.

Bar Induction

Remember that the set of finite sequence of elements of $\mathbb{N}$ is denoted by $\mathbb{N}^*$. Also, given two sequences $s = \langle k_0, \ldots, k_{n-1}\rangle$, $s' = \langle k'_0, \ldots, k'_{n'-1}\rangle \in \mathbb{N}^*$, and $k \in \mathbb{N}$ we write

$$\begin{aligned} s \star s' &= \langle k_0, \ldots, k_{n-1}, k'_0, \ldots, k'_{n'-1}\rangle; \\ s; k &= \langle k_0, \ldots, k_{n-1}, k\rangle. \end{aligned}$$

Each $s \in \mathbb{N}^*$ can be effectively coded as number $\#s \in \mathbb{N}$, where $\#\colon \mathbb{N}^* \to \mathbb{N}$ is a bijection. For $f \in \mathbb{B} \triangleq \mathbb{N}^{\mathbb{N}}$, we let

$$\begin{aligned} \hat{f}(n) &= \langle f(0), \ldots, f(n-1)\rangle \in \mathbb{N}^*, && \text{in particular} \\ \hat{f}(0) &= \langle\rangle \in \mathbb{N}^*, && \text{the empty sequence;} \\ \overline{f}(n) &\triangleq \#(\hat{f}(n)) \in \mathbb{N}. \end{aligned}$$

Finally, for $s \in \mathbb{N}^*$, we write

$$\ulcorner s \urcorner = \mathsf{c}_{\#s}.$$

The λ-representation of the recursive function $-;-$ concatenating a sequence and a natural number is denoted here by $*$ and used intuitively in infix notation. For example, we have $\ulcorner\langle 3,0\rangle\urcorner * \mathsf{c}_5 =_\beta \ulcorner\langle 3,0,5\rangle\urcorner$.

DEFINITION 16.2. For $\mathrm{Q} \subseteq \mathbb{N} \cong \mathbb{N}^*$ considered as a set of sequence numbers, define

(i)	Q is a *bar*	if	$\forall f \in \mathbb{B}, \exists k \in \mathbb{N}\,.\, \hat{f}(k) \in \mathrm{Q}$.
(ii)	Q *climbs upwards*	if	$\forall s \in \mathbb{N}^*\,.\, \big[[\forall k \in \mathbb{N}\,.\, s; k \in \mathrm{Q}] \Rightarrow s \in \mathrm{Q}\big]$.
(iii)	Q is *monotonic*	if	$\forall s \in \mathbb{N}^*, k \in \mathbb{N}\,.\, [s \in \mathrm{Q} \Rightarrow s; k \in \mathrm{Q}]$.
(iv)	$\mathrm{Q}_{\leq}(s)$	$\iff$	$\exists s' \leq s\,.\, \mathrm{Q}(s')$.

Note that Q is monotonic if and only if $\forall s, s' \in \mathbb{N}^*\,.\, [s \leq s' \;\&\; s \in \mathrm{Q} \Rightarrow s' \in \mathrm{Q}]$ and that $\mathrm{Q}_{\leq}$ is monotonic by definition.

Using classical mathematics, the principle of the excluded middle and the axiom of choice, one can prove the following proposition.

PROPOSITION 16.3 (CLASSICAL BAR INDUCTION (BI)). *Let* $\mathrm{Q} \subseteq \mathrm{P} \subseteq \mathbb{N}$. *Suppose that* Q *is a bar and* P *climbs upwards. Then* $\langle\,\rangle \in \mathrm{P}$.

PROOF. Suppose towards a contradiction that $\langle\,\rangle \notin \mathrm{P}$. Then, since P is climbing upwards, for some $s_1 \in \mathbb{N}$ one has $\langle s_1 \rangle \notin \mathrm{P}$. Therefore for some $s_2 \in \mathbb{N}$ one has $\langle s_1, s_2 \rangle \notin \mathrm{P}$. Continuing this way, by the axiom of choice, there exists a function $f \in \mathbb{B}$ such that

$$\begin{aligned} \hat{f}(0) &= \langle\,\rangle \notin \mathrm{P}; \\ \hat{f}(1) &= \langle s_0 \rangle \notin \mathrm{P}; \\ &\vdots \\ \hat{f}(n) &= \langle s_0, \ldots, s_{n-1} \rangle \notin \mathrm{P}. \end{aligned}$$

Therefore, as $\mathrm{Q} \subseteq \mathrm{P}$, one has $\forall n \in \mathbb{N} \,.\, \hat{f}(n) \notin \mathrm{Q}$, contradicting that Q is a bar. □

If Q is also monotonic, then the conclusion can be strengthened.

PROPOSITION 16.4 (MONOTONIC BAR INDUCTION (BI-M)). *Let* $\mathrm{Q} \subseteq \mathrm{P} \subseteq \mathbb{N}^*$. *Suppose that* Q *is a bar,* P *climbs upwards, and* Q *is monotonic. Then* $\mathrm{P} = \mathbb{N}^*$.

PROOF. Let $s = \langle s_0, \ldots, s_{p-1} \rangle \in \mathbb{N}^*$ and assume $s \notin \mathrm{P}$ towards a contradiction. As before the axiom of choice implies that there is a function $f \in \mathbb{B}$ such that $\forall n \,.\, s \star \hat{f}(n) \notin \mathrm{P}$. Let $g \in \mathbb{B}$ be defined by

$$\begin{aligned} g(k) &\triangleq s_k, && \text{if } k < p, \\ g(p+k) &\triangleq f(k), && \text{else.} \end{aligned}$$

Since Q is a bar, one has $\hat{g}(k) \in \mathrm{Q}$ for some $k \in \mathbb{N}$. But then $k < p$, as $\forall n . s \star \hat{f}(n) \notin \mathrm{P}$. So $s{\restriction}_k = \langle s_0, \ldots, s_{k-1} \rangle = \hat{g}(k) \in \mathrm{Q}$, and $s \in \mathrm{Q}$, as Q is monotonic. A contradiction. □

In spite of the classical reasoning, Brouwer has argued on philosophical grounds that bar induction is intuitionistically valid in case Q is decidable.[2] Therefore the following is accepted in fact by him, see Brouwer (1927).

AXIOM 16.5 (DECIDABLE BAR INDUCTION (BI-D)). *Let* $\mathrm{Q} \subseteq \mathrm{P} \subseteq \mathbb{N}$. *Suppose that* Q *is a bar,* P *climbs upwards, and* Q *is decidable. Then* $\langle\,\rangle \in \mathrm{P}$.

Although the proof of Proposition 16.4 was classical, the statement BI-M follows from BI-D and the intuitionistic continuity principle, see Troelstra and van Dalen (1988), p. 231, and hence may be used in intuitionistic reasoning.

[2]This means $\forall s . (\mathrm{Q}(s) \vee \neg \mathrm{Q}(s))$, which intuitionistically is not obvious. If Q is recursive, decidability follows.

Representing Π^1_1-predicates in $\mathcal{H}\omega$

Let P be the Π^1_1-predicate defined by

$$\mathrm{P}(n) \iff \forall f \in \mathbb{B}, \exists m \in \mathbb{N} \,.\, \mathrm{Q}(\overline{f}(m), n), \tag{16.2}$$

where Q is a recursive predicate. Write $\mathrm{Q}^n \triangleq \{s \in \mathbb{N}^* \mid \mathrm{Q}(\#s, n)\}$. Then

$$\mathrm{P}(n) \iff \mathrm{Q}^n \text{ is a bar.}$$

PROPOSITION 16.6. *There exists a* $\Pi \in \Lambda^o$ *such that for all* $M, N \in \Lambda^o$ *one has*

$$\Pi M =_{\mathcal{H}\omega} \Pi N \iff \forall k \in \mathbb{N} \,.\, M\mathsf{c}_k =_{\mathcal{H}\omega} N\mathsf{c}_k.$$

PROOF. See B[1984], Theorem 17.4.9(ii). □

Now using the fixed point theorem, the decidability of Q (hence also of $\mathrm{Q}_{\leq}$), and the combinator Π, define λ-terms $\mathsf{B}^n_0, \mathsf{B}^n_1$ such that

$$\begin{array}{lll} \mathsf{B}^n_i \ulcorner s \urcorner & \twoheadrightarrow_{\beta\Omega} \Omega, & \text{if } \mathrm{Q}^n(s), \\ & \twoheadrightarrow_{\beta\Omega} \Pi(\lambda z.\mathsf{B}^n_i(\ulcorner s \urcorner * z)), & \text{otherwise.} \end{array}$$

Here the i is a dummy parameter that is pushed into infinity, therefore

$$\mathrm{BT}(\mathsf{B}^n_0 \ulcorner s \urcorner) = \mathrm{BT}(\mathsf{B}^n_1 \ulcorner s \urcorner),$$

which is

$$\mathsf{B}^n_0 \ulcorner s \urcorner =_{\mathcal{B}} \mathsf{B}^n_1 \ulcorner s \urcorner.$$

But this equality cannot be proved *a priori* in $\mathcal{H}\omega$. But one has the following.

PROPOSITION 16.7.
(i) $\mathrm{P}(n) \Rightarrow \forall s \in \mathbb{N}^*.\mathcal{H}\omega \vdash \mathsf{B}^n_0 \ulcorner s \urcorner = \mathsf{B}^n_1 \ulcorner s \urcorner.$
(ii) $\mathrm{P}(n) \Rightarrow \mathcal{H}\omega \vdash \Pi(\lambda s.\mathsf{B}^n_0 s) = \Pi(\lambda s.\mathsf{B}^n_1 s).$

PROOF. (i) If $\mathrm{P}(n)$ holds, then Q^n is a bar; therefore also $\mathrm{Q}^n_{\leq}$ is a bar and besides monotonic. Define

$$\mathrm{R}^n \triangleq \{s \in \mathbb{N}^* \mid \mathcal{H}\omega \vdash \mathsf{B}^n_0 \ulcorner s \urcorner = \mathsf{B}^n_1 \ulcorner s \urcorner\}.$$

Then $\mathrm{Q}^n_{\leq} \subseteq \mathrm{R}^n \subseteq \mathbb{N}^*$, with Q^n being a bar and monotonic and R^n climbing upward. Therefore by BI-M the conclusion follows.

(ii) By (i) and Proposition 16.6. □

In this proof the fact that Q^n is a bar seems to be used in an essential way. Therefore, it was conjectured that $\mathcal{H}\omega$ is Π^1_1-complete.

Towards Π^1_1-completeness

A confirmation of Barendregt's conjecture was proposed in Intrigila and Statman (2006), along the following lines.

Step 1 Rather than $\mathsf{B}^n_0, \mathsf{B}^n_1$, a variant $\mathsf{C}^n_0, \mathsf{C}^n_1$ was used, among other things based on a λ-definition of the function $s * k$ satisfying, for $s \in \mathbb{N}^*$ and a variable z:

$$\ulcorner s \urcorner * z \twoheadrightarrow_\beta z R_1 \cdots R_5.$$

Again, one has for all $s \in \mathbb{N}^*$:

$$\mathcal{H}\omega \vdash \mathsf{C}^n_0 \ulcorner s \urcorner = \mathsf{C}^n_1 \ulcorner s \urcorner \iff \mathrm{Q}^n_{\leq}(s) \text{ or } \forall k \in \mathbb{N}\,.\, \mathcal{H}\omega \vdash \mathsf{C}^n_0 \ulcorner s;k \urcorner = \mathsf{C}^n_1 \ulcorner s;k \urcorner. \quad (16.3)$$

This intuitively implies that $\mathrm{Q}^n_{\leq} \subseteq \mathrm{P}^n$ and that P^n climbs upwards. A detailed analysis shows that these terms satisfy the following properties:

- C^n_0 is obtained from C^n_1 by substituting one occurrence of c_0 by c_1;
- for every sequence $s \in \mathbb{N}^*$ such that $\mathsf{C}^n_i \ulcorner s \urcorner$ has a hnf, the above occurrence of c_i does not get erased along the reduction—it rather gets consumed when reaching the head position using the trick $\mathsf{c}_i \mathsf{I} \twoheadrightarrow_\beta \mathsf{I}$. Even in this case, another occurrence of c_i remains in every reduct of C^n_i thanks to a fixed point operator, thus guaranteeing that the two terms remain β-distinguished;
- if $\mathrm{Q}^n_{\leq}(s)$ holds, then $\mathsf{C}^n_0 \ulcorner s \urcorner =_{\mathcal{H}} \Omega =_{\mathcal{H}} \mathsf{C}^n_1 \ulcorner s \urcorner$;
- the equality $\lambda x.\mathsf{C}^n_0 x = \lambda x.\mathsf{C}^n_1 x$ is inconsistent;
- most subterms of $\mathsf{C}^n_0, \mathsf{C}^n_1$ are β-normal forms (this property becomes useful when checking that the two terms have no common β-reduct).

Step 2 Rather than considering the original definition of $\mathcal{H}\omega$, define a system $(\mathcal{H}\omega)^o$ restricted to closed terms.[3] This ensures that most convertibility are obtained by applying the ω-rule rather than β-conversion. One needs to prove that the two systems are equivalent on closed terms $M, N \in \Lambda^o$:

$$(\mathcal{H}\omega)^o \vdash M = N \iff \mathcal{H}\omega \vdash M = N$$

Nota bene: since the ω-rule has infinitely many premises, a derivation in these systems is given by a well-founded countably branching tree whose size can be infinite. Intrigila and Statman associate a countable ordinal with any derivation.

Step 3 Restrict the system $(\mathcal{H}\omega)^o \vdash$ to derivations of a special form, called *cascaded*, yielding a system $(\mathcal{H}\omega)^o \vdash^c$. Prove that this restriction is made without loosing generality: any proof of $(\mathcal{H}\omega)^o \vdash M = N$ can be transformed into a cascaded proof of $(\mathcal{H}\omega)^o \vdash^c M = N$.

[3] That is, one can contract a β-redex of the form $C[(\lambda x.M)N] \in \Lambda^o$ only when $M \in \Lambda^o(x)$ and $N \in \Lambda^o$. Similarly $C[U] \to_\Omega C[\Omega]$ only when $U \in \mathrm{UNS} \cap \Lambda^o$.

Step 4 Check that the cascadization of a proof in system $(\mathcal{H}\omega)^o$ (almost[4]) preserves the associated ordinal.

Step 5 Check that the equivalence (16.3) holds. The direction ($\Rightarrow$) is easy.

To prove the converse ($\Leftarrow$), proceed by induction on the ordinal associated with a cascaded proof of $(\mathcal{H}\omega)^o \vdash^c \mathsf{C}_0^n\ulcorner s\urcorner = \mathsf{C}_1^n\ulcorner s\urcorner$. There are three cases:

1. The ordinal is finite: in this case the equality $\mathsf{C}_0^n\ulcorner s\urcorner = \mathsf{C}_1^n\ulcorner s\urcorner$ holds in $\mathcal{H}$, so one can exploit its characterization in terms of rewriting $\twoheadrightarrow_{\beta\Omega}$ to show that the only possible case is $\mathsf{Q}^n_{\leq}(s)$ whence both terms are unsolvable.
2. The ordinal is infinite. There are two subcases.
   (a) The last applied rule is the ω-rule (whose premises are cascaded). For $k \in \mathbb{N}$, one can find a non-empty, closed, applicative context $[]H_0 \cdots H_n$ such that $\mathsf{C}_0^n\ulcorner s\urcorner\vec{H} \twoheadrightarrow_\beta \mathsf{C}_0^n\ulcorner s; k\urcorner$. Since the proof of $(\mathcal{H}\omega)^o \vdash^c \mathsf{C}_0^n\ulcorner s\urcorner H_0 = \mathsf{C}_1^n\ulcorner s\urcorner H_0$ occurs as a premise, it has (much) smaller ordinal. If one can find a cascaded proof of $(\mathcal{H}\omega)^o \vdash^c \mathsf{C}_0^n\ulcorner s\urcorner\vec{H} = \mathsf{C}_1^n\ulcorner s\urcorner\vec{H}$ having an ordinal which is still smaller, the case follows from bar-induction.
   (b) The derivation actually has the shape of a cascade. In this case the proof of $(\mathcal{H}\omega)^o \vdash^c \mathsf{C}_0^n\ulcorner s\urcorner = \mathsf{C}_1^n\ulcorner s\urcorner$ must rely on the equality between the subterms $\lambda x.\mathsf{C}_0^n x = \lambda x.\mathsf{C}_1^n x$. But this is impossible as the equation is inconsistent,[5] so the whole case is impossible.

An accurate analysis of the proof in Intrigila and Statman (2006) led us to identify two substantial problems that compromise its correctness:

1. In order to show that any proof in $(\mathcal{H}\omega)^o$ can be cascadized (Step 3), the authors cascadize its 'head', by induction hypothesis they obtain a cascaded proof of its 'tail' and they claim to obtain a cascaded proof by composing the two pieces. Unfortunately, their notion of a cascaded proof is not compositional.

2. In Step 5(2a), it is claimed that by first applying the Böhm-out context and then the cascadization process one obtains a proof having a smaller ordinal. This is far from clear, since the cascadization can actually increase the ordinal greatly.

We tried different approaches to fix these problems, but none was conclusive. Therefore, we believe the original problem should be considered still open.

PROBLEM 37 (BARENDREGT ET AL. (1978)). Is the λ-theory $\mathcal{H}\omega$ Π_1^1-complete?

To conclude, we wonder whether the proof of Π_1^1-completeness of $\boldsymbol{\lambda}\omega$ given in Intrigila and Statman (2009) is affected by similar problems. Although not constituting original research, checking their proof and proposing a self-contained, and more modular, presentation would be a non-trivial technical exercise. We specify 'self-contained' since

[4]The ordinal may in fact increase, but only by a natural number k.
[5]This is one of the conditions mentioned in Step 1.

the current version rely on previous results in Intrigila and Statman (2004), that should also be carefully verified. By a 'more modular' presentation we intend that one should extract some lemmas containing the invariants informally used in lengthy proofs of intermediate results, and apply them where needed. In particular, a carefully written proof should define explicitly the notion of *extended trace* and split into lemmas or claims the monolithic proof of Proposition 2.6 in Intrigila and Statman (2009). We believe that this would be an interesting topic for a master thesis in (advanced) λ-calculus. If someone succeeds in this mission, we would be welcome to include the output as a chapter in a future edition of the present book.

16.4 Illative Combinatory Logic

As mentioned on Page 1, around 1928 Church invented the λ-calculus as a foundation for mathematics, containing logic and based on functions rather than sets. Possibly, he did this in order to study the notion of computability and to show that some well-defined functions may not be computable. One of the ideas was to introduce quantification like $\forall x.A$ as a combination of an operator $\mathbb{\Pi}$ acting on an abstraction $\lambda x.A$:

$$\forall x.A = \mathbb{\Pi}(\lambda x.A).$$

Another attractive aspect of such a theory was that both a logical and computational foundation could be given, the latter not being efficiently possible in e.g. set theory.[6] These ideas were first published in Church (1932), that soon was shown to be inconsistent, basically by an oversight. This was corrected in Church (1933). However, also this improved system turned out to contain a hidden inconsistency, as shown by Church's students Kleene and Rosser (1935). Their *tour de force* proof used Gödel's technique[7] of arithmetization and internally coding λ-calculus in itself[8].

After this, Church abandoned the goal of constructing a foundation for logic and mathematics. From his original system he stripped away the inferential part and did keep what is now known as the λ-calculus. Before doing something with this system and even before publishing it Church wanted to be sure that no inconsistency would arise. In the joint work Church and Rosser (1936)[9] it was shown that the system described was consistent. Then, in Church (1936), the Thesis was stated that through the notion of λ-definability one is able to capture exactly the intuitively computable functions. At the

[6]True, one may represent computability in set theory via coded Turing Machine computations, but that is not a convenient computational model.

[7]After Hitler came to power in Germany, Gödel went to Princeton for the 1933-1934 academic year and taught there his incompleteness theorems.

[8]A much simpler proof of inconsistency of a similar system was given in Curry (1942), presented below in Proposition 16.9.

[9]In this paper the λ-calculus was described in a somewhat hard to read style, referring to "Rules I, II, III in Church (1933) [being β-reduction, β-expansion, and α-conversion] as modified by Kleene (1934) [restricting the λK-calculus to the λI-calculus]". Only in Church (1941) a systematic clear-cut description of the λI-calculus and also λK-calculus was given.

same time he showed that—after coding—the notion 'M has a β-normal form' was not λ-definable, hence by Church's Thesis not computable.

Curry, on the other hand, started a program aiming at finding a modified system that would be adequate (sufficiently strong and consistent) as a foundation for logic and mathematics. The systems that arose from these investigations were called by him 'Illative Combinatory Logic' (ICL)—the term 'Illative' coming from the irregular latin verb '*inferre*', meaning 'to infer'. Many systems that were devised turned out to be either too weak or too strong, see Curry et al. (1972).

DEFINITION 16.8. A system of *Illative Combinatory Logic*[10] (ICL) $\mathcal{I}(\vec{\mathsf{P}})$, with constants $\mathsf{P}_1, \ldots, \mathsf{P}_n$, consists of the following.

(i) A term is an element of $\Lambda(\vec{\mathsf{P}})$.

(ii) A *statement* of $\mathcal{I}(\vec{\mathsf{P}})$ is just an element of $\Lambda(\vec{\mathsf{P}})$.

(iii) A *basis* is a set of statements.

(iv) Derivability of a statement X from a basis Γ, in notation $\Gamma \vdash X$, is defined by the general rules $\mathcal{I}_0$

$$\begin{array}{|rcl|} X \in \Gamma & \Rightarrow & \Gamma \vdash X; \\ \Gamma \vdash X, \quad X =_{\beta\eta} Y & \Rightarrow & \Gamma \vdash Y. \end{array}$$

plus some specific rules.

The independent pioneers of ICL, Curry and Church, were confronted with inconsistent theories. A simple theory $\mathcal{I}\mathsf{P}_0^*$, with rules corresponding to 'implication elimination' (Modus Ponens) and 'implication introduction' (cf. the *deduction theorem*) turned out to be inconsistent.[11]

PROPOSITION 16.9 (CURRY'S PARADOX.). *Define the following ICL system* $\mathcal{I}\mathsf{P}_0^*$, *with just one constant* P. *Write* $X \supset Y = \mathsf{P}XY$. *Next to the basic ICL rules* $\mathcal{I}_0$, *the specific rules of* $\mathcal{I}\mathsf{P}_0^*$ *are*

$$\begin{array}{|rcl|} \Gamma \vdash X \supset Y, \quad \Gamma \vdash X & \Rightarrow & \Gamma \vdash Y; \\ \Gamma, X \vdash Y & \Rightarrow & \Gamma \vdash (X \supset Y). \end{array}$$

Then $\mathcal{I}\mathsf{P}_0^*$ *is inconsistent: every statement* Y *can be derived.*

PROOF. Using the fixed point combinator, construct a term X such that $X =_\beta (X \supset Y)$. Then assuming X, one can derive $X \supset Y$, hence Y, by Modus Ponens. It follows that $X \supset Y$ and as $(X \supset Y) =_\beta X$, also X. Therefore Y by Modus Ponens.[12] □

The proof of the inconsistency can be prevented by introducing a constant H that restricts the use of the deduction theorem, as suggested by the Curry school: Curry et al. (1972), Bunder (1974). See also Seldin (2009) for more on the history of ICL.

[10] In spite of the name 'Illative Combinatory Logic' we work with λ-terms.

[11] Curry coined the phrase "Combinatory completeness and deductive completeness are incompatible".

[12] The proof of Löb's theorem (that immediately implies Gödel's second incompleteness theorem), Löb (1955), follows exactly this pattern.

DEFINITION 16.10. Define the ICL system $\mathcal{I}\mathsf{P}$, with constants $\{\mathsf{P}, \mathsf{H}\}$, as follows.

$$\begin{array}{|rrcl|} \Gamma \vdash X \supset Y, & \Gamma \vdash X & \Rightarrow & \Gamma \vdash Y; \\ \Gamma, X \vdash Y, & \Gamma \vdash \mathsf{H}Y & \Rightarrow & \Gamma \vdash (X \supset Y); \\ \Gamma, X \vdash \mathsf{H}Y, & \Gamma \vdash \mathsf{H}Y & \Rightarrow & \Gamma \vdash \mathsf{H}(X \supset Y). \end{array}$$

Here $\mathsf{H}Y$ has to be interpreted that 'Y is member of H' is a class of 'legal' propositions. In general, a statement like AM is to be interpreted that A represents a class (or type) and that M belongs to this class.[13] In collaboration with Curry's last PhD student Martin Bunder this ICL system was found adequate for a natural interpretation of propositional logic, Barendregt et al. (1993).

DEFINITION 16.11. Define the ICL system $\mathcal{I}\Xi$ over $\Lambda(\Xi, \mathsf{L})$, with derivability of X in $\mathcal{I}\Xi$ from Γ, in notation $\Gamma \vdash_{\mathcal{I}\Xi} X$, or simply $\Gamma \vdash X$, is defined by the following specific rules. Write $\mathsf{H} \triangleq \mathsf{L} \circ \mathsf{K}$.

$$\begin{array}{|rrcll|} \Gamma \vdash \Xi XY, & \Gamma \vdash XV & \Rightarrow & \Gamma \vdash YV; & \\ \Gamma, Xx \vdash Yx, & \Gamma \vdash \mathsf{L}X & \Rightarrow & \Gamma \vdash \Xi XY, & \text{if } x \notin \mathrm{FV}(\Gamma, X, Y); \\ \Gamma, Xx \vdash \mathsf{H}(Yx), & \Gamma \vdash \mathsf{L}X & \Rightarrow & \Gamma \vdash \mathsf{H}(\Xi XY), & \text{if } x \notin \mathrm{FV}(\Gamma, X, Y). \end{array}$$

There is an interpretation of the $\forall, \supset$-fragment of first-order (many-sorted) predicate logic into $\mathcal{I}\Xi$. As usual, the interpretation depends on the signature Σ of the (structures and) first-order logic considered. To fix ideas, consider

$$\Sigma = \langle A_1, A_2, f, g, P, c \rangle,$$

where these symbols denote the following:

$$\begin{array}{rl} A_1, A_2 & \text{non-empty sets;} \\ f\colon A_1 \to A_2 & \text{a unary function;} \\ g\colon A_1 \to A_2 \to A_3 & \text{a binary function;} \\ P \subseteq A_1 & \text{a unary relation;} \\ c \in A_1 & \text{a constant.} \end{array}$$

This signature Σ is translated as a basis Γ_Σ in $\mathcal{I}\Xi$ as follows.

$$\Gamma_\Sigma = \mathsf{L}A_1, \mathsf{L}A_2, \mathsf{F}A_1A_2f, \mathsf{F}A_1(\mathsf{F}A_2A_3)g, \mathsf{F}A_1\mathsf{H}P, A_1c, A_2a_2.$$

---

[13]One also can say that in AM the A represents a *predicate* that holds for the *subject* M. In type theory this later became $\vdash M : A$. For this reason the inference

$$\vdash AM, \quad M \twoheadrightarrow N \quad \Rightarrow \quad \vdash AN,$$

which in type theory becomes

$$\vdash M : A, \quad M \twoheadrightarrow N \quad \Rightarrow \quad \vdash N : A,$$

is called 'subject reduction property'.

The assumption $A_2 a_2$ is necessary in order to obtain first-order logic with non-empty domains. Without the assumption of inhabited domains one obtains the so-called *free-logic*, developed in Peremans (1949) and Mostowski (1951), in which less formulas are valid and provable. For example,

$$\forall x \in A_2.Qx \supset \exists x \in A_2.Qx$$

is invalid and underivable in free-logic,[14] but valid and derivable in ordinary logic.

DEFINITION 16.12. In order to translate first-order sentences in the language with signature Σ, consider as an example

$$B \triangleq \forall x_1 \in A_1[(\forall x_2 \in A_2.P(fx_1x_2)) \supset P(x_1)],$$

where we use the λ-calculus notation 'fx_1x_2' for what is usually denoted as '$f(x_1, x_2)$'. This statement B is translated into $\mathcal{I}\Xi$ as $[B]_{\mathcal{I}\Xi}$ which is

$$[B]_{\mathcal{I}\Xi} \triangleq \Xi A_1\big(\lambda x_1.\Xi A_2\big(\lambda x_2.\Xi(\mathsf{K}(P(fx_1x_2)))(\mathsf{K}(Px_1))\big)\big).$$

A set of statements $\Gamma = \{B_1, \ldots, B_n\}$ in the signature Σ is translated as

$$[\Gamma]^{\Sigma}_{\mathcal{I}\Xi} \triangleq \Gamma_{\Sigma} \cup \{[B_1]_{\mathcal{I}\Xi}, \ldots, [B_n]_{\mathcal{I}\Xi}\}.$$

PROPOSITION 16.13. *The interpretation* $[\ ]_{\mathcal{I}\Xi}$ *of the* $\forall, \supset$*-fragment of first-order (many-sorted) predicate logic into* $\mathcal{I}\Xi$ *is sound and complete: for all* $\Gamma = B_1, \ldots, B_n, A$ *in the language with signature* Σ *one has*

$$\Gamma \vdash A \iff [\Gamma]^{\Sigma}_{\mathcal{I}\Xi} \vdash_{\mathcal{I}\Xi} [A]_{\mathcal{I}\Xi}$$

PROOF. See Barendregt et al. (1993). □

REMARK 16.14. (i) There is also another ICL, called $\mathcal{I}\mathsf{G}$, with a formulae-as-types interpretation. Soundness now becomes

$$\Gamma \vdash A \quad \Rightarrow \quad [\Gamma]^{\Sigma}_{\mathcal{I}\mathsf{G}} \vdash_{\mathcal{I}\mathsf{G}} [A]_{\mathcal{I}\mathsf{G}}M, \text{ for some } M \in \Lambda.$$

This is also proved in Barendregt et al. (1993).

(ii) Also in this paper it was shown that both $[\ ]_{\mathcal{I}\Xi}$ and $[\ ]_{\mathcal{I}\mathsf{G}}$ can be seen as a common translation $[\ ]^R$, for some $R \in \Lambda^o$, with

$$\begin{array}{rcl} [\ ]^{\mathsf{K}} & = & [\ ]_{\mathcal{I}\Xi}; \\ [\ ]^{\mathsf{I}} & = & [\ ]_{\mathcal{I}\mathsf{G}}. \end{array}$$

(iii) Completeness for the propositions-as-types interpretation is shown in Dekkers et al. (1998b)

$$\Gamma \vdash A \iff [\Gamma]^{\Sigma}_{\mathcal{I}\mathsf{G}} \vdash_{\mathcal{I}\mathsf{G}} [A]_{\mathcal{I}\mathsf{G}}M, \text{ for some } M \in \Lambda.$$

In Dekkers et al. (1998a) completeness for the two interpretations was given in a uniform way, making use of (iii).

[14]There are also statements without $\exists$ differentiating ordinary logic from free-logic.

REMARK 16.15. (i) In Bunder (1969) the axiom LH was added to systems of ICL, obtaining $\mathcal{I}\Xi_2$, in order to show that second-order logic can be interpreted soundly in the resulting ICL. Being able to quantify over statements gives the power of second-order logic and enables the definition of the usual logical connectives $\bot, \neg, \&, \vee, \exists$.

(ii) If there is hope to show completeness for the canonical interpretation in $\mathcal{I}\Xi_2$, then this system should be consistent. The completeness of $\mathcal{I}\Xi$ and $\mathcal{I}\mathsf{G}$ implies consistency of these systems. The mentioned proofs of these results were rather syntactic and did not carry over to $\mathcal{I}\Xi_2$.

(iii) Consistency of $\mathcal{I}\Xi_2$ was shown in Czajka (2013, 2015) in a semantic way, making use of appropriate Kripke models.[15]

PROBLEM 38. Can completeness of $\mathcal{I}\Xi_2$ be proved using these semantic means? The results in Czajka (2015) provide partial evidence for this.

[15]For this, Czajka used a slightly different version of H.

Part VII

Appendix

Appendix A

Mathematical background

This appendix is devoted to introduce the mathematical tools used in this book giving the reader some working knowledge of category theory, domain theory, and universal algebra[1]. Before that, we propose some books on these topics that might interest the reader. Moreover, a grammar is given for the format of λ-terms that is used in Curry and Feys (1958) and B[1984].

Some books

1. **Category Theory**
    - "Basic category theory for computer scientists", Pierce (1999).
    - "Categories, types and structures: an introduction to category theory for the working computer scientist", by Asperti and Longo (1991).
    - "Categories for the working mathematician", Mac Lane (2013).
2. **Domain Theory**
    - "Continuous Lattices", by Gierz et al. (2012).
    - "Semantic Domains", by Gunter and Scott (1990).
    - "Domains and λ-calculi", by Amadio and Curien (1998).
3. **Universal Algebra**
    - "A course in universal algebra — the millennium edition", by Burris and Sankappanavar (2012).
    - "Lattice theory", by Birkhoff (1967).
    - "The combinatory program", by Engeler (2012).
    - "Algorithmic properties of structures", by Engeler (1993).

[1]No such short introduction is given for the topics logic and computability theory, as it is assumed that the reader interested in λ-calculus is already familiar with these.

4. **Logic**

   - "Introduction to mathematical logic", Mendelson (2009).
   - "Mathematical logic", Shoenfield (2018).

5. **Computability Theory**

   - "Classical Recursion theory, Volume I", Odifreddi (1989).
   - "Recursively enumerable sets and degrees", Soare (1999).

A.1 The lean notation for λ-terms

The well-known abbreviations for λ-terms, like using association to the left for ambiguously denoted applications like xyz, now will be treated as a formal language $\hat{\Lambda}$, that is in a natural bijective correspondence with the language of 'official' terms Λ.

DEFINITION A.1. (i) The language Λ of λ-terms has as alphabet $\Sigma = \{\mathtt{x}, ', \lambda, (\,,)\}$.

(ii) The language Λ over Σ is defined by the following context-free grammar.

$$\begin{array}{|lcl|} \hline S & ::= & V \mid L \mid A \\ V & ::= & \mathtt{x} \mid V' \\ L & ::= & (\lambda V\, S) \\ A & ::= & (S\, S) \\ \hline \end{array}$$

(iii) Λ is the language generated by the non-terminal S, the 'official' λ-terms.
(iv) $\mathcal{V}$ is the sublanguage of Λ generated by V, the variables.
(v) $\mathcal{L}$ is the sublanguage of Λ generated by L, the abstractions.
(vi) $\mathcal{A}$ is the sublanguage of Λ generated by A, the applications.

LEMMA A.2. (i) $\mathcal{V} = \{\mathtt{x}, \mathtt{x}', \mathtt{x}'', \ldots\}$.
(ii) $\mathcal{L} = \{(\lambda x M) \mid x \in \mathcal{V}, M \in \Lambda\}$.
(iii) $\mathcal{A} = \{(MN) \mid M, N \in \Lambda\}$.
(iv) Λ *is the disjoint union of* $\mathcal{V}$, $\mathcal{L}$, *and* $\mathcal{A}$.

In this grammar the S combinator is

$$\mathsf{S} = (\lambda x(\lambda y(\lambda z((xz)(yz))))).$$

This makes λ-terms hard to read. Much simpler is the version in 'lean' notation

$$\mathsf{S} = \lambda xyz.xz(yz).$$

This notation—well-known by the book Curry and Feys (1958)—is introduced as an abbreviation in B[1984], as well as in Section 1.1 of the present book.

NOTATION A.3 (IMPRECISE). For $n \in \mathbb{N}$, define

$$\begin{array}{rcll} MN_1 \cdots N_n & \triangleq & (\cdots((MN_1)N_2)\cdots N_n), & \text{association to the left;} \\ \lambda x_1 \ldots x_n.M & \triangleq & \lambda x_1(\lambda x_2 \cdots (\lambda x_n(M))\cdots), & \text{association to the right.} \end{array}$$

As informal abbreviations these expressions may be intuitively understood. But as expressions of a formal language, to be parsed by a computer, they are not sufficiently specified. For example, if $N_1 = xx$, $N_2 = y$, $N_3 = \lambda z.z$, one should have

$$MN_1N_2N_3 = M(xx)y(\lambda z.z).$$

Moreover, if $M = w$ or $M = \lambda w.w$ this becomes, respectively,

$$\begin{array}{rcl} M_1 & = & w(xx)y(\lambda z.z); \\ M_2 & = & (\lambda w.w)(xx)y(\lambda z.z). \end{array}$$

So there is a need for rules determining when parentheses are to be placed.

A context-free grammar for the 'lean' λ-terms $\hat{\Lambda}$ is given by the following syntax.

DEFINITION A.4. (i) The alphabet of lean terms is $\hat{\Sigma} = \Sigma \cup \{.\} = \{\mathtt{x}, {}', \lambda, (\,,), \,.\,\}$.

(ii) The language $\hat{\Lambda}$ over $\hat{\Sigma}$ is defined by the following context-free grammar.

$$\begin{array}{rcl} \hat{S} & ::= & V \mid \hat{L} \mid \hat{A} \\ V & ::= & \mathtt{x} \mid V' \\ \hat{L} & ::= & \lambda W.V \mid \lambda W.\hat{A} \\ W & ::= & \mathtt{x} \mid V\,W \\ \hat{A} & ::= & FR \\ F & ::= & V \mid (\hat{L}) \mid \hat{A} \\ R & ::= & V \mid (\hat{L}) \mid (\hat{A}) \end{array}$$

(iii) $\hat{\Lambda}$ is the language generated by the non-terminal $\hat{S}$, the 'lean' λ-terms.
(iv) $\mathcal{V}$ is the sublanguage of $\hat{\Lambda}$ generated by V, the variables.
(v) $\hat{\mathcal{L}}$ is the sublanguage of $\hat{\Lambda}$ generated by $\hat{L}$, the abstractions.
(vi) $\hat{\mathcal{A}}$ is the sublanguage of $\hat{\Lambda}$ generated by $\hat{A}$, the applications.

DEFINITION A.5. (i) On $\hat{\Lambda}$ define the binding operation $\lambda\vec{x}\odot$ as follows.

$$\begin{array}{rcll} \lambda\vec{x} \odot M & = & \lambda\vec{x}.M, & \text{if } M \in \mathcal{V} \cup \hat{\mathcal{A}}; \\ \lambda\vec{x} \odot M & = & \lambda\vec{x} \frown \vec{y}.M, & \text{if } M = \lambda\vec{y}.N \in \hat{\mathcal{L}} \text{ and } \frown \text{ denotes} \\ & & & \text{concatenation of sequences.} \end{array}$$

(ii) Define $M \boxdot N$, for $M, N \in \hat{\Lambda}$, by the following table (where var, app, abs represent membership to $\mathcal{V}, \hat{\mathcal{A}}, \hat{\mathcal{L}}$, respectively):

M \ N	var	app	abs
var	MN	$M(N)$	$M(N)$
app	MN	$M(N)$	$M(N)$
abs	$(M)N$	$(M)(N)$	$(M)(N)$

Table A.1: Definition of $M \boxdot N$.

PROPOSITION A.6. (i) $\hat{\mathcal{L}} = \{\lambda x \odot M \mid x \in \mathcal{V}, M \in \hat{\Lambda}\}$. *Moreover, for each* $P \in \hat{\mathcal{L}}$ *there are unique* $x \in \mathcal{V}$ *and* $M \in \hat{\Lambda}$ *such that* $P = \lambda x \odot M$.

(ii) $\hat{\mathcal{A}} = \{M \boxdot N \mid M, N \in \hat{\Lambda}\}$. *Moreover, for each* $P \in \hat{\mathcal{A}}$ *there are unique* $M, N \in \hat{\Lambda}$ *such that* $P = M \boxdot N$.

(iii) $\hat{\Lambda}$ *is the disjoint union of* $\mathcal{V}$, $\hat{\mathcal{L}}$, *and* $\hat{\mathcal{A}}$.

DEFINITION A.7. (i) Define $T \colon \Lambda \to \hat{\Lambda}$ as follows.

$$\begin{aligned} T(x) &\triangleq x; \\ T((MN)) &\triangleq T(M) \boxdot T(N); \\ T((\lambda x M)) &\triangleq \lambda x \odot T(M). \end{aligned}$$

(ii) Define $\hat{T} \colon \hat{\Lambda} \to \Lambda$ as follows.

$$\begin{aligned} \hat{T}(x) &\triangleq x; \\ \hat{T}(M \boxdot N) &\triangleq (\hat{T}(M)\hat{T}(N)); \\ \hat{T}(\lambda x \odot M) &\triangleq (\lambda x \hat{T}(M)). \end{aligned}$$

PROPOSITION A.8. (i) *For all* $M \in \Lambda$ *one has* $\hat{T}(T(M)) = M$.

(ii) *For all* $M \in \hat{\Lambda}$ *one has* $T(\hat{T}(M)) = M$.

PROOF. (i) By induction on the generation of $M \in \Lambda$ distinguishing cases.

Case $M \in \mathcal{V}$. Then $M = x$ and $\hat{T}(T(M)) = \hat{T}(x) = x = M$.

Case $M \in \mathcal{A}$. Then $M = (PQ)$ and

$$\begin{aligned} \hat{T}(T(M)) &= \hat{T}(T(PQ)), \\ &= \hat{T}(T(P) \boxdot T(Q)), \\ &= (\hat{T}(T(P))\hat{T}(T(Q))), \\ &= (PQ), && \text{by the IH,} \\ &= M. \end{aligned}$$

Case $M \in \mathcal{L}$. Then $M = (\lambda x N)$ for some $x \in \mathcal{V}$ and $N \in \Lambda$. Then

$$\begin{aligned} \hat{T}(T(M)) &= \hat{T}(T((\lambda x N))), \\ &= \hat{T}(\lambda x \odot T(N)), \\ &= (\lambda x \hat{T}(T(N))), \\ &= (\lambda x N), && \text{by the IH,} \\ &= M. \end{aligned}$$

(ii) By induction on the generation of $M \in \hat{\Lambda}$ distinguishing cases.

Case $M \in \mathcal{V}$. Then $M = x$ and $T(\hat{T}(M)) = T(x) = x = M$.

Case $M \in \hat{\mathcal{A}}$. Then $M = P \boxdot Q$, for some $P, Q \in \hat{\Lambda}$, and

$$\begin{aligned} T(\hat{T}(M)) &= T(\hat{T}(P \boxdot Q)), \\ &= T((\hat{T}(P)\hat{T}(Q))), \\ &= T(\hat{T}(P)) \boxdot T(\hat{T}(Q)), \\ &= P \boxdot Q, && \text{by the IH,} \\ &= M. \end{aligned}$$

Case $M \in \hat{\mathcal{L}}$. Then $M = \lambda y \odot N$, for some unique $y \in \mathcal{V}$ and $N \in \hat{\Lambda}$. Then

$$\begin{aligned} T(\hat{T}(M)) &= T(\hat{T}(\lambda y \odot N)), \\ &= T((\lambda y \hat{T}(N))), \\ &= \lambda y \odot T(\hat{T}(N)), \\ &= \lambda y \odot N, && \text{by the IH,} \\ &= M. \end{aligned}$$

□

Now Notation A.3 can be improved as follows.

NOTATION A.9. Define for $M, \vec{N} \in \hat{\Lambda}$ and $n \in \mathbb{N}$

$$\begin{aligned} MN_1 \cdots N_n &\triangleq (\cdots((M \boxdot N_1) \boxdot N_2) \cdots \boxdot N_n), && \text{association to the left;} \\ \lambda\vec{x}.M &\triangleq \lambda\vec{x} \odot M, && \text{the right association.} \end{aligned}$$

EXERCISE A.10. (i) Define what is a notion of reduction on $\hat{\Lambda}$.

(ii) Define in a natural way the notion of substitution and reduction $\to_{\hat{\beta}}$ on $\hat{\Lambda}$, such that (iii) and (iv) hold.

(iii) Show that for $M, N \in \Lambda$ one has

$$M \to_\beta N \iff T(M) \to_{\hat{\beta}} T(N).$$

(iv) Conclude that for $M, N \in \hat{\Lambda}$ one has

$$M \to_{\hat{\beta}} N \iff \hat{T}(M) \to_\beta \hat{T}(N).$$

A.2 A summary of category theory

In this section we discuss the notions of category, products, currying and Cartesian closed category (CCC, for short). We also recall the definition of a Seely category and the coKleisli construction allowing to produce a CCC starting from a Seely category.

Categories

We start by defining what a category is.

DEFINITION A.11. A *category* $\mathbf{C}$ is given by:

- a collection of *objects*;
- for each pair of objects A and B, a collection $\mathbf{C}(A, B)$ of *morphisms* (*arrows*);
- for every object A, a morphism $\mathrm{Id}_A \in \mathbf{C}(A, A)$ behaving as the *identity* on A, i.e. for all $f \in \mathbf{C}(A, B)$:
$$\mathrm{Id}_B \circ f = f = f \circ \mathrm{Id}_A$$
- for each triple A, B, C of objects, a *composition* operation
$$\circ : \mathbf{C}(B, C) \times \mathbf{C}(A, B) \to \mathbf{C}(A, C)$$
satisfying associativity, i.e. for all morphisms f, g, h of the appropriate type:
$$f \circ (g \circ h) = (f \circ g) \circ h.$$

A category $\mathbf{C}$ is called *large* or *small* depending on whether its collection of objects is a proper class or a set, respectively. We mostly work with large categories that are however *locally small*, in the sense that $\mathbf{C}(A, B)$ is a set, for every pair of objects A, B. Under this hypothesis, $\mathbf{C}(A, B)$ is also called a *hom-set*.

NOTATION. (i) *When the category* $\mathbf{C}$ *is clear from the context, we often write* $f : A \to B$ *to indicate that* f *is a morphism in* $\mathbf{C}(A, B)$.

(ii) *It is sometimes convenient to indicate composition in diagrammatic order, in which case* $f; g$ *is used. In other words* $f; g \triangleq g \circ f$.

EXAMPLES A.12. (i) The category **Set** has sets as objects and set-theoretic functions as morphisms. Identity morphisms and composition are defined in the usual way.

(ii) The category **Rel** has sets as objects and relations as morphisms. Here composition of $\mathrm{R} \in \mathbf{Rel}(B, C)$ and $\mathrm{S} \in \mathbf{Rel}(A, B)$ is given by their relational product

$$\mathrm{R} \circ \mathrm{S} = \{(a, c) \in A \times C \mid \exists b \in B \,.\, (a, b) \in \mathrm{R} \;\&\; (b, c) \in \mathrm{S}\},$$

whence the identity is given by $\mathrm{Id}_A = \{(a, a) \mid a \in A\}$.

(iii) Given a category $\mathbf{C}$, the *opposite category* $\mathbf{C}^{\mathrm{op}}$ is obtained by reversing all arrows: $\mathbf{C}^{\mathrm{op}}(A, B) \triangleq \mathbf{C}(B, A)$. The composition of arrows in $\mathbf{C}^{\mathrm{op}}$ becomes $f \circ_{\mathbf{C}^{\mathrm{op}}} g \triangleq g \circ f$.

In Exercise A.39 and Definition A.63 we will see other examples of categories, based on partial orders and continuous functions.

DEFINITION A.13. A morphism $f : A \to B$ is called:

(i) a *monomorphism* if, for all $g, h : A' \to A$, we have

$$f \circ g = f \circ h \quad \Rightarrow \quad g = h$$

(ii) an *epimorphism* if, for all $g, h : B \to B'$, we have

$$g \circ f = h \circ f \quad \Rightarrow \quad g = h$$

(iii) an *isomorphism* if there exists a morphism $g : B \to A$, called the *inverse* of f, such that $g \circ f = \mathrm{Id}_A$ and $f \circ g = \mathrm{Id}_B$. In this case we say that (f, g) is an *iso-pair* and the objects A and B are *isomorphic*, written $A \cong B$. Also, g is often denoted by f^{-1}.

EXAMPLES A.14. (i) In **Set** monomorphisms are injective functions, epimorphisms are surjective functions, and isomorphisms are bijective functions.

(ii) A group $\mathcal{G}$ can be seen as a one-object category **C** where all morphisms are isomorphisms. Each element $e \in \mathcal{G}$ becomes a morphism $e : * \to *$, where $*$ denotes the sole object of **C**. The multiplication operation of $\mathcal{G}$ defines the composition in **C**.

DEFINITION A.15. (i) A *product* of two objects A_1, A_2 is given by an object $A_1 \times A_2$, together with two projections $\pi_i : A_1 \times A_2 \to A_i$ (for $i = 1, 2$) such that, for every object C and pair of arrows $f : C \to A_1$ and $g : C \to A_2$, there exists a unique morphism $\langle f, g\rangle : C \to A_1 \times A_2$ making the following diagram commute:

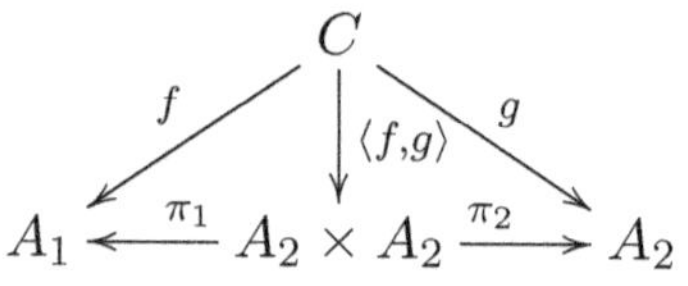

(ii) A *terminal object* in a category is an object $\mathbb{1}$ such that, for every object A, there exists a unique morphism $!_A : A \to \mathbb{1}$.

(iii) A category **C** is *Cartesian* if it has all (finite) products, i.e. if it has a terminal object and every pair of objects admits a product.

(iv) In a Cartesian category, for every pair of morphisms $f : A \to C$ and $g : B \to D$, the *product map* $f \times g : A \times B \to C \times D$ is defined by setting $f \times g \triangleq \langle f \circ \pi_1, g \circ \pi_2\rangle$.

EXAMPLES A.16. (i) Show that the category **Set** is Cartesian.

(ii) Show that the category **Rel** is Cartesian. [Hint: the categorical product is given by the disjoint union.]

(iii) Show that the product of two objects is defined up to isomorphism.

(iv) Show that, for every object A of a Cartesian category, the objects $\mathbb{1} \times A$, A and $A \times \mathbb{1}$ are isomorphic.

(v) Show that the projections $\pi_i : A \times A \to A$ (for $i = 1, 2$) are epimorphisms.

Functors and natural isomorphisms

A functor is a morphism between categories preserving the identity and composition.

DEFINITION A.17. (i) Given two categories $\mathbf{C}, \mathbf{D}$, a *functor* $F : \mathbf{C} \to \mathbf{D}$ consists of

- a map F_0 from the objects of $\mathbf{C}$ to the objects of $\mathbf{D}$;
- for each pair of objects A, B of $\mathbf{C}$, a map $F_{A,B} : \mathbf{C}(A, B) \to \mathbf{D}(F_0(A), F_0(B))$;

satisfying the axioms

$$\begin{aligned} F_{A,C}(g \circ f) &= F_{B,C}(g) \circ F_{A,B}(f) \\ F_{A,A}(\mathrm{Id}_A) &= \mathrm{Id}_{F_0(A)} \end{aligned}$$

for all $f \in \mathbf{C}(A, B)$ and $g \in \mathbf{C}(B, C)$.

(ii) Given functors $F : \mathbf{C} \to \mathbf{D}$ and $G : \mathbf{D} \to \mathbf{E}$ their *composition* $G \circ F : \mathbf{C} \to \mathbf{E}$ is defined as follows:

- $(G \circ F)_0 = G_0 \circ F_0$;
- $(G \circ F)_{A,B} = G_{F_0(A),F_0(B)} \circ F_{A,B}$.

(iii) A *binary functor* (*bifunctor*, for short), is a functor whose domain is the product of two categories, as in

$$F : \mathbf{C}_1 \times \mathbf{C}_2 \to \mathbf{D}$$

For a functor F it is customary to omit the indices and write F for F_0 and $F_{A,B}$.

REMARK A.18. It is easy to check that functors preserve isomorphisms. If (f, g) is an iso-pair in $\mathbf{C}$, and $F : \mathbf{C} \to \mathbf{D}$ is a functor, then $(F(f), F(g))$ is an iso-pair in $\mathbf{D}$.

A natural transformation is a morphism between functors.

DEFINITION A.19. (i) Let $\mathbf{C}, \mathbf{D}$ be categories and $F, G : \mathbf{C} \to \mathbf{D}$ be functors. A *natural transformation* $\nu : F \to G$ is given by a family of morphisms $\nu_A : F(A) \to G(A)$, for every object A of $\mathbf{C}$, such that the naturality diagram below commutes, for all $f : A \to B$,

$$\begin{array}{ccc} F(A) & \xrightarrow{\nu_A} & G(A) \\ \downarrow{\scriptstyle F(f)} & & \downarrow{\scriptstyle G(f)} \\ F(B) & \xrightarrow{\nu_B} & G(B) \end{array}$$

(ii) The *composition* of natural transformations $\nu : F \to G$ and $\mu : G \to H$ is defined componentwise: $(\mu \circ \nu)_A \triangleq \mu_A \circ \nu_A$.

(iii) A *natural isomorphism* ν is a natural transformation such that ν_A is an isomorphism, for every object A.

EXAMPLES A.20. (i) The category **Cat** has small categories as objects, and functors as morphisms.

(ii) Given categories $\mathbf{C}$ and $\mathbf{D}$, the functor category $\mathbf{D}^{\mathbf{C}}$ has functors as objects and natural transformations as morphisms.

Symmetric monoidal closed categories

A monoidal category is a category equipped with some notion of 'tensor product'. Unlike categorical products, tensor products are not unique up to isomorphism and do not necessarily have projections. For the missing axioms in the definition below, and the notion of braided category, we refer to Melliès (2003).

DEFINITION A.21. (i) A *monoidal category* is a category $\mathbf{C}$ equipped with a bifunctor

$$- \otimes - : \mathbf{C} \times \mathbf{C} \to \mathbf{C},$$

called the *tensor product*, associative up to a natural isomorphism

$$\alpha_{A,B,C} : (A \otimes B) \otimes C \to A \otimes (B \otimes C)$$

and with an object 1, which is a unit up to natural isomorphisms

$$\lambda_A : 1 \otimes A \to A, \qquad \rho_A : A \otimes 1 \to A.$$

The structure maps α, λ, ρ must satisfy the usual axioms.

(ii) A *monoidal closed category* is a monoidal category $(\mathbf{C}, \otimes, 1)$ such that, for all objects A, B, there exist an object $A \multimap B$ and a morphism $\mathrm{Eval}_{A,B} : A \otimes (A \multimap B) \to B$ satisfying the following universal property. For every morphism

$$f : A \otimes X \to B$$

there exists a unique morphism $h : X \to (A \multimap B)$ making the next diagram commute

$$\begin{array}{ccc} A \otimes X & & \\ {\scriptstyle \mathrm{Id}_A \otimes h} \downarrow & \searrow^{f} & \\ A \otimes (A \multimap B) & \xrightarrow[\mathrm{Eval}_{A,B}]{} & B \end{array}$$

(iii) A *symmetric monoidal closed category* (*SMCC*) is a monoidal closed category $(\mathbf{C}, \otimes, 1)$ having a natural isomorphism (called *symmetry*)

$$\sigma_{A,B} : A \otimes B \to B \otimes A$$

that satisfies $\sigma_{B,A} = \sigma^{-1}_{A,B}$ as well as the usual conditions on a braiding.

EXAMPLES A.22. (i) The category of modules over a commutative ring $\mathcal{R}$, is a monoidal category, with the tensor product of modules serving as monoidal product and the ring $\mathcal{R}$ (thought of as a module over itself) serving as unit.

(ii) The category **Set** is a monoidal category with the Cartesian product and singletons serving as unit.

(iii) The category **Rel** is an SMCC. The object $A \multimap B$ representing the hom-set $\mathbf{Rel}(A, B)$ and the monoidal product $A \otimes B$ are given by the Cartesian product of sets.

REMARK A.23. Every Cartesian category can be seen as a monoidal category, whose monoidal structure is given by the categorical product (the unit is the terminal object).

DEFINITION A.24. A *Cartesian closed category* $\mathbf{C}$ (*CCC*, for short) is a category with finite products which is closed with respect to its Cartesian monoidal structure $(\mathbf{C}, \times, \mathbb{1})$.

In a CCC, the object internalizing $\mathbf{C}(A, B)$ is often called *exponential object* and is denoted by $A \Rightarrow B$, rather than $A \multimap B$.

The coKleisli construction

A monad is a categorical generalization of a monoid, therefore it has a multiplication and a unit. Kleisli categories over a monad are important concepts in category theory. For our purposes, we rather need their duals namely coKleisli categories over comonads. This duality has a precise meaning, a comonad over a category $\mathbf{C}$ is a monad over $\mathbf{C}^{\mathrm{op}}$.

DEFINITION A.25. (i) A *comonad* over a category $\mathbf{C}$ is a triple $(!, \mathrm{der}, \mathrm{dig})$ where

- $! : \mathbf{C} \to \mathbf{C}$ is an endofunctor;
- $\mathrm{der} : ! \to \mathrm{Id}_{\mathbf{C}}$ (*counit*) and $\mathrm{dig} : ! \to !^2$ (*comultiplication*) are natural transformations, such that the following diagrams commute:

$$\begin{array}{ccc} !A & \xrightarrow{\mathrm{dig}_A} & !!A \\ {\scriptstyle \mathrm{dig}_A}\downarrow & \diagdown\!\!\diagdown & \downarrow{\scriptstyle \mathrm{der}_{!A}} \\ !!A & \xrightarrow[!\mathrm{der}_A]{} & !A \end{array} \qquad\qquad \begin{array}{ccc} !A & \xrightarrow{\mathrm{dig}_A} & !!A \\ {\scriptstyle \mathrm{dig}_A}\downarrow & & \downarrow{\scriptstyle !\mathrm{dig}_A} \\ !!A & \xrightarrow[\mathrm{dig}_{!A}]{} & !!!A \end{array}$$

The counit and comultiplication of the comonad are often called *dereliction* and *digging*, respectively.

(ii) For all $g : !A \to B$, define a morphism $g^! : !A \to !B$ by setting

$$g^! \triangleq !f \circ \mathrm{dig}_{!A}$$

(iii) The *coKleisli category* $\mathbf{C}_!$ of a comonad $(!, \mathrm{der}, \mathrm{dig})$ over $\mathbf{C}$ is defined as follows:

- $\mathbf{C}_!$ has the same objects as $\mathbf{C}$;
- $\mathbf{C}_!(A, B) \triangleq \mathbf{C}(!A, B)$, for all objects A, B;
- the composition of $\mathbf{C}_!$ is given by

$$f \circ_! g \triangleq f \circ g^!$$

EXERCISE A.26. (i) Infer the definitions of monad and Kleisli category of a monad, knowing that they are duals to the notions of comonad and coKleisli, respectively.

(ii) The Kleisli category of the power set monad $\mathscr{P}(\cdot)$ on **Set** has sets as objects and functions $A \to \mathscr{P}(B)$ as morphisms. These morphisms may be identified with relations $\mathrm{R} \subseteq A \times B$. Check that composition in the Kleisli category corresponds to relational composition. Conclude that **Rel** coincides with the Kleisli category of $\mathscr{P}(\cdot)$ over **Set**.

Seely categories arise as categorical models of multiplicative exponential linear logic. For coherence with that tradition, in the following we exceptionally use & and $\top$ to denote the Cartesian product and terminal object, respectively.

DEFINITION A.27. A *Seely*[2] *category* is a symmetric monoidal closed category $(\mathbf{C}, \otimes, 1)$ with finite products $(\mathbf{C}, \&, \top)$ and equipped with

1. a comonad $(!, \mathrm{der}, \mathrm{dig})$;
2. two natural isomorphisms (called *Seely's isomorphisms*)
$$\mathrm{m}^2_{A,B} : !A \otimes !B \cong !(A \,\&\, B), \qquad \mathrm{m}^0 : 1 \cong !\top$$
making
$$(!, \mathrm{m}) : (\mathbf{C}, \&, \top) \to (\mathbf{C}, \otimes, 1)$$
a symmetric monoidal functor;
3. the following coherence diagram commutes
$$\begin{array}{ccc} !A \otimes !B & \xrightarrow{\quad \mathrm{m} \quad} & !(A \,\&\, B) \\ & & \downarrow{\scriptstyle \mathrm{dig}_{A\&B}} \\ \downarrow{\scriptstyle \mathrm{dig}_A \otimes \mathrm{dig}_B} & & !!(A \,\&\, B) \\ & & \downarrow{\scriptstyle !\langle !\pi_1, !\pi_2\rangle} \\ !!A \otimes !!B & \xrightarrow[\quad \mathrm{m} \quad]{} & !(!A \,\&\, !B) \end{array}$$

EXERCISE A.28. (i) Show that the finite multisets functor $\mathscr{M}_{\mathrm{f}}(\cdot)$ is a comonad over **Rel**.
(ii) Show that **Rel** is a Seely category.
(iii) Check that $\mathbf{Rel}_{\mathscr{M}_{\mathrm{f}}(\cdot)}$ coincides with the category **MRel** from Definition 14.3.

THEOREM A.29 (SEELY (1989)). *Given a Seely category* $\mathbf{C}$ *with a comonad* $!(\cdot)$, *its coKleisli category* $\mathbf{C}_!$ *is Cartesian closed.*

PROOF SKETCH. The product objects and pairing are the same as in $\mathbf{C}$, while projections are given by $\pi_i \circ \mathrm{der}$. In fact, we have (for $i = 1, 2$):
$$\begin{array}{rcl} (\pi_i \circ \mathrm{der}) \circ_! \langle f_1, f_2\rangle & = & \pi_i \circ \mathrm{der} \circ \langle f_1, f_2\rangle^! \\ & = & \pi_i \circ \langle f_1, f_2\rangle \\ & = & f_i. \end{array}$$

Now, define $A \Rightarrow B \triangleq (!A) \multimap B$. The natural isomorphisms are obtained as follows
$$\begin{array}{rcl} \mathbf{C}_!(A \,\&\, B, C) & = & \mathbf{C}(!(A \,\&\, B), C) \;\cong\; \mathbf{C}(!A \otimes !B, C) \\ & \cong & \mathbf{C}(!A, !B \multimap C) \;=\; \mathbf{C}(!A, B \Rightarrow C) = \mathbf{C}_!(A, B \Rightarrow C). \end{array}$$

In other words, the Cartesian closedness of $\mathbf{C}_!$ follows from the monoidal closed structure of $\mathbf{C}$ in presence of Seely's isomorphisms. □

[2] To be precise, this definition correspond to the so-called 'new Seely' category, but we do not believe that adding a couple of axioms deserves a change of terminology.

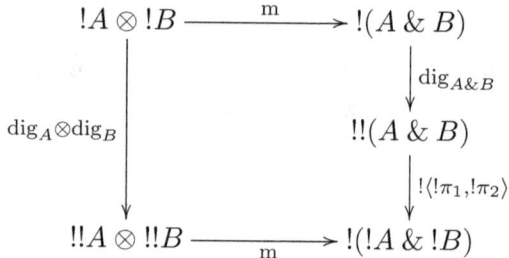

A.3 A summary of domain theory

In this section we review the notions of partially ordered sets, complete partial orders, algebraicity, (Scott-)continuity, Scott-domains and Scott-topology. We also define the category **Alg** of ω-algebraic complete lattices and recall some of its properties.

Complete partial orders and continuity

We start with some basic definitions.

DEFINITION A.30. (i) A *partially ordered set (poset)* $\mathcal{D}$ is a pair $(D, \leq)$, where D is a set and $\leq$ a *partial ordering* on D, namely a transitive, reflexive and antisymmetric relation. The set D is called the *underlying set* of $\mathcal{D}$, its elements are denoted by $a, b, c, \ldots$.

(ii) Two elements c, d of a poset $\mathcal{D}$ are called *compatible* if they have an *upper bound*, i.e., if there exists $e \in D$ such that $c \leq e$ and $d \leq e$; they are *incompatible* otherwise.

(iii) The *least upper bound* (*lub*, or *supremum*) of $c, d \in D$, if it exists, is denoted $c \vee d$.

(iv) Similarly, if the lub of a subset $A \subseteq D$ exists, then it is denoted $\bigvee A$.

DEFINITION A.31. Let $\mathcal{D} = (D, \leq)$ be a poset and $A \subseteq D$.

(i) The set A is a *chain* if it is *totally ordered* by $\leq$, i.e.

$$\forall a, b \in A . [\, a \leq b \text{ or } b \leq a \,]$$

(ii) The set A is *upward closed* if, for all $a \in A$ and $d \in D$, $a \leq d$ implies $d \in A$.

(iii) The *upward closure* of A, written $A\!\uparrow$, is the least upward closed set containing A. Concretely, we have $A\!\uparrow \triangleq \{d \in D \mid \exists a \in A . a \leq d\}$.

(iv) We say that A is *directed* if it is not empty and every two elements in A have an upper bound belonging to A, i.e.

$$\forall a, b \in A, \exists d \in A . [a \leq d \;\&\; b \leq d]$$

We recall here the statement of the Zorn's lemma, also known as Kuratowski–Zorn's lemma[3]. The proof is omitted since the result is quite famous.

THEOREM A.32 (ZORN'S LEMMA). *Given a poset $\mathcal{D}$, if every chain $A \subseteq D$ has an upper bound in A, then D contains a maximal element.*

We often consider posets satisfying additional properties, like having a least element as well as the suprema of directed subsets.

DEFINITION A.33. A *complete partial order* (*cpo*) is a triple $\mathcal{D} = (D, \leq, \bot)$ such that

(i) $(D, \leq)$ is a poset;

(ii) $\bot \in D$ satisfies $\bot \leq d$, for all $d \in D$;

(iii) every directed set $A \subseteq D$ admits a least upper bound, denoted by $\bigvee A$.

The element $\bot$ (*bottom*) is the least element of D. If such a least element exists, then it is unique, whence we say that a poset $(D, \leq)$ is a cpo whenever $(D, \leq, \bot)$ is.

---

[3]As explained by Campbell (1978), this result was proved by Kuratowski in 1922 and independently by Zorn in 1935.

EXAMPLES A.34. The following are cpo's.

(i) The set $\{\bot, \top\}$ ordered by $\bot \leq \top$.

(ii) The set $\{\bot, \mathtt{tt}, \mathtt{ff}\}$ partially ordered by $\bot \leq \mathtt{tt}$ and $\bot \leq \mathtt{ff}$.

(iii) In fact, all finite posets having a bottom element are cpo's.

(iv) Given a set X and a constant $\bot \notin X$, write $X_\bot \triangleq X \cup \{\bot\}$. Define a partial order $\leq$ on $X_\bot$ by setting $x \leq y \iff x = \bot$. Then $(X_\bot, \leq)$ is a cpo, called *flat domain*.

EXERCISE A.35. (i) Show that $(\mathbb{N}, \leq)$ is a poset, not a cpo.

(ii) Show that $(\{\{n\}\!\uparrow \mid n \in \mathbb{N}\}, \subseteq)$ is a cpo.

NOTATION. (i) *It is sometimes convenient to annotate* $\leq, \bot, \vee$ *and* $\bigvee$ *with a subscript* $\mathcal{D}$*, to specify that we are working in the cpo* $\mathcal{D}$*. For instance, we may write* $\leq_{\mathcal{D}}$ *or* $\bot_{\mathcal{D}}$*.*

(ii) *When considering a partial order* $\sqsubseteq$*, we rather use* $\sqcup$ *and* $\bigsqcup$ *for the suprema.*

DEFINITION A.36. Let $\mathcal{D} = (D, \leq)$ and $\mathcal{D}' = (D', \sqsubseteq)$ be two posets.

(i) The *pointwise ordering* of two functions $f, g : D \to D'$ is defined as follows:

$$f \sqsubseteq_{\mathcal{D}\to\mathcal{D}'} g \iff \forall d \in D \,.\, f(d) \sqsubseteq g(d)$$

(ii) A function $f : D \to D'$ is called *monotonic* whenever

$$\forall c, d \in D \,.\, [\; c \leq d \quad \Rightarrow \quad f(c) \sqsubseteq f(d) \;]$$

(iii) If $\mathcal{D}$ and $\mathcal{D}'$ are cpo's, then a function $f : D \to D'$ is called *continuous* if it is monotonic and, for all directed subsets $A \subseteq D$, we have

$$f(\bigvee A) = \bigsqcup f(A) \;\; (\triangleq \bigsqcup \{f(a) \mid a \in A\})$$

(iv) A continuous function f as above is *strict* whenever $f(\bot_{\mathcal{D}}) = \bot_{\mathcal{D}'}$.

We now introduce the notion of compact elements, intuitively representing the 'finite' elements of a cpo.

DEFINITION A.37. Let $\mathcal{D}$ be a cpo.

(i) An element $d \in D$ is *compact* if the following implication holds, for every directed subset $A \subseteq D$ and $d \in D$:

$$d \leq \bigvee A \quad \Rightarrow \quad \exists a \in A \,.\, d \leq a$$

(ii) We write $\mathcal{K}(\mathcal{D})$ for the set of compact elements of $\mathcal{D}$.

(iii) For $d \in D$, define the set of *approximants* of d by $\mathcal{K}(d) \triangleq \{c \in \mathcal{K}(\mathcal{D}) \mid c \leq d\}$.

(iv) The cpo $\mathcal{D}$ is called *algebraic* if, for all $d \in D$, $\mathcal{K}(d)$ is directed and

$$d = \bigvee \mathcal{K}(d)$$

(v) The cpo $\mathcal{D}$ is called ω*-algebraic* if it is algebraic and $\mathcal{K}(\mathcal{D})$ is denumerable.

Note that $\bot$ is always compact. Moreover, $\mathcal{K}(\mathcal{D})$ is closed under lub $\vee$.

PROPOSITION A.38. *Let $\mathcal{D}, \mathcal{D}'$ be algebraic cpo's.*

(i) *Let $f : D \to D'$ be continuous and $d \in D$. We have*

$$f(d) = \bigvee\{f(c) \mid c \in \mathcal{K}(\mathcal{D})\}$$

(ii) *For all continuous functions $f, g : D \to D'$, we have*

$$f\restriction_{\mathcal{K}(\mathcal{D})} = g\restriction_{\mathcal{K}(\mathcal{D})} \iff f = g.$$

PROOF. (i) Assume that f is continuous. It is enough to show that for every $d \in D$ and $c' \in \mathcal{K}(\mathcal{D}')$ such that $c' \leq_{\mathcal{D}'} f(d)$, there exists $c \in \mathcal{K}(\mathcal{D})$ satisfying $c \leq_{\mathcal{D}} d$ and $c' \leq_{\mathcal{D}'} f(c)$. Since D is algebraic, we have $\mathcal{K}(d)$ directed and $d = \bigvee \mathcal{K}(d)$. By continuity of f, we obtain

$$c' \leq_{\mathcal{D}'} f(d) = f\Big(\bigvee \mathcal{K}(d)\Big) = \bigvee f(\mathcal{K}(d)).$$

We conclude by compactness of c'.

(ii) ($\Rightarrow$) By (i).

($\Leftarrow$) Immediate. □

EXERCISE A.39. (i) Given two cpo's $\mathcal{D}, \mathcal{D}'$, show that the set of continuous functions from D to D', pointwisely ordered, constitutes a complete partial order.

(ii) Show that the Cartesian product $D \times D'$, ordered componentwise, is a cpo.

(iii) Conclude that the category **Cpo** of cpo's and continuous function is a CCC.

(iv) Show that, nevertheless, the category **ACpo** of algebraic cpo's and continuous functions is not a Cartesian closed. [Hint: the cpo defined in (i) might not be algebraic.]

There is a canonical way of extending a preorder with a least element to an algebraic cpo, which is reminiscent of the completion for constructing real numbers from rationals.

DEFINITION A.40. (i) A *preordered set* is a pair $\mathcal{P} = (P, \preceq)$ such that P is a set and $\preceq$ a *preorder* on P, namely a reflexive and transitive binary relation.

(ii) A subset $A \subseteq P$ is *downward closed* if $c \in A$ and $d \preceq c$ imply $d \in A$.

(iii) The *downward closure* of $A \subseteq P$, written $A\!\downarrow$, is the least downward closed set containing A. Given $a \in P$, we simply write $a\!\downarrow$ for $\{a\}\!\downarrow$.

(iv) An *ideal I over $\mathcal{P}$* is a directed (hence, non-empty), downward closed subset of P. An ideal I over $\mathcal{P}$ having the form $I = a\!\downarrow$, for some $a \in P$, is called *principal*.

(v) Given a preordered set $\mathcal{P}$, we denote by $\mathcal{I}(\mathcal{P})$ the set of all ideals over $\mathcal{P}$ and say that $\mathcal{I}(\mathcal{P})$ is the *ideal completion* of $\mathcal{P}$.

PROPOSITION A.41. (i) *If $\mathcal{P}$ is a preordered set with $\bot$ then $\mathcal{I}(\mathcal{P})$, partially ordered by inclusion, is an algebraic cpo whose compact elements are exactly the principle ideals.*

(ii) *If $\mathcal{D} = (D, \leq, \bot)$ is an algebraic cpo, then $\mathcal{D} \cong \mathcal{I}(\mathcal{K}(\mathcal{D}))$ in the category* **Cpo**.

PROOF. (i) Clearly, $(\mathcal{I}(\mathcal{P}), \subseteq)$ is a poset having $\bot\!\downarrow$ as bottom element. If A is a directed set of ideals then $\bigcup A$ is an ideal, whence it belongs to $\mathcal{I}(\mathcal{P})$. The rest follows from the facts that principal ideals are compact and every $I = \bigcup\{a\!\downarrow \mid a\!\downarrow \subseteq I\}$. Note that the directedness of $\{a\!\downarrow \mid a\!\downarrow \subseteq I\}$ follows from the directedness of I.

(ii) The two inverse functions are $x \mapsto \{c \in \mathcal{K}(\mathcal{D}) \mid c \leq x\}$ and $I \mapsto \bigvee I$. □

EXERCISE A.42. Show that the set $\mathscr{B}$ of Böhm-like trees is isomorphic to $\mathcal{I}(\mathscr{A}, \leq_\bot)$.

Scott topology

A *topological space* is given by a set X and a collection $\mathcal{O}$ of subsets $O \subseteq X$, such that $\mathcal{O}$ contains both the empty set $\emptyset$ and the whole space X, and it is closed under finite intersection and arbitrary union. The members of $\mathcal{O}$ are called *open sets* and the collection $\mathcal{O}$ is called a topology on X.

DEFINITION A.43. We say that a topological space $(X, \mathcal{O})$ is:

(i) *discrete* if every subset of X is open.

(ii) *compact* if every open cover has a finite subcover, i.e., for every $\mathcal{C} \subseteq \mathcal{O}$ such that $X = \bigcup \mathcal{C}$ there exists a finite subset $\mathcal{F} \subseteq \mathcal{C}$ such that $X = \bigcup \mathcal{F}$.

A subset $A \subseteq X$ is called *closed* if its complement $X - A$ is an open set, and *clopen* if it is both closed and open.

DEFINITION A.44. Let $(X, \mathcal{O})$ be a topological space.

(i) We associate a *specialization preorder* $\leq_{\mathcal{O}}$ by setting

$$a \leq_{\mathcal{O}} b \iff \forall O \in \mathcal{O} . [a \in O \quad \Rightarrow \quad b \in O]$$

In other words, $a \leq_{\mathcal{O}} b$ holds if every neighborhood of a is also a neighborhood of b.

(ii) We say that the space X is

(1) T_0, if $\leq_{\mathcal{O}}$ is a partial order (in this case $\leq_{\mathcal{O}}$ is called the *specialization order*);

(2) T_1 (or Frechet) if, for all distinct $a, b \in X$, there is an open set O such that $a \in O$ whilst $b \notin O$.

(3) T_2 (or Hausdorff) if, for all distinct $a, b \in X$, there exist two open sets O_1 and O_2 with $a \in O_1, b \in O_2$ and $O_1 \cap O_2 = \emptyset$.

(4) *totally disconnected* if, for all distinct $a, b \in X$, there is a clopen set $A \subseteq X$ such that $a \in A \subseteq X - \{b\}$.

REMARK A.45. (i) Any continuous map between T_0-spaces is necessarily monotonic.

(ii) A topological space $(X, \mathcal{O})$ is T_1 if and only if the associated specialization order is discrete, i.e. $a \leq_{\mathcal{O}} b \iff a = b$.

(iii) For a compact Hausdorff space, total disconnectedness is equivalent to require that the clopen sets constitute a base.

In Section 15.1 we work with *Boolean spaces*, namely totally disconnected Hausdorff spaces. By Remark A.45(iii), a Boolean space is a compact Hausdorff space where every open is the union of the clopen sets that it contains.

DEFINITION A.46. Let $\mathcal{D}$ be a cpo.

(i) The *Scott topology* on $\mathcal{D}$ is defined by specifying its open sets, namely those subsets $O \subseteq D$ that are upward closed and satisfy the condition

$$\forall A \subseteq O . A \text{ directed } \& \bigvee A \in O \quad \Rightarrow \quad A \cap O \neq \emptyset$$

(ii) We denote by $\mathcal{O}_S(\mathcal{D})$ the collection of all Scott-open subsets of D.

(iii) For $a \in D$, we let $O_a \triangleq \{b \in D \mid b \not\leq a\}$.

LEMMA A.47. *For all $a \in D$, O_a is a Scott-open.*

PROOF. Easy. □

LEMMA A.48. *A subset $C \subseteq D$ is Scott-closed if and only if*

1. *C is downward closed;*

2. *for all $A \subseteq C$ directed, if $\bigvee A$ exists then $\bigvee A \in C$.*

PROOF. Straightforward from Definition A.46(i). □

LEMMA A.49. *Given a cpo $\mathcal{D}$, the specialization preorder of $\mathcal{O}_S(\mathcal{D})$ is $\leq_{\mathcal{D}}$.*

PROOF. By definition of Scott topology, we have that $\leq_{\mathcal{D}} \subseteq \leq_{\mathcal{O}_S(\mathcal{D})}$. Conversely, let $a \leq_{\mathcal{O}_S(\mathcal{D})} b$ and suppose $a \not\leq_{\mathcal{D}} b$ towards a contradiction. By Lemma A.47, we get $a \in O_b$. By definition of $\leq_{\mathcal{O}_S(\mathcal{D})}$ we derive $b \in O_b$, thus contradicting reflexivity. □

COROLLARY A.50. *The Scott topology on a cpo $\mathcal{D}$ is T_0 (in general, not T_1).*

PROPOSITION A.51. *Let $\mathcal{D}, \mathcal{E}$ be cpo's. Then the Scott continuous functions from $\mathcal{O}_S(\mathcal{D})$ to $\mathcal{O}_S(\mathcal{E})$ are exactly the functions that are continuous in the sense of Definition A.36*(iii).

PROOF. Let $f : \mathcal{D} \to \mathcal{E}$ be Scott continuous and $A \subseteq D$ be directed. By Remark A.45(i), f is monotonic. Now suppose, by the way of contradiction, that $f(\bigvee A) \not\leq \bigvee f(A)$, i.e.

$$\bigvee A \in f^{-1}(O_{\bigvee f(A)})$$

By Lemma A.47 and topological continuity $f^{-1}(O_{\bigvee f(A)})$ is Scott-open, so there is $b \in A$ such that $f(b) \in O_{\bigvee f(A)}$ and this contradicts $f(b) \leq \bigvee f(A)$. The converse is easy. □

PROPOSITION A.52. *If $\mathcal{D}$ is an algebraic cpo, then*

$$\{d\!\uparrow \mid d \in \mathcal{K}(\mathcal{D})\}$$

constitutes a base of $\mathcal{O}_S(\mathcal{D})$.

PROOF. For all $d \in \mathcal{K}(\mathcal{D})$, the set $d\!\uparrow$ is Scott-open by definition of compactness. We need to show:

1. $\forall d_1, d_2 \in \mathcal{K}(\mathcal{D}) \,.\, d_1\!\uparrow \cap\, d_2\!\uparrow \neq \emptyset \quad \Rightarrow \quad \exists d_3 \in \mathcal{K}(\mathcal{D}) \,.\, d_3\!\uparrow \subseteq d_1\!\uparrow \cap\, d_2\!\uparrow$;

2. $O \in \mathcal{O}_S(\mathcal{D})$ & $a \in O \quad \Rightarrow \quad \exists d \in \mathcal{K}(\mathcal{D}) \,.\, a \in d\!\uparrow \subseteq O$.

To prove (1), take $a \in d_1\!\uparrow \cap\, d_2\!\uparrow$. This entails that $d_1, d_2 \in \{d_3 \in \mathcal{K}(\mathcal{D}) \mid d_3 \leq a\}$, so we conclude by directedness. Property (2) follows easily from the definition of Scott-open and by algebraicity. □

The category of ω-algebraic lattices

In the 1930's Birkhoff introduced complete lattices to study the combinations of subalgebras. We present here the order-theoretic definition of lattices, for an algebraic description see the next section.

DEFINITION A.53. (i) A poset $\mathcal{D} = (D, \sqsubseteq)$ is called a *lattice* if, for all $a, b \in D$, both the supremum $a \sqcup b$ and the infimum $a \sqcap b$ exist (in D).

(ii) A lattice $\mathcal{D}$ is *complete* if both $\bigsqcup A$ and $\bigsqcap A$ exists for every subset $A \subseteq D$. In particular, $\bot_{\mathcal{D}} = \bigsqcup \emptyset$ and $\top^{\mathcal{D}} = \bigsqcap \emptyset$ are the bottom and the top elements, respectively.

(iii) A complete lattice is *algebraic* (resp. ω-*algebraic*) if it satisfies the same property as a cpo. See Definition A.37(iv)-(v).

EXAMPLES A.54. (i) The reals $\mathbb{R}$ completed with a top $+\infty$ and a bottom element $-\infty$, is a complete lattice w.r.t. the usual ordering.

(ii) The collection $\mathcal{O}$ of open sets of a topological space is a complete lattice w.r.t. $\subseteq$.

(iii) All finite lattices are algebraic.

(iv) Given a set A, the set $\mathscr{P}(A)$ ordered by inclusion is an algebraic complete lattice, with $\mathcal{K}(\mathscr{P}(A)) = \mathscr{P}_{\mathrm{f}}(A)$. Therefore, if A is countable then $\mathscr{P}(A)$ is moreover ω-algebraic.

EXERCISE A.55. Show that in a complete lattice $\mathcal{D}$, $d \in D$ is compact if and only if

$$\forall A \subseteq D \,.\left[d \sqsubseteq \bigsqcup A \quad \Rightarrow \quad \exists A' \in \mathscr{P}_{\mathrm{f}}(A) \,.\, d \sqsubseteq \bigsqcup A'\right]$$

DEFINITION A.56. Let $\mathcal{D}$ be a complete lattice. We say that a subset $A \subseteq D$ is:

(i) an *antichain* (also called *discrete set*) if its elements are pairwise incomparable:

$$\forall a, a' \in A \,.\, [a \sqsubseteq a' \quad \Rightarrow \quad a = a'];$$

(ii) a *strong antichain* if it does not contain the top element and the only possible common upper bound of two distinct $a, a' \in A$ is the top element;

(iii) *dense* if each pair of distinct comparable elements have an element in the middle:

$$\forall a, a' \in A \,.\, a \sqsubset a' \quad \Rightarrow \quad \exists b \in A \,.\, a \sqsubset b \sqsubset a'.$$

We now show some construction for building ω-algebraic lattices starting from existing ones. We consider fixed two arbitrary ω-algebraic lattices $\mathcal{D}, \mathcal{D}'$.

LEMMA A.57. *The structure $\mathcal{D} \times \mathcal{D}'$ having as underlying set the Cartesian product $D \times D'$ and partially ordered by*

$$(d_1, d_1') \sqsubseteq_{\mathcal{D} \times \mathcal{D}'} (d_2, d_2') \iff d_1 \sqsubseteq_{\mathcal{D}} d_2 \;\&\; d_1' \sqsubseteq_{\mathcal{D}'} d_2'$$

is an ω-algebraic complete lattice.

PROOF. Given a subset $A \subseteq D \times D'$, define $A_1 \triangleq \{d \in D \mid \exists d' \in D' \,.\, (d, d') \in A\} \subseteq D$ and $A_2 \subseteq D'$ symmetrically. It is easy to check that $(\bigsqcup A_1, \bigsqcup A_2)$ gives the supremum of A. The dual definition provides the infimum of A, from which it follows that $\mathcal{D} \times \mathcal{D}'$ is a complete lattice. It is ω-algebraic since $\mathcal{K}(\mathcal{D} \times \mathcal{D}') = \mathcal{K}(\mathcal{D}) \times \mathcal{K}(\mathcal{D}')$ is easily verifiable. □

Note that $(\bot_{\mathcal{D}}, \bot_{\mathcal{D}'})$ and $(\top^{\mathcal{D}}, \top^{\mathcal{D}'})$ are respectively the bottom and the top of $\mathcal{D}\times\mathcal{D}'$.

LEMMA A.58. *The structure*

$$[\mathcal{D} \to \mathcal{D}'] \;\triangleq\; \{f : \mathcal{D} \to \mathcal{D}' \mid f \textit{ is continuous }\}$$

pointwisely ordered by $f \sqsubseteq_{[\mathcal{D}\to\mathcal{D}']} g \iff \forall d \in \mathcal{D}\,.\, f(d) \sqsubseteq_{\mathcal{D}'} g(d)$ *is a complete lattice.*

PROOF. Given a subset $A \subseteq [\mathcal{D} \to \mathcal{D}']$, it is easy to check that

$$\begin{array}{rcl}(\bigsqcup A)(d) & \triangleq & \bigsqcup_{f\in A} f(d);\\ (\bigsqcap A)(d) & \triangleq & \bigsqcap_{f\in A} f(d).\end{array}$$

constitute the supremum and infimum of A, respectively. □

We still need to show that $[\mathcal{D} \to \mathcal{D}']$ is ω-algebraic. With this purpose in mind, we describe its compact elements and discuss some of their properties.

DEFINITION A.59. For all $e \in D, e' \in D'$, define the *step function* $(e \mapsto e')$ by

$$(e \mapsto e')(d) \triangleq \begin{cases} e', & \text{if } e \sqsubseteq_{\mathcal{D}} d,\\ \bot_{\mathcal{D}'}, & \text{otherwise.}\end{cases}$$

LEMMA A.60. *Consider* $f \in [\mathcal{D} \to \mathcal{D}']$.
(i) $\forall d \in \mathcal{D}, d' \in \mathcal{D}'\,.\, d' \sqsubseteq_{\mathcal{D}'} f(d) \iff (d \mapsto d') \sqsubseteq_{[\mathcal{D}\to\mathcal{D}']} f$.
(ii) $\forall c \in \mathcal{K}(\mathcal{D}), c' \in \mathcal{K}(\mathcal{D}')\,.\,(c \mapsto c') \in \mathcal{K}([\mathcal{D} \to \mathcal{D}'])$.
(iii) $f = \bigsqcup\{(c \mapsto c') \mid c \in \mathcal{K}(\mathcal{D}), c' \in \mathcal{K}(\mathcal{D}')\,.\, c' \sqsubseteq_{\mathcal{D}'} f(c)\}$.
(iv) $f \in \mathcal{K}([\mathcal{D} \to \mathcal{D}']) \iff f = (c_1 \mapsto c_1')\sqcup\cdots\sqcup(c_n \mapsto c_n')$ *for some* $\vec{c} \in \mathcal{K}(\mathcal{D}), \vec{c}\,' \in \mathcal{K}(\mathcal{D}')$.

PROOF. (i) Easy.
(ii) From $c \in \mathcal{K}(\mathcal{D})$ it follows that $(c \mapsto c')$ is continuous. Assume $(c \mapsto c') \sqsubseteq_{\mathcal{D}\to\mathcal{D}'} A$, for some A. Then

$$c' = (c \mapsto c')(c) \sqsubseteq_{\mathcal{D}'} \bigsqcup\{f(c) \mid f \in A\}$$

Since c' is compact, there exists $f \in A$ such that $c' \sqsubseteq f(c)$. By (i), $(c \mapsto c') \sqsubseteq_{\mathcal{D}\to\mathcal{D}'} f$.
(iii) It is enough to check that, for all $d \in \mathcal{D}$:

$$\begin{array}{rcl} f(d) & = & f(\bigsqcup\{c \mid c \in \mathcal{K}(\mathcal{D})\,.\, c \sqsubseteq_{\mathcal{D}} d\})\\ & = & \bigsqcup\{f(c) \mid c \in \mathcal{K}(\mathcal{D})\,.\, c \sqsubseteq_{\mathcal{D}} d\}\\ & = & \bigsqcup\{c' \mid c \in \mathcal{K}(\mathcal{D}), c' \in \mathcal{K}(\mathcal{D}')\,.\, c \sqsubseteq_{\mathcal{D}} d \;\&\; c' \sqsubseteq_{\mathcal{D}'} f(c)\}\\ & = & \bigsqcup\{(c \mapsto c')(d) \mid c \in \mathcal{K}(\mathcal{D}), c' \in \mathcal{K}(\mathcal{D}')\,.\, c \sqsubseteq_{\mathcal{D}} d \;\&\; c' \sqsubseteq_{\mathcal{D}'} f(c)\}\\ & = & \bigsqcup\{(c \mapsto c') \mid c \in \mathcal{K}(\mathcal{D}), c' \in \mathcal{K}(\mathcal{D}')\,.\, c' \sqsubseteq_{\mathcal{D}'} f(c)\}(d)\end{array}$$

(iv) ($\Rightarrow$) This implication follows by (iii) using Exercise A.55.
($\Leftarrow$) Since compact elements are closed under finite suprema. □

LEMMA A.61. *For all* $n \in \mathbb{N}$, $c_0, \ldots, c_n \in \mathcal{K}(\mathcal{D})$ *and* $c'_0, \ldots, c'_n \in \mathcal{K}(\mathcal{D}')$, *the following are equivalent:*

1. $(c_0 \mapsto c'_0) \sqsubseteq_{[\mathcal{D} \to \mathcal{D}']} (c_1 \mapsto c'_1) \sqcup \cdots \sqcup (c_n \mapsto c'_n)$;
2. $\exists I \subseteq \{1, \ldots, n\} . \left[\bigsqcup_{i \in I} c_i \sqsubseteq_{\mathcal{D}} c_0 \ \& \ c'_0 \sqsubseteq_{\mathcal{D}'} \bigsqcup_{i \in I} c'_i \right]$.

Moreover, in (1 ⇒ 2), we have that $c'_0 \neq \bot_{\mathcal{D}'}$ *entails* $I \neq \emptyset$.

PROOF. Easy. □

LEMMA A.62. *The complete lattice* $[\mathcal{D} \to \mathcal{D}']$ *is* ω*-algebraic.*

PROOF. By Lemmas A.58 and A.60(iii), it is an algebraic lattice. It remains to be shown that $\mathcal{K}([\mathcal{D} \to \mathcal{D}'])$ is denumerable. Since $\mathcal{D}, \mathcal{D}'$ are ω-algebraic, the set $S \cong \mathcal{K}(\mathcal{D}) \times \mathcal{K}(\mathcal{D}')$ of step functions is denumerable. By Lemma A.60(iv) and the idempotence of $\sqcup$, we have $\mathcal{K}([\mathcal{D} \to \mathcal{D}']) \cong \mathscr{P}_{\mathrm{f}}(S)$ which is also denumerable. □

DEFINITION A.63. The category ω**Alg** has ω-algebraic complete lattices as objects and continuous maps as morphisms.

Gathering all the properties shown in this section, we obtain the following result.

THEOREM A.64. *The category* ω**Alg** *is a (well-pointed) CCC.*

A.4 A summary of universal algebra

We briefly recall some concepts of universal algebra that are useful in Chapter 15.

DEFINITION A.65. (i) A *signature* Σ is constituted by a set of operator symbols together with a function assigning a finite *arity* to each operator $f \in \Sigma$.

(ii) Given $n \in \mathbb{N}$, we let Σ_n be the set of operator symbols from Σ having arity n.

(iii) The elements of Σ_0 are called *nullary operators* or *constants*.

(iv) An *algebra* $\mathcal{A}$ over a signature Σ is determined by a non-empty set A, called the *underlying set* of $\mathcal{A}$, together with an operation $f^{\mathcal{A}} : A^n \to A$, for every $f \in \Sigma_n$.

(v) An algebra $\mathcal{A}$ is *trivial* if its underlying set has cardinality 1.

We will simply speak of an algebra $\mathcal{A}$ whenever the associated signature Σ is not important, or can be inferred from the context. Given an arbitrary subset X of the underlying set of $\mathcal{A}$, it might happen to be closed under the operations in Σ, in which case we have a subalgebra, or not because it lacks certain elements. We now describe the canonical way of extending X in order to satisfy this closure property.

DEFINITION A.66. Let $\mathcal{A}$ be an algebra over a signature Σ.

(i) An algebra $\mathcal{B}$ over the same signature Σ is a *subalgebra* of $\mathcal{A}$ if $B \subseteq A$ and

$$f^{\mathcal{B}}(b_1, \dots, b_n) = f^{\mathcal{A}}(b_1, \dots, b_n),$$

for all $f \in \Sigma_n$ and $b_1, \dots, b_n \in B$. In this case B is called a *subuniverse* of $\mathcal{A}$ and $f^{\mathcal{B}}$ the *restriction* of $f^{\mathcal{A}}$ to B.

(ii) Given $X \subseteq A$, the *subuniverse of $\mathcal{A}$ generated by X* is given by

$$\mathrm{Sg}^{\mathcal{A}}(X) \triangleq \bigcap \{B \mid X \subseteq B \ \&\ B \text{ is a subuniverse of } \mathcal{A}\}$$

(iii) We say that X *generates* $\mathcal{A}$ (and that X is a set of *generators* of $\mathcal{A}$) if $\mathrm{Sg}^{\mathcal{A}}(X) = A$.

(iv) The algebra $\mathcal{A}$ is *finitely generated* if it has a finite set of generators.

We present here the concepts of congruence, quotient algebra, and homomorphism since they are closely related. The reader should go over this material keeping in mind that λ-theories are simply congruences of the (open) term algebra $\mathcal{A}(\boldsymbol{\lambda})$. This remark should also explain the similarities with the notations used in Section 3.1.

DEFINITION A.67. Consider an algebra $\mathcal{A}$ over a signature Σ.

(i) Given a binary relation $\vartheta \subseteq A \times A$ and $a, b \in A$, we write $a \,\vartheta\, b$ for $(a, b) \in \vartheta$.

(ii) A relation ϑ as above is called *compatible* if, for all $f \in \Sigma_n$ and $a_i, b_i \in A$, we have:

$$a_1 \,\vartheta\, b_1 \ \&\ \cdots \ \&\ a_n \,\vartheta\, b_n \quad \Rightarrow \quad f^{\mathcal{A}}(a_1, \dots, a_n) \,\vartheta\, f^{\mathcal{A}}(b_1, \dots, b_n)$$

(iii) A compatible relation ϑ on $\mathcal{A}$ is called a *congruence* if it is an equivalence relation. The *equivalence class* of an element $a \in A$ modulo ϑ is denoted by $[a]_\vartheta \triangleq \{b \mid a \,\vartheta\, b\}$.

(iv) We denote by $\mathrm{Con}(\mathcal{A})$ the set of all congruences on $\mathcal{A}$.

(v) The following congruences are called *trivial* and deserve a special notation:

$$\begin{array}{rcl} \Delta^{\mathcal{A}} & \triangleq & \{(a,a) \mid a \in A\}; \\ \nabla^{\mathcal{A}} & \triangleq & A \times A. \end{array}$$

(vi) A congruence ϑ on $\mathcal{A}$ is *non-trivial* if it is different from both $\nabla^{\mathcal{A}}$ and $\Delta^{\mathcal{A}}$.

REMARK A.68. $(\mathrm{Con}(\mathcal{A}), \leq)$, where $\leq$ stands for set-theoretical inclusion, is a complete lattice having $\Delta^{\mathcal{A}}$ as bottom and $\nabla^{\mathcal{A}}$ as top elements. The meet $\vartheta \wedge \varphi$ of two congruences ϑ, φ is their intersection, while their join is the least equivalence relation including $\vartheta \cup \varphi$.

The compatibility property in the definition of a congruence ϑ allows to transfer the algebraic structure of $\mathcal{A}$ to the set of equivalence classes A/ϑ.

DEFINITION A.69. Let $\mathcal{A}$ be an algebra and consider $\vartheta, \varphi \in \mathrm{Con}(\mathcal{A})$.

(i) The *quotient algebra* of $\mathcal{A}$ modulo φ is the algebra $\mathcal{A}/\varphi$ having as underlying set A/φ and satisfying

$$f^{\mathcal{A}/\varphi}([a_1]_\varphi, \ldots, [a_n]_\varphi) = [f^{\mathcal{A}}(a_1, \ldots, a_n)]_\varphi,$$

for all $f \in \Sigma_n$ and $a_1, \ldots, a_n \in A$.

(ii) Assume $\varphi \subseteq \vartheta$, then $\vartheta/\varphi \triangleq \{([a]_\varphi, [b]_\varphi) \mid a \,\vartheta\, b\}$ is a congruence on $\mathcal{A}/\varphi$.

(iii) The *relative product* of φ, ϑ is given by:

$$\varphi \circ \vartheta \triangleq \{(a,c) \mid \exists b \in A \,.\, a \,\vartheta\, b \,\varphi\, c\}.$$

REMARK A.70. (i) The quotient algebras of $\mathcal{A}$ have the same signature as $\mathcal{A}$.

(ii) For all $\varphi, \vartheta \in \mathrm{Con}(\mathcal{A})$, we have $\varphi \cup \vartheta \subseteq \varphi \circ \vartheta \subseteq \varphi \vee \vartheta$.

(iii) In general $\varphi \circ \vartheta$ is still a compatible relation on $\mathcal{A}$, but not necessarily a congruence (as transitivity or symmetry might fail).

(iv) It is easy to check that

$$\vartheta \vee \varphi = \bigcup_{n \in \mathbb{N}} (\vartheta \circ^n \varphi)$$

An easy way of inducing a congruence is through the kernel of a homomorphism—in fact the λ-theory induced by a model $\mathcal{M}$ is given by the kernel of its interpretation function (see Section 3.2).

DEFINITION A.71. Let $\mathcal{A}, \mathcal{B}$ be two algebras over the same signature Σ.

(i) A function $h : A \to B$ is a *homomorphism* from $\mathcal{A}$ to $\mathcal{B}$, written $h : \mathcal{A} \to \mathcal{B}$, if

$$h(f^{\mathcal{A}}(a_1, \ldots, a_n)) = f^{\mathcal{B}}(h(a_1), \ldots, h(a_n))$$

for every $f \in \Sigma_n$ and $a_1, \ldots, a_n \in A$. The *kernel* of h is the congruence defined by

$$\ker(h) \triangleq \{(a_1, a_2) \in A \times A \mid h(a_1) = h(a_2)\}$$

(ii) A homomorphism $h : \mathcal{A} \to \mathcal{B}$ is an *epimorphism* if it is surjective, and in this case $\mathcal{B}$ is called a *homomorphic image* of $\mathcal{A}$.

(iii) A homomorphism $h : \mathcal{A} \to \mathcal{B}$ is an *isomorphism* if it is both injective and surjective. In this case we write $\mathcal{A} \cong \mathcal{B}$ and say that $\mathcal{A}$ and $\mathcal{B}$ are *isomorphic.*

(iv) The *(direct) product* of $\mathcal{A}$ and $\mathcal{B}$ is the algebra $\mathcal{A} \times \mathcal{B}$ having $A \times B$ as underlying set, and satisfying for all $f \in \Sigma_n$ and $a_i \in A$, $b_i \in B$ $(1 \leq i \leq n)$:

$$f^{\mathcal{A}\times\mathcal{B}}((a_1, b_1), \ldots, (a_n, b_n)) = (f^{\mathcal{A}}(a_1, \ldots, a_n), f^{\mathcal{B}}(b_1, \ldots, b_n))$$

We denote by π_1, π_2 the corresponding projections.

REMARK A.72. (i) It is easy to check that the projections are epimorphisms.

(ii) In general neither $\mathcal{A}_1$ nor $\mathcal{A}_2$ is embeddable in $\mathcal{A}_1 \times \mathcal{A}_2$, however $\mathcal{A}_1$ and $\mathcal{A}_2$ are homomorphic images of $\mathcal{A}_1 \times \mathcal{A}_2$.

We recall some important properties that an algebra might satisfy.

DEFINITION A.73. An algebra $\mathcal{A}$ is

(i) a *discriminator algebra* if it has a quaternary term $t(x, y, w, z)$ satisfying

$$t(x, y, w, z) = \begin{cases} w, & \text{if } x = y, \\ z, & \text{otherwise;} \end{cases}$$

(ii) *congruence distributive* if its congruence lattice $\mathrm{Con}(\mathcal{A})$ is distributive;

(iii) *congruence n-permutable* (for $n \geq 2$) if $\varphi \vee \psi = \varphi \circ^n \psi$, for all $\varphi, \psi \in \mathrm{Con}(\mathrm{A})$;

(iv) *congruence permutable* if it is congruence n-permutable, for all $n \in \mathbb{N}$. As shown in Smith (1976), this is equivalent to requiring that $\mathcal{A}$ is congruence 2-permutable.

Some algebraic structures

We present some examples of algebraic structures that are omnipresent in mathematics.

EXAMPLES A.74. (i) A *monoid* $\mathcal{A}$ is an algebra $(A, \cdot)$ satisfying the identities:

$$x \cdot (y \cdot z) = (x \cdot y) \cdot z \qquad \text{(associativity)}$$
$$x \cdot 1 = x = 1 \cdot x \qquad \text{(identity)}$$

(ii) A *group* $\mathcal{G}$ is an algebra $(G, \cdot, (\cdot)^{-1}, 1)$ such that $(G, \cdot)$ is a monoid and each element x has an inverse x^{-1}, i.e., $x \cdot x^{-1} = 1 = x^{-1} \cdot x$.

(iii) A group $\mathcal{G}$ is *commutative*, or *Abelian*, if $x \cdot y = y \cdot x$ holds.

(iv) A *ring* $\mathcal{R}$ is an algebra $(R, +, \cdot, -, 0)$ such that $(R, +, -, 0)$ is an Abelian group, $(R, \cdot)$ is a monoid and multiplication is distributive with respect to addition, i.e.:

$$\begin{aligned} x \cdot (y + z) &= (x \cdot y) + (x \cdot z) \quad \text{(left distributivity).} \\ (x + y) \cdot z &= (x \cdot z) + (y \cdot z) \quad \text{(right distributivity).} \end{aligned}$$

(v) A *ring $\mathcal{R}$ with unity* is a ring $\mathcal{R}$ having a multiplicative identity 1.

Other significative examples for computer scientists include combinatory algebras, λ-algebras and λ-models. See Section 3.2.

In this book we often work with lattices. At the end of Section A.3 we have seen their order-theoretic definition. From a more algebraic perspective, lattices are rather defined by specifying the properties of their operations $\vee$ and $\wedge$.

DEFINITION A.75. (i) A *lattice* $\mathcal{L} = (L, \vee, \wedge)$ consists of a non-empty set L, together with two binary, commutative and associative operations $\vee, \wedge$, respectively called *join* and *meet*, satisfying the following *absorption laws*:

$$\begin{array}{rcl} x & = & x \vee (x \wedge y) \\ x & = & x \wedge (x \vee y) \end{array}$$

It can be proved that join and meet are *idempotent*: $x \vee x = x$ and $x \wedge x = x$.

(ii) A lattice $\mathcal{L}$ is *distributive* if the following axiom is satisfied, for all $x, y, z \in L$:

$$x \wedge (y \vee z) = (x \wedge y) \vee (x \wedge z).$$

(iii) A lattice $\mathcal{L}$ is called *bounded* if the join and the meet possess *absorbing elements*, respectively denoted 1 and 0, i.e. $x \vee 1 = 1$ and $x \wedge 0 = 0$.

Every lattice $\mathcal{L} = (L, \vee, \wedge)$ can be seen as an instance of Definition A.53(i), by setting

$$a \leq b \iff a \wedge b = a.$$

If $\mathcal{L} = (L, \leq)$ is bounded then 0 and 1 are respectively the bottom and top element.

EXAMPLES A.76. (i) The main example for our purposes is the lattice $\lambda\mathcal{T}$ of λ-theories.

(ii) More generally, given an equational theory T, the set $\mathcal{L}(T)$ of all equational theories containing T is a lattice ordered by $\subseteq$.

(iii) The lattice $\mathrm{Con}(\mathcal{A})$ of congruences of an algebra $\mathcal{A}$.

LEMMA A.77. *A distributive lattice $\mathcal{L}$ satisfies*

$$x \vee (y \wedge z) = (x \vee y) \wedge (x \vee z)$$

PROOF. We have
$$\begin{array}{rcll} x \vee (y \wedge z) & = & (x \vee (x \wedge z)) \vee (y \wedge z), & \text{by absorption,} \\ & = & x \vee ((x \wedge z) \vee (y \wedge z)), & \text{by associativity,} \\ & = & x \vee ((z \wedge x) \vee (z \wedge y)), & \text{by commutativity,} \\ & = & x \vee (z \wedge (x \vee y)), & \text{by distributivity,} \\ & = & x \vee ((x \vee y) \wedge z), & \text{by commutativity,} \\ & = & (x \wedge (x \vee y)) \vee ((x \vee y) \wedge z), & \text{by absorption,} \\ & = & ((x \vee y) \wedge x) \vee ((x \vee y) \wedge z), & \text{by commutativity,} \\ & = & (x \vee y) \wedge (x \vee z), & \text{by distributivity.} \end{array}$$
□

In the following, we consider a bounded lattice $\mathcal{L}$.

An element $x \in L$ is an *atom* if it is a minimal element in $L - \{0\}$. Dually, we say that x is a *coatom* if it is a maximal element in $L - \{1\}$.

NOTATION. *Given* $x \in L$, *we write*

$$L_x \triangleq \{y \in L - \{0\} \mid x \wedge y = 0\}.$$

DEFINITION A.78. A bounded lattice $\mathcal{L}$ is called:
(i) *lower semicomplemented* whenever $L_x \neq \emptyset$, for all $x \neq 1$;
(ii) *pseudocomplemented* if each $L_x \cup \{0\}$ has a greatest element, which is called the *pseudocomplement* of x;
(iii) *complemented* if, for every $x \in L$, there exists y such that $x \wedge y = 0$ and $x \vee y = 1$;
(iv) *atomic* if, for every $x \in L$, there exists an atom $y \leq x$;
(v) *coatomic* if, for every $x \in L$, there exists a coatom y such that $x \leq y$;
(vi) *Boolean* if it is distributive and complemented.

A lattice $\mathcal{L}$ is called a *lattice of equational theories* if it is isomorphic to $\mathcal{L}(T)$ for some equational theory T (using the notation from Examples A.76(ii)). Such lattices are always algebraic, coatomic and possess a compact top element. No stronger property was known before Lampe's discovery that any lattice of equational theories obeys the Zipper condition (see Lampe (1986)).

DEFINITION A.79. A bounded lattice $\mathcal{L}$ satisfies the *Zipper condition* if, for every set I and for every $x_i, y, z \in L$ $(i \in I)$, we have:

$$\bigvee_{i \in I} x_i = 1 \ \& \ (\forall i \in I)\ x_i \wedge y = z \quad \Rightarrow \quad y = z.$$

DEFINITION A.80. A *Boolean algebra* is an algebra $(B, \vee, \wedge, (\cdot)^-, 0, 1)$ such that
(i) $(B, \vee, \wedge, 0, 1)$ is a bounded distributive lattice;
(ii) for all $b \in B$, $b \wedge b^- = 0$ and $b \vee b^- = 1$ hold.
The element b^- is called the *complement* of b.

EXAMPLES A.81. (i) Given a non-empty set X, its powerset $\mathscr{P}(X)$ is a Boolean algebra w.r.t. union, intersection and complementation. Notice that $0 \triangleq \emptyset$ and $1 \triangleq X$.
(ii) Another example is given by the clopen subsets of a topological space, using the same operations as above.
(iii) The Boolean algebra of truth values $\{0, 1\}$ is given by $0 < 1$, $1^- \triangleq 0$ and $0^- \triangleq 1$.
(iv) The trivial algebra $(\{\star\}, \vee, \wedge, (\cdot)^-, \star, \star)$ is a Boolean algebra.

THEOREM A.82 (STONE). *Every finite Boolean algebra is isomorphic to the Boolean algebra of all subsets of some finite set* X.

PROOF. See Burris and Sankappanavar (2012), Corollary 1.10. □

DEFINITION A.83. Let $\mathcal{B}$ be a Boolean algebra and $I \subseteq B$.
(i) We say that I is a *ideal* of $\mathcal{B}$ if
(1) $0 \in I$;
(2) $a, b \in I \quad \Rightarrow \quad a \vee b \in I$;
(3) $a \in I \ \& \ b \leq a \quad \Rightarrow \quad b \in I$.
(ii) An ideal I of $\mathcal{B}$ is a *maximal ideal* if I is maximal w.r.t. the property $1 \notin I$. I.e., I is a maximal ideal whenever $1 \notin I$ and for all $a, b \in B$, $a \wedge b \in I \iff a \in I$ or $b \in I$.

Varieties

Researchers in universal algebra are often interested in the study of classes of algebras over the same signature and closed under some constructions. For instance, we say that a class $\mathbb{C}$ is closed under subalgebras if, for every $\mathcal{A} \in \mathbb{C}$, all subalgebras of $\mathcal{A}$ belong to $\mathbb{C}$. Similarly for homomorphic images, direct products and the like.

DEFINITION A.84. A non-empty class $\mathbb{V}$ of algebras over the same signature Σ is:
(i) an *equational class* if it is axiomatizable by a set of equations.
(ii) a *variety* if it is closed under subalgebras, homomorphic images and direct products. A variety $\mathbb{V}$ is a *subvariety* of a variety $\mathbb{V}'$ whenever $\mathbb{V} \subseteq \mathbb{V}'$ holds.

Birkhoff proved that that conditions (i) and (ii) in the above definition are equivalent.

THEOREM A.85 (BIRKHOFF (1935)). *A class of algebras over the signature Σ is a variety if and only if it is an equational class.*

PROOF. See McKenzie et al. (1987), Theorem 4.131. □

EXAMPLES A.86. (i) Boolean algebras form an equational class, whence a variety.
(ii) The main example for this book is the variety of combinatory algebras.
(iii) Also λ-algebras constitute a variety, whilst λ-models do not.

DEFINITION A.87. Let $\mathbb{V}$ be a variety.
(i) We say that an algebra $\mathcal{A} \in \mathbb{V}$ is the *free algebra over the set X of generators* if $\mathcal{A}$ is generated by X and, for every $\mathcal{B} \in \mathbb{V}$ and $g : X \to \mathcal{B}$, there exists a unique homomorphism $f : \mathcal{A} \to \mathcal{B}$ that *extends* g, i.e., $f(x) = g(x)$ for all $x \in X$.
(ii) A free algebra in the class of all Σ-algebras is called *absolutely free*.

The properties of algebras introduced in Definition A.73 are inherited by varieties in the obvious way (see below).

DEFINITION A.88. An variety $\mathbb{V}$ is
(i) a *discriminator variety* if it is generated by a class $\mathbb{K}$ of discriminator algebras;
(ii) *congruence distributive* if every algebra in the variety is congruence distributive;
(iii) *congruence n-permutable* if every algebra in the variety is congruence n-permutable;
(iv) *congruence permutable* if every algebra in the variety is congruence 2-permutable.

REMARK A.89. If a variety is congruence permutable then all its elements are n-congruence permutable, for all $n \in \mathbb{N}$. On the contrary, there are known examples of congruence 3-permutable varieties that fail to be congruence 2-permutable, such as the varieties of right complemented semigroups (Hagemann and Mitschke (1973)) and the one of implication algebras (Mitschke (1976)).

DEFINITION A.90. (i) A first-order formula Φ is a *universal formula* if it is in prenex form and all the quantifiers are universal.
(ii) A class of algebras is *universal* if it can be axiomatized by universal formulas.

The next result provides a well-known characterization of universal classes in terms of their properties of closure. It is the analogue of Theorem A.85 for universal classes.

THEOREM A.91. *A class of algebras over the signature Σ is universal if and only if it is closed under isomorphisms, subalgebras and ultraproducts.*

PROOF. See Burris and Sankappanavar (2012), Theorem 2.20. □

We do not enter in the definition of ultraproducts since they are beyond the scope of this book, the interested reader can find more details in Burris and Sankappanavar (2012), Chapter IV, Section 6. We mention that they have been used by Bucciarelli et al. (2012a) to construct minimal λ-theories represented by some classes of λ-models.

Bibliography

Abramsky, S. (1991). Domain theory in logical form. *Ann. Pure Appl. Log.*, 51(1-2): 1–77.

Abramsky, S. and McCusker, G. (1997). Lecture notes on game semantics. Available at: `http://www.cs.ox.ac.uk/samson.abramsky/mdorf97.ps.gz`. Proceedings of Marktoberdorf 1997 Summer School.

Abramsky, S. and McCusker, G. (1999). Full abstraction for Idealized Algol with passive expressions. *Theor. Comput. Sci.*, 227(1-2):3–42.

Abramsky, S., Jagadeesan, R., and Malacaria, P. (2000). Full abstraction for PCF. *Inf. Comput.*, 163(2):409–470.

Accattoli, B. and Dal Lago, U. (2016). (Leftmost-outermost) beta reduction is invariant, indeed. *Log. Methods Comput. Sci.*, 12(1).

Accattoli, B. and Guerrieri, G. (2016). Open call-by-value. In Igarashi, A., editor, *Programming Languages and Systems - 14th Asian Symposium, APLAS 2016, Hanoi, Vietnam, November 21-23, 2016, Proceedings*, volume 10017 of *Lecture Notes in Computer Science*, pages 206–226.

Accattoli, B. and Paolini, L. (2012). Call-by-value solvability, revisited. In Schrijvers, T. and Thiemann, P., editors, *Functional and Logic Programming - 11th International Symposium, FLOPS 2012, Kobe, Japan, May 23-25, 2012. Proceedings*, volume 7294 of *Lecture Notes in Computer Science*, pages 4–16. Springer.

Accattoli, B., Barenbaum, P., and Mazza, D. (2014). Distilling abstract machines. In Jeuring, J. and Chakravarty, M. M. T., editors, *Proceedings of the 19th ACM SIGPLAN international conference on Functional programming, Gothenburg, Sweden, September 1-3, 2014*, pages 363–376. ACM.

Aczel, P. (1978). A General Church-Rosser Theorem. University of Manchester. Unpublished.

Alessi, F. (1993). The p model. Internal Report, Udine University.

Alessi, F. and Lusin, S. (2002). Simple easy terms. *Electron. Notes Theor. Comput. Sci.*, 70(1):1–18.

Alessi, F., Dezani-Ciancaglini, M., and Honsell, F. (2001). Filter models and easy terms. In *ICTCS '01: Proc. of the 7th Italian Conference on Theoretical Computer Science. London, United Kingdom*, pages 17–37. Springer-Verlag.

Alessi, F., Barbanera, F., and Dezani-Ciancaglini, M. (2003). Tailoring filter models. In Berardi, S., Coppo, M., and Damiani, F., editors, *Types for Proofs and Programs, International Workshop, TYPES 2003, Torino, Italy, April 30 - May 4, 2003, Revised Selected Papers*, volume 3085 of *Lecture Notes in Computer Science*, pages 17–33. Springer.

Alessi, F., Dezani-Ciancaglini, M., and Honsell, F. (2004). Inverse limit models as filter models. In Kesner, D., van Raamsdonk, F., and Wells, J., editors, *International Workshop on Higher-Order Rewriting*, pages 3–25, Aachen. RWTH Aachen.

Amadio, R. and Curien, P.-L. (1998). *Domains and lambda-calculi*, volume 46 of *Cambridge Tracts in Theoretical Computer Science*. Cambridge University Press.

Ariola, Z. M. and Klop, J. W. (1994). Cyclic lambda graph rewriting. In *Proceedings of the Ninth Annual Symposium on Logic in Computer Science (LICS '94), Paris, France, July 4-7, 1994*, pages 416–425. IEEE Computer Society.

Asperti, A. and Guerrini, S. (1998). *The Optimal Implementation of Functional Programming Languages*. Cambridge Tracts in Theoretical Computer Science. Cambridge University Press.

Asperti, A. and Laneve, C. (1995). Paths, computations and labels in the lambda-calculus. *Theor. Comput. Sci.*, 142(2):277–297.

Asperti, A. and Longo, G. (1991). *Categories, types and structures: an introduction to category theory for the working computer scientist*. MIT Press, Cambridge, MA.

Asperti, A. and Mairson, H. (1998). Parallel beta reduction is not elementary recursive. In *Conference Record of POPL '98: The 25th ACM SIGPLAN-SIGACT Symposium on Principles of Programming Languages*, San Diego, CA.

Asperti, A. and Mairson, H. G. (2001). Parallel beta reduction is not elementary recursive. *Inf. Comput.*, 170(1):49–80.

Asperti, A., Giovannetti, C., and Naletto, A. (1996). The Bologna Optimal Higher-Order Machine. *Journal of Functional Programming*, 6(6).

Avigad, J., Donnelly, K., Gray, D., and Raff, P. (2007). A formally verified proof of the prime number theorem. *ACM Transactions on Computational Logic (TOCL)*, 9(1):2.

Baeten, J. and Boerboom, B. (1979). Omega can be anything it should not be. In *Proc. Koninklijke Netherlandse Akademie van Wetenschappen, Serie A, Indag. Matematicae 41*, pages 111–120.

van Bakel, S., Barbanera, F., Dezani-Ciancaglini, M., and de Vries, F.-J. (2002). Intersection types for lambda-trees. *Theor. Comput. Sci.*, 272(1-2):3–40.

Balabonski, T., Barenbaum, P., Bonelli, E., and Kesner, D. (2017). Foundations of strong call-by-need. *Proc. ACM Program. Lang.*, 1(ICFP):20:1–20:29.

Barbarossa, D. and Manzonetto, G. (2020). Taylor subsumes Scott, Berry, Kahn and Plotkin. *PACMPL*, 4(POPL):1:1–1:23.

Barendregt, H. P. (1971). *Some extensional term models for λ-calculi and combinatory logics.* Ph.D. thesis, Utrecht University. Available at `https://hdl.handle.net/2066/27329`. Republished as Barendregt (2020b).

Barendregt, H. P. (1974). Pairing without conventional restraints. *Zeitschrift für mathematische Logik und Grundlagen der Mathematik*, 20:289–306.

Barendregt, H. P. (1975). RTA list, Problem #97: Is the word problem for the S-combinator decidable? See `https://www.win.tue.nl/rtaloop/problems/97.html`.

Barendregt, H. P. (1977). The type free lambda calculus. In Barwise, J., editor, *Handbook of Mathematical Logic*, volume 90 of *Studies in Logic and the Foundations of Mathematics*, pages 1091–1132. North-Holland, Amsterdam.

Barendregt, H. P. (1984). *The lambda-calculus, its syntax and semantics.* Number 103 in Studies in Logic and the Foundations of Mathematics. North-Holland, revised edition.

Barendregt, H. P. (1992). Representing 'undefined' in lambda calculus. *J. Funct. Program.*, 2(3):367–374.

Barendregt, H. P. (1993). Constructive proofs of the range property in lambda calculus. *Theoretical Computer Science*, 121(1–2):59–69. A collection of contributions in honour of Corrado Böhm on the occasion of his 70th birthday.

Barendregt, H. P. (2001). Discriminating coded lambda terms. In Anderson, A. and Zeleny, M., editors, *Logic, Meaning and Computation. Essays in Memory of Alonzo Church*, pages 275–285. Kluwer.

Barendregt, H. P. (2008). Towards the range property for the lambda theory $\mathcal{H}$. *Theoretical Computer Science*, 398:12–15.

Barendregt, H. P. (2013). Foundations of mathematics from the perspective of computer verification. In *Mathematics, Computer Science and Logic—A Never Ending Story*, pages 1–49. Springer.

Barendregt, H. P. (2020a). Gems of Corrado Böhm. *Log. Methods Comput. Sci.*, 16(3).

Barendregt, H. P. (2020b). *Some extensional term models for λ-calculi and combinatory logics — A 2020 republication.* Ph.D. thesis, Utrecht University.

Barendregt, H. P. and Klop, J. W. (2009). Applications of infinitary lambda calculus. *Information and Computation*, 207(5):559–582.

Barendregt, H. P. and Longo, G. (1980). Equality of lambda terms in the model $\mathbb{T}^\omega$. In Hindley and Seldin, editors, *Essays on Combinatory Logic, Lambda-Calculus, and Formalism*, pages 303–339. Academic Press, San Diego.

Barendregt, H. P., Bergstra, J. A., Klop, J. W., and Volken, H. (1976). Degrees, reductions and representability in the lambda calculus. Technical Report 22, Department of Mathematics, Utrecht University. The Blue Preprint. The report is available via the website `http://repository.ubn.ru.nl/handle/2066/20381`.

Barendregt, H. P., Bergstra, J. A., Klop, J. W., and Volken, H. (1978). Degrees of sensible lambda theories. *Journal of Symbolic Logic*, 43(1):45–55.

Barendregt, H. P., Coppo, M., and Dezani-Ciancaglini, M. (1983). A filter lambda model and the completeness of type assignment. *J. Symb. Log.*, 48(4):931–940.

Barendregt, H. P., Bunder, M. W., and Dekkers, W. (1993). Systems of illative combinatory logic complete for first-order propositional and predicate calculus. *The Journal of Symbolic Logic*, 58(3):769–788.

Barendregt, H. P., Dekkers, W., and Statman, R. (2013a). *Lambda Calculus with Types.* Perspectives in logic. Cambridge University Press.

Barendregt, H. P., Manzonetto, G., and Plasmeijer, R. (2013b). The imperative and functional programming paradigm. In Cooper, S. B. and Van Leeuwen, J., editors, *Alan Turing: His work and impact*, pages 121–126. Elsevier.

Barendregt, H. P., Endrullis, J., Klop, J. W., and Waldmann, J. (2018). Dance of the starlings. In *Raymond Smullyan on Self Reference*, pages 67–111. Springer.

Bastonero, O. (1996). *Modèles fortement stables du λ-calcul et résultats d'incomplétude.* Thèse de doctorat, Université de Paris 7.

Bastonero, O. and Gouy, X. (1999). Strong stability and the incompleteness of stable models for lambda-calculus. *Ann. Pure Appl. Log.*, 100(1-3):247–277.

Batenburg, A. and Velmans, J. (1983). *Invertibility properties of two λ-algebras.* Doctoraalskriptie, University of Utrecht.

Bellot, P. (1985). A new proof for Craig's theorem. *J. Symb. Log.*, 50(2):395–396.

Ben-Yelles, C. (1979). *Type-assignment in the lambda-calculus; syntax and semantics.* Ph.D. thesis, University of Wales Swansea.

Berarducci, A. (1996). Infinite λ-calculus and non-sensible models. In Dekker, M., editor, *Logic and Algebra (Pontignano, 1994)*, volume 180, pages 339–377, New York.

Berarducci, A. and Intrigila, B. (1993). Some new results on easy lambda-terms. *Theor. Comput. Sci.*, 121(1&2):71–88.

Bergstra, J. A. and Klop, J. W. (1979). Church-Rosser strategies in the lambda calculus. *Theor. Comput. Sci.*, 9:27–38.

Bergstra, J. A. and Klop, J. W. (1980). Invertible terms in the lambda calculus. *Theor. Comput. Sci.*, 11:19–37.

Berline, C. (2000). From computation to foundations via functions and application: The λ-calculus and its webbed models. *Theoretical Computer Science*, 249(1):810–161.

Berline, C. and Salibra, A. (2006). Easiness in graph models. *Theoretical Computer Science*, 354(1):4–23.

Berline, C., Manzonetto, G., and Salibra, A. (2009). Effective lambda-models versus recursively enumerable lambda-theories. *Mathematical Structures in Computer Science*, 19(5):897–942.

Berry, G. (1978). Séquentialité de l'évaluation formelle de λ-expressions. In Robinet, B., editor, *Program Transformations; Proc. 3-e Colloque International sur la Programmation*, pages 67–80.

Berry, G. (1979). *Modèles complètement adéquats et stable des λ-calculs typés.* Thèse de doctorat, Université de Paris 7. In French.

Berry, G. and Curien, P.-L. (1982). Sequential algorithms on concrete data structures. *Theor. Comput. Sci.*, 20:265–321.

Bertini, Y. (2005). *The notion of indefinite in lambda calculus.* Thèse de doctorat, Université de Savoie. URL `https://tel.archives-ouvertes.fr/tel-00415825`.

Bigelow, D. and Burris, S. N. (1990). Boolean algebras of factor congruences. *Acta Sci. Math.*, 54:11–20.

Birkhoff, G. (1935). On the structure of abstract algebras. In *Proc. Cambridge Philos. Soc.*, volume 31, pages 433–454.

Birkhoff, G. (1967). *Lattice Theory.* American Mathematical Society, Providence, 3rd edition.

Blanchette, J. and Mahboubi, A., editors (2022). *Handbook of Proof Assistants and Their Applications in Mathematics and Computer Science.* Springer.

Blanchette, J. C., Kaliszyk, C., Paulson, L. C., and Urban, J. (2016). Hammering towards QED. *Journal of Formalized Reasoning*, 9(1):101–148.

Böhm, C. (1968). Alcune proprietà delle forme β-η-normali nel λ-K-calcolo. *Pubblicazioni dell'istituto per le applicazioni del calcolo*, 696:1–19. Lavoro eseguito all'INAC.

Böhm, C., editor (1975). *Lambda-Calculus and Computer Science Theory, Proceedings of the Symposium Held in Rome, Italy, March 25-27, 1975*, volume 37 of *Lecture Notes in Computer Science*. Springer.

Böhm, C. and Dezani-Ciancaglini, M. (1974). Combinatorial problems, combinator equations and normal forms. In Loeckx, J., editor, *Automata, Languages and Programming, 2nd Colloquium, University of Saarbrücken, Germany, July 29 - August 2, 1974, Proceedings*, volume 14 of *Lecture Notes in Computer Science*, pages 185–199. Springer.

Böhm, C. and Micali, S. (1980). Minimal forms in lambda-calculus computations. *J. Symb. Log.*, 45(1):165–171.

Böhm, C. and Tronci, E. (1991). About systems of equations, X-separability, and left-invertibility in the lambda-calculus. *Information and Computation*, 90(1):1–32.

Böhm, C., Dezani-Ciancaglini, M., Peretti, P., and Ronchi Della Rocca, S. (1979). A discrimination algorithm inside $\lambda\beta$-calculus. *Theoretical Computer Science*, 8(3):271–291.

Boudes, P. (2011). Non-uniform (hyper/multi)coherence spaces. *Math. Struct. Comput. Sci.*, 21(1):1–40.

Boudol, G. (1994). Lambda-calculi for (strict) parallel functions. *Inf. Comput.*, 108(1): 51–127.

Breuvart, F. (2014). On the characterization of models of $\mathcal{H}^*$. In Henzinger, T. A. and Miller, D., editors, *Joint Meeting of the Twenty-Third EACSL Annual Conference on Computer Science Logic (CSL) and the Twenty-Ninth Annual ACM/IEEE Symposium on Logic in Computer Science (LICS), CSL-LICS '14, Vienna, Austria, July 14 - 18, 2014*, pages 24:1–24:10. ACM.

Breuvart, F. (2016). On the characterization of models of $\mathcal{H}^*$: the semantical aspect. *Log. Methods Comput. Sci.*, 12(2).

Breuvart, F., Manzonetto, G., Polonsky, A., and Ruoppolo, D. (2016). New results on Morris's observational theory: The benefits of separating the inseparable. In Kesner, D. and Pientka, B., editors, *1st International Conference on Formal Structures for Computation and Deduction, FSCD 2016, June 22-26, 2016, Porto, Portugal*, volume 52 of *LIPIcs*, pages 15:1–15:18. Schloss Dagstuhl - Leibniz-Zentrum fuer Informatik.

Breuvart, F., Manzonetto, G., and Ruoppolo, D. (2018). Relational graph models at work. *Logical Methods in Computer Science*, 14(3).

Brouwer, L. E. (1927). Über definitionsbereiche von-funktionen. *Mathematische annalen*, 97(1):60–75.

Brown, M. and Palsberg, J. (2016). Breaking through the normalization barrier: a self-interpreter for F-omega. In Bodík, R. and Majumdar, R., editors, *Proceedings of the 43rd Annual ACM SIGPLAN-SIGACT Symposium on Principles of Programming Languages, POPL 2016*, pages 5–17. ACM.

de Bruijn, N. G. (1970). The mathematical language AUTOMATH, its usage, and some of its extensions. In *Symposium on automatic demonstration*, pages 29–61. Springer.

Bucciarelli, A. and Ehrhard, T. (1991). Sequentiality and strong stability. In *Proceedings of the Sixth Annual Symposium on Logic in Computer Science (LICS '91), Amsterdam, The Netherlands, July 15-18, 1991*, pages 138–145. IEEE Computer Society.

Bucciarelli, A. and Ehrhard, T. (2001). On phase semantics and denotational semantics: the exponentials. *Ann. Pure Appl. Log.*, 109(3):205–241.

Bucciarelli, A. and Salibra, A. (2004). The sensible graph theories of lambda calculus. In *19th IEEE Symposium on Logic in Computer Science (LICS 2004), 14-17 July 2004, Turku, Finland, Proceedings*, pages 276–285. IEEE Computer Society.

Bucciarelli, A., Ehrhard, T., and Manzonetto, G. (2007). Not enough points is enough. In Duparc, J. and Henzinger, T. A., editors, *Computer Science Logic, 21st International Workshop, CSL 2007, 16th Annual Conference of the EACSL, Lausanne, Switzerland, September 11-15, 2007, Proceedings*, volume 4646 of *Lecture Notes in Computer Science*, pages 298–312. Springer.

Bucciarelli, A., Carraro, A., and Salibra, A. (2012a). Minimal lambda-theories by ultraproducts. In Kesner, D. and Viana, P., editors, *Proceedings Seventh Workshop on Logical and Semantic Frameworks, with Applications, LSFA 2012, Rio de Janeiro, Brazil, September 29-30, 2012*, volume 113 of *EPTCS*, pages 61–76.

Bucciarelli, A., Ehrhard, T., and Manzonetto, G. (2012b). A relational semantics for parallelism and non-determinism in a functional setting. *Ann. Pure Appl. Log.*, 163 (7):918–934.

Bucciarelli, A., Carraro, A., Favro, G., and Salibra, A. (2014a). A graph-easy class of mute lambda-terms. In Bistarelli, S. and Formisano, A., editors, *Proceedings of the 15th Italian Conference on Theoretical Computer Science, Perugia, Italy, September 17-19, 2014*, volume 1231 of *CEUR Workshop Proceedings*, pages 59–71. CEUR-WS.org.

Bucciarelli, A., Kesner, D., and Ronchi Della Rocca, S. (2014b). The inhabitation problem for non-idempotent intersection types. In Díaz, J., Lanese, I., and Sangiorgi, D., editors, *Theoretical Computer Science - 8th IFIP TC 1/WG 2.2 International Conference, TCS 2014, Rome, Italy, September 1-3, 2014. Proceedings*, volume 8705 of *Lecture Notes in Computer Science*, pages 341–354. Springer.

Bucciarelli, A., Kesner, D., and Ventura, D. (2017). Non-idempotent intersection types for the lambda-calculus. *Log. J. IGPL*, 25(4):431–464.

Bucciarelli, A., Kesner, D., and Ronchi Della Rocca, S. (2018). Inhabitation for non-idempotent intersection types. *Log. Methods Comput. Sci.*, 14(3).

Bunder, M. W. (1969). *Set theory based on combinatory logic.* Ph.D. thesis, University of Amsterdam.

Bunder, M. W. (1974). Propositional and predicate calculuses based on combinatory logic. *Notre Dame Journal of Formal Logic*, 15(1):25–34.

Burris, S. N. and Sankappanavar, H. P. (2012). *A course in universal algebra - The millennium edition.* Springer-Verlag, Berlin. Revision of the 1981 book.

Burris, S. N. and Werner, H. (1979). Sheaf constructions and their elementary properties. *Transactions of the American Mathematical Society*, 248:269–309.

Campbell, P. J. (1978). The origin of "Zorn's lemma". *Historia Mathematica*, 5(1):77–89.

Cardone, F. and Hindley, J. R. (2009). Lambda-calculus and combinators in the 20th century. In Gabbay, D. M. and Woods, J., editors, *Logic from Russell to Church*, volume 5 of *Handbook of the History of Logic*, pages 723–817. Elsevier.

Carraro, A. and Guerrieri, G. (2014). A semantical and operational account of call-by-value solvability. In Muscholl, A., editor, *Foundations of Software Science and Computation Structures - 17th International Conference, FOSSACS 2014, Held as Part of the European Joint Conferences on Theory and Practice of Software, ETAPS 2014, Grenoble, France, April 5-13, 2014, Proceedings*, volume 8412 of *Lecture Notes in Computer Science*, pages 103–118. Springer.

Carraro, A. and Salibra, A. (2009). Reflexive Scott domains are not complete for the extensional lambda calculus. In *Proceedings of the 24th Annual IEEE Symposium on Logic in Computer Science, LICS 2009, 11-14 August 2009, Los Angeles, CA, USA*, pages 91–100. IEEE Computer Society.

Carraro, A. and Salibra, A. (2012). On the equational consistency of order-theoretic models of the lambda-calculus. In Cégielski, P. and Durand, A., editors, *Computer Science Logic (CSL'12) - 26th International Workshop/21st Annual Conference of the EACSL, CSL 2012, September 3-6, 2012, Fontainebleau, France*, volume 16 of *LIPIcs*, pages 152–166. Schloss Dagstuhl - Leibniz-Zentrum für Informatik.

Carraro, A. and Salibra, A. (2013). Ordered models of the lambda calculus. *Log. Methods Comput. Sci.*, 9(4).

Carraro, A., Ehrhard, T., and Salibra, A. (2010). Exponentials with infinite multiplicities. In Dawar, A. and Veith, H., editors, *Computer Science Logic, 24th International Workshop, CSL 2010, 19th Annual Conference of the EACSL, Brno, Czech Republic, August 23-27, 2010. Proceedings*, volume 6247 of *Lecture Notes in Computer Science*, pages 170–184. Springer.

Castellan, S., Clairambault, P., and Winskel, G. (2015). The parallel intensionally fully abstract games model of PCF. In *30th Annual ACM/IEEE Symposium on Logic in Computer Science, LICS 2015, Kyoto, Japan, July 6-10, 2015*, pages 232–243. IEEE Computer Society.

Castellan, S., Clairambault, P., Paquet, H., and Winskel, G. (2018). The concurrent game semantics of probabilistic PCF. In Dawar, A. and Grädel, E., editors, *Proceedings of the 33rd Annual ACM/IEEE Symposium on Logic in Computer Science, LICS 2018, Oxford, UK, July 09-12, 2018*, pages 215–224. ACM.

Cheilaris, P., Ramirez, J., and Zachos, S. (2011). Checking in linear time if an S-term normalizes. In *8th Panhellenic Logic Symposium.*

Chen, L. and Rong, Y. (2010). Digital topological method for computing genus and the Betti numbers. *Topology and its Applications*, 157(12):1931–1936.

Church, A. (1932). A set of postulates for the foundation of logic. *Annals of Mathematics*, 33(2):346–366.

Church, A. (1933). A set of postulates for the foundation of logic (second paper). *Annals of Mathematics*, 34:839–864.

Church, A. (1935). A proof of freedom from contradiction. *Proceedings of the National Academy of Sciences of the United States of America*, 21(5):275.

Church, A. (1936). An unsolvable problem of elementary number theory. *American Journal of Mathematics*, 58:354–363.

Church, A. (1941). *The calculi of lambda conversion.* Princeton University Press.

Church, A. and Rosser, J. B. (1936). Some properties of conversion. *Transactions of the American Mathematical Society*, 39(3):472–482.

Clairambault, P. and de Visme, M. (2020). Full abstraction for the quantum lambda-calculus. *Proc. ACM Program. Lang.*, 4(POPL):63:1–63:28.

Comer, S. (1971). Representations by algebras of sections over Boolean spaces. *Pacific J. Math.*, 38:29–38.

Comon, H., Dauchet, M., Gilleron, R., Jacquemard, F., Lugiez, D., Tison, S., and Tommasi, M. (1997). Tree automata techniques and applications. Available at `https://jacquema.gitlabpages.inria.fr/files/tata.pdf`. Release October 1st, 2002.

Coppo, M. and Dezani-Ciancaglini, M. (1980). An extension of the basic functionality theory for the λ-calculus. *Notre Dame Journal of Formal Logic*, 21(4):685–693.

Coppo, M., Dezani-Ciancaglini, M., and Ronchi Della Rocca, S. (1978). (Semi)-separability of finite sets of terms in Scott's $\mathcal{D}_\infty$-models of the lambda-calculus. In Ausiello, G. and Böhm, C., editors, *Automata, Languages and Programming, Fifth Colloquium, Udine, Italy*, volume 62 of *Lecture Notes in Computer Science*, pages 142–164. Springer.

Coppo, M., Dezani-Ciancaglini, M., and Venneri, B. (1981). Functional characters of solvable terms. *Math. Log. Q.*, 27(2-6):45–58.

Coppo, M., Dezani-Ciancaglini, M., Honsell, F., and Longo, G. (1984). Extended type structure and filter lambda models. In Lolli, G., Longo, G., and Marcja, A., editors, *Logic Colloquim '82*, pages 241–262. Elsevier Science Publishers B.V. (North-Holland).

Coppo, M., Dezani-Ciancaglini, M., and Zacchi, M. (1987). Type theories, normal forms and $\mathcal{D}_\infty$-lambda-models. *Inf. Comput.*, 72(2):85–116.

Corradini, A., Duval, D., Echahed, R., Prost, F., and Ribeiro, L. (2019a). The PBPO graph transformation approach. *J. Log. Algebraic Methods Program.*, 103:213–231.

Corradini, A., Heindel, T., König, B., Nolte, D., and Rensink, A. (2019b). Rewriting abstract structures: Materialization explained categorically. In Bojanczyk, M. and Simpson, A., editors, *Foundations of Software Science and Computation Structures - 22nd International Conference, FOSSACS 2019, Held as Part of the European Joint Conferences on Theory and Practice of Software, ETAPS 2019, Prague, Czech Republic, April 6-11, 2019, Proceedings*, volume 11425 of *Lecture Notes in Computer Science*, pages 169–188. Springer.

Courcelle, B. and Raoult, J.-C. (1980). Completions of ordered magmas. *Fundam. Informaticae*, 3(1):105.

Curry, H. B. (1930). Grundlagen der kombinatorischen logik. *American Journal of Mathematics*, 52:509.

Curry, H. B. (1942). The inconsistency of certain formal logics. *The Journal of Symbolic Logic*, 7(3):115–117.

Curry, H. B. and Feys, R. (1958). *Combinatory logic. Volume I.* Number 1 in Studies in Logic and the Foundations of Mathematics. North-Holland, Amsterdam.

Curry, H. B., Hindley, J. R., and Seldin, J. P. (1972). *Combinatory Logic. Volume II.* North-Holland Publishing Company.

Cvetko-Vah, K. and Salibra, A. (2015). The connection of skew Boolean algebras and discriminator varieties to Church algebras. *Algebra universalis*, 73(3):369–390.

Czajka, Ł. (2013). Higher-order illative combinatory logic. *The Journal of Symbolic Logic*, 78(3):837–872. Revised and corrected in `http://arxiv.org/abs/1202.3672`.

Czajka, Ł. (2015). *Semantic Consistency Proofs for Systems of Illative Combinatory Logic.* PhD thesis, Warsaw University. Available at `https://lukaszcz.github.io/papers/phd.pdf`.

Czajka, Ł. (2020). A new coinductive confluence proof for infinitary lambda calculus. *Log. Methods Comput. Sci.*, 16(1).

van Daalen, D. (1980). *The Language Theory of Automath.* Ph.D. thesis, Technical University Eindhoven. Large parts of this thesis, including the treatment of RuS, have been reproduced in Nederpelt et al. (1994).

Dal Lago, U. and Leventis, T. (2019). On the Taylor expansion of probabilistic lambda-terms. In Geuvers, H., editor, *4th International Conference on Formal Structures for Computation and Deduction, FSCD 2019, June 24-30, 2019, Dortmund, Germany*, volume 131 of *LIPIcs*, pages 13:1–13:16. Schloss Dagstuhl - Leibniz-Zentrum für Informatik.

Dal Lago, U., Masini, A., and Zorzi, M. (2009). On a measurement-free quantum lambda calculus with classical control. *Math. Struct. Comput. Sci.*, 19(2):297–335.

Danos, V. and Ehrhard, T. (2011). Probabilistic coherence spaces as a model of higher-order probabilistic computation. *Inf. Comput.*, 209(6):966–991.

David, R. and Nour, K. (2014). About the range property for $\mathcal{H}$. *Logical Methods in Computer Science*, 10(1).

Davis, M. (1982). Why Gödel didn't have Church's thesis. *Information and control*, 54 (1-2):3–24.

De Benedetti, E. and Ronchi Della Rocca, S. (2015). Call-by-value, elementary time and intersection types. In van Eekelen, M. C. J. D. and Dal Lago, U., editors, *Foundational and Practical Aspects of Resource Analysis - 4th International Workshop, FOPARA 2015, London, UK, April 11, 2015, Revised Selected Papers*, volume 9964 of *Lecture Notes in Computer Science*, pages 40–59.

De Carvalho, D. (2009). Execution time of lambda-terms via denotational semantics and intersection types. *CoRR*, abs/0905.4251.

De Carvalho, D. (2018). Execution time of λ-terms via denotational semantics and intersection types. *Mathematical Structures in Computer Science*, 28(7):1169–1203.

de Groote, P., editor (1995). *The Curry-Howard Isomorphism*, volume 8 of *Cahiers du centre de logique.* Academia, Louvain-la-Neuve.

Dedekind, R. (1965). Was sind und was sollen die zahlen? In *Was sind und was sollen die Zahlen?. Stetigkeit und Irrationale Zahlen*, pages 1–47. Springer.

Dekkers, W., Bunder, M. W., and Barendregt, H. P. (1998a). Completeness of two systems of illative combinatory logic for first-order propositional and predicate calculus. *Archive for Mathematical Logic*, 37(5):327–341.

Dekkers, W., Bunder, M. W., and Barendregt, H. P. (1998b). Completeness of the propositions-as-types interpretation of intuitionistic logic into illative combinatory logic. *J. Symb. Log.*, 63(3):869–890.

Dershowitz, N. (1979). Orderings for term-rewriting systems. In *20th Annual Symposium on Foundations of Computer Science, San Juan, Puerto Rico, 29-31 October 1979*, pages 123–131. IEEE Computer Society.

Dershowitz, N., Kaplan, S., and Plaisted, D. A. (1991). Rewrite, rewrite, rewrite, rewrite, rewrite, ... *Theor. Comput. Sci.*, 83(1):71–96.

Dezani-Ciancaglini, M. (1976). Characterization of normal forms possessing inverse in the $\lambda\beta\eta$-calculus. *Theor. Comput. Sci.*, 2(3):323–337.

Dezani-Ciancaglini, M. (1996). *Logical Semantics for Concurrent Lambda-Calculus.* PhD thesis, Nijmegen University.

Dezani-Ciancaglini, M. and Giovannetti, E. (2001). From Böhm's theorem to observational equivalences: an informal account. *Electr. Notes Theor. Comput. Sci.*, 50(2): 83–116.

Dezani-Ciancaglini, M. and Margaria, I. (1987). Polymorphic types, fixed-point combinators and continuous lambda-models. In Wirsing, M., editor, *Formal Description of Programming Concepts - III: Proceedings of the IFIP TC 2/WG 2.2 Working Conference on Formal Description of Programming Concepts - III, Ebberup, Denmark, 25-28 August 1986*, pages 425–450. North-Holland.

Di Gianantonio, P., Honsell, F., and Plotkin, G. D. (1995). Uncountable limits and the lambda calculus. *Nord. J. Comput.*, 2(2):126–145.

Di Gianantonio, P., Franco, G., and Honsell, F. (1999). Game semantics for untyped $\lambda\beta\eta$-calculus. In Girard, J.-Y., editor, *Typed Lambda Calculi and Applications, 4th International Conference, TLCA'99, L'Aquila, Italy, April 7-9, 1999, Proceedings*, volume 1581 of *Lecture Notes in Computer Science*, pages 114–128. Springer.

Di Pierro, A. (2017). A probabilistic semantics for the pure λ-calculus. In Hung, D. V. and Kapur, D., editors, *Theoretical Aspects of Computing - ICTAC 2017 - 14th International Colloquium, Hanoi, Vietnam, October 23-27, 2017, Proceedings*, volume 10580 of *Lecture Notes in Computer Science*, pages 70–76. Springer.

Díaz-Caro, A., Guillermo, M., Miquel, A., and Valiron, B. (2019). Realizability in the unitary sphere. In *34th Annual ACM/IEEE Symposium on Logic in Computer Science, LICS 2019, Vancouver, BC, Canada, June 24-27, 2019*, pages 1–13. IEEE.

Dierks, V., Erné, M., and Reinhold, J. (1994). Complements in lattices of varieties and equational theories. *Algebra Universalis*, 31:506–515.

Dutle, A., Moscato, M., Titolo, L., Muñoz, C., Anderson, G., and Bobot, F. (2020). Formal analysis of the Compact Position Reporting algorithm. *Formal Aspects of Computing.*

Ehrhard, T. (2002). On Köthe sequence spaces and linear logic. *Math. Struct. Comput. Sci.*, 12(5):579–623.

Ehrhard, T. (2005). Finiteness spaces. *Math. Struct. Comput. Sci.*, 15(4):615–646.

Ehrhard, T. (2012). Collapsing non-idempotent intersection types. In Cégielski, P. and Durand, A., editors, *Computer Science Logic (CSL'12) - 26th International Workshop/21st Annual Conference of the EACSL, CSL 2012, September 3-6, 2012, Fontainebleau, France*, volume 16 of *LIPIcs*, pages 259–273. Schloss Dagstuhl - Leibniz-Zentrum für Informatik.

Ehrhard, T. (2018). An introduction to differential linear logic: proof-nets, models and antiderivatives. *Math. Struct. Comput. Sci.*, 28(7):995–1060.

Ehrhard, T. and Guerrieri, G. (2016). The bang calculus: an untyped lambda-calculus generalizing call-by-name and call-by-value. In Cheney, J. and Vidal, G., editors, *Proceedings of the 18th International Symposium on Principles and Practice of Declarative Programming, Edinburgh, United Kingdom, September 5-7, 2016*, pages 174–187. ACM.

Ehrhard, T. and Regnier, L. (2003). The differential lambda-calculus. *Theor. Comput. Sci.*, 309(1-3):1–41.

Ehrhard, T. and Regnier, L. (2006). Böhm trees, Krivine's machine and the Taylor expansion of lambda-terms. In Beckmann, A., Berger, U., Löwe, B., and Tucker, J. V., editors, *Logical Approaches to Computational Barriers, Second Conference on Computability in Europe, CiE 2006, Swansea, UK, June 30-July 5, 2006, Proceedings*, volume 3988 of *Lecture Notes in Computer Science*, pages 186–197. Springer.

Ehrhard, T. and Regnier, L. (2008). Uniformity and the Taylor expansion of ordinary lambda-terms. *Theor. Comput. Sci.*, 403(2-3):347–372.

Endrullis, J. and de Vrijer, R. (2008). Reduction under substitution. In Voronkov, A., editor, *Rewriting Techniques and Applications, 19th International Conference, RTA 2008, Hagenberg, Austria, July 15-17, 2008, Proceedings*, volume 5117 of *Lecture Notes in Computer Science*, pages 425–440. Springer.

Endrullis, J. and Klop, J. W. (2013). De Bruijn's weak diamond property revisited. *Indagationes Mathematicae*, 24(4):1050–1072.

Endrullis, J. and Polonsky, A. (2011). Infinitary rewriting coinductively. In Danielsson, N. A. and Nordström, B., editors, *18th International Workshop on Types for Proofs and Programs, TYPES 2011, September 8-11, 2011, Bergen, Norway*, volume 19 of *LIPIcs*, pages 16–27. Schloss Dagstuhl - Leibniz-Zentrum für Informatik.

Endrullis, J., de Vrijer, R. C., and Waldmann, J. (2009). Local termination. In Treinen, R., editor, *Rewriting Techniques and Applications, 20th International Conference, RTA 2009, Brasília, Brazil, June 29 - July 1, 2009, Proceedings*, volume 5595 of *Lecture Notes in Computer Science*, pages 270–284. Springer.

Endrullis, J., de Vrijer, R. C., and Waldmann, J. (2010). Local termination: theory and practice. *Log. Methods Comput. Sci.*, 6(3).

Endrullis, J., Hendriks, D., and Klop, J. W. (2012). Highlights in infinitary rewriting and lambda calculus. *Theor. Comput. Sci.*, 464:48–71.

Endrullis, J., Grabmayer, C., Hendriks, D., Klop, J. W., and van Oostrom, V. (2014). Infinitary term rewriting for weakly orthogonal systems: Properties and counterexamples. *Log. Methods Comput. Sci.*, 10(2).

Endrullis, J., Hansen, H. H., Hendriks, D., Polonsky, A., and Silva, A. (2015). A coinductive framework for infinitary rewriting and equational reasoning. In Fernández, M., editor, *26th International Conference on Rewriting Techniques and Applications, RTA 2015, June 29 to July 1, 2015, Warsaw, Poland*, volume 36 of *LIPIcs*, pages 143–159. Schloss Dagstuhl - Leibniz-Zentrum für Informatik.

Endrullis, J., Klop, J. W., and Polonsky, A. (2016). Reduction cycles in lambda calculus and combinatory logic. In van Eijck, J., Iemhoff, R., and Joosten, J. J., editors, *Liber Amicorum Alberti — A Tribute to Albert Visser*. College Publications.

Endrullis, J., Hendriks, D., Klop, J. W., and Polonsky, A. (2017). Clocked lambda calculus. *Math. Struct. Comput. Sci.*, 27(5):782–806.

Endrullis, J., Hansen, H. H., Hendriks, D., Polonsky, A., and Silva, A. (2018). Coinductive foundations of infinitary rewriting and infinitary equational logic. *Log. Methods Comput. Sci.*, 14(1).

Endrullis, J., Klop, J. W., and Overbeek, R. (2020). Decreasing diagrams for confluence and commutation. *Log. Methods Comput. Sci.*, 16(1).

Engeler, E. (1981). Algebras and combinators. *Algebra Universalis*, 13(3):389–392.

Engeler, E. (1993). *Algorithmic Properties of Structures: Selected Papers of Erwin Engeler*. World Scientific.

Engeler, E. (2012). *The combinatory programme*. Springer Science & Business Media.

Érdi, G. (2021). *Retrocomputing in Clash: Haskell for FPGA Hardware Design*. Lulu.

Faggian, C. and Guerrieri, G. (2021). Factorization in call-by-name and call-by-value calculi via linear logic. In Kiefer, S. and Tasson, C., editors, *Foundations of Software Science and Computation Structures - 24th International Conference, FOSSACS 2021, Held as Part of the European Joint Conferences on Theory and Practice of Software, ETAPS 2021, Luxembourg City, Luxembourg, March 27 - April 1, 2021, Proceedings*, volume 12650 of *Lecture Notes in Computer Science*, pages 205–225. Springer.

Faggian, C. and Ronchi Della Rocca, S. (2019). Lambda calculus and probabilistic computation. In *34th Annual ACM/IEEE Symposium on Logic in Computer Science, LICS 2019, Vancouver, BC, Canada, June 24-27, 2019*, pages 1–13. IEEE.

Fiore, M. P. (1995). Order-enrichment for categories of partial maps. *Math. Struct. Comput. Sci.*, 5(4):533–562.

Fiore, M. P., Gambino, N., Hyland, M., and Winskel, G. (2007). The Cartesian closed bicategory of generalised species of structures. *Journal of the London Mathematical Society*, 77(1):203–220.

Folkerts, E. (1995). *Kongruenz von unlösbaren lambda termen.* Ph.D. thesis, University WWU-Münster. In German.

Folkerts, E. (1998). Invertibility in $\lambda\eta$. In *Thirteenth Annual IEEE Symposium on Logic in Computer Science, Indianapolis, Indiana, USA, June 21-24, 1998*, pages 418–429. IEEE Computer Society.

Folkerts, E. (2020). Personal communication.

Fox, A. (2003). Formal Specification and Verification of ARM6. In Basin, D. A. and Wolff, B., editors, *Theorem Proving in Higher Order Logics 2003*, volume 2758 of *Lecture Notes in Computer Science*. Springer.

Franco, G. (2001). *Some intensional model of lambda calculus.* Ph.D. thesis, Università degli studi di Udine.

Frege, G. (1879). *Begriffsschrift und andere Aufsätze.* Georg Olms Verlag, Hildesheim. Zweite Auflage. Mit E. Husserls und H. Scholz' Anmerkungen herausgegeben von Ignacio Angelelli, Nachdruck, 1971.

Gardner, P. (1994). Discovering needed reductions using type theory. In Hagiya, M. and Mitchell, J. C., editors, *Theoretical Aspects of Computer Software, International Conference TACS '94, Sendai, Japan, April 19-22, 1994, Proceedings*, volume 789 of *Lecture Notes in Computer Science*, pages 555–574. Springer.

Gentzen, G. (1935a). Untersuchungen über das logische schließen. I. *Mathematische zeitschrift*, 35.

Gentzen, G. (1935b). Untersuchungen über das logische schließen. II. *Mathematische zeitschrift*, 39.

Gierz, G., Hofmann, K. H., Keimel, K., Lawson, J. D., Mislove, M., and Scott, D. S. (2012). *A compendium of continuous lattices*. Springer Science & Business Media.

Girard, J.-Y. (1987). Linear logic. *Theor. Comput. Sci.*, 50:1–102.

Girard, J.-Y. (1988). Normal functors, power series and λ-calculus. *Ann. Pure Appl. Logic*, 37(2):129–177.

Girard, J.-Y. (1989). Geometry of interaction I. Interpretation of system F. In Ferro, R., Bonotto, C., Valentini, S., and Zanardo, A., editors, *Logic colloquium '88*, volume 127 of *Studies in Logic and The Foundations of Mathematics*, pages 221–260. North-Holland.

Given-Wilson, T. and Jay, B. (2011). A combinatory account of internal structure. *J. Symb. Log.*, 76(3):807–826.

Gödel, K. (1930). Die Vollständigkeit der Axiome des logischen Funktionalkalküls. *Monatshefte für Mathematik und Physik*, 37:349–360.

Gödel, K. (1931). On formally undecidable propositions of principia mathematica and related systems. *Monatshefte für Math. und Physik–Monthly journals on mathematics and physics*, 38:1.

Gödel, K. (1934). On undecidable propositions of formal mathematical systems. Institute for Advanced Study, mimeographed lecture notes by S. C. Kleene and J. B. Rosser. Revised and amplified in (?, pp. 41–74).

Gonthier, G. (2007). The four colour theorem: Engineering of a formal proof. In Kapur, D., editor, *Computer Mathematics, 8th Asian Symposium, ASCM 2007, Singapore, December 15-17, 2007. Revised and Invited Papers*, volume 5081 of *Lecture Notes in Computer Science*, page 333. Springer.

Gonthier, G., Abadi, M., and Lévy, J.-J. (1992a). Linear logic without boxes. In *Proceedings of the Seventh Annual Symposium on Logic in Computer Science (LICS '92), Santa Cruz, California, USA, June 22-25, 1992*, pages 223–234. IEEE Computer Society.

Gonthier, G., Abadi, M., and Lévy, J.-J. (1992b). The geometry of optimal lambda reduction. In Sethi, R., editor, *Conference Record of the Nineteenth Annual ACM SIGPLAN-SIGACT Symposium on Principles of Programming Languages, Albuquerque, New Mexico, USA, January 19-22, 1992*, pages 15–26. ACM Press.

Gonthier, G., Asperti, A., Avigad, J., Bertot, Y., Cohen, C., Garillot, F., Roux, S. L., Mahboubi, A., O'Connor, R., Biha, S. O., Pasca, I., Rideau, L., Solovyev, A., Tassi, E., and Théry, L. (2013). A machine-checked proof of the odd order theorem. In Blazy, S., Paulin-Mohring, C., and Pichardie, D., editors, *Interactive Theorem Proving - 4th International Conference, ITP 2013, Rennes, France, July 22-26, 2013. Proceedings*, volume 7998 of *Lecture Notes in Computer Science*, pages 163–179. Springer.

Gouy, X. (1995). *Etude des théories équationnelles et des propriétés algébriques des modèles stables du λ-calcul.* Thèse de doctorat, Université Paris-Diderot (Paris VII).

Grellois, C. and Melliès, P.-A. (2015). An infinitary model of linear logic. In Pitts, A. M., editor, *Foundations of Software Science and Computation Structures - 18th International Conference, FoSSaCS 2015, Held as Part of the European Joint Conferences on Theory and Practice of Software, ETAPS 2015, London, UK, April 11-18, 2015. Proceedings*, volume 9034 of *Lecture Notes in Computer Science*, pages 41–55. Springer.

Guerrieri, G. and Manzonetto, G. (2018). The bang calculus and the two Girard's translations. In Ehrhard, T., Fernández, M., de Paiva, V., and Tortora de Falco, L., editors, *Proceedings Joint International Workshop on Linearity & Trends in Linear Logic and Applications, Linearity-TLLA@FLoC 2018, Oxford, UK, 7-8 July 2018*, volume 292 of *EPTCS*, pages 15–30.

Guerrieri, G., Paolini, L., and Ronchi Della Rocca, S. (2017). Standardization and conservativity of a refined call-by-value lambda-calculus. *Log. Methods Comput. Sci.*, 13(4).

Gunter, C. A. and Scott, D. S. (1990). Semantic domains. In *Handbook of Theoretical Computer Science, Volume B, ed J. van Leeuwen*, pages 633–674. Elsevier.

Hagemann, J. and Mitschke, A. (1973). On n-permutable congruences. *Algebra Univ.*, 3:8–12.

Hales, T. C., Adams, M., Bauer, G., Dang, D. T., Harrison, J., Hoang, T. L., Kaliszyk, C., Magron, V., McLaughlin, S., Nguyen, T. T., Nguyen, T. Q., Nipkow, T., Obua, S., Pleso, J., Rute, J. M., Solovyev, A., Ta, A. H. T., Tran, T. N., Trieu, D. T., Urban, J., Vu, K. K., and Zumkeller, R. (2017). A formal proof of the Kepler conjecture. *Forum of mathematics, Pi*, 5. Available at `http://arxiv.org/abs/1501.02155`.

Halpern, J. D. (1964). The independence of the axiom of choice from the Boolean prime ideal theorem. *Fund. Math*, 55:57–66.

Halpern, J. D. and Lévy, A. (1971). The Boolean prime ideal theorem does not imply the axiom of choice. In *Axiomatic Set Theory, Proceedings of Symposia in Pure Mathematics*, volume 13, Part 1, pages 83–134. American Mathematical Society, Providence, Rhode Island.

Harmer, R. and McCusker, G. (1999). A fully abstract game semantics for finite nondeterminism. In *14th Annual IEEE Symposium on Logic in Computer Science, Trento, Italy, July 2-5, 1999*, pages 422–430. IEEE Computer Society.

Harrison, J. (2009). Formalizing an analytic proof of the prime number theorem. *Journal of Automated Reasoning*, 43(3):243–261.

Hayashi, S. (1985). Adjunction of semifunctors: Categorical structures in non-extensional lambda calculus. *Theor. Comput. Sci.*, 41:95–104.

Hindley, J. R. (1969). An abstract form of the Church-Rosser Theorem. I. *J. Symb. Log.*, 34(4):545–560.

Hindley, J. R. (2008). M. H. Newman's typability algorithm for lambda-calculus. *J. Log. Comput.*, 18(2):229–238.

Hindley, J. R. and Longo, G. (1980). Lambda calculus models and extensionality. *Z. Math. Logik Grundlag. Math*, 26:289–310.

Hindley, J. R. and Seldin, J. P. (2008). *Lambda-Calculus and Combinators: An Introduction.* Cambridge University Press, Cambridge, 2nd edition.

Hirokawa, S. (1984). Some properties of one-step recurrent terms in λ-calculus. Technical Report 515, Kyoto University.

Honsell, F. (2000). TLCA list, Problem #22: Is there a continuously complete cpo model of the λ-calculus whose theory is precisely $\boldsymbol{\lambda}\eta$ or $\boldsymbol{\lambda}$? See `http://tlca.di.unito.it/opltlca/`. First raised in Honsell and Ronchi Della Rocca (1992).

Honsell, F. and Ronchi Della Rocca, S. (1992). An approximation theorem for topological lambda models and the topological incompleteness of lambda calculus. *J. Comput. Syst. Sci.*, 45(1):49–75.

Howard, W. A. (1980). The formulae-as-types notion of construction. In Hindley, J. R. and Seldin, J. P., editors, *To HB Curry: Essays on combinatory logic, lambda calculus and formalism*, pages 479–490. Academic Press.

Huet, G. P. and Lévy, J.-J. (1991). Computations in orthogonal rewriting systems, II. In Lassez, J.-L. and Plotkin, G. D., editors, *Computational Logic - Essays in Honor of Alan Robinson*, pages 415–443. The MIT Press.

Hyland, J. M. E. (1975a). A survey of some useful partial order relations on terms of the lambda calculus. In Böhm, C., editor, *Lambda-Calculus and Computer Science Theory, Proceedings of the Symposium Held in Rome, March 25-27, 1975*, volume 37 of *Lecture Notes in Computer Science*, pages 83–95. Springer.

Hyland, J. M. E. (1975b). A syntactic characterization of the equality in some models for the λ-calculus. *Journal London Mathematical Society (2)*, 12(3):361–370.

Hyland, J. M. E. (2017). Classical lambda calculus in modern dress. *Mathematical Structures in Computer Science*, 27(5):762–781.

Hyland, J. M. E. and Ong, C.-H. L. (2000). On full abstraction for PCF: I, II, and III. *Inf. Comput.*, 163(2):285–408.

Hyland, M., Nagayama, M., Power, J., and Rosolini, G. (2006). A category theoretic formulation for Engeler-style models of the untyped lambda. *Electr. Notes Theor. Comput. Sci.*, 161:43–57.

Intrigila, B. (1991). A problem on easy terms in calculus. *Fundam. Informaticae*, 15(1): 99–106.

Intrigila, B. (1997). Non-existent Statman's double fixedpoint combinator does not exist, indeed. *Inf. Comput.*, 137(1):35–40.

Intrigila, B. (2000). TLCA list, Problem #25: How many fixed points can a combinator have? See `http://tlca.di.unito.it/opltlca/`. First raised in Intrigila and Biasone (2000).

Intrigila, B. and Biasone, E. (2000). On the number of fixed points of a combinator in lambda calculus. *Mathematical Structures in Computer Science*, 10(5):595–615.

Intrigila, B. and Nesi, M. (2003). On structural properties of eta-expansions of identity. *Inf. Process. Lett.*, 87(6):327–333.

Intrigila, B. and Statman, R. (2004). The omega rule is Π^0_2-hard in the $\lambda\beta$-calculus. In *Symposium on Logic in Computer Science (LICS 2004)*, pages 202–210. IEEE Computer Society.

Intrigila, B. and Statman, R. (2006). Solution of a problem of Barendregt on sensible λ-theories. *Log. Methods Comput. Sci.*, 2(4).

Intrigila, B. and Statman, R. (2007). On Henk Barendregt's favorite open problem. *Reflections on Type Theory, Lambda Calculus, and the Mind.* Essays dedicated to Henk Barendregt on the occasion of his 60th birthday.
Available at `http://www.cs.ru.nl/barendregt60/essays/`.

Intrigila, B. and Statman, R. (2009). The omega rule is Π^1_1-complete in the $\lambda\beta$-calculus. *Logical Methods in Computer Science*, 5(2).

Intrigila, B. and Statman, R. (2015). Lambda theories allowing terms with a finite number of fixed points. *Mathematical Structures in Computer Science*, pages 1–23.

Intrigila, B., Manzonetto, G., and Polonsky, A. (2017). Refutation of Sallé's longstanding conjecture. In Miller, D., editor, *2nd International Conference on Formal Structures for Computation and Deduction, FSCD 2017, September 3-9, 2017, Oxford, UK*, volume 84 of *LIPIcs*, pages 20:1–20:18. Schloss Dagstuhl - Leibniz-Zentrum fuer Informatik.

Intrigila, B., Manzonetto, G., and Polonsky, A. (2019). Degrees of extensionality in the theory of Böhm trees and Sallé's conjecture. *Logical Methods in Computer Science*, 15 (1).

Jacobs, B. and Rutten, J. (1997). A tutorial on (co)algebras and (co)induction. *EATCS Bulletin*, 62:62–222.

Jacopini, G. (1975). A condition for identifying two elements of whatever model of combinatory logic. In Böhm, C., editor, *International Symposium on Lambda-Calculus and Computer Science Theory*, volume 37 of *Lecture Notes in Computer Science*, pages 213–219. Springer.

Jacopini, G. and Venturini Zilli, M. (1985). Easy terms in the lambda-calculus. *Fundamenta Informaticae*, 8(2):225–233.

Jay, B. (2018). Self-quotation in a typed, intensional lambda-calculus. *Electr. Notes Theor. Comput. Sci.*, 336:207–222.

Johnstone, P. T. (1982). *Stone spaces.* Cambridge Studies in Advanced Mathematics. Cambridge University Press.

Johnstone, P. T. (1983). The point of pointless topology. *Bulletin (New Series) of the American Mathematical Society*, 8(1):41–53.

Joly, T. (2003). Encoding of the halting problem into the monster type & applications. In Hofmann, M., editor, *Typed Lambda Calculi and Applications, 6th International Conference, TLCA 2003, Valencia, Spain, June 10-12, 2003, Proceedings*, volume 2701 of *Lecture Notes in Computer Science*, pages 153–166. Springer.

Keimel, K. and Lawson, J. D. (2009). D-completions and the d-topology. *Ann. Pure Appl. Log.*, 159(3):292–306.

Kelly, G. M. (2005). Basic concepts of enriched category theory. In *Reprints in Theory and Applications of Categories*, volume 10, pages vi+137. Cambridge University Press, Cambridge. Reprint of the 1982 original [Cambridge Univ. Press, Cambridge; MR0651714].

Kennaway, R. and de Vries, F.-J. (2003). Infinitary rewriting. In *Term Rewriting Systems*, volume 55 of *Cambridge tracts in Theoretical Computer Science*, pages 668–711. Cambridge University Press.

Kennaway, R., Klop, J. W., Sleep, M. R., and de Vries, F.-J. (1995). Infinitary lambda calculi and Böhm models. In Hsiang, J., editor, *Rewriting Techniques and Applications, 6th International Conference, RTA-95, Kaiserslautern, Germany, April 5-7, 1995, Proceedings*, volume 914 of *Lecture Notes in Computer Science*, pages 257–270. Springer.

Kennaway, R., Klop, J. W., Sleep, M. R., and de Vries, F.-J. (1997). Infinitary lambda calculus. *Theor. Comput. Sci.*, 175(1):93–125.

Kennaway, R., van Oostrom, V., and de Vries, F.-J. (1999). Meaningless terms in rewriting. *J. Funct. Log. Program.*, 1999(1).

Kennaway, R., Severi, P., Sleep, M. R., and de Vries, F.-J. (2005). Infinitary rewriting: From syntax to semantics. In Middeldorp, A., van Oostrom, V., van Raamsdonk, F., and de Vrijer, R. C., editors, *Processes, Terms and Cycles: Steps on the Road to Infinity, Essays Dedicated to Jan Willem Klop, on the Occasion of His 60th Birthday*, volume 3838 of *Lecture Notes in Computer Science*, pages 148–172. Springer.

Ker, A. D., Nickau, H., and Ong, C.-H. L. (1999). A universal innocent game model for the Böhm tree lambda theory. In Flum, J. and Rodríguez-Artalejo, M., editors, *Computer Science Logic, 13th International Workshop, CSL '99, 8th Annual Conference of the EACSL, Madrid, Spain, September 20-25, 1999, Proceedings*, volume 1683 of *Lecture Notes in Computer Science*, pages 405–419. Springer.

Ker, A. D., Nickau, H., and Ong, C.-H. L. (2002). Innocent game models of untyped lambda-calculus. *Theor. Comput. Sci.*, 272(1-2):247–292.

Kerinec, E., Manzonetto, G., and Pagani, M. (2020). Revisiting call-by-value Böhm trees in light of their Taylor expansion. *Log. Methods Comput. Sci.*, 16(3).

Kesner, D. (2016). Reasoning about call-by-need by means of types. In Jacobs, B. and Löding, C., editors, *Foundations of Software Science and Computation Structures - 19th International Conference, FOSSACS 2016, Held as Part of the European Joint Conferences on Theory and Practice of Software, ETAPS 2016, Eindhoven, The Netherlands, April 2-8, 2016, Proceedings*, volume 9634 of *Lecture Notes in Computer Science*, pages 424–441. Springer.

Kesner, D., Ríos, A., and Viso, A. (2018). Call-by-need, neededness and all that. In Baier, C. and Dal Lago, U., editors, *Foundations of Software Science and Computation Structures - 21st International Conference, FOSSACS 2018, Held as Part of the European Joint Conferences on Theory and Practice of Software, ETAPS 2018, Thessaloniki, Greece, April 14-20, 2018, Proceedings*, volume 10803 of *Lecture Notes in Computer Science*, pages 241–257. Springer.

Ketema, J. and Simonsen, J. G. (2009). Infinitary combinatory reduction systems: Confluence. *Log. Methods Comput. Sci.*, 5(4).

Kleene, S. C. (1934). Proof by cases in formal logic. *Annals of mathematics*, pages 529–544.

Kleene, S. C. (1935). A theory of positive integers in formal logic. Part I. *American journal of mathematics*, 57(1):153–173.

Kleene, S. C. (1936a). General recursive functions of natural numbers. *Mathematische Annalen*, 112:727–742.

Kleene, S. C. (1936b). λ-definability and recursiveness. *Duke mathematical journal*, 2 (2):340–353.

Kleene, S. C. (1952). *Introduction to metamathematics*, volume 483. van Nostrand New York.

Kleene, S. C. (1955). Recursive predicates and quantifiers. *Trans. Am. Math. Soc*, 79: 312–340.

Kleene, S. C. (1963). Recursive functionals and quantifiers of finite types II. *Transactions of the American Mathematical Society*, 108(1):106–142.

Kleene, S. C. and Rosser, J. B. (1935). The inconsistency of certain formal logics. *Annals of Mathematics*, pages 630–636.

Klein, G., Elphinstone, K., Heiser, G., Andronick, J., Cock, D., Derrin, P., Elkaduwe, D., Engelhardt, K., Kolanski, R., Norrish, M., Sewell, T., Tuch, H., and Winwood, S. (2009). seL4: formal verification of an OS kernel. In Matthews, J. and Anderson, T., editors, *ACM Symposium on Principles of Operating Systems*, pages 207–220. Big Sky.

Klop, J. W. (1980a). *Combinatory Reduction Systems.* Ph.D. thesis, Mathematisch Centrum. Mathematical Centre Tracts 127.

Klop, J. W. (1980b). Reduction cycles in combinatory logic. In Hindley and Seldin, editors, *Essays on Combinatory Logic, Lambda-Calculus, and Formalism*, pages 193–214. Academic Press, San Diego.

Klop, J. W. (2007). New fixed point combinators from old. In Barendsen, E., Capretta, V., Geuvers, H., and Niqui, M., editors, *Reflections on Type Theory, λ-Calculus, and the Mind. Essays dedicated to Henk Barendregt on the occasion of his 60th birthday*, pages 197–211. Radboud University Nijmegen.

Klop, J. W., van Oostrom, V., and van Raamsdonk, F. (2007). Reduction strategies and acyclicity. In Comon-Lundh, H., Kirchner, C., and Kirchner, H., editors, *Rewriting, Computation and Proof, Essays Dedicated to Jean-Pierre Jouannaud on the Occasion of His 60th Birthday*, volume 4600 of *Lecture Notes in Computer Science*, pages 89–112. Springer.

Koymans, C. P. J. (1982). Models of the lambda calculus. *Inf. Control.*, 52(3):306–332.

Kozen, D. and Silva, A. (2017). Practical coinduction. *Math. Struct. Comput. Sci.*, 27 (7):1132–1152.

Krivine, J.-L. (1993). *Lambda-calculus, types and models.* Ellis Horwood series in computers and their applications. Masson.

Kuper, J. (1997). On the Jacopini technique. *Inf. Comput.*, 138(2):101–123.

Kuper, J. (1999). $\mathsf{Y}\Omega_3$ is almost easy. *The 1998 European Summer Meeting of the Association for Symbolic Logic. Logic Colloquium '98*, 5(1):59–153.

Kurz, A., Petrisan, D., Severi, P., and de Vries, F.-J. (2013). Nominal coalgebraic data types with applications to lambda calculus. *Log. Methods Comput. Sci.*, 9(4).

Laird, J., Manzonetto, G., and McCusker, G. (2013). Constructing differential categories and deconstructing categories of games. *Inf. Comput.*, 222:247–264.

Lambek, J. and Scott, P. (1986). *Introduction to higher order categorical logic*, volume 007 of *Cambridge studies in advanced mathematics*. Cambridge University Press, Cambridge, New York, Melbourne.

Lampe, W. A. (1986). A property of the lattice of equational theories. *Algebra Universalis*, 23:61–69.

Lamping, J. (1989). An algorithm for optimal lambda calculus reduction. In *17th ACM SIGPLAN-SIGACT Symposium on Principles of Programming Languages*, pages 16–30. ACM.

Lassen, S. B. (1999). Bisimulation in untyped lambda calculus: Böhm trees and bisimulation up to context. *Electr. Notes Theor. Comput. Sci.*, 20:346–374.

Lercher, B. (1976). Lambda-calculus terms that reduce to themselves. *Notre Dame J. Formal Log.*, 17(2):291–292.

Leroy, X. (2009). A formally verified compiler back-end. *Journal of Automated Reasoning*, 43(4):363.

Lévy, J.-J. (1975). An algebraic interpretation of equality in some models of the lambda calculus. In Böhm, C., editor, *Lambda Calculus and Computer Science Theory*, volume 37 of *Lecture Notes in Computer Science*, pages 147–165. Springer.

Lévy, J.-J. (1978a). Approximations et arbres de Böhm dans le lambda-calcul. In Robinet, B., editor, *Lambda Calcul et Sémantique formelle des langages de programmation, Actes de la 6ème École de printemps d'Informatique théorique, La Châtre*, LITP-ENSTA, pages 239–257. (In French).

Lévy, J.-J. (1978b). *Réductions correctes et optimales dans le λ-calcul.* Thèse d'état, Université Paris-Diderot (Paris 7). In French.

Lévy, J.-J. (1980). Optimal reductions in the lambda-calculus. In Hindley and Seldin, editors, *Essays on Combinatory Logic, Lambda-Calculus, and Formalism*, pages 159–191. Academic Press, San Diego.

Lévy, J.-J. (1993). Böhm trees and extensionality. Unpublished.

Lévy, J.-J. (2005). Le lambda calcul — notes du cours. In French. `http://pauillac.inria.fr/~levy/courses/X/M1/lambda/dea-spp/jjl.pdf`.

Lévy, J.-J. (2017). Redexes are stable in the λ-calculus. *Math. Struct. Comput. Sci.*, 27 (5):738–750.

Löb, M. H. (1955). Solution of a problem of Leon Henkin. *The Journal of Symbolic Logic*, 20(2):115–118.

Longley, J. (2009). Some programming languages suggested by game models (extended abstract). In Abramsky, S., Mislove, M. W., and Palamidessi, C., editors, *Proceedings of the 25th Conference on Mathematical Foundations of Programming Semantics, MFPS 2009, Oxford, UK, April 3-7, 2009*, volume 249 of *Electronic Notes in Theoretical Computer Science*, pages 117–134. Elsevier.

Longley, J. and Normann, D. (2015). *Higher-Order Computability.* Theory and Applications of Computability. Springer, Berlin, Heidelberg.

Longo, G. (1983). Set-theoretical models of λ-calculus: theories, expansions, isomorphisms. *Ann. Pure Appl. Logic*, 24(2):153–188.

Lusin, S. and Salibra, A. (2003). A note on absolutely unorderable combinatory algebras. *J. Log. Comput.*, 13(4):481–502.

Lusin, S. and Salibra, A. (2004). The lattice of lambda theories. *J. Log. Comput.*, 14(3): 373–394.

Mac Lane, S. (2013). *Categories for the working mathematician*, volume 5. Springer Science & Business Media.

Manzonetto, G. (2008). *Models and theories of λ-calculus.* Ph.D. thesis, Università Ca'Foscari and Université Paris-Diderot.

Manzonetto, G. (2009). A general class of models of $\mathcal{H}^*$. In Královic, R. and Niwinski, D., editors, *Mathematical Foundations of Computer Science 2009, 34th International Symposium, MFCS 2009, Novy Smokovec, High Tatras, Slovakia, August 24-28, 2009. Proceedings*, volume 5734 of *Lecture Notes in Computer Science*, pages 574–586. Springer.

Manzonetto, G. and Ruoppolo, D. (2014). Relational graph models, Taylor expansion and extensionality. *Electr. Notes Theor. Comput. Sci.*, 308:245–272.

Manzonetto, G. and Salibra, A. (2006). Boolean algebras for lambda calculus. In *21th IEEE Symposium on Logic in Computer Science (LICS 2006), 12-15 August 2006, Seattle, WA, USA, Proceedings*, pages 317–326. IEEE Computer Society.

Manzonetto, G. and Salibra, A. (2008). From lambda-calculus to universal algebra and back. In Ochmanski, E. and Tyszkiewicz, J., editors, *Mathematical Foundations of Computer Science 2008, 33rd International Symposium, MFCS 2008, Torun, Poland, August 25-29, 2008, Proceedings*, volume 5162 of *Lecture Notes in Computer Science*, pages 479–490. Springer.

Manzonetto, G. and Salibra, A. (2009). Lattices of equational theories as Church algebras. In Drossos, C., Peppas, P., and Tsinakis, C., editors, *Proc. 7th Panhellenic Logic Symposium*, pages 117–121. Patras University Press.

Manzonetto, G. and Salibra, A. (2010). Applying universal algebra to lambda calculus. *J. Log. Comput.*, 20(4):877–915.

Manzonetto, G., Polonsky, A., Saurin, A., and Simonsen, J. G. (2019). The fixed point property and a technique to harness double fixed point combinators. *Journal of Logic and Computation.*

Margaria, I. and Zacchi, M. (1983). Right and left invertibility in lambda-beta-calculus. *RAIRO Theor. Informatics Appl.*, 17(1):71–88.

Martini, S. (1992). Categorical models for non-extensional lambda-calculi and combinatory logic. *Math. Struct. Comput. Sci.*, 2(3):327–357.

McCarthy, J. (1962). Computer programs for checking the correctness of mathematical proofs. In *Proceedings of a Symposium of Pure Mathematics*, volume V, pages 219–227. American Mathematical Society, Providence.

McKenzie, R. N., McNulty, G., and Taylor, W. (1987). *Algebras, lattices, varieties, Volume I.* Wadsworth Brooks, Monterey, California.

Melliès, P.-A. (2003). Categorical models of linear logic revisited. Available at: `https://hal.archives-ouvertes.fr/hal-00154229`. Preprint.

Melliès, P.-A., Tabareau, N., and Tasson, C. (2018). An explicit formula for the free exponential modality of linear logic. *Math. Struct. Comput. Sci.*, 28(7):1253–1286.

Mendelson, E. (2009). *Introduction to mathematical logic.* Chapman and Hall/CRC.

Meredith, C. A. and Prior, A. N. (1963). Notes on the axiomatics of the propositional calculus. *Notre Dame J. Formal Log.*, 4(3):171–187.

Meyer, A. R. (1982). What is a model of the lambda calculus? *Inf. Control.*, 52(1): 87–122.

Mitschke, A. (1976). Implication algebras are 3-permutable and 3-distributive. *Algebra Univ.*, 1:182–186.

Mogensen, T. Æ. (1992). Efficient self-interpretations in lambda calculus. *J. Funct. Program.*, 2(3):345–363.

Morris, J. (1968). *Lambda calculus models of programming languages.* Ph.D. thesis, Massachusetts Institute of Technology (MIT).

Mostowski, A. (1951). On the rules of proof in the pure functional calculus of the first order. *The Journal of Symbolic Logic*, 16(2):107–111.

Mulder, H. (1986). On a conjecture by J.W. Klop. Unpublished.

Musil, R. (1930-1943). *Der Mann ohne Eigenschaften. The man without qualities.* Rohwolt.

Myhill, J. (1957). Finite automata and the representation of events. Technical Report 57-624, Wright Air Development Command.

Nakajima, R. (1975). Infinite normal forms for the lambda-calculus. In Böhm, C., editor, *Lambda-Calculus and Computer Science Theory, Proceedings of the Symposium Held in Rome, Italy, March 25-27, 1975*, volume 37 of *Lecture Notes in Computer Science*, pages 62–82. Springer.

Nederpelt, R., Geuvers, J., and de Vrijer, R., editors (1994). *Selected Papers on Automath*, volume 133 of *Studies in Logic and the Foundations of Mathematics*. North-Holland, Amsterdam.

Nerode, A. (1958). Linear automata transformation. In *Proceedings of AMS*, volume 9, pages 541–544.

Newman, M. H. A. (1942). On theories with a combinatorial definition of "equivalence". *Annals of mathematics*, 43(2):223–243.

Nickau, H. (1996). *Hereditarily sequential functionals: a game-theoretic approach to sequentiality.* Ph.D. thesis, Universität Gesamthochschule Siegen.

Odifreddi, P. (1989). *Classical Recursion Theory (Volume I).* North–Holland Publishing Co., Amsterdam.

Olimpieri, F. (2021). Intersection type distributors. In *36th Annual ACM/IEEE Symposium on Logic in Computer Science, LICS 2021, Rome, Italy, June 29 - July 2, 2021*, pages 1–15. IEEE.

Ong, C.-H. L. (2017). Quantitative semantics of the lambda calculus: Some generalisations of the relational model. In *32nd Annual ACM/IEEE Symposium on Logic in Computer Science, LICS 2017, Reykjavik, Iceland, June 20-23, 2017*, pages 1–12. IEEE Computer Society.

van Oostrom, V., van de Looij, K.-J., and Zwitserlood, M. (2004). Lambdascope. another optimal implementation of the lambda-calculus. In *Workshop on Algebra and Logic on Programming Systems (ALPS)*, Kyoto.

Overbeek, R. and Endrullis, J. (2020). Patch graph rewriting. In Gadducci, F. and Kehrer, T., editors, *Graph Transformation - 13th International Conference, ICGT 2020, Held as Part of STAF 2020, Bergen, Norway, June 25-26, 2020, Proceedings*, volume 12150 of *Lecture Notes in Computer Science*, pages 128–145. Springer.

Overbeek, R., Endrullis, J., and Rosset, A. (2021). Graph rewriting and relabeling with PBPO^{+}. In Gadducci, F. and Kehrer, T., editors, *Graph Transformation - 14th International Conference, ICGT 2021, Held as Part of STAF 2021, Virtual Event, June 24-25, 2021, Proceedings*, volume 12741 of *Lecture Notes in Computer Science*, pages 60–80. Springer.

Pagani, M., Selinger, P., and Valiron, B. (2014). Applying quantitative semantics to higher-order quantum computing. In Jagannathan, S. and Sewell, P., editors, *The 41st Annual ACM SIGPLAN-SIGACT Symposium on Principles of Programming Languages, POPL '14, San Diego, CA, USA, January 20-21, 2014*, pages 647–658. ACM.

Paolini, L. (2008). Parametric lambda theories. *Theor. Comput. Sci.*, 398(1-3):51–62.

Paolini, L., Piccolo, M., and Ronchi Della Rocca, S. (2017). Essential and relational models. *Mathematical Structures in Computer Science*, 27(5):626–650.

Peremans, W. (1949). Een opmerking over intuitionistische logica. Technical Report ZW 16/49, Stichting Mathematisch Centrum.

Pierce, B. C. (1999). *Basic Category Theory for Computer Scientists.* MIT Press, Cambridge, MA, USA.

Pierce, R. S. (1967). *Modules over commutative regular rings.* Memoirs Amer. Math. Soc.

Pigozzi, D. and Salibra, A. (1995). Lambda abstraction algebras: Representation theorems. *Theor. Comput. Sci.*, 140(1):5–52.

Pigozzi, D. and Salibra, A. (1998). Lambda abstraction algebras: Coordinatizing models of lambda calculus. *Fundam. Informaticae*, 33(2):149–200.

Plotkin, G. D. (1971). A set-theoretical definition of application. Technical Report MIP-R-95, School of artificial intelligence.

Plotkin, G. D. (1974). The lambda-calculus is ω-incomplete. *Journal of Symbolic Logic*, 39(2):313–317.

Plotkin, G. D. (1975). Call-by-name, call-by-value and the lambda-calculus. *Theor. Comput. Sci.*, 1(2):125–159.

Plotkin, G. D. (1977). LCF considered as a programming language. *Theor. Comput. Sci.*, 5(3):223–255.

Plotkin, G. D. (1978). $\mathbb{T}^\omega$ as a universal domain. *J. Comput. System Sci.*, 17:209–236.

Plotkin, G. D. (1982). The category of complete partial orders: a tool for making meanings. Postgraduate lecture notes.

Plotkin, G. D. (1993). Set-theoretical and other elementary models of the λ-calculus. *Theoretical Computer Science*, 121(1):3510–409.

Poincaré, J. H. (1903). *La science et l'hypothèse [Science and hypothesis].* Paris: E. Flammarion.

Polonsky, A. (2011a). Axiomatizing the quote. In Bezem, M., editor, *Computer Science Logic (CSL'11) - 25th International Workshop/20th Annual Conference of the EACSL*, volume 12 of *Leibniz International Proceedings in Informatics (LIPIcs)*, pages 458–469, Dagstuhl, Germany. Schloss Dagstuhl–Leibniz-Zentrum fuer Informatik.

Polonsky, A. (2011b). *Proofs, Types and Lambda Calculus*. Ph.D. thesis, University of Bergen, Norway.

Polonsky, A. (2012). The range property fails for $\mathcal{H}$. *J. Symb. Log.*, 77(4):1195–1210.

Prawitz, D. (2006). *Natural deduction: A proof-theoretical study*. Courier Dover Publications. Reprint of the original published in 1965 with Almqvist & Wicksell.

Rogers, J. H. (1967). *Theory of Recursive Functions and Effective Computability*. McGraw-Hill.

Ronchi Della Rocca, S. (1982). Characterization theorems for a filter lambda model. *Information and Control*, 54(3):201–216.

Ronchi Della Rocca, S. and Paolini, L. (2004). *The Parametric Lambda Calculus: A Metamodel for Computation*. Texts in Theoretical Computer Science. An EATCS Series. Springer Berlin Heidelberg.

Rosser, B. J. (1936). Extensions of some theorems of Gödel and Church. *Journal of Symbolic Logic*, 1(3):87–91.

Rosser, J. B. (1935). A mathematical logic without variables. Part I. *Annals of Mathematics, Series 2*, 36(1):127–150. Also Part 2: Duke Mathematical Journal 1 (1935), pp. 328–355.

Rosser, J. B. (1971). Letter to H. P. Barendregt.

Sabry, A., Valiron, B., and Vizzotto, J. K. (2018). From symmetric pattern-matching to quantum control. In Baier, C. and Dal Lago, U., editors, *Foundations of Software Science and Computation Structures - 21st International Conference, FOSSACS 2018, Held as Part of the European Joint Conferences on Theory and Practice of Software, ETAPS 2018, Thessaloniki, Greece, April 14-20, 2018, Proceedings*, volume 10803 of *Lecture Notes in Computer Science*, pages 348–364. Springer.

Salibra, A. (2000). On the algebraic models of lambda calculus. *Theor. Comput. Sci.*, 249(1):197–240.

Salibra, A. (2001a). Nonmodularity results for lambda calculus. *Fundamenta Informaticae*, 45:379–392.

Salibra, A. (2001b). A continuum of theories of lambda calculus without semantics. In *16th Annual IEEE Symposium on Logic in Computer Science, Boston, Massachusetts, USA, June 16-19, 2001, Proceedings*, pages 334–343. IEEE Computer Society.

Salibra, A. (2001c). Towards lambda calculus order-incompleteness. *Electron. Notes Theor. Comput. Sci.*, 50(2):145–158.

Salibra, A. (2003). Topological incompleteness and order incompleteness of the lambda calculus. *ACM Trans. Comput. Log.*, 4(3):379–401.

Salibra, A. (2005). Personal communication.

Salibra, A. and Goldblatt, R. (1999). A finite equational axiomatization of the functional algebras for the lambda calculus. *Inf. Comput.*, 148(1):71–130.

Salibra, A., Ledda, A., Paoli, F., and Kowalski, T. (2013). Boolean-like algebras. *Algebra Universalis*, 69:113–138.

Salvati, S., Manzonetto, G., Gehrke, M., and Barendregt, H. P. (2012). Loader and Urzyczyn are logically related. In Czumaj, A., Mehlhorn, K., Pitts, A. M., and Wattenhofer, R., editors, *Automata, Languages, and Programming - 39th International Colloquium, ICALP 2012, Warwick, UK, July 9-13, 2012, Proceedings, Part II*, volume 7392 of *Lecture Notes in Computer Science*, pages 364–376. Springer.

Santo, J. E. (2020). The call-by-value lambda-calculus with generalized applications. In Fernández, M. and Muscholl, A., editors, *28th EACSL Annual Conference on Computer Science Logic, CSL 2020, January 13-16, 2020, Barcelona, Spain*, volume 152 of *LIPIcs*, pages 35:1–35:12. Schloss Dagstuhl - Leibniz-Zentrum für Informatik.

Schmidt, D. A. (1986). *Denotational Semantics: Methodology for Language Development.* Allyn and Bacon, Boston.

Schönfinkel, M. (1924). Über die bausteine der mathematischen logik. *Mathematische Annalen*, 92:305–316. (English translation in J. van Heijenoort ed.'s book "From Frege to Gödel, a source book in Mathematical Logic", 1879-1931, Harvard University Press, 1967).

Schroer, D. E. (1965). *The Church-Rosser Theorem.* Ph.D. thesis, Cornell Univ. Informally circulated 1963. 673 pp. Obtainable from University Microfilms Inc., Ann Arbor, Michigan, U.S.A., Publication No. 66-41.

Scott, D. S. (1972). Continuous lattices. In Lawvere, editor, *Toposes, Algebraic Geometry and Logic*, volume 274 of *Lecture Notes in Mathematics*, pages 97–136. Springer.

Scott, D. S. (1974). The language LAMBDA (abstract). *J. symbolic logic*, 39:425–427.

Scott, D. S. (1975). Some philosophical issues concerning theories of combinators. In *International Symposium on Lambda-Calculus and Computer Science Theory*, pages 346–366. Springer.

Scott, D. S. (1976). Data types as lattices. *SIAM J. Comput.*, 5(3):522–587.

Scott, D. S. (1980). Relating theories of the λ-calculus. In Hindley, R. and Seldin, J., editors, *To H.B. Curry: Essays in Combinatory Logic, Lambda Calculus and Formalisms*, pages 403–450. Academic Press.

Seely, R. (1989). Linear logic, ∗-autonomous categories and cofree coalgebras. *AMS Contemporary Mathematics*, 92:371–328.

Sekimoto, S. and Hirokawa, S. (1988). One-step recurrent terms in lambda-beta-calculus. *Theor. Comput. Sci.*, 56:223–231.

Seldin, J. P. (2009). The logic of Church and Curry. In Gabbay, D. and Woods, J., editors, *Handbook of the History of Logic*, pages 5–819. Elsevier.

Selinger, P. (1996). Order-incompleteness and finite lambda models (extended abstract). In *Proceedings, 11th Annual IEEE Symposium on Logic in Computer Science, New Brunswick, New Jersey, USA, July 27-30, 1996*, pages 432–439. IEEE Computer Society.

Selinger, P. (1997). *Functionality, polymorphism, and concurrency: a mathematical investigation of programming paradigms.* Ph.D. thesis, University of Pennsylvania.

Selinger, P. (2002). The lambda calculus is algebraic. *J. Funct. Program.*, 12(6):549–566.

Selinger, P. (2003). Order-incompleteness and finite lambda reduction models. *Theor. Comput. Sci.*, 309(1-3):43–63.

Selinger, P. (2008). Lecture notes on the lambda calculus. *CoRR*, abs/0804.3434. Available at `http://arxiv.org/abs/0804.3434`.

Severi, P. and de Vries, F.-J. (2002). An extensional Böhm model. In Tison, S., editor, *Rewriting Techniques and Applications, 13th International Conference, RTA 2002, Copenhagen, Denmark, July 22-24, 2002, Proceedings*, volume 2378 of *Lecture Notes in Computer Science*, pages 159–173. Springer.

Severi, P. and de Vries, F.-J. (2005a). Continuity and discontinuity in lambda calculus. In Urzyczyn, P., editor, *Typed Lambda Calculi and Applications, 7th International Conference, TLCA 2005, Nara, Japan, April 21-23, 2005, Proceedings*, volume 3461 of *Lecture Notes in Computer Science*, pages 369–385. Springer.

Severi, P. and de Vries, F.-J. (2005b). Order structures on Böhm-like models. In Ong, C.-H. L., editor, *Computer Science Logic, 19th International Workshop, CSL 2005, 14th Annual Conference of the EACSL, Oxford, UK, August 22-25, 2005, Proceedings*, volume 3634 of *Lecture Notes in Computer Science*, pages 103–118. Springer.

Severi, P. and de Vries, F.-J. (2011a). Decomposing the lattice of meaningless sets in the infinitary lambda calculus. In Beklemishev, L. D. and de Queiroz, R. J. G. B., editors, *Logic, Language, Information and Computation - 18th International Workshop, WoLLIC 2011, Philadelphia, PA, USA, May 18-20, 2011. Proceedings*, volume 6642 of *Lecture Notes in Computer Science*, pages 210–227. Springer.

Severi, P. and de Vries, F.-J. (2011b). Weakening the axiom of overlap in infinitary lambda calculus. In Schmidt-Schauß, M., editor, *Proceedings of the 22nd International Conference on Rewriting Techniques and Applications, RTA 2011, May 30 - June 1, 2011, Novi Sad, Serbia*, volume 10 of *LIPIcs*, pages 313–328. Schloss Dagstuhl - Leibniz-Zentrum für Informatik.

Severi, P. and de Vries, F.-J. (2017). The infinitary lambda calculus of the infinite eta Böhm trees. *Mathematical Structures in Computer Science*, 27(5):681–733.

Shoenfield, J. R. (2018). *Mathematical logic.* AK Peters/CRC Press.

Simonsen, J. G. (2006). On modularity in infinitary term rewriting. *Inf. Comput.*, 204 (6):957–988.

Smith, J. D. H. (1976). *Mal'cev Varieties*, volume 554 of *Lecture Notes in Math.* Springer, Berlin.

Smullyan, R. (1985). *To Mock A Mockingbird.* Alfred A. Knopf, New York.

Soare, R. I. (1999). *Recursively enumerable sets and degrees: A study of computable functions and computably generated sets.* Springer Science & Business Media.

Statman, R. (1986). Every countable poset is embeddable in the poset of unsolvable terms. *Theor. Comput. Sci.*, 48(3):95–100.

Statman, R. (1988a). Combinators hereditarily of order one. Technical Report 88-32, Department of Mathematics, Carnegie Mellon University.

Statman, R. (1988b). Combinators hereditarily of order two. Technical Report 88-33, Department of Mathematics, Carnegie Mellon University.

Statman, R. (1989). The word problem for Smullyan's Lark combinator is decidable. *J. Symb. Comput.*, 7(2):103–112.

Statman, R. (1991). There is no hyper–recurrent S, K combinator. Technical Report 91–133, Department of Mathematics, Carnegie Mellon University, Pittsburgh, PA.

Statman, R. (1993a). Some examples of non-existent combinators. *Theor. Comput. Sci.*, 121(1&2):441–448.

Statman, R. (1993b). RTA list, Problem #52: Is there a fixed point combinator Y for which $Y \leftrightarrow^* Y(SI)$?. See `https://www.win.tue.nl/rtaloop/`. First raised in Statman (1993a).

Statman, R. (1993c). RTA list, Problem #53: Are there hyper-recurrent combinators? See `https://www.win.tue.nl/rtaloop/`. First raised in Statman (1993a).

Statman, R. (1993d). RTA list, Problem #54: In combinatory logic, is there a uniform universal generator? See `https://www.win.tue.nl/rtaloop/`. First raised in Statman (1993a).

Statman, R. (2000). On the word problem for combinators. In Bachmair, L., editor, *Rewriting Techniques and Applications, 11th International Conference, RTA 2000, Norwich, UK, July 10-12, 2000, Proceedings*, volume 1833 of *Lecture Notes in Computer Science*, pages 203–213. Springer.

Statman, R. (2001). Marginalia to a theorem of Jacopini. *Fundamenta Informaticae*, 45 (1-2):117–121.

Statman, R. (2005). Two variables are not enough. In Coppo, M., Lodi, E., and Pinna, G. M., editors, *Theoretical Computer Science, 9th Italian Conference, ICTCS 2005, Siena, Italy, October 12-14, 2005, Proceedings*, volume 3701 of *Lecture Notes in Computer Science*, pages 406–409. Springer.

Statman, R. and Barendregt, H. P. (1999). Applications of Plotkin-terms: partitions and morphisms for closed terms. *Journal of Functional Programming*, 9(5):565–575.

Statman, R. and Barendregt, H. P. (2005). Böhm's theorem, Church's delta, numeral systems, and Ershov morphisms. In Middeldorp, A., van Oostrom, V., van Raamsdonk, F., and de Vrijer, R. C., editors, *Processes, Terms and Cycles: Steps on the Road to Infinity, Essays Dedicated to Jan Willem Klop, on the Occasion of His 60th Birthday*, volume 3838 of *Lecture Notes in Computer Science*, pages 40–54. Springer.

Tait, W. W. (1967). Intensional interpretations of functionals of finite type I. *J. Symb. Log.*, 32(2):198–212.

Terese (2003). *Term Rewriting Systems*, volume 55 of *Cambridge Tracts in Theoretical Computer Science*. Cambridge University Press.

Troelstra, A. S. and van Dalen, D. (1988). *Constructivism in Mathematics, an Introduction.* Number 121 and 123 in Studies in Logic and the Foundations of Mathematics. North-Holland.

Turing, A. M. (1937a). On computable numbers, with an application to the Entscheidungsproblem. *Proceedings of the London Mathematical Society 2*, 1(42):230–65.

Turing, A. M. (1937b). Computability and λ-definability. *The Journal of Symbolic Logic*, 2(4):153–163.

Univalent Foundations Program (2013). *Homotopy Type Theory: Univalent Foundations of Mathematics.* Available at `https://homotopytypetheory.org/book`, Institute for Advanced Study.

Urban, J. (2021). ERC project AI4Reason final scientific report. Available at `http://ai4reason.org/PR_CORE_SCIENTIFIC_4.pdf`.

Urzyczyn, P. (1999). The emptiness problem for intersection types. *J. Symb. Log.*, 64 (3):1195–1215.

Vaggione, D. (1996a). $\mathcal{V}$ with factorable congruences and $\mathcal{V} = \mathrm{I}\Gamma^a(\mathcal{V}_{DI})$ imply $\mathcal{V}$ is a discriminator variety. *Acta Sci. Math.*, 62:359–368.

Vaggione, D. (1996b). Varieties in which the Pierce stalks are directly indecomposable. *Journal of Algebra*, 184:424–434.

Venturini Zilli, M. (1978). Head recurrent terms in combinatory logic: A generalization of the notion of head normal form. In Ausiello, G. and Böhm, C., editors, *Automata, Languages and Programming, Fifth Colloquium, Udine, Italy, July 17-21, 1978, Proceedings*, volume 62 of *Lecture Notes in Computer Science*, pages 477–493. Springer.

Vial, P. (2017). Infinitary intersection types as sequences: A new answer to Klop's problem. In *32nd Annual ACM/IEEE Symposium on Logic in Computer Science, LICS 2017, Reykjavik, Iceland, June 20-23, 2017*, pages 1–12. IEEE Computer Society.

Vial, P. (2018). Every λ-term is meaningful for the infinitary relational model. In Dawar, A. and Grädel, E., editors, *Proceedings of the 33rd Annual ACM/IEEE Symposium on Logic in Computer Science, LICS 2018, Oxford, UK, July 09-12, 2018*, pages 899–908. ACM.

Visser, A. (1980). Numerations, λ-calculus, and arithmetic. In Hindley and Seldin, editors, *Essays on Combinatory Logic, Lambda-Calculus, and Formalism*, pages 259–284. Academic Press, San Diego.

Voevodsky, V. (2014). The origins and motivations of univalent foundations. Available at `https://www.ias.edu/ideas/2014/voevodsky-origins`.

de Vrijer, R. C. (1987). *Surjective Pairing and Strong Normalization: Two Themes in Lambda Calculus*. Ph.D. thesis, Universiteit van Amsterdam.

Vuillemin, J. (1973). *Proof techniques for recursive programs*. Ph.D. thesis, Computer Science Department, Standford University, USA.

Wadler, P. (2015). Propositions as types. *Communications of the ACM*, 58(12):75–84.

Wadsworth, C. P. (1976). The relation between computational and denotational properties for Scott's $\mathcal{D}_\infty$-models of the lambda-calculus. *SIAM J. Comput.*, 5(3):488–521.

Waldmann, J. (1997). Nimm zwei. Technical Report IR-432, Vrije Universiteit, Research Group Theoretical Computer Science, Amsterdam.

Waldmann, J. (1998). *The Combinator S*. Ph.D. thesis, Friedrich-Schiller-Universität Jena.

Waldmann, J. (2000). The combinator S. *Inf. Comput.*, 159(1-2):2–21.

Waldmann, J. (2022). Personal communication.

Whitehead, A. N. and Russell, B. (1910). *Principia Mathematica Volume*, volume I. Cambridge University Press.

Wiweger, A. (1984). Pre-adjunctions and λ-algebraic theories. *Colloq. Math*, 48(2): 153–165.

Wupper, H. (2000). Design as the discovery of a mathematical theorem, what designers should know about the art of mathematics. *Journal of Integrated Design and Process Science*, 4(2):1–13.

Zachos, E. K. (1978). *Kombinatorische Logik und S-Terme.* Ph.D. thesis, Berichte des Instituts für Informatik, Eidgenössische Technische Hochschule Zürich.

Zantema, H. (1995). Termination of term rewriting by semantic labelling. *Fundamenta Informaticae*, 24:89–105.

Zantema, H. (2008). Normalization of infinite terms. In Voronkov, A., editor, *Rewriting Techniques and Applications, 19th International Conference, RTA 2008, Hagenberg, Austria, July 15-17, 2008, Proceedings*, volume 5117 of *Lecture Notes in Computer Science*, pages 441–455. Springer.

Zhao, D. and Fan, T. (2010). Dcpo-completion of posets. *Theor. Comput. Sci.*, 411 (22-24):2167–2173.

Zylberajch, C. (1991). *Syntaxe et sémantique de la facilité en λ-calcul.* Thèse de doctorat, Université Paris 7. In French.

Indices

There is an index of definitions, of names, and of symbols.

- The structure of the index of definitions is oriented towards the definiendum rather than its attributes. As an example, the notion of an 'easy λ-term' appears under 'λ-term' rather than 'easy'.
- The index of names contains authors of cited references.
- The index of symbols is subdivided as follows.

  1. Lambda terms: general, operations, classes, relations, theories.
  2. Reductions: general, operations, classes, relations.
  3. Combinatory terms: general, operations, classes, relations.
  4. Sequences: general, operations, classes, relations.
  5. Infinitary terms and trees: general, operations, classes, relations, theories.
  6. Type assignment: general, special types, operations, classes, relations, systems, derivations.
  7. Models: general, special elements, operations, relations, interpretations,
  8. Universal algebra: general, operations, relations, congruences, classes & varieties.
  9. Categories: general, objects, morphisms, functors.
  10. Miscellaneous: functions, operations, classes, relations, interpretations.

  The meaning of the main division is the following. There are objects and relations between objects. Using the operations one constructs new objects. By collecting objects one obtains classes, by collecting valid relations one obtains theories. Categories consist of classes of objects together with arrows (morphisms) and arrows between arrows (functors).

Index of definitions

Index of names

Index of symbols

1. Lambda terms

A. general

B. operations

C. classes

www.ingramcontent.com/pod-product-compliance
Lightning Source LLC
Chambersburg PA
CBHW081837230426
43669CB00018B/2735
* 9 7 8 1 8 4 8 9 0 4 1 5 6 *